Accelerator-Based Atomic Physics Techniques and Applications

Accelerator-Based Atomic Physics Techniques and Applications

Editors

Stephen M. Shafroth

James C. Austin

The University of North Carolina
Department of Physics
Chapel Hill, North Carolina

AIP PRESS

American Institute of Physics **Woodbury, New York**

AIP Press
American Institute of Physics
500 Sunnyside Boulevard
Woodbury, NY 11797-2999

Library of Congress Cataloging-in-Publication Data

Accelerator-based atomic physics : techniques and applications / editors, Stephen M. Shafroth, James C. Austin.
p. cm.
Includes bibliographical references and index.
ISBN 1-56396-484-8
1. Particle accelerators--Technique. 2. Ion sources. 3. Particle beams. 4. Collisions (Nuclear physics) I. Shafroth, Stephen M. II. Austin, James C. (James Christopher)
QC787.P3A278 1997 96-46751
539.7 ' 3--dc21 CIP

10 9 8 7 6 5 4 3 2 1

I would like to dedicate this book to my mother, Janet Durrie Shafroth, who almost lived to see it in print. Her constant encouragement kept me going during the dark days, when it seemed we would never finish.

Stephen M. Shafroth

Contents

Preface

This book grew out of a workshop entitled, "Unique Instruments for Accelerator-Based Atomic Physic Experiments." The workshop preceded the Thirteenth International Conference on the Application of Accelerators in Research and Industry, which was held at Denton, Texas, November 7–10, 1994. At the workshop it became evident that many of the subjects presented would be of interest to a much wider audience. In order to round out the book, we decided to extend the coverage to include chapters by experts in most of the major subfields of accelerator-based atomic physics. The authors were asked to look into the future and tell what they thought would be some of the most interesting developments in their respective fields. As the book has developed we have encouraged the authors to focus on the details that make it possible to do an experiment, information that is often missing from journal articles. We wanted a book that would fill a niche and also be pedagogically useful, a book that could serve as an introduction to the field for graduate students and scientists from other disciplines while providing enough depth and new research to be of interest to scientists working in the field.

We are grateful to Professor Jerry Duggan of the University North Texas, whose idea it was to hold the workshop. The book could not have appeared without the authors' exemplary cooperation. Finally, we wish to acknowledge the support of the Department of Physics and Astronomy of the University of North Carolina at Chapel Hill.

Stephen M. Shafroth
James C. Austin
Editors

Introduction

Stephen M. Shafroth, James C. Austin, and Eugen Merzbacher

Accelerator-Based Atomic Physics Techniques and Applications can be divided into eight broad categories which provide an overview of instrumentation and experimental techniques currently in use, as well as applications to physics and applied science. The goals of the book are to bring together and unify major areas of atomic physics in one volume, with emphasis on shared accelerator-based techniques; to introduce the techniques and applications to scientists in government laboratories, industries, and academe, including graduate students; and to serve as a helpful reference. It is also intended to be useful to scientists in other disciplines such as chemistry and medicine. Each contribution includes a brief section on future prospects for the techniques or instruments covered.

It is no exaggeration to say that the use of particle accelerators has transformed experimental atomic physics in the past three decades. Projectiles of every conceivable kind, in all possible charge states and with almost any desired energy, can now be made to interact with target atoms or ions under a wide range of conditions. The collision dynamics of virtually all types of interactions in the domain of atoms has been or is being studied with suitable particle beams and detection facilities. Both atomic collision theory and atomic structure theory have been and continue to be subjected to extensive tests, and much new physics has challenged both experimental and theoretical (and computational) physicists to deepen their understanding of the fundamental processes involved. There is a natural tendency in the physics literature to emphasize experimental results and to play down the details of the techniques employed to arrive at them. In these contributions from many of the major players in the field, we have attempted to bring together practical and useful information about the techniques, instruments, and applications which make the pursuit of accelerator-based atomic collision physics so worthwhile.

We start with sources of highly charged ions. Information about these sources can be found in the technical literature, but introductory and expository material is scarce. For perspective, an overview is included of the physics of multicharged ions (MCI) produced by these sources and collisions of MCI with electrons, neutral atoms, ions, surfaces, and solids (Winter). Some of the practitioners of ion "sorcery" tell about ECR, electron-cyclotron resonance ion sources (Melin and Girard) and CRYEBIS, the cryogenic electron-beam ion sources (Stöckli).

Moving from sources to beam processing devices, we introduce the reader to merged beams and heavy ion storage rings. Storage rings combined with cooler technology are providing interesting new sources of trapped ions. Heavy-ion storage rings play a major role in accelerator-based atomic physics in Europe, but only one storage ring exists in the U.S., the Indiana University nuclear facility, which so far is confined to light ions. Thus it is valuable to explain in introductory language how heavy-ion storage rings work and what has been accomplished with them as well as what lies ahead in this field (Müller and Wolf). The technique of merging beams of highly charged ions with neutral atomic beams, which has allowed the study of interactions and processes that are otherwise inaccessible, is explained, by reference to the ORNL Ion–Atom Merged beam facility, along with its advantages and disadvantages. It will soon be possible to record electron spectra in a merged beam configuration, and the implications are considered (Havener).

While most accelerator-based atomic physics studies are performed with charged, or even highly charged ion beams, fast neutral atomic beams are also interesting projectiles. They have not been discussed very much in the literature. Of course, they are not easy to steer and focus, and this causes problems, but for some specialized and quite fundamental studies they have proved to be of great importance. At low energies neutral (H and He) beam–atom collisions have shed light on the "cusp"-like feature in the electron spectrum near zero degrees at nearly the same velocity as the neutral projectiles. The detection of these electrons poses unique experimental problems (Berényi). In conjunction with lasers, relativistic neutral beams from LAMPF (Los Alamos Meson Physics Facility) have yielded important spectroscopic information. Some of the techniques for producing neutral beams were developed under the Strategic Defense Initiative, but have recently been declassified. New medical applications and accelerator diagnostics are also noted (Bryant and MacKerrow).

Characteristic X rays arising from ion–atom collisions have been the subject of investigation ever since Chadwick discovered them early in the century. We do not attempt to summarize all the progress which has been made in X-ray physics since then, but prefer to describe in detail certain experimental topics which have been neglected or are relatively new.

Parametric X rays are coherent radiation arising from electron bombardment of crystals. This method of X-ray production is not well known but may provide an intense, nearly monoenergetic, tunable source of short wavelength X rays, somewhat like synchrotron radiation, with medical and biological applications. Synchrotron radiation produces a wide band of wavelengths from which, for many research purposes, a narrow band must be selected by means of a monochromator. With parametric X rays, one needs only to change the angle of incidence of the electron beam with respect to the crystal planes (Shchagin and Maruyama). Continuous bremsstrahlung from electrons passing through matter, the traditional process by which X rays are produced, has not been the subject of as much study as one might think. Invariably, thick target bremsstrahlung X-rays are produced when any accelerated beam stops in matter. In order to better understand this production mechanism for X rays, the technique for producing and detecting bremsstrahlung radiation with thin solid or gas targets while eliminating background radiation is described. The instrumental challenges in these measurements offer instructive lessons about the calibration of the efficiency of X-ray detectors (Quarles).

X-ray microbeams from synchrotron radiation are becoming important for use in materials analysis, and the new and innovative methods of producing them are described (Chevallier and Dhez). High-resolution X-ray spectrometers, semiconductor X-ray detectors, and electron spectrometers are used to study "hollow atoms," which are formed when slow, highly charged ions, like Ar^{17+}, capture electrons above, at, or below surfaces of metals, semiconductors, or insulators (Giardino and Briand).

The science and art of electron spectroscopy has required the development of improved detection methods. A unique multi-angle spectrometer has been designed and used for studying Auger electron angular distributions from collisionally excited atomic target atoms (Sulik and Stolterfoht). The detection of Auger electrons emitted by fast-moving excited projectile ions at zero degrees poses interesting technical problems for the design of electron spectrometers and much new and interesting collision physics has been discovered using zero degree electron spectroscopy (Zouros and Lee).

Molecular ions have now joined atomic ions as interesting projectiles for collision experiments. We start with molecular-ion imaging with fast ion beams, including an introduction to such processes as Coulomb explosion and charge exchange dissociation. Molecular-ion imaging is a powerful tool for the study of the mechanisms leading to molecular dissociation and for the investigation of molecular structure (Zajfman and Heber). An innovative technique for determining the lifetimes of fast moving doubly charged ions in certain metastable states is explained. These doubly charged molecular ions are produced in collisions with a target gas such as Ar. They then quickly dissociate and the two parts spread out in a cone about the beam direction. The detection and analysis of these kinematically constrained cones leads to lifetime determinations of the metastable states in question (Rosner, Gertner, Ben-Itzhak). The angular distribution of Auger electrons relative to the internuclear axis arising from collisionally excited, dissociating, doubly charged, homonuclear diatomic molecules is measured by standard coincidence techniques. These measurements open up new unexplored regimes with similar techniques for heteronuclear diatomic molecules (Wood and Edwards).

The techniques of beam-foil spectroscopy for lifetime measurements of highly excited states of fast ions have evolved beyond the standard movable foil method and reached a new phase with the use of storage rings, Doppler tuned spectrometers, double-foil methods, and position- and energy-sensitive X-ray detectors (Träbert). Resonant laser excitation of fast atomic beams provides a method for making precision lifetime measurements, which are important for new stringent tests of quantum electrodynamics and particle theory. Again, these experiments illustrate the successful marriage between accelerator-related atomic physics and laser technology (Tanner).

The book concludes with five chapters covering diverse timely topics. We learn how thermal positron beams are made using electron LINACS (Hulett).

The instrumentation required to study light emission by atoms sputtered from surfaces under ion bombardment is tied to an explanation of the mechanism at work (Hippler and Ghose). Yet another fruitful example of laser technology (pulsed and CW) in accelerator-related atomic physics comes from Rydberg atom collision studies. These difficult experiments have been dreamed of for many years and are now beginning to lead to a new class of atomic collision physics results. For example, absolute electron capture cross sections for highly charged ions colliding with Rydberg atoms are measured for the first time (DePaola). Fast collisions of ions with ground state atomic targets can be used to produce highly charged recoil ions, which are extracted from the collision area and focused into a slow recoil ion beam. This beam then interacts with atoms and may gain or lose electrons as well as energy. The method of translational energy gain spectroscopy is used to electrostatically analyze the recoil ions after the collision (Kamber).

In a new experimental method characterizing ion–atom collisions with the highest precision yet achieved in a kinematically complete experiment, a cooled supersonic gas jet target is used in conjunction with a high-resolution recoil ion momentum spectrometer and an integrated 4π electron analyzer. Longitudinal (along the beam direction) momentum of the recoil ions and ejected electrons are determined by time-of-flight, while transverse momentum of the recoil ions and electrons are determined using two-dimensional position-sensitive channel plate detectors. This technique, called COLTRIMS, provides a remarkable demonstration of several of the significant advances which have recently been made in accelerator-based atomic physics (Schmidt-Böcking *et al.*). It is a fitting conclusion to this compilation of useful techniques.

CHAPTER 1

Applications of High-Performance Multicharged Ion Sources

H. P. Winter*

This chapter gives a short introduction to the working principles, characteristic features, and some typical applications of modern, high-performance multicharged ion sources (MCIS). ECRIS (Electron Cyclotron Resonance-heated plasma Ion Sources), EBIS (Electron Beam Ion Sources) and EBIT (Electron Beam Ion Traps) conveniently deliver slow MCI beams for various investigations in atomic collision physics. This will be exemplified by recent experimental studies on MCI structure and collisions of MCI with electrons, neutral particles, ions, clusters and solid surfaces.

1.1 INTRODUCTION

Multicharged ions (MCI) are interesting tools, as well as objects of intrinsic interest, in areas as diverse as: MCI collisions with photons, electrons, heavy particles, and solids; properties of thermonuclear and astrophysical plasmas; atomic spectra and structure as tests of general relativity and quantum electrodynamics; interrelation of atomic and nuclear phenomena; and the development of soft X-ray lasers [1]. During the last two decades tremendous progress has been achieved in the production of intense, low-energy MCI beams, which has greatly improved the conditions for experiments in all of the above-mentioned fields. Electron Cyclotron Resonance-heated plasma Ion Sources (ECRIS) can now deliver, e.g., fully stripped Ar at electrical currents of up to several hundred nA. Completely stripped ions up to U^{92+} will soon be produced by Electron Beam Ion Sources (EBIS) and have already been observed in the closely related Electron Beam Ion Traps (EBIT). All these types of multicharged ion sources (MCIS) possess the capability to produce MCI of

*) Institut für Allgemeine Physik, Technische Universität Wien, Wiedner Hauptstraße 8-10, A-1040 Wien, Austria

practically all chemical species, and they can be operated over reasonably long periods under rather stable conditions. Apart from their usefulness as a first stage of an ion accelerator (e.g., for external injection into cyclotrons), these MCIS are especially well suited for experiments with comparably slow MCI beams; i.e., at ion kinetic energies below 1 atomic unit (a.u.), corresponding to 25 keV per atomic mass unit (amu). We will cover briefly the working principles and basic features of ECRIS, EBIS and EBIT devices. For more details the reader is referred to chapters 1.2 (EBIS/EBIT) and 1.3 (ECRIS). By way of example, numerous applications utilizing such devices as sources of slow MCI beams will be dicussed in the following fields:

MCI structure and collisions with electrons EBITs are well suited for photon spectroscopy from excited, ionized, or recombined MCI within the trapping electron beam. Such experiments provide detailed information on the atomic structure of MCI and allow derivation of cross sections and reaction rates for collisional processes.

MCI collisions with neutral and ionized heavy particles MCI impact on neutrals or ions in various collision geometries results in signatures such as emission of electrons or photons and inelastic energy loss of scattered MCI. Detailed information on dynamics and electronic transitions in such collisions can be obtained from these signatures.

MCI collisions with clusters, surfaces and solids Near a solid surface, slow MCI are transiently converted into so-called "hollow atoms", multiply excited neutral particles that last sufficiently long to reveal rather exotic features. Hollow atoms can be studied via electron or photon emission, projectile energy loss, or the distribution of charge states of the scattered projectiles. MCI collisions with clusters bridge the gap between the respective binary and surface collisions. In addition to the well established "kinetic" sputtering process of solids, recently for MCI impact on insulators a novel "potential sputtering" process has been identified, which depends only on the MCI charge state, i.e., on the potential projectile energy.

1.2 WORKING PRINCIPLES OF HIGH-PERFORMANCE MCIS

1.2.1 General remarks on production and transport of slow MCI beams

An ideal MCIS delivers bright beams of MCI (Z^{q+}; atomic number Z, charge state q) from virtually all chemical species and in all charge states q up to the respective fully stripped ("naked") ions ($q = Z$). However, life is generally not so simple, and therefore anyone interested in MCI-related studies has to match his/her experimental needs (and financial means) with the types of MCIS actually available.

MCI may be produced in many different ways, of which successive electron impact ionization has proven to be the most effective (for a general overview of ion source principles and -technology see Brown [2]). MCI production by single electron impact on neutral atoms or molecules involves reaction cross sections that decrease dramatically with increasing q, whereas cross sections for successive removal of single electrons from the increasingly higher-charged ions decrease much slower with q. However, in order to utilize successive ionization processes the respective ions need to be kept within an ionization environment for the necessary ionization times $\tau_{i,q}$, which depend on the relevant ionization- and recombination processes as well as the respective ion confinement conditions. On the other hand, intense ion beams (as defined by their emittance and brightness, see below) can only be efficiently produced and transported toward an experimental setup if the respective ion source provides a sufficiently small initial MCI energy spread (or, in more relaxed terminology, a low MCI temperature). In effect, the genuine art of MCIS builders and operators consists of balancing the contradictory requirements of producing MCI efficiently while keeping them as cold as possible.

The concept of ion beam emittance derives from Liouville's theorem, which states that the density of non-interacting particles subject to conservative forces in the six-dimensional phase space stays constant along each particle trajectory. If we regard an ion beam drifting along the z direction and neglect the ion beam space charge (see below), we may substitute the six-dimensional phase space $\{x, y, z; p_x, p_y, p_z\}$ by its two-dimensional sub-spaces $\{x, p_x\}$ and $\{y, p_y\}$ (Holmes [3]). Further assuming that $p_x, p_y \ll p_z$, we define

$$x' \equiv \frac{p_x}{p_z} = \tan\alpha_x \approx \alpha_x; y' \equiv \frac{p_y}{p_z} = \tan\alpha_y \approx \alpha_y \tag{1.1}$$

where α_x and α_y are angles of a particular ion trajectory relative to the z-axis at given z in the xz and yz planes, respectively. These angles can be measured, e.g., by directing an ion beam onto a slotted plate followed by a movable wire detector. Liouville's theorem simply means that the "areas" $\varepsilon_x(x, x')$ and $\varepsilon_y(y, y')$, which are called (two-dimensional) ion beam *emittances*, will stay constant along a drifting ion beam. For an accelerated or decelerated ion beam this statement remains valid for the relativistic ion momenta, leading to the "normalized emittance" $\varepsilon_n = \beta \cdot \gamma \cdot \varepsilon$ (Holmes [3]), where β and γ have their usual meaning.

Fig. 1.1 shows a typical emittance plot with contours for different fractions of the total ion beam intensity. Since such emittance "areas" have an elliptical shape, it is customary to give the size of the respective emittances in units of $\pi \cdot$ millimeter $\cdot$ milliradian.

Based on this emittance concept, the relevant figure of merit of ion beams is the ion current density in phase space, the so-called *brightness* or *normalized brightness*, respectively:

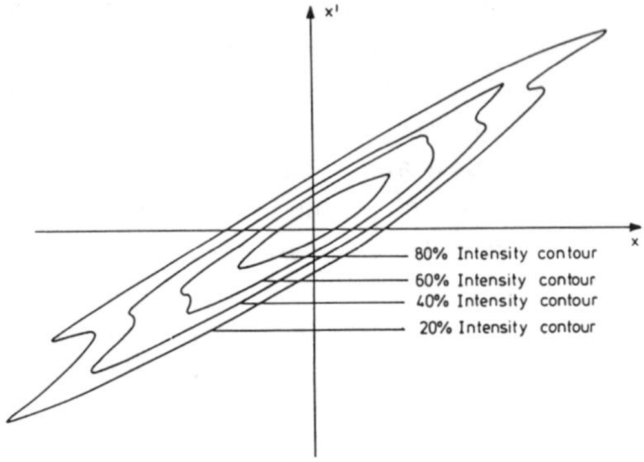

FIGURE 1.1 Typical emittance plot with contours for different total ion beam fractions.

$$B \equiv \frac{I}{\varepsilon_x \cdot \varepsilon_y}; B_n \equiv \frac{I}{\varepsilon_{n,x} \cdot \varepsilon_{n,y}} \tag{1.2}$$

Brightnesses are given in units of Ampere/$(\pi \cdot mm \cdot mrad)^2$. Thus, we can simply state that ion beams possess the higher quality the lower their emittance and the higher their brightness. If ion beams involve non-negligible space charge due to mutual repulsion of the ions by their Coulomb interaction, the above-given preconditions for the validity of Liouville's theorem break down. However, since c.w. or slowly pulsed positive ion beams drifting in reasonably good technical vacua (typically above 10^{-7} mbar) remain space-charge compensated by those electrons which they themselves produce in collisions with background gas molecules, the concepts of emittance and brightness may still be adhered to. Space charge influence along acceleration or deceleration regions, as well as imperfect ion-optical systems, enhance the ion beam emittance and give rise to some characteristic distortions from the ideal elliptical shape, causing a considerably larger "effective emittance" (i.e. the elliptical envelope of a real emittance figure, cf. fig. 1.1). Such distortions will affect ion beam transport in ion-optical systems or experimental geometries. The ion-optical properties of such systems can be characterized by their "acceptance" according to the same criteria as for the ion beam emittance.

1.2.2 ECRIS — Electron Cyclotron Resonance Ion Sources

An ECRIS capable of efficient MCI production consists of the main components shown in fig. 1.2. A discharge chamber filled with the working gas (pressure of typically = 10^{-5} mbar; for ion production from non-volatile substances, see below) is immersed in a "min-B" magnetic field geometry (i.e. the magnetic field strength increases from the center outward), which provides the

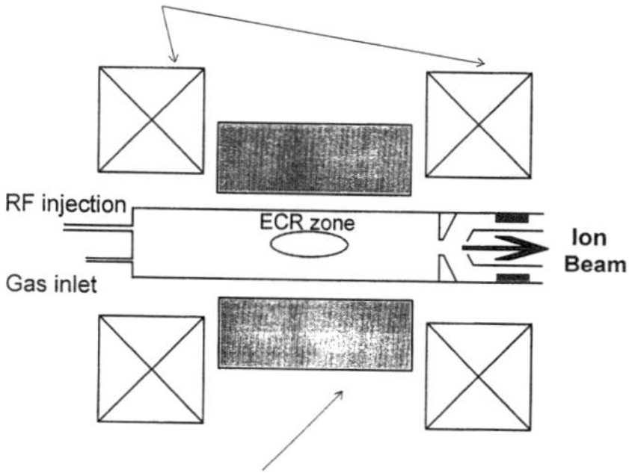

FIGURE 1.2 Main components of an ECRIS for MCI production (cf. text).

necessary ion (plasma) confinement. Such magnetic field configurations can be generated by sets of solenoids (which for larger setups may be superconducting) combined with a multipolar magnetic field produced from coils carrying high current (Joffe bars) or, for more compact constructions, permanent magnets (Sm-Co, Fe-Nd-B). Microwaves with a frequency of up to 28 GHz nowadays are fed into the discharge chamber to produce a plasma, wherein the electrons are preferentially heated by electron cyclotron resonance (ECR) when drifting through the so-called ECR zone (cf. fig. 1.2), where the electron cyclotron frequency $\nu_{ec} = \frac{e \cdot B}{2\pi \cdot m_e}$ matches the applied microwave frequency (e and m_e are electron charge and -mass, respectively, and B is the local magnetic field). Some electrons achieve rather high energy (up to hundreds of keV) and can ionize very efficiently even highly charged ions, which themselves stay rather cold. This provides effective step-by-step electron impact ionization conditions up to high q values and permits extraction of high-quality MCI beams.

Different techniques have been developed for producing MCI from solid elements or compounds; e.g., sputtering or melting from suitable electrodes brought near to the ECR zone or vaporization from a little oven inside the discharge chamber. Stable operation of ECRIS calls for preferential replenishment of the plasma electrons, which in the early ECR types was achieved by a first discharge stage featuring higher gas pressure. More recently, electrons are delivered either by hot filaments or by ion-induced electron emission from oxide-coated inner walls or probes negatively biased with respect to the ECR plasma.

Since its first practical realization in the late sixties by R. Geller and coworkers in Grenoble, France (cf. Bliman et al. [4]), the ECRIS has undergone a rapid development toward its now already high degree of maturity (more technical details will be given by G. Melin in chapter 2). Modern ECRIS are typically operated with ECR frequencies of 10 - 15 GHz, at a power level of several hundred Watts, and can produce, e.g., c.w. beams of 4 μA of Ar^{16+} or 12 μA of U^{33+} (see G. Melin in chapter 2). The kinetic energy spread of the extracted ions is typically between 5 and 10 eV times the charge state q. Emittance measurements for $5 \times q$ keV N^{4+} or Ar^{8+} ions extracted from a single-stage 5 GHz ECRIS (Leitner et al. [5]) delivered values for $\varepsilon_{95\%}$ of well below $100\pi \cdot$ mm $\cdot$ mrad (Brandstötter [6]). Of special interest to experimentalists is the present trend to rather compact, all-permanent magnet ECRIS, which consume astonishingly low microwave power. For example, *Nanogan 10 GHz* (Sortais [7]) can produce 10 $e\mu A$ of Ar^{8+} with only 1 W microwave power, which can be delivered by low-cost, rugged, solid-state microwave transmitters. Vacuum pumping is provided from the extraction region only, thus superseeding a pump on the high voltage terminal.

1.2.3 EBIS - Electron Beam Ion Sources

Whereas ECRIS produce MCI from magnetically confined plasmas, EBIS and EBIT devices utilize high-density electron beams for both ion confinement and ionization. The EBIS principle was introduced around 1965 by E.D. Donets [8]. An appropriate electron beam (see fig. 1.3: typical length about one m, diameter some ten to hundred μm, current densities of some hundred to several thousand A/cm^2) guided inside a series of drift tubes set to variable potentials is focussed by a strong magnetic field of typically 1 T, delivered by a long (usually superconducting) solenoid. This electron beam, by virtue of its strong negative space charge, provides excellent confinement for positive ions as long as suitable positive potential wells are applied on both ends (cf. fig. 1.3). This confinement can last for a few milliseconds up to many seconds, until the full space-charge compensation of the electron beam is reached. Ions brought into the electron beam during the confinement time, either by gas injection or from suitable external beam sources, will be ionized in a step-by-step manner. The lower the surrounding background gas pressure, the more efficiently this ionization proceeds. The background pressure can be kept below 10^{-10} mbar by cryogenic pumping on the inner walls of the superconducting solenoid. As a special feature, at any time during ion confinement only narrow groups of MCI charge states are present. With the EBIS used in short-pulsed fashion, desired MCI are extracted by rapidly lowering one axial potential barrier (extraction times typically 5 - 50 μs, cf. fig. 1.3). In this mode an EBIS is also excellently suited as for use as an MCI injector for heavy ion synchrotrons. On the other hand, it can also be run c.w. in the so-called "leaky mode", where it delivers less highly charged ions than in the (short-) pulsed mode. Existing EBIS are capable of producing up to fully stripped Xe^{54+} ions (about 10^4 per pulse), or fully stripped Ar^{18+} (about 10^8 per second; Donets [9]). However, there is still considerable potential for further development, which requires systematic studies of relevant properties of EBIS.

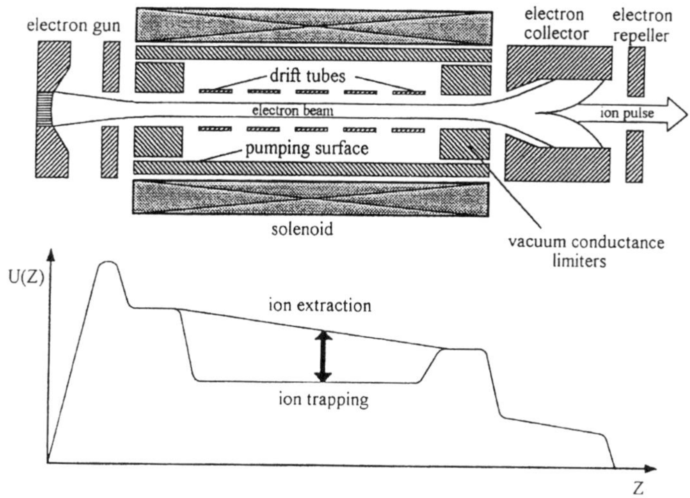

FIGURE 1.3 Main components of an EBIS for MCI production (cf. text).

In particular, the emittance of MCI beams from EBIS (Schmieder [10]) should be studied more systematically. According to Becker [11], typically EBIS emittances of $10\pi \cdot \text{mm} \cdot \text{mrad}$ can be expected, which values would be considerably smaller than for ECRIS (see above). Technical details on the construction and operation of EBIS are presented in chapter 3 by M. Stöckli.

1.2.4 EBIT - Electron Beam Ion Traps

EBIS and EBIT feature as common working principle the step-by-step ionization of ions trapped in the space charge of dense electron beams. There are also characteristic differences. First of all, the EBIT as pioneered by Marrs et al. [12] involves a much shorter electron beam than an EBIS (only a few cm, cf. fig. 1.4), which is easier to realize and probably more stable, giving rise to a comparably more efficient ion confinement. Second, the EBIT has essentially been conceived as a device for studying the radiation from trapped MCI (cf. fig. 1.4a). Currently the most advanced EBIT version (Lawrence Livermore National Laboratory "superEBIT" involving a 200 keV electron beam) permits studies on electron impact excitation and ionization for up to *H-like* U^{91+} (Marrs et al. [13]). For any EBIT, the produced MCI may also be extracted along the electron beam direction, as was first demonstrated by Schneider et al. [14]. In this case, similar ion charge spectra and MCI yields are obtained as from the EBIS (fig. 1.4b). External injection of low-charge ions from non-gaseous elements or compounds can be applied in a similar way as for the EBIS; e.g.; from a metal vapour vacuum arc source ("MEVVA"; Brown [2]).

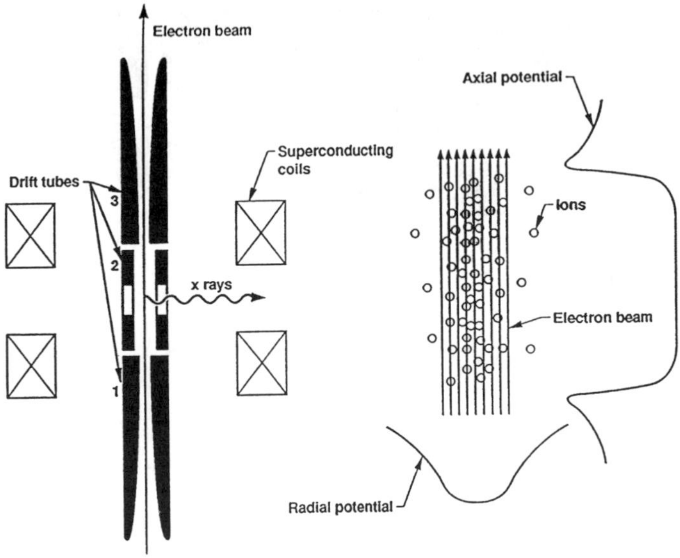

FIGURE 1.4 (a) Basic setup (left side) and ion trapping potentials (right side) for an EBIT. Reprinted from Steven R. Elliot, "Studies of highly charged ions with EBIT and super-Ebit", *Nucl. Instrum. Meth.* **B98** (1995) 114, with kind permission of Elsevier Science - NL, Sara Burgerhartstraat 25, 1055 KV Amsterdam, The Netherlands (courtesy of D. Schneider).

Summing up, the EBIT is a very compact device, comparable in size, technology and costs to a medium-performance electron microscope, which can produce up to fully stripped ions of virtually all elements and thus holds great promise for studies involving very highly charged ions.

1.3 EXPERIMENTAL APPLICATIONS OF SLOW MCI BEAMS

MCI production and related applications have been regularly covered in topical sessions of ion accelerator conferences. In 1982 the Research Institute of Physics in Stockholm, Sweden organized a symposium which was exclusively dedicated to the "Production and Physics of Highly Charged Ions" (Liljeby [16]). This first "HCI conference" initiated a successful biannual series of similar meetings which took place in Oxford, UK (1984), Groningen, The Netherlands (1986), Grenoble, France (1988), Gießen, Germany (1990), Manhattan, USA (1992), and Vienna, Austria (1994). Proceedings of the most recent HCI conference [1] provide a good impression on the healthy growth of atomic collision studies involving MCI, with about two hundred and forty

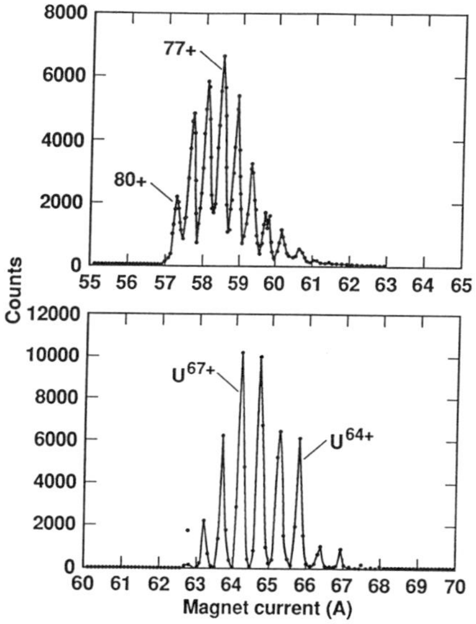

FIGURE 1.4 (b) Charge state spectra of Th and U ions extracted from the LLNL EBIT. Reprinted from [15] with kind permission of Gordon and Breach Publishers, courtesy of D. Schneider).

participants from twenty seven countries discussing practically all aspects of MCI production, properties, and experimental applications. In the following pages we present a number of interesting examples, with special emphasis on experimental techniques.

1.3.1 Studies of MCI structure and electron collisions

Here we can be rather brief, since nowadays the interactions of MCI with electrons and/or photons are studied most conveniently and precisely in heavy ion storage rings (cf. chapter 5 by A. Müller and A. Wolf). Such investigations give excellent access to the detailed electronic structure of MCI as well as their excited state lifetimes (see also chapter 17 by E. Träbert). Heavy ion storage rings with electron coolers permit highly-resolved measurements on electron collisions resulting in excitation, ionization and recombination of stored MCI.

In the following pages we present some other examples where no such techniques are involved.

Total cross sections for electron impact ionization of MCI have been measured by means of the crossed beams method at University of Gießen for about 15 years (Müller et al. [17]). Initially, a fixed perpendicular crossing of MCI- and electron beams was applied, but more recently the "animated beam method", first devised by Defrance et al. [18], was adopted (Müller et

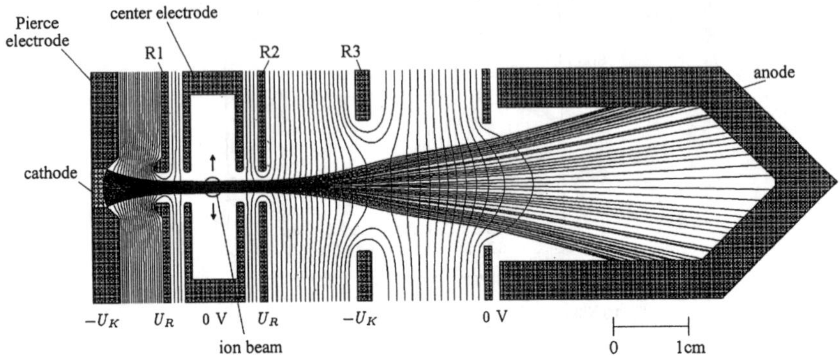

FIGURE 1.5 (a) Longitudinal section of setup by Stenke et al., for crossed animated beams measurements of total electron impact ionization cross sections of MCI. C: cathode of electron gun, R1 - R3: ring electrodes, CE: center electrode; IB: space-fixed MCI beam, A: water-cooled electron collector. Reprinted from [20] with kind permission of Elsevier Science - NL, Sara Burgerhartstraat 25, 1055 KV Amsterdam, The Netherlands (courtesy of E. Salzborn).

al. [19], Stenke et al. [20]). An intense, ribbon-shaped electron beam is mechanically swept with constant velocity across a spatially fixed MCI beam, which in most cases is delivered by an ECRIS. Absolute electron impact ionization cross sections can be determined from the total currents of the MCI and electron beams, the number of ionized ions accumulated through preset time intervals, the velocity of the animated electron beam, and an apparative factor involving the electron and MCI velocities and their overlap geometry (Müller et al. [19]). Fig. 1.5a depicts a section through the most recent version of this experimental setup, and fig. 1.5b presents absolute total cross sections for single ionization of W^{10+} ions at electron impact energies from threshold up to 5 keV (Stenke et al. [21]).

With the availability of intense MCI beams from ECRIS, reliable cross section measurements for elastic and inelastic electron scattering on MCI have become possible. Fig. 1.6 shows an apparatus for studies of MCI excitation, also at low electron impact energy, where backward electron scattering becomes dominant. Recently, with this setup Jalabert et al. [22] have investigated $Ar^{6+}(3s^{2\,1}S$ - $3s3p\,^1P)$ electron impact excitation at collision energies from 27 to 100 eV.

MCI beams from EBIS are generally too weak (see chapter 3) to be usable in crossed-beam studies with electrons as described above. However, electronic structure, electron impact excitation and ionization, and electronic recombination of highly charged ions can be investigated by observation of the related x-ray photon emission from the trapping region of an EBIT (for a recent review cf. Elliott 1995). As an example, electron impact ionization up to

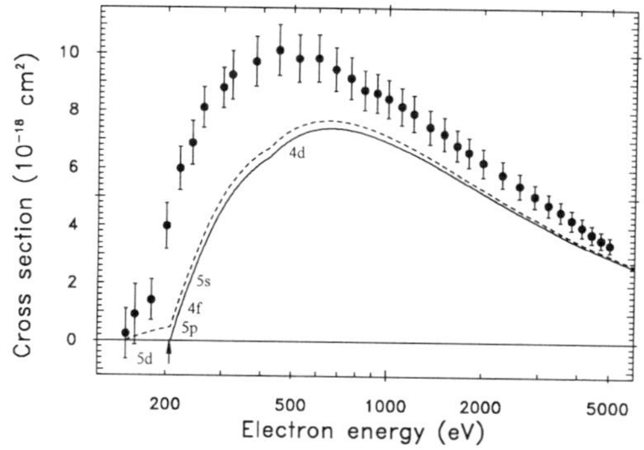

FIGURE 1.5 (b) Total cross section for electron-impact ionization of W^{10+} ions (Stenke et al. [21]). Error bars indicate total experimental errors, and solid curve shows the result of a model calculation. Reprinted from M Stenke, K Aichele, D Hathiramani, G Hofmann, M Steidl, R Völpel and E Salzborn, *J. Phys. B: At. Mol. Opt. Phys.* **28** (1995) 2711, with kind permission of IOP Publishing Limited (courtesy of E. Salzborn).

bare uranium was studied by means of x-ray emission due to K-shell capture into *H-like* U^{91+} and bare U^{92+}, respectively (Marrs et al. [13]). Furthermore, tests on QED predictions have been made via precise Doppler-shift-free measurements for the $2s_{1/2}$ - $3p_{3/2}$ electric dipole transition energies for ions U^{82+} through U^{89+} (Beiersdorfer et al. [23]), and dielectronic recombination of highly charged ions has been studied by shifting the electron beam energy through corresponding resonance values.

High-precision ion mass spectroscopy at the $\Delta M/M = 10^{-9}$ level can nowadays be achieved by means of MCI trapping techniques. Increased mass resolution is possible due to the higher ion cyclotron frequencies $\omega_c = qB/M$, which at given ion mass number M follow the ion charge state q. Moreover, trapping of different ion species with equal q/M can be applied (Carlberg et al. [24]). Fig. 1.7 depicts the "Stockholm-Mainz Ion Levitation Experiment (SMILETRAP)", which involves a Penning MCI trap in a superconducting solenoid for the B field. The "precision trap" is fed via a "pre-trap" from the Stockholm cryogenic EBIS ("CRYSIS"; Beebe et al. [25]). The pre-trap assures higher precision by precooling of MCI before their transfer into the precision trap, wherein the actual spectroscopy studies are conducted. In the upper left part of fig. 1.7, a preliminary measurement with fully stripped Ar^{18+} ions is indicated (Carlberg et al. [24], Kluge [26]).

1.3.2 MCI collisions with neutrals and ions

Experimental investigation of MCI collisions with neutral atoms and molecules is treated in different chapters of this book (see chapter 4 by

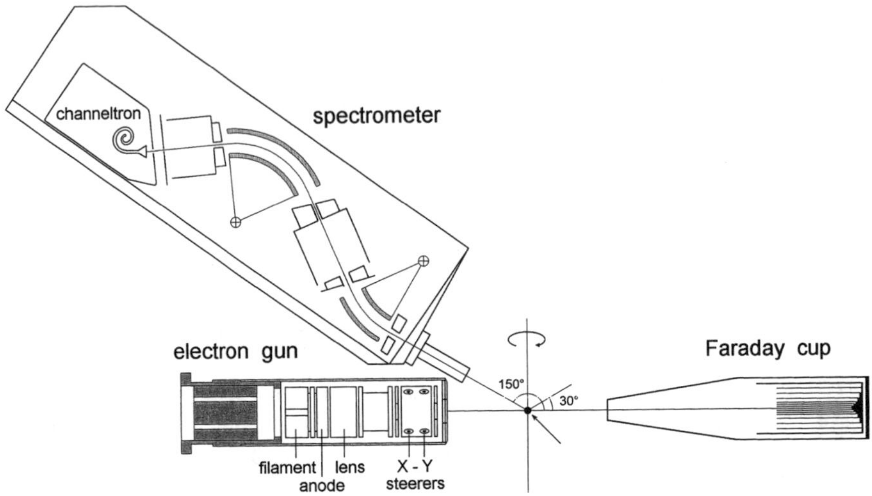

FIGURE 1.6 Experimental setup by Jalabert et al. [22] for studying elastic and inelastic MCI—electron collisions. The MCI beam from an ECRIS of the Grenoble AIM facility (perpendicular to the drawing plane) enters the collision chamber (pressure $< 10^{-8}$ Pa), where it crosses the electron beam (energy from a few eV up to hundreds of eV, electron current some μA) at a variable intersection angle (45° — 135°). Scattered electrons are analyzed by means of a rotable electron spectrometer mounted 30° out of the electron–ion beam plane. Reprinted from D. Jalabert, C. Ristori, M. Maurel, J. C. Rocco, B. H. Huber, "Electron impact excitation of Ar^{6+} ($3s^2\ ^1S \rightarrow 3s3p^1P$)", *Nucl. Instrum. Meth.* **B98** (1995) 132, with kind permission of Elsevier Science - NL, Sara Burgerhartstraat 25, 1055 KV Amsterdam, The Netherlands (courtesy of B.A. Huber).

C.C. Havener on "Merged beams", chapter 5 by A. Müller and A. Wolf on "Heavy ion storage rings", chapter 13 by T. Zouros and D. Lee on "Electron spectroscopy", chapter 23 by H. Schmidt-Böcking on "Recoil ion momentum spectroscopy", and chapter 22 by E. Kamber on "Translational energy spectroscopy"). Since most of these deal primarily with relatively fast MCI, we will confine ourselves to comparably slower MCI (impact velocity ≤ 1 a.u.). Inelastic collisions between such slow singly charged or multicharged ions and neutral target particles (atoms, molecules) can give rise to various electron transfer processes, such as single electron capture (SEC) and for multi-electron targets also double- (DEC) or even multiple-electron capture. "True double electron capture" (TDC) is relevant only at low q, whereas DEC combined with projectile autoionization (ADC, also called transfer ionization) can generally be as important as SEC. Furthermore, both projectile ions and target particles can become excited or even ionized, but at low impact energy these latter processes are less important than electron transfer.

If for respective cross section measurements the initial and final states of

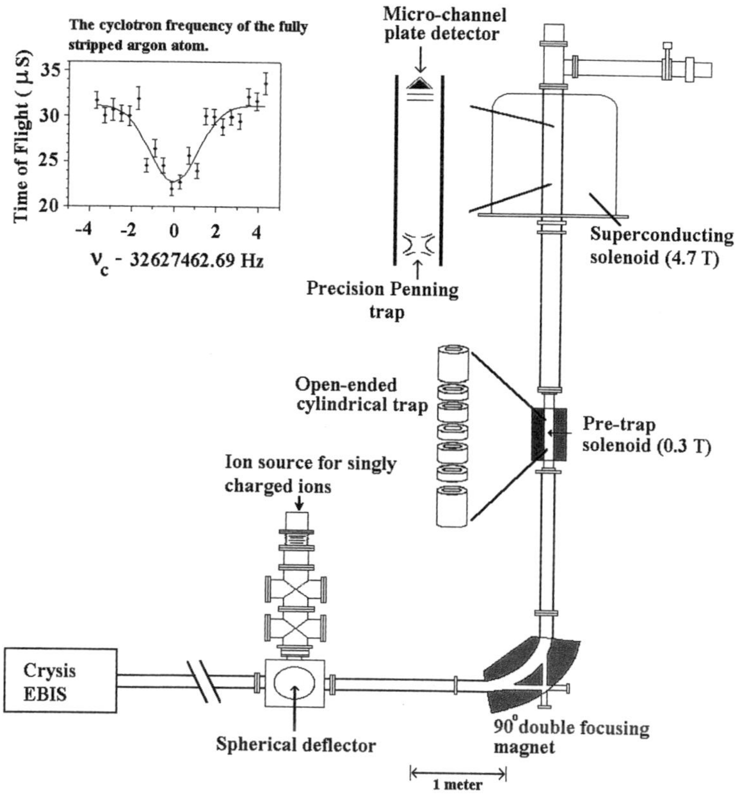

FIGURE 1.7 Schematic view of "SMILETRAP" , showing the Stockholm CRYSIS, the "pre-trap" and the precision trap (cf. text). In the upper left hand corner a cyclotron resonance curve for trapped Ar^{18+} is indicated, with the resonance width yielding a mass resolution of $> 1.2 \times 10^7$ (Carlberg et al. [24]; courtesy of R. Schuch).

colliding particles are well defined and (practically) all scattered reaction products are detected, we speak of "state-to-state selective total cross sections". Variation of the projectile scattering angle gives access to the respective "differential cross sections". It should be emphasized that important cross sections are always related to occurence of electronic transitions at relatively large internuclear distances, for which (except at rather low impact energy) the projectiles are scattered into the near-forward direction.

For total cross section measurements, depending on the extent of state definition, we may further distinguish "initial-state selective" and "final-state selective" apparent total cross sections. More commonly, however, the former and the latter are simply named "total cross sections" and "state-selective cross sections", respectively. At low impact energy, total cross sections for both SEC and DEC strongly depend on the initial states of both reaction partners, so measurements without proper initial state definition are not really meaningful. In particular, metastable fractions of low-q MCI beams need to be taken into account carefully [27]. If the primary MCI species cannot form long-lived excited states, and if neutral target particles are probably in their electronic (and/or vibronic) ground states, the respective initial states are well defined and thenceforth final-state selective measurements suffice to provide complete state-to-state selectivity. However, clear separation between SEC, TDC and ADC will require coincident detection of different final reaction products.

Neutral target particles are provided either from static gas targets, as for inert species like H_2 and He, as dissociated or reactive species like atomic hydrogen (see below), or as slow or fast neutral beams. For total cross section measurements the MCI beam crosses the target gas cell, or neutral target particles can be fed into an ion trapping region.

Atomic hydrogen is commonly produced by means of a hot furnace, with its dissociation degree determined by comparing single and double electron capture reactions. For state-selective measurements involving photon emission spectroscopy (PES) or translational energy spectroscopy (TES), a low energy target particle beam can be crossed with the MCI beam. For such purposes, inert neutral particles can be effused from a nozzle into the interaction region, and reactive beams can be provided by effusion from a hot furnace or from RF or microwave discharges. The merged beams technique applies neutral beams resulting from charge exchange from positive ions, or from photodetachment from negative ions. Furthermore, highly excited Rydberg state atom beams can be obtained via collisional detachment from negative ions.

Reaction products are positive ions, neutral atoms, electrons, and photons, which can also be detected in coincidence. Absolute detection of photons or neutral particles is comparably difficult, especially for low energy. The accuracy of measured cross sections is strongly related to the experimental technique, with errors usually resulting mainly from inaccurately known fluxes and compositions of the primary collision partners, and from insufficiently defined collision geometries and collection and detection efficiencies for the various reaction products.

Total cross section measurements

Absolute total cross sections are particularly important for calibrating the (usually) relative data from corresponding state-selective EC cross sections (see below). Toward low impact energy the collection efficiency of scattered and/or charge-exchanged projectiles deserves special attention and, in particular, can vary for different reaction channels. This difficulty can be avoided with the merged beams technique (cf. chapter 4 by C.C. Havener).

If SEC- and DEC total cross sections are of comparable importance, coincident detection of scattered projectiles and recoiled target particles is necessary (see, e.g., Wu et al. [28]). Coincident detection permits not only separation of SEC, TDC and ADC reactions (see above), but also distinction of the latter from single- and double ionization processes. With highly charged iodine ions (q =10, 15) from an EBIS (NICE), Nakamura et al. [29] have recently measured absolute total cross sections for multiple electron capture from noble gas atoms by determining the charge of scattered projectiles in coincidence with differently charged recoil ions. Multiple ionization of Ar by multicharged projectile ions has been investigated by de Nijs et al. [30] using a reflectron-type spectrometer for separating different recoil ion charge states according to their different times-of-flight.

Trapping of MCI in electric and magnetic fields enables measurement of electron capture rate coefficients for very low impact energies (for a review of respective methods cf. Church 1993). After admission of neutral reaction partners into the MCI trapping region, the resulting decrease of MCI storage time can be related to the relevant charge-changing processes. Additional laser cooling of the trapped ions permits extension toward still lower impact energy.

Measurement of state-selective cross sections

A number of techniques are nowadays available, which in the following paragraphs will be described briefly:

Translational energy spectroscopy (TES). TES (cf. chapter 22 by E. Kamber, and H.B. Gilbody [31]) makes use of the kinetic energy gain or loss of projectile ions scattered into near-forward directions. It is a direct and rather sensitive method since single particle counting can be applied and, in contrast to PES and electron spectroscopy (see below), reaction channels involving final ground or long-lived excited states are also covered. Its state-selectivity is defined by the kinetic-energy resolution of the electrostatic or TOF analyzers that have to be applied for kinetic energy spectroscopy of the inelastically scattered projectiles. The accurate definition of the scattering cone for projectile detection is especially important. TES also permits studies with state-selected, long-lived excited multicharged projectiles, which can be produced with a charge-exchange cell followed by an ion monochromator in front of the main collision region (Huber and Kahlert [32]). TES has recently also been utilized for separating SEC from ADC by slow, highly charged Xe ions ($25 < q < 44$) from the Stockholm EBIS "CRYSIS" in collisions with He atoms (Cederquist et al. [33]). Detailed measurements on various single and double electron capture channels for different projectile scattering angles have been conducted by Barat and Roncin [34], who used the instrument shown in fig. 1.8. The MCI beam crosses an effusive gas target, from which the recoil ions are extracted toward a multichannel plate (MCP) detector with a time-to-amplitude converter for ion charge state determination. Post-collisional

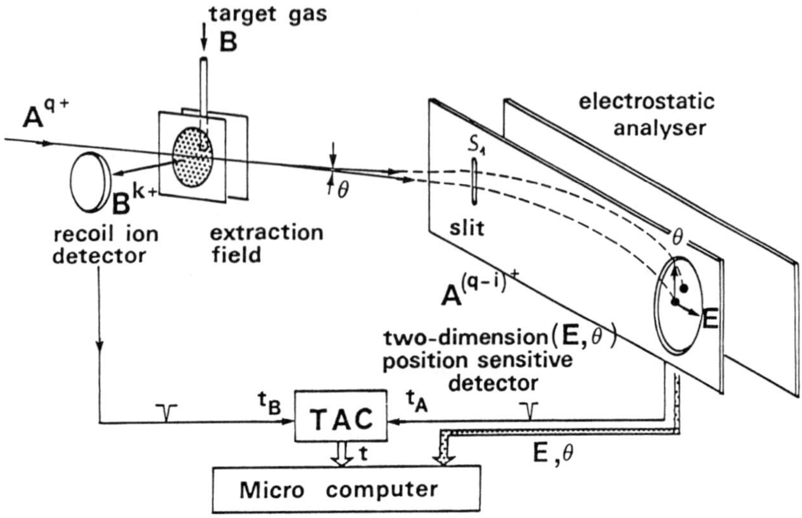

FIGURE 1.8 Apparatus by Barat and Roncin [34] for coincidence measurements between scattered projectile ion energy and recoil ion charge state. Reprinted from M. Barat and P. Roncin, "Multiple electron capture by highly charged ions at keV energies," *J. Phys. B* **25** (1992) 2205, with kind permission of IOP Publishing Limited (courtesy of M. Barat).

projectiles are selected according to their kinetic energy and scattering angle by an electrostatic parallel plate analyzer with a two-dimensional position sensitive MCP detector.

Fig. 1.9 shows illustrative results obtained with this instrument from Gaboriaud et al. [35] for impact of 3 keV F^{6+} or Ne^{7+} ions on He or Ne, respectively. The data have been depicted as three-dimensional profiles vs. the projectile ion energy gain ΔE and the projectile ion scattering angle θ, respectively. The main reaction channels are single electron capture (marked by "SC") and transfer excitation (marked by "TE"). In a TE process a target electron is captured into the MCI and another electron, either in the target or the projectile, is excited into a higher orbital.

Photon emission spectroscopy The final-state composition of projectiles after electron capture or collisional excitation can also be determined from the related photon emission in the visible, VUV or XUV regions. Corresponding absolute cross sections are measured with absolutely-calibrated photon spectrometers (for experimental details see Hoekstra et al. [39]; Folkerts et al. [37]). Fig. 1.10 shows the most recent setup of the KVI Groningen group with visible and VUV spectrometers, which are both calibrated via electron impact excitation of atoms or molecules, for which purpose an electron gun can be placed into the ion beam axis. The target particles are supplied through the capillary of a RF discharge, which is only activated for produc-

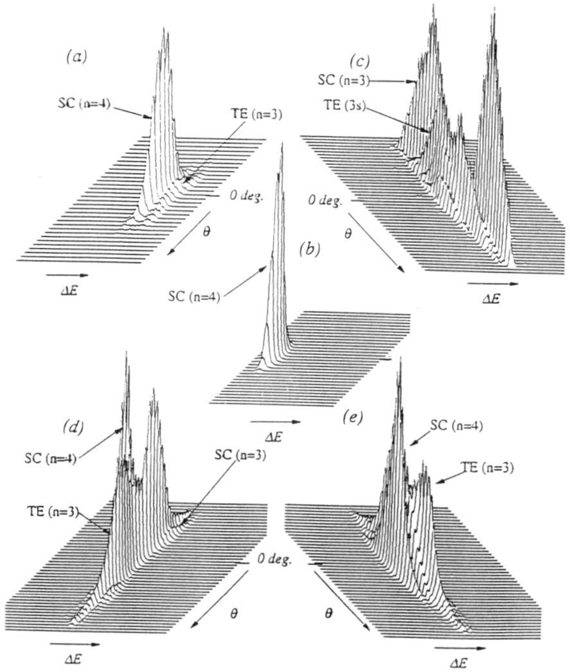

FIGURE 1.9 Projectile ion-recoil ion coincidence data from Gaboriaud et al. [35] for single electron capture ("SC") and transfer excitation (i.e. single electron capture accompanied by projectile or target excitation - "TE") in collisions of 3 keV F^{6+} or Ne^{7+} with He or Ne. Arrow lengths for kinetic energy and scattering angle correspond to 50 eV and 1 degree, respectively. (a - e) refer to the following systems: (a) Ne^{7+} — He collisions; (b) F^{6+} — He collisions; (c) Ne^{7+} — Ne collisions; (d, e) F^{6+} — Ne collisions. Reprinted from M. N. Garboriaud, M. Barat, P. Roncin, and V. Sidis, "Transfer excitation in F^{6+} and Ne^{7+} ions colliding with He and Ne targets," *J. Phys. B* **27** (1994) 4595, with kind permission of IOP Publishing Limited (courtesy of M. Barat).

tion of reactive particles, e.g. atomic hydrogen. Alternatively, an alkali atom beam can be supplied from a hot oven. Selection of hydrogen-like (i.e. energy-degenerate) final states populated by SEC into fully stripped ions is achieved by utilizing their different lifetimes, for which the PES observation field can be shifted with a rotable mirror along the ion beam direction, as indicated in fig. 1.10 for He^{2+} — He collisions (Hoekstra [36]).

The PES technique permits considerably higher state resolution than TES, but is much less sensitive because photon detection is necessary. Due to this fact, the PES technique can only be utilized if sufficiently high MCI fluxes are provided (typical MCI currents of 100 nA are needed), as deliverable from ECRIS. The accuracy of PES data depends mainly on the absolute spectrom-

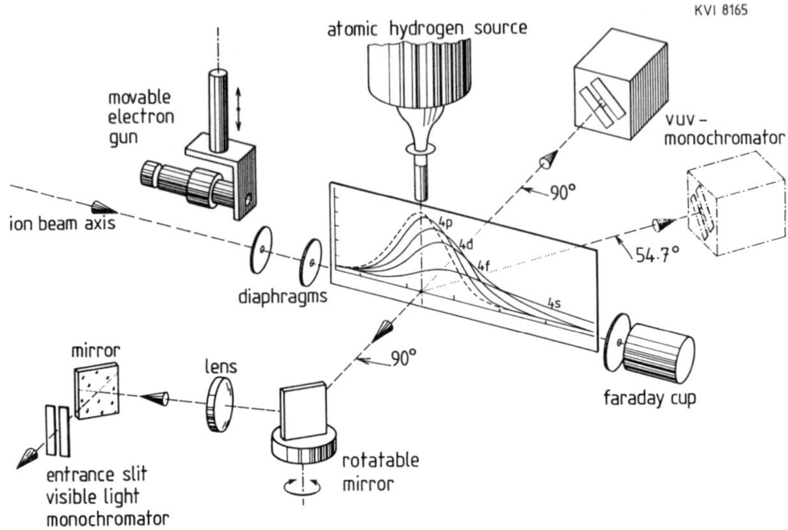

FIGURE 1.10 Schematic view of the PES setup at KVI Groningen. Insert demonstrates selection of energy-degenerate H-like final ion states, which are, e.g., populated in He^{2+} — He collisions, as explained in text (Hoekstra et al. [39]. Reprinted from R. Hoekstra, F. J. Heer and R. Morganstern, "State-selective electron capture in collisions of He^{2+} with H", *J. Phys. B* **27** (1994) 4595, with kind permission of IOP Publishing Limited. Courtesy of R. Morgenstern.

eter/detector calibration and the definition of photon emission characteristics (spatial distribution and polarization) .

Furthermore, a clear correlation of observed photons with originally excited states (influence of cascade population and branching) is mandatory. Unfortunately, PES cannot deliver information from ground state- or long-lived excited projectile states. Generally, TES and PES are complementary methods; if applied together, they can be quite helpful for studying the more complicated collision systems. This has been demonstrated for SEC in C^{3+} — H(1s) collisions (PES by Ciric et al. [40], TES by Wilkie et al. [41]), and He^{2+} — Li(2s) collisions (PES by Hoekstra et al. [38], TES by Aumayr et al. [42]). In the latter case, application of PES was indispensable for achieving selectivity among the H-like energy-degenerated final projectile states (see above). TES and PES have also been applied together by Roncin et al. (1994) for measuring differential cross sections and orientation parameters in B^{3+} — He collisions.

Electron emission spectroscopy Spectroscopy of electrons resulting from the decay of doubly-excited autoionizing states, which are formed by double-electron capture from multi-electron neutrals into MCI, gives access to state-selective information on the involved reaction channels. Important

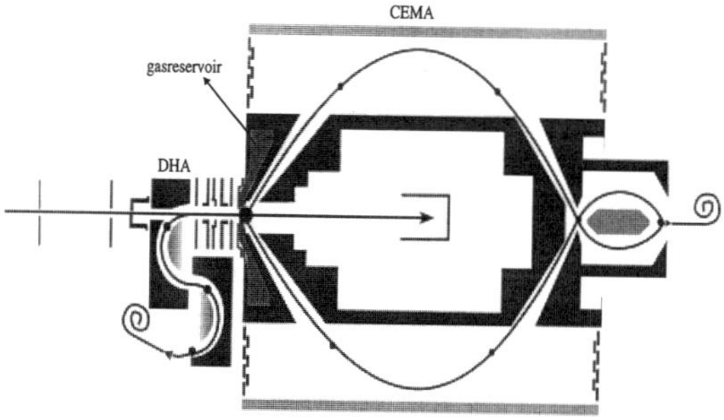

FIGURE 1.11 Electron energy spectrometer used by de Nijs et al. [44] for projectile ion — electron coincidence studies. Courtesy of R. Morgenstern.

studies in this field have been conducted, e.g., at HMI Berlin (Stolterfoht et al. [43]) and KVI Groningen (de Nijs et al. [44]). Fig. 1.11 shows the most recent apparatus of the latter group.

The primary MCI beam crosses a gas target from which the electrons emitted by autoionizing projectiles can enter toward the forward-scattering direction into a cylindrical mirror analyzer (CMA) . The resulting target ions are extracted backward into a double-hemispherical analyzer (DHA) . The charge states of post-collisional projectiles are determined from times-of-flight measured with the emitted electron and the target ion as the start and stop signals, respectively. Coincidence measurements between emitted electrons and post-collisional projectile charge states (see fig. 1.12 for impact of N^{7+} on Ar) clearly identify the reaction channels involved.

An interesting field of study concerns collisions of singly and multiply charged ions with laser-excited atoms (recent work has been reviewed by Schippers [45]).

MCI collisions with atoms in highly excited Rydberg states can result in very large electron capture cross sections. Such Rydberg atoms are produced by laser excitation. DePaola et al. [46] have measured absolute SEC cross sections for impact of Xe^{q+} from the KSU CRYEBIS ($q \leq 40$; see chapter 3 by M. Stöckli) on Rb($10f$) Rydberg atoms. At the lowest impact energies they found SEC cross sections of up to 3×10^{10} cm^2. Moreover, projectile-state-selective SEC from Rb($17p$) Rydberg atoms by 18 and 40 keV Kr^{8+} ions has been studied by Pesnelle et al. [47]. Fig. 1.13 shows the experimental setup, which is served by MCI from the AIM ECRIS in Grenoble. MCI cross at 90° the *Rb* atoms, which are excited into their 17p state by a second-harmonic generated 301.6 nm UV laser beam travelling anticollinear to the MCI beam. Highly excited Kr^{7+} ions, due to SEC from the Rb($17p$) atoms, are deflected by a first electrostatic spectrometer and enter a field-ionizing lens followed by a second electrostatic spectrometer. The field-ionized Kr^{8+} ions are registered according to their region of origin, from which their original principal quantum

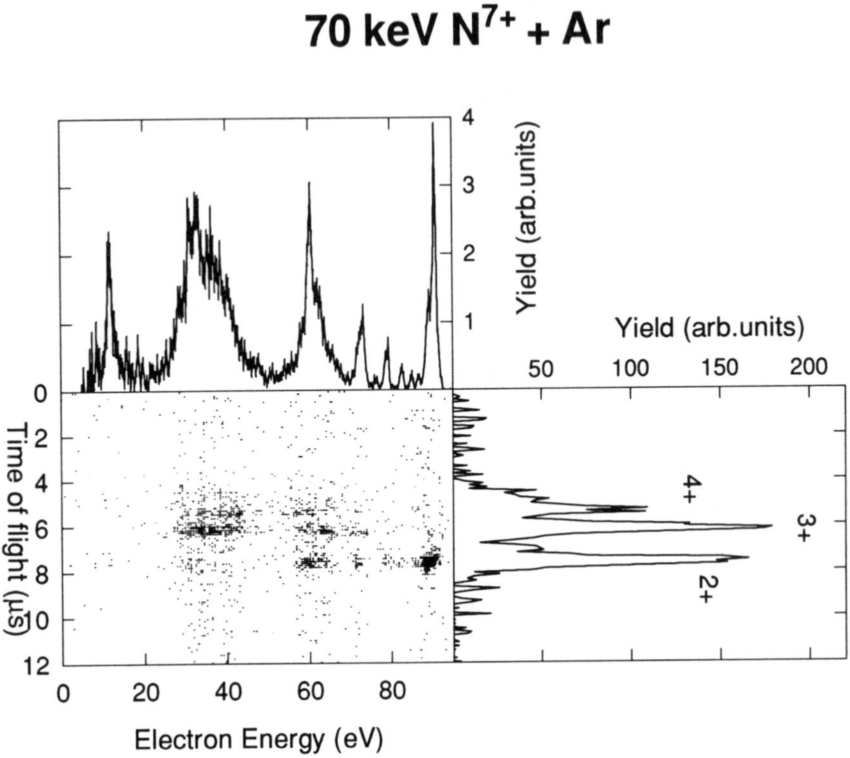

FIGURE 1.12 Ejected electron-projectile charge state coincidence "scattergram" for impact of 70 keV fully stripped N^{7+} on Ar. Reprinted from de Nijs, et al. [44], with kind permission of IOP Publishing; courtesy of R. Morgenstern).

number in the $q = 7$ state can be derived. Fig. 1.14 shows data for 40 keV MCI impact energy vs. the projectile quantum number n, with the relative measurements calibrated at $n = 90$ against a classical-trajectory Monte Carlo (CTMC) calculation.

Charge-changing ion-ion collisions are especially difficult to study, since ion beams provide a rather thin collision target and background signal due to ion collisions with residual gas molecules is a tremendous problem. Fig. 1.15 shows an experimental setup used for such measurements at the University of Giessen. Singly or multiply charged ion beams from two ECRIS cross each other at 17.5°. Post-collisional charge-changed ions are identified by an elaborate arrangement of electrostatic analyzers and deflector plates, by performing coincidence and anti-coincidence measurements for separation of the useful signal from the background. With this apparatus, total cross sections for resonant charge transfer in $^3He^{2+}$ — $^4He^{+}$ collisions at barycentric energies between 4 and 200 keV have been measured recently (Melchert et al. [48], cf.

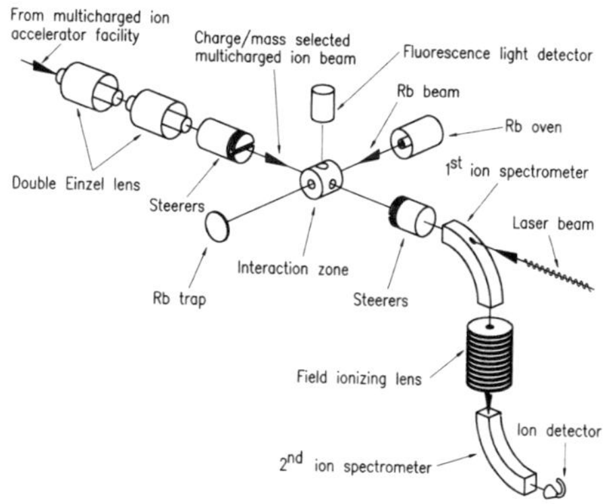

FIGURE 1.13 Schematic diagram of the apparatus of Pesnelle et al. for projectile-state-selective SEC measurements in MCI collisions with Rb Rydberg atoms (A. Pesnelle, J. Pascale, R. Trainham and J. J. Andrä, to be published in Phys. Rev. A).

fig. 1.16).

1.3.3 MCI collisions with clusters and solids

As a last field of typical applications for high-performance MCIS, we now briefly cover some recent studies on MCI interaction with clusters and solid surfaces. Such processes may be regarded, at least for some of the involved mechanisms, as a natural extension of MCI collisions with single neutral particles. This is especially true for collisions with neutral clusters, where electron capture plays a dominant role as it does for neutral atoms or molecules. In addition, such capture reactions can induce instabilities in the post-collisional clusters, due to Coulomb repulsion. This has recently been demonstrated in several attempts to produce multicharged fullerenes $(C_{60})^{q+}$ via multi-electron capture by MCI. Walch et al. [49] achieved $q = 6$ by impact of Ar^{8+} and Xe^{14+}, and Jin et al. [50] even $q = 9$ (the highest fullerene charge so far demonstrated) for impact of up to U^{46+}. In both sets of experiments ECRIS delivered the MCI beams. This was also the case in another recent study by Chandezon et al. [51] on the stability of $(Na_n)^{q+}$ clusters formed by collisions with H^+, O^{5+} or Ar^{8+}. Fig. 1.17 shows a comparison of the minimum sizes "n" of fivefold-charged Na clusters after impact of Ar^{8+} and H^+. For impact of MCI also somewhat smaller clusters can survive the collision, which is explained by less cluster heating in comparison to proton impact, since the MCI produce multicharged clusters at comparably larger impact parameters.

Some features of ion collisions with surfaces are treated in chapter 20 by R. Hippler. However, the situation for impact of slow MCI on a solid surface is rather special because of the large number of different electronic transitions involved. Such collisions can transiently form so-called "hollow atoms" , i.e.

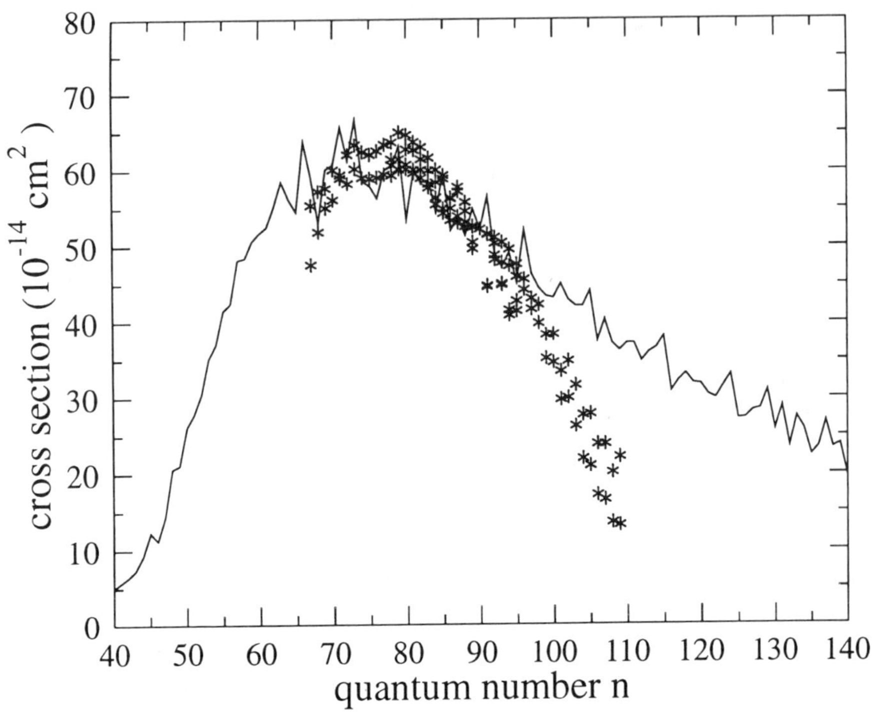

FIGURE 1.14 Experimental cross sections for SEC from Rb(17p) into Kr^{8+} at 40 keV vs. quantum number n defined within the classical field ionization model, *'s. The CTMC capture cross section calculated for each value of n is represented by a continuous line. The experimental cross section has been normalized to the CTMC calculations at n =90° [47] (courtesy of A. Pesnelle).

multiply excited neutral projectiles with most of their electrons in high-n shells while the lower-n shells temporarily remain empty (see fig. 1.18).

The currently-assumed scenario for interaction of slow MCI with metal surfaces has evolved from many experimental studies of the total electron yields, slow electron number statistics and fast Auger electron energy distributions, together with information from scattered projectile charge states and energies, and soft X-ray emission (see below). In a simple-minded approach, this scenario can be divided into three stages: the projectile's approach, its close contact with the surface, and its subsequent penetration into the target bulk (or backscattering into vacuum). The main properties of this scenario have been quantitatively explained with the so-called "classical over-the-barrier" (COB) model by Burgdörfer et al. [53]. According to the COB model, a slow MCI near a metal surface induces a collective response of the metal electrons, which at sufficiently large distance from the surface can be described by the

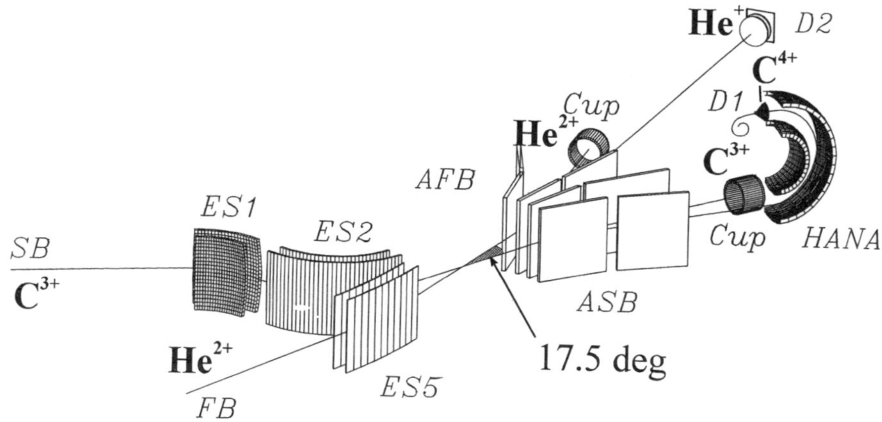

FIGURE 1.15 Apparatus at University of Giessen for measuring absolute cross sections for fast collisions between singly and/or multiply charged ions (here shown for SEC from C^{3+} in collisions with He^{2+}; final reaction products C^{4+} and He^{+} are detected in coincidence (courtesy of E. Salzborn).

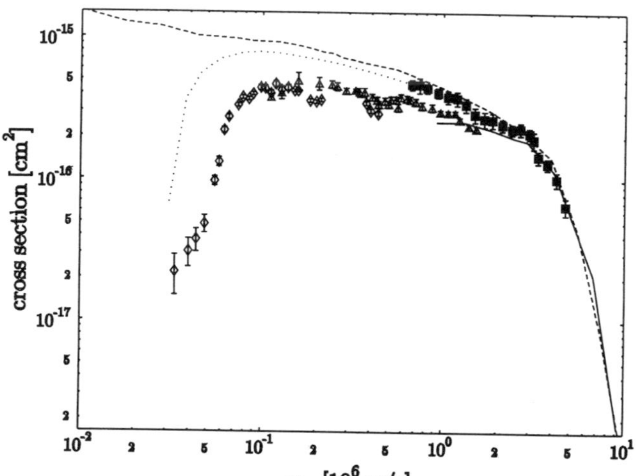

FIGURE 1.16 Total cross sections for electron transfer in He^{2+} — He^{+} collisions (Melchert et al. [48], full squares) in comparison with calculated (lines) and other experimental data. Reprinted from F Melchert, S Krüdener, R Schulze, S Petri, S Pfaff and E Salzborn, *J. Phys. B: At. Mol. Opt. Phys.* **28**, L355 (1995), with kind permission of IOP Publishing Limited (courtesy of E. Salzborn).

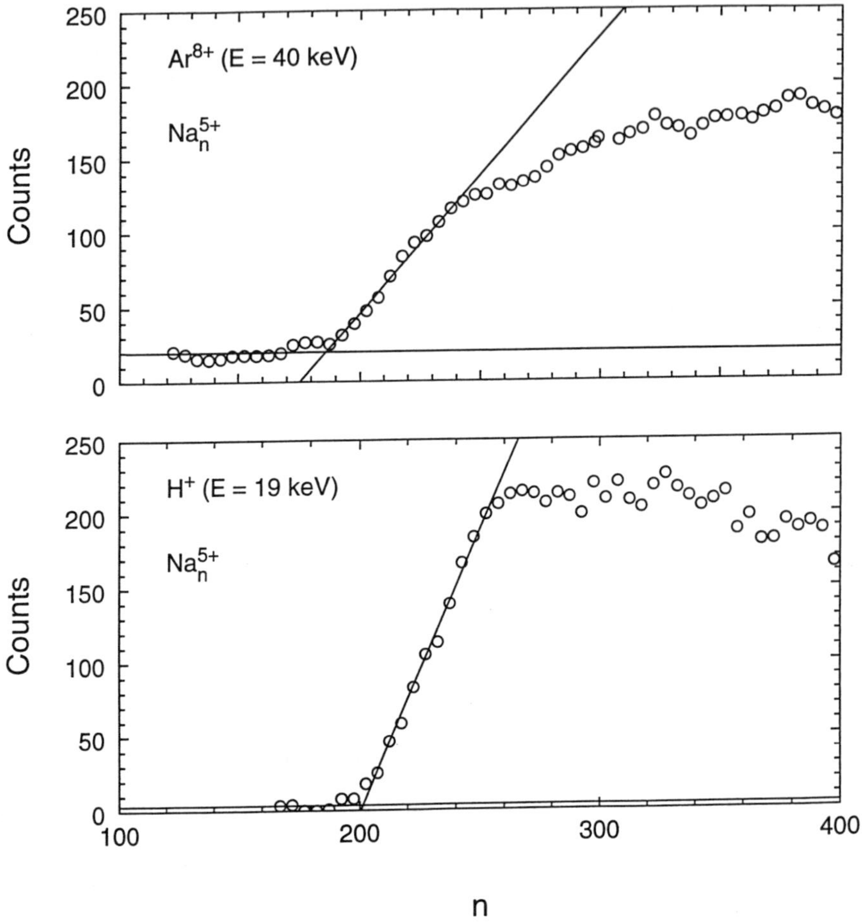

FIGURE 1.17 Size distribution of $(Na_n)^{5+}$ clusters after impact of Ar^{8+} (top) and H^+ (bottom), respectively (Chandezon et al. [51]; courtesy of B.A. Huber).

classical image potential. The latter accelerates the MCI towards the surface and thus sets a lower limit on the effective projectile impact velocity, corresponding to an upper limit for the MCI above-surface interaction time. The image charge acceleration has been demonstrated by several experiments. H. Winter et al. [54] showed, for grazing incidence of Xe^{q+} ($q \leq 33$) from the Stockholm CRYSIS on a clean flat Al(111) surface, that the outgoing trajectory of the neutral Xe atoms is steeper than the incoming MCI trajectory, because the image charge acceleration works only during the latter. In a similar investigation, Meyer et al. [56] applied I^{q+} ($q \leq 25$) and Pb^{q+} ($q \leq 36$)

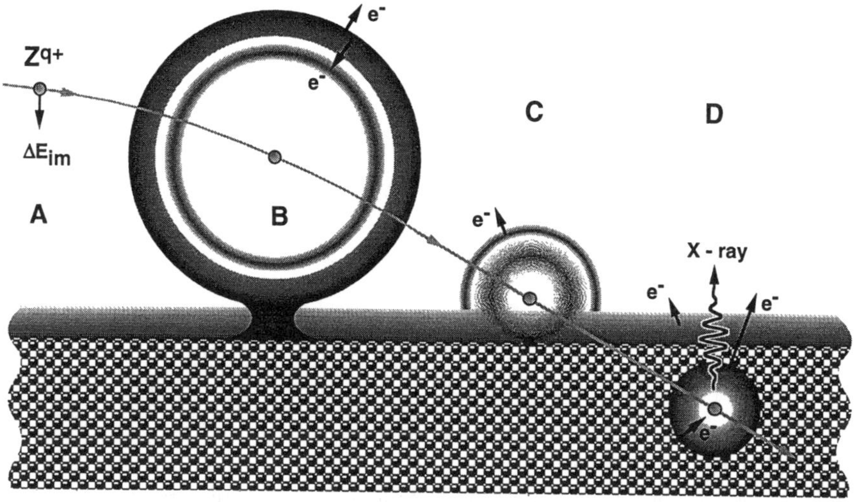

FIGURE 1.18 Development of hollow atom during slow MCI impact on a metal surface [52]. A..approaching MCI, B..hollow atom above surface, C..removal of outer electrons, D..final relaxation at/below surface.

from an ECRIS on Au(110). In a quite different approach, Aumayr et al. [55] showed that slow electron emission due to MCI impact (see below) saturates at low impact velocity. Their experiment involved highly charged Th^{q+} ions ($q \leq 80$) from the Lawrence Livermore National Laboratory EBIT (Schneider et al. [14]), and was subject to somewhat higher experimental errors than the former described scattering measurements.

Fig. 1.19 compares some of the above mentioned experimental results with predictions from a recently refined version of the COB model (Lemell et al. [57]), which gives even better agreement with the available measurements.

The image interaction shifts up the projectile electron states and enhances the height of the potential barrier between the MCI and the surface. However, inside a critical distance that depends only on the metal work function and the MCI charge state, electrons from the solid can be captured into highly excited projectile states (resonant neutralization). This capture goes on until complete neutralization of the MCI is achieved, corresponding to formation of the "primary hollow atom" (cf. fig. 1.18, stage B). However, the hollow atom immediately starts to autoionize. With increasing approach its levels are further up-shifted because of the image interaction as well as the screening of the projectile charge by already captured electrons. Both autoionization and up-shifting give rise to slow electron emission from the projectile and, in parallel, increasingly tightly bound projectile states are populated as soon as they fall into resonance with occupied electron states in the surface. On the other hand, electrons in up-shifted states can be recaptured into the metal via

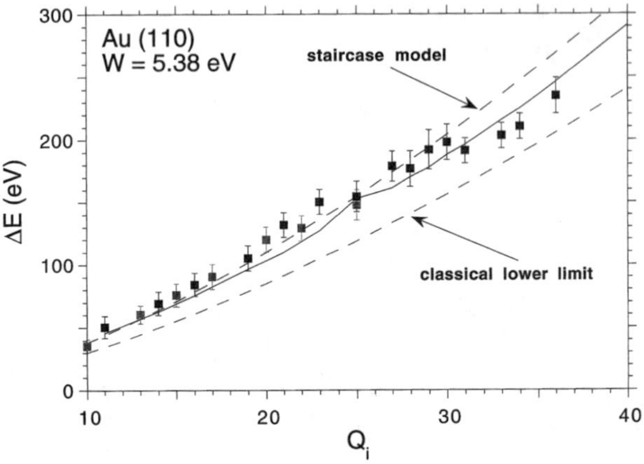

FIGURE 1.19 Comparison of COB calculations for the image charge acceleration of a MCI nearing a metal surface (dashed lines) with available experimental data (Lemell et al. [57]).

resonance ionization. This interplay of various electronic transitions continues until close contact with the surface, causing both electron emission and gradual shifting of electrons bound to the projectile into lower n-shells (stage C). All the involved projectile states are filled and again depopulated within a few femtoseconds, and are thus of a highly transient nature. Emitted electrons are rapidly replaced, so that the hollow atom stays neutral until its close contact with the surface, after which relaxation of the then-formed "secondary hollow atom" can proceed below the surface via three competing processes.

At low impact velocity, mainly Auger neutralization fills still-present projectile inner shell vacancies, whereas at higher impact velocity they are directly filled in close collisions with target atoms via quasimolecular transitions [52]. These sub-surface Auger neutralization processes account for most of the observed fast (projectile) Auger electrons. Fig. 1.20 shows Auger electron spectra obtained for impact of N^{6+} on Si(100) (top) and LiF(100) (bottom) (Limburg et al. [58]). As a particularly interesting feature, characteristic differences on the low energy side of these electron spectra point to a smaller probability for formation of hollow atoms above an insulator surface (here, LiF) as compared to semiconductor- (Si) or metal surfaces.

In competition with the just discussed fast Auger electron emission, for MCI with higher Z the transient inner shell vacancies can also be filled via characteristic X-ray emission (see chapter 11 of this book by J.-P. Briand).

In comparison to fast Auger electron emission, a much larger number of slow electrons will be emitted, primarily during stages B and C of the hollow atom formation. These slow electrons provide the bulk of the total electron yield resulting from slow MCI-surface interaction, which can be directly mea-

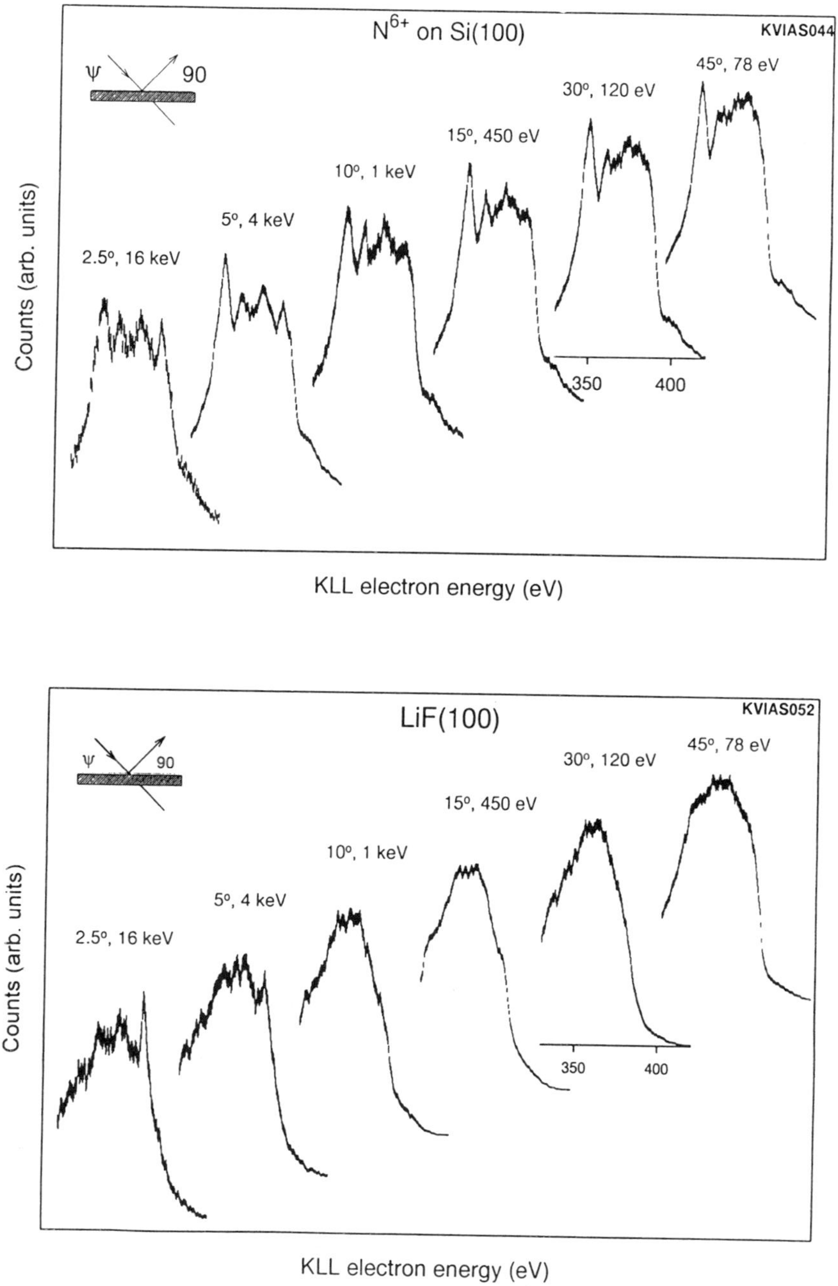

FIGURE 1.20 Comparison of fast Auger electron spectra resulting from impact of N^{6+} on Si(100) (top) and LiF(100), respectively (Limburg *et al.* [58]).

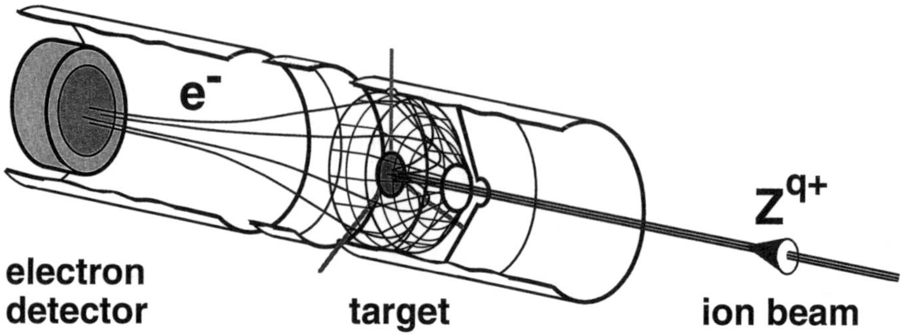

FIGURE 1.21 Setup for measuring electron emission number statistics (ES). Electrons from the target are accelerated by ca. 20 kV into a surface barrier detector, which delivers a pulse height spectrum in correspondence with the respective ES (Lakits et al. [60]).

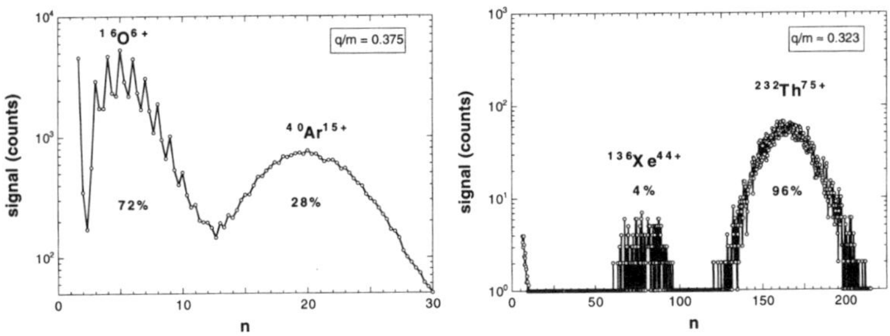

FIGURE 1.22 ES for impact of "mixed" MCI beams from EBIT on clean gold [61]. Left side: O^{6+}/Ar^{15+} beam ($q/m = 0.375$); right side: Xe^{44+}/Th^{75+} beam ($q/m = 0.323$).

sured from currents of impinging MCI and ejected electrons. Unfortunately, such measurements become practically impossible for higher q because of increasingly small MCI fluxes. An elegant way to circumvent this difficulty has been devised by Lakits et al. [59,60]: Instead of ion and electron currents one measures the electron emission number statistics (ES), which can be achieved with rather small MCI fluxes as delivered, e.g., by EBIT [55]. Since the mean value of the electron number statistics is the total electron yield, no absolute measurement of the incident MCI flux is needed. Fig. 1.21 shows an appropriate experimental setup for measuring electron number statistics resulting from MCI-surface collisions. As a further advantage, different MCI beam fractions with equal q/m can be conveniently exposed, as exemplified in fig. 1.22.

Finally, for impact of MCI on insulator surfaces (polycrystalline LiF, NaF), recently a novel "potential sputtering" process has been identified (see fig. 1.23 from Neidhart et al. [62]). It involves no impact energy threshold, in

22. Jalabert D, Ristori C, Maurel M, Rocco J C and Huber B A *Nucl. Instrum. Meth. Phys. Res. B* **98**, 132 (1995).
23. Beiersdorfer P, Knapp D. Marrs R E, Elliott S R and Chen M H, *Phys. Rev. Lett.* **71**, 3939 (1993).
24. Carlberg C et al., *IEEE Trans. Instrum. Meas.* **44**. 553 (1995).
25. Beebe E, Liljeby L, Pikin A, Björkhage M, Engström A and Paal A, *Rev. Sci. Instrum.* **65**, 1718 (1994).
26. Kluge H-J, *Nucl. Instrum. Meth. Phys. Res. B* **98**, 500 (1995).
27. Aumayr F and Winter HP, *Physica Scripta T* **28**, 96 (1989).
28. Wu W, Giese J P, Ben-Itzhak I, Cocke C L et al, *Phys. Rev. A* **48**, 36217 (1993).
29. Nakamura N, Currell F J, Danjo A, Kimura M, Matsumoto A et al, *J. Phys. B: At. Mol. Opt. Phys.***28**, 2959 (1995).
30. de Nijs G, Folkerts H O, Hoekstra R and Morgenstern R, *J. Phys. B: At. Mol. Opt. Phys.* **29**, 85 (1995).
31. Gilbody H B, *Adv. At. Mol. Opt. Phys.* **32**, 149 (1994).
32. Huber B A and Kahlert H J, *J. Phys. B: At. Mol. Phys.* **16**, 4655 (1983).
33. Cederquist H, Biedermann C, Selberg N and Hvelplund P. *Phys. Rev. A* **51**, 2191 (1995).
34. Barat M and Roncin B, *J. Phys. B: At. Mol. Opt. Phys.* **25**, 2205 (1992).
35. Gaboriaud M N, Barat M, Roncin P and Sidis V, *J. Phys. B: At. Mol. Opt. Phys.* **27**, 4595 (1994).
36. Hoekstra R, *Comm. At. Mol. Phys.* **30**, 361 (1995).
37. Folkerts H O, Bliek F W, Meng L, Olson R E, Morgenstern R et al, *J. Phys. B: At. Mol. Opt. Phys.* **27**, 2475 (1994).
38. Hoekstra R, Wolfrum E, Beijers J P M, de Heer F J, Winter HP and Morgenstern R, *J. Phys. B: At. Mol. Opt. Phys.* **25**, 2587 (1992).
39. Hoekstra R, de Heer F J and Morgenstern R, *J. Phys. B: At. Mol. Opt. Phys.* **24**, 4025 (1991).
40. Ciric D, Brazuk A, Dijkkamp D, de Heer F J and Winter HP, *J.Phys. B: At. Mol. Phys.* **18**, 3629 (1985).
41. Wilkie F G, McCullough R W and Gilbody H B, *J. Phys. B: At. Mol. Phys.* **19**, 239 (1986).
42. Aumayr F, Schweinzer J and Winter HP, *J. Phys. B: At. Mol. Opt. Phys.* **22**, 1027 (1989).
43. Stolterfoht N, Sommer K, Swenson J K, Havener C C and Meyer F W, *Phys. Rev. A* **42**, 5396 (1990).
44. de Nijs G, Hoekstra R and Morgenstern R, *J. Phys. B: At. Mol. Opt. Phys.* **27**, 2557 (1994).
45. Schippers S, *Nucl. Instrum. Meth. Phys. Res. B* **98**, 177 (1995).
46. DePaola B D, Huang M-T, Winecki S, Stöckli M, Kanai Y et al, *Phys. Rev. A* **52**, 2136 (1995)
47. Pesnelle A, Trainham R, Pascale J, Monnand E and Andrä H J, *Phys. Rev. Lett.* **74**, 4169 (1995).
48. Melchert F, Krüdener S, Schulze R, Petri S, Pfaff S and Salzborn E, *J. Phys.*

B: At. Mol. Opt. Phys. **28**, L355 (1995).
49. Walch B, Cocke C L, Voelpel R and Salzborn E, *Phys. Rev. Lett.* **72**, 1439 (1994).
50. Jin J, Khemliche H, Prior M H and Xie Z, *Phys. Rev. A* **53** 615 (1996).
51. Chandezon F, Guet C, Huber B A, Jalabert D, Maurel M et al, *Phys. Rev. Lett.* **74**, 3784 (1995).
52. Aumayr F, *Invited Lectures, XIX Int. Conf. on the Physics of Electronic and Atomic Collisions (July 26 - August 1, 1995, Whistler/Canada)*, AIP Conference Proceedings 360, eds. L. Dube et al, p. 631 (New York: American Institute of Physics).
53. Burgdörfer J, Lerner P and Meyer F W, *Phys. Rev. A* **44**, 5674 (1991).
54. Winter H, Auth C, Schuch R and Beebe E, *Phys. Rev. Lett.* **71**, 1939 (1993).
55. Aumayr F, Kurz H, Schneider D, Briere M A, McDonald J W, Cunningham C E and Winter HP, *Phys. Rev. Lett.* **71**, 1943 (1993).
56. Meyer F W, Folkerts L, Folkerts H O and Schippers S, *Nucl. Instrum. Meth. Phys. Res. B* **98**, 441 (1995).
57. Lemell C, Winter HP, Aumayr F, Burgdörfer J and Meyer F W, *Phys. Rev. A* **53**, 880 (1996).
58. Limburg J, Schippers S, Hoekstra R, Morgenstern R, Kurz H, Aumayr F and Winter HP, *Phys. Rev. Lett.* **75**, 217 (1995).
59. Lakits G, Aumayr F and Winter HP, *Rev. Sci. Instrum.* **60**, 3151 (1989).
60. Lakits G, Aumayr F, Heim M and Winter HP, *Phys. Rev. A* **42**, 5780 (1990).
61. Aumayr F, Kurz H, Winter HP, Schneider D, Briere M A and McDonald J W, Rev. Sci. Instrum. **64**, 3499 1993.
62. Neidhart T, Pichler F, Aumayr F, Winter HP, Schmid M and Varga P, *Phys. Rev. Lett.* **74**, 5280 (1995).
63. Neidhart T, Pichler F, Aumayr F, Winter HP, Schmid M and Varga P, *Nucl. Instrum. Meth. Phys. Res. B* **98**, 465 (1995).

CHAPTER 2

ECR Ion Sources

Gérard Melin and Alain Girard*

ECR ion sources (ECRIS), an outgrowth of fusion plasma research, keep regularly improving and expanding since the pioneer time of the first facilities about twenty years ago. For multiply charged ions, both gases and metals, ECRIS are known to be highly reliable, low-cost-technology devices that can deliver intermediate charge states at high intensity in the continuous mode of operation (as high as 0.5 eμA of U^{43+} !). The basic principles of ECRIS operation are here presented, as well as the underlying atomic physics and plasma physics and the way they are modeled. The design, main features and performances of existing ECRIS are then examined. Finally the present trends of development, and what could be a new, advanced concept for an ECRIS, are discussed.

2.1 INTRODUCTION

The Electron Cyclotron Resonance Ion Sources (ECRIS) grew out of fusion plasma machines, the so-called open-ended mirror machines [1] initially built for their plasma confinement properties. It is amazing to notice that today's ECRIS, which are small mirror machines, are optimized for their ion loss rates, e.g. the ion currents to be extracted. This remark reveals the basic compromise that ECR sources of multiply charged ions (MCI) have to face: good confinement and high losses, both optimized! This contradictory situation is discussed below.

ECRIS for MCI were invented and developed by R.Geller and his coworkers about twenty years ago [2]. Accelerators, especially cyclotrons, together with the Grenoble group contributed to their present status, but an increasing number of small-scale facilities for atomic physics experiments now utilize ECRIS and are interested in their development. ECR sources of monocharged

*) Département de Recherche Fondamentale sur la Matière Condensée
Service des Ions, des Atomes et des Agrégats – C.E.A. Grenoble
17, rue des Martyrs – 38054 Grenoble Cedex 9 FRANCE

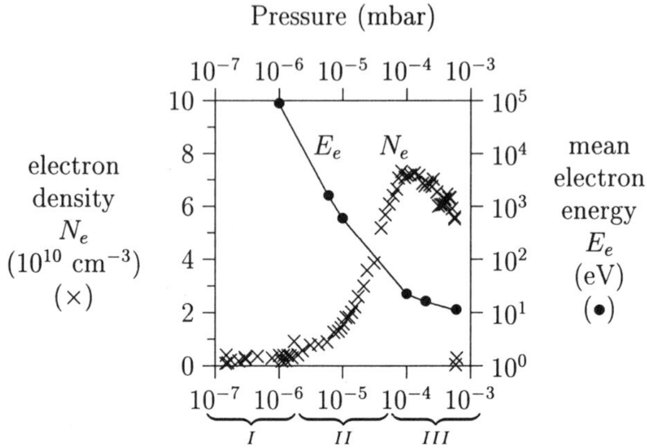

FIGURE 2.1 Typical evolution of electron density and mean electron energy of the plasma streaming out of ECRIS, as a function of gas pressure in the source.

ions also exist and have many applications, especially for material processing. Although the present review is devoted to MCI production using ECRIS, the domain of ion production of ECRIS is much broader, ranging from high-intensity monocharged ion sources to low-intensity highly charged ion sources. Here a rapid overview of ECRIS possibilities is given.

Figure 2.1 illustrates qualitatively the behavior of ECRIS. The mean electron energy E_e and the electron density N_e of the plasma streaming out of the source are shown as a function of the gas pressure. Here the values of E_e and N_e are only indicative. For the sake of simplicity, the magnetic field necessary to initiate the ECR discharge and to confine the plasma, as well as the rf power, do not appear. At low pressure, in the zone labelled I in fig. 2.1, the high electron energy makes it possible to ionize the atom inner shells, including the K shell. This is the domain of low-intensity highly charged ion sources for accelerators; the ion currents typically vary between 1 nA and 1 mA. The reader will understand that in zone I, aside from the discharge start-up the magnetic field plays another important role, the confinement of electrons and ions. Obviously the electron confinement allows the electron density to increase and therefore reduces the ionization times. It also helps the electron heating because of multiple electron/rf wave interactions. These aspects will be analyzed in the course of this chapter. Ion confinement is also important: increasing the density of ions having charge state q, by good ion confinement, evidently improves the production of charge state $q + 1$ ions.

The intermediate-pressure zone, labelled II in fig. 2.1, is that of medium charge ions. The extracted ion currents lie in between a few tens of μA and a few tens of mA for ion charge number of about 2 to 5. This is the domain of ion implanter sources for material and surface treatments—hardening, ion implantation for microelectronics, etc.—more generally utilized in low energy electrostatic accelerators.

Finally, the high-pressure zone (10^{-4} to 10^{-3} mbar), labelled III in fig. 2.1, is the domain of monocharged ion sources of intensity as high as a few amperes: proton sources, plasma sources for material processing (surface cleaning, etching, metal deposition by sputtering, etc.). Beyond this pressure range the ECR discharge collapses, the rate of collisions with neutrals being too high and the electron energy too low for efficient ionization.

Whatever the working zone of fig. 2.1, ECRIS have advantages over other ion sources. They can work in continuous operation over a wide range of ion species with good stability, reliability and longevity. They have no cathode, no filament—almost no maintenance. They are easy to handle, operate and repair. Their rather simple technology (the magnetic fields can be also obtained by permanent magnets) and reasonable cost make them quite attractive.

This chapter essentially concentrates on ECRIS as MCI sources. In ECRIS the production of high charge state ions is made through the confinement of a hot electron plasma, and not directly controlled as in electron beam ion sources (EBIS). This situation, where both plasma physics, atomic physics, and (to a lesser extent) surface physics get simultaneously involved, makes ECRIS behavior and data interpretation somewhat confusing, as many parameters play a role in the ion production. This contributed to rendering them difficult to understand (not to say mysterious), but much progress has been made both in the understanding of ECRIS physics and, consequently, in improving their performance. However, as explained below there are still some ECRIS parameters whose control remains difficult and not well-understood, which sometimes gives rise to differences of understanding, different recipes, among the numerous users of ECRIS.

2.2 BASIC PRINCIPLES OF ECRIS OPERATION

The ECRIS is an open-ended magnetic trap that confines a hot electron plasma. Electrons heated by interaction with rf waves are magnetically confined. They move back and forth along the magnetic field lines, and thus undergo a large number of oscillations through the configuration, while the ions get progressively ionized by electron impact.

Thus, basically the system has the property of confining a plasma. Its main ingredients are:

(i) the magnetic configuration,

(ii) the heating of electrons by rf waves at the cyclotron frequency, i.e. at the magnetic field $|B|_{ecr}$ such that

$$\omega_{\rm ce} = \frac{e\,|B|_{ecr}}{m_e} = \omega_{\rm rf} \;, \tag{2.1}$$

(iii) the necessary sources of electrons, which allow the electron density to build up in the trap.

As compared to EBIS, ECRIS do not have direct control of parameters such as electron energy and ion confinement; these are only indirectly controlled by external knobs, limited in number and not univocal, such as rf power, pressure, magnetic field, and rf frequency. But the main advantage of ECRIS with respect to EBIS is the multi-use of the same electrons for enhancing ionization, which results from the electron confinement.

The extracted ion current I_q of charge state q at one end of an ECRIS is equal to the ion loss rate of the ECRIS trapped plasma:

$$I_q \simeq \frac{1}{2}\frac{n_q qeV_x}{\tau_q} \quad , \tag{2.2}$$

here the coefficient $\frac{1}{2}$ means that the magnetic configuration has two symmetrical ends, usually called extraction and injection; n_q is the density of the ion of charge state q; τ_q the confinement time of this ion; V_x is the part of the hot plasma volume that maps along the magnetic field lines into the extraction area. Assuming plasma charge neutrality

$$n_e = \sum_q n_q q \quad , \tag{2.3}$$

n_e being the electron density, one may expect that, approximately,

$$I_q \propto n_e V_x \quad . \tag{2.4}$$

The formula (2.2) that gives the intensity (I_q) of a given ion charge state (q) delivered by the ECRIS brings up in itself some contradiction:
(i) on one hand high charge state (q) ions require long τ_q because of the ionization times, which in turn means that the losses (I_q) cannot be very large,
(ii) on the other hand high losses (high I_q) evidently would not allow the plasma density (n_q and n_e) to build up, which would seriously reduce the ionization efficiency in order to produce high charge state (q) ions.

This situation results from the self-consistent aspect of a plasma confinement machine. Fortunately, owing to the large number of parameters involved in ECRIS operation, some compromise is possible and by means of the knobs mentioned above, one can match the ionization time to the ion confinement time for the desired ion. This tuning is always necessary in ECRIS, and is different for every ion.

2.2.1 Magnetic configuration

R.Geller [2] and his coworkers definitively [3] contributed to ECRIS development as follows:

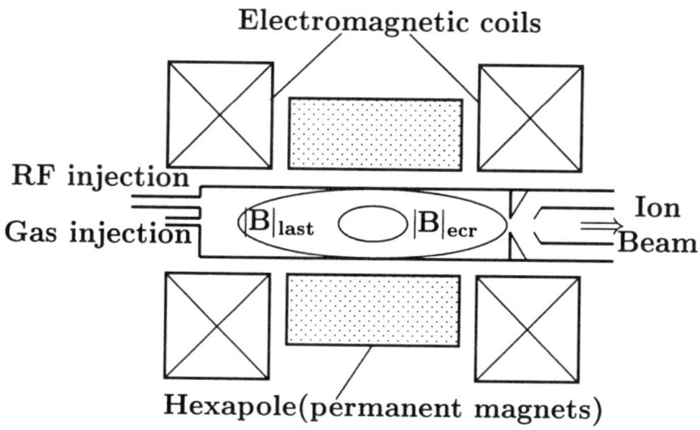

FIGURE 2.2 Sketch of an ECR ion source.

(i) a hot electron mirror machine with hexapolar minimum-B is utilized in order to produce multiply charged ions.
(ii) the resonance surface, as defined by the formula (2.1) has to be closed in order to obtain high charge state ions.
(iii) the source is miniaturized by using permanent magnets for the hexapolar minimum-B (the Supermafios hexapole [2], made of six conductors, had a power consumption as large as 2.5 MW !).

Most ECRIS keep this initial design, sketched in fig. 2.2. The figure shows a mirror field created by two sets of coils combined with a radial hexapolar field, giving magnetic flux tubes of the shape of the one shown in fig. 2.3a. The resonance surface defined by (2.1), which is egg-shaped with azimuthal hexapolar modulations (except at the axial mirror field midplane), lies at the center and confines the rf-heated electrons as shown in fig. 2.3b [4]. The hot electron plasma volume V_p is approximately limited by this surface, as experimentally proved by using a movable probe while monitoring the X-ray flux [5]. In ECRIS, as in mirror machines, electrons spiral around the magnetic field lines and are confined in the radial, or perpendicular ($\perp$), direction. Axial, or parallel ($\|$) confinement occurs because the magnetic field increases at each end of the field lines. The electron confinement results from the conservation of: (i) its magnetic moment μ_e, and (ii) its total energy E_e. Outside the ECR zone of heating, during its motion along the field line,

$$\mu_e = \frac{\frac{1}{2} m_e v_\perp^2}{|B|} \quad , \quad E_e = \frac{1}{2} m_e \left(v_\perp^2 + v_\|^2 \right) \quad .$$

Here $v_\perp$ and $v_\|$ are the perpendicular and parallel components of the electron velocity. Therefore the electrons are reflected within the configuration, i.e. when $v_\| = 0$, if their "pitch angle", defined as $\tan^{-1}(v_\perp / v_\|)$, at the minimum magnetic field B_{min}, exceeds the so-called "loss cone" angle,

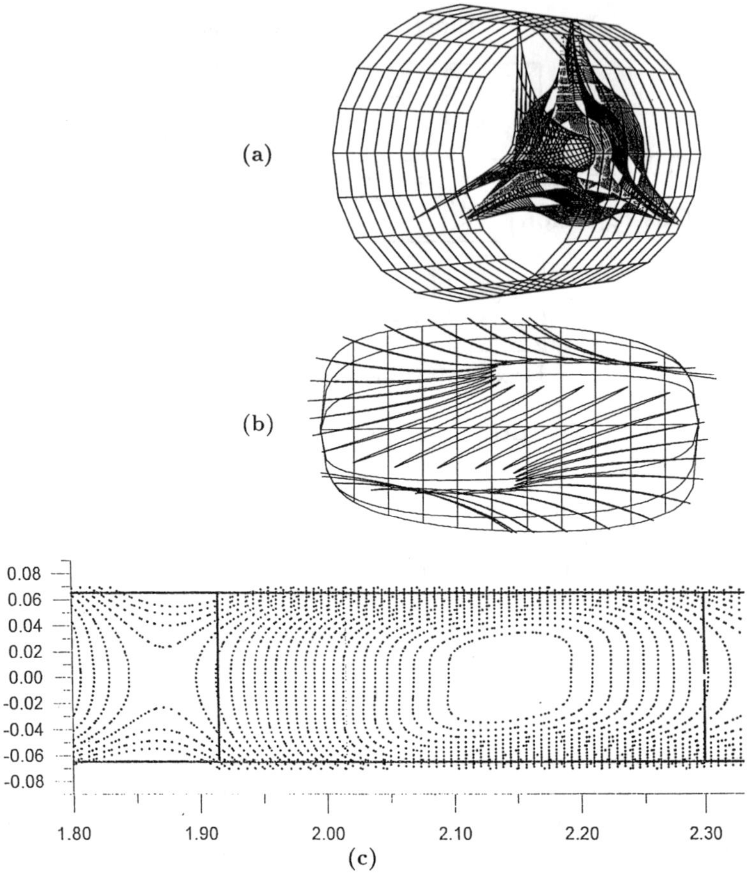

FIGURE 2.3 (a) Magnetic flux tubes created by an axial mirror field and a radial hexapolar field within an ECRIS plasma chamber. The field lines start from coaxial circular contours at the midplane, simulating the equal-plasma density contours. (b) Guiding center trajectories of electrons confined within the resonance surface. Reprinted from J. Vámosi, S. Biri, "TrapCAD – a tool to design and study magnetic traps of ECR ion sources" [4], with kind permission of Elsevier Science - NL, Sara Burgerhartstraat 25, 1055 KV Amsterdam, The Netherlands. (c) Axial cutaway of magnetic isobars of a 14 GHz ECRIS. The $|B|$ value varies from 0.5 Tesla (center surface) to 1 Tesla ($|B|_{last}$ surface at the walls). Distances are in meters.

$\sin^{-1}(B_{min}/B_{max})^{1/2}$ [1].

The equal-$|B|$ surfaces of ECRIS magnetic configuration (mirror field plus multipolar minimum-B), are closed and nested inside each other from the center of the configuration (fig. 2.3c), the resonance surface (equal $|B|_{ecr}$) being one of these surfaces. Regarding the plasma equilibrium, i.e. the confinement capability of this configuration, it can be demonstrated [6] that the equal $|B|$ surfaces (magnetic isobars) are also equal plasma-pressure surfaces (plasma isobars). Thus the plasma is confined only within closed magnetic isobars: when such a magnetic isobar surface touches the walls, the plasma is no longer confined. Therefore if $|B|_{last}$ is the highest $|B|$ value corresponding to a closed equal $|B|$ surface not touching the walls, the higher $|B|_{last}$, the better the confinement. This criterion appears to be satisfied in the high-performance ECRIS. It should be clearly stressed that the magnetic isobars are not magnetic field line envelopes. The electrons follow the field lines while moving through the magnetic configuration, this allows collection of particles (electrons and ions) at the ends of the machine. The field lines are not closed and hit the walls (fig. 2.3a). The plasma pressure is not constant along the field lines—it is anisotropic, owing to the cyclotron motion of particles. Although the relationship between the plasma pressure and the magnetic pressure $|B|^2/2\mu_0$ is not straightforward, it can be shown that the anisotropic plasma pressure depends only upon $|B|$ [7].

The other property of the ECRIS magnetic configuration, the inward curvature of the magnetic field lines (fig. 2.3) was proved early to be necessary for magnetohydrodynamic (MHD) stability with respect to interchange modes: these surface instabilities reduce the plasma density build-up [1].

2.2.2 Electron cyclotron heating

In considering the cyclotron heating of electrons in a magnetic configuration like the one defined above, one has to deal with two different aspects, which are separately analyzed in this section: the particle behavior during the heating process, and the rf waves, used for heating, that propagate in ECRIS plasmas.

Single particle approximation and consequences

Electron cyclotron heating plays an intrinsic role in electron confinement. In the single-particle approximation, an electron moving back and forth along the magnetic field lines receives kicks (+ or −) of transverse velocity ($v_\perp$) from the rf electric field while crossing the resonances (fig. 2.4) [8]. Its phase in the cyclotron motion, with respect to the rf wave, becomes random. The theory predicts that electron heating will occur, even though it occurs only to a limited extent at high energy [9]. Consequently, the electron heating can be modeled as a diffusion process in velocity space. The electron velocity distribution function (EVDF) of the ECRIS plasma may be calculated from a Fokker-Planck equation [10], assuming the distribution to be the result of the combined effect of this diffusion and of Coulomb collisions. These collisions lead to electron losses into the loss cone through pitch angle scattering (partial transfer of $v_\perp$ to $v_\parallel$ at each collision). Solutions to the Fokker-Planck equation [10] during electron heating show that the EVDF is non-Maxwellian and has

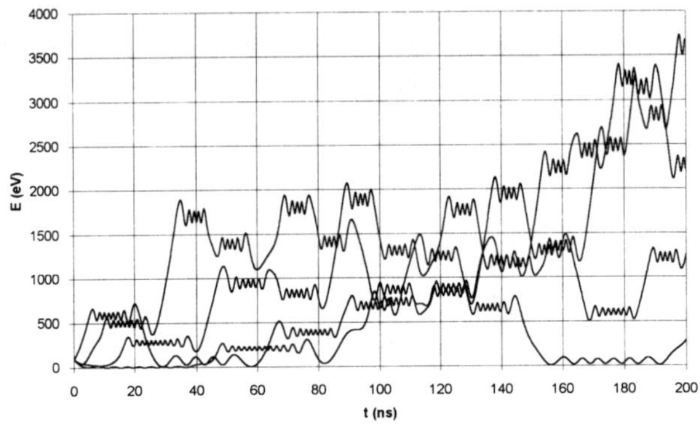

FIGURE 2.4 Examples of electron cyclotron heating: electron energy versus time of four electrons starting with the same initial energy and different phases, showing the resonance crossings and the stochastic heating. From [8].

a high energy tail. Experimental measurements qualitatively confirm this prediction (see below).

rf waves in ECRIS

Despite the electron heating, in ECRIS plasmas the cold plasma and Appleton-Hartree (electron waves only) approximations are still valid, and the main ECRIS plasma waves are [11]:

- in parallel or quasi-parallel propagation (wave vector **k** parallel to **B**); this includes the whistler wave (electric field circularly polarized, perpendicular to **k**), and a high frequency wave, sometimes called the maser wave.
- in perpendicular or quasi-perpendicular propagation (wave vector **k** perpendicular to **B**); this includes the O wave (electric field linearly polarized, perpendicular to **k**, and parallel to **B**), and the X wave (electric field elliptically polarized in the plane perpendicular to **B**). Figure 2.5 gives an illustration of the dispersion relations $\omega = \mathrm{f}(\mathrm{k}_{||}, \mathrm{k}_{\perp})$, i.e. the conditions of propagation of these waves. The waves experience resonances (ω_{ce}, ω_{uh}) and cut-offs (ω_{pe}, ω_R, ω_L), with ω_{uh}, ω_R, ω_L given as follows:

$$\omega_{uh} = \left(\omega_{ce}^2 + \omega_{pe}^2\right)^{1/2} \quad , \tag{2.5}$$

$$\omega_{R/L} = \frac{\left(\omega_{ce}^2 + 4\omega_{pe}^2\right)^{1/2} \pm \omega_{ce}}{2} \quad . \tag{2.6}$$

At a resonance, the wavelength goes to zero, and the rf power can be absorbed. However, hot plasma effects not taken into account here are necessary to

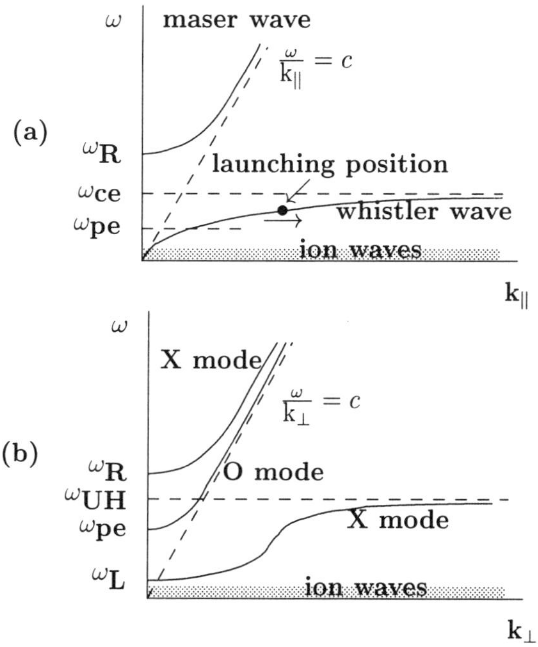

FIGURE 2.5 Dispersion relations of electron plasma waves: (a)parallel propagation $\mathbf{k}||\mathbf{B}$, (b)perpendicular propagation $\mathbf{k}\perp\mathbf{B}$.

describe the absorption. At a cutoff the wave may convert into other waves, be reflected, or be partially absorbed. In ECRIS the preferred wave to use is the whistler wave , having a right-handed circularly polarized electric field that rotates in the electron gyromagnetic direction; this wave carries the rf power from the launching antenna (fig. 2.5), usually at high magnetic field, to the cyclotron resonance $\omega_{rf} = \omega_{ce}$ where it normally couples to the electrons. The fraction of the rf power not absorbed at the cyclotron resonance, owing to reflections either on cut-offs [11] or on the chamber walls, may also be absorbed as an X wave at the so-called upper hybrid resonance, $\omega_{rf} = \omega_{uh}$.

The frequency $\omega_{pe}/2\pi$ is the so-called electron plasma frequency given by the formula,

$$\omega_{pe} = \left(\frac{n_e e^2}{m_e \epsilon_0}\right)^{1/2} \simeq 2\pi \times 9000.\sqrt{n_e} \;\; (2\pi\text{Hz, cm}^{-3}), \tag{2.7}$$

and the electron density corresponding to $\omega_{pe} = \omega_{rf}$ is referred to as the *cut-off* electron density. The limitation $\omega_{pe} \leq \omega_{rf}$ suggested by the plasma shielding of rf waves [2] (correct in unmagnetized plasmas), is only the cut-off of the

O wave, a priori not excited in ECRIS. However this cut-off has been often referred to to give the scaling for ECRIS maximum performance [2], using formula (2.4),

$$(I_q)_{\max} \propto \omega_{\mathrm{rf}}^2 , \tag{2.8}$$

which has not yet been proved.
In the ECRIS plasma two limiting cases have to be considered, depending upon the vacuum wavelength λ_0 of the wave being launched for heating,:
• $\lambda_0 <$ plasma dimensions:
Assuming that the plasma parameters (electron density n_e and magnetic field **B**) are known, one can calculate the dispersion relation of the plasma wave and make the ray tracing of the wave into the plasma from the antenna, in order to evaluate the electric field and the absorption.
• $\lambda_0 \simeq$ plasma dimensions:
Electromagnetic (em) cavity effects dominate because of boundary conditions, and the em field is not localized and has a spatial distribution. Although many em modes may exist [12] (multimode cavity), they may be selected depending upon the rf coupling system [13]. Most of the existing ECRIS have to deal with eigenmodes, and usually eigenmodes of cylindrical cavities: these modes $\mathrm{TE_{m,n,p}}$ or $\mathrm{TM_{m,n,p}}$ (the indices stand for the ψ azimuthal, r radial, and z axial or parallel coordinates, respectively), are actually standing waves formed by oblique waves with respect to the parallel direction, as their dispersion relation in vacuum can always be written in the form,

$$k_0^2 = k_\perp^2(\mathrm{m,n}) + k_\parallel^2(\mathrm{p}) ,$$

where k_0 is the vacuum wave number. Thus the excitation of plasma waves with the $k_\perp$ and $k_\parallel$ components of the selected eigenmode, has to be considered accordingly. Usually this excitation is quasi-parallel, because $k_\parallel \gg k_\perp$, owing to the dimensions of the ECRIS plasma chambers, and the rf power is also coupled to the whistler wave. However the global calculation of the absorption is difficult.

From plasma parameter measurements [14] one can account for as much as about one half of the total rf input power. The fraction of the rf power not absorbed by the plasma may be either absorbed by the chamber walls or radiated through the extraction hole.

2.2.3 Electron sources

Supplying cold electrons to the ECRIS magnetic trap is a necessary condition to obtain an electron density large enough for efficient ionization. Remarkable improvements have been achieved by using different methods, either from external sources or from internal sources [15] [16].
• external sources: (i) 1st stage, an auxiliary ECR discharge, which in the early sources was considered to be important for the singly charged ion supply to

the main stage [2]; recently the electron supply proved to be the essential role of the 1st stage [17] [18]. (ii) the plasma cathode developed at Riken [15] is actually a 1st stage with an electrode for low-energy electron extraction and injection into the main stage. (iii) a low voltage electron gun was used on the AECR source [19]; the effect was quite impressive in the performance, but the gun cathode had a limited lifetime.

• internal sources: (i) multi-ionization of atoms is of course a permanent source of new cold electrons; however it cannot sustain by itself the total electron density. (ii) convenient coatings on the ECRIS walls having a high yield for secondary electrons, e.g. ThO_2 [2], SiO_2 [19], or Al_2O_3 [20], prove to be very efficient; however under the impact of plasma particles, these coatings need to be frequently regenerated, except for Al_2O_3. (iii) a negatively biased disk or probe reduces plasma electron losses and/or provides new electrons from secondary emission of impinging plasma particles [17] [21]. Recent data [16] show that the internal sources by wall coating may have the additional interesting effect of lowering the plasma potential (see below).

The permanent refueling of cold electrons in the ECRIS by these electron sources compensates for the electron losses caused by the various loss mechanisms of the confined plasma.

2.3 ATOMIC PHYSICS AND PLASMA EFFECTS – ECRIS MODELING

As described below in ECRIS modeling, the production of a given ion charge state in ECRIS obeys a balance equation between (i) the rates of production and losses of this ion allowed by atomic physics processes (ionization, charge exchange, etc...), and (ii) the rate of losses caused by the imperfect ion confinement of the magnetic configuration, which is nonetheless necessary if one wants to extract some ions of this charge state!. Roughly speaking, atomic physics processes balance plasma physics effects. Both are described in this section.

2.3.1 Main atomic physics processes in ECRIS

The step-by-step ionization of atoms by electron impact, or successive ionization, is by far the dominant ionization process in ECRIS. The rate of ionization by electrons of an ion from charge state $q-1$ to charge state q is calculated from the ionization cross-sections after averaging over the plasma EVDF, usually considered as a maxwellian distribution of temperature T_e for the sake of simplicity. The cross-sections are given by the Lotz formula [22]. The result of the integration over the EVDF gives the rates of ionization, as shown for example on fig. 2.6 for argon gas. The coefficients of the Lotz formula, however, are known only for few elements, the usual and noble gases, with a reasonable accuracy. Another mechanism of single ionization may occur in

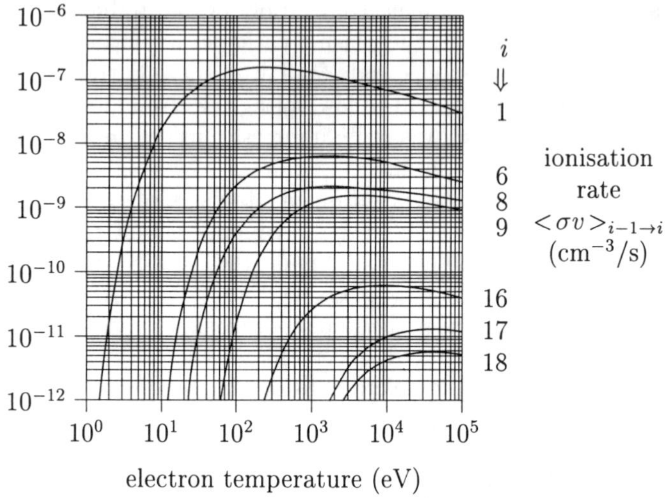

FIGURE 2.6 A few argon ionization rates as a function of the electron temperature.

two steps, excitation and subsequent decay by auto-ionization. Experimental evidence of the sequential ionization is found with metallic ions by rf pulsing, while monitoring in time the peak intensity of each charge state [23].

The multiple-step ionization $q - n$ to q, sometimes called shake-off ionization, may substantially reduce the ionization times in a specific electron energy range; however, once averaged over the EVDF, this mechanism remains negligible. Orders of magnitude of these ionization cross-sections can be found in the comprehensive data compilation of H.Tawara and T.Kato [24].

Another mechanism of ionization by electron impact in ECRIS is the Auger effect. High energy electrons may directly produce a K shell vacancy in the atom, which by Auger effect may be followed by a double vacancy in the L shell, soon followed by twice more holes in the M shell and so on (the Auger cascade effect [25]). This remarkable effect is observed by X ray spectroscopy of the ECRIS plasma, for example while monitoring the K_α and K_β lines of krypton at different input rf powers [5]:

• at very low rf power (a few watts), the characteristic K line of neutral or quasi-neutral krypton is already observed, which means that electrons energetic enough (~15 keV) already exist.

• the K line FWHM increases with rf power. This indicates that all states of krypton ions with one K vacancy and many L vacancies (K^1L^n states) are produced.

Although remarkable, this mechanism would be efficient only for high-Z atoms, and cannot account for a significant part of the ionization.

The recombination process is the reverse of ionization and consists of a capture by an ion of a free electron. The excess energy is removed by the emission of one photon (radiative recombination, binary collision of one electron and a multicharged ion), or several photons (dielectronic recombination, absorption by an ion of an energetic electron, elevating a bound electron to an excited state). Although important only at high charge states, the latter mechanism, dielectronic recombination, dominates over the former; however, for plasma densities so far involved in ECRIS ($n_e \leq 10^{12}\text{cm}^{-3}$), even dielectronic recombination is not very important.

The charge exchange effect, usually considered as the limiting mechanism for high charge states, may be seriously reduced by the burn-up effect of neutrals in the ECRIS plasma, which means that dimension effects may be important: large dimension ECRIS may allow the formation of a central core plasma where the density of neutrals is so low that charge-exchange processes become negligible. Thus, charge exchange is only important to some extent in today's ECRIS. Regarding charge-exchange effects between different ions, either between charge states and/or species, a few experimental data exist, but globally they are not so important.

2.3.2 Plasma effects in ECRIS

ECRIS are subject to specific laws of plasma confinement, which determine the rates of losses, i.e. the currents of ions and electrons, that can be extracted. They are summarized in this section.

Tail electrons, or high energy electrons, definitely exist in ECRIS, as proved by different measurements (bremsstrahlung X rays, electron cyclotron emission, diamagnetism) [26] [27] [14]. As a result of electron cyclotron heating, the EVDF that develops in ECRIS can be divided in two electron populations, both considered to be maxwellian for the sake of simplicity: bulk electrons of temperature from a few hundreds eV to several keV (fig. 2.7), which are dominant in density (fig. 2.8), and tail electrons with a temperature of several tens of keV (fig. 2.9). However, in the coming sections only one population of electrons is considered for simplicity, bulk maxwellian electrons of temperature T_e and density n_e.

Ambipolar potential

In ECRIS a positive potential ϕ (a few tens of volts) with respect to the walls is usually observed. This ϕ potential has to be compared with that of a sheath between a wall and a plasma [28]. The colder electrons of ECRIS leave the plasma faster than the ions. The potential ϕ appears in order to accelerate the ions and to retard the electrons, so that the total current at the walls is zero. The plasma potential ϕ may be experimentally measured from the ion beam energy with mass spectrometry as shown in fig. 2.10 [16]. The plasma potential ϕ strongly depends upon the ion charge state distribution in the ECRIS: the high charge of ions partially compensates for their low mobility,

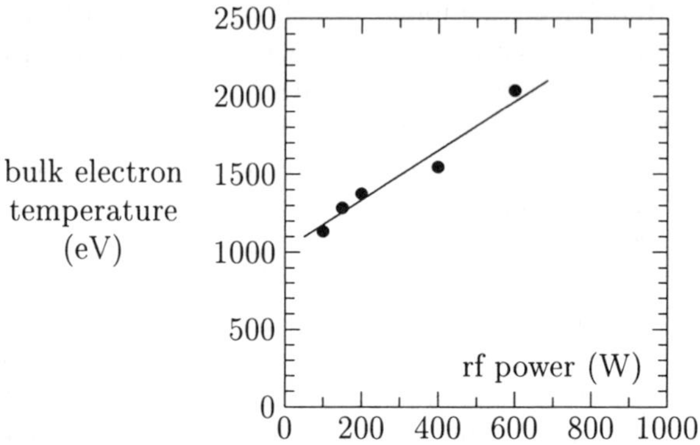

FIGURE 2.7 Bulk electron temperature versus rf power from [14].

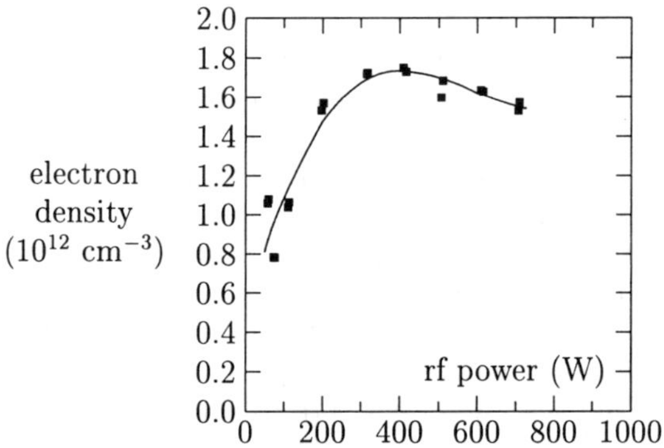

FIGURE 2.8 Total electron density versus rf power from [14].

so that at the same electron density ECRIS delivering higher charge state ions have a lower plasma potential. A low plasma potential is also expected to have some desirable effects: a reduction in the sputtering of the wall material in the source [29], and a better emittance. Thus, control of the plasma potential may become an important issue for ECRIS.

Actually the plasma potential along the axis of the ECRIS confinement region is a positive potential hill ϕ with a small negative well $\Delta\phi$ located at the center, i.e. on the top of this hill (fig. 2.11). This small dip of potential $\Delta\phi$ (perhaps only a few volts at the most), could be created by the high energy electrons magnetically confined, and is going to be very important for the ion confinement, however its existence has not been confirmed experimentally.

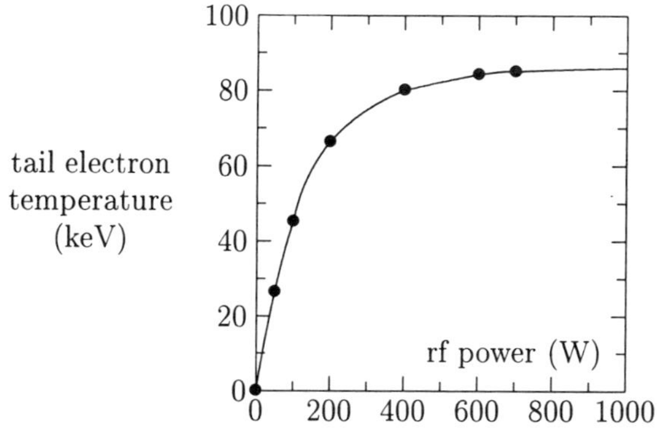

FIGURE 2.9 Tail electron temperature versus rf power from [14].

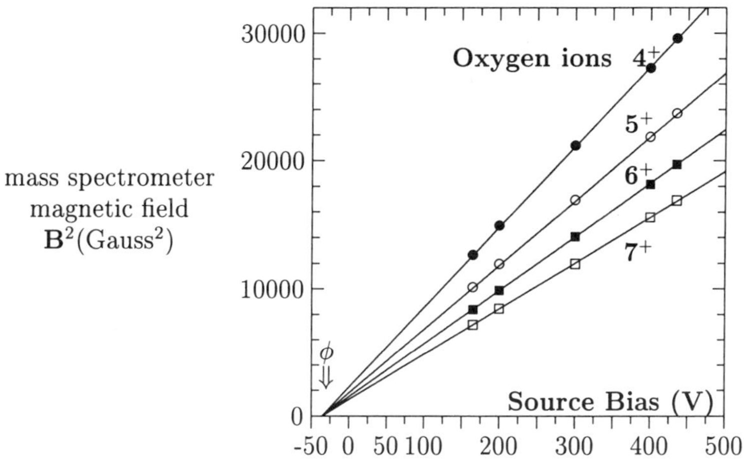

FIGURE 2.10 Magnetic field squared of mass spectrometer versus source bias giving by least square fit the source plasma potential ϕ for extracted oxygen ions of different charge states, from [16].

Electron and ion confinement

Without complete solutions of the Fokker-Planck equation [10], which would give the EVDF in mirror geometry with electron cyclotron heating and Coulomb collisions, averaged parameters of the electron confinement can be calculated under some approximations: (i) isotropic maxwellian EVDF, considering only the perpendicular velocity component (resulting from the heating

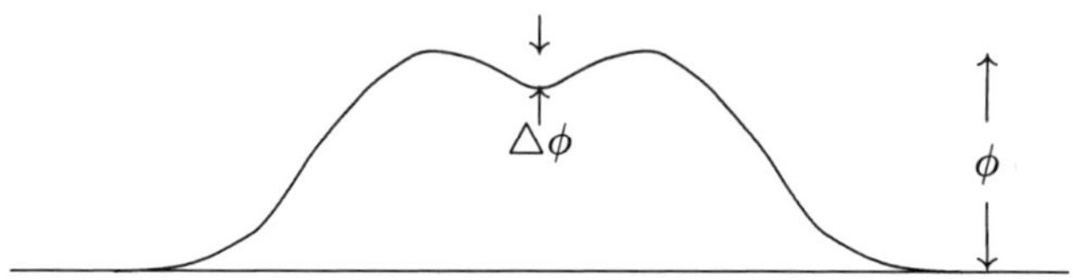

FIGURE 2.11 Postulated plasma potential along the axis of the confinement region of an ECR ion source from [45] (abscissa is the axial distance). Thanks to the University of California, Lawrence Livermore National Laboratory, and the Department of Energy, under whose auspices this work was performed. Reproduced with permission. The U.S. Government retains non-exclusive, royalty-free license in and to this material.

effect); (ii) electrons of energy lower than that of the ambipolar potential, $e\phi$, cannot be lost.

The electron particle confinement time τ_e is defined as the ratio of the total electron density and the number of electrons lost per unit time and per unit volume by Coulomb scattering into the the loss cone at the rate given by Spitzer [30]. Utilizing the Spitzer 90° collision frequency [30], an upper estimate of τ_e for ECRIS can be obtained in the limit $e\phi/kT_e \ll 1$:

$$\tau_e \simeq 4.6\,10^3 \frac{T_e^{3/2}}{n_e\, q_{\text{eff}}} \sqrt{\frac{\phi}{T_e}} \quad (\text{s, eV, cm}^{-3}\text{, V})\ , \tag{2.9}$$

Here, $q_{\text{eff}} = \sum n_q q^2/n_e$ is the mean effective ion charge. Here only the dominant electron-ion collisions are taken into account.

The electron energy confinement time τ_{T_e} is similarly defined as the ratio of the total electron energy content and the electron energy lost per unit time and per unit volume by Coulomb scattering into the loss cone. With the same approximations, again in the limit $e\phi/kT_e \ll 1$, an upper estimate of τ_{T_e} is given by

$$\tau_{T_e} \simeq 5.2\,10^3 \frac{T_e^{3/2}}{n_e\, q_{\text{eff}}} \quad (\text{s, eV, cm}^{-3})\ , \tag{2.10}$$

The difference between the two formulas (2.9) and (2.10), i.e. the dependence of the plasma potential ϕ, simply means that the low energy electrons, close to $e\phi$ in energy, make an important contribution to the total particle losses, but their contribution to the total energy loss is small.

The formulas (2.9) and (2.10) do not depend upon the mirror ratio $R = B_{max}/B_{min}$ as they used the Spitzer 90° collision frequency, which means that the electrons in the $(v_\perp, v_\parallel)$ space are supposed to be localized near the $v_\perp$ axis and that they experience 90° scattering deflections which put them on the $v_\parallel$ axis, the loss cone angle being zero. Previous Fokker-Planck computer calculations of collisional losses of light ions for fusion plasma confinement have shown that the mirror ratio dependence of the confinement time of these light ions can be fitted by a function $\log_{10} R$ [31]. Perhaps such a function could be included in formulas (2.9) and (2.10) in order to account for the mirror ratio effect even for heavy ions, but this has not been calculated. Nonetheless as given here these formulas are approximate upper limits to the corresponding confinement times.

As explained below in the section ECRIS modeling, a rough power balance of the ECR discharge in steady state can be written as

$$\frac{P_{\text{abs}}}{V_p} \simeq \frac{n_e k T_e}{\tau_{T_e}} ,$$

where P_{abs} is the absorbed rf power and V_p is the plasma volume. Thus the classical Coulomb scattering losses allows an electron density scaling,

$$n_e \propto \sqrt{\frac{P_{\text{abs}}}{V_p}}\, T_e^{1/4} , \tag{2.11}$$

which, using formula (2.4), may explain the observed extracted ion current behavior as a function of the rf power (fig. 2.12). However kinetic instabilities [32] that make the ECR discharge unstable, may reduce the actual electron energy confinement time to a value lower than the classical value given by formula (2.10). This may cause a performance saturation [10] [33].

The ion confinement time τ_q in the ECRIS plasma strongly depends upon the ion collisionality, i.e. the number of ion-ion collisions that the ion experiences during each bounce through the ECRIS plasma. This collisionality is given by the ratio of the bounce time τ_b,

$$\tau_b \simeq \frac{l}{v_{T_i}} \quad \text{with} \quad v_{T_i} = \sqrt{\frac{2kT_i}{A_i\, m_p}} \quad \text{and} \quad A_i = \frac{m_i}{m_p} ,$$

where l is the characteristic plasma length, m_i and m_p are, respectively, the ion and proton masses, to the ion-ion scattering time τ_{ii}, which is given by the Spitzer 90° scattering time [30] for the ion of charge state q and mass ratio A_i as follows,

$$\frac{1}{\tau_{ii}} = \sum_j \frac{1}{\tau_{ij}} \simeq 4\pi e^4 \frac{n_e\, q^2 q_{\text{eff}} \ln\Lambda}{A_i^{1/2} m_p^{1/2} k T_i^{3/2}} \quad \text{(cgs units)},$$

in practical units, with $\ln\Lambda \simeq 10$

$$\tau_{ii} \simeq 5.74 \times 10^5 \frac{A_i^{1/2} T_i^{3/2}}{n_e\, q^2 q_{\text{eff}}} \quad (\text{s, eV, cm}^{-3}).$$

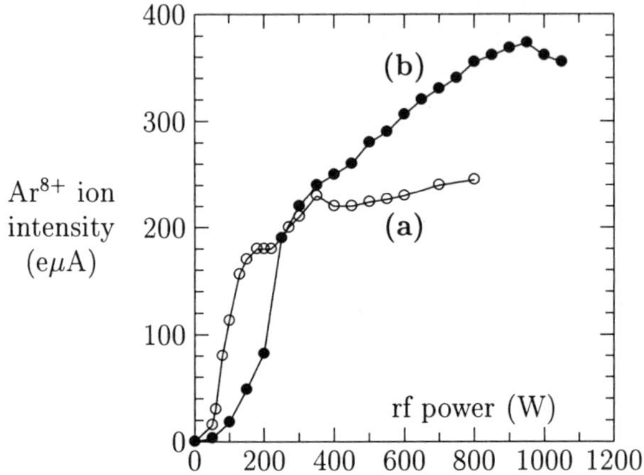

FIGURE 2.12 Ar^{8+} ion current versus input rf power at different operating conditions of the Caprice source [33]. (a) the ion current is maximized at the input rf power of 100 W by slightly changing the pressure and the magnetic field prior to the rf power excursion, (b) the ion current is maximized at 500 W prior to the rf power excursion.

The ion temperature T_i is considered to be the same for all charge states and species owing to the large ion-ion collision frequency. Depending upon the ion collisionality, the following expressions for the ion confinement time have to be considered [34],

• $\tau_{ii} \gg \tau_b$, i.e. in non-collisional confinement, the Pastukhov formula prevails [35],

$$\tau_P = \frac{\sqrt{\pi}}{2} \ln(2R+2) \frac{qe \bigtriangleup \phi}{kT_i} \frac{R+1}{R} \frac{\exp\left(\frac{qe\Delta\phi}{kT_i}\right)}{1+\frac{qe\Delta\phi}{2kT_i}} \tau_{ii} \quad . \tag{2.12}$$

• $\tau_{ii} \sim \tau_b$, i.e. in moderately collisional confinement, the flow formula prevails,

$$\tau_f \simeq Rl\sqrt{\frac{A_i m_p}{kT_i}} \exp\left(\frac{qe\,\Delta\phi}{kT_i}\right) \quad . \tag{2.13}$$

• $\tau_{ii} \ll \tau_b$, i.e. in highly collisional confinement, the diffusion formula prevails,

$$\tau_D \simeq \frac{a^2}{D_i} \quad \text{with} \quad D_i = \frac{kT_i}{m_i\,\nu_{ii}} \simeq v_{T_i}^2 \tau_{ii} \quad , \tag{2.14}$$

where a is the smallest characteristic plasma dimension (plasma diameter).

Practically, the final expressions for the ion confinement τ_q may be taken as follows:

• if $\tau_{ii} \geq \tau_b$, the ion *feels* the magnetic field as well as the potential well $\Delta\phi$, and

$$\tau_q \simeq \tau_P + \tau_f \quad .$$

Most of the ions of today ECRIS are likely in a regime where $\tau_q \simeq \tau_f$.
• if $\tau_{ii} \ll \tau_b$, the ion does not *feel* the magnetic field and,

$$\tau_q \simeq \tau_D \quad .$$

The question whether this formula should also include the effect of the potential well $\Delta\phi$ is still open, but this regime only applies to very highly charged ions.

ECRIS in pulsed mode: afterglow effect

The ECRIS in pulsed mode can deliver higher intensities of highly charged ions during the afterglow plasma [5] [36]. Turning the rf power off causes the electron population to collapse rapidly because of (i) a decrease of the electron temperature T_e (the electron heating does not exist any longer), and (ii) a reduction of some electron sources, such as the secondary emission from hot electrons impinging on the walls. The confined highly charged ions are compelled to leave the plasma because of charge neutrality. Thus the ion loss rate and the extracted ion intensity may increase for a short time, owing to a different compromise between intensity and charge state as discussed earlier.

Impressive extracted intensities have been obtained, as much as 100eμA of Bi^{28+} [37] and Pb^{28+} [36], 30eμA of Au^{27+} [33] (fig. 2.13), i.e. a few pμA of these metallic ions in pulses a few ms long at a frequency of a few tens of Hz. The shape of the afterglow pulse strongly depends upon the operating conditions of the source at the rf power turn-off.

Gas mixing

In order to obtain higher charge states, a carrier gas is generally used in ECRIS with the main element (gas or metal) to be ionized. The density of this carrier gas is dominant with respect to that of the other element, and the carrier gas mass is always lower [2] [38]. Various interpretations have been proposed, none of them fully satisfactory: (i) a dilution effect lowering the mean ion charge (q_{eff} defined earlier) and reducing the electron loss rates according to the formulas (2.9) and (2.10) [2]; (ii) an increase of the electron density because of the better ionization efficiency of the carrier gas, the conditions (n_e, T_e) of the ECRIS plasma with carrier gas being more favorable to the desired multicharged ion production [39]; (iii) the ion cooling which results from the mass effect in ion-ion collisions between the two species [40] [41]: it lowers the ion temperature T_i and increases the ion confinement time of highly charged states, as shown by the formulas (2.12) to (2.14). Oxygen, nitrogen

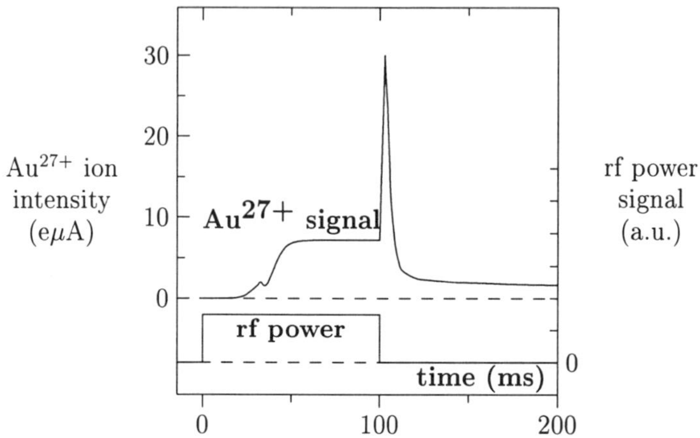

FIGURE 2.13 Au^{27+} ion current versus time showing afterglow pulse at rf power turn-off. From [33].

and helium are the usual gases for this experimental technique. However gas mixing does not always work out, especially in high-performance sources [42].

A remarkable experiment referred to as the *oxygen anomaly* [41] indicates that at least a mass effect is present in gas mixing: a mixture consisting of three oxygen isotopes, masses $A = 18$ (32.9%), $A = 17$ (51.2%) and $A = 16$ (15.9%) is utilized. The anomaly is that the extracted current ratio I_q^{O18}/I_q^{O17} increases with the charge state q: at high charge state this current ratio (0.78 at $q = 7$) is larger than the initial mixture ratio (0.64), whereas at low charge state this ratio (0.58 at $q = 3$) is lower than the initial mixture ratio. When introducing ^{4}He gas into this oxygen mixture, the anomaly is reduced. The explanation is still elusive, but both mass and charge state play a role, likely through the ion confinement time.

2.3.3 ECRIS modeling

ECRIS modeling has been undertaken since as far back as the first ECRIS [43] [44] [45], the charge state distribution (CSD) of the extracted currents being often the only piece of data! It is remarkable that despite the large number of unknown inputs, the fitting of experimental CSD by codes has been often relatively good, although over a limited range of variation of the parameters. This is likely because many internal variables are distribution-averaged (in space, velocity, etc.) so that the ion loss rates, i.e. the extracted ion currents, do not vary rapidly when changing the external knobs, pressure, magnetic field, rf power, etc., except for bursting phenomena such as kinetic instabilities. Still the relationships between these external knobs and the

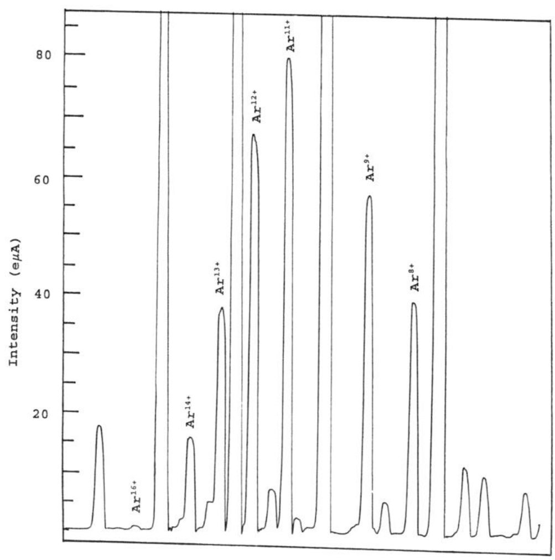

FIGURE 2.14 Argon charge state distribution of currents delivered by the 14 GHz Caprice source tuned on Ar^{16+} from [52]. Abcissa is proportional to $\left(\frac{m}{q}\right)^{1/2}$. The main argon ion peaks are indicated.

internal variables n_e, T_e, n_q $(q = 1, q_{max})$, n_0 (neutral density), T_i, τ_q, τ_e, τ_{T_e}, ϕ, $\triangle\phi$, etc ... , are not straightforward, and simple scaling laws that would summarize the physics of ECRIS seem quite difficult if not impossible to establish.

Atomic data are the best known inputs, at least for most gases and medium-Z elements. Like the first codes, the recent ones [6] [34] [46] describe the time evolution of the particle densities, thus using a fluid approximation (integrated parameters) in a 0 D model (no spatial dimension). Maxwellian velocity distribution functions are usually considered for simplicity; however a theoretical work using actual solutions of the EVDF equation with the rf heating term has been recently undertaken [10]. Some effects are ignored or not correctly described: gas mixing, electron sources, plasma–wall interactions, diamagnetic effect of hot electrons, etc. None of these codes is self-consistent.

The plasma physics description given above seems to be correct for τ_e and τ_q, although the ion confinement time τ_q may be not as optimistic as that given by the above formulas. For the atomic physics description the step-by-step ionization by electron impact and the charge exchange with neutrals are the only two mechanisms considered. The evolution as a function of the charge state q of these three dominant mechanisms, plasma confinement, step-by-step ionization, and charge exchange, explains the peaked shape of the observed CSD (for instance, see argon CSD in fig. 2.14), which is generally fairly well simulated in steady state $(\frac{d}{dt} = 0)$. The main equations of ECRIS modeling are as follows:

• neutral density inside the plasma (n_0),

$$\frac{dn_0}{dt} = \frac{\frac{Q_0}{kT_0} - n_0 S_p}{V} - n_e <\sigma v>^{ion}_{0\to1} n_0 - n_0 \sum_{q=2}^{q_{max}} <\sigma v>^{cx}_{q\to q-1} n_q \ . \qquad (2.15)$$

Q_0 is the total input neutral flow rate and S_p the pumping speed of the pumping system in the plasma chamber of volume V (Q_0 and S_p are usually expressed, respectively, in torr·liter/s and liter/s as practical units). T_0 is the temperature of neutrals, equal to the room temperature.

• electron population (n_e, T_e),

$$\frac{dn_e}{dt} = S_e + n_e \sum_{q=0}^{q_{max}-1} <\sigma v>^{ion}_{q\to q+1} n_q - \frac{n_e}{\tau_e} \ , \qquad (2.16)$$

S_e (number of electrons per unit plasma volume) stands for the external electron sources.

$$\frac{d\,(n_e T_e)}{dt} = \frac{P_{abs}}{V_p} - \frac{n_e T_e}{\tau_{T_e}} - n_e \nu_{ei} \,(T_e - T_i) - P_{ion} \ . \qquad (2.17)$$

On the right-hand side of this electron power balance equation, the first two terms are by far dominant. The third term is the electron-ion power transfer, which results from the electron-ion collision frequency ν_{ei} and which is the ion heating term. Owing to the large difference between ion and electron masses, this collisional heating is not very efficient: T_i is only of the order of 1 eV and $T_i \ll T_e$. In practical units ν_{ei} is written,

$$\nu_{ei} \simeq 3.2\,10^{-9} \frac{n_e\, q_{\text{eff}} \ln \Lambda}{A_i\, T_e^{3/2}} \quad (\text{s}^{-1}, \text{cm}^{-3}, \text{eV}).$$

P_{ion} is the power consumed for ionization.

• ion population (n_q, $q = 1, q_{max}$, T_i),

$$\frac{dn_1}{dt} = S_1 + n_e <\sigma v>^{ion}_{0\to1} n_0 - n_e <\sigma v>^{ion}_{1\to2} n_1$$
$$+ n_0 \sum_{q=2}^{q_{max}} <\sigma v>^{cx}_{q\to q-1} n_q + n_0 <\sigma v>^{cx}_{2\to1} n_2 - \frac{n_1}{\tau_1} \ . \qquad (2.18)$$

Here, S_1 stands for an external monocharged ion source (for example, the first stage of the source).

$$\frac{dn_q}{dt} = + n_e <\sigma v>^{ion}_{q-1\to q} n_{q-1} - n_e <\sigma v>^{ion}_{q\to q+1} n_q$$
$$+ n_0 <\sigma v>^{cx}_{q+1\to q} n_{q+1} - n_0 <\sigma v>^{cx}_{q\to q-1} n_q - \frac{n_q}{\tau_q} \ , \qquad (2.19)$$

$$\frac{dn_{q_{max}}}{dt} = + \, n_e <\sigma v>^{ion}_{q_{max}-1\to q_{max}} \, n_{q_{max}-1} \; - \; \frac{n_{q_{max}}}{\tau_{q_{max}}}$$
$$- \; n_0 <\sigma v>^{cx}_{q_{max}\to q_{max}-1} \, n_{q_{max}} \quad , \tag{2.20}$$

$$\sum_{q=1}^{q_{max}} \frac{d\,(n_q T_i)}{dt} = n_e \nu_{ei} \,(T_e - T_i) - \sum_{q=1}^{q_{max}} \frac{n_q T_i}{\tau_{T_i}} \quad . \tag{2.21}$$

τ_{T_i} would be the ion energy confinement time.
In steady state ($\frac{d}{dt} = 0$) a few simplified relationships can be obtained,

- particle balance from the sum of ion density equations (2.18) to (2.20)

$$\frac{\frac{Q_0}{kT_0} - n_0 S_p}{V} = \sum_{q=1}^{q_{max}} \frac{n_q}{\tau_q} - S_1 = n_0 (n_e <\sigma v>^{ion}_{0\to 1} + \sum_{q=2}^{q_{max}} <\sigma v>^{cx}_{q\to q-1} \, n_q) \quad ,$$

which can be written with a good approximation as,

$$\frac{\frac{Q_0}{kT_0} - n_0 S_p}{V} = \sum_{q=1}^{q_{max}} \frac{n_q}{\tau_q} - S_1 \simeq n_0 n_e <\sigma v>^{ion}_{0\to 1} \quad . \tag{2.22}$$

- charge balance from the same equations (2.18) to (2.20), multiplying each equation by the corresponding ion charge before summation

$$\frac{n_e}{\tau_e} + S_1 = \sum_{q=1}^{q_{max}} \frac{q\, n_q}{\tau_q} + S_e \quad . \tag{2.23}$$

- power balance from equation (2.17), as already done above to get the scaling (2.11). Keeping only the dominant terms of equation (2.17) and using expression (2.10) for τ_{T_e} yields the ECR discharge law for classical coulomb losses

$$\frac{n_e^2 q_{\text{eff}}}{T_e^{1/2}} \left(\frac{V_p}{P_{abs}} \right) = \text{constant} \quad . \tag{2.24}$$

- ion temperature: a rough estimate of T_i can be obtained from relations (2.21) and (2.22), while neglecting S_1 and considering that

$$\sum_{q=1}^{q_{max}} \frac{n_q T_i}{\tau_{T_i}} \approx T_i \sum_{q=1}^{q_{max}} \frac{n_q}{\tau_q} \quad .$$

Thus T_i is approximately given by

$$T_i \approx T_e \frac{\nu_{ei}}{n_0 <\sigma v>^{ion}_{0\to 1}} \quad . \tag{2.25}$$

• peak of CSD: from equation (2.19), neglecting charge exchange, it is possible to obtain the derivative of the ion current,

$$\frac{(\Delta I)_{q-1}}{I_{q-1}} \approx \frac{n_q}{n_{q-1}} - 1 \simeq \frac{\tau_q - \tau_{ion}}{\tau_{ion}(1 + \tau_q n_e < \sigma v >^{ion}_{q \to q+1})} ,$$

where $\tau_{ion} = (n_e < \sigma v >^{ion}_{q-1 \to q} - n_e < \sigma v >^{ion}_{q \to q+1})^{-1}$ is the ionization time for charge state q. Therefore the peak of the charge state distribution of the extracted ion currents, i.e. the most abundant extracted current, may be interpreted as the crossover between two times as a function of charge state q, the ionization time and the confinement time. If q_M is the charge state of the most abundant current of the CSD (see for example figure 2.14):
-at $q < q_M$, $(\Delta I)_{q-1} > 0$, $\tau_q > \tau_{ion}$, the ion confinement time is larger than the ionization time.
-at $q > q_M$, $(\Delta I)_{q-1} < 0$, $\tau_q < \tau_{ion}$, the ion confinement time is lower than the ionization time.

2.4 DESIGN AND PERFORMANCE OF EXISTING ECRIS

None of the ECRIS in operation today can pretend to have the full benefit of the understanding of the physics and theory that has been presented above. The importance of several parameters is still unclear and puzzling when designing a source. The choice of some parameters often results from a cost compromise rather than from a physics issue, for instance, the dimensions of the source: the radial magnetic field at the chamber wall, usually created by permanent magnets, has often dictated the diameter of the chamber. The design of existing sources is sometimes already old, and new principles or concepts do not always have direct applications. Although the development of new ideas is very active, performance improves at a slower pace, because: (i) the technology requirements of new methods are not always easy to integrate, and some advantages of the ECRIS technology, such as simplicity and compactness, may become drawbacks; (ii) most laboratories involved in ECRIS development are dependent on accelerator facilities, and are often compelled by priority to supply the accelerators highly charged ion beams rather than working on source development. This section surveys the main existing sources, design and specific aspects as well as performance; they are sorted into three categories: classical, superconducting, and permanent-magnet-only sources.

More than 50 ECRIS are now in operation. Table 2.1 gives the main parameters of a few sources, briefly described below.

TABLE 2.1 Operating parameters of a few ECRIS (design parameters only for SERSE): rf frequency F, maximum axial field B_z^{max}, maximum axial mirror ratio $R_z = B_z^{max}/B_z^{min}$, hexapole magnetic field at wall B_r^{wall}, axial mirror to mirror distance L_z, plasma chamber diameter d, power consumed by the axial coils P_{coils}.

source	F(GHz)	B_z^{max} (T)	R_z	B_r^{wall} (T)	L_z (cm)	d (cm)	P_{coils} (kW)
AECR	14	1.05	2.6	0.64	~40	7.0	<90
Minimafios	18	1.44	2.7	0.72	35	7.0	160
Caprice	14	1.45	3.5	1.05	21	6.6	60/70
ECR4	14	1.10	2.8	0.90	~20	6.6	50
Riken	10	~0.60	1.8	–	50	10	–
SCECR	6.4	1.60	8.8	0.57	50	14	~0
SERSE	14	2.70	7.0	1.40	48	13	~0

TABLE 2.2 A few representative (best known data) ion currents (eμA) from different sources for gases, cw regime.

ion	AECR	Minimafios	Caprice	ECR4	Riken	SCECR
O^{6+}	>500	800	1130	1000	150	840
O^{7+}	>150	180	190	–	60	185
Ar^{8+}	–	–	680	600	–	370
Ar^{11+}	>160	140	185	–	80	170
Ar^{13+}	59	45	34	–	12	56
Ar^{14+}	33	20	17	15	8	30
Ar^{16+}	4	1.4	1	–	0.8	2.1
Ar^{17+}	–	0.02	–	–	–	–
Ar^{18+}	–	0.001	–	–	–	–
Kr^{17+}	100	32	70	50	31	100
Kr^{19+}	77.5	22	25	20	16	65
Kr^{20+}	57.5	15	15	8.5	8.5	43
Kr^{21+}	40.5	9	7	–	–	27
Kr^{23+}	18.5	3	2	–	–	7
Kr^{24+}	11	2	–	–	–	5.1
Xe^{22+}	48	–	40	30	–	78
Xe^{25+}	68	–	32	17	–	55
Xe^{27+}	37.5	–	12	–	–	34
Xe^{29+}	–	–	5	3	–	4.4
Xe^{30+}	5	–	–	–	–	1.4

2.4.1 Classical sources

Most sources utilize electromagnetic coils for the axial field and permanent magnets for the radial hexapolar field. A few representative ion beam intensities are given in table 2.2 for some gases and in table 2.3 for some metals. Various techniques for metallic ions are described in references [47]; ovens were first developed at LBL [48]. The data given in tables 2.2 and 2.3

TABLE 2.3 A few representative (best known data) ion currents (eμA) from different sources for metals, cw regime.

ion	AECR	Caprice	ECR4
Fe^{18+}	–	1	–
Au^{27+}	–	10	–
Au^{31+}	–	1.5	–
Pb^{27+}	–	20	25
Pb^{31+}	–	10	8
Pb^{36+}	–	0.3	–
Pb^{37+}	–	0.1	–
Bi^{33+}	12.1	–	–
Bi^{34+}	10.2	–	–
Bi^{36+}	6.5	–	–
Bi^{40+}	1.8	–	–
U^{27+}	–	20	10
U^{29+}	–	15	–
U^{33+}	12	5	–
U^{37+}	7.1	0.9	–
U^{39+}	3.7	0.17	–
U^{43+}	0.5	–	–

are simply indicative: various techniques have been used and any comparison between these data would be meaningless, since the working conditions (extraction aperture, extraction voltage, electron sources, two-frequency heating, etc.) may be quite different.

Figure 2.15 shows a sketch of the 14 GHz AECR built at LBL [19] [49]. The engineering design of this source differs from others essentially by the radial pumping slots in the main chamber, but whether this explains its good performance is not yet clear. Its efficient internal electron sources, and perhaps consequently its low plasma potential recently put in evidence [16] are more likely its main advantages. The peak performance is high despite its relatively low magnetic field, which makes further progress possible. The two-frequency heating effect that was recently tested [50] has actually boosted the performances (see the argon, bismuth and uranium data in tables 2.2 and 2.3).

The 18 GHz Minimafios source has a higher magnetic field, and generally worked with gas mixing and a biased disk. Like the AECR source, its long axial dimension allowed a higher ion confinement time, which may also explain its good performance for highly charged ions. This source has been now shut down [17] [26].

The 14 GHz Caprice source shown in fig. 2.16 is the last version of the first compact sources, as initially designed by B.Jacquot [3] [51]. Aside from the

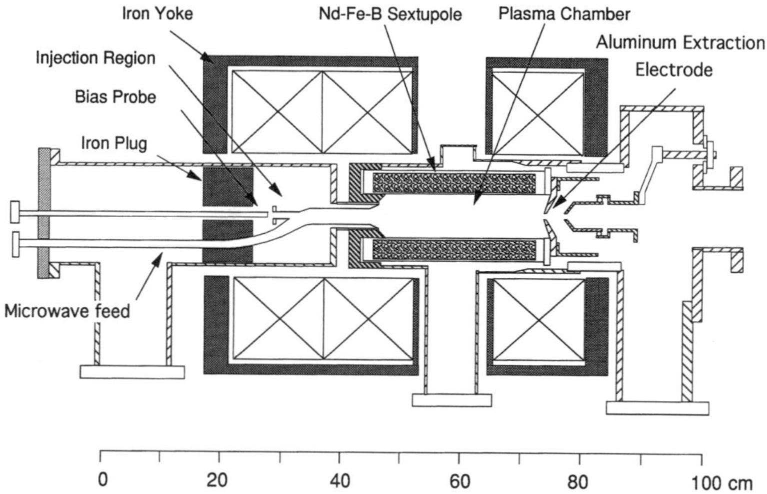

FIGURE 2.15 Elevation view of the 14 GHz AECR source (LBL) showing the bias probe used as an electron source at injection [19] [49].

source compactness, the rf power coaxial coupling in the plasma chamber was really its main original feature and a breakthrough for efficient rf coupling; since then this system has been successfully applied in many sources. The Caprice performance is still improving [33]. This high-magnetic-field source offers an outstanding compromise between the many desirable characteristics of an ECRIS: high charge states, high intensities especially at medium charge states, high stability, reliability, simplicity, and modularity in the design (easy access for changing parts). The best performance is obtained with an aluminum chamber [52] (see argon and uranium data in tables 2.2 and 2.3).

The 14 GHz ECR4 source built at GANIL [53] is also a compact source. It follows the main features of Caprice, but with a stronger iron yoke. The recent version ECR4M, which is able to work at 18 GHz [54], has interesting performance at 14 GHz (see argon data in table 2.2).

The 10 GHz Riken ECR ion source, likely one of the largest-volume classical sources, is worth noticing as to the various studies of electron sources being performed [15] [20] [21].

The measured emittances of these classical sources usually vary between 50π and 150π mm·mrad (for about 80% of the beam intensity) for charge states below ten. At higher charge states the emittance gets better, because of the increased ion velocity, but also because the high-charge-state ions actually originate from the plasma center along the main axis, owing to the electrostatic ion confinement.

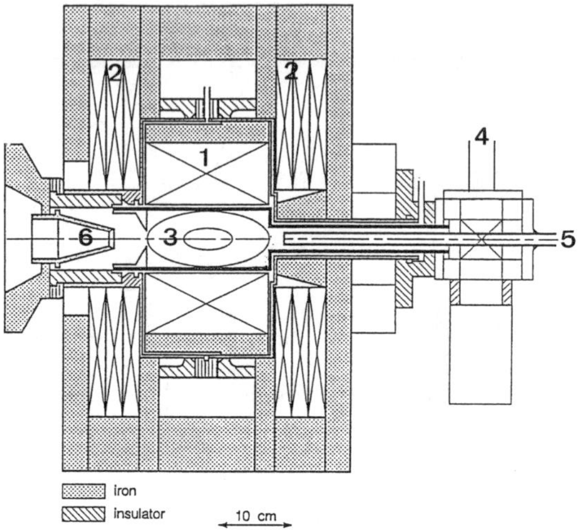

FIGURE 2.16 Cutaway of the 14 GHz compact high magnetic field Caprice source: 1-permanent magnet hexapole, 2-solenoid coils, 3-closed $|B|$ surfaces, 4-rf power input, 5-gas inlet, electric oven feedthroughs, 6-ion extraction [33].

2.4.2 Superconducting sources

Until recently this demanding technology did not reach the performance of the best classical sources; it was considered to be the solution for reducing the huge power consumption of the Supermafios hexapole [2]. Today four superconducting sources have been built. The Ecrevette and Ecrevis sources of Louvain-la-neuve [55] have already been shut down. The Jülich 14.4 GHz source [56] has been running successfully for many years at a field of 0.7 T. Its performance is better than other sources for very highly charged ions, likely owing to the large size of the facility. Good results of superconducting sources were only recently given by the SCECR source (fig. 2.17) at NSCL/MSU. The SCECR works at 6.4 GHz (see tables 2.1, 2.2 and 2.3) in the so-called high-B mode [57], a high axial magnetic field, steep axial field gradients, a configuration actually similar to that of the compact sources Caprice and ECR4, but the radial hexapolar field. Interestingly this source has good performance for very highly charged states, and its next operation step at 14 GHz with a higher hexapole field is expected to be really promising, and hopefully a breakthrough. It is likely that higher performance of the SCECR at 14 GHz will be obtained by using more intense cold electron sources than in the today operation, because of the large size plasma volume and the higher electron density.

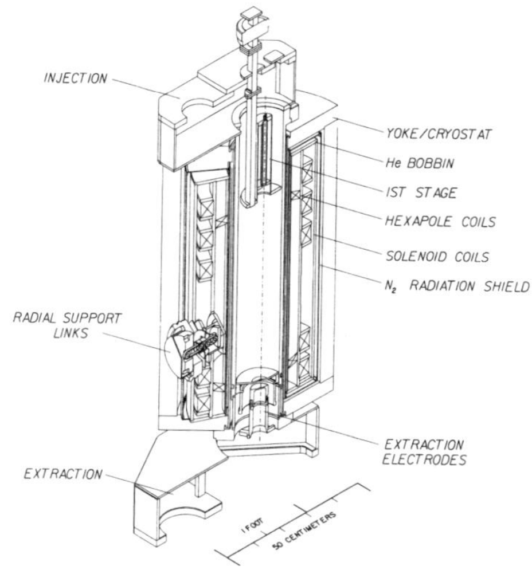

FIGURE 2.17 Cutaway of the 6.4 GHz high magnetic field superconducting SCECR source (NSCL/MSU).

2.4.3 Permanent magnet only sources

Although limited in performance and delivering medium charge states, these sources are worth mentioning: both axial and radial magnetic fields are obtained only with permanent magnets. The disadvantage of the lack of axial field tuning is compensated by miniaturization and reduced power consumption—only the rf generator consumes power. This advantage, together with their simplicity and flexibility, make them quite attractive for high voltage platform accelerators. The 8 GHz Neomafios source built by the Grenoble group [58] has been extensively used at the Riken Institute in Japan with as many as 47 different elements [59]. A 10 GHz version of Neomafios [60] with improved performance is now working at the Osaka University RCNP [61], and can deliver Ar^{11+} 5eμA – Ar^{8+} 100eμA beams. Nanogan 10 GHz, a quite compact source built by the Ganil group [62] for on-line production of radioactive ions, delivers Ar^{11+} 3eμA – Ar^{8+} 60eμA beams; it is quite remarkable for its low weight and small size, as well as its low rf power consumption.

Owing to high quality permanent magnets (FeNdB) new compact sources are being designed, and interesting performance sources in this field may be expected soon [63]. Another challenge is the use of the small-size–low-cost 2.45 GHz rf generators for ECRIS at high voltage. Several prototypes are already in operation with modest performance.

2.5 PROSPECTS FOR DEVELOPMENT OF ECRIS

This section summarizes the current thinking on the main issues for further advances of ECRIS performance, surveys the new concepts, and discusses ECRIS now under construction.

2.5.1 Main issues for future ECRIS

Despite the compromise between intensity and charge state in the ECRIS, there are still opportunities for improved performance. The previous sections suggest the dominant trends and issues for future ECRIS: improved magnetic configuration, reinforced electron cyclotron heating, and enhanced electron sources to supply enough cold electrons to the plasma. Higher charge states and higher intensities clearly require improved particle confinement times τ_e and τ_q, so as to build high-electron-density plasmas efficient for ionization and confining highly charged ions.

Accordingly, this leads to higher magnetic fields (both radial and axial), higher mirror ratios, and also to higher rf frequencies in order to fit the resonance condition (2.1). The evidence that higher magnetic fields lead to higher performance is almost general, but the most impressive example is given by the SCECR source at MSU [57]. The question of whether there is an intrinsic effect of the rf frequency itself, such as a shielding effect of rf waves by the plasma [2], is still not clear, but the current idea is that the gains originally credited to a frequency scaling like that of formula (2.8) may have been due instead to the increase of magnetic field strength.

The impressive results obtained with two-frequency heating in the AECR source at LBL [50] have suggested the Multiple Frequency ECR source (MF-ECR) concept [64], an extension to several frequencies of the two-frequency heating concept, in a magnetic configuration with high magnetic fields and high mirror ratios. Closed ECR heating surfaces, nested in each other, could be simultaneously produced at 10, 14 and 18 GHz (these frequencies being commercially available), and would drastically reinforce electron heating and plasma confinement, thus improving source performance.

Finally, more intense electron sources are desirable in order to reach a higher electron density: aluminum as a plasma chamber material is a good candidate for doing so; it forms a resistant oxide layer that has a high secondary emission coefficient [64]. This technique has given excellent results in the Caprice source [52]. An intense electron source is also reported to improve the plasma stability and to reduce the plasma potential ϕ [16].

Together with a higher electron density, a larger-size plasma is expected to allow a better ion confinement. This, in turn, should reduce the neutral density within the plasma by neutral burn-up, and hopefully also reduce charge exchange losses, today the dominant limiting mechanism of high-charge-state ion production.

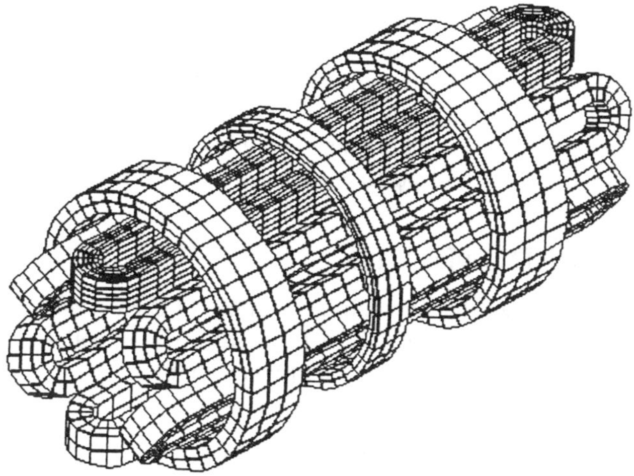

FIGURE 2.18 Hexapole and solenoidal coils of the superconducting SERSE source (axial length 0.8 m) [65].

Electron temperature control does not seem to be a critical issue so far. However, the respective roles of the two quasi-maxwellian, bulk and tail, components of the electron population need to be understood.

Regarding the design requirements of the above issues, only superconducting ECRIS (axial and hexapolar coils) can reach the necessary multi-Tesla fields. Optimized rf coupling systems [13] will be also necessary for multiple frequency heating.

2.5.2 New ECRIS under construction and new concepts

The SERSE project [65], a superconducting source being built by the CEA/DRFMC-Grenoble in a joint venture with the INFN/LNS-Catania, is expected to confirm the main issues described above. The SERSE source is designed to reach the highest magnetic fields so far used in ECRIS, 1.4 T radially at the chamber wall and 2.7 T axially at the mirror throat. The other parameters are given in table 2.1. A sketch of the magnet system is shown in fig. 2.18. The central coil works with reverse field. At 14 GHz the expected performance of SERSE is 1 eμA of Si^{14+}, Ar^{17+}, Kr^{29+}, and Xe^{40+} ion beams. Further operation of SERSE at 30 GHz with gyrotron tubes is also planned.

A new 18 GHz ECRIS at Riken [66] has just started. It is a high-magnetic-field source, 1.4 T axially and radially, of classical technology, coils and permanent magnets, although the hexapole design is quite specific. The source dimensions are similar to those of Minimafios. The source should be flexible enough to tailor the mirror ratio and the axial magnetic field gradient without changing the peak field strength. Preliminary results look quite promising.

A new concept of ECRIS recently proposed is under study [67]. Basically this concept is defined as a "volume" ECR zone source, whereas today ECRIS are "surface" ECR zone sources. In a "volume" ECR zone source, the whole plasma volume lies in a magnetic well with a flat bottom and is everywhere in resonance with the rf waves. This concept would offer noticeable improvements in the power absorption and particle confinement.

ACKNOWLEDGMENTS

The authors acknowledge the help and advice of their Grenoble colleagues, as well as that of A.G.Drentje of KVI Groningen whose collaboration is very much appreciated, while preparing this review chapter.

REFERENCES

1. T.C.Simonen, *Proc. IEEE* **69**, 935 (1981).
2. R.Geller, *Europhys. News* **22**, 8 (1991); R.Geller, *Annu. Rev. Nucl. Part. Sci.* **40**, 15 (1990); R.Geller, *IEEE Trans. Nucl. Sci.* **NS-23**, 904 (1976); R.Geller, *ibid.* **NS-26**, 2120 (1979).
3. B.Jacquot, F.Bourg, R.Geller, *Nucl. Instrum. Meth.* **A254**, 13 (1987); B.Jacquot, M.Pontonnier, *ibid.* **A287**, 341 (1990) and *ibid.* **A295**, 5 (1990).
4. J.Vámosi and S.Biri, *Nucl. Instrum. Meth.* **B94**, 297 (1994).
5. G.Melin et al., *Rev. Sci. Instrum.* **61**, 236 (1990).
6. A.Girard et al., *Proc. ECRIS 12*, Report N° INS-J-182, Riken, 164 (1995).
7. J.B.Taylor, *Phys. Fluids* **6**, 1529 (1963).
8. S.Biri, A.Valek and J.Vámosi, *Proc. ECRIS 12*, Report N° INS-J-182, Riken, 207 (1995).
9. F.Jaeger, A.J.Lichtenberg and M.A.Lieberman, *Plasma Physics* **14**, 1073 (1972); M.A.Lieberman and A.J.Lichtenberg, *ibid.* **15**, 125 (1973); A.J.Lichtenberg, M.J.Schwartz and D.T.Tuma, *ibid.* **11**, 101 (1969).
10. A.Girard, *Proc. ECRIS 11*, Report N° KVI 996, Groningen, 86 (1993).
11. T.H.Stix, *Waves in Plasmas*, American Institute of Physics, New York (1992).
12. C.M.Lyneis, *Proc. 13th Int. Conf. Cycl. Appl.*, Vancouver, World Scientific, editors G.Dutto and M.K.Craddock, 301 (1992).
13. G.Melin et al., *Proc. ECRIS 12*, Report N° INS-J-182, Riken, 105 (1995).
14. A.Girard et al., *Rev. Sci. Instrum.* **65**, 1714 (1994).
15. T.Nakagawa et al., *Jpn. J. Appl. Phys.* **32**, 1335 (1993).
16. Z.Q.Xie and C.M.Lyneis, *Proc. ECRIS 11*, Report N° KVI 996 Groningen, 106 (1993); *Rev. Sci. Instrum.* **65**, 2947 (1994).
17. G.Melin et al., *Proc. ECRIS 10*, Report N° ORNL CONF-9011136, 1 (1990).
18. S.Gammino, J.Sibring, A.G.Drentje, *Rev. Sci. Instrum.* **63**, 2872 (1992).
19. Z.Q.Xie, C.M.Lyneis, R.S.Lam, and S.A.Lundgren, *Rev. Sci. Instrum.* **62**, 775 (1991).
20. T.Nakagawa, *Jpn. J. Appl. Phys.* **30**, 930 (1991).

21. T.Nakagawa and T.Kageyama, *Jpn. J. Appl. Phys.* **30**, 1588 (1991); T.Nakagawa et al., *ibid.* **31**,1129 (1992).
22. W.Lotz, *Z. Physik* **206**, 205 (1967); *ibid.* **216**, 241 (1968); *ibid.* **220**, 466 (1969); *ibid.* **232**, 101 (1970).
23. R.Harkewicz et al., *Proc. ECRIS 12*, Report N° INS-J-182, Riken, 19 (1995).
24. H.Tawara and T.Kato, *Atomic Data and Nuclear Data Tables* **36**, 167 (1987).
25. M.O.Krause and T.A.Carlson, *Phys. Rev.* **158**, 18 (1967).
26. C.Barué, P.Briand, A.Girard, G.Melin, and G.Briffod, *Rev. Sci. Instrum.* **63**, 2844 (1992).
27. C.Barué, M.Lamoureux, P.Briand, A.Girard, G.Melin, *J. Appl. Phys.* **76**, 2662 (1994).
28. F.F.Chen, *Introduction to Plasma Physics and Controlled Fusion*, 2nd ed., Plenum Press, New York (1984).
29. D.Meyer et al., *Proc. ECRIS 12*, Report N° INS-J-182, Riken, 34 (1995).
30. L.Spitzer, *Physics of fully ionized gases*, Wiley Interscience Publishers, New York (1967).
31. J.W.Shearer, Report N° UCID-19577 LLNL (1982).
32. R.C.Gardner, *Ph.D. thesis*, Report N° MIT/PFC/RR-86-23 (1986).
33. D.Hitz, G.Melin, M.Pontonnier, T.K.Nguyen, *Proc. ECRIS 11*, Report N° KVI 996 Groningen, 91 (1993).
34. D.R.Whaley, W.D.Getty, *Phys. Fluids B* **2**, 1195 (1990).
35. V.P.Pastukhov, *Nucl. Fusion* **14**, 3 (1974).
36. P.Sortais, *Rev. Sci. Instrum.* **63**, 2801 (1992).
37. P.Briand, R.Geller, G.Melin, *Nucl. Instrum. Meth.* **A294**, 673 (1990).
38. A.G.Drentje, *Nucl. Instrum. Meth.* **B9**, 526 (1985).
39. M.Delaunay, *Rev. Sci. Instrum.* **63**, 2861 (1992).
40. T.A.Antaya, *J. Phys. Colloque C1* **50**, 707 (1989).
41. A.G.Drentje and J.Sijbring, *Proc. ECRIS 10*, Report N° ORNL CONF-9011136, 17 (1990); A.G.Drentje, *Rev. Sci. Instrum.* **63**, 2875 (1992); A.G.Drentje, A.Girard, D.Hitz and G.Melin, *Rev. Sci. Instrum.* **67**, 953 (1996).
42. Z.Q.Xie, private communication (1994).
43. Y.Jongen, Reports N° LC-8001 and N° LC-8003, Université Catholique de Louvain (1980).
44. S.Bliman and N.Chang-Tung, *J. Phys.* **42**, 1247 (1981).
45. H.I.West, Report N° UCRL-53391 LLNL (1982).
46. J.H.Booske, F.Aldabe, R.F.Ellis and W.D.Getty, *J. Appl. Phys.* **64**, 1055 (1988); S.Pesĭć and M.Vuković, *Phys. Rev. A* **42**, 3571 (1990); G.D.Shirkov, *Plasma Sources Sci. Technol.* **2**, 250 (1993).
47. R.Geller, P.Ludwig and G.Melin, *Rev. Sci. Instrum.* **63**, 2795 (1992); R.Harkewicz, J.Stacy, J.Greene and R.Pardo, *ibid.* **65**, 1104 (1994).
48. C.M.Lyneis, *Proc. 11th Int. Conf. Cycl. Appl.*, Ionics, Tokyo, 707 (1986); D.J.Clark and C.M.Lyneis, *J. Phys. Colloque C1* **50**, 759 (1989).
49. C.M.Lyneis, Z.Q.Xie, D.J.Clark, R.S.Lam and S.A.Lundgren, *Proc. ECRIS 10*, Report N° ORNL CONF-9011136, 47 (1990).

50. Z.Q.Xie and C.M.Lyneis, *Rev. Sci. Instrum.* **66**, 4218 (1995).
51. B.Jacquot and M.Pontonnier, *Proc. ECRIS 10*, Report N° ORNL CONF-9011136, 133 (1990); D.Hitz, B.Jacquot, R.Geller, M.Pontonnier and F.Bourg, *ibid.*, 173 (1990).
52. D.Hitz et al., *Proc. ECRIS 12*, Report N° INS-J-182, Riken, 126 (1995); *Rev. Sci. Instrum.* **67**, 883 (1996). 1995.
53. P.Sortais et al., *Proc. ECRIS 10*, Report N° ORNL CONF-9011136, 35 (1990).
54. R.Leroy et al., *Proc. ECRIS 12*, Report N° INS-J-182, Riken, 57 (1995).
55. Y.Jongen, C.Pirart, G.Ryckewaert, *Proc. ECRIS 4*, internal report CENG, 3.1 (1982).
56. H.Beuscher, *Rev. Sci. Instrum.* **61**, 262 (1990).
57. T.A.Antaya, S.Gammino, *Rev. Sci. Instrum.* **65**, 1723 (1994); T.A.Antaya et al., *Proc. 3rd Int. Conf. Radioac. Nucl. Beams*, East Lansing, May 1993.
58. P.Sortais et al., *Proc. ECRIS 8*, Report N° NSCL # MSUCP-47 East Lansing, 334 (1987).
59. T.Nakagawa et al., *Proc. 9th Symp. Acc. Sci. Techn.*, Tsukuba, 4 (1993).
60. P.Ludwig, R.Geller, and G.Melin, *Rev. Sci. Instrum.* **63**, 2892 (1992).
61. M.Tanaka et al., *Phys. Rev. A* **50**, 1184 (1994).
62. P.Sortais et al., *Proc. ECRIS 11*, Report N° KVI 996 Groningen, 97 (1993).
63. P.Sortais et al., *Proc. ECRIS 12*, Report N° INS-J-182, Riken, 44 (1995).
64. C.M.Lyneis and Z.Q.Xie, *Proc. ECRIS 12*, Report N° INS-J-182, Riken, 119 (1995).
65. G.Ciavola and S.Gammino, P.Briand, G.Melin and P.Seyfert, *Rev. Sci. Instrum.* **65**, 1057 (1994); see also *Proc. EPAC 94*, London, **2**, 1421 (1994).
66. T.Nakagawa et al., *Proc. 9th Symp. Acc. Sci. Techn.*, Tsukuba, 19 (1993).
67. G.D.Alton, *Rev. Sci. Instrum.* **65**, 775 (1994); *Proc. ECRIS 12*, Report N° INS-J-182, Riken, 100 (1995).

CHAPTER 3

The Operation of Electron Beam Ion Sources for Atomic Physics

Martin P. Stöckli*

For over ten years Electron Beam Ion Sources (EBISes) provided beams of highly charged ions allowing the atomic physics community to start the exploration of the collision physics of highly charged ions with the highest charge states of any low energy ions available. A detailed discussion of the EBIS ionization process should allow the reader to determine the highest charge state and the maximum yield he can expect when planning an experiment with an EBIS. A detailed discussion of a series of EBIS operations and the resulting ion beam characteristics demonstrate the amazing range of requirements, which can be met with a properly selected mode of operation, reducing the required amount of beam time for many experiments or targets.

3.1 INTRODUCTION

Interest in the nature of highly charged ions was stimulated by the detection of radiation indicating their existence inside natural as well as man-made plasmas [1]. However, the observed radiation does not allow for a detailed understanding of the processes which govern those ions which, as charged particles, are normally trapped inside the plasma. Detailed understanding called for experiments in which the ions interact under controlled conditions with other matter where some, if not all, reaction products can be observed. The most stringent conditions can be applied with an ion source producing the ions, which can be sorted according to their energy per charge and mass per charge during their transport to the target, where the interaction of interest should take place and can be studied in detail with a variety of detectors.

Most ion sources use electrons to knock electrons out of the atoms or molecules they encounter [2], which requires 4-25 eV depending on the specie.

*) J.R. Macdonald Laboratory, Department of Physics, Kansas State University, Manhattan, KS 66506-2601 USA.

The number of ionizing collisions is proportional to the ionization cross section, the electron density, and the density of atoms or molecules, and therefore, singly charged ions are the dominant charge state produced by ion sources optimized for the total current output. If the ion confinement is increased with a magnetic field and/or by reducing the gas flow, the increased confinement time increases the probability for a successive ionization of the ion in a collision with another electron, enhancing the production of multiply charged ions. Duoplasmatrons have been developed to produce up to Xe^{6+} in dc operation and up to Xe^{11+} in pulsed operation [3]. Penning ion sources have been developed to produce Xe^{13+} with a cold and Xe^{16+} with a hot cathode [3]. However, such high charge states can be produced only with heavy species as the inner shell electrons of lighter elements are too strongly bound for the ionizing electrons found in traditional ion sources. For example, the production of completely stripped (bare) ions with nuclear charge Z requires approximately $Z^2 \cdot 13.6$ eV, which for heavy elements is out of reach for traditional ion sources. This requirement led to the invention of highly charged ion sources twenty some years ago. In an Electron Cyclotron Resonance ion source (ECRIS) microwave power heats the electrons where a few of them can reach energies of several MeV's [4–8]. On the other hand the Electron Beam Ion Source (EBIS), invented by Donets, uses an electron beam with monoenergetic electrons with an energy given by the applied high voltages [7–17]. The Livermore Super Electron Beam Ion Trap (SuperEBIT) [18] recently produced bare uranium with an electron beam energy of 198 keV [19], substantially above the required ionization energy of 131.8 keV [20], and therefore mankind is now in a position where low energy ions of any natural element can be produced with any desired charge state under controlled conditions in a small, but correspondingly equipped laboratory.

The ionization probability decreases rapidly with charge states, especially when inner shells require ionization. The confinement time required to reach the desired charge state increases correspondingly, and therefore the ion production yield per time unit, and hence the extracted time-averaged ion beam currents, decrease rapidly with charge state and may be as small as nA, pA or even fA. However, many atomic physics experiments have been successfully completed with count rates as little as 100 ions per second which represents roughly a fA beam of highly charged ions (e.g. [19,21]). The feasibility of experiments is determined by the number of ions which are within the phase space accepted by the target, which is enhanced by the small emittance and the small energy spread from a properly tuned EBIS. These two features enhance also the resolution of any magnetic analysis, yielding ion beams with minimal, if any, contributions from other species with similar mass-to-charge ratios. Such contaminants can be suppressed further by selecting a confinement time which favors the desired ion beam and simultaneously reduces the production of the contaminant ions. In addition, ion beams from EBISes contain essentially no ions in metastable excited states, as such ions have normally plenty of time to decay before they are expelled due to low collision rates in a normal EBIS operation. Another important factor is the time structure of the ion beam which ideally should match the time structure of the acceptance by the target. Electron Beam Ion Sources are Electron Beam Ion Traps designed for the production of high yields of highly charged ions to be spilled, and by controlling the spillage one can generate almost any desired time structure.

Almost thirty EBISes and EBIT's have been built over the last twenty years and most of them have been used successfully as injectors for accelerators or as ion sources for atomic physics [7,9–17,22,23]. However, the broad range of goals and the diversity of EBIS engineers resulted in a variety of devices with correspondingly diverse characteristics, some of which are reflected in the name of the device. CRYEBIS [24–27], CRYSIS [28], Kryo-EBIS [10] and KRION [9,23,29–32] for example, reflect the cryogenic nature opposed to EBISes with the traditional warm bore. Cryogenic EBISes use the liquid helium to generate the desired high magnetic field with a superconducting solenoid. However, a magnetic field is not absolutely required as demonstrated by the X-EBIS [33,34] which uses no magnetic field at all. SuperEBIS [35], Mini-EBIS [36], and MicroEBIT [37] reflect the ambition of the designer to optimize performance, economy, or size. Most EBISes are designed with a trapping length of roughly 1 m to provide sufficient ion yields. On the other hand EBITs are designed with 2-3 cm long traps, which are sufficient to study the beam electrons interacting with trapped ions by observing the radiation emitted in a radial direction. The short trap produces smaller ion yields when operated as an EBIS, but the reduced complexity accelerated the development of higher electron beam energies, culminating with the production of bare U. There are several papers available which discuss the design considerations for the construction of a specific EBIS [7,9–11,24,38]. EBISes and EBIT's have only one truly basic common feature, the electron beam, which can be described by its energy, its current density and its total current. These three parameters, together with the residual gas pressure which can be maintained during operation, essentially determine the performance and the limits of all EBISes and EBIT's. In this paper we will demonstrate how these basic parameters can be determined and used to predict the performance for other charge states and ion specie. In addition we will discuss how the ion expulsion method as well as the seed particle injection method affect the performance as well as other characteristics which are important for certain experiments. Sometimes only minor modifications are required to operate an EBIS with another method and to improve the desired quality or yield for a certain experiment. We hope this paper assists the EBIS operator, as well as the EBIS user, to benefit from the many options an EBIS offers to enhance the ion beam characteristics desired for their specific experiment.

3.2 EVOLUTION OF CHARGE STATES

3.2.1 Direct Electron Impact Single Ionization Approximation

As mentioned above, the basic, common feature of all EBISes is an electron gun providing an electron beam which is compressed to a small diameter by a combination of electric and magnetic fields [10,38]. Atoms and molecules passing through the dense electron beam may be hit and ionized by an electron and so become trapped in the space charge of the electron beam. In this fashion the ions end up in a dense shower of energetic electrons which continue the ionization process and increase the ion charge state q in successive collisions. Any electron bound to the ion can be removed as long as the beam electron can supply the required binding energy. The probability for directly removing an electron rises rapidly after the electron beam exceeds the corresponding binding energy, reaches a maximum for 2-3 times the binding energy, and decreases slowly with further increases of the beam energy. The probability for removing the qth electron, the electron impact ionization cross section $\sigma_i(q-1,q)$, is the sum of the probabilities for removing any of the remaining electrons. Several approximations have been developed for direct electron impact single ionization, of which the Lotz approximation [39] appears to be the most popular, because it is simple and reliable [40,41]. These cross sections are shown in Fig. 3.1 for all possible Ar ions and for 6 different electron beam

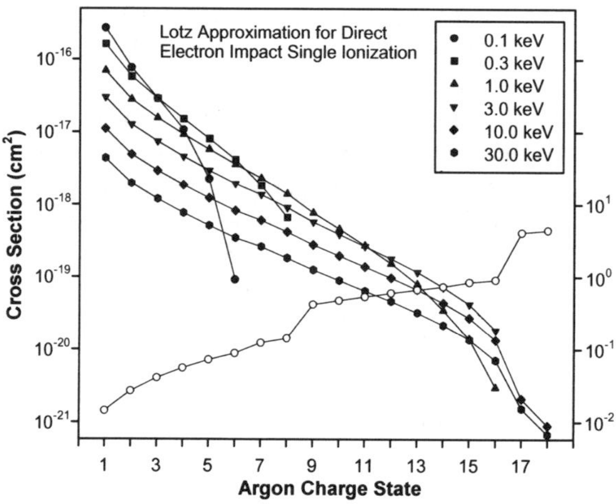

FIGURE 3.1 Ionization cross sections and ionization energies of argon versus the ionic charge state. The ionization energies are shown with open symbols. The full symbols show the cross section for direct electron impact single ionization for six impact energies using the Lotz approximation.

energies. It shows that a 0.1 keV beam removes the first few electrons most efficiently. However, the energy of the most efficient electron beam changes gradually with increasing charge state due to the maximum cross section at 2-3 times the binding energies.

The binding energy of the weakest bound of the remaining electrons, the ionization energy I_q, is also shown in Fig. 3.1. It determines the highest charge state which can be produced with a certain electron beam energy. For example, the ionization energy of 124 eV makes Ar^{6+} the terminal charge state for a 0.1 keV beam, the ionization energy of 423 eV makes the Ne-like Ar^{8+} the terminal charge state for a 0.3 keV beam, and the ionization energy of 4.12 keV makes He-like Ar^{16+} the terminal charge state for 1 and 3 keV beams. Any electron beam in excess of 4.43 keV can remove the last two electrons, but it requires for each electron roughly 10,000 times more beam electrons than for the removal of the first electron from the neutral atom.

Multiple ionization has a higher threshold given by the sum of the corresponding ionization energies and is less likely than single ionization. The probability decreases rapidly with the number of removed electrons [42]. However, its relative importance increases with the number of electrons on the atom, for example double ionization of neutral atoms accounts 5% for Si (Z=14), 7% for Se (Z=34), and 23% for Pb (Z=82) [43]. However, these fractions decrease with increasing charge state as there are less electrons available, being less than 3% for Kr^{10+} (Z=36) [44].

The evolution of charge states can be described with a series of rate equations which account for all gains and losses in the population of each charge state. The population gain of charge state q is dominated by the single ionization of the lower charge state q-1, where the losses are normally dominated by the ionization producing the next higher charge state q+1. With this approximation [40] the population n_q can be calculated from

$$\frac{dn_q}{d(jt/e)} = n_{q-1}(t) \cdot \sigma_i(q-1,q) - n_q(t) \cdot \sigma_i(q,q+1) \tag{3.1}$$

The populations evolve with the number of beam electrons colliding with the ions, the product of the electron beam current density j and the evolved time t, the so-called ionization factor jt. Based on those approximations, Fig. 3.2 shows the evolution of charge states for argon ionized by a 10 keV electron beam. If the electron beam has a current density of 1000 A/cm^2, Ar^{8+} becomes the dominant charge state after 1 ms, Ar^{16+} after 25 ms, Ar^{17+} after 0.1 s and bare Ar after 0.2 s.

When the population n_q of charge state q reaches its maximum N_q with an ionization factor jT_q, Equation 3.1 can be rewritten as

$$N_q/n_{q-1}\,(j \cdot T_q) = \sigma_i(q-1,q)/\sigma_i(q,q+1) \tag{3.2}$$

which shows that the population fractions are governed by the ratio of the corresponding ionization cross sections. For example, the fact that every charge state has a time span where it is dominant is due to the ionization cross section which decreases monotonically with charge state. The first few charge states reach high maximum populations in part due to the ionization cross section which decreases rapidly for the first few charge states. The maximum population of approximately 35% for all intermediate Ar charge states is caused by the ionization which becomes 1.4 times less likely with the removal of every electron. The maximum population increases gradually from Ar^{12+} through Ar^{15+} which is due to the ionization which becomes more difficult with the depletion of the L-shell. The soaring maximum population of Ar^{16+} is due to K electrons which are 6.4 times more difficult to remove than the last L-electron. Ar^{17+} reaches also an high population due to the second K-electron which is 2.4 time more difficult to remove than the first one. Ar^{18+}, like any other terminal charge state, can reach 100% at least in theory. The population enhancement of the Ne-like Ar^{8+} is barely noticeable because the 10 keV beam energy is much higher than the ionization energy of 423 eV. However, an enhancement can be expected for a 1 keV beam due to the change of slope observed in Fig. 3.1. Such enhancements are most significant for the He-like and H-like ions as there are only two electrons in the K-shell. Other closed shell configurations yield a significant population enhancement only if the ionization cross section decreases substantially due a significant increase in the ionization energy and a beam energy substantially below the optimum for the ionization of the inner shell. If one dials an electron beam energy below the

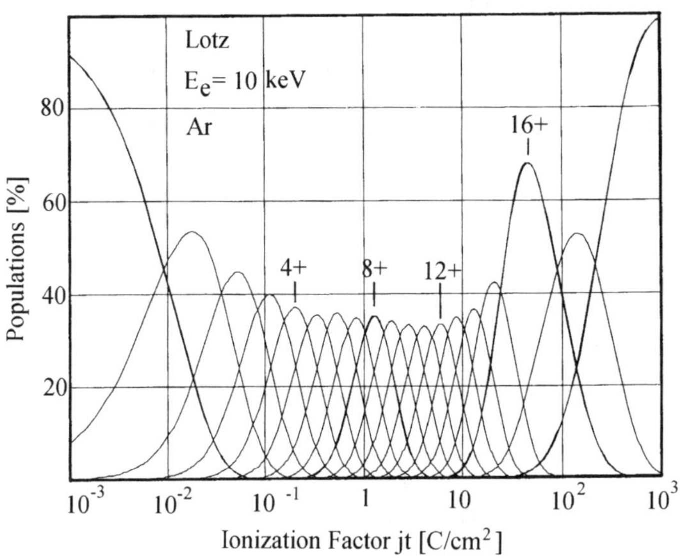

FIGURE 3.2 Evolution of charge states for Ar ionized by a 10 keV electron beam as a function of the ionization factor jt.

ionization energy of an ion with a partly filled shell, such as Ar^{6+}, the ionization rates of the lower charge states, such as Ar^{4+} and Ar^{5+}, become very small and the production very inefficient. However, the maximum population in excess of 30% for higher charge states is much higher than the populations observed with any other ion source, which is due to the fact that only an EBIS allows for an accurate control of the confinement time.

The ionization factor required for the maximum population N_Q of charge state Q can be estimated from the ionization cross sections without calculating the entire evolution of charge states:

$$\frac{j \cdot T_Q}{e} = \sum_{q=1}^{Q} \left(\frac{1}{\sigma_i(q-1,q)} \right) \tag{3.3}$$

This formula indicates that the ionization factor or confinement time required to reach the desired charge state is dominated by the smallest of all applicable cross sections, the ionization cross section for the last of the desired number of electrons to be removed. Even under optimal conditions, the production of Ar^{17+} uses approximately 70% of the time to remove the K-electron. Even the production of intermediate Ar charge states under optimum conditions uses roughly 35% of the time to remove the last electron. Hence the production can be optimized, or the required confinement time can be minimized, by selecting an electron beam energy which maximizes the ionization probability for the last electron to be removed, which is roughly 2-3 times its ionization energy I_q.

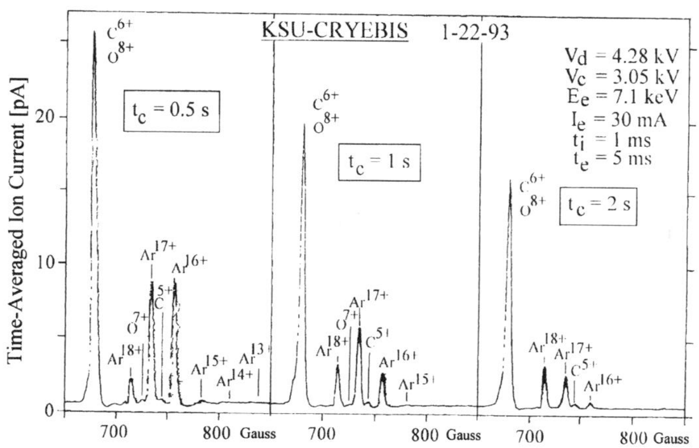

FIGURE 3.3 Measured time-averaged ion beam currents versus the field in the analyzing magnet. The beams were produced by the KSU-CRYEBIS with confinement times (t_c) of 0.5, 1, and 2 s, 1 ms injection time (t_i), 5 ms expulsion time (t_e), with 4.28 kV dam voltage (V_d), 3.05 kV cathode voltage (V_c), and a 30 mA (I_e), 7.1 keV (E_e) electron beam.

Reaching the maximum population N_Q is important for an operation which requires a bunch of ions with charge state Q with a regular, fixed frequency, for example when supplying a synchrotron. However, the number of ions per second, or the time-averaged ion current, is the figure of merit when supplying a target without a time structure, and therefore one should optimize the ratio $n_Q(t_c)/t_t$, where t_t is the time required for the total production cycle. The production cycle becomes dominated by the confinement time t_c for charge states which require a confinement in excess of 1 ms. The ratio $n_q(t_c)/t_c$ reaches its maximum value faster than $n_q(t_c)$, in times which are approximately 20% shorter than T_q for intermediate charge states Ar^{11+} through Ar^{14+}. The reductions are larger for charge states which reach higher maximum fractions N_Q, reaching approximately 50% for Ar^{17+}, in agreement with a measurement [45]. The ratio $n_q(t_c)/t_c$ yields also a maximum for terminal charge states contrary to $n_q(t_c)$ which has no well defined optimum.

Figure 3.3 shows time-averaged ion current distributions for Ar ionized with three different confinement times. The number of ions produced per second can be obtained by dividing the time-averaged ion beam currents through the corresponding ion charge $q \cdot e$, where e is the elementary charge. These numbers can be used to determine the fractions populated by the different charge states as shown in Fig. 3.4. The figure shows that the measured population fractions can be predicted fairly well by selecting an appropriate ionization factor in the evolution of charge states calculated with the proper electron beam energy. The good agreement is due to the Lotz approximation which predicts well the ratio of the ionization cross sections. That is in agreement with

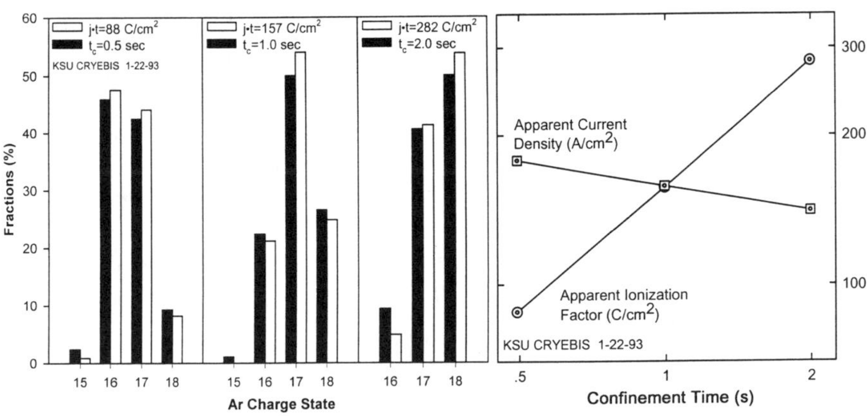

FIGURE 3.4 Left: Charge state distributions predicted by the evolution of charge states for a 7.1 keV electron beam for 88, 157, and 282 C/cm^2 compared with the normalized charge state distributions from Fig. 3.3. Right: The apparent ionization factor and the apparent electron beam current density from Fig. 3.3 versus confinement time.

experiments which found that the Lotz cross sections are fairly reliable for Li-like ions [46]. Even in the extreme case of He-like and H-like uranium, Lotz underestimates the cross section only by 50% [19].

Matching the relative ratios among the 2-3 dominant populations with the ratios predicted by the evolution of charge states allows one to determine the ionization factor with high resolution. The evaluated ionization factors jt and the selected confinement times t_c can be used to determine the electron beam current density with high resolution, a parameter which otherwise is almost impossible to measure for most EBISes. If the charge state distribution is measured for several confinement times without changing any other parameter, one expects the ionization factor to increase proportionally with the confinement time. However, Fig. 3.4 shows that the ionization factor does not increase perfectly linear with confinement time, and therefore yields an electron beam density which decreases slowly with confinement time. This inconsistency obviously limits the accuracy and reliability of any electron beam density determined in such a fashion which is reflected in the term apparent electron beam density.

The charge state evolution of xenon, which has 54 electrons in its neutral state, is shown in Fig. 3.5 for an electron beam energy of 8.9 keV. In this case the charge states below 40+ cannot account for more than 24% of the total ion population because the M-shell has more electrons with similar ionization energies. On the other hand this case does exhibit a substantially enhanced maximum fraction of 49% for the Ne-like Xe^{44+} because the 8.9 keV electron beam energy is much higher than the 3.3 keV required to ionize Xe43+, but is not much higher than the ionization energy of 7.7 keV required to ionize Xe^{44+}

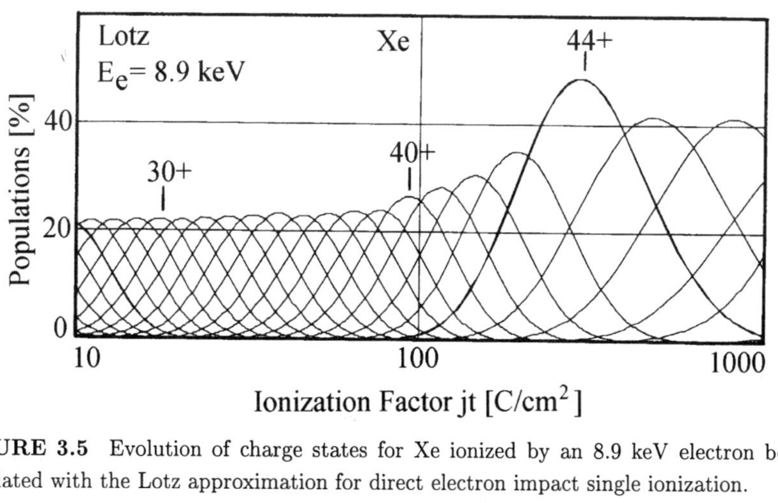

FIGURE 3.5 Evolution of charge states for Xe ionized by an 8.9 keV electron beam calculated with the Lotz approximation for direct electron impact single ionization.

[20]. This enhancement can be seen in Fig. 3.6, which shows the production of highly charged Xe ions for five different confinement times, where Xe^{44+} accounts for 46% of all Xe ions after a 1 s confinement.

3.2.2 Real-World Limits

A more detailed analysis of Fig. 3.6 shows the limits of the simple ionization model described by Equation 3.1. For example, despite the fact that the confinement time was the only parameter changed in Fig. 3.6, Fig. 3.7 shows an ionization factor which increases much less than linear with confinement time and therefore the apparent current density drops rapidly with confinement time. Both figures show how the situation deteriorates with increasing confinement time. For example, the confinement of 1 s and 2 s produce essentially the same charge state distribution and hence fit essentially the same ionization factor. The 5 s confinement produces a charge state distribution with slightly smaller fractions of the higher charge states and therefore indicates an ionization factor below the one for 2 s. Equation 3.1 seems to fail for long confinement times required for the production of high charge states which features small ionization rates.

When smaller ionization rates are involved it becomes important to account for the loss due to recombination, which, according to Fig. 3.6 and 3.7, is able to match the ionization rate. The ions, for example, can capture an electron

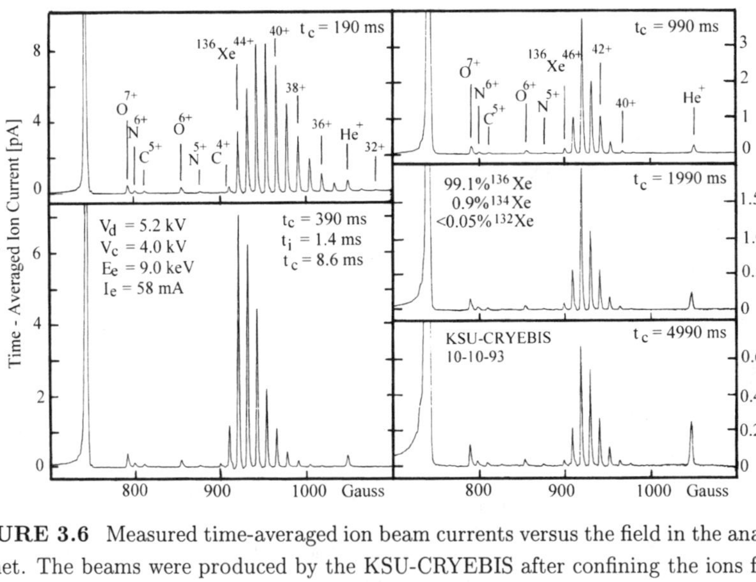

FIGURE 3.6 Measured time-averaged ion beam currents versus the field in the analyzing magnet. The beams were produced by the KSU-CRYEBIS after confining the ions for 0.2, 0.4, 1, 2, or 5 s (t_c) in a 58 mA, 9 keV electron beam.

from the beam in a dielectronic recombination or in a radiative recombination. The loss rates are obviously proportional to the number of colliding electrons, $j \cdot dt$, the ion population $n_q(t)$, and the probability given by the corresponding cross section σ_{dr} and σ_{rr}, and therefore easily can be included in Eq. 3.1. In a dielectronic recombination, the electron transfers its excessive energy to an electron bound by the ion, which is a likely process if the transferred energy matches the difference between the levels available to the bound electrons. As this process occurs only at very well defined energies, this loss rate can be avoided by selecting an electron beam energy which does not match any dielectronic resonance [47]. This happens often unnoticed when adjusting the electron beam energy to optimize the yield of the desired charge state.

In a radiative recombination the electron must emit its excessive energy with an X ray, which causes the probability to decrease with the inverse of the electron energy and therefore is unlikely for an electron beam energy optimized for ionization. However, the radiative recombination rate increases roughly with $(q+Z)^2$ [48], whereas the ionization rate decreases rapidly with increasing ionic charge q and nuclear charge Z. Therefore the radiative recombination loss rates should be included in Eq. 3.1 when calculating the evolution of high charge states of heavy ions. For example, the radiative recombination rate matches approximately the ionization rate for helium- and hydrogen-like Dy^{64+} and Dy^{65+} in a 150 keV beam [49]. On the other hand, the evolutions of charge states shown in Fig. 3.2 and 3.5 are fairly reliable because Ar^{17+} in a 7.1 keV electron beam encounters an ionization rate which is roughly

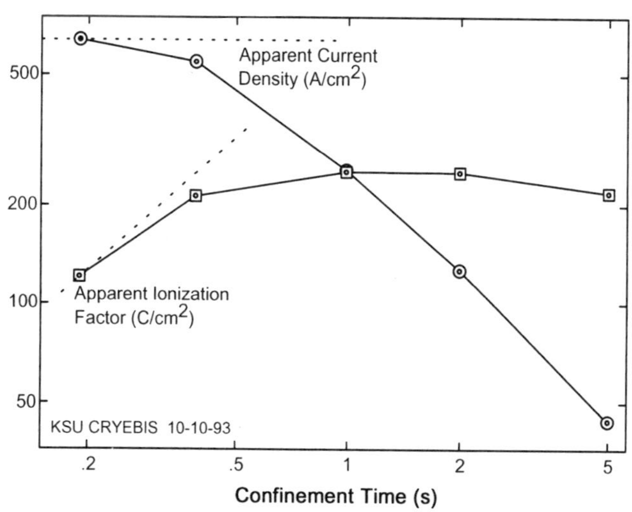

FIGURE 3.7 Apparent ionization factor and apparent electron beam current density versus confinement time. The ionization factors were obtained by matching the ratio between the dominant charge states in Fig. 3.6 with the ratio predicted by Fig. 3.5.

40 times larger than the radiative recombination rate and Xe^{43+} encounters in a 9 keV electron beam an ionization rate which is roughly 8 times larger than the radiative recombination rate [50].

More important are the electrons captured when the ions collide with neutral atoms or molecules which are always present in a real world. This recombination has been studied extensively, and it has been shown that the capture cross section $\sigma_c(q, Z_i, Z_n)$ for low velocities [51] can be estimated reliably with the molecular classical overbarrier model [52]. It is roughly

$$\sigma_c(q, Z_i, Z_n) \approx \sigma_c(q) \approx q \cdot 10^{-15} cm^2 \tag{3.4}$$

and becomes increasingly important with increasing charge state. The recombination with neutrals produces a loss rate of

$$\frac{dn_q}{dt} = -n_q(t) \cdot n_n \cdot v \cdot \sigma_c \tag{3.5}$$

where n_n is the density of the neutral particles, v is the relative velocity between the colliding particles, which is normally dominated by the energy of the ions. Although the capture cross section is well known, the loss rate normally cannot be predicted accurately as one lacks reliable numbers for the density of the neutral particles and the collision velocity v. However, if the term from Eq. 3.5 is included in Eq. 3.1, one could fit the product of $n_n \cdot v$ to match an observed equilibrium charge state distribution (like in Fig. 3.6), which is cumbersome and of questionable merit. If a substantial fraction of ions is in a terminal charge state q, the product $n_n \cdot v$ can be reliably calculated from the balance between the recombination rate of the population n_q and the ionization rate of the population n_{q-1}.

$$n_{q-1} \cdot \sigma_i(q-1, q, Z_i) = n_q \cdot [\sigma_{rr}(q, Z_i) + \sigma_{dr}(q, Z_i) + n_n \cdot v \cdot \sigma_c(q, Z_i, Z_n)] \tag{3.6}$$

where n_q and n_{q-1} are the equilibrium populations observed for very long confinement times. If one measures the ion velocity, which will be discussed later, Eq. 3.6 allows one to determine the average neutral density in the ionization region under real operational conditions which allows for some rather interesting diagnostics. For example, one can measure the increase of the neutral density when the confinement time is increased excessively which is expected to be responsible for the decrease of the ionization factor for the 5 s confinement time in Fig. 3.7. Another interesting example is the decreasing neutral density due to decreasing desorption rates following the initial tuning of high charge states which is often used to explain the observed yield increases of an undisturbed EBIS. The cause of those phenomena has often been identified through correlation with readings from ion gauges, which sample a very different environment and hence allow no quantitative confirmation.

The ionization rates drop rapidly with increasing charge state which causes the recombination to become increasingly important with confinement time. Consequently an equilibrium is reached where the ionization rates are balanced by the recombination rates putting an upper limit on the apparent ionization factor. This upper limit depends on the ratio of the electron beam current density and the operational neutral density, and therefore depends on the source tuning and more importantly on the specific EBIS design. The production of the highest charge state requires obviously an EBIS with a high current density and low operational residual gas pressure which lead to the development of cryogenic EBISes. Most cryogenic EBISes use liquid helium for a superconducting solenoid to produce the desired high current densities, and at the same time for cryopumping the trapping region. So far no evidence has been presented which would indicate substantial loss rates in an EBIS caused by recombination with low charged ions or secondary electrons.

3.2.3 Contributions from Multi-electron Processes

The charge state distribution calculated for 253 C/cm^2 (Fig. 3.8) predicts the fraction of the dominant Xe^{44+} rather well, but overestimates the higher charge states and underestimates the lower charge states found for a

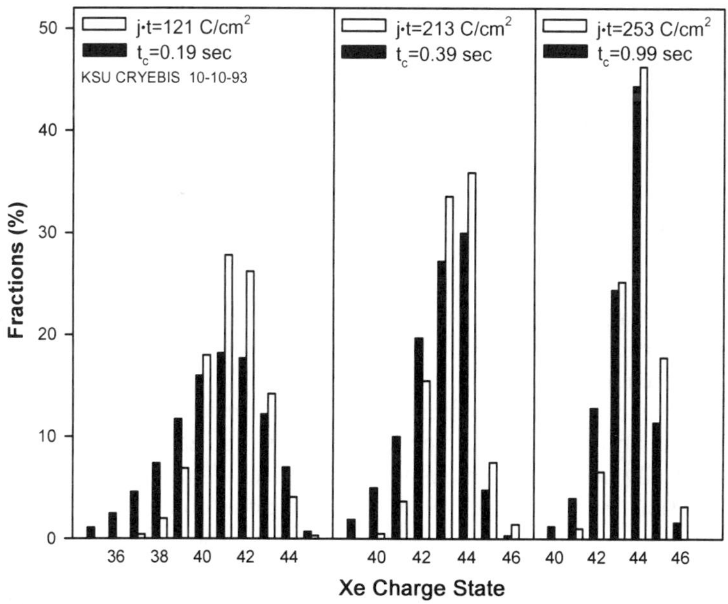

FIGURE 3.8 Charge state distributions predicted by Fig. 3.5 for 121, 213, and 253 C/cm^2 compared with the normalized charge state distributions for confinement times of 0.2, 0.4, and 1 s from Fig. 3.6.

1 s confinement time, which is expected for recombination as an important contributor, because it feeds the lower charge states instead of the higher charge states. However, the dominant Xe^{41+} population found for a confinement time of 0.2 s, where recombination is much less important, is over estimated by a factor of 1.5 by the prediction for 121 C/cm^2. The 253 C/cm^2 ionization factor was determined with the population of Xe^{43+} and Xe^{44+} which have 0-1 M-electrons left, whereas the 121 C/cm^2 ionization factor was determined with Xe^{41+} and Xe^{42+} which have 2-3 M-electrons left. When several electrons are available, multi-electron processes can become important with the formation of multiple excited states, which can autoionize and contribute indirectly to the ionization rates [42,53]. For example, if the electron beam has sufficient energy (4.4-4.9 keV for Xe), an L-electron can be excited forming a doubly-excited state which either decays radiatively or autoionizes. Its probability is dominated by the L-excitation cross section which depends on the number of L-electrons, but does not change substantially if there is an electron in the M-shell [54]. As the direct ionization decreases rapidly with the number of remaining M-electrons, the L-excitation-autoionization becomes more important and exceeds the direct ionization by a factor of 4 to 5 for Na-like ions [55]. In addition, the impacting electron can be Resonantly captured, after Exciting an L-electron, forming a doubly-excited state, which can Double Autoionize (REDA). If the electron beam is tuned on a resonance close to the thresholds for L-shell excitation, REDA can contribute substantially to the ionization rate of Xe^{43+}, depending on the energy spread of the electron beam [54,56]. These indirect processes cause the total ionization cross section of ions with few M-electrons to decrease less rapidly than predicted by direct ionization, which explains the lower fractions observed for 0.2 s confinement time. These findings agree with measurements by Donets, who converted the evolution of Xe charge states into total ionization cross sections [31]. Detailed calculations for ions with a full L-shell and 1-2 M-electrons yield a total ionization cross section with a maximum close to the thresholds for L-excitation [54], which suggests that the production can be enhanced with an energy roughly 1.5 times higher than the ionization energy of the desired ion. Indirect processes are less important for inner shells, where they can contribute 15-20% to the direct ionization of Li-like ions [57].

Indirect ionization [42,53–58] and multiple ionization [58–60] are important for the production of many electron ions requiring ionization factors which can be substantially below predictions based on direct electron impact ionization. The predictions can be improved with corrections obtained from measured charge state evolutions [61] until better approximations for the total ionization cross sections become available. However, the direct ionization gives fairly reliable predictions for the production of few electron ions because the confinement time is dominated by the direct removal of the last few electrons. Electron beam current densities should be determined from the ionization of He-like or H-like ions, and, if the impact energy is kept below the threshold for inner shell excitation, with Li-like ions.

3.3 ION CONFINEMENT

The ions are confined predominantly by the space charge of the electron beam which is the basis for defining the ion capacity of an EBIS. The charge of the electron beam Q^- inside the trapping length l is in practical units

$$Q^-[nC] = 52.8 \cdot 1[m] \cdot I_e[A]/\sqrt{E_e(keV)} \tag{3.7}$$

where I_e is the total electron beam current and E_e its energy. It is commonly assumed that the trapping is lost unless the charge Q^+ from all ions present in the ionization region is less than the Q^-, which defines the neutralization factor

$$Q^+/Q^- = e \cdot \left(\sum_{i,q} q \cdot n_q(Z_i) \right) /Q^- \tag{3.8}$$

where Q^+ must be calculated from the populations $n_q(Z_i)$ of all present species Z_i and charge states q. However, if all the secondary electrons, knocked out from the ions, remain trapped, the net space charge does not change. Such a situation must have been responsible for the production of 1.6 nC of N^{6+} and N^{7+} after a 0.4 s confinement in a 0.75 m long, 6 mA, 3.1 keV electron beam corresponding to a neutralization factor of 12 [62]. However, Eq. 3.8 holds for most other reported operations which normally involve higher electron beam currents [9,10,63–67]. In addition it has been found that the neutralization factor decreases with confinement time [64–66], especially for short confinement times [9].

Only a very few electrons are close enough to knock electrons out of the ions trapped inside the electron beam. But the trapped ions encounter also collisions with electrons which are not close enough to ionize, but interact with their Coulomb attraction [68]. Only a small fraction of energy can be transferred in a single collision with a large impact parameter due to the conservation of energy and momentum. However, the high frequency of large impact parameter collisions produces a steady flux of energy which is transfered from the electron beam to the ions with charge state q and atomic mass number A. The energy transfer rate dE_i/dt is approximately [18]

$$dE_i/dt[eV/s] \approx \frac{7 \cdot q^2 \cdot j[A/cm^2]}{E_e[keV] \cdot A} \tag{3.9}$$

which is roughly 7 keV/s for the Xe^{44+} confined in the 638 A/cm^2 beam shown in Fig. 3.6. The 7 keV Xe^{44+} ions can be confined in axial direction by raising the potential of the drift tubes at both ends of the trap to 160 V above the center drift tubes, which poses no difficulty. The confinement in radial direction is predominantly due to space charge of the electrons and the ions and therefore requires a detailed discussion. If the electron beam has a uniform current density within its diameter r_o, the potential difference between the edge and the center of the beam, the beam trapping potential U_b, is

$$U_b = U(r_o) - U(0) = 479 \cdot I_e[A]/\sqrt{E_e[keV]} \tag{3.10}$$

The beam used in Fig. 3.6 generates a beam trapping potential of only 9 V and if the Xe^{44+} ions increased their energy according to Eq. 3.9 they would leave the electron beam within 0.06 s. The ions with charge state q and an energy in excess of $q \cdot U_b$ leave the electron beam only for a fraction of time and keep returning inside the beam due to conservation of angular momentum [11]. Hence the ionization and therefore the evolution of charge states continue but with a rate reduced by the fraction of time the average ion spends outside the electron beam. This is another reason for the apparent electron beam density to decrease with confinement time which can become evident for short confinement times long before the evolution of charge states is affected by recombination rates.

The heating problem can be reduced by minimizing the heat absorbed during the required confinement time t_c, and therefore we have to maximize the ratio between the ionization rate and the heating rate, for which the electron beam energy is the only free parameter. This ratio reaches a maximum for high energies where the direct ionization cross section decays also with E_e^{-1}. An electron beam energy of 3-4 times the ionization energy is recommended as no substantial improvements can be expected for energies much above the maximum of the ionization cross section.

If the electron beam is coaxial with the surrounding drift tube, its space charge depresses the potential on the axis by the radial trapping potential U_r

$$U_r = U(R_o) - U(0) = 958 \cdot (ln(R_o/r_o) + 1/2) \cdot I_e[A]/\sqrt{E_e[keV]} \tag{3.11}$$

where R_o is the inner radius of the drift tube. The beam used in Fig. 3.6 generates a radial trapping potential of 128 V based on a uniform current density of 638 A/cm^2 and the 0.66 cm inner diameter of the drift tube. If the ion energy increases according to Eq. 3.9, the Xe^{44+} ions with the mean energy would impact on the drift tube after 0.8 s where they would neutralize and disappear from the ion inventory. The problem is aggravated by the fact that the trapping potential is normally reduced by the space charge of the trapped ions. The real potential should be calculated from the net space charge which rarely is done as the electron distribution as well as the ion distributions can only be approximated with crude assumptions due to a lack of detailed measurements. Fig. 3.9 shows the total number of Xe ions observed in Fig. 3.6 as a function of confinement time. The number of ions per second decreases proportionally with the repetition rate which in this case is the inverse of the cycle time. The number of Xe ions per expulsion remains almost constant, decreasing approximately 3% per second. This yield behavior is exactly what one would expect for a batch production where a trap can only be filled to a certain capacity and no ions are lost. However, there are 3800 electrons for each Xe ion inside the trap, which indicates a rather low neutralization factor in the few percent range. Figure 3.3, where each Ar ion is matched by 1200 electrons in the confinement region, indicates a loss rate of less than 6% per second, which agrees with a previously reported result [45]. However, other EBISes operating with higher neutralization factors reported substantially higher ion loss rates which increase rapidly with confinement time [9,64,66].

The ion heating is an intrinsic part of the electron ion collisions and the longer than expected trapping times can be explained in part with the magnetic field, which we neglected, but which in fact increases the radial confinement. However, more important are the ion-ion collisions which allow the highly charged ions to dissipate some of their energy. According to Eq. 3.9 low charged ions encounter substantially less heating. When they collide with the highly charged ions energy is exchanged in an effort to equilibrate the temperature of all ions. That reduces the net heating rate of the highly charged ions but increases the heating rate of the low charged ions. However, since the confinement of the ions is proportional to their charge, the low charged ions

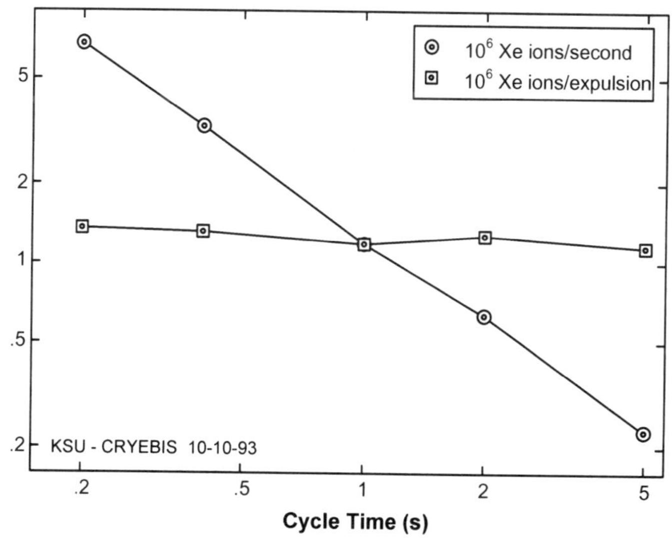

FIGURE 3.9 Number of expelled Xe ions per second and per expulsion from Fig. 3.6 versus cycle time. The almost constant number of ions per expulsion is typical for saturated batch mode productions.

are able to overcome the trapping potential long before the highly charged ions and thus remove some of the electron beam induced heat when they escape [69]. Continuity of this cooling process requires a steady flux of light atoms or ions which enter the trap with a temperature below the temperature of the heavy ions, which, for example, is provided by the residual gas [10,45,69–71]. Some groups enhance the cooling by adding a small flow of Ti [72], Ne [19] or He [73] to increase the life time [72] or production [19,73] of highly charged ions. However, such a flow must be adjusted very carefully because the benefit of cooling is compromised by the heavy ions capturing electrons from the cooling gas, which increases the recombination rates and hence increases the required confinement times. Figure 3.9 shows a rather small loss rate of heavy ions in the KSU-CRYEBIS which indicates that the ions generated from the residual gas provide sufficient cooling. In addition Figs. 3.7 and 3.8 indicate limitations due to recombination with the residual gas and hence the addition of a cooling gas would be counterproductive under the present conditions of the KSU-CRYEBIS. The enhanced recombination rate essentially can be eliminated with the injection of low charged, light ions [66,73] or low charged ions of the production specie [73,74] which requires careful potential adjustments to assure that the energy of the cooling ions inside the trap is substantially below the trapped ions of interest.

3.4 ION EXPULSION

3.4.1 Steady State Operations

The radial confinement is determined predominantly by the net space charge of the partly compensated electron beam and therefore cannot always be controlled accurately. The axial confinement is accomplished with elevated voltages applied to drift tubes at both ends of the confinement region and therefore easily can be controlled. Ions escape predominantly over the lowest potential barrier and therefore the ion beam formation is directed by keeping the voltage of the downstream barrier below the voltage of the upstream barrier as well as below the radial trapping potential U_r. Figure 3.10 shows how the ion beam can be formed in many different ways depending on the desired charge states, yields and other characteristics. Our discussion focuses first on steady state operations where the ions escape continuously from a steady state trap. Figure 3.10 starts with the "cw" [75], "time of flight" ("TOF") [76–80], or "continuous" [7,81] mode which uses no downstream voltage barrier [76,81]. Neutral particles are initially ionized close to the upstream barrier and become radially trapped inside the electron beam. The energetic electrons continue to bombard and ionize the ions, and at the same time accelerate them towards the downstream end of the ionization region [68,77]. The ion confinement time is given by the time-of-flight required by the ions to reach the open downstream end if the downstream movement of the ions is not inhibited by the space charge from other slow moving ions [76,78], or by unintentional potential barriers in the form of apertures with reduced inner diameters [79], non uniform magnetic fields [81], and misalignments between electron beam and drift tubes [9]. Such problems can be reduced and the operation improved with a potential tilted towards the exit by using a conical drift tube [81] or many drift tubes connected to a potential divider [76,78], where the latter solution allows one to optimize the desired charge state by adjusting the potential tilt for the desired confinement time. However, the range of confinement times is rather limited, which benefits only the production of high yields of low charged ions. The second potential distribution in Fig. 3.10 is used for the "continuous" [82–87], "DC" [36,77,88–91] or "leaky" [11,92–94] mode of operation for which the downstream end of the ionization region is kept at a few to a few tens of Volts above the voltage applied to the center of the trap. This potential

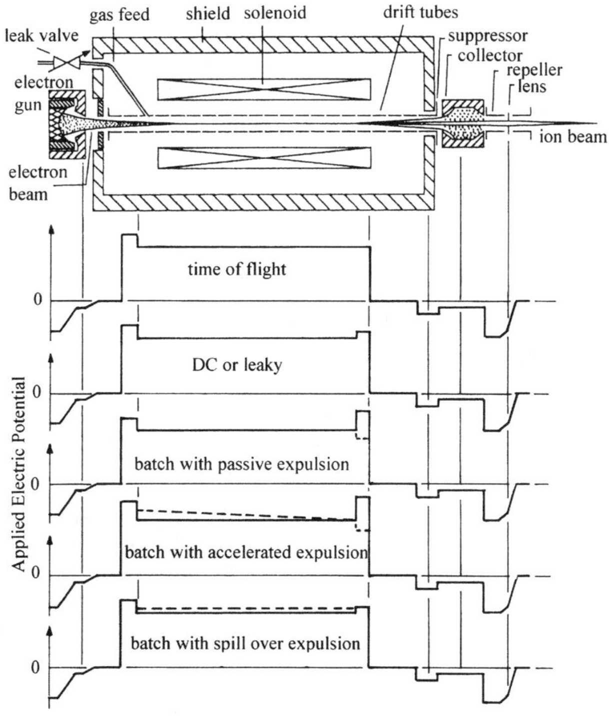

FIGURE 3.10 Top: schematic cross section through a shielded electron beam ion source with an external electron gun. Bottom: potential distributions applied to the drift tubes for the (time-of-flight) "TOF" mode, the "DC" or "leaky" mode, and the "batch" modes with a "passive" expulsion, an "accelerated" expulsion, and a "spill-over" expulsion. The potential distributions during confinement time are indicated with solid lines, whereas the dashed lines indicate the potential distributions changed for the expulsion time.

barrier returns the low energy ions to the ionization region and thus confines the ions in excess of their time-of-flight. During their confinement the ions are heated by the electron beam and heated or cooled by the other trapped ions until they escape after reaching an energy per charge slightly exceeding the downstream potential barrier, which obviously favors the escape of the low charged ions. If the downstream potential barrier is increased by a combination of lowering the voltage applied to the center of the trap [11,82,90–92] and reducing the neutral gas density [71], which lowers the ions' space charge compensation, the average confinement time will increase, lowering the total ion output current, but increasing the ion output currents of high charge states. The Frankfurt Kryo-EBIS can be tuned with a 50 mA, 3 keV electron beam to produce Ar^{14+} as the dominant charge state in a DC operation [84]. The corresponding charge state distribution is compared in Fig. 3.11 with the distribution predicted for 7.2 C/cm^2 assuming equal confinement times for all ions just like for pulsed injections discussed in Section 3.5. The prediction exceeds the measured Ar^{14+} fraction by a factor of 2.3 which is not surprising because the ions surely cover a broad range of confinement times due to the reduced confinement of the low charge states and due to the energy exchange among the ions. If one assumes the confinement times to be equally distributed between zero and a maximum value t_{max}, just like for continuous injections discussed in Section 3.5, the prediction for 11.8 C/cm^2 exceeds

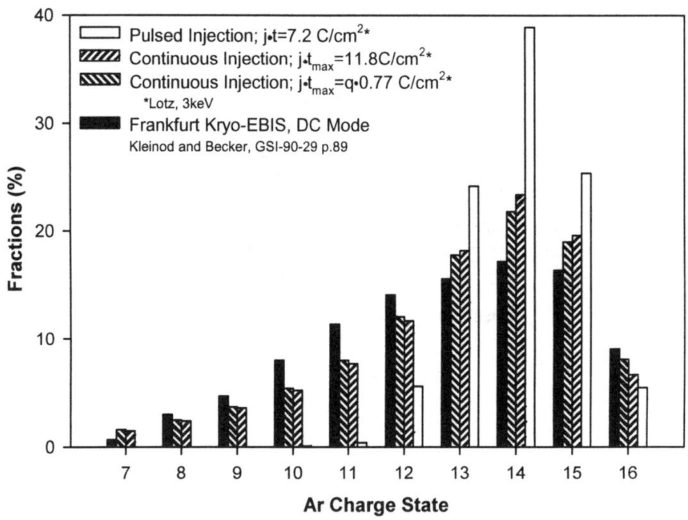

FIGURE 3.11 Ar charge state distribution obtained with the Frankfurt Kryo-EBIS in a DC-mode compared with the distributions expected for 7.2 C/cm^2 with one fixed confinement time (pulsed injection), as well as for 11.8 C/cm^2 and $q \cdot 0.77$ C/cm^2 with the confinement times equally distributed between 0 and t_{max} (continuous injection).

the measured Ar^{14+} fraction by a factor of 1.36. Figure 3.11 shows that this assumption still underestimates the fractions of the higher as well as the fractions of the lower charge states indicating longer average confinement times for the higher charge states and shorter average confinement times for the lower charge states. If one assumes the confinement times to be equally distributed between zero and a maximum value t_{max}, which is proportional to the charge state q, the prediction for q·0.77 C/cm^2 exceeds the measured Ar^{14+} fraction by 1.27, a marginal improvement. The inability to predict the charge state distribution with a simple assumption underlines the complex nature of the DC operation where the confinement times depend on the effective trapping potentials, which in turn depend on the evolution of charge state and energy of each ion, which in turn depend on all other ions due to ion-ion interactions. Models which include the loss through continuous ion expulsion [50,89–93] can explain some features, but never have been developed to have real predictive power. Based on the 11.8 C/cm^2 from Fig. 3.11 and the estimated 1000 A/cm^2 [84], one can deduct average confinement times in the range of 10 ms. However, there is no easy way to verify this estimate as DC operations have no reliable control of the confinement time hence do not allow for a reliable estimate of the current density. The DC mode is recommended for the production of ions with intermediate charge states which require confinement times longer than that obtained with an TOF mode, but not requiring very high ionization factors.

The small radius of the source volume, the gradual heating, and the confinement increasing with charge state keep each ion from acquiring an energy per charge substantially above the downstream potential barrier and therefore all ions of interest escape with an energy per charge equal to or very slightly above the lowest potential in the center of the electron beam in the exit barrier. The elevated voltage of the exit barrier prohibits the accumulation of ions and therefore the potential in the center of this drift tube is given by the drift tube voltage lowered by the space charge of the uncompensated electron beam (Eq. 3.11), which remains constant as long as the drift tube voltage, the electron beam current and the electron beam diameter remain unchanged. Accordingly DC modes can deliver ion beams with an energy which is very stable and has a very narrow spread. An energy spread of 10 eV/q was quoted for the Kryo-EBIS [77], where 0.8 eV/q or less was determined for NICE [95–97] and as little as 0.3 to 0.4 eV/q for the Mini-EBIS operating with moderate intensities [89,98].

Steady state modes of operation are sometimes favored because they produce monotonous ion beams with very small variations of the instantaneous ion beam current, which is favored by most atomic physics experiments. However, it is the simplicity which makes steady state operations popular. Required are only a few high voltage supplies available off-the-shelf at low prices featuring high levels of stability and regulation and very low PARD (periodic and random deviations). The tuning is also amazingly simple. After setting up the desired electron beam one dials the downstream barrier voltage according to the radial trapping potential U_r and the desired ion energy and applies a 20-50 V higher voltage to the upstream barrier. After adjusting the leak valve for a small flow of the seed gas and powering the trapping region with a voltage of roughly 10 V below the downstream barrier a small current should appear on the diagnostic device tuned to the desired charge state. From then on one can optimize the desired ion current by carefully adjusting the flow of the seed gas and the voltage of the trapping region. Fine tuning the ion transport and the electron beam can bring further enhancements.

Tuning the seed gas flow and the voltage applied to the trapping region balances the requirement for a deep trap to achieve the desired confinement times and the requirement for a shallow trap to help the highly charged ions escape. These two contradicting requirements are encountered in all steady state operations of sources for multiply and highly charged ions [99,100].

3.4.2 Batch Mode Operations with Fast Expulsions

The contradicting requirements discussed above can be separated in time with a time dependent operation which first strongly confines the ions to prevent their escape and then terminates the confinement completely to let all ions escape. Such a "pulsed" [24,30,63,77,81,82,84,87,90,91,98,101], "cyclic" [7,25,102], "containment" [70,79,80,83,85,88], "confinement" [86], or "batch" [77,99,103] mode provides direct control of the confinement time, and therefore can yield the narrow charge state distributions and the high dominant fractions predicted by the evolution of charge states, and accordingly higher yields of highly charged ions. However, batch modes are substantially more complicated and costly because they require the switching and varying of high voltages in a very controlled fashion depending on the desired time structure of the high yield beams of highly charged ions. Figure 3.12 shows as an example the switchyard of the KSU-CRYEBIS, where a central clock controls the length of the various time periods and coordinates the channels of a custom-made, multi-channel, low-voltage function generator. The low voltages are multiplied with high voltage amplifiers to obtain the required voltage levels. High voltage supplies are used for drift tubes which maintain a constant voltage level throughout the entire operation. Figure 3.10 shows the potential distributions used for the passive expulsion for which the confining downstream barrier is dropped rapidly below the trap potential to let the ions escape as soon as they reach the downstream end of the trap. The ions escape with an energy given by the potential of the trap and the energy acquired from the electron beam heating, yielding an energy spread which, in principle, can be as large as the radial trapping potential U_r (Eq. 3.11). The length of the ion beam pulse depends on the length of the trap and the distribution of the longitudinal ion velocities and grows with the distance from the source due to the energy spread of the ions. Fig. 3.10 shows also the potential distributions used for an accelerated ion expulsion where the potential of the trap is tilted during expulsion to accelerate the ions towards the exit. As the added energy increases with the distance from the downstream end, the accelerated expulsion produces an approximate time focus in a certain downstream location with a shorter length of the ion pulse and a higher peak intensity. The location of the approximate time focus can be controlled with the slope of the expulsion potential. However, the length of the pulse will grow rapidly behind the time focus due to the ion energy spread which is enlarged by the tilted potential. Instantaneous and accelerated expulsions produce ion pulses with a length of tens of μs [25,63,73,86,101,104] desired for targets with short acceptance windows such as synchrotrons or targets excited by pulsed lasers, where the energy spread is not of primary concern. Ion energy spreads of 50-70 eV per charge have been reported [7,25,102].

3.4.3 Batch Mode Operations with Slow Expulsions

The energy spread of the ions can be substantially reduced with the spill-over expulsion shown in Fig. 3.10 for which the trap voltage is increased to let the ions spill over the downstream barrier, the dam, as soon as their energy per charge matches or exceeds the remaining downstream potential barrier. The ions escape with an energy equal to or slightly above the lowest potential in the center of the dam which remains unchanged. The trap potential $U_t(t)$ is commonly increased linearly in time

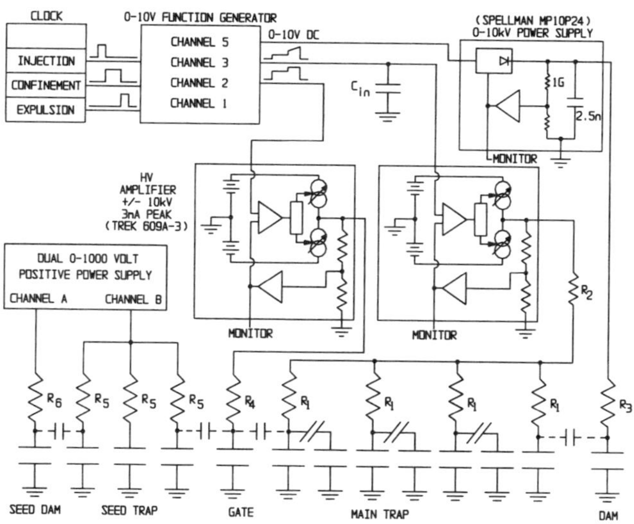

FIGURE 3.12 Schematic of the high voltage switchyard used to operate the KSU-CRYEBIS in a batch mode.

$$U_t(t) = V_o + V_r \cdot t/t_e \tag{3.12}$$

where V_o is the starting voltage and V_r the voltage range covered during the expulsion period t_e. The expulsion of all ions requires the final voltage $V_o + V_r$ to match or exceed V_d, the voltage applied to the dam. Reduced energy spread of the entire expulsion requires the voltage range V_r to cover approximately U_r, the maximum ion energy per charge which can be trapped by the electron beam. The remaining energy spread per charge of the escaped ions depends on U_b, the trapping voltage of the electron beam, and the number of times each ion can sample a certain remaining barrier potential which depends on the ramp rate $dU_t(t)/dt$. Therefore very low ramp rates should yield ion energy spreads close to the values obtained in the DC mode. The number of ions escaping per time unit dN_q/dt is given for each charge state q by the ramp rate $dU_t(t)/dt$ and the number of such ions in the range of ion energy which can escape in the corresponding range dU_t. Hence, the ions with charge state q yield an instantaneous ion current $i_q(t)$ of

$$i_q(t) = q \cdot e \cdot dN_q(t)/dt = e \cdot q^2 \cdot dN_q/dE_t \cdot dU_t(t)/dt \tag{3.13}$$

If the ramp rate $dU_t(t)/dt$ is slow enough to produce a minimal energy spread of the escaped ions and the expulsion time is short enough to maintain the distribution of charge states, the instantaneous ion current can be used to determine the energy distribution of the trapped ions dN_q/dE_t. Equation 3.13 shows also how the instantaneous ion current can be reduced and the duty cycle correspondingly increased when the expulsion period is extended and the ramp rate is reduced correspondingly [65,105,106]. Figure 3.13 shows a series of expulsions with different ramp rates V_r/t_e. Fast ramps like in Fig. 3.13a can produce short ion beam pulses without a substantial reduction of the ion energy spread, similar to the passive expulsion discussed in 1.4.2. When the ramp rate is reduced like in Fig. 3.13b, the instantaneous ion current represents the ion energy distribution inside the trap. The correspondingly reduced energy spread of the escaped ions improves the ion beam transmission and therefore increases the magnetically analyzed time-averaged ion beam current I_a as shown in Fig. 3.13.

Experiments with steady-state targets often have to limit the ion beam intensity to keep the peak instantaneous counting rate within the acceptable range of their data taking system. Such experiments prefer high duty cycles to improve the data accumulation rate which grows proportionally with the time-averaged data rate. An appropriate figure of merit is the effective duty cycle, f_{eff}, defined as

$$f_{eff} = I_a/i_p \tag{3.14}$$

where I_a is the time-averaged ion beam current and i_p is the maximum or peak value of the instantaneous ion beam current. For example, the ramp rate reduction allowed us to improve the effective duty cycle from 3% in Fig. 3.13ai to 10% in Fig. 3.13bi. If the voltage range V_r is reduced below U_r and the starting voltage V_o is correspondingly increased, as in Fig. 3.13c, one often encounters another ion beam peak at the beginning of the expulsion period, which, in this case, is caused by the high ramp rate before the desired ramp rate is established. This double-peaked expulsion increases the effective duty cycle to 21% but reduces, in this case, the transmitted time-averaged ion beam current.

If the effective duty cycle is increased with a reduction of the trap voltage ramp rate and a corresponding increase of the expulsion period, the increased cycle time reduces the expulsion frequency and hence the time-average ion beam current. However, experiments with time-independent targets prefer a high time-averaged ion beam current simultaneously with the high effective duty cycle, as it often allows for a reduction of the accepted phase space, and hence for an improved experimental resolution. Consequently, such

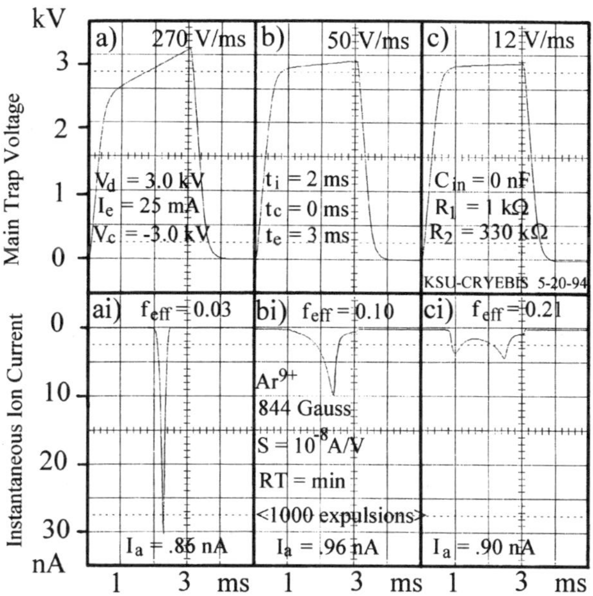

FIGURE 3.13 Top: The high voltage filtered with 331 kohm ($R_1 + R_2$) and applied to the main trap with ramp rates of 270, 50, and 12 V/ms for one cycle of 5 ms. Bottom: The magnetically analyzed (844 Gauss) instantaneous Ar^{9+} beam current was recorded with a current preamplifier sensitivity S of 10^{-8} A/V and minimum rise time RT as the average of <1000 expulsions>. Lowering the ramp rate increases the effective duty cycle f_{eff} but also affects the time-averaged beam current I_a.

experiments are best served with an ion beam which optimizes the merit current I_m, defined as the product of the time-averaged ion beam current and the effective duty cycle

$$I_m = I_a \cdot f_{eff} = I_a^2 / i_p \tag{3.15}$$

An example is shown in Fig. 3.14a with the production of Ar^{8+} yielding a time-averaged ion current of 248 pA with a 20% effective duty cycle and a 49 pA merit current. If the expulsion period is increased from 0.9 to 1.4 ms and the expulsion delayed and extended by a few tenths of a ms, the time-averaged ion current drops to 206 pA, the effective duty cycle to 17% and the merit current to 35 pA and therefore seems to be undesirable. However, if the ion beam current is measured approximately 3 m further down stream with enhanced resolution, the 3.8 mm wide slits let pass a maximum time-averaged Ar^{8+} current of only 30 pA for the fast expulsion but 76 pA for the slower expulsion which reverses the ranking. This example demonstrates the benefit of tuning an ion source with the yield accepted by the target or, if impossible, with a sensor which has a similar acceptance.

The reason for the lower Ar^{8+} current passing through the narrow slit can be found in Fig. 3.15 which shows a full-width-half-maximum (FWHM) of 1.7 Gauss for the slower expulsion of Ar8+ which puts an upper limit of 12 eV/q on the energy spread of the ion beam. The fast expulsion, however, yields a FWHM of 5.3 Gauss indicating an energy distribution with two peaks separated by 4.1 Gauss or 28 eV/q which cannot pass simultaneously through the slits. It is interesting to note the ratio between the two peaks to change

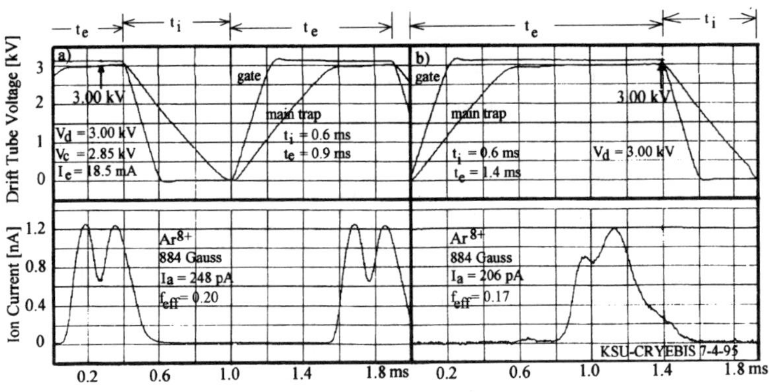

FIGURE 3.14 Top: The high voltage applied to the main trap for the production of Ar^{8+} with a fast (0.9 ms) expulsion and a moderate (1.4 ms) expulsion which delays the expulsion for a few tenths of a ms. Bottom: The magnetically analyzed (884 Gauss) instantaneous Ar^{8+} beam current yields a higher time-averaged current for the fast expulsion due to a higher repetition rate.

with charge state. The double-peaked energy distribution does not seem to be consistent with Fig. 3.14 where the main trap voltage seems to barely exceed the 3.00 kV dam voltage. A part of the problem is identified in Fig. 3.16 which shows the high voltage applied to the main trap for a typical operation with a 2 ms long injection, 3 ms long confinement, and 5 ms long expulsion period. The lower part of the Figure shows the voltage deviations observed on the dam which are generated during the fast switching due to the capacitive coupling between those electrodes. The rapid increase of the trap voltage causes an increase of the dam voltage and hence any ion escaping at that time leaves the trap with a correspondingly higher energy. If the dam is powered with a high voltage supply, the small slew rate requires a rather extended recovery time, but the voltage deviations can be reduced with additional capacitive loads. However, a capacitive load is counterproductive with bipolar high voltage amplifiers which can sink current and feature high slew rates and therefore are able to recover much faster. Powering the dam with such an amplifier benefits the production of low charge states which require minimal confinement time and should be expelled rapidly. The cycling also can be accelerated by reducing the voltage step between the different time periods, for example by injecting at a voltage level slightly below the expulsion level, which requires minor modifications of the KSU-CRYEBIS.

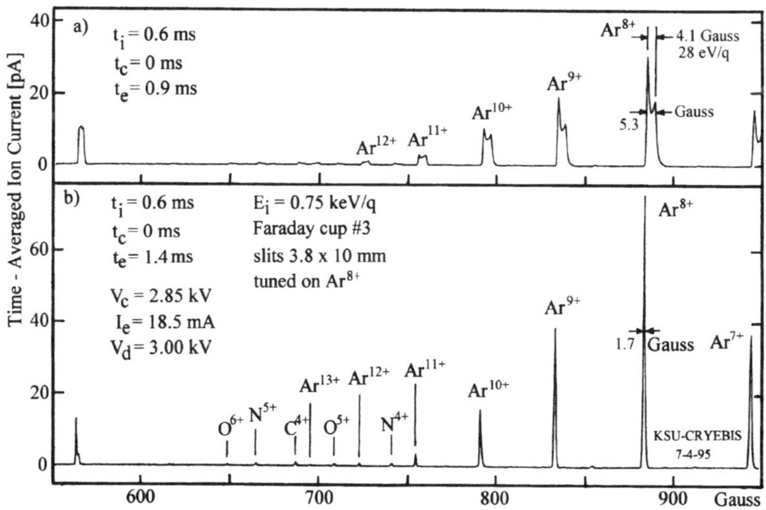

FIGURE 3.15 The time-averaged ion beam currents from Fig. 3.14 measured further downstream with a higher resolution indicates a double-peaked energy distribution of the Ar ions from the fast expulsion in Fig. 3.14a. The moderate expulsion from Fig. 3.14b indicates a much narrower energy distribution and hence a higher current for a properly tuned analyzing magnet.

The situation is different for the production of the higher charge states as the required confinement time dwarfs the switching time, which allows for delaying the expulsion for a few ms to minimize the effect of the dam voltage deviations generated by the switching of capacitive coupled drift tubes. In this case the ion energy spread can be lowered by powering the dam with a high voltage supply which normally features lower PARD than corresponding high voltage amplifiers. However, long confinement times coupled with high effective duty cycles require long expulsion times and correspondingly small ramp rates V_r/t_e, another challenge for high voltage amplifiers which tend to overshoot and ring especially with capacitive loads, as shown in Fig. 3.17a. If the overshoot causes the trap voltage to exceed the dam voltage, a substantial fraction of ions can escape without a well defined energy. If the escape of ions is minimized during the overshoot by lowering the expulsion start voltage, no ions can escape during a substantial fraction of the expulsion time period due to the time needed for the linear ramp to reach the appropriate voltage level, which is responsible for the small duty cycles as shown in Fig. 3.17ai. The effect of the overshoot can be reduced with a filter, for example with

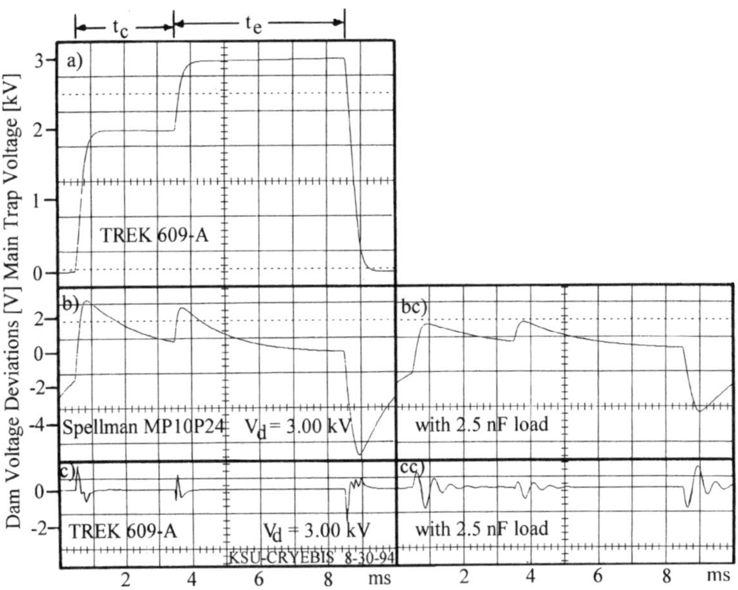

FIGURE 3.16 Top: The high voltage applied to the main trap with a TREK 609A-3 high voltage amplifier during a typical cycle of 10 ms. Stray capacitance couples the main trap to the dam. Bottom: Induced deviations from the mean dam voltage V_d held at 3 kV with a Spellman MP10P24 high voltage supply or with a TREK 609A-3 high voltage amplifier. Right: The induced deviations after adding an 2.5 nF capacitor to the dam drift tube.

a resistor between the high voltage amplifier and the drift tubes, as shown in Figs. 3.12 and 3.13. However, the overshoot changes with the voltage step size and some other factors, and therefore any counter measure requires adjustments for optimum suppression. Adjustments can be handled much easier on the low voltage level, for example by lowering the slew rate of the control voltage [100]. Control voltages with controlled slew rates allow one to keep the amplifier's error signal in the controlled range which can eliminate the overshoot. Figure 3.12 shows a simple, but rather efficient solution with

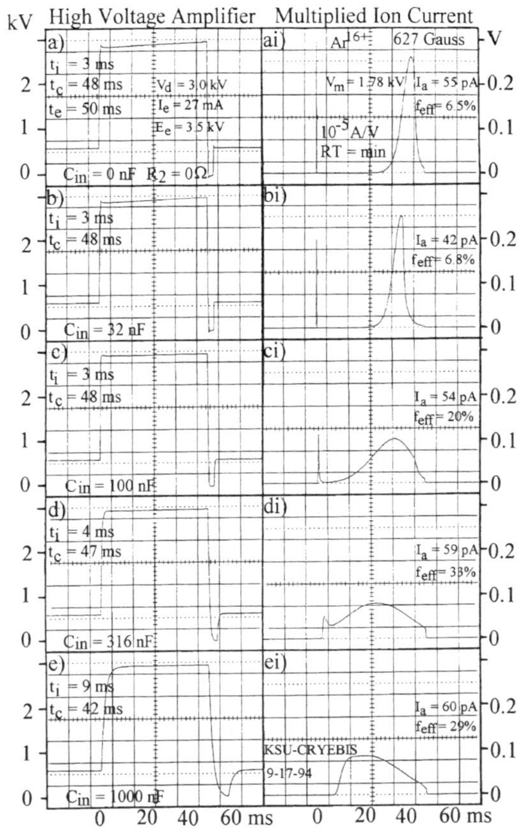

FIGURE 3.17 The unfiltered (R_2=0) high voltage applied to the main trap with the control voltage being filtered with different capacitors C_{in} for one cycle of 0.1 s. Right: The magnetically analyzed (627 Gauss) instantaneous Ar^{16+} beam current was multiplied with a channeltron held at 1.78 kV (V_m) and recorded with a current preamplifier sensitivity of 10^{-5} A/V. Filtering the control voltage with more than 100 nF allows for the complete elimination of the spike at the beginning of the expulsion period caused by the overshooting HV amplifier.

an adjustable capacitor C_{in} loading the control voltage circuit which allows for a high initial slew rate before decreasing rapidly when approaching the desired voltage level. Figure 3.17 shows how the initial ion spill can be reduced by increasing the control circuit capacitor, with the desired value depending on several factors, in particular the control circuit and the size of the voltage step. In the case of Fig. 3.17, 316 nF completely eliminates the initial ion spill as the broader initial peak appears approximately 2 ms after switching and is caused by the higher initial ramp rate before reaching the constant ramp rate. This initial peak can be delayed further and/or blended out by increasing the capacitor C_{in} and/or adjusting the expulsion voltages. However, the capacitor C_{in} reduces also the slew rate when lowering the voltage for the next injection cycle and therefore requires correspondingly longer injection periods. High yield productions of low charge states require rapid cycling and therefore a reduction of the filter capacitor C_{in} to the minimum level without overshoot. Figure 3.14 shows the production of Ar^{8+} where C_{in} was selected to eliminate just the overshoot and therefore some ringing remains as shown by the multiple peak structure of the instantaneous ion beam current.

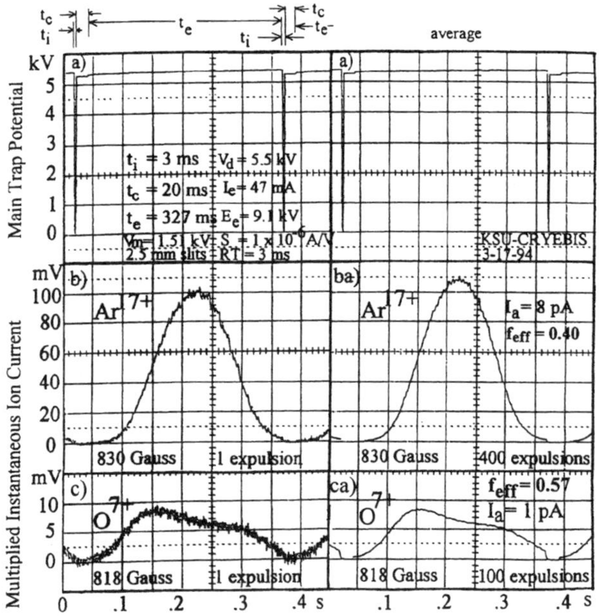

FIGURE 3.18 Top: The voltage applied to the main trap tuned for a leaky batch mode production of Ar^{17+}. Bottom: The instantaneous ion beam current multiplied by a channeltron showing Ar^{17+} as well as O^{7+} from the residual gas. The single expulsion on the left exhibits a small noise level and hence minor details of the waveform can be seen only in the average of many expulsions on the right.

3.4.4 Leaky Batch Mode Production

The optimum confinement time for the production of high charge states is roughly defined with Eq. 3.3. However, for very slow expulsions the definition of the confinement time becomes fuzzy as the ions are still confined during a fraction of the expulsion period. Figure 3.18 shows the production of Ar^{17+} where the nominal confinement time is reduced to 20 ms while the nominal expulsion time is extended to 327 ms. This production yields a substantial time-averaged Ar^{17+} beam current of 8 pA with a high effective duty cycle of 40%. Ions leak out of the trap during 93% of time not far away from the 100% obtained with the steady state leaky mode. However, the escaping ions are not replaced until the trap is empty, and therefore is in fact a batch operation. The 20 ms nominal confinement time is sufficient for at least a 10 fold ionization of Ar but insufficient for the production of an appreciable amount of Ar^{17+}, and therefore the Ar^{17+} ion beam current does not appear until 20 ms after the start of the expulsion time. Approximately 100 ms after the start of the confinement time the Ar^{17+} current starts to grow rapidly due to the rapidly growing fraction of the Ar^{17+} population. Roughly 200 ms after the start of the confinement time, the Ar^{17+} beam intensity reaches a maximum and then falls off to almost zero at the end of the expulsion period. The decline is due to the declining number of ions remaining in the trap and therefore, in this case, has little to do with evolution of charge states or ion energy distribution. It is interesting to note that the O^{7+} current starts right at the beginning of the expulsion period and peaks before the Ar^{17+} current due to the lower

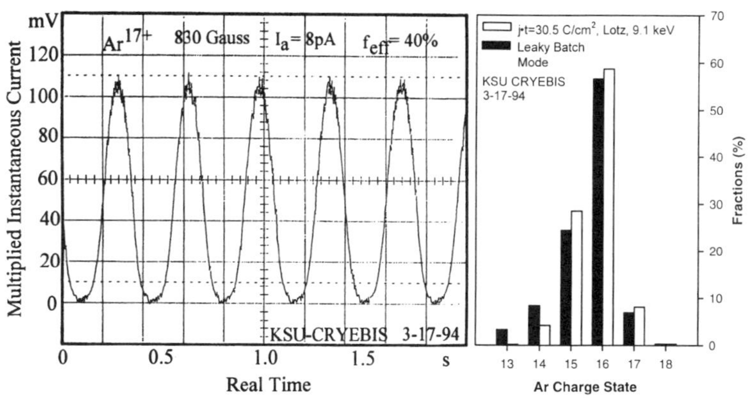

FIGURE 3.19 Left: A 2 s long observation of the instantaneous Ar^{17+} current multiplied by a channeltron. The leaky batch mode production from Fig. 3.18 delivers Ar^{17+} ions for more than 90% of time close to the 100% obtained with a leaky mode. Right: The Ar charge state population predicted for 30.5 C/cm^2 compared with the population fractions determined from a measured charge state distribution resembles the batch mode nature.

charge state. However, the O^{7+} current maintains some intensity up to the end of the expulsion period indicating the presence of residual gas ions during the entire cycle which are believed to be crucial for the cooling of the highly charged ions. Figure 3.19 shows the measured charge state distribution to match the prediction for an ionization factor of 30.5 C/cm^2, which needs to be interpreted with caution as the actual ion confinement time varies by a factor of 17. Consequently, the measured charge state distribution is slightly broader than predicted. Figure 3.19 shows also the true signature of this "leaky batch" mode, which is the almost continuous ion beam current with oscillating intensity, interrupted only briefly when preparing the trap with the next batch of ions. The strong variation of the ion current is due to the strongly varying number of ions available with the desired charge state. If the initial confinement time would be increased to increase the initial number of the desired charge state, the ion beam intensity would still vary strongly due to the evolution of charge state as shown in Fig. 3.20. If, for example, the expulsion time represents 93% of the total cycle time, as in Fig. 3.18, Fig. 3.20 predicts that the evolution of charge state will contribute at least a factor of 5 to the intensity variation. The intensity variations shown in Fig. 3.20 are actually enhanced by the depopulation of the trap. The leaky batch mode production obviously does not take full advantage of the possibility to line up large fractions of ions in the desired charge state in part by selecting time cycles which favor the expulsion of the desired charge state rather then losing it through ionization to the next higher charge state.

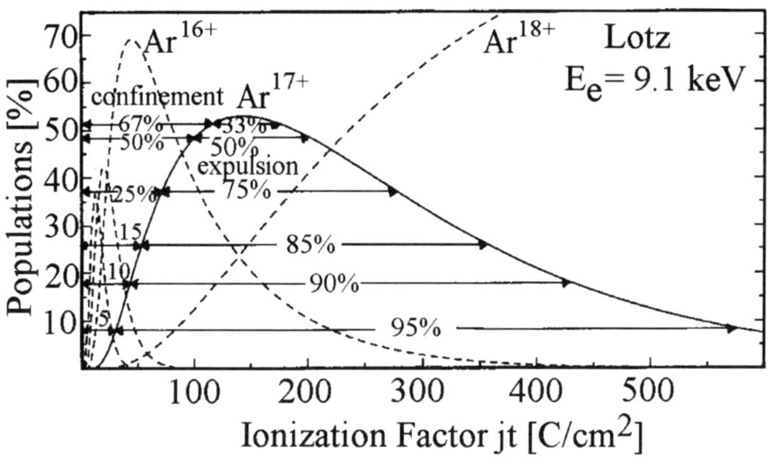

FIGURE 3.20 The evolution of Ar^{16+}, Ar^{17+} and Ar^{18+} predicted with the Lotz approximation for direct electron impact single ionization for 9.1 keV electrons. The figure shows the fraction of Ar^{17+} to change substantially for any expulsion time which exceeds the confinement time.

3.4.5 DC Batch Mode Production

Figure 3.20 indicates that the population of a charge state varies little as long as the expulsion period does not exceed the confinement period. If, for example, the confinement period is twice as long as the expulsion period, the optimum time-averaged (over the expulsion period) Ar population reaches 97 to 99% of the maximum population depending on the charge state. Even if the expulsion period matches the confinement period, the optimum time-averaged Ar population still reaches 92 to 97% of the maximum population. A further increase of the expulsion period rapidly diminishes the time-averaged populations and hence the ion yields, which sets an upper limit for an efficient production. However, even if the duty cycle is limited to 50%, the effective duty cycle can be increased if the peak instantaneous ion current is reduced while the instantaneous ion current at the beginning and at the end of the expulsion period is increased, which requires a nonlinear ramp of the trap voltage during expulsion. If, for example, the ramp rate $dU_t(t)/dt$ is proportional to the inverse of the ion energy distribution dE_t/dN_q, Eq. 3.13 predicts an instantaneous ion current which remains constant during the entire expulsion period. The reality is more complicated as the depopulation of the trap changes the trap's space charge potential and therefore the optimum expulsion ramp $U_t(t)$ depends on the population of the desired charge state as well as on the total ion population remaining inside the trap.

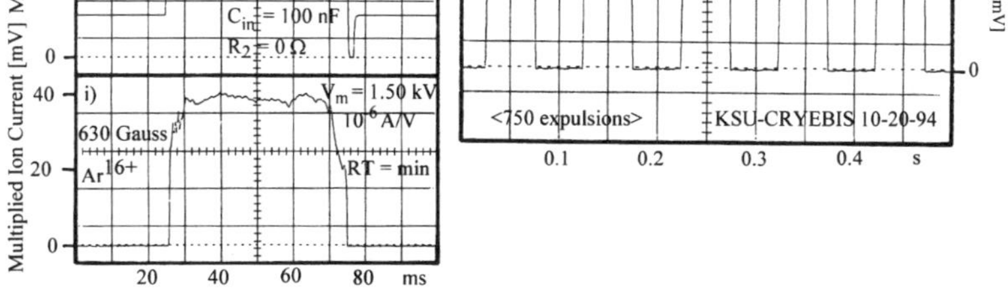

FIGURE 3.21 Left: The voltage applied to the main trap tuned for a DC batch mode production of Ar^{16+} shown with increased resolution on top. The voltage profile was tuned with an arbitrary wave form generator to keep the intensity of the expelled Ar^{16+} ions constant during the entire expulsion period as shown on the bottom. Right: The instantaneous Ar^{16+} current multiplied with a channeltron and averaged over 750 periods of 0.5 s resembles a square wave which minimizes the peak instantaneous current and hence maximizes the effective duty cycle with an expulsion time which does not exceed the confinement time.

Figure 3.21 shows the production of Ar^{16+} expelled with a nonlinear voltage ramp which keeps the ion beam current almost constant during the entire expulsion period. The effective duty cycle of 45% is close to the 50% one could expect ideally for the selected time periods. This expulsion ramp was generated by manually programming an arbitrary wave form generator, increasing the ramp rate at times of low instantaneous ion current and lowering the ramp rate at times of high instantaneous ion current, which is a tedious and time consuming process. The residual gas pressure decreases usually after the initial tuning which slowly lowers the neutralization factor and slowly increases the net trapping potential. Therefore, one expects the instantaneous ion beam current at the beginning of the expulsion period to slowly decrease while the instantaneous ion beam current at the end of the expulsion period slowly increases. In addition, the residual gas pressure usually also changes with the temperature of the environment which undergoes a typical daily cycle, also affecting the instantaneous ion current distribution. The "DC batch" mode, with a constant current expulsion, is very sensitive and therefore requires regular fine tuning of the expulsion ramp to maintain the optimum effective duty cycle.

A computer can, in a split-second, analyze the measured instantaneous ion current and calculate the changes of the ramp rate as a function of time required for a constant ion expulsion rate. Iterations are needed for a proper correction of the net space charge potential which changes during the expulsion time with the depopulation of the trap. Fine tuning should be accomplished with the average instantaneous current measured over many expulsions to minimize the effect of random fluctuations. Once implemented, a computer controlled system easily can fine-tune the expulsion ramp at regular intervals to maintain the optimum effective duty cycle.

As indicated in Fig. 3.20, the evolution of charge states continues during the expulsion period, and therefore, for a 50% duty cycle, one should select a confinement time which is roughly 1/3 shorter than the confinement time which optimizes the yield with a rapid expulsion. A DC batch mode production with such a reduced confinement time should yield almost the same ion yield per expulsion as obtained with a production optimized with a rapid expulsion. However, the total cycle time is roughly 1/3 longer than a batch production with a rapid expulsion and therefore the time-averaged ion yields match only 3/4 of the time-averaged yields obtained with rapid expulsions.

Figure 3.21 shows the true signature of the DC batch mode which is an almost constant ion beam current during the entire length of the expulsion period. The DC-batch mode production steps up the complexity of the required controls, but delivers high, effective duty cycles of highly charged ions with optimum yields because it benefits almost fully from the unique EBIS feature, the direct control of the confinement time.

3.4.6 The Future: Multi-Trap and Flexi-Trap Operations

Efficient batch mode productions with higher duty cycles can be accomplished if the conflicting requirements for confinement and expulsion are separated in space rather than in time. If one, as proposed by Becker [77], subdivides the ionization region into two separate traps, atoms or low charged ions can be injected into the upstream trap where they can be confined strongly while being ionized. After the selected confinement time, the ions are moved into the downstream trap, from where they are expelled over the downstream barrier while the next batch is ionized in the upstream trap. If the expulsion is adjusted to last for the entire selected confinement time, one obtains a steady flux of escaping ions equivalent to a 100% duty cycle [77]. If the expulsion is accomplished with the voltage profile used for the DC batch mode, one obtains a monotonous ion current with an effective duty cycle close to 100%, which may only briefly fluctuate when restocking the downstream trap [99]. Each ion batch reaches its maximum batch charge Q^+ at the end of the ionization cycle when the batch is transferred into the downstream trap, and therefore both traps should have the same capacity to minimize unwanted ion losses when operating with high neutralization factors. The minimum capacity Q_d^- of the downstream trap is determined by its length l_d and the maximum local electron beam energy, which does not have to exceed $E_i/q - eV_c$, given by the desired ion energy per charge and the cathod voltage V_c. The minimum capacity Q_u^- of the upstream trap is determined by its length l_u and the electron beam energy E_e which optimizes the production of the desired charge state q. Hence the maximum yields can be expected with traps with a length ratio of

$$l_u/l_d = \sqrt{(E_i/q - eV_c)/E_e} \tag{3.16}$$

Depending on the desired charge state, specie, and ion energy, the ratio may be larger or smaller than 1. Most EBISes contain a series of drift tubes and therefore may allow for the movement of the gate into the best position. If this flexibility is not available, equal length may be the most reasonable compromise. Hence one can expect approximately only half of the ion yield per batch obtained with a 50% DC batch mode but with twice the frequency, yielding almost the same time-averaged ion current but with a 100% effective duty cycle.

Higher time-averaged ion yields could be expected if the actual expulsion time were reduced to 1/3 by subdividing the ionization region into 3 traps. If the electron beam current density is approximately uniform over the ionization region, the first trap would bring the ions already close to the desired final charge state where the second trap predominantly enhances the population of the desired charge state. In an optimized production of Ar^{17+}, for example, the center trap would increase the mean charge state by approximately 0.5, a 3% increase. Therefore the upstream and the center trap should be of equal length to minimize unwanted ion losses when operating with high neutralization factors. The reduced total cycle time and the slightly improved time-averaged populations promise an increase of the time-averaged ion yields by 13-17%. However, a fraction of the expected increase is lost to the additional gate which reduces the total ion capacity. Whereas a single trap can be operated with one single high voltage amplifier, each additional trap requires two additional high voltage amplifiers to control the voltages of the additional trap and gate. The increasing complexity and the decreasing gain do not encourage operations beyond three traps.

Multiple traps permitting continuous expulsion can also support the production of many other time structures which may be desired for certain experiments. For example, targets excited with pulsed lasers or experiments which use the time-of-flight to determine certain reaction parameters, would benefit from smaller ion bunches delivered with a higher frequency. Such ion beam bunching could be accomplished by varying the voltage of the downstream dam with the desired time pattern, originally proposed by Arianer in combination with a DC mode [105]. However, the ion beam bunching combined with a multiple trap DC batch mode provides higher yields of the high charge states with a small ion energy spread.

The stacking of user devices such as ion traps or synchrotrons with an EBIS occurs often with efficiencies which are limited by the length of the ion pulse obtained with fast expulsions, passive or accelerated. In most cases the length of the ion pulse is limited by the range of transit times of the low energy ions traveling through the trap before they reach the downstream end. An accelerated expulsion cannot correct for the axial ion velocity distribution and therefore is unable to produce an exact time focus. On the other hand, the ions could be produced in a long trap with minimal losses due to a limited neutralization factor, and then shortly before expulsion, the ions could be compressed into a short trap by gradually moving the upstream dam or gate. Once this flexible trap is fully compressed, the ions are quickly released with a fast expulsion, passive or accelerated. A careful implementation of the compression is required to limit the increase of the ion energies and to avoid excessive losses. The losses, however, would affect predominantly the unwanted low charged ions. Losses of the desired highly charged ions could be minimized by stepping up the electron beam current for the brief compression and expulsion period. The increased ion energies and the shorter length of the compressed trap should yield a substantial reduction of the transit times and accordingly shorter ion beam pulses.

3.5 INJECTION OF SEED PARTICLES

3.5.1 Direct Gas Injection

The production yields of highly charged ions depend substantially on the injection method of the seed particles. Steady state operations discussed in Section 3.4.1 require a steady supply of seed particles which essentially can be supplied only with direct gas injection, where a gas with the desired atomic specie, the seed gas, is leaked into the space of the dense electron beam which ionizes and traps the seed particles. As discussed, the TOF mode works most efficiently if the gas leak is a short distance downstream of the upstream barrier. However, a DC mode may be slightly more efficient if the gas is injected into the center of the solenoid, the location of the smallest beam diameter and hence the highest trapping potential U_r. Ions born in the location of the highest trapping potential start their confinement with the lowest possible energy which increases their average confinement time. This should produce slightly narrower charge state distributions and hence higher yields of the desired charge state.

The direct gas injection also can be used for batch mode productions with a single trap. The desired ion yields also may be slightly higher with a central injection point as the lower initial ion energy improves the confinement and hence the ionization efficiency for long confinement times. The ionic output from a batch operated EBIS in the range from pA to 100 nA corresponds to 10^{-8} to 10^{-14} cc/s gas at standard pressure and temperature. If 0.1% of seed particles are successfully ionized and expelled, the seed gas flow needs to be regulated in the range of 10^{-5} to 10^{-11} cc/s, which requires an ultra-high-vacuum leak valve with ultra fine control. However, with a continuous direct gas injection, the trap accumulates ions with individual confinement times equally distributed between the confinement time of the trap and 0. Therefore, the expelled ions exhibit a charge state distribution substantially broader than predicted by the evolution of charge states for a single confinement time and accordingly smaller maximum populations [107]. For example, Fig. 3.11 shows the population of Ar^{14+} expected from a continuous gas injection to be 40% lower than from an initial gas injection. The ions born in the second half of the confinement period are unlikely to reach the desired charge state but contribute substantially to the neutralization of the trap and therefore limit the yield of the desired charge state. Fast operating, pulsed valves are commercially available, but the small conductance between an externally mounted, pulsed valve and the injection point limits the slew rate of the effective gas pressure. Hence pulsed systems have only been installed where the main solenoid provides sufficient access [63,108]. Differentially-pumped, pulsed gas injection systems can produce higher slew rates but consume substantiatly more gas. For example, the sytem on the RIKEN Mini-EBIS delivers 0.16 s long (FWHM) gas pulses [108], which have shown to enhance the high charge states [109].

3.5.2 Electrically Switched Gas Injection

The speed limitations of neutral gas transports can be circumvented by controlling the particle flux after ionization, as ions can be rapidly controlled by switching voltages. Invented by Donets [30], the electrically-switched injection is shown schematically in Fig. 3.22, where the solid line indicates the upstream barrier to be upstream of the gas injection site, so that the generated ions can enter the confinement region during the injection time. The dashed line shows the upstream barrier after being moved downstream of the gas injection site at the beginning of the confinement time to prevent low charged ions from entering the confinement region. The low charged ions generated at the injection site during the confinement time are drained in upstream direction and crash most likely into the electron gun. Top performance of an EBIS as well as an electron gun demand a low gas pressure which puts an upper limit on the leak rate of the seed gas. The limited leak rate of the seed gas requires therefore a certain injection time to accumulate sufficient ions in the confinement region needed for a high yield production. The required injection time increases the cycle time and hence reduces the time-averaged yields. For long confinement

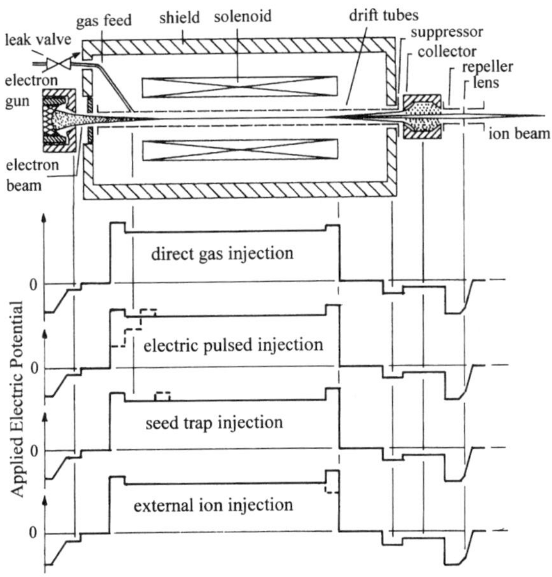

FIGURE 3.22 Top: Schematic cross section through a shielded electron beam ion source with an external electron gun. Bottom: Potential distributions applied to the drift tubes for the direct gas injection, the electric pulsed injection, the pulsed injection from a seed trap, and the pulsed ion injection from an external ion source. The potential distributions during injection time are indicated with solid lines, whereas the dashed lines indicate the potential distributions changed for the confinement time.

times, this limitation also could be reduced with a pulsed gas injection system which provides a peak pressure during the injection time.

Electrically switched gas injections have been installed at the end of the solenoid to minimize the capacity lost from the ionization region and to avoid the freezing in the cryogenic central region [90,110,111]. However, the injection from this location seems to be inefficient [67,112], most likely due to the high magnetic field gradient in the fringing fields which interferes with the axial transport of the ions. Therefore gas injections are normally installed in the high field region [30,86,113].

3.5.3 The Operation of a Seed Trap

The requirement for a high seed gas pressure during the injection period can be relaxed if the low charged ions are accumulated over a period of time in a separate trap, the seed trap. Figure 3.22 shows a schematic gate separating the seed trap from the main trap. The leak rate of the seed gas can be adjusted so that the seed trap fills up during the time required for the ions in the main trap to reach the desired charge state and to be expelled. Only a very short injection time is needed to lower the gate and to move the accumulated ions from the seed trap to the main trap. As the accumulation period stretches over the confinement- and expulsion-times, the charge states of the ions in

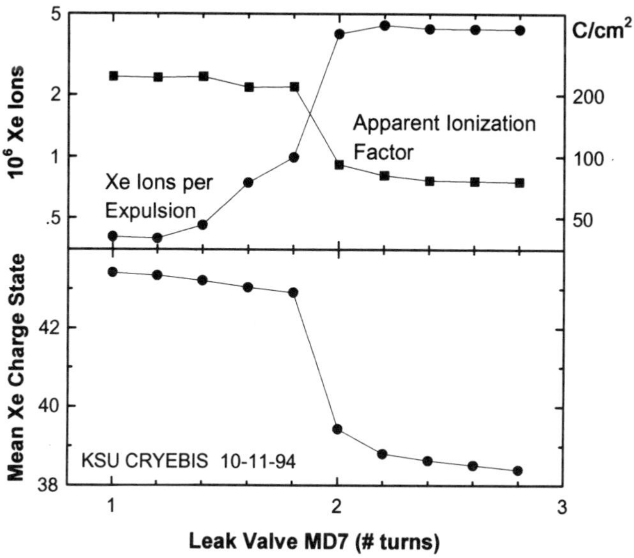

FIGURE 3.23 Number of Xe ions expelled in one expulsion, apparent ionization factor and mean Xe charge state as a function of setting of the leak valve which feeds the seed trap in the fringing field.

the seed trap evolve according to energy and average density of the electron beam inside the seed trap integrated over the range of confinement times. As discussed in Section 3.4.6, a seed trap inside the high field ionization region should have roughly the same length as the main trap to balance the maximum neutralization factors of the two traps. However, the length of main trap could be increased for the production of highly charged ions which require a high electron beam energy while compensating the capacity of the shortened seed trap with a lower electron beam energy. For this reason the seed trap of the KSU-CRYEBIS is normally kept at a low voltage which requires the voltage of the main trap to be cycled between the low injection voltage and the high expulsion voltage as shown in Fig. 3.13, 3.14, 3.16, 3.17, 3.18, and 3.21. Locating the seed trap in the fringing fields of the magnetic solenoid is very tempting due to the lower seed trap capacity required for the slower evolution of charge states by the lower electron beam current density, as well as the gain in the capacity of the main trap. However, the capacity of a seed trap in the fringing fields is difficult to predict due to the rapidly changing electron beam diameter.

Figure 3.23 shows the yields of Xe ions obtained with the seed trap installed in the fringing field of the KSU-CRYEBIS as a function of the seed gas leak rate. With an electron beam tuned to optimize the production of Xe^{44+} with a small gas flow, the number of Xe ions increases rapidly from 0.4 to $4 \cdot 10^6$ Xe ions. Then, however, the number of expelled Xe ions does not increase any further when increasing the leak rate. At the same time the apparent ionization factor drops by a factor of 2.4 and the mean charge state drops

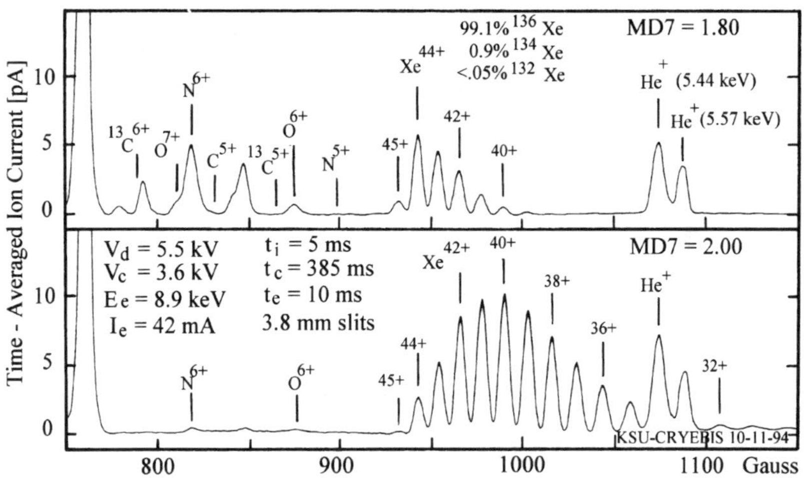

FIGURE 3.24 Xe charge state distributions for leak valve settings of 1.80 and 2.00 differ substantially.

by 3.5. The change is very obvious in the charge state distributions shown in Fig. 3.24, which also shows an almost complete elimination of the residual gases with the higher Xe leak rate. We believe that the valve opening of 2 turns is sufficient to fill the seed trap in 0.4 s so that the ions from the residual gas and any additional Xe are boiled off. The electron beam optimized with little compensation in the seed trap area loses some of its compression as the compensation is increased. CRYEBIS I yielded a decreasing apparent electron beam density when the valve was closed [25] which could be explained with the presumption that its operation was optimized with a high gas flow. The effect of the seed trap on the electron beam current density in the ionization region can be eliminated by locating the seed trap between the ionization region and the electron beam collector [74,114]. The efficiency of the injection from the seed trap in the fringing field of the KSU-CRYEBIS is questioned by the belief that there are roughly 240 electrons inside the seed trap for each of the expelled $4 \cdot 10^6$ Xe ions [67]. The small injection efficiency or the lower than expected seed trap capacity can be partly compensated through multiple injections, as done by Donets in his ion-by-ion cooling operation [74].

The voltage difference between the seed trap and the main trap plays a crucial role during the injection period. If the seed trap voltage is slightly above the main trap voltage, the ions gain energy when they enter the main trap which speeds up the expulsion and cycling when producing low charge states which require no nominal confinement time. However, for the production of high charge states, which require long confinement times, the ion energy should be minimized to prolong the confinement within the electron beam. Sometimes it appears beneficial to retard the ions by applying a slightly higher voltage to the main trap which removes some of the energy the ions acquired from the electron beam inside the seed trap.

3.5.4 External Ion Injection

The contradicting requirements regarding the neutral density for an efficient injection and an efficient production of high charge states can be eliminated with the injection of low charged ions from another ion source. Essentially all EBISes use a high field solenoid to compress their electron beam, and therefore require the ions to be injected on the solenoid axis, entering through the collector on the same trajectories on which the ions are expelled after being ionized. The trajectories are merged with an electrical switchyard, switching back and forth between injection and expulsion period. The low charged ions from the external ion source are slowed down to essentially zero energy when they enter the confinement region and are trapped by raising the downstream barrier as shown in Fig. 3.22. Singly charged ions from gaseous or low vapor pressure substances are injected from plasmatron [104,114] or PIG [115] ion sources, which require substantial differential pumping to maintain the low pressure required for a successful EBIS operation [115]. A very interesting and promising additional application of an external ion source is the injection of low charged ions over the down stream barrier during the confinement time. The low charged ion beam is mostly reflected at the upstream barrier and most ions will exit the EBIS after a single return pass without becoming trapped, which allows for cooling the highly charged ions [66] or for slow injections [104] without adding a compromising neutral gas.

3.5.5 Production of Metal Ion Beams

The production of ion beams from nonvolatile materials is a challenge for all ion sources because of the difficulty to inject the material into the ion source cavity with a steady, controlled rate. External ion injection delegates this problem to the external ion source for which there are a variety of methods available [116]. Plasmatron and PIG ion sources can produce metal beams from the particles sputtered from certain electrodes, and therefore have a limited life time. A hollow cathode ion source substantially extends the life-time for metal ion beams [117]. MEVVA are popular external ion sources for EBITs [94,118–120] as they generate the pulsed metal beams with a vacuum arc adding no detrimental gas load. They are not as popular with EBISes due to their limited lifetime and shot-to-shot variations.

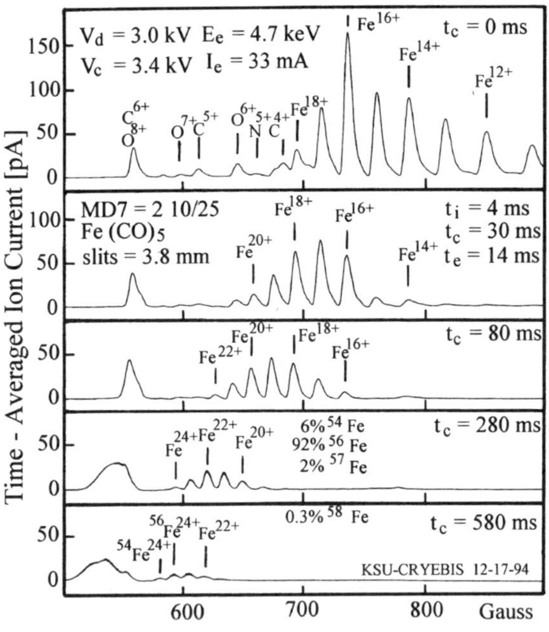

FIGURE 3.25 Charge state distributions for 5 different confinement times obtained with the injection of iron pentacarbonyl.

However, cost and complexity of external ion sources can be avoided if metal ions are generated directly inside the EBIS. Laser ablation from an internal target has been successfully tested on SILFEC [121] but has not been used in a regular production EBIS. A much simpler approach is the sputtering from a wire probe tested on the EBIT [122]. The sputtering starts before the wire probe penetrates the electron beam and increases with decreasing distance from the electron beam. Evidence suggests that the sputtering is produced by the ions surrounding the electron beam. It has been shown to work very efficiently with samples smaller than 0.1 μg and therefore is ideally suited for rare materials. However, the sputter rate is slow and therefore is promising only for the production of high charge states with a continuous injection into the main trap or a seed trap.

Volatile metal organics, substances where metal ions are bonded with C, O, N, and H, have been used for a long time to produce metal ion beams with many different ion sources [116], because they eliminate the lifetime problems caused by the sputtering. The drawback is the large fraction of C, O, N, and H found inside the plasma which limit the production of the specie of interest. For multiply and highly charged ion sources the intense organic beams can easily interfere with small beams of multiple charged metal ions [123,124]. However, an EBIS is ideally suited for this method because the light organic ions can be separated efficiently from the heavier ions of interest. Figure 3.25 shows the production of Fe beams with the KSU-CRYEBIS by injecting iron pentacarbonyl into the seed trap. For every Fe atom there are five C and five O atoms injected, but Fig. 3.25 shows for short confinement times two Fe ions for each C and O ion, most likely because the light C and O ions are boiled out of the seed trap. Although the not-fully-ionized C and O beams are clearly present for the shortest confinement time, they rapidly disappear with increasing confinement time because they get boiled out of the trap or completely ionized with the increasing ionization factor. Figure 3.25 shows clearly how the partly ionized light ions disappear from the low magnetic rigidity range before it is invaded by the higher charge states of the heavier ions, which minimizes ion beam overlaps and impurities.

3.5.6 Concluding Remarks

The production of highly charged ions with an EBIS is governed by the operational electron beam current density which depends on the EBIS and its specific operation. Having determined the electron beam current density, the evolution of charge state, normally calculated with the Lotz approximation for direct electron impact single ionization, predicts the confinement time required for the desired charge state. This prediction is fairly reliable for few electron ions but overestimates the confinement times required for the production of many electron ions because multi-electron processes are neglected. However, multi-electron processes can also substantially lower the fraction of ions which populate the desired charge state.

The operational electron beam energy should exceed the required ionization energy by roughly 20-50% to assure an efficient production of the desired charge state. Roughly 50% excess energy should maximize the production of Na-like and Mg-like ions, whereas 100-200% excess energy should maximize the production of higher charge states. The highest charge states may benefit from even higher excess energies which lower the heating rate more than the ionization rate. However, the electron beam energy affects also the trapping capacity and often the electron beam current density, and therefore the optimal electron beam energy must be established through a careful tuning process.

Assuming a moderate neutralization of the trapping capacity, and the confinement time and population fraction predicted by the evolution of charge state, one can estimate the number of ions in the desired charge state expelled with each batch, as well as the time-averaged ion beam current. However, the expected ions yields differ also substantially for different modes of operation as discussed in the previous sections.

The broad range of desired characteristics makes the delivery of ion beams for atomic physics a very interesting challenge. Experiments with pulsed targets, for example targets pumped with pulsed lasers, prefer short ion bunches with the highest possible peak ion current with a frequency matching the target frequency. Accordingly one needs to optimize the peak ion current, which is very similar to the synchrotron operation, which has been widely discussed in the past. However, most targets for atomic physics are in a steady state and prefer a steady supply of ions with minimal variations of the intensity. Accordingly one should optimize the effective duty cycle simultaneously with the time-averaged ion beam current which normally benefits from shorter confinement times as well as from short injection times. The duty cycle can be optimized and the intensity variations minimized with EBISes operated in a steady state mode, or a batch mode with a slow expulsion, which is discussed above in detail.

Steady state modes should produce higher yields with smaller energy spread for ionic charge states which require a confinement time which is not substantially longer than the typical switching time. However, batch modes, with slow expulsions if desired, should produce higher yields of charge states which require longer confinement times, as a larger fraction of ions reaches the desired charge state during expulsion, in agreement with the previous experience. However, the discussed expected yields are also affected by the achieved neutralization factor and by the achieved apparent electron current density which are presently not well understood and may differ for different modes of operation.

The highest charge states which can be produced with the KSU-CRYEBIS are limited presently by the maximum electron beam energy of 10 keV which can be maintained during safe and reliable operations, a limit which will be raised in the future. The ion yields which can be produced with the KSU-CRYEBIS are limited presently by the inefficient injection of low charged ions from the seed trap, another target for future upgrades. However, if required by the experiment, one can increase the ion yields by increasing the electron beam current, which we normally avoid as it increases the lHe consumption which is normally less than 0.5 l/h. EBISes which are cooled with He liquifiers pay no penalty for operation with higher currents and therefore produce routinely larger ion beams such as the ion bunches in excess of 2 nC per expulsion produced with CRYSIS when operating with an electron beam of 200 mA [66]. Naturally, not all EBISes produce the highest charge states and highest ion yields, but they still provide unique opportunities for atomic physics due to the production of ion beams with very small energy spread and emittance reported to be roughly 10^{-7} m·rad [7,25,77,102,106,107,125].

ACKNOWLEDGMENTS

I would like to thank my mentors, foremost J. Arianer, E.D. Donets, R. Becker, and C.L. Cocke, for their guidance on this interesting and challenging journey. More importantly, I wish to extend my appreciation to the numerous colleagues, friends, students, and support personnel who helped me with the design, operation, and understanding of our EBIS as well as with the preparation of this manuscript, especially J. Peterson for typsetting, D. Richard for proofreading, and S. Schauf for drafting. I am indebted to R. Becker and O. Delferriere, who made their programs available for the calculation of the evolution of charge states presented in this paper. Many of the discussed developments were accomplished with the CRYogenic Electron Beam Ion Source at Kansas State University (KSU-CRYEBIS), which is supported by the Division of Chemical Sciences, Office of Basic Energy Sciences, Office of Energy Research, U. S. Department of Energy.

REFERENCES

1. Feldman, U., *X-Ray and Atomic Inner-Shell Physics, 1982*, New York, 1982, pp. 447-462; Kallne, E., and Kallne, J., *X-Ray and Atomic Inner-Shell Physics*, 1982, New York, 1982, pp. 463-479; Fawcett, B.C., Supplement to *Z. Phys. D***21**, S1-5 (1991).
2. Brown, I.G., editor, *The Physics and Technology of Ion Sources*, New York: John Wiley & Son, 1989.
3. Wolf, B., editor, *Handbook of Ion Sources*, New York: CRC Press, 1995.
4. Jongen, Y., and Lyneis, C. M., in (2), pp. 207-228.
5. Wolf, B.H., in (3), pp. 121-148.
6. Melin, G., this book, pp. 33-66.
7. Arianer, J., Geller, R., *Ann. Rev. Nucl. Part. Sci.* **31**, 19-51 (1981).
8. Sortais, P., *Nucl. Instr. and Meth. B***98**, 508-516 (1995).

9. Donets, E.D., in (2), pp. 245-279.
10. Becker, R., in (3), pp. 157-182.
11. Schmieder, R.W., *Physics of Highly-Ionized Atoms*, New York: Plenum, 1989, pp. 321-376.
12. Wolf, B. H., and Klein, H., *Workshop on EBIS and Related Topics*, Darmstadt: GSI-P-3-77, 1977.
13. Arianer, J., and Olivier, M., *II EBIS Workshop*, Saclay-Orsay: Laboratoire National Saturne II and Institut de Physique Nucleaire, 1981.
14. Kostroun, V.O., and Schmieder, R.W., *Third International EBIS Workshop*, Ithaca, NY, Cornell University, 1985.
15. Hershcovitch, A., *International Symposium on Electron Beam Ion Sources and their Applications*, New York: American Institute of Physics, conference proceedings No. 188, 1989.
16. Donets, E.D., and Yudin, I.P., *The 5th International Symposium on Electron Beam Ion Sources and their Applications*, Dubna: EBIS-5 Publishing Group, 1992.
17. Liljeby, L., editor, *Proceedings of the 6th International Symposium on Electron Beam Ion Sources and their Applications*, Stockholm: Physica Scripta, to be published.
18. Marrs, R.E., *Experimental Methods in the Physical Sciences, Vol. 29A,* Academic Press, 1995, Ch. 14, pp. 391-420.
19. Marrs, R.E., Elliott, S.R., and Knapp, D. A., *Phys. Rev. Lett.* **72**, 4082-4085 (1994).
20. Scofield, J., LLNL, Livermore, CA 94550, private communication.
21. Tawara, H., Iwai, T., Kaneko, Y., Kimura, M., Kobayashi, N., Matsumoto, A., Ohtani, S., Okuno, K., Takagi, S., and Tsurubuchi, S., *J. Phys. B***18**, 337-350 (1985).
22. Stockli, M.P., Supplement to *Z. Phys. D***21**, S111-115 (1991).
23. Donets, E.D., *VIth International Conference on the Physics of Highly Charged Ions*, New York: American Institute of Physics, conference proceedings No. 274, 1993, pp. 663-670.
24. Arianer, J., Cabrespine, A., and Goldstein, C., *Nucl. Instr. and Meth.* **193**, 401-413 (1982).
25. Arianer, J. Cabrespine, A., Goldstein, C., Junquera, T., Courtois, A., Deschamps, G., and Olivier, M., *Nucl. Instr. and Meth.* **198**, 175-187 (1982).
26. Arianer, J. Goldstein, C., Laurent, H., and Malard, M., *IEEE Trans. on Nucl. Sci. NS*-**30**, 2737-2739 (1983).
27. Stockli, M.P., Arianer, J., Cocke, C.L., Richard, P., *Nucl. Instr. and Meth. B***40/41**, 1020-1023 (1989).
28. Liljeby, L., and Engstrom, A., in (15), pp. 27-31.
29. Donets, E.D., and Pikin, A.I., *Sov. Phys. Tech. Phys.* **20**, 1477-1482 (1976).
30. Donets, E.D., *IEEE Trans. on Nucl. Sci. NS*-**23**, 897-903 (1976).
31. Donets, E.D., *Phys. Scripta T***3**, 11-18 (1983).
32. Ovsyannikov, V.P., Alfeev, V.S., Drobin, V.M., Kovalenko, A.D., and Romanov, Yu.I., *Rev. Sci. Instrum.* **65**, 1133-1134 (1994).

33. Kleinod, M., Becker, R., Kester, O., Lakatos, A., Thomae, H., Zipfel, B., and Klein, H., *Rev. Sci. Instrum.* **65**, 1069-1071 (1994).
34. Thomae, H., Becker, R., Bongers, H., and Klenoid, M., *Nucl. Instr. and Meth.* *B***98**, 577-580 (1995).
35. Schmieder, R.W., Bisson, C.L., Haney, S., Toly, N. VanHook, A.R., Weeks, J., *Rev. Sci. Instrum.* **61**, 259-261 (1990).
36. Okuno, K., in (15), pp. 33-43.
37. Khodja, H., and Briand, J.-P., in (17).
38. Becker, R., Donets, E.D., Kleinod, M., Margolis, H.S., Silver, J.D., *Rev. Sci. Instrum.* **67**, 983-985 (1996).
39. Lotz, W., *Astrophys. J., Suppl.* **14**, 207-238 (1967); Lotz, W., *Z. Phys.* **206**, 205-211 (1967); Lotz, W., *Z. Phys.* **216**, 241-247 (1968).
40. Tawara, H., and Imamura, H., in (12), pp. 77-79.
41. Itikawa, Y., and Kato, T., Institute of Plasma Physics, Nagoya University, Nagoya 464, Japan, Report IPPJ-AM-17, 1981.
42. Tawara, H., in (15), pp. 427-443.
43. Freund, R.S., Wetzel, R.C., Shul, R.J., Hayes, T.R., *Phys. Rev.* *A***41**, 3575-3595 (1990).
44. Oualim, E.M., Duponchelle, M., and Defrance, P., *Nucl. Instr. and Meth.* *B***98**, 150-153 (1995).
45. Cocke, C.L., Ali, R., Bhalla, C.P., Stockli, M., and Schulz, M., *Nucl. Instr. and Meth.* *B***53**, 432-437 (1991).
46. e.g. Wong, K.L., Beiersdorfer, P., Chen, M.H., Marrs, R.E., Reed, K.J., Scofield, J.H., Vogel, D.A., and Zasadzinski, R., *Phys. Rev.* *A***48**, 2850-2859 (1993).
47. e.g. Ali, R., Bhalla, C. P., Cocke, C. L., and Stockli, M., *Phys. Rev. Lett.* **64**, 633-366 (1990); Ali, R., Bhalla, C.P., Cocke, C.L., Schulz, M., and Stockli, M., *Phys. Rev.* *A***44**, 223-231 (1991).
48. Kim, Y.S., and Pratt, R.H., *Phys. Rev.* *A***27**, 2913-2924 (1983).
49. Penetrante, B.M., Bardsley, J.N., Levine, M.A., Knapp, D.A., and Marrs, R.E., *Phys. Rev.* *A***43**, 4873-4882 (1991).
50. Penetrante, B.M., Bardsley, J.N., DeWitt, D., Clark, M., Schneider, D., *Phys. Rev.* *A***43**, 4861-4872 (1991).
51. e.g. Ali, R., Cocke, C.L., Raphaelian, M.L.A., and Stockli, M., *Phys. Rev.* *A***49**, 3586-3596 (1994).
52. Niehaus, A.J., *Phys.* *B***19**, 2925-2937 (1986).
53. Gregory, D.C., *Electronic and Atomic Collisions*, Amsterdam: Elsevier Science Publishers, 1986, pp. 205-214.
54. Chen, M.H., and Reed, K.J., *Phys. Rev.* *A***47**, 1874-1877 (1993).
55. Reed, K.J., Chen, M.H., and Moores, D.L., *Phys. Rev.* *A***44**, 4336-4344 (1991).
56. Reed, K.J., Chen, M.H., and Moores, D.L., *Phys. Rev.* *A***42**, 5315-5318 (1990).
57. Chen, M.H., and Reed, K.J., *Phys. Rev.* *A***45**, 4525-4529 (1992).
58. Gregory, D.C., Huq, M.S., Meyer, F.W., Swenson, D.R., Sataka, M., Chantrenne, S., *Phys. Rev.* *A***41**, 106-115 (1990).
59. Muller, A., *Phys. Lett.* *A***113**, 415-419 (1986).

60. Stenke, M., Hathiramani, D., Hofmann, G. Shevelko, V.P. Steidl, M., Völpel, R., and Salzborn, E., *Nucl. Instr. and Meth. B***98**, 138-141 (1995).
61. Becker, R., in (15), pp. 359-370.
62. Faure, J., Feinberg, B., Courtois, A., and Gobin, R., *Nucl. Instr. and Meth.* **219**, 243-246, (1984).
63. Brown, I.G., and Feinberg, B., *Nucl. Instr. and Meth.* **220**, 251-263 (1984).
64. Faure, J., Antoine, P., Ciret, J.C., Degueurce, L., Gros, P., Courtois, A., Gastineau, B., Gobin, R., Leaux, P., Leroy, P.A., Penicaud, J.P., in (15), pp. 102-112.
65. Ovsyannikov, V.P., in (16), pp. 59-63.
66. Liljeby, L., Beebe, E.N., and Pikin, A.I., in (17).
67. Stockli, M.P., Gibson, P.E., Cocke, C. L., DePaola, B.D., Fry, D., Kravis, S., Parks, D., Richard, P., Tipping, T.N., Walch, B., and Winecki, S., *Rev. Sci. Instrum.* **67**, 1162-1164 (1996).
68. Becker, R., in (13), pp. 185-196.
69. Levine, M.A., Marrs, R.E., Henderson, J.R., Knapp, D. A., Schneider, M.B., *Phys. Scripta T***22**, 157-163 (1988).
70. Bebee, E.N., in (15), pp. 166-177.
71. Becker, R., Kleinod, M., Thomae, H., and Donets, E.D., *VIth International Conference on the Physics of Highly Charged Ions*, New York: American Institute of Physics, Conference Proceedings No. 274, 1993, pp. 686-689.
72. Schneider, M.B., Levine, M.A., Bennett, C.L., Henderson, J.R., Knapp, D.A., and Marrs, R.E., in (15), pp. 158-165.
73. Visentin, B., Faure, J., Gobin, R., Leroy, P.A, *Nucl. Instr. and Meth. A***313**, 23-25 (1992).
74. Donets, E.D., *Rev. Sci. Instrum.* **61,** 225-229 (1990).
75. Becker, R., Klein, H., and Schmidt, W., *IEEE Trans. on Nucl. Sci. NS-***19**, 125-130 (1972).
76. Kleinod, M., Becker, R., Klein, H., and Schmidt, W., *IEEE Trans. on Nucl. Sci. NS-***23**, 1023-1026 (1976).
77. Becker, R., Kleinod, M., and Klein, H., *Nucl. Instr. and Meth. B***24/25**, 838-840 (1987).
78. Kleinod, M., and Becker, R., in (12) pp. 38-41.
79. Becker, R., and Klein, H., *IEEE Trans. on Nucl. Sci. NS-***23**, 1017-1022 (1976).
80. Becker, R., Kleinod, M., and Klein, H., in (13), pp. 48-58.
81. Clausnitzer, G., Klinger, H., Muller, A., and Salzborn, E., *Nucl. Instr. and Meth.* **128**, 1-7 (1975); Clausnitzer, G., Klinger, H., Muller, A., and Salzborn, E., *IEEE Trans. on Nucl. Sci. NS-***23**, 1027-1030 (1976).
82. Imamura, H., Kaneko, Y., Iwai, T., Ohtani, S., Okuno, K., Kobayashi, N., Tsurubuchi, S., Kimura, M., and Tawara, H., *Nucl. Instr. and Meth.* **188**, 233-238 (1981).
83. Kleinod, M., Becker, R., and Klein, H., in (15), pp. 15-23.
84. Kleinod, M., and Becker, R., in *Topical Problems of Accelerator and Applied Heavy Ion Physics*, Gesellschaft fur Schwerionenforschung, Darmstadt, Report GSI-90-29, 97-99 (1990).

85. Kleinod, M., Becker, R., Dietrich, J., Klein, H., Lakatos, A., Pfisterer, B., Thomae, H., in (16), pp. 64-70.
86. Beebe, E.N., and Kostroun, V.O., *Rev. Sci. Instrum.* **63**, 3399-3411 (1992).
87. Beebe, E., Liljeby, L., Engstrom, A., and Bjorkhage, M., in (16), pp. 24-37.
88. Kleinod, M., Becker, R., and Klein, H., in (14), pp. 17-25.
89. Okuno, K., *Jap. J. of Appl. Phys.* **28**, 1124-1131 (1989).
90. Okuno, K., *Nucl. Instr. and Meth. B***53**, 387-394 (1991).
91. Faure, J., in (15), pp. 3-10.
92. Schmieder, R.W., Bisson, C.L., Haney, S., Toly, N., Van Hook, A.R., Weeks, J., in (15), pp. 45-58.
93. Schmieder, R.W., and Bisson, C.L., *Rev. Sci. Instrum.* **61**, 256-258 (1990).
94. Schneider, D., DeWitt, D., Clark, M.W., Schuch, R., Cocke, C.L., Schmieder, R., Reed, K.J., Chen, M.H., Marrs, R.E., Levine, M., and Fortner, R., *Phys. Rev. A***42**, 3889-3895 (1990).
95. Ohtani, S., Kaneko, Y., Kimura, M., Kobayashi, N., Iwai, T., Matsumoto, A., Okuno, K., Takagi, S., Tawara, H., and Tsurubuchi, S., *J. Phys. B***15**, L533-L535 (1982).
96. Okuno, K., Tawara, H., Iwai, T., Kaneko, Y., Kimura, M., Kobayashi, N., Matsumoto, A., Ohtani, S., Takagi, S., and Tsurubuchi, S., *Phys. Rev. A***28**, 127-134 (1983).
97. Ohtani, S., *Phys. Scripta T***3**, 110-114 (1983).
98. Okuno K., Soejima, K., and Kaneko, Y., in (16) pp. 38-58.
99. Stockli, M.P., *Rev. Sci. Instrum.* **67**, 892-4 (1996).
100. Sortais, P., *Rev. Sci. Instrum.* **67**, 867-872 (1996).
101. Beebe, E., Liljeby, L., Engstrom, A., and Bjorkhage, M., *Phys. Scripta* **47**, 470-474 (1993).
102. Arianer, J., *Nucl. Instr. and Meth. B***9**, 516-521 (1985).
103. Stockli, M., Ali, R.M., Cocke, C.L., Raphaelian, M.L.A., Richard, P., and Tipping, T.N., *Rev. Sci. Instrum.* **63**, 2822-2824 (1992).
104. Beebe, E., Liljeby, L., Pikin, A., Bjorkhage, M., Engstrom, A., and Paal, A., *Rev. Sci. Instr.* **65**, 1718-1722 (1994)
105. Arianer, J., and Goldstein, C., *IEEE Trans on Nucl. Sci. NS-***23**, 979-986 (1976).
106. Arianer, J., Cabrespine, A., and Goldstein, C., *IEEE Trans on Nucl. Sci. NS-***26**, 3713-3715 (1979).
107. Gastineau, B., Faure, J., and Cortois, A., *Nucl. Instr. and Meth. B***9**, 538-543 (1985).
108. Kravis, S. D., Watanabe, N., Awaya, Y., Kimura, M., Kojima, T., Okuno, K., and Oura, M., in (17).
109. Watanabe, Kravis, S., Awaya, Y., Kojima, T. M., Okuno, K., and Oura, M., in RIKEN *Accel. Prog. Rep.* **29**, 178 (1996).
110. Arianer, J., *IEEE Trans on Nucl. Sci. NS-***28**, 1018-1023 (1981).
111. Stockli, M. P., Cocke C. L., and Richard, P. in (15), pp. 114-125.
112. Goldstein, C., Laurent, H., and Malard, M., in (14), pp. 1-15.
113. Kostroun, V.O., *Nucl. Instr. and Meth. B***10/11**, 771-774 (1985).

114. Faure, J., Feinberg, B., Courtois, A., and Gobin, R., *Nucl. Instr. and Meth.* **219**, 449-455 (1984).
115. Assoufid, L., and Kostroun, V.O., *Nucl. Instr. and Meth. B***69**, 492-502 (1992).
116. Wolf, B. H., in (3), pp. 331-363; Wolf, B. H., *Rev. Sci. Instrum.* **67**, 965-970 (1996).
117. Visentin, B., Harrault, F., Gobin, R., and Leroy, P.A., *Rev. Sci. Instrum.* **65**, 1129-1131 (1994).
118. Silver, J. D., Varney, A. J., Margolis, H. S., Baird, P. E. G., Grant, I. P., Groves, P. D., Hallett, W. A., Handford, A. T., Hirst, P. J., Holmes, A. R., Howie, D. J. H., Hunt, R. A., Nobbs, K. A., Roberts, M., Studholme, W., Wark, J. S., Williams, M. T., Levine, M. A., Dietrich, D. D., Graham, W. G., Williams, I. D., O'Neil, R., and Rose, S. J., *Rev. Sci. Instrum.* **65**, 1172-1174 (1994).
119. Margolis, H.S., Varney, A.J., Jarjis, R.A., and Silver, J.D., *Nucl. Instr. and Meth. B***98**, 562-565 (1995).
120. Gillaspy, J. D. in (17).
121. Collart, C., Liebe, A., Nicol, P., Steinegger, A., in (13), pp. 170-174.
122. Elliott, S. R., and Marrs, R. E., *Nucl. Instr. and Meth. B***100**, 529-535 (1995).
123. Koivisto, H., Arje, J., and Nurmia, M., *Nucl. Instr. and Meth. B***94**, 291-296 (1994).
124. Waldman, H. and Martin, B., *Nucl. Instr. and Meth. B***98**, 532-535 (1995).
125. Visentin, B., Courtois, A., Delferriere, O., Faure, J., Gobin, R., Harrault, F., and Leroy, P.A., in (16) pp. 11-23.

CHAPTER 4

Merged Beams with Multicharged Ions

C. C. Havener*

The feasibility of using a merged-beams technique to measure interactions at low collision energies was first successfully demonstrated in 1966 and involved measurements of symmetric charge exchange for 5 eV/amu collisions of Ar^+ + Ar. Soon after, the technique was expanded to study collisions with a variety of atomic ions and neutrals, molecular ions, and electrons. Merged-beams measurements were found to be technically difficult and were generally limited to ions with low charge states (where beams of sufficient intensity existed) and to collision processes characterized by large cross sections. The recent advent of advanced ion sources, e.g., the ECR source, has allowed the merged-beams technique to be further expanded to study collisions with multicharged ions. At ORNL an ion-atom merged-beams technique has recently been used to study electron capture with multicharged ions to near-thermal collision energies. Details of the apparatus will be used to illustrate the merged-beams technique. Recent electron capture cross section measurements will be presented to illustrate the importance of the merged-beams technique in the study of low energy electron capture.

4.1 EARLY MERGED BEAMS

The use of merged beams to access very low (thermal to eV) heavy particle interaction energies is not new. There has always been a need to access energies above that available from thermal sources which have an upper limit of about 0.3 eV and below that available from conventional ion sources which have a lower limit of about 100 eV. Special low energy ion sources incorporate a "cool" source of ions extracted from, e.g, a laser produced plasma [1] or ions from recoils [2]. Such sources are able to produce beams typically as low as $q \times 10$ eV, where q is the charge state of the ion. However, beam-gas or crossed-beam

*) Oak Ridge National Laboratory, Oak Ridge, Tennessee 37831-6372.

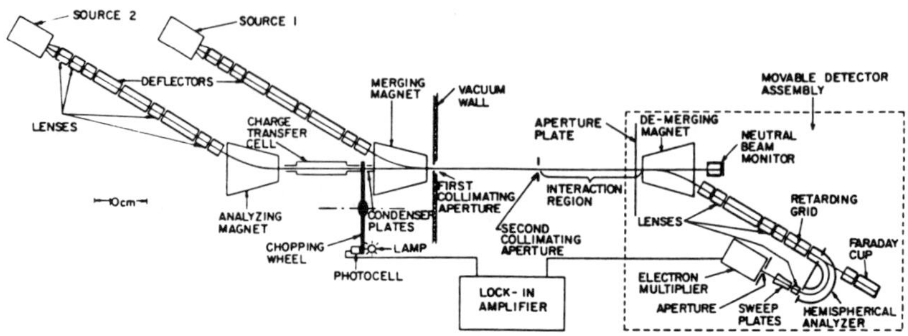

FIGURE 4.1 Schematic of an early merged-beams apparatus [5] developed at the Space Science Laboratory to study ion-neutral collisions. This apparatus demonstrated the feasibility of merged beams and was used in numerous collision studies.

techniques which utilize these and other sources convert a significant portion of the beam energy into collision energy and the techniques are then limited to the quality and intensity of the beams. In merged-beams, relatively fast (keV) beams are merged producing low energy collisions in a moving center-of-mass frame. It was realized [3,4] almost thirty years ago that these collisions in the center-of-mass frame could be performed with an energy resolution significantly lower than the energy spread of each beam.

The experimental feasibility of the technique was first demonstrated [5] at the Space Science Laboratory, General Dynamics/Convair, San Diego, California, where the cross section for the symmetric charge exchange of Ar^+ + Ar → Ar + Ar^+ at a 100 eV (5 eV/amu) collision energy was successfully measured using the ion and neutral primary beams at energies of $E_1 = 2000$ eV and $E_2 = 3465$ eV, respectively. A 1.5 eV spread in the primary beams led to an estimated uncertainty in the collision energy of 0.21 eV. Their apparatus is shown in Fig. 4.1 and illustrates the technical difficulty in merged beams. Note that two Ar^+ ion sources were needed during operation. The neutral Ar beam was produced by passing one Ar^+ beam through a charge exchange cell. To separate the product Ar^+ from the primary beams and also from several sources of ion and detector backgrounds, a de-merging magnet, retarding grid assembly, hemispherical analyzer, and a beam chopping scheme were needed. Further details can be found in the literature [5].

Another early merged-beams apparatus [6] was developed to look at symmetric resonant charge transfer in H^+ + H or D^+ + D collisions. A clever "overtaking" beam technique was used in which a fraction of the ion beam was first neutralized by a charge exchange cell and then the remaining ions electrostatically accelerated in order to overtake the neutral beam, thus producing the desired low energy collisions. Cross sections were successfully measured in the energy range between 2.5 and 50 eV/amu and were used to investigate whether isotope effects existed for these collisions.

From the success of these early experiments, the basic merged-beams technique showed promise. There was an intense interest in the collision physics made possible by merged beams and several apparati were developed mostly during the late 1960s and 1970s which studied ion-neutral collisions for symmetric [7,5,8] and non-symmetric [9] atomic and molecular [10–14] species. Neutral-neutral collisions [20,21] and ion-ion collisions for symmetric [15] and non-symmetric [16–18] atomic reactants and molecular ions [19] were also studied. Merged-beam technology developed and measurements were performed with excited-state beams [8,9], with self-merged beams [6,8], and with detailed energy and angular scattering analysis [12,13,18] of the products. For electron capture in $Na^+ + O^-$ collisions, observation of an optically allowed transition to the product $Na(^2S)$ ground state led to state selective measurements [17] with merged beams. The earlier experiments were technically rather difficult and were typically limited to collision systems with relatively large cross sections using beams with low charge states.

With the development of advanced ion sources such as the electron-cyclotron-resonance (ECR) ion source, intense beams of multicharged ions became available and a new opportunity arose to study collisions in merged beams. A merged-beams apparatus [22] was developed at Oak Ridge National Laboratory and used in conjunction with the ORNL ECR ion source [23]. Total electron capture measurements were first reported [22] in 1989 for the $O^{5+} + H(D)$ system from 0.9 to 800 eV/amu. Improvements in the apparatus and ECR ion source technology now permit measurements from 20 meV/amu to 5000 keV/amu for a variety of multicharged ions on atomic H or D. While special beam trapping techniques (e.g., octopole ion beam guide) in combination with advanced multicharged ion sources have recently extended more conventional beam-gas techniques to measurements [24] of total cross section to 0.1 eV/amu energy and state selective measurements [25] to tens of eV/amu, these have been limited so far to measurements with noble or molecular gases. The merged-beams technique is still unique in its ability to measure absolute cross sections for fundamental collision systems to thermal energies.

It is important to point out that the merged-beams technique was also used in electron-ion collisions in early experiments with atomic [26,27] and molecular ions [28,29] and later in more successful measurements [30] of dissociative recombination of the molecular ions H_2^+ and H_3^+. A merged-beams geometry has recently been used to measure ion excitation by electron impact for singly charged [31] and for multicharged ions [32]. With the advent of ion-storage rings there have been numerous merged-beams measurements of electron-impact ionization of ions, dielectronic recombination of ions, and dissociative recombination of molecular ions. Because of beam expansion techniques [33], the low energy limit and resolution in an ion-storage ring experiment surpass the performance of a single-pass configuration with electrons;

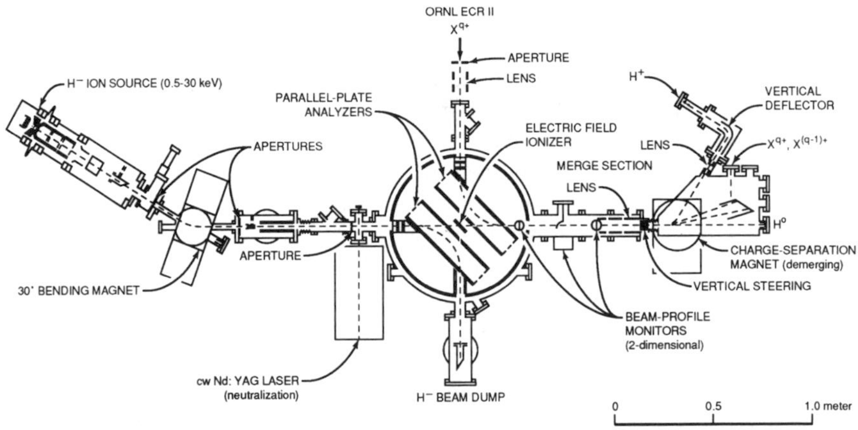

FIGURE 4.2 Schematic of the current ORNL ion-atom merged-beams apparatus.

however, no such technology exists for ion-neutral collisions. These multipass electron-ion merged-beams experiments are discussed in another chapter.

4.2 MERGED-BEAMS TECHNIQUE

Figure 4.2 is a schematic of the current ORNL ion-atom merged-beams apparatus which illustrates the merged-beams technique for the generic electron capture reaction

$$X^{q+} + H \rightarrow X^{(q-1)+} + H^{+} \tag{4.1}$$

for some multicharged ion, X^{q+} colliding with atomic H. An intense multicharged ion beam (1-10 μA) with energies up to q x 30 keV is provided by the ORNL $2\omega_c$ CAPRICE ECR ion source [34] and merged electrostatically with a neutral H (or D) beam. A ground state H beam of up to 100 nA is produced by photodetachment of a 2-3-μA beam of H^- in a 100 W cw Nd:YAG laser cavity ($\lambda = 1.06\ \mu$m). The 6-10 keV H^- beam is extracted from a duoplasmatron, analyzed by a 30^0 bending magnet, focused and steered horizontally and vertically so that subsequent to neutralization, a nearly parallel neutral beam will travel down the merge path, interact with the multicharged ion beam, and then be collected in the neutral detector. Production of a neutral beam by photodetachment has several advantages over the more common technique of using a charge exchange or detachment gas cell. These advantages, as described by Van Zyl *et al.* [35], include compatibility with ultrahigh vacuum, generation of a ground state beam, and the ability to perform *in situ* absolute calibration of the neutral beam detector. The neutral beam passes through an electric field of up to 30 kV/cm (see Fig. 4.2) which removes by ionization a significant fraction of the small excited state component (currently $< 0.01\%$) present in the neutral H beam. This excited state component is produced by collisional detachment of H^- with background gas.

The merged beams interact in a field-free region for a distance of 47.1 cm, after which the primary beams are magnetically separated from each other and the product or "signal" H^+ ions. The merge path is kept at ultrahigh vacuum conditions (1 x 10^{-10} Torr) to minimize the H^+ background generated by stripping of the H beam as it travels through the background gas. To maximize the angular acceptance of H^+, a cylindrical Einzel lens is placed toward the end of the merge path. Due to a limited magnetic dispersion, the $X^{(q-1)+}$ product of the reaction is not measured separately but is collected with the X^{q+} beam in a large Faraday cup. Since only the H^+ signal is measured, the apparatus actually measures electron loss, a sum of electron capture and ionization. Ionization of H at the low energies of interest here, though, is negligible compared to the electron capture process. The neutral-beam intensity is measured by secondary-electron emission from a stainless-steel plate, and the signal H^+ is recorded by a channel electron multiplier (CEM) operated in pulse counting mode. The H^+ undergo a second (vertical) dispersion out of the plane of magnetic analysis in order to attenuate transmission of photons to the CEM. The photons are produced by ion impact on the Faraday cup.

The following discussion is meant to illustrate several aspects of merged-beams and is not meant to completely describe the experimental procedure. Such details are apparatus dependent and can be found elsewhere [22].

4.2.1 Collision energy

It may be useful to begin with a discussion of collision energy. Conventional beam-gas collision studies usually define collision energy as simply the beam energy of the incident ion beam, X^{q+}, and is given by qeV where q is the charge on the ion, e the electronic charge, and V is the ion source potential. Since the velocity of the target is on average zero, the relative collision velocity is just the velocity, v, of X^{q+} with mass m, and is given by the relation $E = qeV = \frac{1}{2}mv^2$. In order to produce pure multicharged ion beams for some m/q combinations,

FIGURE 4.3 Schematic of two beams with mass m_1 and m_2, travelling at different velocities v_1 and v_2, intersecting at angle θ.

however, isotopes often must be used. There can be some confusion, then, in calculating the collision velocity v, if the isotope is not specified in the tabulation of the data. Though not usually done, the collision energy is less ambiguous if reported as E/m, thus removing the mass term.

This confusion gets worse when considering merged beams where the target is not a stationary gas but a moving beam. Figure 4.3 shows two intesecting beams with energy E_1, E_2, with mass m_1, m_2 and travelling at velocities $\vec{v_1}, \vec{v_2}$ intersecting at angle θ. The interaction energy in the center-of-mass frame W is defined [5] to be

$$W = \frac{\mu}{2} \mid \vec{v_1} - \vec{v_2} \mid^2, \tag{4.2}$$

where μ is the reduced mass given by $\frac{m_1 m_2}{m_1+m_2}$.

The relative collision energy in the center-of-mass frame, E_{rel}, is defined by

$$E_{rel}(eV/amu) = W/\mu = 1/2(v_1^2 + v_2^2 - 2v_1v_2cos(\theta)), \tag{4.3}$$

which can be written in terms of the lab frame energies of the two beams, E_1, E_2, as

$$E_{rel}(\frac{eV}{amu}) = \frac{E_1}{m_1} + \frac{E_2}{m_2} - 2\sqrt{\frac{E_1 E_2}{m_1 m_2}} cos(\theta). \tag{4.4}$$

The relative collision energy, which has been defined as the interaction energy divided by reduced mass, is only dependent on the square of the relative collision velocity and can be directly compared to the E/m energy identification above, as defined for beam-gas measurements.

Figure 4.4 shows the collision energy in the center-of-mass frame, E_{rel}, obtained for three different neutral-beam lab frame energies, using as an example the O^{5+} + H collision system. Collisions with D, a heavier isotope of H, are also included. To change the energy E_{rel}, the energy of the O^{5+} can be varied while the velocity of the H or D beam is held constant. It is clear from the figure that a wide range of relative energies ($\sim$ 0 - 800 ev/amu) can be accessed for this collision system using this one apparatus. For an 8-keV D beam, velocity matching occurs with a O^{5+} beam energy of 64 keV. Therefore, for an O^{5+} beam energy < 64 keV, the D travels faster than the O^{5+}, and for

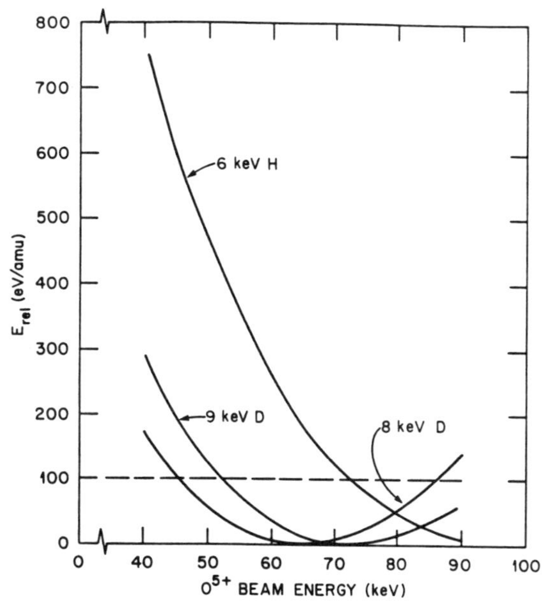

FIGURE 4.4 Relative energy, E_{rel}, obtainable for 6-keV H, 8-keV D, and 9-keV D neutral beams as a function of the O^{5+} beam energy.

an energy > 64 keV, the D travels slower than the O^{5+}. When other H- and D-beam energies are considered, one value of E_{rel} may be obtained under several laboratory conditions. For example, as seen in Fig. 4.4, an interaction energy of 100 eV/amu may be obtained in several ways: D faster, D slower, and H faster than the O^{5+} beam. Using beams at different lab energies to measure the same relative collision energy provides important diagnostic checks.

4.2.2 Energy resolution / Low energy limit

An important advantage of the merged-beams technique is the ability to perform collisions with good energy resolution even at the low (< 1 eV/amu) collision energies. Measurements are performed with primary beams that have relatively large energy uncertainties; however, these uncertainties are "deamplified" in the center-of-mass frame. This deamplification can be illustrated using Fig. 4.4, where the slope of the curves, $\triangle E_{rel}/\triangle E_2$, gives the relationship between the uncertainty in the relative collision energy, E_{rel}, and the uncertainty in the O^{5+} beam energy, E_2. As E_{rel} decreases, the slope decreases, thereby increasing this deamplification effect at the low collision energies.

The uncertainty, ΔE_{rel}, in E_{rel} due to the uncertainties (spreads), ΔE_1, ΔE_2, in the D and O^{5+} beam energy, E_1, E_2, is calculated using Eq. 4 and

TABLE 4.1 Uncertainties in E_{rel} due to uncertainties (spreads) in the beam energies and merge angles.

E_{rel}	D	O^{5+}	ΔE_{rel} (eV/amu)			
			for ΔE_1	for ΔE_2	for $\Delta\theta$	for $\Delta\theta$
(eV/amu)	E_1(keV)	E_2(keV)	$\Delta E_1 = 6eV$	ΔE_2=46 eV	$\Delta\theta = 0.1^0$	$\Delta\theta = 0.35^0$
100	8	45.0	0.47	0.54	0.011	0.13
10	8	57.3	0.15	0.15	0.011	0.14
1	8	61.5	0.05	0.047	0.012	0.15
0.1	8	62.9	0.015	0.014	0.012	0.15
0.01	8	63.33	0.005	0.004	0.012	0.15

presented individually in Table I as a function of collision energy. Different collision energies are obtained by varying the O^{5+} beam energy while the D beam energy is held constant at 8 keV. An energy uncertainty (spread) of 6 eV is considered [36] a conservative upper limit for a duoplasmatron operated in a source gas "starvation" mode. An energy spread of 9.2 qeV is estimated for beams extracted from the ECR source by taking the quadrature sum of the energy spread due to the source [37] (6.5 V) and to the voltage ripple of the source power supply (6.5 V). ΔE_{rel} is also calculated for two values of $\Delta\theta$ corresponding to the range of typical uncertainties (spreads) in merge angle.

The total uncertainty in collision energy is obtained by taking a quadrature sum of the ΔE_{rel} due to ΔE_1, ΔE_2, and an appropriate $\Delta\theta$. Note that at the lowest collision energies, the spread in angle is the major source of collision energy uncertainty. This estimated spread in merge angle is comparable to the measured finite merge angle and together they determine the low energy limit of the apparatus.

4.2.3 Angular collection

Another important aspect of the merged-beams technique is the potentially large angular collection of the reaction products. The low-energy electron capture collisions under study are exoergic, and since both products are positively charged, significant angular scattering can occur [38] in the center-of-mass frame. Therefore, experiments must be designed with as large an angular collection of products as possible. In this respect the merged-beams technique is ideal since the angular scattering in the center-of-mass frame becomes significantly compressed in the laboratory frame, the frame where the product ions are collected. This angular compression can be evaluated by examining the kinematic transformation in more detail.

After the collision the exoergicity of the reaction, Q, is added to the relative energy of the colliding particles. Hence, the final energy in the center-of-mass frame is

$$W^f = \frac{1}{2}\mu v_r^{f^2} = \frac{1}{2}\mu v_r^{i^2} + Q. \tag{4.5}$$

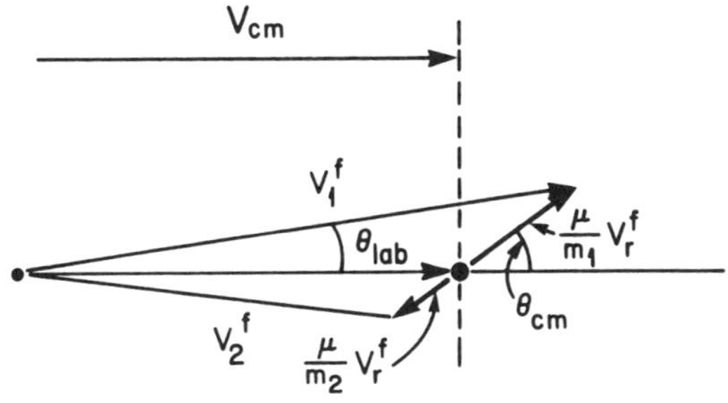

FIGURE 4.5 Vector (Newton) diagram after the collision. The symbols v_1^f and v_2^f correspond to the final velocities of the projectiles with masses m_1 and m_2, respectively, in the laboratory frame. v_r^f is their final relative velocity.

This additional energy is shared between the two product ions in accordance with the conservation of linear momentum. After the collision, the final velocities v_1^f and v_2^f are given by

$$\vec{v_i^f} = \vec{V_{c.m.}} + \frac{\mu}{m_i}\vec{v_r^f}. \tag{4.6}$$

Figure 4.5 shows the vector (Newton) diagram for a collision where the lighter projectile m_1 is travelling faster than m_2. The faster of the two collision partners gains in velocity, while the slower loses. The angle $\theta_{c.m.}$ and θ_{lab} is shown for the lighter projectile m_1.

It is evident that the larger the center-of-mass velocity $\vec{V_{c.m.}}$, the greater the angular compression in the laboratory frame. As can be seen from Fig. 4.5, the laboratory scattering angle θ_{lab} is related to the angle of scattering in the center of mass $\theta_{c.m.}$ by

$$tan(\theta_{lab}) = \frac{(\mu/m_1)v_r^f sin\theta_{c.m.}}{V_{c.m.} + (\mu/m_1)v_r^f cos\theta_{c.m.}}. \tag{4.7}$$

Using Eq. (5)-(7) and an estimate [39] of 2.3^0 for the angular acceptance in the laboratory, the angular acceptance in the center-of-mass frame was calculated for the collision system O^{5+} + H(D) which has an average exoergicity of 8.2 eV [1] For measurements performed with a 6-keV-H neutral beam, up to 58^0 scattering of the signal (H^+ products) can be collected at a collision energy of 1 eV/amu. For measurements at higher $V_{c.m.}$ with 8 keV, up to 90^0 scattering (all of the products) in the center-of-mass frame can be collected at collision energies below 2 eV/amu.

---

1) At these energies, capture takes place predominantly into the n=4 shell of O^{4+}, which, if one assumes a statistical population of the n=4 sublevels, has an average exoergicity of 8.2 eV.

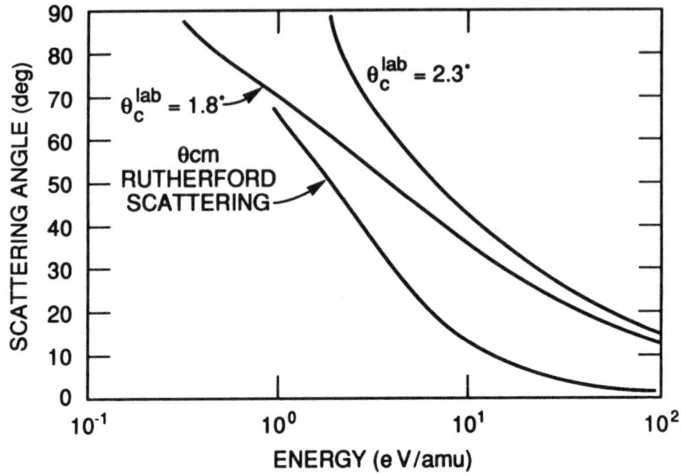

FIGURE 4.6 Estimated angular collection of the reaction products in the center-of-mass frame for the O^{5+} + D collision system for an angular acceptance in the lab of $\theta_c^{lab} = 2.3^0$ (present apparatus) and $\theta_c^{lab} = 1.8^0$ (original apparatus [22]). Also shown is the "half-coulomb" Rutherford estimate (see text).

When the angular acceptance in the center-of-mass frame is less than 90^0, the effect of angular scattering must be considered. Usually, the only estimate readily available is that given by "half-Coulomb" Rutherford minimum scattering [38] which is shown in Fig. 4.6 for an 8-keV-beam for the O^{5+} collision system. Also shown is the angular acceptance in the center-of-mass frame for two different angular acceptances in the laboratory frame.

Due to a discrepancy between early merged-beams measurements [22] for O^{5+} and fully-quantal coupled-channel molecular orbital calculations by Gargaud [40], detailed differential cross-section calculations [41] were performed for this system. They predict significantly larger scattering than the Rutherford "half-Coulomb" estimate, requiring an angular collection in the lab of at least 2.3^0 for a collision energy between 1 and 10 eV/amu. This explained why the early merged-beams measurements [22] with an angular acceptance of only 1.8^0 led to a loss of signal.

The angular acceptance of the present apparatus was designed to be at least 2.3^0 in the laboratory frame, in order to guarantee no significant loss of signal for this system. Indeed, O^{5+} measurements [39] with the current apparatus agree well with the calculations of Gargaud.

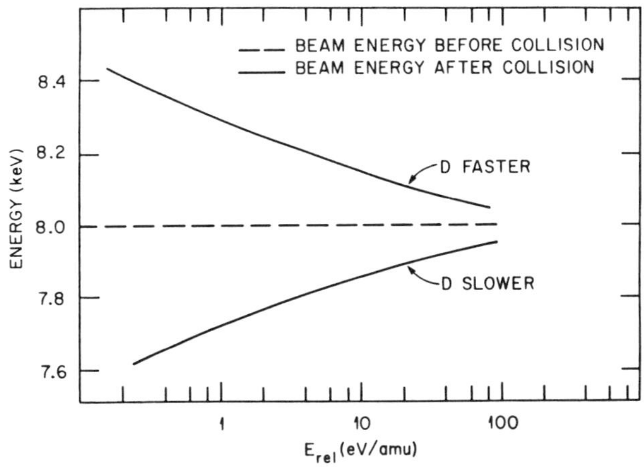

FIGURE 4.7 Resultant energy in the laboratory frame of an initial 8-keV D beam after interaction with O^{5+}. 8.2 eV was used as the exoergicity of the reaction.

4.2.4 Center-of-mass dynamics amplified in the laboratory frame

Another aspect of the merged-beams technique is that the exoergicity(or endoergicity) Q of the reaction is amplified when transformed from the center-of-mass to the laboratory frame. This can be illustrated using the O^{5+} + H collision system as an example. Figure 4.7 shows the maximum shift in energy for an 8-keV D beam as a function of E_{rel}, due to the 8.2-eV exoergicity. The maximum shift is calculated using Eq. (6) and assuming that scattering is restricted to the forward angle ($\theta = 0$) for D faster than O^{5+} (upper curve) and to the backward angle ($\theta = 180\,\text{deg}$) for D slower (lower curve). Especially at low relative energies, these shifts can be large and, as was experimentally observed [22], could be correllated to the exoergicity of the reaction.

For some merged-beams apparrati which were capable of product energy dispersion, this amplification effect leads to a large enough difference in the lab beam energies to facilitate signal/background separation (e.g., Ref. [10]). Even in merged-beams experiments investigating symmetric resonant reactions [5,7] where the change in energy is zero, energy analysis of the product could still be used to separate the unshifted signal from an endoergic background reaction.

As mentioned above, at low relative energies significant angular scattering can also take place which complicates the velocity transformation from the lab to the center-of-mass frame. To extract detailed information about the reaction process is difficult; however, there has been some success. For transfer ionization in He^{2+} + H^- collisions, the exoergicity of the reaction was measured [18] using magnetic and time-of-flight analysis of the product.

For several merged-beams measurements of molecular ion - neutral collision systems [12,13] the observed product energy distribution was asymmetrical in the laboratory frame which, due to differences in angular scattering, indicate direct rather than long-lived complexes present in the exoergic (or endoergic) reactions studied. Studies of the width of the energy distribution as a function of energy revealed additional information about the product scattering [42].

To take further advantage of this amplification effect, a merged-beams apparatus was developed [14] to perform detailed energy analysis of the products. The apparatus used a high-resolution analyzer which consisted of two 90^0-deflection, spherical electrostatic sector analyzers in tandem, which, together with deconvolution analysis, was able to provide details of the reaction process under study. For example, for $H_2^+ + H_2 \rightarrow H_3^+ + H$ collisions near 0 eV/amu, most of the reactant kinetic and exoergic energy was found [14] to go into internal excitation of the products. This leads to a decrease in the H_3^+ production cross section at 1 eV/amu due to H_3^+ dissociation. In another measurement on the same apparatus a similar deconvolution analysis [43] was performed for $D_2^+ + N \rightarrow ND^+ + D$ collisions.

4.2.5 Beam purity

While producing "pure" primary beams may not always be possible, detection or control of impurities or excited states in the beams is always desirable and even sometimes feasible. As has been previously observed [22] at ORNL in merged-beams studies of O^{5+} + H(D) at collisions energies below 100 eV/amu, a very small (0.02%) Rydberg population in the H beam can result in significant beam-beam signal compared to that due to capture from the ground state. The excited states of H are created by collisional stripping of the H^- beam on background gas. Some of the excited H have trajectories which allow the atoms to merge with the multicharged ions and, if electron loss occurs, produce a beam-beam H^+ signal. The cross section for electron removal from an excited state of H with an electron in, e.g., an n shell of 20 is typically a factor of 10^5 greater than the capture cross section from the ground state [44]. Thus, contributions from excited states can be a significant fraction of the beam-beam signal, and have been studied [44] in detail using this apparatus. To correct for the signal due to excited states, the beam-beam signal is currently measured with the laser on and then off, and the difference, then, corresponds to the signal due to capture from the H ground state. To reduce the Rydberg population, the H beam is made to pass through an electric field ionizer with fields typically 30 kV/cm (see Fig. 4.2), which ionizes the higher n levels of H (down to n=12), causing them to be swept out of the neutral beam. This reduction in excited states decreases the required correction to the ground state cross section to only a few percent (if any).

A multicharged ion beam from the ECR ion source/magnetic analyzer can sometimes have as a contaminant another ion with the same m/q. The amount can be estimated from the intensities of neighboring charge states of the contaminant and, if the cross section is known, the contribution to the cross section can be subtracted out. If the contamination is significant, then the ion of interest is preferably produced from an isotope, which, with a different m/q, may elimimate contaminants. In addition, some ion beams contain a fraction of metastable ions. These metastable components can be present during the measurement and for some systems affect the result. Li-like ions are thought to have no long-lived states[2] and thus are good candidates for detailed comparison with theory.

4.2.6 Cross-section measurement

Cross sections are determined by measuring the signal produced by the beam-beam interaction over the merged length L. For ion-atom merged beams the electron capture cross section is determined from the H^+ signal rate using

$$\sigma = \frac{R}{\varepsilon}\frac{\gamma q e^2}{I_1 I_2}\frac{v_1 v_2}{v_r}\frac{1}{L\Omega}, \tag{4.8}$$

where R, I_1, and I_2 correspond to the measured H^+ signal count rate, effective current produced by neutral particles, and O^{5+} current, respectively. The true signal rate is given by R/ε, with ε being the efficiency of the CEM and electronics for detecting H^+(or D^+) at velocity v_1, and the true neutral "current" is I_1/γ, with γ being the measured effective secondary-electron emission coefficient for the neutral-particle detector. The quantity q is the charge of the ion, and e is the electronic charge. v_r is the relative velocity of the two beams. Ω is the overlap integral [5] (effective form factor) and is a measure of the spatial overlap of the two beams along the merged path L.

Signal/Noise

In a merged-beams experiment the (beam-beam) product signal is orders of magnitude smaller than that obtained in a typical beam-gas experiment. In a beam-gas experiment the signal rate is proportional to the incident beam current, the cross section, and the target thickness Π, given by product of the target density and the effective length of the gas cell. For a 1 cm gas cell maintained at 1 mTorr, $\Pi = 3.3 \times 10^{13}$ cm^{-2}.

For a merged-beams experiment the target thickness that one beam generates for the other beam can be defined by

$$\Pi_{b-b} = \frac{I}{qe}\frac{v_r}{v_1 v_2}L\Omega. \tag{4.9}$$

---

2) There has been recent speculation [45] that doubly excited O^{5+} ions exist whose cascade to the ground state may be on the order of a μsec.

TABLE 4.2 Operating parameters for the recent N^{2+} + D measurements at E_{rel} = 21.8 eV/amu

Primary beams	
D @ 7.266 keV	47 nA (particle)
	$\gamma = 1.11$, neutralization efficiency of D^{-}= 5%
N^{2+} @ 42.990 keV	7 μA (electrical)
Signal	588 Hz
Background from D stripping	16 kHz
Background from photons	400 Hz
Signal-to-noise ratio:	3.7×10^{-2} Torr
$v_1\ v_2\ /\ v_r$	9.912×10^{8}
Ω	9.52 cm^{-2}
σ_{21}	2.3×10^{-16} cm^2
Time for 10% standard deviation	20 seconds
Merge path length, L	47.1 cm
Vacuum in merge section	$1 - 2 \times 10^{-10}$ Torr
Angular collection (lab frame)	2.3^0

For measurements of O^{5+} + D at 25 eV/amu and typical operating parameters [22] of the original ORNL ion-atom apparatus, the multicharged ion target thickness seen by the neutral D beam is 4.4 x 10^5 cm^{-2}. As one can see from the equation, Π_{b-b} is proportional to v_r, so as the collision energy decreases, the signal decreases as $\sqrt{E_{rel}}$.

The most recent measurements with N^{2+} were performed with the operating parameters shown in Table 4.2. The current operating parameters take advantage of several improvements to the apparatus made during the last few years. They include a shorter merge path, L, to take advantage of the better overlaps at the beginning of the merge path, increased angular acceptance, and new laser cavity optics, providing a factor of ten higher neutralization efficiency. Also the $2\omega_c$ CAPRICE ECR ion source provides higher velocity, more intense, and skinnier ion beams than the previous ECR source. These improvements result in an increase in Π_{b-b}, e.g. for the collision system presented in Table 4.2, $\Pi_{b-b} = 1.0 \times 10^7$ cm^{-2} which is an increase of a factor of twenty over the target thickness of the original O^{5+} experiment [22]. With the more intense D beams now avalable, the time to take measurements at 10% standard deviation is significantly reduced.

As can be seen in Table 4.2, there are background processes which complicate the measurement of beam-beam signal. When the neutral H or D beam strips on residual gas in the merge section ($H + H_2 \rightarrow H^+$), the H^+ produced is also detected by the CEM detector. To minimize this background, the merge path is kept at ultrahigh vacuum (10^{-10} Torr). Nevertheless, the resultant gas target thickness, Π, for the merge path of 47.1 cm is 1.6 x 10^8 cm^{-2}, which is larger than typical values for Π_{b-b}. Luckily the cross section for electron capture ($\sim 10^{-15} - 10^{-14}$ cm^2) is greater than the cross section for the H stripping on background gas ($\sim 10^{-16}$ cm^2). Another source of background is photons created by the neutralization of multicharged ions as they are collected in the Faraday cup. Only after double dispersion of the H^+ products and gold blacking of possible reflecting surfaces could the photons reaching the CEM be reduced to an acceptable level (< 100 Hz/μA). The signal/noise ratio generally decreases for lower energies, due mainly to the fact that the Π_{b-b} is proportional to v_r while the background rates essentially remain the same.

To separate the signal from background, a two-beam modulation technique [22] was used on the ORNL apparatus. Both beams were switched on and off at rates of 500 Hz - 2 kHz by applying 100-200 V between parallel deflecting plates placed near where the ion beams were extracted from their respective sources. No dependence on chopping frequency was observed except possibly at the very lowest freqency where background pressure modulation by the beams may have had an effect on the signal, and at the very highest frequencies where spurious effects due to switching the beams on and off may have been significant. Spurious signals due to spatial beam intermodulation can be both serious and subtle [46]. Merged-beams experiments are particularly susceptible to such effects due to the extended interaction length. Local electric fields produced by beam space charge are enhanced for multiply charged ion beams and can affect the trajectories of the background and signal H^+. Any loss of background due to spatial displacement has serious consequences on the determination of the beam-beam signal. The demerging scheme used has, for this reason, a low dispersion to maximize the H^+ transmission to the detector. The chopping scheme incorporates various refinements [22] and is capable of separating a signal from background nearly 10^5 larger.

As mentioned previously, other merged-beams apparati using singly charged ions have shown that, if the product ion beam is energy analyzed with sufficient resolution, the energy difference between the signal and background can be used to significantly reduce background, even to the point where only one beam requires modulation [12]. Applying a voltage to the whole merge region so that the signal ions are born at a different potential can also result in further reduction to the point that no beam modulation is required [14].

Beam current measurement, I_1, I_2

Faraday cups are used to measure the primary ion beams currents. To ensure proper ion collection, a positive voltage is applied via a battery between the cup and the ammeter. However, care must be taken to ensure that any leakage from the battery is not significant with respect to the beam currents measured. While this may not have an effect during measurement of the full primary beams, it may affect current measurements on the order of nA that are encountered during beam-beam overlaps measurements.

Measurement of the neutral beam flux can be accomplished using secondary-electron emission from a simple stainless-steel disk. In the ORNL apparatus a guard ring biased at +200 V draws electrons from the disc. The ratio of electrons emitted to neutral H or D striking the disc is γ, the secondary emission coefficient. The effective neutral current I_N then, is given by $\frac{I_1}{\gamma}$, where I_1 is the measured electron current leaving the disc. A procedure for *in situ* absolute calibration [22] of the neutral detector involves alternately switching the laser on and off and equating the loss of H^- to the H produced and comparing this to the secondary electron emission current. Typically, γ was measured to be in the range 0.9 to 1.3, with increases of a few percent after venting the apparatus to atmosphere with dry N_2, and a slight incident-beam velocity dependence.

v_1, v_2, v_r, *and L determination*

The velocities v_1, v_2, and therefore v_r are determined from the acceleration potential of the two ion sources. These potentials V_1 and V_2 correspond to the net potential difference across which the ions are accelerated. In an ion source, the ions are extracted from a plasma which may be at a nonzero potential with respect to the voltage supplied to the source. This plasma potential is due to the difference in the mobilities of the electrons and positive ions. Previous measurements [47] indicate that 10 eV is a reasonable estimate for this shift for negative ion extraction [36] of a duoplasmatron ion source operated in source gas "starvation" mode.

An estimate of the plasma potential of 20 ±10 V (per charge) for the ECR source was obtained by measuring the voltage applied to the merger (which is a parallel plate analyzer) vs the voltage applied to the ECR source and extrapolating to zero ECR voltage. These plasma potential shifts are small and of little consequence for energies above 1 eV/amu.

The merge length L is determined by considering positions along the merge path from which collision products (signal) can be transmitted to the detector. Ion-trajectory calculations [22] indicate that for the ORNL ion-atom merged-beams apparatus the merge path begins when the beams exit the plane of the last plate of the parallel-plate analyzer. The merge path ends at a position along the merge path inside the Einzel lens (see Fig. 4.2) where the negative voltage at which the H^+ is created results in a loss of transmission to the CEM detector.

Ω *determination*

In order to determine an accurate cross section from the measured signal, it is necessary to accurately measure the overlap of the two beams. This is certainly one of the technical challenges of the merged-beams technique. Rigorously, the overlap of the beams should be determined by integrating the product of the current densities j_1 and j_2 of the two beams at every point along the merged-path. Properly normalized we define Ω as

$$\Omega = \frac{\int\int\int j_1(x,y,z)j_2(x,y,z)dxdydz}{\int\int\int j_1(x,y,z)dxdydz \int\int\int j_2(x,y,z)dxdydz} \tag{4.10}$$

where z is the direction of propagation of the beams and xy defines a plane perpendicular to the beams. Since Ω is nonzero only where beams overlap, Ω is called the overlap integral [5]. Obviously some approximation must be made in order to make the measurement of Ω experimentally tractable. The particular method chosen is often selected on the basis of the spatial distribution and stability of the beams, suitability in vacuum, and whether beams can be selectively turned on/off. An early scheme [48] to determine the full three-dimensional profiles used square apertures formed by a pair of crossed slits scanned over the entire interaction region. Faster, two-dimensional profile monitoring [49] using a fluorescent screen and digitized video techniques at several (adjustable) positions along the merge path is currently in operation on an electron-ion merged-beams apparatus [50]. An alternate scheme [51] to measure two-dimensional profiles consists of pinhole apertures on rotating paddles at fixed positions along the merge path. Other schemes, including those used on the ORNL apparatus, use two orthogonal one-dimensional scans of each beam, I_i, giving, in the notation of Eq. (10), at $z = z_j$,

$$I_i(y) = \int j_i(x,y,z=z_j)dx, \tag{4.11}$$

$$I_i(x) = \int j_i(x,y,z=z_j)dy. \tag{4.12}$$

Assuming that the numerator of Eq. (4.10) is separable in x and y and substituting what is actually measured [as shown in Eqs. (4.11) and (4.12)], Eq. (4.10) becomes

$$\Omega(z=z_j) = \frac{\int I_1(y)I_2(y)dy \int I_1(x)I_2(x)dx}{\int I_1(y)dy \int I_1(x)dx \int I_2(y)dy \int I_2(x)dx}. \tag{4.13}$$

The integrals in the denominator are just the total integrated beam currents. Schemes to measure such beam profiles include differential measurements of the beam intensity as beams are intercepted by knife edges [22] or rotating paddles with knife edges [52].

Currently the ORNL merged-beams apparatus uses both a transmission measurement as slits are translated into the beam and a commercial rotating wire beam profile monitor. The principle of operation of the rotating wire scanners is as follows: A helical shaped 0.25-mm wire is rotated at 19 Hz in the beam in such a way that the beam "sees" a wire scanning horizontally

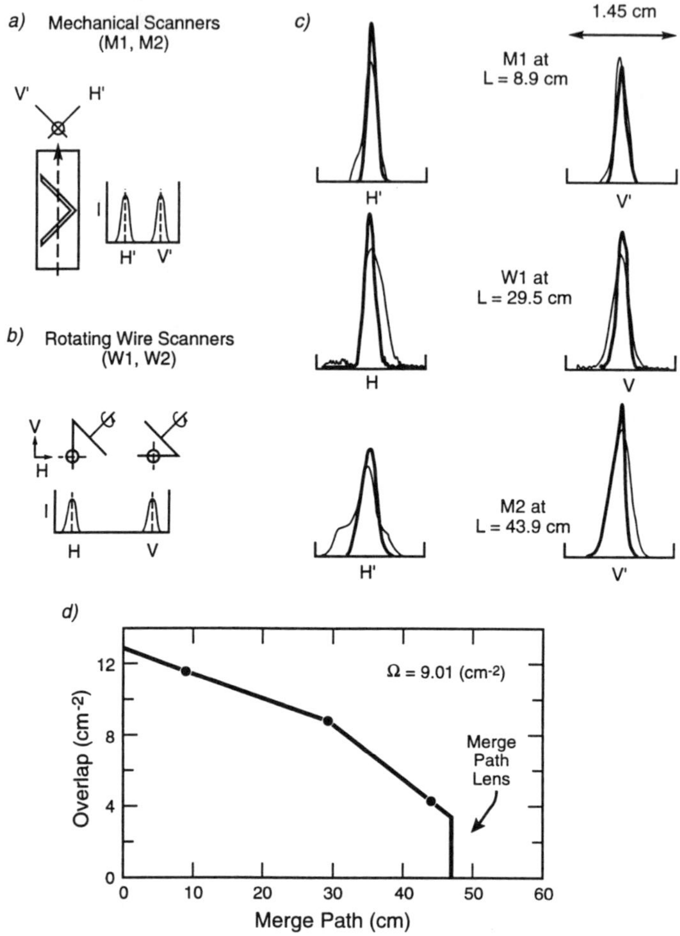

FIGURE 4.8 Two-dimensional beam profiles taken at the three positions along the merge path for the recent measurements [80] for Si^{4+} + D. In (c) the thick lines denote the D profiles while the thin lines denote the Si^{4+} profiles.

and then vertically for a distance of ± 1.0 cm from the center of the scanner. An analoq signal produced from the secondary electrons ejected from the wire is in direct proportion to the horizontal and vertical profiles of the beam and corresponds to the integration (along the wire) as given in Eqs. (4.11) and (4.12). Properly synchronized with the rotation of the wire, the profiles of each beam can be displayed in real time on an oscilloscope. This facilitates tuning beams with minimum divergence and maximum overlap with relative ease. The beam profiles are measured by recording these signals using a waveform digitizer. The beam overlap integral is then determined using a finite-sum approximation to Eq.(4.13).

Figure 4.8 shows two-dimensional beam profiles taken at three positions along the merge path for recent measurements [80] for Si^{4+} + D. At each position there are horizontal and vertical profiles of the ion and neutral beams, the latter being the narrowest, with a FWHM of 1.5 to 2 mm. The Si^{4+} is generally larger, with FWHM of 2 to 3 mm, making the overlap of the two beams stable. The first and third profiles, labelled M_1 and M_2, are taken with respect to a 45 deg rotated frame of reference, thereby indicating the elliptical nature of the beam, if any. Figure 4.8 shows the values of the three beam overlaps calculated from the measured profiles plotted as a function of postition along the merged path. Values for the remainder of the merge path are estimated by linear interpolation or extrapolation as shown in the figure. An Einzel lens is located 51 cm into the merged path to aid in collection of the products. Any products created inside the lens will be born at a different potential and will not be transmitted through the magnetic and electrostatic dispersion elements to the detector (see Fig.4.2).

4.3 ORNL MERGED-BEAMS MEASUREMENTS

The ORNL ion-atom merged-beams apparatus has been successful in performing benchmark electron capture cross section measurements [22,39,44,53–56] for various multicharged ions on H and D from 20 meV/amu to 5000 eV/amu, over five orders of magnitude in energy. In this energy range the relative nuclear motion between collision partners is slow compared to the orbital motion of active electrons in the system. Electrons of the temporary quasimolecule formed in the collision have sufficient time to adjust to the changing interatomic field as the nuclei approach and separate. Simple scaling laws which are successful at higher energies ($v \sim 1 a.u.$) do not exist at eV/amu energies, as verified experimentally for total cross sections by Phaneuf *et al.* [1]. Calculations [57] for the ions C^{4+}, O^{4+}, and Si^{4+} show that the total cross sections are essentially the same ($\sim q$ x $10^{-15} cm^2$) at keV/amu energies, while at lower energies the cross sections are electron core dependent and vary widely for the different 4+ ions, up to a factor of 10 at 1 eV/amu. Fully-quantal coupled-channel molecular-orbital calculations (e.g., Ref. [57]) are considered most accurate at these energies but are difficult to perform and have only recently been tested in detail.

Merged-beams measurements have provided critical comparison with theory. As a recent example we consider the C^{3+} + H system, which, because of its importance, has been studied in detail [1,56,58–65,71,66] for several years. Carbon ions are one of the major impurity ions in magnetic fusion plasma devices. Energy loss from high temperature plasmas is enhanced by the radiation from impurity ions that have undergone capture to excited states. Low energy measurements of charge transfer for C^{3+} are also important for astrophysics, since the cross section is predicted to sharply increase at low (near-thermal)

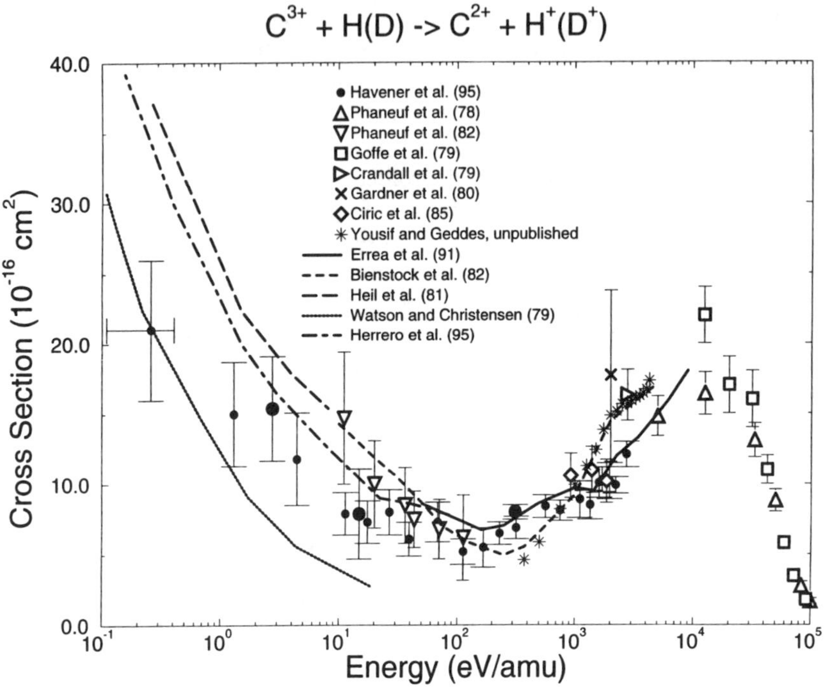

FIGURE 4.9 ORNL ion-atom merged-beams measurements [56] (filled circles) for C^{3+} + H(D) compared to other experiments and theory. All vertical error bars denote relative uncertainties estimated at a 90% confidence level (see text). The measurements at 2.7, 14.8, and 312 eV/amu are performed with ^{13}C instead of ^{12}C. All calculations except those of Herrero *et al.* below 4.4 eV/amu were performed with H. The merged-beams measurements below 600 eV/amu were performed with D.

energies.

Figure 4.9 shows merged-beams measurements [56] of absolute total cross sections for electron capture by C^{3+} in collisions with ground state hydrogen and deuterium in the energy range 0.3-3000 eV/amu. Relative uncertainties in the measured signal are mostly due to statistical uncertainties in separating the beam-beam signal from the backgrounds. Occasionally spatial instabilities in the beams whose effect on the beam-beam overlap was beyond our ability to monitor added an additional component which was combined in quadrature. Absolute uncertainties in the cross section were estimated to be 12% at the 90% confidence level and must be added in quadrature to the relative uncertainties to obtain total uncertainties. A detailed discussion can be found in Ref. [22].

C^{3+} is Li-like and no long lived metastable components of the beam are known to exist [67] that would obscure detailed comparison between theory and experiment. In general, good agreement is obtained with published experimental measurements at the lower [1] (10-110 eV/amu) and higher [58–61] (> 1 keV/amu) collision energies, where a beam-gas (hydrogen furnace) technique was used. The improved accuracy and the large energy range of the merged-beams measurements indicate an energy dependence different than was suggested by interpolating previous published and unpublished measurements and by the theoretical calculation of Bienstock *et al.* [63] which attempted to reconcile the previous low- and high-energy total capture cross section data. In order to span such a large energy range their calculations use a coupled channel method at the lower energies and a distorted wave method at the higher energies. As can be seen in the figure, the unpublished total electron capture measurements by Yousif and Geddes show excellent agreement with the calculations of Bienstock *et al.* and have led to the latter's use in normalization of state-selective translational energy spectroscopy (TES) measurements [68], recently reviewed by Gilbody [69].

The lack of more extensive total cross section measurements for the C^{3+} system has previously hampered critical comparison of experimental and theoretical state-selective cross sections. Capture to specific states was investigated by both photon emission spectroscopy (PES) [70] and TES. However, even though the two measurements agree well within their limited energy overlap for capture to the dominant channel $(1s^2 2s3s)^3S$, the PES measurements compare best with the calculations of Errea *et al.* [62] at the higher energies ($>$ 1 keV/amu) while the TES measurements agree with the calculations of Bienstock *et al.* at the lower energies ($<$ 700 eV/amu). Comparison between measurements and theory is obscured by the fact that the TES measurements are normalized to the total capture cross sections of Bienstock *et al.* and cannot fully resolve different product states. In addition, the PES measurements have uncertainties of 30%. A more extensive discussion can be found in the literature [56,69,71,62].

The merged-beams measurements above 100 eV/amu show excellent agreement with the more recent 22-state molecular orbital calculation by Errea *et al.* [62], even reproducing the slight structure in the cross section at collision energies between 1000 and 2000 eV/amu. When the merged-beams measurements are compared to the unpublished total electron capture measurements by Yousif and Geddes, there seems to be a discrepancy. It should be noted, however, that when the total uncertainties of the measurements are considered, the discrepancy is only slight. The new merged-beams measurements do suggest that the normalization of the TES measurements to the total cross section calculation of Bienstock *et al.* should be reconsidered for the C^{3+} system.

At energies below 100 eV/amu the merged-beams measurements agree well with the recent calculations of Herrero *et al.* [66] who used the same molecular potentials and couplings as Errea *et al.*. Only at the lowest energy of 0.26

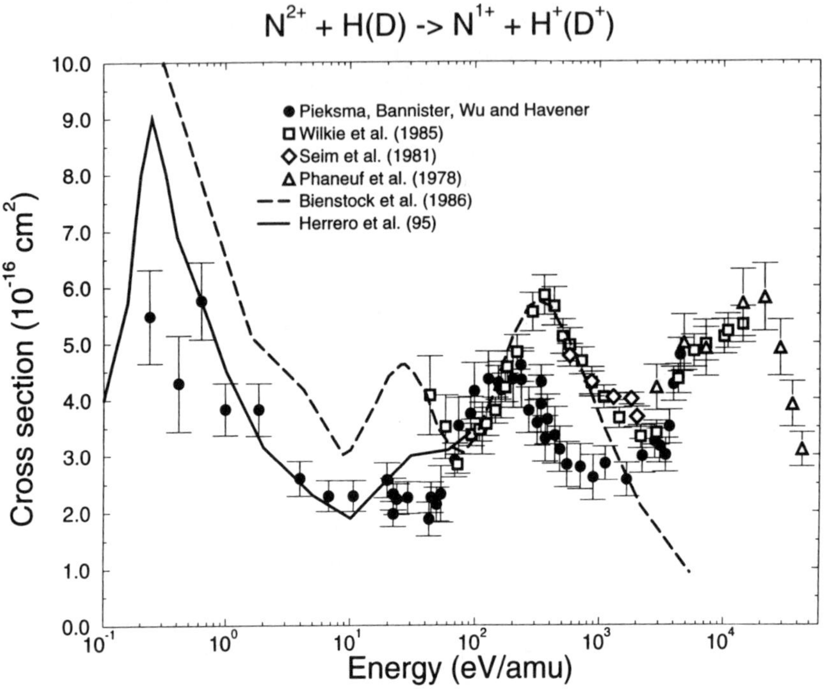

FIGURE 4.10 Preliminary ORNL ion-atom merged-beams measurements (filled circles) for N^{2+} + H(D) compared to other experiments and theory.

eV/amu do the measurements deviate significantly from the predicted cross section. At energies below 4.4 eV/amu, Herrero *et al.* performed calculations with both H and D to investigate the possible dynamical (trajectory) effect of different reduced masses on the cross section. For this collision system, though, the cross section for collisions with H are predicted to be at most 10% larger than for collisions with D at energies > 0.1 eV/amu. Calculations by Watson and Christensen [65] and Opradolce *et al.* [71] (not shown) included only contributions from singlet states and lie consistently below the merged-beams measurements.

Another system which is currently under study is N^{2+} + H. Nitrogen has been extensively observed in the spectra of gaseous nebulae. The interpretation of these observations requires an understanding of the ionization balance in the nebulae. Models which describe the ionization and recombination processes require accurate atomic data, including electron transfer with atomic hydrogen.

Preliminary merged-beams measurements [72] for N^{2+} are shown in Fig. 4.10. As can be seen in the figure, at high energies the merged-beams measurements agree with previous beam-gas (hydrogen furnace) measurements [58], but below 1 keV/amu there is a striking difference between the two measurements. This discrepancy is currently not understood but may be due, e.g., to either vibrationally excited H_2 in the hydrogen furnace, or a large metastable component in the N^{2+} beam. However, no evidence for metastables was observed [72] when the N^{2+} beam was directed into the ORNL electron-ion crossed beams apparatus [73] to look for electron-impact ionization below threshold. In addition, the previous studies with the hydrogen furnace have shown that when the relative metastable component of the N^{2+} beam was varied, no difference in the cross section with H was detected [74].

Previous TES measurements [74] show that throughout the whole energy range, capture to the $N^+(2s2p^3)^3D^0$ state is the dominant channel, while at keV/amu energies there is a growing contribution from capture to $N^+(2s2p^3)^3P^0$. The molecular-orbital coupled-channel calculations of Bienstock *et al.* [81] include only capture to the dominant channel, so the excellent agreement between the calculations and the previous measurements of Wilke *et al.* [74] just below 1 keV/amu may be somewhat fortuitous (see Fig. 4.10). At lower energies our measurements agree with the more recent total capture calculations of Herreo *et al.* [75] and give support for their significant lower rate coefficient (1.5 times lower) calculated for state-selective capture to $^3D^0$.

Only recently has the merged-beams apparatus been able to accurately explore collision energies less than 1 eV/amu. At these energies the ion-induced dipole potential between the reactants is important and is given by

$$V(r) = -\frac{\alpha q^2}{2r^4} \text{ (a.u.)}, \tag{4.14}$$

where r is the internuclear distance, α is the polarizability of the neutral atom ($\alpha = 4.5$ a.u. for both H and D), and q is the charge of the ion. At low collision energies, the force due to this potential, and the resulting acceleration of the particles towards each other, will significantly modify the reactant trajectories. This results in the incident trajectories accessing smaller internuclear distances than the initial impact parameter. At low enough energies trajectory effects may dominate the electron transfer process. This is the basis of the Langevin model [76] which provides some physical intuition for the process. According to this simple model, for impact parameters less than some critical impact parameter, the reactant trajectories follow collapsing orbits during which electron capture is assumed to occur. The Langevin model predicts that the cross section scales as $\sigma \propto q/v\sqrt{\alpha/\mu}$, where v is the collision velocity and μ the reduced mass. Both the simple Langevin model and fully quantal molecular-orbital calculations [77] predict that these trajectory effects can lead to strong enchancements in the cross section. At thermal energies shape resonances have been predicted [77] to contribute an additional enhancement in the cross section. Since the ion-induced dipole potential scales as q^2, these effects may become stronger for ions with higher charge, and can appear [78]

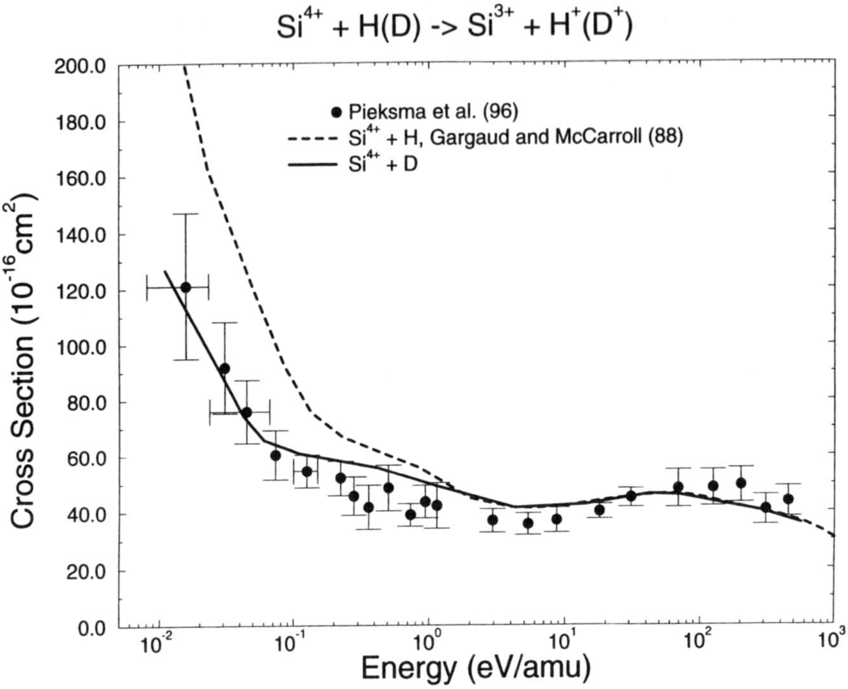

FIGURE 4.11 ORNL ion-atom merged-beams measurements [80] (filled circles) for Si^{4+} + D compared to theory.

at higher collision energies.

Merged-beams measurements for Si^{4+} + D are presented in Fig. 4.11. These are the first cross section measurements in the thermal energy regime for multicharged ions on D . Measurements with H could not be performed below 1 eV/amu due to the limited range of acceleration voltage of the ECR ion source. Such low energies with good resolution were possible due to the small merge angles $< 0.1^0$ and low divergence of the two beams that was achieved. The divergence of the Si^{4+} beam was $< 0.25^0$ and the D beam was $< 0.1^0$.

State-of-the-art molecular-orbital coupled-channel calculations [57] for the Si^{4+} + H/D systems are also shown in the figure and can be compared to the measurements on an absolute scale. Excellent agreement is found above 1 eV/amu, where strong rotational coupling between capture to the Si^{3+} (3d) and (4s) configurations exists. At energies below 1 eV/amu, measurements agree with the calculations with D, not with the calculations for H. This is the first time such isotope effects are observed. Indeed at the lowest energies, the cross sections for H compared to D are shown to be a factor of 1.9 larger. This large isotope effect is in contradiction to the Langevin model which predicts a factor of only 1.4 (i.e., the ratio of the reduced masses). The failure of the model to predict the magnitude of the isotope effect is due to the fact that it does not take into account the quantum-mechanical transition probability for capture, and this probability is different for collisions involving H and D. This probability for capture depends on, among other things, the radial velocity. At low enough energies, the radial velocity is determined by the ion-induced dipole attraction and the mass of the reactants.

At low energies, the measurements exhibit a $1/v$ dependence, which is also predicted by the molecular-orbital calculations and the Langevin model. Both the existence of a suprisingly large kinematic isotope effect and an observed sharp $1/v$ low-energy increase of the cross section are attributed to trajectory effects due to an attractive ion-induced dipole.

4.4 THE FUTURE OF THE ORNL MERGED BEAMS

Merged-beams measurements continue to be a technical challenge and the new collison physics that can be explored depends on, among other things, ion source technology. For example, the new capabilites of the ORNL ECR ion source, which now include the ability to produce intense metallic beams, will be used to study low energy electron capture by low to moderately charged ions of refractory metals. These ions are of interest to divertor and edge plasma of present and future magnetic fusion devices. Shape resonances due to ro-vibrational states of the transient quasimolecule formed in the collision have been predicted for some systems [77] at energies below 1 eV/amu. A new PC-based data acquisiton system will allow the apparatus to continuously scan over and thereby better resolve any structure that may appear in the cross section.

There is also an interesting possibility to perform collisions with specially prepared beams. The merged-beams technique can be used to study collisions with excited states, as has been done before in a limited way for H in a band of excited states [44]. Electron capture cross sections with H in excited states are enormous and scale with the geometric size of the atom. Plans to produce beams of metastable He are being considered. With major modification the ORNL merged-beams apparatus can be adapted to study collisions with O which play important roles in astrophysical plasmas as well as in the earth's and other planetary atmospheres.

The ion-atom merged-beams technique is suited not only for measurements of electron capture at low energies, but also for measurement of the ionization of H by multicharged ions at keV/amu energies. While the total ionization cross section has been previously investigated [79], there are still uncertainties about what mechanisms are important near threshold. The total ionization cross section is currently being measured by observing the beam-beam generated electrons. Also an electron spectrometer is being developed to measure the energy and angular distribution of the emitted electrons. There is a real advantage to performing measurements in merged beams, since electrons emitted at near zero energy in the center-of-mass frame are measured in the lab frame with energies corresponding to nearly 10 eV.

ACKNOWLEDGMENTS

This research was sponsored by the Division of Chemical Sciences, Office of Basic Energy Sciences of the U. S. Department of Energy under contract No. DE-AC05-96OR22464 with Lockheed Martin Energy Research Corporation. The author acknowledges all those who were involved in the development of the ORNL ion-atom merged-beams apparatus, especially Ron Phaneuf while he was in Oak Ridge. The author would also like to thank S. Trujillo for permission to use the illustration of their early merged-beams apparatus and Fred Meyer for his helpful comments on the manuscript.

REFERENCES

1. R. A. Phaneuf, I. Alvarez, F. W. Meyer, and D. H.Crandall, *Phys. Rev. A* **26**, 1892 (1982).
2. S. Yaltkaya, E. Y. Kamber, and S. M. Ferguson, *Phys. Rev. A* **48**, 382 (1993).
3. S. M. Trujillo, R. H. Neynaber, L. Marino, and E. W. Rothe, in *Proceedings of the IVth International Conference on the Physics of Electronic and Atomic Collisions* (Science Bookcrafters, Inc., New York, 1965), p. 260.
4. V. A. Belyaev, B. G. Brezhnev, and E. M. Erastov, *JETP Lett.* **3**, 207 (1966).
5. S. M. Trujillo, R. H. Neynaber, and E. W. Rothe, *Rev. Sci. Instrum.* **37**, 1655 (1966).
6. V. A. Belyaev, B. G. Brezhnev, and E. M. Erastov, *JETP* **25**, 777 (1967).
7. R. D. Rundell *et al Phys. Rev. A* **19**, 33 (1979).
8. P. M. Koch and J. E. Bayfield, *Phys. Rev. Lett.* **34**, 448 (1975).
9. M. Burniaux *et al.*, *J. Phys. B: Atom. Molec. Phys.* **12**, 221 (1977).
10. R. H. Neynaber and S. M. Trujillo, *Phys. Rev.* **167**, 63 (1968).
11. R. H. Neynaber, G. D. Magnuson, S. M. Trujillo, and B. F. Meyers, *Phys. Rev. A* **5**, 285 (1972).
12. P. K. Rol, E. A. Entemann, and K. L. Wendell, *J. Chem. Phys.* **61**, 2050 (1974).
13. K. L. Wendell and P. K. Rol, *J. Chem. Phys.* **61**, 2059 (1974).
14. W. R. Gentry, D. J. McClure, and C. H. Douglass, *Rev. Sci. Instrum.* **46**, 367 (1975).
15. J. Moseley, W. Aberth, and J. R. Peterson *Phys. Rev. Lett.* **24**, 435 (1970).
16. W. Aberth, J.R.Peterson, D.C. Lorents, and C. J. Cook, *Phys. Rev. Lett.* **20**, 979 (1968).
17. J. Weiner, *Phys. Rev. A* **4**, 1824 (1971).
18. C. H. Cherkani, S. Szucs, H. Hus, and R. Brouillard, *J. Phys. B: At. Mol. Opt. Phys.* **24**, 2367 (1991).

19. J. R. Peterson, W. H. Aberth, J. T. Moseley, and J. R. Sheridan, *Phys. Rev. A* **3**, 1651 (1971).
20. G. Ringer and W. R. Gentry, *J. Chem. Phys.* **71**, 1902 (1979).
21. D. P. Wang, S. Y. Tang, and R. H. Neynaber, *J. Phys. B: At. Mol. Opt. Phys.* **18**, 1513 (1985).
22. C. C. Havener *et al.*, *Phys. Rev. A* **39** 1725 (1989).
23. F. W. Meyer *Nucl. Instrum. Methods Phys. Res., Sect. B* **9**, 532 (1985).
24. H. Saitoh *et al.*, in *Proceedings of the Nineteenth ICPEAC, Whistler, BC*, p. 279, abstracts.
25. R. Hoekstra, in *AIP Conference Proceedings of the XIXth Int. Conf. on Physics of Electronic and Atomic Collisions, Whistler*, edited by L. J. Dubè *et al.* (American Institute of Physics, New York, 1995) **360**, p. 547
26. B. Davis and A. H. Barnes, *Phys. Rev.* **34** 152 (1929); **37** 1368 (1931).
27. P. Mahadevan, *Proc. R. Soc. A* **327**, 317 (1972).
28. G. Hagen, in *5th Int. Conf. on Physics of Electronic and Atomic Collisions*, edited by I. P. Flaks and E. S. Solvyov (Leningrad:Nauka, Leningrad, 1967), p. 165, abstracts.
29. L. P. Theard, in *Proc. 6th Int. Conf. on Physics of Electronic and Atomic Collisions, Cambridge, Mass.*, edited by I. Amdur (Cambridge:MIT Press, 1969), p. 1042, abstracts.
30. D. Auerbach *et al.*, *J. At. Mol. Phys. B* **10**, 3797 (1977).
31. S. J. Smith, K-F. Man, R. J. Mawhorter, I. D. Williams, and A. Chutjian, *Phys. Rev. Lett.* **67**, 30 (1991).
32. E. K. Wahlin *et al.*, *Phys. Rev. Lett.* **66**, 157 (1991).
33. J. R. Mowat *et al.*, *Phys. Rev. Lett.* **74**, 50 (1995).
34. B. Jacquot and M. Pontonnier, *Nucl. Instrum. Methods* **A295**, 5 (1990).
35. B. V. Zyl, N. G. Utterback, and R. C. Amme, *Rev. Sci. Instrum.* **47**, 814 (1976).
36. L. E. Collins and R. H. Gobbett, *Nucl. Instrum. Methods* **35**, 277 (1965).
37. T. A. Antaya, *J. de Physique* **50**, cl-707 (1989).
38. R. E. Olson and M. Kimura, *J. Phys. B: At. Mol. Phys.* **15**, 4231 (1982).
39. C. C. Havener, F. W. Meyer, and R. A. Phaneuf, in *Proceedings XVII Int. Conf. on the Physics of Atomic and Electronic Collisions, Brisbane,* edited by W. R. MacGillivray *et al.* (IOP Publishing Ltd., 1992), p. 381, invited papers.
40. M. Gargaud, 1987 Doctoral Thesis, L'Université de Bordeau I.
41. L. R. Andersson, M. Gargaud, and R. McCarroll, *J. Phys. B: At. Mol. Opt. Phys.* **24**, 2073 (1991).
42. R. H. Neynaber and S. Y. Tang, *J. Chem. Phys. B:* **17**, 3565 (1984).
43. D. J. McClure, C. H. Douglass, and W. R. Gentry, *J. Chem. Phys.* **66**, 2079 (1977).
44. C. C. Havener *et al.*, in *AIP Conference Proceedings of the VIth International Conference on the Physics of Highly Charged Ions*, edited by P. Richard et al. (American Institute of Physics, New York, 1993), **274**, p. 43.
45. E. W. Bell, 1993 Ph. D Thesis, University of Colorado, Boulder, Colorado.
46. P. O. Taylor, R. A. Phaneuf, and G. H. Dunn, *Phys. Rev. A* **22**, 435 (1980).

47. C. Lejeune, *Nucl. Instrum. Methods* **116**, 417 (1974).
48. S. K. Sethuraman, J. R. Gibson, and J. L. Moruzzi, *J. Phys. E: Sci. Instrum.* **10**, 14 (1977).
49. J. L. Forand *et al.*, *Rev. Sci. Instrum.* **61**, 3372 (1990).
50. M. E. Bannister, X. Q. Guo, T. M. Kojima, and G. H. Dunn, *Phys. Rev. Lett.* **21**, 3336 (1994).
51. D. E. Nitz, M. W. Geis, K. A. Smith, and R. D. Rundel, *Rev. Sci. Instrum.* **47**, 306 (1976).
52. C. J. Keyser, H. R. Froelich, J. B. A. Mitchell, and J. W. McGowan, *J. Phys. E: Sci. Instrum.* **12**, 316 (1979).
53. M. S. Huq, C. C. Havener, and R. A. Phaneuf, *Phys. Rev. A* **40**, 1811 (1989).
54. C. C. Havener, M. P. Nesnidal, M. R. Porter, and R. A. Phaneuf, *Nucl. Instrum. Methods* **56/57**, 95 (1991).
55. L. Folkerts *et al.*, *Phys. Rev. A* **51**, 3685 (1995).
56. C. C. Havener, A. Müller, P. A. Zeijlmans van Emmichoven, and R. A. Phaneuf, *Phys. Rev. A* **51**, 2982 (1995).
57. M. Gargaud and R. McCarroll, *J. Phys. B: At. Mol. Opt. Phys.* **21**, 513 (1988).
58. R. A. Phaneuf, F. W. Meyer, and R. H. McKnight, *Phys. Rev. A* **17**, 534 (1978).
59. T. V. Goffe, M. B. Shah, and H. B. Gilbody, *J. Phys. B: At. Mol. Opt. Phys.* **12**, 3763 (1979).
60. D. H. Crandall, R. A. Phaneuf, and F. W. Meyer, *Phys. Rev. A* **19**, 504 (1979).
61. L. D. Gardner *et al.*, *Phys. Rev. A* **21**, 1397 (1980).
62. L. F. Errea, B. Herrero, L. Méndez, and A. Riera, *J. Phys. B: At. Mol. Opt. Phys.* **24**, 4061 (1991).
63. S. Bienstock, T. G. Heil, C. Bottcher, and A. Dalgarno, *Phys. Rev. A* **25**, 2850 (1982).
64. T. G. Heil, S. E. Butler, and A. Dalgarno, *Phys. Rev. A* **23**, 1100 (1981).
65. W. D. Watson and R. B. Christensen, *Astrophys. J.* **231**, 627 (1979).
66. B. Herrero, I. L. Cooper, A. S. Dickinson, and D. R. Flower, *J. Phys. B: At. Mol. Phys.* **28**, 4607 (1995).
67. R. F. Welton, T. F. Moran, and E. W. Thomas, *J. Phys. B: At. Mol. Phys.* **24**, 3815 (1991).
68. F. G. Wilkie, R. W. McCullough, and H. B. Gilbody, *J. Phys. B: At. Mol. Opt. Phys.* **19**, 239 (1986).
69. H. B. Gilbody, *Adv. At. Mol. Opt. Phys.* **32**, 149 (1994).
70. D. Ćirić *et al.*, *J. Phys. B: At. Mol. Opt. Phys.* **18**, 3629 (1985).
71. L. Opradolce, L. Benmeuraiem, R. McCarroll, and R. D. Piacentini, *J. Phys. B: At. Mol. Opt. Phys.* **21**, 503 (1988).
72. M. Pieksma, M. E. Bannister, W. Wu, and C. C. Havener, to be submitted to *Physical Review A*.
73. M. E. Bannister *et al.*, *Phys. Rev. A* **52**, 413 (1995).
74. F. G. Wilke *et al.*, *J. Phys. B: At. Mol. Opt. Phys.* **18**, 479 (1985).
75. B. Herrero, I. L. Cooper, A. S. Dickinson, and D. R. Flower, *J. Phys. B: At. Mol. Opt. Phys.* **28**, 711 (1995).

76. G. Giomousis and D. P. Stevenson, *J. Chem Phys.* **29**, 294 (1958).
77. N. Shimakura and M. Kimura, *Phys. Rev. A* **44**, 1659 (1991).
78. P. C. Stancil and B. Zygelman, *Phys. Rev. Lett.* **75**, 1495 (1995).
79. M. Pieksma *et al.*, *Phys. Rev. Lett.* **73** 46 (1994).
80. M. Pieksma, M. Gargaud, R. McCarroll and C. C. Havener, sub. to *Physical Review A*
81. S. Bienstock, A. Dalgarno and T. G. Heil, *Phys. Rev. A*, **33** 2078 (1986).

CHAPTER 5

Heavy Ion Storage Rings

Alfred Müller* and Andreas Wolf†

Within the last decade nine heavy ion storage rings equipped with electron cooling devices were completed and the related experimental programs started with great success. Unique new results were obtained—particularly in the field of atomic physics—by employing the outstanding possibilities of cooler rings for preparation and manipulation of high-quality beams of ions. In this chapter, the new experimental possibilities provided by heavy ion storage rings are reviewed. Several groups of experiments which have become feasible only by the construction and use of heavy ion storage rings are discussed.

5.1 INTRODUCTION

Previous interest in circular particle accelerators and storage rings was mainly stimulated by the increasing demand for high collision energies in nuclear and elementary particle physics. While the energy frontier continues to be extended, other aspects of storage rings have also become important. For both nuclear and atomic physicists the preparation, storage and manipulation of beams of "expensive" particles has become a prime goal, and the gain in beam quality and intensity facilitated by phase-space cooling techniques has become obvious. The remarkable progress in phase-space cooling of heavy particles, which culminated in the antiproton accumulators at CERN and Fermilab, together with the example of the Low-Energy Antiproton Ring (LEAR) at CERN (operating since 1982), have promoted the construction of many new "cooler" storage rings for hadrons and heavy ions all over the world. Within a relatively short time span of only five years the new rings have become operational and have already provided a wealth of outstanding new experimental results.

---

*) Institut für Kernphysik, Universität Giessen, D–35392 Giessen, Germany
†) Physikalisches Institut der Universität Heidelberg and Max-Planck-Institut für Kernphysik, D–69117 Heidelberg, Germany

Storage ring facilities which are at least partially dedicated to atomic physics studies comprise the TSR (Test Storage Ring) of the Max Planck Institute for Nuclear Physics in Heidelberg, ASTRID (Aarhus STorage RIng Denmark) of the University of Aarhus, ESR (Experimental Storage Ring) of the Gesellschaft für Schwerionenforschung (GSI) in Darmstadt and CRYRING (combination of a CRYogenic electron beam ion source—CRYEBIS—with a storage RING) of the Manne-Siegbahn Laboratory at the University of Stockholm. Important contributions to the field of atomic collisions have also been made at LEAR of CERN in Geneva, at TARN II (a follow-up of the Test Accumulation Ring for the NUMATRON accelerator facility) of the University of Tokyo, and at the Cooler ring of the Indiana University Cyclotron Facility (IUCF) in Bloomington. In this Chapter mainly ASTRID, CRYRING, ESR and TSR will be considered since they are the storage rings where a major fraction of the available beam time is devoted to atomic physics.

A substantial body of literature has now accumulated on the topic of ion storage rings, the related beam cooling methods (in particular electron cooling), and experiments with stored ion beams. The electron cooling technique was pioneered by G. I. Budker and his group at Novosibirsk, and a number of ideas later realized at ion storage rings can be found already in early reviews [1,2] originating from that work. For more details concerning the techniques of ion storage and cooling and the physics related to the new facilities the reader is referred to reviews by Poth [3,4], Parkhomchuk and Skrinsky [5], Meshkov [6], Pollock [7], and Schuch [8], to proceedings of symposia on physics with stored (and trapped) ions [9–13] and on cooling techniques [14–16], as well as to references contained therein. Atomic physics experiments at storage rings have been reviewed recently by Schuch [17], Habs [18], Bosch [19], Müller [20], Larsson [21], and Mokler and Stöhlker [22].

5.2 CONSTRUCTION AND PERFORMANCE OF STORAGE RINGS

The main purpose of the storage rings discussed here is the trapping of beams of energetic charged particles by recirculation on a closed orbit. Small traps for low energy atoms and ions are outside the scope of this contribution. Trapping is the essential function that opens up attractive possibilities for accumulating intense ion beams and efficiently manipulating and observing such beams over extended periods of time. It allows for the application of phase-space cooling, for the preparation or relaxation of excited species, and for repeated interaction of the beam particles with radiation or internal targets of different types. In the following subsections we will first discuss the general arrangement and the typical layout and experimental instrumentation of ion storage rings and then address specific aspects of the operation of these machines. All storage rings presently operating with electron coolers are listed in Table 5.1 and some of their main properties are given.

TABLE 5.1 Ion storage rings with electron cooling devices in operation. For each ring the maximum rigidity $B\rho$ of particle beams, the circumference C, and the first year of operation are listed.

Ring name	Laboratory	Location	$B\rho$ (Tm)	C (m)	Year
LEAR (Low Energy Antiproton Ring)	CERN[a]	Geneva (Switzerland)	6.7	78.0	1982[b]
Cooler	IUCF[c]	Bloomington, IN (USA)	3.6	86.8	1987
TSR (Test Storage Ring)	MPI[d]	Heidelberg (Germany)	1.7	55.4	1988
CELSIUS[e]	TSL[f]	Uppsala (Sweden)	7.0	82.0	1988
TARN II[g]	INS[h]	Tokyo (Japan)	5.8	77.8	1989
ASTRID (Aarhus STorage RIng Denmark)	ISA[i]	Aarhus (Denmark)	2.0	40.0	1990
ESR (Experimental Storage Ring)	GSI[j]	Darmstadt (Germany)	10.0	108.4	1990
CRYRING (CRYSIS[k] synchrotron-storage RING)	MSL[l]	Stockholm (Sweden)	1.4	51.6	1992
COSY (COoler SYnchrotron)	FZ[m]	Jülich (Germany)	12.0	184.0	1993

[a] Conseil Européen pour la Recherche Nucléaire
[b] Electron cooling since 1987
[c] Indiana University Cyclotron Facility
[d] Max-Planck-Institut für Kernphysik
[e] Cooling with Electrons and Storing of Ions from the Uppsala Synchrocyclotron
[f] The Svedberg Laboratory
[g] Test Accelerator Ring for the NUMATRON facility
[h] Institute for Nuclear Study
[i] Institute for Synchrotron Radiation, Univ. of Aarhus
[j] Gesellschaft für Schwerionenforschung
[k] CRYogenic (electron-beam) Ion Source
[l] Manne Siegbahn Laboratory, Univ. of Stockholm
[m] Forschungs-Zentrum Jülich

5.2.1 General layout

The general arrangement of an ion storage ring is similar to that of a synchrotron accelerator, which provides a periodic structure of bending magnets and magnetic focussing elements arranged in a closed loop. In fact, the heavy ion storage rings employed for atomic physics experiments can also be operated in a synchrotron mode, i.e., ions trapped in the ring can be accelerated or decelerated by repeated passage through rf cavities, ramping the field strength in the bending and focusing magnets and the rf frequency synchronously to each other and to the changing particle energy/momentum. The ion storage rings addressed here have circumferences of about 50–100 m and shapes of the closed ion orbit varying from nearly quadratic (as in the cases of TSR and ASTRID) to hexagonal (for the ESR) and nearly circular (in the case of CRYRING). The ion orbit is located in an evacuated beam pipe which, together with the apertures of the magnetic elements, defines the acceptance of the ring. Stable orbits within the acceptance are usually possible only for ions in a definite charge state (although the large acceptances of some rings permit simultaneous storage of ions in different adjacent charge states); in most cases, charge-changing collisions of the ions in the residual gas lead to beam loss and hence, to obtain sufficient storage times, the beam pipe must be kept under ultrahigh vacuum. Typical storage times range between a few seconds and many hours, as discussed below. The time taken for each single turn of an ion in a storage ring can vary over a wide range, between 0.5 μs for 500-MeV/u U^{92+} ions in the ESR and 150 μs for 200-keV $C_{60}{}^+$ ions in ASTRID; the common typical values are of the order of a few microseconds. The two examples also characterize the range of beam velocities used, extending from $\approx 2 \times 10^{-5}$ to $\approx 3/4$ of the speed of light; slightly relativistic beams at about 10% of the speed of light are typical. Usually ion storage rings are operated with continuous coasting beams with electrical ion currents as high as 2–3 mA for U^{92+} in the ESR and, for example, up to 18 mA for C^{6+} ions in the TSR. Using the rf cavity of the ring at a constant frequency equal to (some integral multiple of) the revolution frequency, it is also possible to produce a bunched circulating beam at a stationary energy.

The product $B\rho$ of the magnetic field strength B in the dipole magnets and the radius ρ of the orbit within the dipoles represents the magnetic rigidity of the stored ion beam, which is proportional to the ion momentum and inversely proportional to their charge-to-mass ratio. The magnetic rigidity for the maximum magnetic field strength limits the momentum or (for a given ion charge and velocity) the mass of the ions that can be stored; hence, it represents an important characteristic of ion storage rings.

Considering the long distances travelled by the stored ions at about a million revolutions per second, one clearly understands that the stable transport of the beam through the periodic focusing structure is a fundamental issue in the design. In fact the essential beam transport elements (dipole and quadrupole magnets and eventually higher-order multipole lenses) consume an appreciable fraction of the ring circumference. Free space between these components is occupied by the beam injection system, the rf cavities, diagnostic installations, and the electron cooling device, discussed below. Additional experimental installations, such as a gas target facility or an additional electron target, are often subject to severe spatial limitations. On the other hand, cooling and diagnostic devices themselves often open up possibilities for experiments.

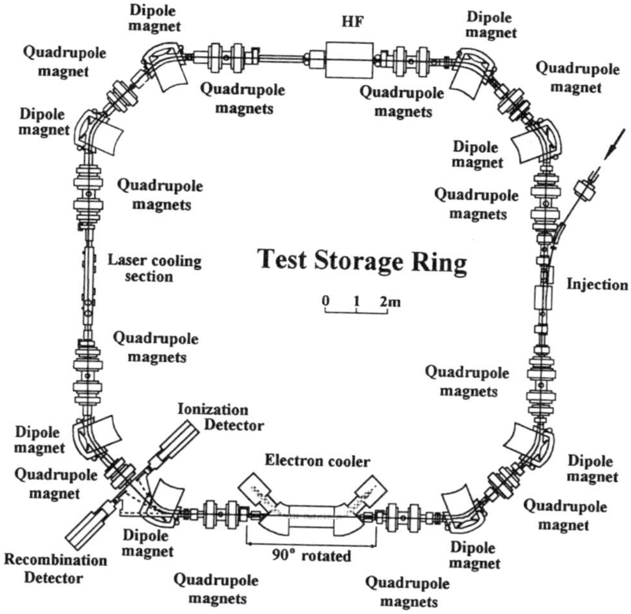

FIGURE 5.1 Schematic of the heavy-ion storage ring TSR in Heidelberg

5.2.2 Typical storage-ring instrumentation

Before summarizing the basic characteristics of the rings used for atomic physics investigations, let us illustrate the layout and the typical instrumentation of an ion storage ring, using as an example the heavy-ion Test Storage Ring TSR [23] at Heidelberg, schematically shown in Fig. 5.1. This machine, with a circumference of 55.4 m, has been added to the existing heavy-ion accelerator facility at the Max-Planck-Institute for Nuclear Physics, which comprises a 12-MV tandem accelerator and an rf post-accelerator with an integral gap voltage of 25 MV. Three of the four straight sections between the indicated dipole and quadrupole magnets are occupied by elements for the ion injection, the electron cooling device, the rf cavity of the ring, and diagnostic equipment; the fourth straight section is dedicated to experiments and presently occupied by installations for laser cooling and laser spectroscopy. The electron cooling device itself also serves as the central piece of equipment for various experiments addressing charge-changing processes in electron–ion collisions. Behind the first dipole magnet downbeam from the electron cooling device, neutral ions can be detected in the straight-through port of the magnet chamber. In addition, ions leaving the electron–ion overlap region in the electron cooler with a charge state smaller than that of the circulating particles can be collected with movable detectors on the outward side of the circulating beam; ions with higher charge state or lower mass (products of molecular fragmentation) can be collected on the inward side (see Fig. 5.1). Thus it is possible to observe recombination, ionization and dissociation in ion–electron collisions occuring in the electron cooler. Similar detectors are present at all other storage rings as well. They are used to monitor atoms and ions which drop out of the main beam because of charge- or mass-changing processes which, apart from ion–electron collisions, also occur by in-beam scattering, decay processes, ion–gas collisions, photoionization and -detachment, and so on. All the principal storage rings for atomic physics have excellent vacuum conditions, allowing the storage of ions for long periods of time.

In the laser cooling section and along the beam overlap region in the electron cooler (see Fig. 5.1), laser beams are applied parallel or antiparallel to the circulating ion beam; similar arrangements are used at the other heavy-ion storage rings. Furthermore, the TSR employs an inductive accelerator unit [24], which is used in laser-cooling and electron-cooling experiments, and a transverse beam profile monitor detecting ions produced in the residual gas by the circulating ion beam (see Sec. 5.2.7). The diagnostic and experimental tools that are employed at storage rings also comprise detectors for low energy photons, X-ray detectors, gas targets, and (last but not least) the analysis of ion circulation frequencies monitored by observation of the Schottky noise of the ion beam, which provides a powerful tool to detect small variations of the ion momentum.

In addition to the magnetic lattice of the ring itself, which defines the maximum magnetic rigidity of the stored ions (1.7 Tm at the TSR), also the injection device imposes limits on the ion species which can be stored, and their energies. The tandem/rf accelerator facility available as injector for the TSR allows the ring to be filled with heavy ions of a large number of elements in many charge states, ranging up to bare chlorine (Cl^{17+}) and highly stripped ions such as Cu^{26+} and Au^{50+}, but including also singly charged atomic and molecular ions (e.g., $^9Be^+$, CH^+). The ions are mostly stored at the energy delivered by the injector (typically 6–12 MeV/u for highly charged heavy ions), but can also be accelerated (up to $\approx$ 30 MeV/u for bare ions) and decelerated in the ring.

Among the principal atomic-physics rings discussed here, ASTRID, CRYRING, and TSR form a group of devices similar in size. ASTRID [25], with a circumference of 52 m and a maximum $B\rho$ of 2.0 Tm, presently uses a 200-kV electrostatic accelerator as an injector and so far has been used mostly for positive ions in low charge states and for negative ions at comparatively low beam velocities. The experimental instrumentation mainly includes lasers and related photon detectors, ion counters, Schottky-noise pickups, and an electron cooler/target. With the relatively high magnetic rigidity of ASTRID and the availability of a versatile very-low-energy injector, there is a unique capability to work with quite heavy species such as fullerene ions. Another unique property of ASTRID is that this storage ring is used part-time as a synchrotron light source, storing relativistic electrons.

CRYRING [26], with a circumference of 40 m and a maximum $B\rho$ of 1.4 Tm, uses an rf-quadrupole (RFQ) accelerator as an injector, providing ions with energies up to 300 keV/u. For light-ion injection a plasmatron ion source provides singly charged atomic and molecular ions. Heavy ions with the high charge states required for efficient preacceleration and for acceptance by the ring are produced by a cryogenic electron beam ion source (CRYEBIS). Usually, after injection the ions are rapidly accelerated to energies of several MeV/u. The instrumentation of CRYRING comprises an electron cooling device offering low electron temperatures (see Sec. 5.2.4), a novel transverse beam profile monitor (see Sec. 5.2.7), laser equipment, and versatile ion detectors; an internal gas target is under construction.

The ESR [27], operated by the GSI in Darmstadt, differs from the other storage rings by its size and by the superior performance of its injector. This ring has a circumference of 108 m and a maximum $B\rho$ of 10 Tm. The injector of the ESR is the heavy-ion synchrotron SIS with maximum energies up to 2 GeV/u for lighter ions and 1 GeV/u for uranium ions. The ESR can accept ion energies up to 560 MeV/u for all ions. These energies are sufficient to produce completely stripped ions of even the heaviest elements; U^{92+} ions have been successfully stored in the ESR with electrical currents as high as $\approx$ 2–3 mA. Thus the ESR provides a unique opportunity to perform experiments with very highly charged few-electron systems. The facility is equipped with an electron cooler capable of very high electron accelaration voltages (up to $\approx$ 300 kV), with lasers and photon counters, an internal gas target, various particle counters, and X-ray detectors. The ESR can also be loaded with nuclear fragments (including radioactive species) produced when high-energy ions from the SIS collide with a fixed target; the fragment separator (FRS) situated between the SIS and the ESR allows individual isotopes to be selected out of the reaction products according to their charge and mass.

Of course, each existing ring can be used to store any given ion in any given charge state provided the magnetic rigidity of the ion beam is below the maximum allowed by the ring. Thus, in principle, the ESR could run with protons and ASTRID would be capable of storing completely stripped uranium ions. However, apart from the effort of producing these ions, the possible storage times would be quite limited due to the energy constraint set by the maximum magnetic dipole fields. (At the corresponding low velocities U^{92+} ions capture electrons from the residual gas with large cross sections.) On the other hand, storing protons or fullerenes in the ESR might be seen as wasting the specific capabilities of the whole accelerator facility available at the GSI. So each of the rings provides a niche for specific experiments for which the existing environment is best suited.

Aside from the available ion injectors and the maximum magnetic rigidity of the rings, the layout of bending and focussing magnets (the ring 'lattice') can also make differences in an experiment. CRYRING, for example, with its nearly circular shape, particularly allows smooth bending of the circulating ion beam. This may be an advantage in the quest for crystalline ion beams (see Sec. 5.2.4). It also makes a slight difference with regard to the motional electric fields seen by the ions in the bending magnets.

5.2.3 Storage times of ions in rings

Ions can only be accumulated and manipulated in a storage ring when their storage times, represented by the mean lifetime τ, are sufficiently long. The storage times are limited by a number of factors which are of varying importance for different species of ions. Slow, highly charged ions for example have huge cross sections for electron transfer (capture) in collisions with neutral species. For ions of lower charge, having some remaining electrons with relatively small binding energies, collisional stripping (electron loss) limits the lifetime. Due to these processes, the residual gas pressure in the ring becomes the dominant factor determining the storage time at heavy-ion rings, even though these machines are operated at ultra-high vacuum in the range of 10^{-10} to 10^{-12} mbar.

In a wider view, the processes which may limit the storage time in heavy-ion rings are:

- Charge changing collisions (electron capture or electron loss) of the ions in the residual gas; ions with the "wrong" charge usually drop out of the ring.
- Scattering of ions from residual gas particles (Rutherford scattering from nuclei) and related energy loss (mostly from electronic stopping). Here, it is useful to distinguish multiple (small-angle) scattering, which mostly can be compensated for by continuous phase-space cooling of the ions, from single (large-angle) scattering, which drives ions out of the ring acceptance after only one scattering event. Energy loss can usually be compensated for by cooling the ion beam.
- Ion–ion collisions within the stored beam, which may comprise scattering processes leading to similar consequences as residual-gas collisions, or which may even comprise charge-changing processes; the effect of these intra-beam collisions depends strongly on the ion beam intensity and also on the ion 'temperature' in the beam's comoving frame of reference.
- Recombination of ions with electrons in the electron cooler.
- Beam instabilities caused by resonant excitation of the ion motion in the periodic focusing structure or by self-fields of the ion beam at high intensities.
- Removal of weakly bound electrons by field ionization, field detachment, radiative detachment, etc.

TABLE 5.2 Typical mean lifetimes τ of stored ion beams. Average background gas pressures are in the 10^{-11} mbar range. The keywords indicating the dominant loss mechanisms refer to angular scattering in ion-atom/molecule collisions (single/multiple scattering), capture of electrons from the residual gas (capture), radiative recombination in the electron cooler (RR), electron loss in collisions with residual gas particles (stripping, detachment), and fragmentation of molecular ions.

Ion	Energy	Ring	Lifetime	Loss mechanism	Reference
p	21 MeV	TSR	36 h	single scattering	[29][a]
p	21 MeV	TSR	3 h	multiple scattering	[29][b]
D^-	2 MeV/u	ASTRID	2 s	detachment	[30]
$C_{60}{}^+$	58 keV	ASTRID	11 s	fragmentation	[31]
H_3^+	11 MeV/u	CRYRING	2 s	fragmentation	[32]
C^{6+}	6.1 MeV/u	TSR	2 h	RR	[29][a]
O^{7+}	8.9 MeV/u	TSR	420 s	stripping	[33]
Cu^{26+}	8.9 MeV/u	TSR	122 s	capture	[29][a]
U^{92+}	300 MeV/u	ESR	1 h	RR/capture	[19]

[a] electron cooling on
[b] electron cooling off

Besides these omnipresent loss processes, experiments carried out with the ions may introduce additional beam loss; thus, any target in the ring (gas, photons, or charged particles), introduced to study certain collision processes, can cause loss of ions due to scattering and charge-changing collisions.

Beam decay constants characteristic of various loss processes have been discussed and compared to observations at the TSR by Habs et al. [28]. An updated summary of beam lifetimes (time interval during which the beam intensity drops to 1/e of the initial intensity) observed at ion storage rings with different species and beam energies is given in Table 5.2. Only for bare ion species with low charge (such as protons) is the beam lifetime limited by scattering in the residual gas. For most of the other atomic ions, charge changing processes (electron capture for bare and highly charged non-bare ions, electron loss for medium- and low-charged non-bare ions) are found to be the limiting factors, and fragmentation (dissociative excitation and ionization) is the dominant loss mechanism for molecular ions. With respect to electron capture (and to a lower degree also with respect to Rutherford scattering) the residual gas composition is critical, since heavy gas components such as argon have much higher cross sections for the relevant processes than hydrogen which, fortunately, frequently dominates in ultrahigh vacuum systems. Thus, e.g., the lifetime of Cu^{26+} ions given in Table 5.2 is limited by the $\approx 1\%$-fraction of Ar in the residual gas. The intrabeam collisions mentioned among our items have not so far been found to present serious limitations of the beam lifetime; however, intrabeam scattering has a large influence on the equilibrium ion 'temperature' in the comoving reference frame of the beam during phase space cooling.

5.2.4 Ion beam cooling

In line with Liouville's theorem, the phase space volume occupied by an ensemble of ions can be reduced (or increased) by manipulations using non-conservative forces. Accordingly, the phase space volume of a particle beam (the beam emittance) can be reduced by beam cooling. The name 'cooling' reflects the idea that the statistical motion of the ions, seen in a reference frame where the average ion velocity vanishes, can be associated with an ion temperature T_i.

An example of a process causing natural beam cooling is synchrotron radiation emission in combination with rf acceleration. Imposed beam-cooling methods are stochastic, electron, and laser cooling. Stochastic cooling of hot beams of antiprotons was the experimental key to the observation of the W vector boson. The principle of stochastic cooling [34] is based on the detection of ion position and momentum deviations at one location of the ring and a corrective action, derived from the dectected signal, applied downbeam such that the delay in the correction-signal transmission is synchronized with the ion time-of-flight between the detection and the correction electrodes. Technically, this scheme of active feedback is realized by fast broad-band amplification of signals representing the fluctuations of the beam charge or of the average lateral position of the ions within the pick-up device. The damping times of the thermal motion obtained by this cooling method usually are of the order of several minutes; more favorable conditions with damping times of some seconds are obtained for hot, fast, low-current ion beams (such as secondary particles produced by shooting an accelerated beam on a target). Stochastic cooling has not been employed yet for atomic physics experiments in storage rings, but a stochastic cooling system is foreseen [36] for fragments of nuclear reactions injected into the ESR via the fragment separator (see Sec. 5.2.2).

The principle of electron cooling was invented by Budker [37] and first introduced to the NAP-M proton storage ring in Novosibirsk [38]. A cold beam of electrons with an average velocity matching the ion velocity is merged for a distance of about 1 to 2 m with the circulating ion beam. Coulomb collisions between electrons and ions lead to a friction force which slows the motion of those ions traveling with non-zero velocity in the co-moving reference frame of the electron beam. The typical damping time constants for protons are of the order of a second. The ultimate limit of the resulting "cooling" is reached at thermal equilibrium, i.e., when the ion temperature T_i is equal to the electron temperature T_e. The ion velocity spread is then proportional to $(T_e\, m_e/m_i)^{1/2}$ where m_e and m_i are the electron and ion rest masses, respectively.

Electron beams for cooling [39] are based on the technology of electrostat-

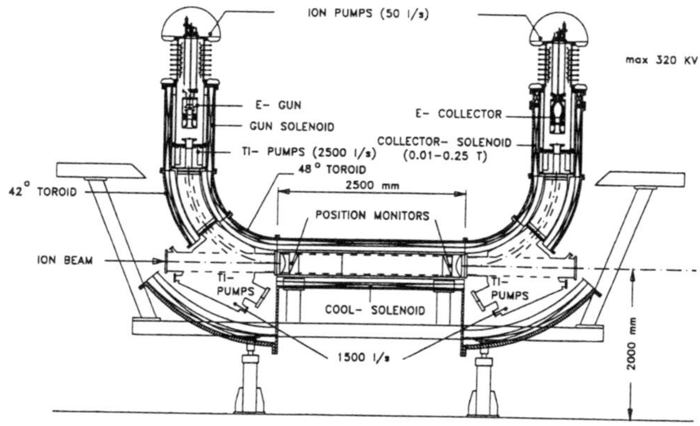

FIGURE 5.2 Schematic of the ESR electron cooling device [35]

ically accelerated, magnetically guided electron beams, using a longitudinal field of moderate strength (0.03–0.2 T) to confine the transverse electron motion. As an example for this arrangement, a schematic of the electron cooling device at the ESR is shown in Fig. 5.2. In the electron beams produced by such devices, the 'temperature' represented by the motion of the individual electrons in the co-moving reference frame of the beam can be kept very low. The temperatures in the transverse and the longitudinal degrees of freedom differ strongly and in particular the longitudinal temperature $T_{\|}$ reaches very small values with thermal energies $kT_{\|}$ of the order of 10^{-4} eV. The individual electron velocities $\vec{v}$ can be assumed to scatter around an average longitudinal velocity v_0 according to the anisotropic Maxwellian distribution function

$$f(\vec{v}, v_0) = \frac{m_e}{2\pi kT_\perp} \exp\left(-\frac{m_e v_\perp^2}{2kT_\perp}\right) \sqrt{\frac{m_e}{2\pi kT_{\|}}} \exp\left(-\frac{m_e(v_{\|} - v_0)^2}{2kT_{\|}}\right) , \quad (5.1)$$

which characterizes the electron beam and its potential for cooling ions and for providing good energy resolution as a target in electron–ion collision studies. Whereas in the earlier electron cooling devices the transverse temperature was given by its value at the electron source (usually a thermocathode leading to $kT_\perp = kT_c \approx 0.1$ eV), recent work showed that it is often advantageous to lower the transverse temperature at the expense of proportionally reducing the electron density; this can be accomplished by adiabatic expansion of the electron beam, lowering the strength of the magnetic guiding field on the way from the electron source to the electron–ion interaction region [40,41]. The transverse temperature decreases in this arrangement because the ratio $v_\perp^2/B$ of the squared transverse electron velocity and the flux density of the magnetic guiding field is an adiabatic invariant and hence, for a field strength B_c at the cathode, $T_\perp = (B/B_c)T_c$; at the same time, the electron beam cross section increases by a factor of B_c/B as the electrons are guided by the diverging magnetic flux lines in the transition region between the high and the low magnetic field strength. The lowest temperatures to-date appear to have been achieved at CRYRING with thermal energies of $kT_\perp = 10$ meV and $kT_\parallel = 0.06$ meV reported for an adiabatically expanded electron cooler beam [40].

The cooling time τ_c depends on the electron density n_e ($\tau_c \propto n_e^{-1}$), the ion charge q and its mass ($\tau_c \propto m_i/q^2$), and the electron and ion temperatures [$\tau_c \propto \max(T_e, T_i)^{3/2}$]. With magnetically guided electrons at a density of 10^7 cm^{-3} (acting only over a few percent of the storage ring circumference) it takes typically only millisecond exposures to cool a beam of U^{92+} ions, and protons are typically cooled within a few seconds. Longitudinal ion momentum spreads $\Delta p/p$ of the order of 10^{-4} to 10^{-5} are obtained. The typical diameters of the cooled ion beam at ring positions suitable for experiments are around 1–2 mm with an angular spread of 0.1–0.3 mrad, and emittances as small as 0.05 π mm mrad are achieved. The scaling relations for the cooling time given above represent only approximate expressions resulting from elementary plasma physics [42]. Electron cooling was studied in much detail since many years [4], and in recent measurements at ion storage rings in particular the ion-charge scaling was found to deviate from the above relation [43,44]; it seems to agree better with new predictions obtained by theoretical calculations for strongly coupled plasmas [45]. Thus, studies of the electron cooling process at ion storage rings continue to be of great interest.

While electron cooling provides ion beams of exceptional quality, the forces damping any relative motion of the ions with respect to the electrons can also be disturbing, considering the use of the electron cooler as an electron target. When the electron energy is detuned from cooling, the ion beam is dragged behind and the average relative electron–ion velocity diminishes with time. This can either be accounted for in the final data analysis by using experimental or theoretical information on the velocity dependence of the drag force, or it can be avoided by intermittent cooling of the ion beam during the experiment. A typical data-taking scheme making use of intermittent cooling at the TSR is based on switching between cooling and measurements at given relative electron–ion velocities with 20 ms periods [46].

Even much lower temperatures than those obtained by electron cooling can be achieved in laser-cooling schemes, where photons from a directed laser beam are absorbed and isotropically reemitted by the ions. Thus, a net momentum is transferred to a velocity class of ions (characterized by a certain Doppler shift) matching the resonance condition for photoabsorption. The population of the velocity class is depleted and transferred to an adjacent velocity class. By tuning the laser wavelength across the whole Doppler profile of the ion beam, the velocity distribution is "broomed" to one narrow pile. The remaining longitudinal temperatures of the ions can be below 1 mK, corresponding to longitudinal ion momentum spreads $\Delta p/p$ of the order of 10^{-7}.

One motivation of laser cooling is the generation of crystalline ion beams

[47,48], as reviewed by Habs and Grimm [49]. Laser cooling experiments were performed at the TSR on Li^+ ions [50] and later on Be^+ [51], and at ASTRID on Li^+ [52] and on Mg^+ [53]. For details of these experiments, achieving primarily cooling of the longitudinal degree of freedom of the ions, we refer to the review in Ref. [49]. Laser cooling schemes are also under development for transverse ion beam cooling; however, the remarkable achievements obtained in this field are so far applicable only to the few ion species which can resonantly absorb laser photons. Highly charged ions usually do not provide appropriate excited states in the limited available laser wavelength range, although this range can be somewhat extended in present-day experiments by making use of Doppler shifts in laser interaction with nearly relativistic ions. Cooling highly charged ions by collisions with laser-cooled ions in a low charge state is being discussed [49] in view of expanding the applicability of laser cooling.

5.2.5 Ion beam accumulation

Advanced accumulation and cooling techniques at heavy ion storage rings open up the possibility of producing and storing ion beams of unprecedented brilliance. The injection into a heavy ion storage ring is usually accomplished by time-dependent deflection fields, using either fast switching electric or magnetic kickers, or more slowly ramping deflectors. In the filling procedure, a first obvious intensity limit is set by storing a beam pulse with a geometrical length corresponding to the circumference of the ring (single-turn injection). In this case the emittance and the momentum spread of the stored beam, as well as its current, are the same as for the beam arriving from the accelerator. However, this limit of the beam current can often be overcome, using *stacking techniques.* They can be employed if the usable phase space volume of the ring (the machine acceptance) is not exhausted after single-turn injection; the injection is simply repeated, steering the additional incoming ions into the remaining empty regions of the phase space. The current and the occupied phase-space volume of the stored beam thus increase in proportion to each other. A multiturn stacking technique, filling up the available transverse phase space, and longitudinal stacking (also called rf stacking), filling up the momentum acceptance of the machine, are used separately or in combination at some ion storage rings; their performance has been discussed in detail for the SIS/ESR facility [54] and for the TSR [55]. For example, multiturn stacking at the TSR is accomplished by sweeping the beam at the injection position in the horizontal plane during the filling procedure, which lasts for ≈ 30 turns. The result is a circulating beam with an increased intensity (a factor of ≈ 30) but with a very large emittance.

By the combination of stacking and phase-space cooling, even higher beam currents can be achieved. Applying phase-space cooling once all the machine acceptance has been filled up by stacking, new free phase space can be made available for the injection of additional ions, while the stored ions are kept at higher density in a small phase-space region. Repeated filling (either by single-turn injection or by stacking) in connection with a beam cooling method thus makes it possible to *accumulate* ions in the storage ring. For a given incoming ion flux, the intensity of the stored beam reached in a steady state is given by the rate at which the filling can be repeated (essentially determined by the phase-space cooling time) and the loss rate of the stored ions. In particular, electron cooling with its high cooling rates has been used successfully for ion accumulation. Thus, combining multiturn stacking, rf stacking, and electron cooling, a beam current of 18 mA (3×10^{10} stored ions) for C^{6+} ions at 6.1 MeV/u, and a current of 1.1 mA (7×10^{8} ions) for S^{16+} ions at the same specific energy, were achieved at the TSR [56]. Similarly, the intensity of fully stripped U ions at 300 MeV/u accumulated in the ESR using rf stacking and electron cooling has reached nearly 3 mA (1×10^{8} ions) [27,35].

5.2.6 Limits of ion beam intensity and quality

The beam intensities that can be achieved by accumulation in ion storage rings often reach values at which the stationary and fluctuating self-fields of the stored beam become important for the ion motion and eventually cause resonant excitation of the ions, which may lead to rapid beam loss. Beam instabilities of this kind in fact limit the maximum currents accumulated so far at ion storage rings for lighter fully stripped ions, which are abundantly available at injection and have long storage lifetimes; reported currents close to these intensity limits are, e.g., 18 mA for C^{6+} ions at 6.1 MeV/u in the TSR [56,57], and ≈ 7 mA for ions with $Z \leq 30$ at ≈ 250 MeV/u in the ESR [27]. The most important feature causing instability appears to be a coherent transverse oscillation of the ion beam, as supported by the observation that a properly tuned active feedback system acting on the transverse ion motion shifts up the intensity limits somewhat [57,27]. For more details on the stability and diagnostics of intense cooled heavy ion beams we refer to a recent review [58].

Both the emittance and the momentum spread of cooled stored ion beams increase with the beam intensity, since multiple scattering in intrabeam collisions of the ions represents the dominant heating mechanism. For the emittance ϵ in equilibrium with electron cooling as a function of the ion current I, the increase approximately follows a power law $\epsilon \propto I^a$, where values between ≈ 0.43 [59] and ≈ 0.66 [60] were reported for the exponent a. Corresponding values reported for the scaling of the equilibrium momentum spread, $\Delta p/p \propto I^b$, are $b \approx 0.21$ [59] and $b \approx 0.33$ [60]. Reported absolute equilibrium values of ϵ and $\Delta p/p$, which can be considered as typical, are $\epsilon \approx 0.1\ \pi$ mm mrad and $\Delta p/p \approx 1 \times 10^{-4}$ at a beam current of 10 μA and at lower energies near 6 MeV/u [59]. At higher energies near 300 MeV/u and at similar electrical beam currents, the reported equilibrium values are lower

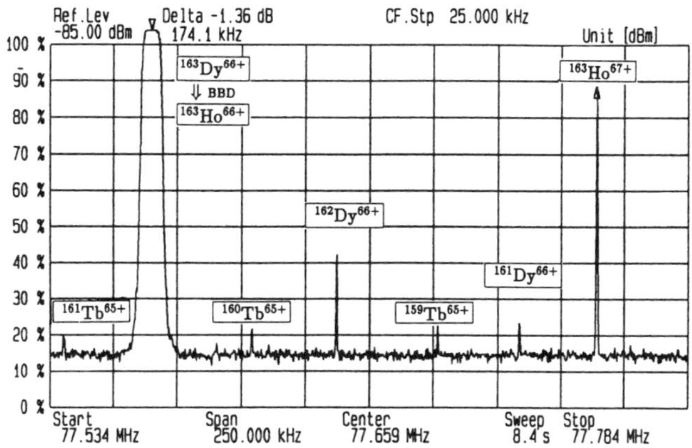

FIGURE 5.3 Schottky noise spectrum near the 43rd harmonic of the revolution frequency ($\approx$ 1.79 MHz) showing peaks for different bare ions stored and cooled simultaneously in the ESR [27]. One division of the horizontal scale corresponds to a relative frequency difference of $\approx 3.2 \times 10^{-4}$. The electron cooler imposes the same beam velocity on all ions; different revolution frequencies result from the variation of the closed-orbit circumference with the charge-to-mass ratio. Starting from the primary stored $^{163}Dy^{66+}$ ions (broad cut-off peak), various nuclear fragments are produced by collisions in the internal gas jet of the ESR. In addition, the $^{163}Dy^{66+}$ ions spontaneously convert to $^{163}Ho^{66+}$ by bound-state β decay [61]; these hydrogen-like ions rapidly loose their electron in the gas jet, forming $^{163}Ho^{67+}$ and giving rise to the clearly observable peak on the right.

($\epsilon \approx 0.01\ \pi$ mm mrad, $\Delta p/p \approx 2 \times 10^{-5}$ [60]). In common usage $\Delta p/p$ denotes the full width at half maximum of the momentum distribution and ϵ the phase space volume occupied by $\approx 86\%$ of the beam particles.

Considering simple theoretical scalings of intrabeam scattering and electron cooling rates, one expects the equilibrium emittances and the squared relative momentum spread to scale linearly with the charge-to-mass ratio of the ions for a fixed electrical ion current; moreover, the intrabeam scattering rates for a given current vary inversely proportional to the fourth power of the beam momentum. The physics of beam instabilities and heating processes is a subject of intense research at ion storage rings, pursued with the goal of further improving the brightness limits.

5.2.7 Ion beam diagnostics

The behavior and quality of an ion beam circulating in a storage ring can be monitored by a number of techniques. The revolution frequency and, usually with limited accuracy, also the intensity of a stored coasting ion beam can be determined from spectra of the Schottky noise picked up by metallic probes at certain positions of the ring. The Schottky noise originates from fluctua-

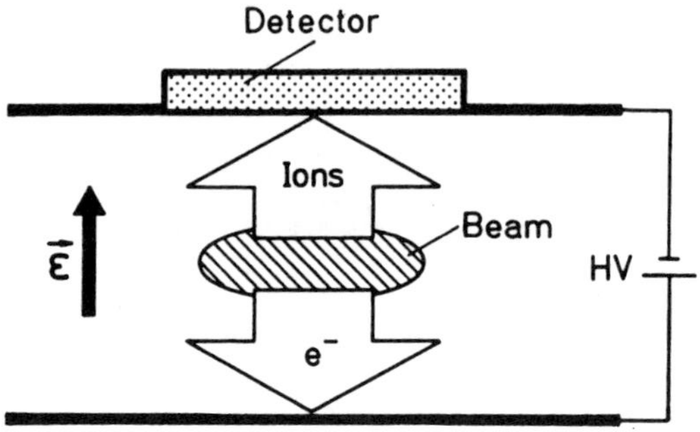

FIGURE 5.4 (a) Schematic of the detection method used in the beam profile monitors [65] at the TSR. The cross section of the circulating ion beam is indicated by a shaded area. Secondary ions produced in the residual gas by the energetic circulating beam are accelerated by the electric field $\vec{\mathcal{E}}$ of a strength of $\approx$ 1 kV/cm between two biased parallel metal plates and detected by a position sensitive detector providing a projected image of the stored beam at that location. (b) Measured beam profile of C^{6+} ions of 6.1 MeV/u stored in the TSR (current $\approx$ 20 μA) after multiturn injection (upper graph) and after electron cooling for 1.5 s (lower graph) [66]. [Fig. 5.4(b) reprinted with permission, from Proceedings of the Fourth European Particle Accelerator Conference, V. Suller and Ch. Petit-Jean-Genaz, eds., ©1994 World Scientific Publishing Co.]

tions of the beam density or current, due to the random spatial distribution of the individual ions. The frequency spectrum from a dc beam in a linear transport system would resemble continuous white noise. For a stored ion beam on a closed orbit, however, the noise signal repeats periodically with the revolution frequency. Hence, its spectral power is concentrated in bands near integral multiples of this frequency. For low ion intensity, the shape of any of these bands represents the distribution of the revolution frequencies in the beam. For ions with a given charge-to-mass ratio, the average frequency of the noise band indicates the average momentum of the beam particles and its width indicates the momentum spread of the beam [62]. For high intensities the relation between the revolution frequency distribution and the noise spectrum is complicated by the Coulomb interaction between the beam ions [58]. Nevertheless, measurements of the Schottky spectrum represent one of the most sensitive and powerful tools for detecting stored ions and analyzing their longitudinal momentum. An example of a Schottky spectrum obtained with several heavy nuclei stored in the ESR is shown in Fig. 5.3.

Whereas some information on the intensity of the circulating beam can also be obtained from Schottky spectra, accurate non-destructive beam current measurements are performed by dc current transformers [63], which have a resolution somewhat below 1 μA. Similar electromagnetic measurements are possible for lower circulating ion currents down to about 10 nA if the stored beam is bunched.

For monitoring the transverse beam profile, two approaches are commonly used. First, charge-changed particles from the circulating beam itself—produced in collisions with free electrons from the electron cooler or with residual gas particles—are imaged by collecting them on a position-sensitive detector as they drop out of the circulating beam. If the detected charge-changed particles are neutral, the interpretation of the image of the ion beam obtained this way is particularly easy, and the result is very reliable (see, e.g., Ref. [26]). If the charge-changed particles are still ions the effects of the magnetic fields of focusing elements and bending magnets have to be accounted for. In the second approach, one detects recoil ions formed in the residual gas when the ion beam passes; most of the recoil ions from ion–atom and ion–molecule collisions have very low energies and can be imaged easily onto a detector by using appropriate (usually homogeneous) magnetic [64] or electric [65] guiding fields. The detection method used at the TSR is illustrated by Fig. 5.4(a). Beam profiles demonstrating the transverse electron cooling of stored C^{6+} ions are shown in Fig. 5.4(b). Here, a beam diameter of the order of 1 mm is reached and can be monitored at a resolution [65] of 0.2–0.3 mm.

5.3 STORAGE RING EXPERIMENTS: SPECIAL TECHNIQUES AND EXAMPLES

5.3.1 Lifetime experiments

The most direct application of a storage ring to atomic and nuclear physics is based on its capability to trap ions for a period of time. Thus, it allows the observation of relaxation and decay of the stored species and the measurement of decay times ranging from microseconds to tens of days. This has been used in a number of storage ring experiments in which either the reduction of the parent species or the accumulation of decay products was observed.

An example is the measurement of the lifetime of the metastable $Be^-(2s2p^2\ ^4P_{3/2})$ state at ASTRID [67]. The negative ions were produced by charge exchange in a Na-vapor cell and μs-pulses were injected into the ring with an energy between 10 and 120 keV. Metastable Be^- ions are subject to autodetachment. Resulting Be atoms can be collected by a detector placed behind a dipole magnet in the extension of one straight section of the ring. The counting rate of Be atoms is shown as a function of the storage time in Fig. 5.5. Within 600 μs the detachment yield drops by 6 orders of magnitude with a measured lifetime of 42.9 μs. Careful study showed that the measured decay constant did not depend on ring or beam parameters, apart from an ex-

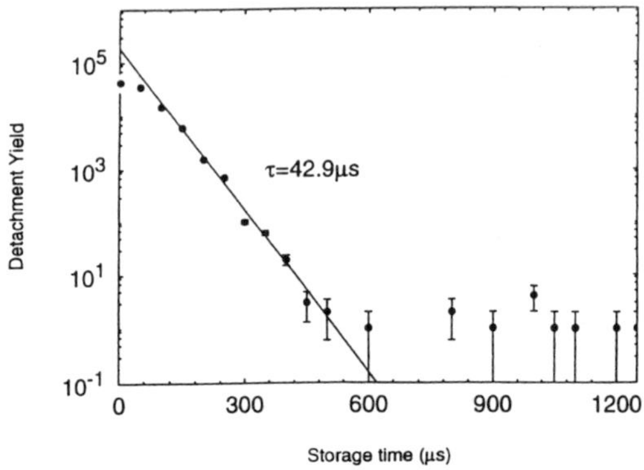

FIGURE 5.5 Autodetachment yield vs time for a 50-keV Be^- beam stored in ASTRID [67]. The solid line is an exponential fit from which the lifetime can be obtained. The first few revolutions are not considered in the fit since the counting rates were too high there. Beyond 600 μs only electronic detector noise is seen. (Reprinted with permission from Physical Review Letters.)

pected influence of the magnetic bending fields, which were strong enough to mix magnetic substates from different fine-structure components. By measuring with different ion energies it was possible to extrapolate the data to zero energy and hence zero magnetic field. The corrected lifetime determined by this technique is (45 ± 5) μs. The same group studied lifetimes of metastable states of other negative ions using a similar approach [68].

Lifetime measurements have also been carried out at the ESR and the TSR. At the ESR, in an experiment on the borderline between atomic and nuclear physics, the half life for bound-state β decay of bare $^{163}Dy^{66+}$ into hydrogen-like $^{163}Ho^{66+}$ was determined to be (47^{+5}_{-4}) d [69], observing $^{163}Ho^{67+}$ ions produced by electron loss of $^{163}Ho^{66+}$ ions. (The Schottky noise spectrum displayed in Fig. 5.3 was obtained during this experiment.) The β^- decay of neutral ^{163}Dy atoms is forbidden, the nuclear decay energy not being sufficient to release an electron into the continuum; hence, neutral ^{163}Dy atoms are absolutely stable. However, for bare ^{163}Dy nuclei, the electron originating from the nuclear β^- decay can be captured into a vacant atomic shell, which reduces the final-state energy of the reaction by the binding energy of the electron. The binding energy of an electron in one of the inner shells of the resulting $^{163}Ho^{66+}$ ion thus makes the β^- decay of $^{163}Dy^{66+}$ to become energetically allowed, and the related decay constant was measured for the first time in the ESR experiment. As a nuclear decay process influenced by the atomic charge state, bound-state β decay is of considerable applied interest in the areas of astrophysics and cosmochronology [70]. In a stellar plasma during nucleosynthesis by the s-process, the temperature is typically about 4×10^8 K. At the corresponding thermal energy near 30 keV, atomic charge

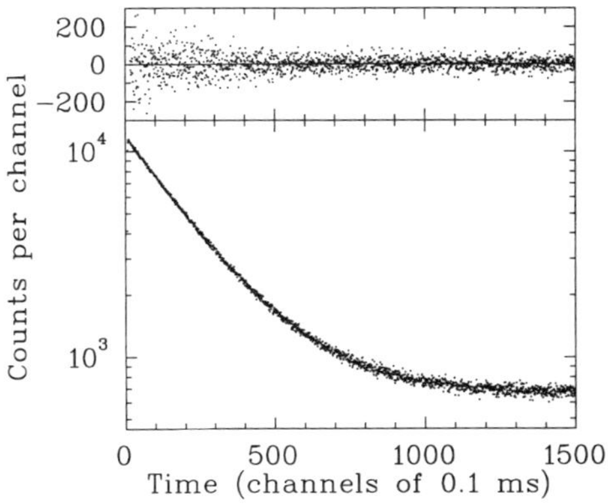

FIGURE 5.6 Decay curve of metastable $C^{4+}(1s2s\,^3S_1)$ ions in the TSR [71], measured by observing the dielectronic recombination rate of the metastable ions with free electrons from the electron cooler. The electron kinetic energy relative to the ions was set to the peak of the $1s2p(^1P)6l$ dielectronic resonance near 2.6 eV. An exponential fit (residues shown in the upper graph) yields a radiative decay rate of $(48.57 \pm 0.11)\,\mathrm{s}^{-1}$ for the $M1$ transition $1s2s\,^3S_1$-$1s^2\,^3S_0$, while the metastable-ion loss rate due to other processes amounts to only $0.016\ \mathrm{s}^{-1}$.

states are very high. Bound-state β decay therefore has significant influence on the cosmic relative abundances of certain chemical elements.

Lifetimes of long-lived metastable states in multiply charged helium-like ions were measured with high precision at the TSR. This experiment yielded the magnetic-dipole decay rate of metastable boron, carbon and nitrogen ions; the most precise result was obtained for the $C^{4+}(1s2s\ ^3S_1)$ level with a measured lifetime of (20.589±0.045) ms [71]. Here, dielectronic recombination resonance rates were recorded in order to obtain a signal proportional to the number of ions in the $2\,^3S$ parent state as a function of time (see Fig. 5.6).

The correction for the relativistic time dilatation in measurements of intrinsic lifetimes can be quite substantial at the velocities applied in ion storage rings. For example, the lifetime for bound-state β decay measured in the laboratory frame of the ESR differs from the intrinsic lifetime of $^{163}\mathrm{Dy}^{66+}$ (observed in a comoving frame) by a Lorentz factor γ of 1.316. All the lifetimes quoted here are intrinsic and suitable corrections based on the ion velocity in the storage ring were applied when necessary.

5.3.2 Particle-mass measurements

Similar to the capabilities of an ion trap, a storage ring offers the potential of precise mass measurements. In principle, a measurement of the $^{238}\mathrm{U}^{92+}$ and $^{238}\mathrm{U}^{91+}$ ion masses can be used to determine the binding energy of the electron in the hydrogen-like state. With a mass precision of $m_i/\Delta m_i = 2 \times 10^{11}$ the Lamb-shift contribution to this binding energy (≈ 460 eV) would be experimentally accessible with an uncertainty of about 1 eV, which coincides with the present accuracy limit of theoretical calculations. Storage rings (as well as ion traps) presently have not reached a precision of this order for species such as $^{238}U^{91+}$. However, interest in mass measurements begins already at a much lower level of precision. An experimental program aiming at the determination of atomic masses is being pursued at the ESR. Nuclei far from stability can be produced very efficiently by fragmentation of high-energy ions. Utilizing the GSI fragment-separator facility (see Sec. 5.2.2) selected individual isotopes resulting from such reactions are injected into the ESR, accumulated, and cooled. With the obtained circulating radioactive beam of relatively high brillance, a mass determination can be performed by Schottky noise diagnostics [72] (see also Fig. 5.3). A mass resolution of the order of $\Delta m_i/m_i \approx 10^{-6}$ has been demonstrated [73]. One factor limiting the precision in such mass measurements is the imperfect knowledge of the circumference of the ion beam trajectory in the ring. A deviation of 1 mm out of 108 m results in a relative uncertainty of 10^{-5} for absolute mass measurements. However, by comparison with known masses of particles circulating in the ring at the same time, it is possible to achieve a precision in the 10^{-7} range.

5.3.3 Laser experiments

A laser provides a precision tool for exciting, probing and manipulating atoms or ions. The spectral bandwidth of relatively simple pulsed dye laser systems is, e.g., ≈ 0.03 meV (and can be much smaller); the resulting energy resolution is orders of magnitude better than that provided by the cold electron beam of the cooling device in a storage ring. Lasers are therefore suitable not only for ultimate beam cooling, as discussed above, but also for precise probing of excitation energies by inducing electronic transitions between atomic levels. They can also be used for precise beam diagnostics based on the observation of Doppler line shifts and Doppler broadening of induced transitions.

With regard to laser studies on atomic and molecular ions, cooler rings provide a confinement technique suitable for fast beams, complementary to the traps available for slow, neutral and charged particles. Also for laser experiments, the storage-ring arrangement in connection with phase-space cooling and versatile diagnostic techniques offers several advantages over single-pass ion beams, in particular a high degree of control over the ion beam, the availability of accumulated 'rare' species (highly charged ions for example) at high intensities, and the possibility of reaching long interaction times with the laser radiation due to the ion confinement in the ring. In view of the high and very well defined ion beam velocities, a special feature of laser studies at storage rings is the wide range of Doppler shifts at low Doppler spreads. For example, considering the relativistic Doppler shift for ions moving at a velocity of 2/3 of the speed of light, as available at the storage ring ESR, the frequency acting in their co-moving reference frame can differ from that of the incident laser light in the laboratory frame by factors of 0.45 for parallel laser and ion beams and 2.2 for antiparallel beams.

Laser experiments at storage rings comprise the following main projects:

- Laser cooling of fast, dense ion beams. (A short discussion was given at the end of Sec. 5.2.4.)

- Laser spectroscopy of fast ion beams. A particularly interesting aspect of such experiments is the precise verification of the transverse Doppler shift, providing an Ives-Stillwell-type test of the special theory of relativity. Measurements using $^7Li^+$ ions, in progress at the TSR, have already [74] reached a precision exceeding that of earlier laser-spectroscopic relativity tests of this type.

- Laser spectroscopy of low-lying levels in highly charged ions. Fluorescence after laser excitation was observed on the magnetic dipole transition between the two hyperfine levels of the $1s$ ground state of hydrogen-like Bi^{82+} and the level splitting was determined to be (5.0835 ± 0.0008) eV [75] (the corresponding splitting in atomic hydrogen amounts to 4×10^{-6} eV). A complete theoretical evaluation of the hyperfine splitting of the Bi^{82+} ground state is presently not available. Uncertainties of the calculations are on the 1% level while the possible error in the experiment is on the level of 0.02%.

- Laser-stimulated radiative recombination. Here a pulsed high-power laser induces transitions of free electrons in the electron cooler from their initial state in the continuum to a bound atomic level. In these experiments [76], at least three beams are merged at a time: the circulating ion beam, the electron beam of the cooler device, and the laser beam; the enhancement of the recombination rate by the presence of the laser pulse is detected. By tuning the laser wavelength, the energy distribution of the electrons before the transition to a bound state is probed with a sub-meV energy resolution. This process has been observed at the TSR [77,78], the ESR [79], and CRYRING [80]; it enables detailed studies on the recombination in particular of highly charged ions moving very slowly relative to the electron beam (see also Sec. 5.3.5). An additional laser was used to induce a subsequent transition from the bound level populated by laser-stimulated recombination to another bound level [79]; the angular-momentum splitting of transitions between Rydberg states of multiply charged, beryllium-like ions could be measured by this method with a precision of $\approx 10^{-5}$ of the transition energy [81].

Only a short enumeration of these experiments can be given here. The ongoing activities of laser groups at the different storage rings are documented, e.g., by papers in Ref. [13].

5.3.4 X-ray measurements

Cooler rings provide high quality ion beams which in many cases are much more intense than comparable beams obtained directly from an accelerator. In combination with an internal target (usually a gas jet or thin fibres) luminosities of the order of 10^{34} $cm^{-2}\,s^{-1}$ appear to be possible, and 10^{32} $cm^{-2}\,s^{-1}$ have been reported [82]. This number depends on the storage lifetime obtained for a beam which is degraded by collisions in the internal target. Deterioration of beam quality by small-angle scattering can be compensated for by continuous electron cooling. Thus, it is possible to study ion–atom or ion–molecule collisions with excellent beams and very clean experimental conditions. Moreover, the high intensity of the stored beams also allows one to work with very dilute targets such as an electron beam or even an ion beam. Many of the experiments carried out so far at heavy ion storage rings have made use of the fact that the electron beam of the cooling device can be used as a target with a luminosity and an energy resolution which is not accessible in any other experimental arrangement. Examples of electron–ion collision studies will be given below; we here first turn to the aspect of X-ray spectroscopy.

Spectroscopic studies with few-electron highly charged ions are particularly interesting as these ions allow very sensitive studies of relativistic and quantum electrodynamical effects on the atomic structure. Thus, the 1s Lamb shift of the hydrogen atom is 3.56×10^{-5} eV, while that of hydrogen-like uranium (U^{91+}) amounts to 457.6 eV [83]. A precise measurement of the latter quantity can provide a critical test of quantum electrodynamics in strong fields. Two scenarios are pursued towards this goal. One is to observe X rays resulting from radiative recombination of completely stripped U ions into the ground state of hydrogen-like U together with the characteristic radiation resulting from the deexcitation of hydrogen-like U ions formed by recombination into excited states. These experiments employ the electron cooling device of a storage ring [84,85]. They give direct information on the total binding energy of the 1s ground state of U^{91+} [86]. The other scenario is based on observing the emitted radiation after the capture of an electron by fully stripped U in the internal gas target [87,88].

The only storage ring facility where these X-ray studies can be carried out with the heaviest ions is the ESR, where both scenarios described above are being successfully pursued. The X rays detected by Ge(i) detectors are measured in coincidence with hydrogen-like product ions and are thus unambiguously related to events where one electron is transferred to the completely stripped projectile. Deceleration of the ions in the ring before the measurement greatly reduces uncertainties related to Doppler spreads and Doppler shifts [88]. The lowest uncertainties are quoted for the experiments which make use of electron–ion recombination in the cooling device [86]; here, the additional information on the ion velocity (which is identical to the cooler electron velocity) reduces the uncertainties resulting from Doppler corrections.

5.3.5 Electron–ion collisions

Studies of electron–ion interactions have already been mentioned in connection with X-ray spectroscopy and laser physics. The combination of high-quality cooled ion beams with the electron beam of the cooling device provides ideal conditions to study inelastic electron–ion collision processes with very good energy resolution. An electron beam constitutes a rather dilute target with densities of the order of only 10^7–10^8 cm^{-3} in a cooling device; this disadvantage, however, is partly made up for by the long interaction length of typically 2 m. The recirculation of ions in the ring for millions of turns per second then leads to luminosities which allow one to measure electron–ion collision cross sections down to about 10^{-21} cm^2 or less. The technique used for detecting collision products which changed their charge state in the cooling device either by recombination with electrons or by electron-impact ionization was already described in Sec. 5.2.2, and the properties of the electron beam were discussed in Sec. 5.2.4. All the experiments relying on the electron cooler as a target have greatly profited from the idea of cooling the electron beam by adiabatic expansion in a decreasing magnetic guiding field [40,41] (see Sec. 5.2.4). Except for the ESR all the rings under discussion here have implemented the expansion technique and further progress is expected soon.

Recombination of atomic ions during electron cooling

During electron cooling, the velocity of a coasting stored ion beam automatically adjusts itself to the electron beam velocity, and the ions (with a velocity spread becoming much smaller than that of the electrons near the thermal equilibrium between the two beams) find themselves practically at rest in a surrounding electron 'gas'. At the typical electron densities given above, one should expect the measured recombination rates to represent mainly radiative recombination, due to spontaneous transitions of slow free electrons to bound levels of the recombination products (see Ref. [76] and references therein). Although estimates for three-body recombination and the associated collisional-radiative cascade at the electron densities and temperatures found in electron coolers yielded significant rates for highly charged ions [89], the *stationary* plasma situation assumed in these calculations does not well represent the conditions during electron cooling, where one passage of an ion through the cooling device typically takes only ≈ 30 ns. Thus, also for highly charged ions, the radiative recombination rates predicted for merged beams in a storage ring were widely considered to be valid. Nevertheless, several measurements of low-energy recombination rates for multiply charged ions, performed since 1989, show large enhancement factors of the experimental rates in comparison with the predictions for radiative recombination. The first results showing such phenomena were obtained at relatively high electron densities ($\gtrsim 10^9$ cm^{-3}) in a single-pass arrangement with merged, velocity matched electron and ion beams [90–92]. During electron cooling of multiply charged ions in a storage ring an enhancement of the low-energy recombination rate was found soon after at the TSR [93,78,94] at even lower electron densities ($\gtrsim 10^7$ cm^{-3}). Stronger enhancement factors in the low-density regime were revealed by more recent storage-ring experiments at LEAR [95] and at CRYRING [96]. The detailed measurements of low-energy recombination spectra at the single-pass setup and the ion storage rings clearly indicate that the rate enhancement rapidly decreases when the average relative velocity between electrons and ions becomes larger than roughly the longitudinal velocity spread of the electrons; at relative velocities exceeding the transverse electron velocity spread agreement with the theoretical rates of radiative recombination is found. Further studies of this phenomenon at ion storage rings (also helped by the method of laser-stimulated recombination, see Sec. 5.3.3) are necessary to understand the interaction of highly charged ions with a surrounding 'gas' of cold electrons.

Dielectronic recombination of atomic ions

Dielectronic recombination (DR) is a resonant two-step process where in the first step a free electron and an ion recombine with simultaneous excitation of a bound core electron and in a second step the new ion charge state is stabilized by the emission of photons. The principle of DR measurements using merged beams of electrons and ions involves observing the rate of recombined ions as a function of the average relative velocity between the two beams. At ion storage rings, the high intensity of the circulating beam is particularly advantageous for such experiments. Already in the first DR experiment at a storage ring [33], the small DR cross sections of hydrogen-like ions (O^{7+}) were measured with high accuracy. Since this work, starting in 1989, many DR measurements were carried out at storage rings on ions ranging from He^{+} to U^{89+}; a recent review of these measurements can be found in Ref. [97]. The use of heavy-ion storage rings and the method of adiabatic magnetic expansion of the electron-cooler beam have provided the technical basis for measurements of recombination cross sections and rates with unprecedented quality in both statistics and energy resolution. Electron–ion resonances showing widths as small as 25 meV have been observed.

Out of the results of recent storage-ring experiments with multiply charged ions, we here discuss in particular those on lithium- and sodium-like ions, which have recently been studied in great detail at the TSR. The wide energy ranges investigated cover transitions of K- and L-shell electrons in Si^{11+} and Cl^{14+} ions [99] and of L- and M-shell electrons in Cl^{6+}, Fe^{15+}, and Se^{23+} ions [98], respectively. As an example, Fig. 5.7 shows measured and calculated DR rates for Fe^{15+} ions in the center-of-mass energy range of 0–40 eV, covering the $3s \rightarrow 3p, 3d, 4\ell$ transitions in the ion core. For the magnetically expanded electron beam, the transverse thermal energy of the electrons amounted to

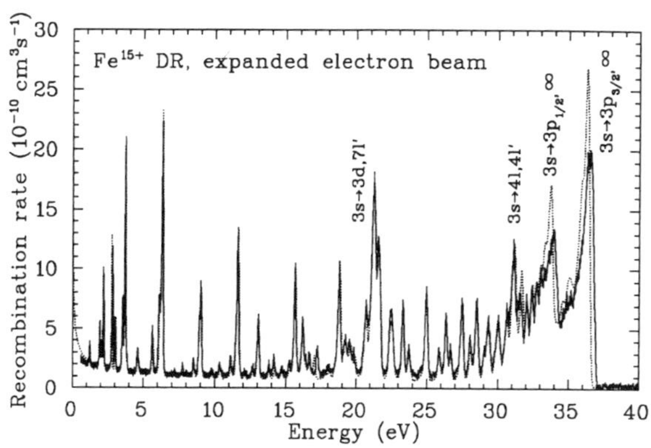

FIGURE 5.7 Recombination rate coefficient $\alpha = \langle v\sigma \rangle$ of sodium-like $Fe^{15+}(1s^2 2s^2 2p^6 3s)$ ions in the energy range of the $3\ell \to 3\ell'$ ($\Delta n = 0$) DR resonances, measured at the Heidelberg heavy-ion storage ring TSR [98]. The electron energy is given in the electron–ion center-of-mass frame. The rate α is a convolution of the recombination cross section $\sigma(v)$ with the distribution function of the electron–ion relative velocity v. The theoretical calculation (dotted line) is convoluted with a velocity distribution [cf. Eq. (5.1)] specified by $kT_\perp = 15$ meV and $kT_\| = 0.15$ meV.

only 15 meV, and a longitudinal temperature of only 0.15 meV was obtained in this experiment.

As the parent ion charge state is increased along an isoelectronic sequence, the lowest excitation energies rapidly increase. Energy splittings between electronic levels in general become very big for the heaviest few-electron ions. Since the energy resolution of electron–ion collision experiments at storage rings does not explicitly depend on the ion charge state, measurements with highly charged ions become increasingly attractive with regard to the observation of fine-structure splittings. Many such splittings and other features of the DR cross section can be resolved for highly charged ions which would never be seen for ions of low charge states. An example is the fine structure of the first excited ^2P terms in the lithium isoelectronic sequence. The energy splitting increases from 4.2×10^{-5} eV [100] for the Li atom to 4.178 keV for the lithium-like U^{89+} ion [101]. Similar effects occur in other isoelectronic sequences.

Presently, the ESR of GSI is the only facility where comfortable intensities of the heaviest few-electron ions are available for merged beams experiments. First experiments on the recombination of such ions have been carried out with Au^{76+}, Bi^{80+}, and U^{89+} [102,103]. Apart from the clarification of the Z-dependence of electron–ion collision cross sections and rates, another new and important aspect is coming up with the capability to study the highest-Z few-electron systems. Relativistic effects in the wave functions and on

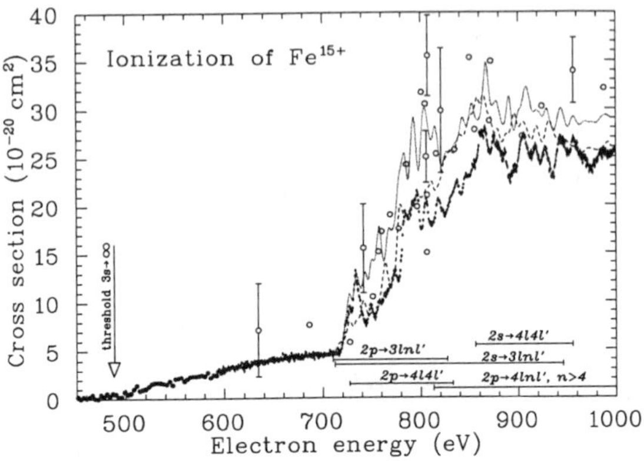

FIGURE 5.8 Ionization cross sections of Fe^{15+} ions. The solid dots are from a storage ring measurement [104] and the open circles from a crossed-beams experiment [105]. Theoretical calculations by Chen et al. [106] (dashed line) and Badnell and Pindzola [104] (dotted line) are indicated. The ground-state ionization energy (489 eV) is marked by an arrow. Ranges of intermediate excited states are displayed by horizontal bars. The threshold of excitation-autoionization is 710 eV.

transition probabilities can be tested and, moreover, a new precision tool to study quantum electrodynamic contributions to atomic binding energies may evolve from the present studies of DR. Lamb shifts in the heaviest one-electron systems are up to several hundreds of eV. Considering that energy spreads well below 0.1 eV are possible for DR experiments, even with 60-GeV uranium ions, the potential for a high-precision spectroscopy by DR provides a fascinating new prospect for the field [103].

Detachment and ionization of ions

The lightest ion studied with respect to electron removal is D^-. For the low atomic numbers, correlation effects between the electrons active in the collisions are of particular importance. Interest in the detachment of D^- by electron impact arises, for example, from the peculiar situation in the detachment threshold region where two slow outgoing electrons interact in the vicinity of a neutral residual atom. With a measurement of the detachment cross section at low energies, new insight into the Coulomb three-body problem can be expected when very precise data become available for energies barely above threshold. At ASTRID the detachment of D^- ions by electron impact was studied [30] in a center-of-mass energy range from 0 to 20 eV. Although the measured data are in very reasonable overall agreement with model calculations, the precision at very low energies is not sufficient yet for a rigorous test of theory.

A new experimental program dealing with electron impact ionization of highly charged ions has been successfully started at the TSR [104]. Similar to the procedures for the measurement of recombination rates, the ionized ions are collected by a detector behind the first dipole magnet in the ring downbeam the electron cooler and the counting rate is measured as a function of the electron energy. Results were obtained on resonant indirect ionization mechanisms for a number of lithium-like and sodium-like ions. Experimental results are shown in Fig. 5.8 for Fe^{15+}. The measurements clearly reveal the rich structure in the cross section function caused by indirect resonant and non-resonant inner-shell excitations. Obviously, indirect ionization mechanisms dominate the ionization cross section of Fe^{15+} ions at energies above 800 eV.

Recombination of molecular ions with electrons

Recombination of molecular ions with electrons proceeds by the non-radiative process of dissociative recombination, where the excess energy released by the binding of the free electron is consumed by molecular dissociation [107]. Earlier studies on the energy-dependent cross section of dissociative recombination, using low-energy colliding beams techniques, often suffered from the fact that rotational and vibrational excitation of molecules is hard to avoid in an ion source and the lifetimes are too long compared to flight times of these ions in bench-top experiments. In a storage ring it is no problem in principle to wait long enough so that at least most of the excited states have relaxed and a beam of vibrationally cold molecular ions is available. Such ions are necessary for meaningful experiments since the cross sections of dissociative recombination are very sensitive to the vibrational state of the molecular ion, and since otherwise it is nearly impossible to give a quantitative characterization of the distribution of excited states in the beam. Also, structure in dissociative recombination cross sections is smeared out unless vibrationally cold ions are used.

The first storage-ring measurements on dissociative recombination of electrons and molecular ions were reported in 1993, concerning HeH^+ [108], ${H_3}^+$ [32], and HD^+ [109]. Meanwhile, such measurements were extended to other light molecular ions [110], and more elaborate detectors were used to analyze the composition and even the internal excitation of the dissociating fragments. Details of such measurements are discussed by Zajfman and Heber in Chapter 14 of this book.

5.4 CONCLUSION

Storage rings are excellent tools to accumulate and prepare bright beams of heavy ions, including "expensive" particles such as ions in very high charge states up to U^{92+}, ions of rare isotopes, and radioactive nuclei. The efficient use of such beams is made possible by the available cooling techniques and by the recirculation of ions in the ring for up to several million times per second. Thus, relatively small numbers of the order of 10^8 stored highly charged ions can provide effective electrical currents of the order of 1–10 mA. The useful storage time of a beam of completely stripped uranium ions can be from many minutes up to an hour, depending on the charge-changing collisions in the ring and in the electron bath of the cooler device. Protons can be stored for days.

In many ring experiments the electron cooling device—primarily designed for the phase space compression of an ion beam circulating in the ring—was employed as an electron target for high resolution electron–ion collision studies. Storage rings are also well suited for laser experiments. Laser cooling, laser excitation of ions, and laser-stimulated electron–ion recombination are major topics in the research programs at several ring facilities. Molecular ions, negative ions, and even ions of C_{60} and C_{70} molecules have been stored in cooler rings, and relaxation and collision processes are being studied with unprecedented experimental precision.

With appropriate targets, high luminosities can be obtained within the ring. Improvement of beam quality by cooling also facilitates efficient deceleration or additional acceleration of ion beams. Further use of the ions in high-energy collider facilities, where head-on collisions between accelerated stored particles can be studied, was one of the original motivations to build cooler rings.

Storage rings have considerably expanded the possibilities to carry out basic research on fundamental issues of physics. Tests of the theory of relativity are performed, the influence of the electronic state of the atomic shell on nuclear decay properties is studied, precision mass spectrometry of short-lived atomic species is possible, and the use of the heaviest few-electron ions in ring experiments has opened up new possibilities to study the structure and dynamic behavior of atoms in the relativistic and strong-field regime. Quantum electrodynamic effects become very sizable in this regime and experiments carried out with very highly charged ions promise to provide high-precision tests for theoretical calculations of the Lamb shift.

ACKNOWLEDGMENTS

Experimental work at ion storage rings depends on the efforts of many people involved in the construction and operation of these facilities and represents the joint activity of many collaborators. We want to thank in particular our collaborators and colleagues at the Max Planck Institute for Nuclear Physics in Heidelberg and at the GSI, Darmstadt for fruitful cooperative work over many years. We also acknowledge the financial support of our work by these institutions, as well as by the German Bundesministerium für Bildung, Wissenschaft, Forschung und Technologie (BMBF), and by the Human Capital and Mobility program of the European Community.

ANNOTATED SUGGESTED READINGS

- J. D. Lawson, *The Physics of Charged Particle Beams,* University Press, Oxford, 1978
  This book presents a broad view of the physics underlying the formation and the 'inner life' of charged particle beams. The discussion of beam optics, space charge and collision effects, and finally waves and instabilities in beams is very enlightening as it emphasizes the fundamental ideas and the relations to other fields.

- CERN Accelerator School, *General Accelerator Physics,* edited by S. Turner, CERN 94-01 (two volumes), Geneva, 1994; *Advanced Accelerator Physics,* edited by S. Turner, CERN 87-03 (two volumes), Geneva, 1987
  These reports contain writeups of the lectures given at courses of the CERN Accelerator School and provide detailed technical information about the construction and operation of charged particle accelerators and storage rings. They are suitable for students and newcomers to the field who are interested in understanding the technical basis of storage ring experiments.

- H. Poth, *Phys. Rep.* **196**, 135 (1990)
  This article gives comprehensive information on electron cooling in storage rings. After an elementary introduction, many details are discussed on a more demanding level.

- R. E. Pollock, *Annu. Rev. Nucl. Part. Sci.* **41**, 357 (1991)
  Pollock's review of storage rings equipped with electron cooling is easy to read and contains almost no formulas. On about 30 pages it provides comparisons of the different rings and cooling devices both in tables and in figures. Emphasis is on the application of rings in nuclear physics.

- R. Schuch, *Cooler Storage Rings: New Tools for Atomic Physics* in: *Fundamental Processes and Applications of Atoms and Ions*, edited by C. D. Lin, World Scientific Publishers, Singapore, 1993
  This article is similar in nature to the present one although Schuch provides more details on storage ring operation than the present format would allow. Formulas and figures help to go more deeply into the matter. The text is suitable for students and newcomers.

- M. Larsson, *Rep. Prog. Phys.* **58**, 1267 (1995)
  This review on *Atomic and Molecular Physics with Storage Rings* gives less information on storage ring techniques but concentrates on the atomic physics research programs at storage rings. Emphasis is on laser supported experiments and studies of collisions involving molecular ions. The text is easy to read and provides a very substantial list of references.

The authors thank Mats Larsson for providing his manuscript prior to publication.

- P. H. Mokler and T. Stöhlker, *Adv. At. Mol. and Opt. Phys.* (1996), in print
  This review on *The Physics of Highly-Charged Heavy Ions Revealed by Storage/Cooler Rings* concentrates on the structure and the collisions of very highly charged ions which are available at the accelerator facilities of the GSI. It also provides information on experiments and theoretical considerations not directly related with storage rings. The text is complementary to the other suggested readings and addresses a similar audience. The authors thank Paul Mokler for providing the manuscript prior to publication.

REFERENCES

1. G. I. Budker and A. N. Skrinsky, *Sov. Phys. Usp.* **21**, 277 (1978).
2. Ya. Derbenev and A. N. Skrinsky, in *Physics Reviews*, Vol. 3, ed. by I. M. Khalatnikov (Harwood Acad. Publ., Chur, 1981), p. 165.
3. H. Poth, *Nature* **345**, 399 (1990)
4. H. Poth, *Phys. Rep.* **196**, 135 (1990)
5. V. V. Parkhomchuk and A. N. Skrinsky, *Rep. Prog. Phys.* **54** (1991) 919.
6. I. N. Meshkov, *Phys. Part. Nucl.* **25**, 631 (1994).
7. R. E. Pollock, *Annu. Rev. Nucl. Part. Sci.* **41**, 357 (1991).
8. R. Schuch, in: *Fundamental Processes and Applications of Atoms and Ions*, ed. by C. D. Lin (World Scientific Publishers, Singapore, 1993).
9. *Nuclear Physics with Stored, Cooled Beams*, ed. by P. Schwandt and H. O. Meyer, AIP Conference Proceedings No. 128 (AIP, New York, 1985).
10. *Proceedings of the Workshop and Symposium on the Physics of Low-Energy Stored and Trapped Particles (PSTP–87),* Stockholm, 1987, ed. by A. Barany, A. Kerek, M. Larsson, S. Mannervik, and L.-O. Norlin, special issue of *Physica Scripta,* **T22** (1988).
11. *Cooler Rings and their Applications,* Proceedings of the 19th INS Symposium, Tokyo, 1990, ed. by T. Katayama and A. Noda (World Scientific, Singapore, 1991).
12. *Trapped Carged Particles and Related Fundamental Issues,* Proceedings of the Nobel Symposium 91, Lysekil, 1994, ed. by I. Bergström, C. Carlberg, and R. Schuch, special issue of *Physica Scripta,* **T59** (1995).
13. *Proceedings of the First Euroconference on Atomic Physics with Stored Highly Charged Ions*, ed. by A. Wolf, H. J. Kluge, and H. Orth, special issue of *Hyperfine Interact.,* **99** (1996).
14. *Proceedings of the Workshop on Electron Cooling and Related Applications (ECOOL 84)*, Karlsruhe, 1984, ed. by H. Poth, KfK 3846 (Kernforschungszentrum, Karlruhe, 1984).

15. *Proceedings of the Workshop on Electron Cooling and New Cooling Techniques*, Legnaro, 1990, ed. by R. Calabrese and L. Tecchio (World Scientific Publishers, Singapore, 1991).
16. *Proceedings of the Workshop on Beam Cooling and Related Applications*, Montreux, 1993, ed. by J. Bosser, CERN 94–03 (CERN, Geneva, 1994).
17. R. Schuch, in *X-Ray and Inner-Shell Processes*, ed. by T. A. Carlson, M. O. Krause, and S. T. Manson, AIP Conference Proceedings 215 (American Institute of Physics, New York, 1990), p. 174.
18. D. Habs, *Nucl. Phys.* **A553**, 337c (1993).
19. F. Bosch, in *The Physics of Electronic and Atomic Collisions*, ed. by T. Andersen, B. Fastrup, F. Folkmann, H. Knudsen, and N. Andersen (AIP Press, New York, 1993), p. 432
20. A. Müller, *Nucl. Instrum. Methods* **B87**, 34 (1994).
21. M. Larsson, *Rep. Prog. Phys.* **58**, 1267 (1995).
22. P. H. Mokler and T. Stöhlker, *Adv. At. Mol. and Opt. Phys.* (1996) in print.
23. E. Jaeschke, G. Bisoffi, M. Blum, A. Friedrich, C. Geyer, M. Grieser, B. Holzer, H. W. Heyng, D. Habs, M. Jung, D. Krämer, A. Noda, W. Ott, R. E. Pollock, R. Repnow, F. Schmitt, and M. Steck, *Part. Accel.* **32**, 97 (1990).
24. C. Ellert, D. Habs, E. Jaeschke, T. Kambara, M. Music, D. Schwalm, P. Sigray, and A. Wolf, *Nucl. Instrum. Methods* **A314**, 399 (1992).
25. S. P. Møller, in *Conference Record of the 1991 IEEE Particle Accelerator Conference*, San Francisco, 1991, ed. by L. Lizana and J. Chew, IEEE Catalog No. 91CH3030-7 (IEEE, New York, 1991), p. 1741.
26. K. Abrahamsson, G. Andler, L. Bagge, E. Beebe, P. Carlé, H. Danared, M. Engström, C. J. Herrlander, J. Hilke, J. Jeansson, A. Källberg, S. Leontein, L. Liljeby, A. Nilsson, A. Paál, A. Pikin, K.-G. Rensfelt, U. Rosengård, A. Simonsson, J. Starker, M. af Ugglas, H. Meuth, and A. Schnase, in *Proceedings of the Fourth European Particle Accelerator Conference*, London, 1994, ed. by V. Suller and Ch. Petit-Jean-Genaz (World Scientific Publishers, Singapore, 1994), p. 380.
27. B. Franzke, K. Beckert, F. Bosch, H. Eickhoff, B. Franczak, A. Gruber, O. Klepper, F. Nolden, P. Raabe, H. Reich, P. Spädtke, M. Steck, J. Struckmeier, in *Proceedings of the 1993 Particle Accelerator Conference*, Washington D.C., 1993, ed. by S. T. Corneliussen, IEEE Catalog No. 93CH3279-7 (IEEE, Piscataway, 1993), p. 1645.
28. D. Habs, W. Baumann, J. Berger, P. Blatt, A. Faulstich, P. Krause, G. Kilgus, R. Neumann, W. Petrich, R. Stockstad, D. Schwalm, E. Szmola, K. Welti, A. Wolf, S. Zwickler, E. Jaeschke, D. Krämer, G. Bisoffi, M. Blum, A. Friedrich, C. Geyer, M. Grieser, H.-W. Heyng, B. Holzer, R. Ihde, M. Jung, K. Matl, W. Ott, B. Povh, R. Repnow, M. Steck, E. Steffens, D. Dutta, T. Kühl, D. Marx, S. Schröder, M. Gerhard, R. Grieser, G. Huber, R. Klein, M. Krieg, N. Schmidt, R. Schuch, J.F. Babb, L. Spruch, W. Arnold, and A. Noda, *Nucl. Instrum. Methods* **B43**, 390 (1989).
29. M. Grieser, D. Habs, R. v. Hahn, C. M. Kleffner, R. Repnow, M. Stampfer, E. Jaeschke, and M. Steck, in same proceedings as Ref. [25], p. 2817.

30. L. H. Andersen, D. Mathur, H. T. Schmidt, and L. Vejby-Christensen, *Phys. Rev. Lett.* **74**, 892 (1995)
31. P. Hvelplund, in same proceedings as Ref. [19], p. 709.
32. M. Larsson, H. Danared, J. R. Mowat, P. Sigray, G. Sundström, L. Broström, A. Filewich, A. Källberg, S. Mannervik, K. G. Rensfeldt, and S. Datz, *Phys. Rev. Lett.* **70**, 430 (1993)
33. G. Kilgus, J. Berger, P. Blatt, M. Grieser, D. Habs, B. Hochadel, E. Jaeschke, D. Krämer, R. Neumann, G. Neureither, W. Ott, D. Schwalm, M. Steck, R. Stokstad, E. Szmola, A. Wolf, R. Schuch, A. Müller, M. Wagner, *Phys. Rev. Lett.* **64**, 737 (1990).
34. D. Möhl, G. Petrucci, L. Thorndal, and S. van der Meer, *Phys. Rep.* **58**, 75 (1980).
35. B. Franzke, private communication.
36. F. Nolden, G. Berthes, D. Böhne, B. Franzke, and A. Schwinn, in *Proceedings of the European Particle Accelerator Conference*, Rome 1988, ed. by S. Tazzari (World Scientific Publishers, Singapore, 1989), p. 579.
37. G. I. Budker, *Atomnaya Energiya* **22**, 346 (1967); *Phys. Abstracts* **71**, 47 (1968).
38. G. I. Budker, N. S. Dikanskij, V. I. Kudelainen, I. N. Meshkov, V. V. Parkhomchuk, D. V. Skrinskij, and B. N. Sukhina, *Part. Accel.* **7**, 197 (1976).
39. A. Wolf, *Physica Scripta* **T22**, 55 (1988).
40. H. Danared, G. Andler, L. Bagge, C. J. Herrlander, J. Hilke, J. Jeansson, A. Källberg, A. Nilsson, A. Paál, K.-G. Rensfelt, U. Rosengård, J. Starker, and M. af Ugglas, *Phys. Rev. Lett.* **72**, 3775 (1994).
41. S. Pastuszka, U. Schramm, M. Grieser, C. Broude, R. Grimm, D. Habs, J. Kenntner, H.-J. Miesner, T. Schüßler, U. Schramm, D. Schwalm, and A. Wolf, *Nucl. Instrum. Methods* **A369**, 11 (1996).
42. L. Spitzer, *Physics of Fully Ionized Gases* (Wiley, New York, 1962), chapter 5.
43. M. Steck, K. Beckert, F. Bosch, H. Eickhoff, B. Franzke, O. Klepper, R. Moshammer, F. Nolden, H. Reich, and P. Spädtke, in Ref. [16], p. 395.
44. A. Wolf, C. Ellert, M. Grieser, D. Habs, B. Hochadel, R. Repnow, and D. Schwalm, in Ref. [16], p. 416.
45. G. Zwicknagel, C. Toepffer, and P.-G. Reinhard, *Hyperfine Interact.* **99**, 285 (1996).
46. G. Kilgus, D. Habs, D. Schwalm, A. Wolf, N. R. Badnell, and A. Müller, *Phys. Rev. A* **46**, 5730 (1992).
47. J. P. Schiffer and P. Kienle, *Z. Phys. A* **321**, 181 (1985).
48. R. Hasse, *Phys. Rev. Lett.* **67**, 600 (1991).
49. D. Habs and R. Grimm, *Annu. Rev. Nucl. Part. Sci.* **45**, 391 (1996).
50. S. Schröder, R. Klein, N. Boos, M. Gerhard, R. Grieser, G. Huber, A. Karafillidis, M. Krieg, N. Schmidt, T. Kühl, R. Neumann, V. Balykin, M. Grieser, D. Habs, E. Jaeschke, D. Krämer, M. Kristensen, M. Music, W. Petrich, D. Schwalm, P. Sigray, M. Steck, B. Wanner, and A. Wolf, *Phys. Rev. Lett.* **64**, 2901 (1990).

51. W. Petrich, M. Grieser, R. Grimm, A. Gruber, D. Habs, H.-J. Miesner, D. Schwalm, B. Wanner, H. Wernøe, A. Wolf, R. Grieser, G. Huber, R. Klein, T. Kühl, R. Neumann, and S. Schröder, *Phys. Rev. A* **48**, 2127 (1993).
52. J. S. Hangst, M. Kristensen, J. S. Nielsen, O. Poulsen, J. P. Schiffer, and P. Shi, *Phys. Rev. Lett.* **67**, 1238 (1991).
53. J. Hangst, A. Labrador, V. Lebedev, N. Madsen, J. S. Nielsen, O. Poulsen, P. Shi, and J. P. Schiffer, *Phys. Rev. Lett.* **74**, 86 (1995); J. Hangst, J. S. Nielsen, O. Poulsen, P. Shi, and J. P. Schiffer, *Phys. Rev. Lett.* **74**, 4432 (1995).
54. H. Eickhoff, K. Blasche, U. Blell, B. Franzke, and J. Pinkow, in same proceedings as Ref. [27], p. 357.
55. G. Bisoffi, M. Grieser, E. Jaeschke, D. Krämer, and A. Noda, *Nucl. Instrum. Methods* **A287**, 320 (1990).
56. M. Grieser, M. Blum, D. Habs, R. v. Hahn, B. Hochadel, E. Jaeschke, C. M. Kleffner, M. Stampfer, M. Steck, and A. Noda, in Ref. [11], p. 190.
57. M. Grieser, D. Habs, R. v. Hahn, C. Ellert, C. M. Kleffner, J. Liebmann, R. Repnow, D. Schwalm, M. Stampfer, A. Wolf, E. Jaeschke, and S. Papureanu, in *Proceedings of the Third European Particle Accelerator Conference*, Berlin, 1992, ed. by H. Henke, H. Homeyer, and Ch. Petit-Jean-Genaz (Editions Frontières, Gif-sur-Yvette, 1994), p. 152.
58. I. Hofmann and G. Kalisch, in Ref. [16], p. 69.
59. B. Hochadel, F. Albrecht, M. Grieser, D. Habs, D. Schwalm, and A. Wolf, in Ref. [16], p. 198.
60. M. Steck, K. Beckert, H. Eickhoff, B. Franzke, F. Nolden, and P. Spädtke, in same proceedings as Ref. [27], p. 1738.
61. M. Jung, F. Bosch, K. Beckert, B. Franzke, A. Gruber, O. Klepper, W. Koenig, C. Kozhuharov, R. Mann, R. Moshammer, F. Nolden, U. Schaaf, G. Soff, P. Spädtke, M. Steck, Th. Stöhlker, and K. Sümmerer, *Nucl. Phys.* **A553**, 309c (1993).
62. D. Boussard, in *Proceedings of the CERN Accelerator School*, Oxford, 1985, ed. by S. Turner, CERN 87–03, (CERN, Geneva, 1987), p. 416.
63. K. Unser, *IEEE Trans. Nucl. Sci.* **NS-28**, 2344 (1981).
64. T. Quinteros, R. Schuch, M. Pajek, P. Sigray, H. Cederquist, H. Danared, L. Bagge, A. Filevich, J. Jeansson, A. Källberg, and A. Paál, *Nucl. Instrum. Methods* **A333**, 288 (1993).
65. B. Hochadel, F. Albrecht, M. Grieser, D. Habs, D. Schwalm, E. Szmola, and A. Wolf, *Nucl. Instrum. Methods* **A343**, 401 (1994).
66. M. Grieser, F. Albrecht, D. Habs, R. v. Hahn, B. Hochadel, C. M. Kleffner, J. Liebmann, J. Kenntner, H.-J. Miesner, S. Pastuszka, R. Repnow, U. Schramm, D. Schwalm, and A. Wolf, in same proceedings as Ref. [26], p. 518.
67. P. Balling, L. H. Andersen, T. Andersen, H. K. Haugen, P. Hvelplund, and K. Taulbjerg, *Phys. Rev. Lett.* **69**, 1024 (1992).
68. H. K. Haugen, L. H. Andersen, T. Andersen, P. Balling, N. Hertel, P. Hvelplund, and S. P. Möller, *Phys. Rev. A* **46**, R1 (1992).
69. M. Jung, F. Bosch, K. Beckert, H. Eickhoff, H. Folger, B. Franzke, A. Gruber, P. Kienle, O. Klepper, W. Koenig, C. Kozhuharov, R. Mann, R. Moshammer,

F. Nolden, U. Schaaf, G. Soff, P. Spädtke, M. Steck, Th. Stöhlker, and K. Sümmerer, *Phys. Rev. Lett.* **69**, 2164 (1992).
70. F. Käppeler, H. Beer, and K. Wisshak, *Rep. Prog. Phys.* **52**, 945 (1989).
71. H. T. Schmidt, P. Forck, M. Grieser, D. Habs, J. Kenntner, G. Miersch, R. Repnow, U. Schramm, T. Schüßler, D. Schwalm, and A. Wolf, *Phys. Rev. Lett.* **72**, 1616 (1994).
72. H. Geissel, K. Beckert, F. Bosch, H. Eickhoff, B. Franczak, B. Franzke, M. Jung, O. Klepper, R. Moshammer, G. Münzenberg, F. Nickel, F. Nolden, U. Schaaf, C. Scheidenberger, P. Spädtke, M. Steck, K. Sümmerer, and A, Magel, *Phys. Rev. Lett.* **68**, 3412 (1992).
73. B. Franzke, K. Beckert, H. Eickhoff, F. Nolden, H. Reich, U. Schaaf, B. Schlitt, A. Schwinn, M. Steck, and Th. Winkler, *Physica Scripta* **T59**, 176 (1995).
74. R. Grieser, R. Klein, G. Huber, S. Dickopf, I. Klaft, P. Knobloch, P. Merz, F. Albrecht, M. Grieser, D. Habs, D. Schwalm, and T. Kühl, *Appl. Phys. B* **59**, 127 (1994).
75. I. Klaft, S. Borneis, T. Engel, B. Fricke, R. Grieser, G. Huber, T. Kühl, D. Marx, R. Neumann, S. Schröder, P. Seelig, and L. Völker, *Phys. Rev. Lett.* **73**, 2425 (1994).
76. A. Wolf, in *Recombination of Atomic Ions*, ed. by W.G. Graham, W. Fritsch, Y. Hahn, and J.A. Tanis, NATO ASI, Ser. B, Vol. 296, (Plenum, New York, 1992), p. 209.
77. U. Schramm, J. Berger, M. Grieser, D. Habs, E. Jaeschke, G. Kilgus, D. Schwalm, A. Wolf, R. Neumann and R. Schuch, *Phys. Rev. Lett.* **67**, 22 (1991).
78. A. Wolf, D. Habs, A. Lampert, R. Neumann, U. Schramm, T. Schüßler and D. Schwalm, in *Atomic Physics 13*, ed. by H. Walther, T. W. Hänsch, and B. Neizert, AIP Conf. Proceedings, vol. 275 (Am. Inst. Phys., New York, 1993), p. 228.
79. S. Borneis, F. Bosch, T. Engel, M. Jung, I. Klaft, T. Kühl, D. Marx R. Moshammer, R. Neumann, S. Schröder, P. Seelig and L. Völker, *Phys. Rev. Lett.* **72**, 207 (1994).
80. S. Asp, R. Schuch, D. R. DeWitt, C. Biedermann, Gao Hui, W. Zong, G. Andler, and E. Justiniano, *Nucl. Instrum. Methods* (to be published).
81. T. Schüßler, U. Schramm, T. Rüter, C. Broude, M. Grieser, D. Habs, D. Schwalm, and A. Wolf, *Phys. Rev. Lett.* **75**, 802 (1995).
82. C. Ekström, in Ref. [11], p. 228.
83. G. Soff, *Z. Phys. D* **21**, 7 (1991).
84. H. F. Beyer, K. D. Finlayson, D. Liesen, P. Indelicato, C. T. Chantler, R. D. Deslattes, J. Schweppe, F. Bosch, M. Jung, O. Klepper, W. Koenig, R. Moshammer, K. Beckert, H. Eickhoff, B. Franzke, A. Gruber, F. Nolden, P. Spädtke, and M. Steck, *J. Phys. B* **26**, 1557 (1993).
85. D. Liesen, H. F. Beyer, K. D. Finlayson, F. Bosch, M. Jung, O. Klepper, R. Moshammer, K. Beckert, H. Eickhoff, B. Franzke, F. Nolden, P. Spädtke, M. Steck, G. Menzel, and R. D. Deslattes, *Z. Physik D* **30**, 307 (1994).
86. H. F. Beyer, D. Liesen, K. D. Finlayson, F. Bosch, M. Jung, O. Klepper, R. Moshammer, K. Beckert, H. Eickhoff, B. Franzke, F. Nolden, P. Spädtke, M.

Steck, G. Menzel, and R. D. Deslattes, *Phys. Lett. A* **184**, 435 (1994); GSI Nachrichten, GSI 07–94 (1994), p. 1.

87. T. Stöhlker, P. H. Mokler, K. Beckert, F. Bosch, H. Eickhoff, B. Franzke, M. Jung, T. Kandler, O. Klepper, C. Kozhuharov, R. Moshammer, F. Nolden, H. Reich, P. Rymuza, P. Spädtke, and M. Steck, *Phys. Rev. Lett.* **71**, 2184 (1993).
88. Th. Stöhlker, GSI Nachrichten, GSI 05–95 (1995), p. 4.
89. H. F. Beyer, D. Liesen, and O. Guzman, *Part. Accel.* **24**, 163 (1989).
90. A. Müller, S. Schennach, M. Wagner, J. Haselbauer, O. Uwira, W. Spies, E. Jennewein, R. Becker, M. Kleinod, U. Pröbstel, N. Angert, J. Klabunde, P. H. Mokler, P. Spädtke, and B. Wolf, *Physica Scripta* **T37**, 62 (1991).
91. A. Frank, A. Müller, J. Haselbauer, S. Schennach, W. Spies, O. Uwira, M. Wagner, R. Becker, E. Jennewein, M. Kleinod, N. Angert, and P. H. Mokler, in *Atomic Physics of Highly-Charged Ions,* ed. by P. Richard, M. Stöckli, C. L. Cocke, and C. D. Lin, AIP Conf. Proceedings, vol. 274 (Am. Inst. Phys., New York, 1993), p. 532.
92. O. Uwira, A. Müller, W. Spies, A. Frank, J. Linkemann, L. Empacher, P.H. Mokler, R. Becker, M. Kleinod, and S. Ricz, *Nucl. Instrum. Methods* **B98**, 162 (1995).
93. A. Wolf, J. Berger, M. Bock, D. Habs, B. Hochadel, G. Kilgus, G. Neureither, U. Schramm, D. Schwalm, E. Szmola, A. Müller, M. Wagner, and R. Schuch *Z. Phys. D (Supplement)* **21**, S 69 (1991).
94. O. Uwira, A. Müller, W. Spies, J. Linkemann, A. Frank, T. Cramer, L Empacher, R. Becker, M. Kleinod, P. H. Mokler, J. Kenntner, A. Wolf, U. Schramm, D. Schwalm, and D. Habs, *Hyperfine Interact.* **99**, 295 (1996).
95. S. Baird, J. Bosser, C. Carli, M. Chanel, P. Lefèvre, R. Ley, R. Maccaferri, S. Maury, I. Meshkov, D. Möhl, G. Molinari, F. Motsch, H. Mulder, G. Tranquille, and F. Varenne, *Phys. Lett.* **B361**, 184 (1995).
96. H. Gao, D. R. DeWitt, R. Schuch, W. Zong, S. Asp, and M. Pajek, *Phys. Rev. Lett.* **75**, 4381 (1995).
97. A. Müller, in *Atomic and Plasma-Material Interaction Data for Fusion*, Vol. 6, ed. by R. K. Janev (International Atomic Energy Agency, Vienna, 1995), p. 59.
98. J. Linkemann, J. Kenntner, A. Müller, A. Wolf, D. Habs, D. Schwalm, W. Spies, O. Uwira, A. Frank, A. Liedtke, G. Hofmann, E. Salzborn, N. R. Badnell, and M. S. Pindzola, *Nucl. Instrum. Methods* **B98**, 154 (1995).
99. J. Kenntner, J. Linkemann, C. Broude, D. Habs, G. Hofmann, A. Müller, E. Salzborn, D. Schwalm, and A. Wolf, *Nucl. Instrum. Methods* **B98**, 142 (1995).
100. S. Bashkin and J. O. Stoner, jr., *Atomic Energy Levels & Grotrian Diagrams*, vol. 1 (North Holland, Amsterdam, and American Elsevier, New York, 1975).
101. Y.-K. Kim, D. H. Baik, P. Indelicato, and J. P. Desclaux, *Phys. Rev. A* **44**, 148 (1991).
102. W. Spies, A. Müller, J. Linkemann, A. Frank, M. Wagner, C. Kozhuharov, B. Franzke, K. Beckert, F. Bosch, H. Eickhoff, M. Jung, O. Klepper, W. König,

P. H. Mokler, R. Moshammer, F. Nolden, U. Schaaf, P. Spädtke, M. Steck, P. Zimmerer, N. Grün, W. Scheid, M. S. Pindzola, and N. R. Badnell, *Phys. Rev. Lett.* **69**, 2768 (1992).

103. W. Spies, O. Uwira, A. Müller, J. Linkemann, L. Empacher, A. Frank, C. Kozhuharov, P. H. Mokler, F. Bosch, O. Klepper, B. Franzke, and M. Steck, *Nucl. Instrum. Methods* **B98**, 158 (1995).
104. J. Linkemann, A. Müller, J. Kenntner, D. Habs, D. Schwalm, A. Wolf, N. R. Badnell, and M. S. Pindzola, *Phys. Rev. Lett.* **74**, 4173 (1995).
105. D. C. Gregory, L.-J. Wang, F. W. Meyer, and K. Rinn, *Phys. Rev. A* **35**, 3256 (1987).
106. M. H. Chen, K. J. Reed, and D. L. Moores, *Phys. Rev. Lett.* **64**, 1350 (1990).
107. J. B. A. Mitchell, Phys. Rep. **186**, 215 (1990).
108. T. Tanabe, I. Katayama, N. Inoue, K. Chida, Y. Arakaki, T. Watanabe, M. Yoshizawa, S. Ohtani, and K. Noda, *Phys. Rev. Lett.* **70**, 422 (1993).
109. P. Forck, M. Grieser, D. Habs, A. Lampert, R. Repnow, D. Schwalm, A. Wolf, and D. Zajfman, *Phys. Rev. Lett.* **70**, 426 (1993).
110. *Dissociative Recombination: Theory, Experiment and Applications III,* ed. by D. Zajfman, J. B. A. Mitchell, D. Schwalm, and B. R. Rowe (World Scientific, Singapore, 1996).

CHAPTER 6

Relativistic Neutral Hydrogen Beams

H. C. Bryant* and Edward P. MacKerrow†

High emittance beams of hydrogen atoms moving at speeds approaching the speed of light are presently possible. The most practicable method for producing such beams is by the photodetachment or foil stripping of relativistic beams of the negative ion of hydrogen, H^-. The former case, for which the relativistic kinematics of the interaction of a laser beam with an atomic beam must be carefully considered, produces ground state atoms only; the latter case produces excited as well as ground state hydrogen atoms. Stripping to the hydrogen ground state can also be accomplished using static, typically motional, electric fields, which in general degrade the beam emittance. More novel methods of neutral beam production include laser-induced recombination in merged electron-proton merged beams and capture from pair production; these more difficult methods can be used for anti hydrogen production. Very high-energy hydrogen atoms may be present in the cosmic rays. Laser techniques can be used for beam phase space tailoring as well as for improved accelerator diagnostics, based on the atomic physics of H^-.

6.1 INTRODUCTION

The possibility of the production of high energy neutral beams is interesting for a number of reasons. On the practical side, the neutralization of a beam allows for its passage through magnetic fields that would create awkward problems for a charged beam, as in the injection process for accelerators or in the use of neutral beams to probe orbiting missiles. For the study of fundamental science, such beams present clearly many opportunities.

*) University of New Mexico, Department of Physics and Astronomy, Albuquerque New Mexico 87131

†) Los Alamos National Laboratory, Chemical Science and Technology Division, Los Alamos New Mexico 87545

An atom traveling near the speed of light is uniquely isolated from other atoms in the lab because of the very different kinematics that distinguish all of its interactions. This uniqueness can be used to study atom-field interactions, and, moreover, the atom can be used as a probe. Careful checks, for example, of the relativistic transformations from lab to atom frame, using the exquisitely precise atomic energy levels, could detect small, but critical, discrepancies in our understanding of spacetime. The field transformations make it possible to subject the atom to field strengths that one would be very hard-pressed indeed to produce in any other way. The absorption spectrum of the relativistic atoms can be probed in a tunable manner, with fixed wavelength lasers, using the relativistic Doppler shift. The tunable wavelength range, in the atom's rest frame, using this technique is much larger than that of modern day tunable lasers. Photoabsorption measurements of this type can in turn be used as an ultra-sensitive diagnostics for accelerator operation.

6.1.1 The acceleration of H^-

The acceleration of negative ions leads us into this chapter because it gives rise to the possibility of neutral beams traveling at near luminal speeds. The need of particle physicists for efficient high-energy accelerators, using H^- in the first stage, makes possible the production of relativistic hydrogen atoms for fundamental studies in atomic physics.

How, then, are relativistic neutral beams produced? One must begin with negative ion beams and neutralize them after acceleration, rather than accelerating positive ions and neutralizing them by electron attachment processes. The large momentum transfers associated with electron attachment make electron pick-up reactions very unlikely [1]. Exceptions to this rule would be the use of the positive molecular ion of hydrogen, or the use of various forms of merged beams in which the nucleus combines with an electron moving with near the same velocity, as discussed below. Negative ion sources will not be dealt with in this chapter, except to note that the art is still an active area for research [2], and we shall discuss accelerators only briefly below. After the beam of negative ions is accelerated to the requisite speed, there are essentially three ways to neutralize it: neutralization by photodetachment, stripping by thin foils (including gas jets or plasmas), and stripping in strong motional electric fields produced by passing the beam through a magnet.

High energy accelerators of protons [3], such as the Tevatron at FERMILAB, generally begin the acceleration with an H^- beam in a linac which is then stripped by a thin foil in the process of injection into a booster ring [4]. The foil is placed at the point of tangency to the ring so that the curvature of the particle's trajectory conveniently changes sign in the transition magnetic field. The flexibility afforded by negative ions is used also at LAMPF [5] where both H^- and H^+ are accelerated at different phases of the same traveling electromagnetic wave in the linac. The accelerated H^- is used also in the loading, by stripping to protons, of the Los Alamos Proton Storage Ring, PSR. At the meson factory, TRIUMF, the internally circulating H^- beam is extracted as protons by a stripping foil. The acceleration of H^- allows for the ease of production of multiple beams by stripping only portions of the beam, which because of the different charge states, follow different paths in the magnetic fields of bending magnets. Emittance reduction can be obtained by passing the H^- beam through a hole in a foil and then through a bending magnet so that the stripped ions, occupying undesirable portions of the phase space, can be dumped or put to less critical use.

6.1.2 H_2^+

Recently it has been suggested that H_2^+ beams should be considered for the first stages in high energy accelerators [6]. Such an alternative to negative ions could be attractive because of the greater ease of stripping in foils and a greater resistance to dissociation by the motional electric fields applied by bending magnets. Neutral hydrogen beams could be derived from this ion by stripping in very thin foils or by photodissociation. Practical applications of this technique await reliable estimates and measurements of the lifetime against dissociation of this ion in electric fields and its cross section for dissociation in thin foils [7]. We will not discuss these possibilities further in this chapter; many interesting technical and scientific questions remain to be answered.

6.2 PRODUCTION BY PHOTODETACHMENT

The production of a neutral beam from a negative ion beam by laser photodetachment [8] is an excellent method if one needs a pulsed beam of atoms in their ground state. Present technology does not allow for continuous and complete photodetachment of most accelerator ion beams, even though this is possible in principle using a laser with a high enough average power. Below we discuss the physics and technique of photodetachment for a relativistic H^- beam specifically, although the principles, of course, have much more general application.

6.2.1 Relativistic kinematics

To the present moment, the use of special relativity has been completely adequate for the working out the kinematics of accelerated beams and for the understanding of the relationship between laboratory measurements and behavior in the atom's frame. That is, all checks that have been made so far are consistent with special relativity [9]. Nevertheless, we shall introduce the kinematics in the more general form due to H. P. Robertson [10], so that potential breakdowns in the relations can be easily spotted.

In matrix form the Robertson transformation,α_{ik}, can be written as

$$\alpha_{ik} = \begin{pmatrix} \gamma g_1 & 0 & 0 & -i\beta\gamma g_4 \\ 0 & g_2 & 0 & 0 \\ 0 & 0 & g_2 & 0 \\ i\beta\gamma g_1 & 0 & 0 & \gamma g_4 \end{pmatrix}, \tag{6.1}$$

where α_{ik} relates four-vectors in the moving frame, x'_i, to those in the rest frame, x_i, by

$$x_i = \sum_{k=1}^{4} \alpha_{ik} x'_k, \tag{6.2}$$

where, for example, x_i is the four-vector $(ix_0, \vec{r}) = (ict, \vec{r})$, with elements (ict, x, y, z). The relative velocity between frames is $\vec{\beta} = \vec{v}/c$; in this case we have chosen the coordinate systems K and K' so that $\vec{v} = v\hat{x}$, which defines the axis labeled by the subscript "1". When the g_i are all equal to unity, then we recover the Lorentz transformation, and we shall for the time being, make them so. The relativistic parameter, γ, is defined in the usual sense as $\gamma = (1 - \beta^2)^{-1/2}$.

If the axes in K and K' remain parallel, but the velocity $\vec{v}$ of the frame K' in frame K is an arbitrary direction, the Lorentz transformation is [11]

$$\begin{aligned} x'_0 &= \gamma\left(x_0 - \vec{\beta}\cdot\vec{r}\right), \\ \vec{r}' &= \vec{r} + \frac{(\gamma - 1)}{\beta^2}\left(\vec{\beta}\cdot\vec{r}\right)\vec{\beta} - \gamma\vec{\beta}x_0. \end{aligned} \tag{6.3}$$

All quantities which transform under Lorentz transformations, Eq. (6.3), in the same manner as space-time are referred to as "four-vectors". An example is the four-vector composed of the frequency, ω, and wavevector, $\vec{k}$, for an electromagnetic plane wave. This can be seen from the fact that the phase of the plane wave, ϕ, is invariant between two moving frames

$$\phi = \omega t - \vec{k}\cdot\vec{r} = \omega' t' - \vec{k}'\cdot\vec{r}'. \tag{6.4}$$

The phase can be determining by counting the number of crests that have passed an observer, and since this is merely a counting operation it must be independent of coordinate frame. If we express t and $\vec{r}$ in terms of t' and $\vec{r}'$ using Eq. (6.3), and then equate the coefficients of t' and $\vec{r}'$ on both sides of the equation we obtain the relativistic Doppler shift

$$\omega' = \gamma\omega\left(1 - \beta\cos\theta\right), \tag{6.5}$$

for the frequency of light. This equation is the "cornerstone" of relativistic photodetachment. We have used the fact that for a light wave, $\left|\vec{k}\right| = \omega/c$.

The angle θ is defined by $\cos\theta = \hat{k}\cdot\vec{\beta}$, and the transformation of this angle between frames is

$$\tan\theta' = \frac{\sin\theta}{\gamma\left(\cos\theta - \beta\right)}. \tag{6.6}$$

The Doppler shift is very useful in delivering a tunable photon energy to the ion's rest frame by merely changing the angle between a laser beam and the ion beam.

The light intensity transforms between the two frames in a slightly different manner. This can be seen by examining the transformation of the electric field, $\vec{E}$, and magnetic field, $\vec{B}$, [11]

$$\begin{aligned}\vec{E}' &= \gamma\left(\vec{E} + \vec{\beta}\times\vec{B}\right) - \frac{\gamma^2}{(\gamma+1)}\vec{\beta}\left(\vec{\beta}\cdot\vec{E}\right),\\ \vec{B}' &= \gamma\left(\vec{B} - \vec{\beta}\times\vec{E}\right) - \frac{\gamma^2}{(\gamma+1)}\vec{\beta}\left(\vec{\beta}\cdot\vec{B}\right).\end{aligned} \tag{6.7}$$

For a plane wave, $\vec{B} = \hat{k}\times\vec{E}$, and using the above definition of θ, we get

$$\begin{aligned}\vec{E}' &= \gamma\left(1 - \beta\cos\theta\right)\vec{E},\\ \vec{B}' &= \gamma\left(1 - \beta\cos\theta\right)\vec{B}.\end{aligned} \tag{6.8}$$

The intensity (irradiance) of the light field, I, is given by the time average of the Poynting vector, $\vec{S}$,

$$I = \langle S\rangle = \frac{c}{4\pi}\langle\left|\vec{E}\times\vec{B}\right|\rangle, \tag{6.9}$$

which, in the rest frame, becomes, using Eq. (6.8),

$$\begin{aligned}I' &= \gamma^2\left(1 - \beta\cos\theta\right)^2 I,\\ &= \left(\frac{\omega'}{\omega}\right)^2 I.\end{aligned} \tag{6.10}$$

The photon flux [$\mathrm{s}^{-1}\cdot\,\mathrm{cm}^{-2}$] of the light beam is an important quantity in photoneutralization rate equations. The flux transforms like the photon frequency, since it is the intensity divided by the photon energy:

$$F' = F\gamma\left(1 - \beta\cos\theta\right). \tag{6.11}$$

6.2.2 Photodetachment yield

In order to calculate photodetachment production rates from crossed beam experiments we must carefully consider the distribution of photons in the laser beam, the phase-space of the ion beam, and the overlap region between the two beams.

The concept of luminosity [12] is only useful in *approximating* reaction rates from crossed beam experiments. More accurate methods for estimating reaction rates will be described after a short discussion on luminosity.

Luminosity, in the context of crossed beam experiments, relates the reaction rate, R, to the detachment cross section, σ, in a very simple way

$$R \approx L\sigma. \tag{6.12}$$

Here the luminosity, L, is given by [12]

$$L = \left[(\vec{v_1} - \vec{v_2})^2 - \frac{|\vec{v_1} \times \vec{v_2}|^2}{c^2} \right]^{1/2} \int dV n_1 n_2. \tag{6.13}$$

The velocity for each beam is $\vec{v_i}$ and the number density for each beam is n_i. The integral is carried out over the interaction region. The luminosity gives a measure of how many particles from each beam are simultaneously present per unit time, per unit cross sectional area of the beam. The units of luminosity are [cm$^{-2}\cdot$ s^{-1}].

For a photon beam, of distribution $n_{h\nu}$, intersecting an ion beam of distribution n_{ion}, we can set $\vec{v_1} = \vec{c}$ and $\vec{v_2} = c\vec{\beta}$ in Eq. (6.13), defined such that $\vec{v_1} \cdot \vec{v_2} = \beta c^2 \cos\theta$, to obtain the luminosity

$$L = c\,(1 - \beta\cos\theta) \int dV n_{h\nu} n_{ion}. \tag{6.14}$$

A simple example is the calculation for the luminosity of two intersecting cylindrical beams of uniform densities. Of course the distributions of the individual beams must be considered in real life applications, as we will address later. Since, for uniform beams, the number densities are constant, the integration in Eq. (6.14) results in the volume of the interaction region. The easiest way to obtain the interaction volume for two intersecting cylinders is to slice the interaction region up into parallelpipeds. With the y axis chosen to be perpendicular to the plane defined by the laser beam and the ion beam, the differential volume element is

$$dV = \frac{4\sqrt{(R_<^2 - y^2)\,(R_>^2 - y^2)}}{\sin\theta} dy, \tag{6.15}$$

where $R_<(R_>)$ is the smaller or (larger) radius of either beam. The luminosity is then given by

$$L_u\,(R_<, R_>) = \frac{8cn_{h\nu}n_{ion}\,(1 - \beta\cos\theta)}{\sin\theta} \int_0^{R_<} dy \sqrt{R_<^2 - y^2}\sqrt{R_>^2 - y^2}, \tag{6.16}$$

In general this integral must be carried out numerically.

We can obtain an analytical expression for the luminosity if the two beams have the same radius, R,

$$L_u\left(R_< = R_> \equiv R\right) = \frac{16 I_{ion} F R\left(1 - \beta\cos\theta\right)}{3\pi\beta ec\sin\theta}, \tag{6.17}$$

where have written the photon number density in terms of the photon flux density, $F = n_{h\nu}c$, and the ion number density in terms of the ion current, $I_{ion} = \pi R^2 n_{ion} ce\beta$. Here e is the electronic charge and c is the speed of light. At a head on, or parallel intersection, the luminosity is singular. This is due to the infinite interaction volume for colinear beams.

Another simplifying case is when one of the beams is much larger than the other. We can then expand the integrand in Eq. (6.16) to obtain an analytical expression for the luminosity. The result is

$$\begin{aligned} L_u\left(R_> \gg R_<\right) &= \left(\frac{\pi c\left(1-\beta\cos\theta\right) n_{ion} n_{h\nu}}{4\sin\theta}\right)\frac{R_<^2}{R_>}\left(8R_>^2 - R_<^2\right), \\ &\approx \left(\frac{2\pi c\left(1-\beta\cos\theta\right) n_{ion} n_{h\nu}}{\sin\theta}\right) R_<^3. \end{aligned} \tag{6.18}$$

The above calculations for the luminosity can be used with the simple rate equation, Eq. (6.12), when the number of photodetachments is small. When this is not the case Eq. (6.12) predicts an infinite number of photodetachments as the luminosity approaches infinity. This unphysical result is due to the "thin-target" approximation; we have assumed that the ion density remains constant as the ions pass through the photon target. Only for single photon processes, where the probability of interaction is small, does Eq. (6.12) remain valid.

In the case when the probability of interaction is high, but for which the interaction mechanism is not intensity dependent, we can write

$$Rate = \frac{I}{e}\left(1 - \exp\left(\frac{-L\sigma e}{I}\right)\right), \tag{6.19}$$

where I is the ion beam current that passes through the laser beam and e is the electronic charge. We can see that for either a large cross section or high photon flux that the detachment rate saturates to the expected value, I/e, which is merely the number of ions that passed through the laser pulse.

The concept of luminosity, however, is inadequate to unfold the physics of an interaction when the cross section is dependent on the photon flux density. Examples of flux-dependent cross sections are multiphoton cross sections and cross sections which have included the effects of the ponderomotive potential. In addition to this complication, the particle beam is most often appreciably attenuated, or completely consumed, within the photon pulse ("thick target"). In order to address realistic photoneutralization we will first discuss ponderomotive effects in photodetachment.

6.2.3 Ponderomotive effects

Photodetachment of ions using a strong laser results in the detached electron being "bathed" in the time–dependent electromagnetic field of the laser. The escaping electron acquires a "quiver" energy in the laser field referred to as the "ponderomotive energy" [13–15]. The ponderomotive energy of the electron inside the laser field is

$$\begin{aligned} E_P &= \frac{e^2 \langle \vec{E}\,(\vec{r},\omega,t)\rangle^2}{4m_e c\omega^2}, \\ &= \frac{e^2}{8\pi^2 m_e c^3 \epsilon_0} \langle I\,(\vec{r},\omega,t)\rangle \lambda^2, \end{aligned} \tag{6.20}$$

where the angle brackets indicate the time-average of the laser field.

A useful numerical form for the ponderomotive energy is

$$E_P\,[eV] = \left(9.4\times10^{-14}\right) I\left[W/cm^2\right]\lambda^2\left[\mu m^2\right]. \tag{6.21}$$

The ponderomotive energy is a relativistic invariant as can be seen from Eq. (6.10). When the ponderomotive energy is greater than the photo-ejected electron's kinetic energy the detachment channel closes. This can be seen from energy considerations; the ejected electron energy must be greater than the average oscillatory kinetic energy of the electron in the laser field. When the n-photon channel closes the $(n+1)$ channel remains open as the lowest order MPD process. This is one reason why direct observation of ponderomotive shifts is difficult; the n-photon ponderomotive shift in the threshold may be buried by the higher order contributions.

6.2.4 Photodetachment using realistic beams

The intensity of a realistic laser pulse increases steadily from the wings of the pulse to the peak intensity at the core. The possibility of ion stabilization [16–18] inside the laser pulse must be carefully considered in the context of the photon distribution. As the ion enters the wings of the laser pulse the detachment rate increases as the intensity increases. At a cutoff intensity the ponderomotive energy is too high for the ejected electron to escape the atom into the surronding laser field; the photodetachment rate then goes to zero and remains at zero until the laser intensity decreases on the ion's exit from the pulse. Analogous to the "eye of the hurricane" the core of the laser pulse can be a region of stability for the ion. The size and shape for the "ponderomotive-eye" are functions of the photon energy, peak laser power, and the laser focal size.

As an ion traverses the wings of the laser pulse it may photodetach before it can reach the ponderomotive–eye. This is the case for large focal volumes and slow moving ions. Saturation of this type is also possible for relativistic ions when the intersection angle, θ, is large – resulting in a long interaction time in the wings of the laser pulse. Of course, if the laser intensity is weak then there may not be a region inside the laser pulse where ponderomotive channel closing occurs – no ponderomotive eye exists.

Not all ions will undergo the same intensity history in passing through the laser focus. Ions that pass through the center of the laser focus will experience the peak laser field whereas ions that pass through the tails of the laser focus will experience a much weaker field. The wings of the laser focus will give different detachment rates as compared with the high-intensity core.

Model calculations [19] for the photoneutralization of an 800 MeV H^- beam, typical of LAMPF, using a Q-swithced Nd:YAG laser are shown in Fig. (6.1). This calculation takes into account ponderomotive and saturation effects. The geometry assumed in this calculation is one that would be used for complete spatial neutralization of the ion beam – the laser and ion beam are assumed to have the same rms radius. Fig. (6.1) shows that at large Doppler shifts, "head-on" intersection angles, the detachment fraction saturates at a much lower pulse energy than for more perpendicular intersections. The enhanced photoneutralization at "head-on" angles is due to the much larger interaction volume.

As the laser pulse energy is increased, saturation of the photodetachment signal occurs more readily at lower rest frame photon energies. This effect is shown in Fig.(6.2) for the same laser and ion beam parameters used in Fig.(6.1).

The above calculations for photoneutralization assume a Gaussian-distributed ion beam and laser beam. The predictions for complete photoneutralization using a reasonable Nd:YAG laser are promising for accelerator applications that do not require the entire time structure of the ion beam to be neutralized. With clever designs, using multiple pass interaction chambers, photoneutralizers may someday be utilized in accelerator switchyard applications to deliver atomic beams in specifically prepared quantum states.

6.2.5 Distribution of detached electrons

When electrons are detached from a negative ion beam, the center of mass angular distribution is reflected in not only the laboratory angular distribution, but also in the lab energy distribution. The angular distribution of electrons in the center of mass, taking, for example, the case of linearly-polarized laser light parallel to the ion's velocity vector, $\vec{v}$, would be [20]

$$\frac{d\sigma}{d\Omega} = \frac{\sigma}{4\pi}\left[1 + \sum_{k=1}^{N} \beta_{2k} P_{2k}\left(\cos\theta_c\right)\right], \tag{6.22}$$

where σ is the total cross section, N is the number of absorbed photons, P_{2k} is a Legendre polynomial of order $2k$, θ_c is the angle between the polarization

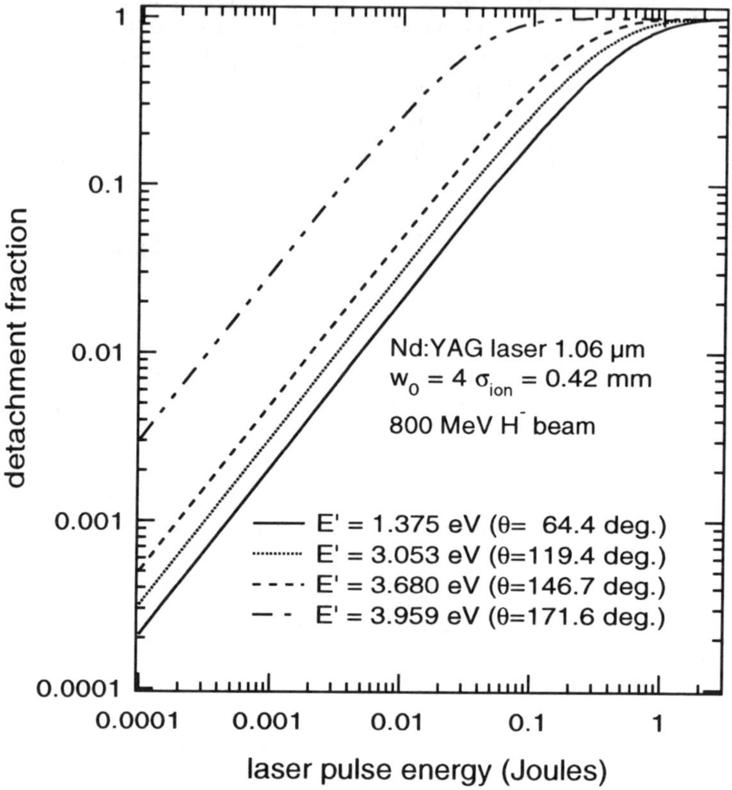

FIGURE 6.1 The detachment fraction versus laser pulse energy for various rest frame photon energies, E', as predicted by MacKerrow and Bryant [19]. This calculation takes into account ponderomotive and saturation effects. The laser waist, w_0, was chosen to have a rms radius equal to that of the ion beam's. The enhancement in photodetachment at "head-on" angles, large θ, is the result of the increased interaction volume for near co-linear beams.

direction and the electron's center of mass velocity vector, and β_{2k} are the asymmetry parameters, not to be confused with the velocity parameter for a particle. For example, if $x = \cos\theta_c$, we have

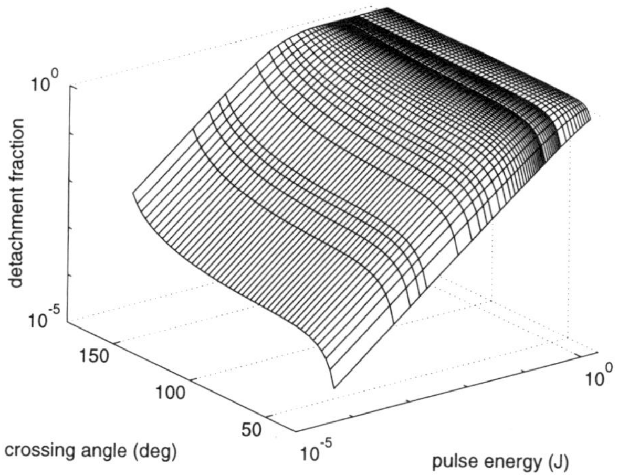

FIGURE 6.2 Photodetachment calculations [19] based upon the same laser and ion beam parameters used in Fig.(6.1). As the laser pulse energy is increased to approximately 1 Joule, saturation occurs for all intersection angles which result in a rest frame photon energy above the detachment threshold, 0.754 eV.

$$\begin{aligned} P_2 &= \frac{1}{2}\left(3x^2 - 1\right) \\ P_4 &= \frac{1}{8}\left(35x^4 - 30x^2 + 3\right) \\ P_6 &= \frac{1}{16}\left(231x^6 - 315x^4 + 105x^2 - 5\right) \\ P_8 &= \frac{1}{128}\left(6435x^8 - 1201x^6 + 6930x^4 - 1260x^2 + 35\right) \end{aligned} \tag{6.23}$$

To find the angular distribution in the laboratory we need to compute

$$\frac{d\sigma}{d\Omega_{LAB}} = \frac{d\sigma}{d\Omega_{CM}}\frac{d\Omega_{CM}}{d\Omega_{LAB}}. \tag{6.24}$$

Writing

$$\frac{d\Omega_{CM}}{d\Omega_{LAB}} = \frac{d\left(\cos\theta_c\right)}{d\left(\cos\theta_L\right)}, \tag{6.25}$$

and

$$\tan\theta_L = \frac{\sin\theta_c}{\gamma\left(\cos\theta_c + \frac{\beta}{\beta_e} + \frac{\beta\beta_e}{2}\right)}, \tag{6.26}$$

where $\beta = v/c$, $\gamma = 1/\sqrt{1-\beta^2}$, for the average beam particle in the lab and $\beta_e = v_e/c$, for the center of mass velocity of the electron, we find

$$\frac{d\Omega_{CM}}{d\Omega_{LAB}} = \frac{\sin^3\theta_c}{\sin^3\theta_L}\left[1 + \left(\frac{\beta}{\beta_e} + \frac{\beta\beta_e}{2}\right)\cos\theta_c\right]^{-1}. \qquad (6.27)$$

This form gives us a "rainbow" at

$$\cos\left(\theta_{cRB}\right) = -\left[\frac{\beta}{\beta_e} + \frac{\beta\beta_e}{2}\right]^{-1}. \qquad (6.28)$$

The energy distribution in the lab is given by

$$\frac{d\sigma}{dT} = \frac{d\sigma}{d\Omega_{CM}}\frac{d\Omega_{CM}}{dT}. \qquad (6.29)$$

Even for the most common case one might encounter, when only one photon is absorbed, and we are restricted to the second Legendre polynomial, one finds that β_2 ranges between 2 and -1 [21]. For one photon absorption where the laser is polarized normal to the plane of the laser beam and the ion beam, the rms spread in the lab angular distribution (in radians) of electrons from the beam along the laser polarization axis is

$$\delta_{rms} = \frac{\beta_1}{\gamma\beta}\sqrt{\frac{5+2\beta_2}{15}}. \qquad (6.30)$$

The corresponding spread in electron kinetic energy in the lab is given by

$$\delta T_{rms} = \gamma\beta\beta_1 m_e c^2\sqrt{\frac{5-\beta_2}{15}}. \qquad (6.31)$$

For example, for single photodetachment from an 800 MeV H^- beam using the fourth harmonic of the Nd:YAG laser at a lab angle of 56.45 degrees, so that the center of mass photon energy is 12.650 eV and the residual hydrogen atom is left in the H^0(n=2) state, we find the average 435 keV ejected electron energy in the lab has angular and energy spreads of 1.34 mrad and 0.97 keV for β_2 = 2, and 0.77 mrad and 1.37 keV for β_2 = -1, respectively.

6.2.6 Polarized beams by photodetachment

Ady Hershcovitch has pointed out that in principle it is possible to "edit", out of a beam of negative hydrogen ions, polarization states of the nucleus by operating with the center of mass photon energy near the detachment threshold in a magnetic field [22]. The hyperfine energy splitting in the final state causes the photodetachment threshold energy to depend on the orientation of the nucleus. He proposes that the editing can best be done in a storage ring so the ions can be exposed to the proper laser frequency for an extended period. Because of the very small energy differences, the center of mass photon energy must be as narrow as 150 MHz. One should note, however, that the best measurement of the electron affinity of hydrogen to date with a resolution of 100 MHz did not resolve the hyperfine structure in the threshold [24]. In the case of the storage ring, the circulating hydride ions would eventually become a pure state of nuclear polarization; at this point they would be stripped to form a nuclear-polarized beam.

6.3 PRODUCTION BY STRIPPING WITH THIN FOILS

Neutral atomic beams may be easily produced by passing a negative ion beam through matter where electron-detaching collisions neutralize it. The matter could be in the form of a differentially-pumped gas cell, or even a plasma, but more commonly, for simplicity, in the form of a thin foil such as carbon, aluminum oxide (Al_2O_3) or formvar ($C_5H_7O_2$). Stripping by matter, especially thin foils, has the advantage of simplicity; its drawbacks are that the process is incomplete. If the stripper is too thin, too many negative ions will survive, whereas if the stripper is too thick, too many of the neutrals will be stripped down to charged states. Even at the optimum thickness the transmitted beam will be a mixture of charged and neutrals. Moreover, some fraction of the neutrals will find themselves in excited states, which as we shall see below, has unpleasant consequences in some applications.

6.3.1 Theoretical background of foil stripping

The theory of H^- stripping has been treated recently [25]. It is found that a complete description requires collective contribution to the ion–matter interaction; nevertheless, for most design purposes it seems that the process can be modeled in terms of successive incoherent collisions with the individual atoms composing the foil, as is done for example in the paper by Mohagheghi et al. [26]. Theoretical studies of the relative production rates of the first excited state, H^0(n=2), in collisional electron detachment processes from H^- indicate that this state is produced more copiously than expected from calculations for other projectiles in the isoelectronic sequence [28,29].

6.3.2 Optimal foil thickness and production of excited states

In a recent paper [31], absolute measurements are reported on the yields of neutral hydrogen from foil stripping at 800 MeV of an H^- beam. Fig. (6.3) and Fig. (6.4) show the results of these measurements for self-supported carbon foils. There are two points to be noted in Fig. (6.4) : the yields of neutral hydrogen in the H^0(n=1) state and the first excited state, H^0(n=2), peak at a carbon thickness of approximately 40 $\mu g/cm^2$ whereas the yields of all the higher-lying states peak at 80 μg/cm^2. This remarkable result, which is also seen in the work of Mohagheghi et al. [26], implies that the production of the lower states is essentially a one step process, whereas the production of the higher states must have important contributions from secondary excitations of the two lowest states. The practical benefit is that one can minimize the production of excited states by using foils that are optimized for the ground state or perhaps even a little thinner. In accord with theoretical expectation [30] as well as observation [32] the yield of states above H^0(n=2) follows a n^{-3} law.

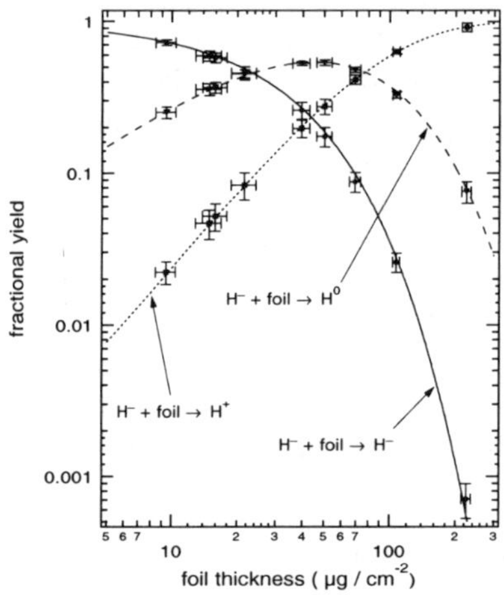

FIGURE 6.3 Measurements of Gulley et al. [31] on the overall branching ratios for the interaction of 800 MeV H^- ions with thin carbon foils. The solid curves are the best fit of an empirical relation derived by Gulley et al. to the measured yields. The branching ratios into individual hydrogenic states are shown in Fig. (6.4).

6.3.3 Effects of external fields on the stripping process

Does the presence of a magnetic field in region of the stripping foil affect the character of the resulting neutral beam? Recent experimental evidence points toward an effect on the distribution of Stark states in the excited manifolds of neutral hydrogen [33]. Although the distribution among principal quantum numbers appears to be unaffected by the presence of a field, the Stark state that is most resistant to electric-field-stripping seems to be enhanced relative to those Stark substates that are most easily stripped. The experimental data were obtained for only one field strength and only the H^0(n=4) manifold has been carefully scrutinized. A full report of this observation is currently in preparation [34].

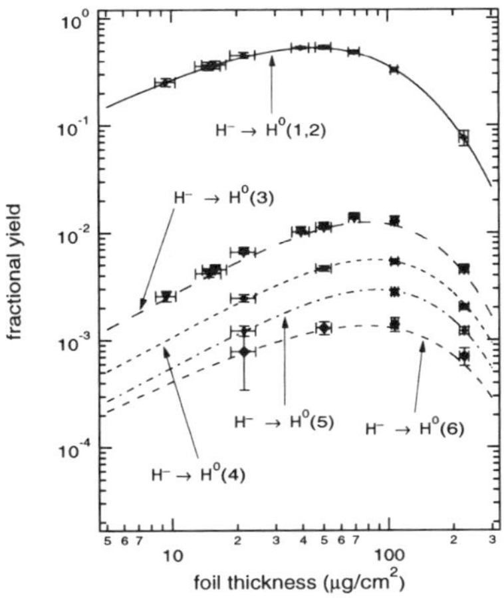

FIGURE 6.4 Measurements of Gulley et al. for the foil stripping of 800 MeV H^- ions by thin carbon foils into various hydrogenic states. The experiment was unable to separate the H^0(n=1) and H^0(n=2) states, but did resolve the higher energy states.

6.4 PRODUCTION BY FIELD STRIPPING

6.4.1 Lifetime of H^- in an electric field

Recently the lifetimes of H^- moving with 800 MeV in magnetic fields have been remeasured [35] in fields ranging up to 1.3 T. Although magnetic fields themselves do not affect the lifetime of the ion, the motional electric fields from even modest laboratory fields can be substantial for relativistic ions. The motional electric field due to a transverse magnetic field can be found using Eq. (6.8) to be

$$\left|\vec{E}_m\right| = \gamma\beta c\left|\vec{B}_t\right|, \tag{6.32}$$

so that an 800 MeV H^- ion moving transversely to a 1 Tesla field would experience an electric field in its rest frame of 4.68 MV/cm. The lifetime of the H^- ion can be well parameterized by the formula

$$\tau = \frac{a}{\left|\vec{E}_m\right|}\exp\left(\frac{b}{\left|\vec{E}_m\right|}\right). \tag{6.33}$$

The experimentally estimated values, recent as well as earlier measurements [36,37], are given in Table (6.1).

Parameter	Jason et al.	Keating et al.
a $[10^{-6} s \cdot V \cdot m^{-1}]$	2.47(10)	3.073(10)
b $[10^{9} V \cdot m^{-1}]$	4.49(1)	4.448(7)

TABLE 6.1 The experimentally determined parameters to be used with Eq.(6.33).

For exceptionally large electric fields (in excess of 1 MV/cm) this simple parameterization must be modified according to the theory of Leonard Scherk [38]. The paper by Keating et al. [35] should be consulted for more details.

Fig. (6.5) reproduces figure 4 of Keating et al. [35]. Here the lifetime of H^- in its ground state is plotted versus the applied rest frame electric field.

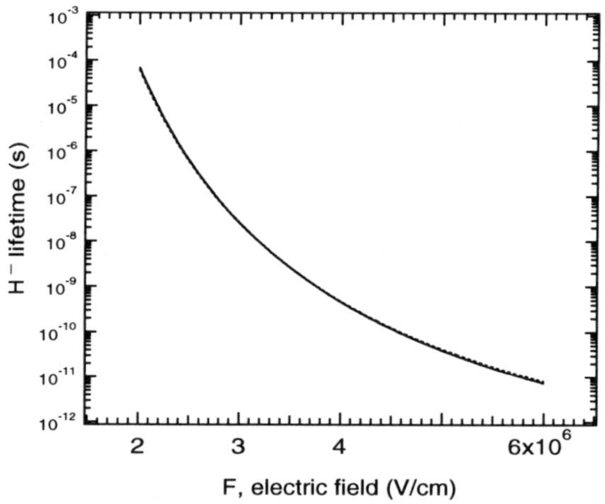

FIGURE 6.5 Plot of the lifetime against electron detachment of $H^-(1s^2\ ^1S)$ versus rest frame electric field. The plot contains two curves which lie on top of each for this range of electric field strength. One curve is from Jason et al. [36] and the other is from Keating et al. [35], both curves are derived from experimental measurements.

6.4.2 Effects on beam emittance

A magnetic field is presently in use at the proton storage ring (PSR) at Los Alamos to produce a neutral hydrogen beam by electron-detachment from an H^- beam. Although this method is very convenient, one must pay a price in emittance growth when one uses it.

According to Richard Hutson [39], the stripper magnet at the PSR rises to its maximum value of 1.8 T in about one cm and introduces an angular spread in the resulting 800 MeV H^0 beam of approximately 0.37 mrad rms. This smearing must be added in quadrature to the angular divergence in the plane perpendicular to the magnetic field in determining its effect on the beam emittance.

The problem of smearing arises because of the non-negligible deflection of the parent ions by the magnetic field before they penetrate into the field where their stripping rate is close to instantaneous. Let us look at the formulas:

The angular deflection θ of an ion of momentum p that has penetrated a distance l along the z axis into a transverse magnetic field $\vec{B}$ is

$$\theta = \frac{e}{p}\int_0^l \vec{B}\cdot \vec{dz}. \tag{6.34}$$

The fraction of the incident beam neutralized as a function of distance is given by

$$\frac{df}{dz} = -\frac{f}{\beta\gamma c\tau\,(B)}. \tag{6.35}$$

The fraction per unit deflection angle is

$$\frac{df}{d\theta} = \frac{df}{dz}\frac{dz}{d\theta}. \tag{6.36}$$

Combining these relations we find

$$\frac{df}{d\theta} = \frac{p}{\beta\gamma ce\tau\,(B)}\exp\left[-\int_0^\theta \frac{pd\theta}{\beta\gamma ce\tau\,(B)}\right]. \tag{6.37}$$

This formula can be simplified by use of the expression for the lifetime τ, Eq. (6.33), so that

$$\frac{df}{d\theta} = \frac{p}{ea}\exp\left(-\frac{b}{\left|\vec{E}_m\right|}\right)\exp\left[-\frac{p}{ea}\int_0^\theta d\theta'\exp\left(-\frac{b}{\left|\vec{E}_m\right|}\right)\right]. \tag{6.38}$$

The root mean square of the angular distribution can be calculated in a straight-forward way from the above formulas.

Richard Hutson [39] has suggested that the angular divergence introduced by the magnetic stripping process may be mitigated by use of an alternating magnetic field, a "wiggler" magnet. To our knowledge, this suggestion, although promising, has not yet been implemented. Hutson [39] also briefly describes a pulsed helical field magnet constructed by Dodge Warren of Los Alamos that may also be of interest of neutral beam designers.

6.5 PRODUCTION BY NOVEL METHODS

6.5.1 Laser-induced recombination

Although merged beams of positive ions with electrons can recombine to produce a neutral beam by radiative recombination, the yield can be greatly increased by laser-stimulating the radiative transition. The production of energetic neutral hydrogen by laser-stimulated radiative recombination has been observed and reported by two experimental teams [40,41]. The first team, U. Schramm et al., working at the ion storage ring TSR at Heidelberg were able to demonstrate a gain of about 25 over ordinary radiative recombination for 21 MeV protons with a merged, cooled electron beam overlapped by a laser beam tuned to photon energy slightly larger than the bound H^0(n=2) free transition energy. The second team, F. B. Yousif et al., working with a 330 keV proton beam reported a gain over spontaneous radiative recombination of 4790 ± 2830 when they used a 15.3 W CO_2 laser beam to stimulate transitions to H^0(n=11) and H^0(n=12). In principle, this method should be applicable to very high proton energies, so that extreme relativistic beams of hydrogen atoms could be prepared into selected quantum states. Of course, using antiprotons and cooled positrons could also allow for the production of neutral beams of antimatter using the same technique.

6.5.2 Capture from pair-production

Recently high energy antihydrogen atoms [42] have been observed from the capture of positrons produced in positron-electron pairs by peripheral collisions of antiprotons with heavy nuclei. A European team [43] led by Walter Oelert, report having produced 11 antihydrogens including 2 ± 1 background events by directing a jet of xenon gas across the antiproton beam at LEAR at CERN. Positrons from electron-positron pairs made by occasional close encounters with the xenon nuclei then have a small chance of capture by the antiproton. This technique, first suggested by Charles Munger, Stanley J.Brodsky, and Ivan Schmidt [44], is in principle capable of producing energetic hydrogen and antihydrogen beams, but with extremely low efficiency; the 11 observed events required 5×10^{12} antiprotons for production!

6.5.3 Cosmic sources

One might speculate that a fraction of the abundant cosmic H^- ions could be accelerated and then deprived of one of their electrons by one of the mechanisms discussed above. This process could lead to extremely high energy neutral atoms albeit widely dispersed in energy and position. Indeed, this possibility has been discussed [45].

6.6 BEAM COOLING BY RESONANCE PRODUCTION

In the production of neutral atomic beams by photodetachment one can use narrow resonant features in the photodetachment spectrum of the parent ion not only for diagnostic purposes, as discussed further below, but also for producing a neutral beam whose momentum spread and phase space is smaller than the parent population [46]. This procedure is a form of cooling analogous to the technique of evaporative cooling used in atom traps. In our discussion below we shall deal with the H^- ion; similar situations should also be found in heavier ion beams, of course. We shall first describe an example of how the momentum distribution can be narrowed, and then how the phase space could be tailored.

6.6.1 The Feshbach resonance for momentum spread narrowing

The most prominent needle-like structure in the photodetachment spectrum of H^- is the singlet P Feshbach resonance lying just 27 meV below the threshold for leaving the residual H^0 in its first excited state. In order to excite this resonance, which was first observed at LAMPF [47], the center of mass photon energy must be 10.9264(6) eV [49] . The energy width of this resonance is of the order of 30 μeV [50] and the photodetachment cross section at its peak is about 50 square Bohrs [51] ($1.4\text{x}10^{-15}$ cm^2). The use of this resonance for cooling arises from its exquisitely well-defined energy and strength relative to the neighboring photodetachment continuum, 0.3 square Bohrs ($8.4\text{x}10^{-18}$ cm^2). If a well-collimated laser beam, tuned to the resonance, is used as the photodetaching agent, only those H^- ions with the appropriate Doppler shifts will undergo detachment, so that the population of the resulting neutral beam will be cooler than the parent beam. Whether this cooling is transverse or longitudinal depends upon the relative direction of the laser beam. The spread in the center of mass photon energy of a laser beam can be found as follows: To first order the spread δE in the rest frame photon energy E depends on three variables, the intersection angle θ, the beam momentum, p, and the energy in the lab, E_L:

$$\delta E \approx \frac{\partial E}{\partial \theta}\delta\theta + \frac{\partial E}{\partial E_L}\delta E_L + \frac{\partial E}{\partial p}\delta p. \tag{6.39}$$

In general this is not accurate since there can be a correlation that we neglect in what follows, between p and θ in regions where the ion beam dispersion is non-zero. For most ion beams the contribution to the rest frame photon energy spread from the laser linewidth in the laboratory frame is small; we will therefore neglect δE_L from here on. With these approximations we can calculate the rms rest frame photon energy spread using Eq. (6.39) and Eq. (6.5)

$$\delta E_{RMS} = E\left[\left(\frac{\beta\sin\theta}{1-\beta\cos\theta}\right)^2 \delta\theta^2 + \beta^2\left(\frac{\beta-\cos\theta}{1-\beta\cos\theta}\right)^2\left(\frac{\delta p}{p}\right)^2\right]^{1/2}. \tag{6.40}$$

For head-on photon-ion collisions, $\theta \approx 180°$, there is no dependence on θ. Since a successful photodetachment, for a modest range of laser intensities, is likely to occur only if the spread is within the narrowly defined energy of the resonance, we can restrict the laboratory momentum distribution to lie in the range given by $\delta p/p = \Gamma/(\beta E)$, where Γ is the center of mass width of the resonance. Thus, for 800 MeV, producing a neutral beam by sending a properly-tuned laser beam directly into the ion beam would result in a neutral (ground state) hydrogen beam whose momentum spread $\delta p/p$ would be no larger than

$$\frac{\Gamma}{\beta E} = \frac{3 \times 10^{-5} eV}{(0.842)(10.926 eV)} = 3.3 \times 10^{-6}. \tag{6.41}$$

A similar constraint could be put on the transverse momentum spread by constraining the angular spread. We note that when $\cos\theta = \beta$, the contribution to the energy spread from momentum spread vanishes. For 800 MeV, for example, $\theta_m = 32.7°$. Unfortunately, this "magic" angle is a backward angle, where the energy of the photon is red-shifted. It should also be noted that the angular contribution, from $\delta\theta$, is maximum at this angle. The center of mass photon energy at the magic angle is just $E = E_L/\gamma$ (from Eq. (6.5). Never-the-less, advances in laser technology may eventually allow for precise, tunable lasers in the vacuum ultraviolet (the required laboratory photon energy for an 800 Mev H^- ion beam is about 20.2 eV). In any case, the constraint on $\delta\theta$ in order to excite the resonance at the magic angle is $\Gamma/(\beta\gamma E)$ for an 800 MeV beam. At angles other than the two discussed so far, photodetachment through the resonance simply restricts the beam to lie in a small region on the p vs. θ plane. It is interesting to note that, although an autodetaching resonance, similar to the Feshbach resonance, has recently been produced by two photon absorption [52], multiphoton excitation of the Feshbach resonance, say with three photons, would be useless here, since single photon detachments would be overwhelming.

As discussed below, a study of the resonance shapes can be an important diagnostic tool for the parent ion beam. Of particular relevance to cooling is the paper by Holtkamp and Quick [53].

6.6.2 Phase space tailoring

Because negative ions, the parents of neutral atomic beams, are exceptionally frangible, the transverse phase space, or emittance, can be tailored by "supercollimation", using apertures just thick enough to strip the ions without introducing undesirable angular spreads in the stripped beam. In some cases the stripped beam can then be put to less critical uses, as was done on beam-line X at LAMPF. In simplest terms, the beam which survives unstripped after passing through two successive apertures of areas A_1 and A_2

separated by a distance d, with no beam-focusing elements interposed, will populate an area in transverse phase space given approximately by

$$\int dx'dxdy'dy = \frac{A_1 A_2}{d^2}. \tag{6.42}$$

In addition to the possibilities for beam tailoring that such a scheme presents, beam line accidents can be minimized with proper design. When the apertures themselves are designed to spoil the vacuum if the beam striking them is excessive, for example, unauthorized or accidentally large beam currents can be quickly suppressed [54].

Because the extra electron is so loosely bound, collimation and beam tailoring by means of photodetachment appears to be a valuable tool for beam design. One can imagine using laser beams with shaped focal regions that would strip the outer portions of an ion beam and allow the center region of the beam to pass unscathed. (A "photon aperture") One can even envisage the control of the intensity distribution across a beam in a time dependent way, for example in radiation therapy, as proposed by Carol Johnstone [56].

Partial stripping, as has been demonstrated [57], does not require a laser; bright incoherent light sources, with an appreciable amount of their Doppler-shifted spectrum with wavelengths less than 1.6 μm, can strip a useful amount of beam. For example, suppose we pass a fast H^- beam through an optical cavity pumped with xenon flash lamps so that the effective photon density in the cavity is that of a blackbody at 9000 K, corresponding to an energy density of 5 μJ/ cm^3. Taking the average photon energy to be 1 eV, this works out to be 3 $\times 10^{13}$ photons/cm^3. Taking the probability of stripping to be the cross section times the area density of photons in our "photon target", the effective cross section to be 10 megabarns, and, for example, the length of the cavity to be 10 cm, we find the probability for stripping to be 0.3 percent. This arrangement would be a good way to generate an intermittent H^0 beam of 75 μA from a parent beam of 25 mA, without any disturbance to the parent beam.

6.7 DIAGNOSTICS

The prospect of a laser diagnostic for accelerators is very exciting. A laser diagnostic can in principle monitor H^- beam parameters, such as momentum spread, in a non-invasive manner. Laser diagnostics can also be used for beam profile monitors by scanning the ion beam and measuring the photodetached electron signal. A profile monitor of this type has promise since the wire scanners used today scatter beam and cause beam spill on high current machines resulting in radioactive contamination.

6.7.1 Measurement of the absolute beam momentum

The relativistic Doppler shift can be written in terms of the beam momentum, $\vec{p}$, as

$$E\left(E_L, \left|\vec{p}\right|, \theta\right) = \frac{E_L}{E_0}\left[\sqrt{p^2c^2 + E_0^2} - pc\cos\theta\right], \tag{6.43}$$

where E_0 is the H^- rest mass energy (939.294 MeV/ c^2), E_L (E) is the laser photon energy in the lab frame (rest frame) , and c is the speed of light. If the center frequency of a known resonance is measured, E, at an angle θ, then the beam momentum, p can be found. The angle of intersection, θ, must also be measured; this is usually done with an angle encoder on the laser mirror turn-table. In practice this angle is obtained by measuring more than one resonance. Two measurements are required since the angle θ is given by two parameters N and N_0

$$\theta = \frac{N - N_0}{K}, \tag{6.44}$$

where N is the encoder reading for the center of the resonance and N_0 is the encoder reading for the beam direction (called the "encoder zero"). The parameter K is the number of encoder counts per degree, or the encoder resolution. Previous experiments [55] used a 12 bit shaft encoder ($K = 11.3$ deg^{-1}) and a 14 bit shaft encoder ($K = 552$ deg^{-1}). The angular resolution of these two encoders are 1.53 mrad per step and 31.9 μrad per step respectively. Of course higher resolution angular encoders (> 22 bit) also exist.

Beam energy measurements are usually made on the $H^-(^1P^o)$ Feshbach resonance. This narrow resonance is measured on both sides of the ion beam using the Nd:YAG laser fourth harmonic (0.266 μm). The Doppler shift formula, Eq. (6.5), is then fit to the two data points to determine N_0 and p. This method requires that the mirror system pass through the ion beam. Another less invasive method would be to use two different laser wavelengths from the Nd:YAG laser (0.355 μm and 0.266 μm) and measure a Feshbach resonance angle for each wavelength staying on one side of the ion beam only.

Measuring the position of more than one spectroscopic feature, such as the shape [48] and Feshbach resonances, without letting the mirror system cross the ion beam, is another method that can be used for determining the two unknown parameters.

If a thin foil can be temporally inserted into the beam, then energy measurement precision is much improved. The foil interaction will populate many different hydrogenic states. The laser can then be used to promote hydrogenic transitions to states that are easily field ionized in an electron spectrometer. This allows many angles to be measured for a more accurate curve fit for the parameters N_0 and p. This method has been used at LAMPF where the beam energy was measured to be 797.26 ± 0.12 MeV [27].

6.7.2 Measurement of the beam momentum spread, $\delta p/p$

Since the $H^-(^1P^o)$ Feshbach resonance has a narrow spectral width (predicted to be about 30 μeV [23]), the measured width of the resonance using photodetachment will be the experimental resolution and can therefore give the momentum spread of the ion beam. The experimental resolution, to first order is

$$\delta E = \gamma\beta E_L \left[\left(\frac{1-\beta\cos\theta}{\beta} \right)^2 \left(\frac{\delta E_L}{E_L} \right)^2 + (\sin\theta\delta\theta)^2 + (\beta - \cos\theta)^2 \left(\frac{\delta p}{p} \right)^2 \right]^{1/2} . \tag{6.45}$$

The resolution associated with photodetachment is normally dominated by either the H^- beam momentum spread or the combined angular divergence of the laser beam and ion beam. From Eq.(6.45) we can see that the rest frame energy resolution depends upon the intersection angle used for photodetachment. The angular dependence of the energy resolution is shown in Fig.(6.6).

The H^- photodetachment spectrum has been measured at 800 MeV and parameterized [48]. The shape and Feshbach resonances are shown in Fig. (6.7) for an 800 MeV H^- beam at various momentum spreads.

As can be seen from Fig. (6.7) the momentum spread of the beam can be monitored by measuring the width of a resonance using the Doppler shift technique. Note that if the momentum spread is too large, the Feshbach resonance will be washed out. At this point the width of the shape resonance starts to broaden with increasing momentum spread and it can then be used to monitor the momentum spread of the beam.

The Feshbach resonance has been used to successfully monitor changes in the momentum spread resulting from detuning of a single Rf module on the LAMPF linear accelerator [64].

Another method of using the Feshbach resonance as a momentum spread diagnostic has been presented by Holtkamp and Quick [53]. They proposed sending a tunable laser beam onto the H^- beam in a head on fashion (θ =180°) and detecting H^0 atoms resulting from the decay of the Feshbach resonance. Working at θ =180° is advantageous since the linewidth of the Feshbach resonance is most sensitive to $\delta p/p$ at this angle. However, this technique requires special beam line geometry, is extremely sensitive to any movement of the beam, and may have a high H^0 background due to gas stripping. The method also requires tuning, and measuring, the laser wavelength continuously.

6.7.3 Angular divergence

The angular divergence contribution to the photon energy spread has two components: the laser beam divergence and the ion beam divergence. Typical

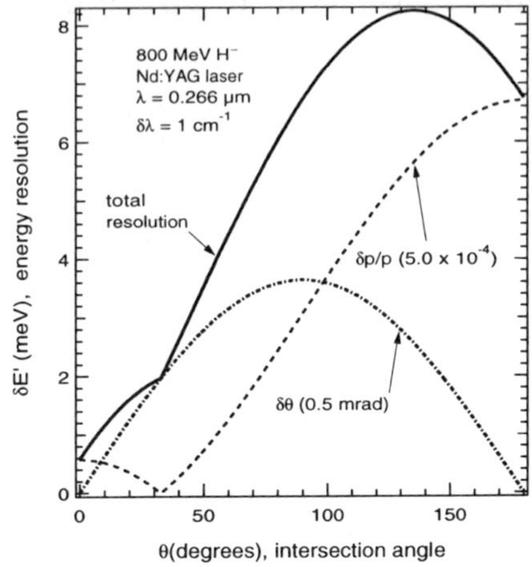

FIGURE 6.6 The energy resolution for laser photodetachment as a function of the intersection angle. For "standard" lasers the laser linewidth in the laboratory frame is negligible as compared with the angular uncertainty and momentum uncertainty. At the "magic" angle of $\cos\theta = \beta$ the contribution to the total rest frame energy uncertainty is zero. At this intersection angle laser photodetachment is insensitive to the ion beam momentum spread.

divergences for an unfocussed laser beam are approximately 0.5 mrad (half-angle) and typical ion beam divergences range from 100 μrad to 5 μrad. If the laser beam is focused onto the ion beam, for example in multiphoton experiments to achieve high intensity, the divergence of the laser beam will contribute the most to the rest frame energy uncertainty.

6.7.4 Transverse emittance diagnostics

It may be possible to measure the $x - x'$, $y - y'$ phase space of a relativistic ion beam using very small, inexpensive, diode lasers. If this method works the result would be a non-invasive beam diagnostic that could monitor accelerator performance continuously. The concept was first considered by Swenson and MacKerrow in 1993 [63]. A small diode laser would scan across the transverse dimension of the ion beam just before a bending magnet. This laser would photodetach a small amount of the H^- beam. The resulting neutrals would then pass un-deflected through the dipole field to an array detector, such as a microchannel plate detector. The correlation between the transverse location of the diode laser and the image on the array would determine the angular spread, x', as a function of the laser position, x - in other words the transverse beam emittance. The measurement would also give the current distribution of the ion beam, similar to a wire scanner, but without the scattered radiation problem associated with secondary emission wire scanners.

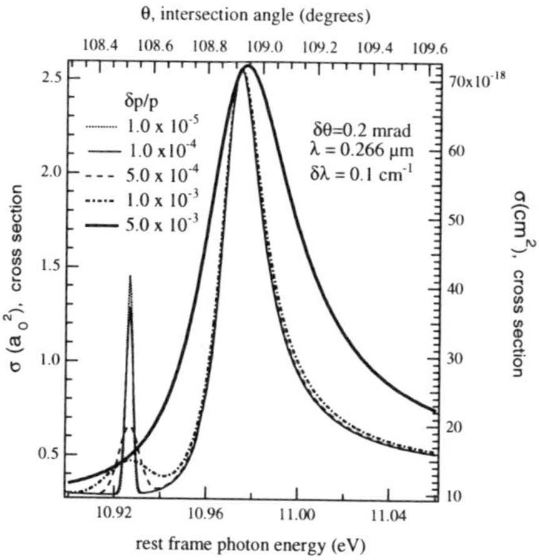

FIGURE 6.7 The H^- Feshbach and shape resonance, decay into H^0(n=2), shown at different values of $\delta p/p$. The lower energy Feshbach resonance is more sensitive to $\delta p/p$ than the higher energy shape resonance. This plot assumes a Nd:YAG laser operating at the 4th harmonic (0.266 μm) with an angular divergence of 0.2 mrad.

6.7.5 Laser resonance fluorescence

A solution to the problem of the non-invasive determination of the direction and divergence of a relativistic neutral hydrogen beam was suggested by Rohringer [60] in 1977. The idea is that some fraction (estimated at about 6%) [28] of the neutral beam is formed in $H^0(2s)$ states in the stripping process, described above. Hydrogen atoms in the $H^0(2s)$ state can then be photoexcited to a higher state, say the $H^0(3p)$, whose radiation to the ground state $H^0(1s)$ will then be detected. The exciting laser beam, precisely collimated to a divergence less than that expected in the particle beam, will be tuned to the resonant frequency corresponding to one of the $H^0(2s)$-$H^0(3p)$ absorption lines when directed at the beam at the "magic" angle, that is, $\cos\theta_m = \beta$. At this angle, as discussed above, to first order, the center of mass laser frequency is independent of small variations in the beam momentum. Thus, the beam direction, and its divergence, can be determined from the angles of two laser beams in orthogonal planes, tuned to one of the magic angle $H^0(2s)$-$H^0(3p)$ resonant frequencies. The angles of the two laser beams in orthogonal planes would be scanned and the strength of the induced resonant radiation intensity from the beam would allow continuous monitoring of the direction and the angular divergence of the neutral beam. The feasibility of this approach has been demonstrated, with just one laser beam, at Argonne National Laboratory with a 50 MeV H^- beam, where a laser-induced fluorescence signal was clearly visible above a particle-beam-induced background [58]. From the angular width of the fluorescence signal, these authors were able to deduce a

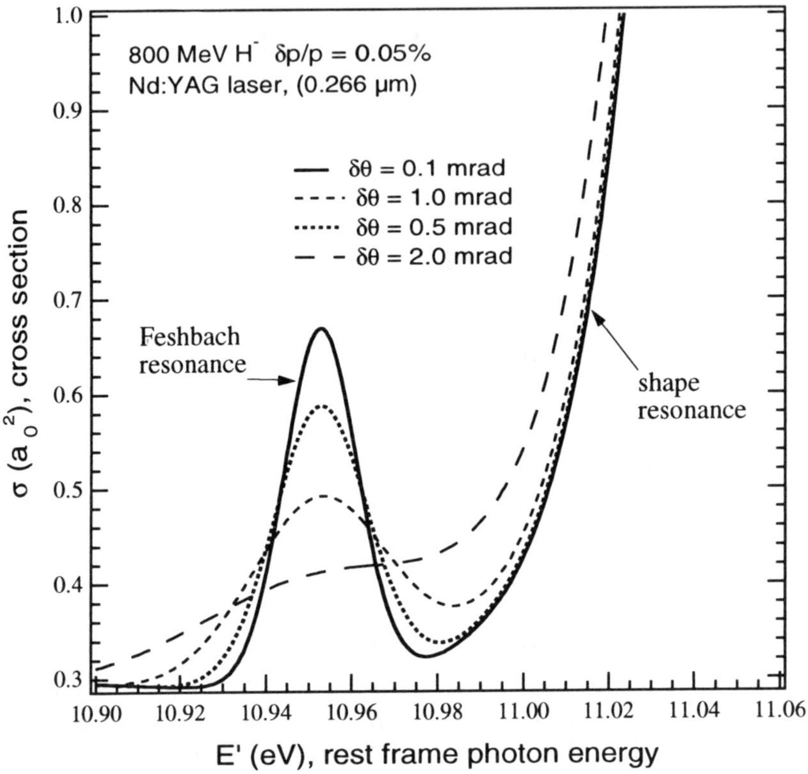

FIGURE 6.8 The H^- Feshbach and shape resonance, decay into H^0(n=2), shown at different values of laser angular divergence and a fixed 800 MeV H^- momentum spread of 5×10^{-4}.

divergence of the neutral beam of 0.267 mrad FWHM. This technique should in principle be capable of much higher angular resolution if the need arose. In a lucid examination of neutral particle beam sensing [59], David Pegg discusses the details of the physics of resonant fluorescence and also examines the possibility of detecting the resonant absorption, say from the $H^0(2s)$ to the $H^0(3p)$, by subsequent ionization of the $H^0(3p)$ state before it has a chance to radiate. Either the electrons or protons released can be detected with greater efficiency and probably less background than the fluorescent L_β or L_α radiation.

6.7.6 Wire shadows

The angular divergence of a high-energy neutral beam can be determined by the "shadow" cast by a thin wire passing through the beam [61]. Some understanding of the problems of determining the direction of a neutral beam by the wire shadow method with a precision of 0.4 μrad has been obtained [62].

6.8 CONCLUSION

We have sketched various techniques presently available for the production of relativistic neutral hydrogen beams. The most straight-forward way to produce such beams is by stripping beams of H^- ions which have been accelerated to the desired energy of the neutral beam. We have discussed the advantages and disadvantages of the ways such beams of H^- may be neutralized. Since the highest energies that can be obtained by acceleration of the hydride ion are limited by the small electron affinity of hydrogen in collusion with the fields required for the acceleration process, if one wishes to go to higher energy, another strategy is required. Fortunately, there are other methods, which may allow for the production of higher energy beams. Disassociation of H_2^+ should allow for somewhat higher energies, and finally various forms of merged beam methods would seem to be feasible to even higher energies. In fact, as has been recently demonstrated, relativistic beams of antihydrogen can be produced, albeit presently exceedingly weak ones, by the merging of pair-produced positrons with antiprotons. We have also discussed diagnostic methods, and how neutral beams may be tailored to have smaller divergences and energy spreads than their parent beams.

Only time will tell what interesting new physics the availability of high quality relativistic hydrogen (and antihydrogen) beams will help to reveal.

ACKNOWLEDGMENTS

The authors would like to acknowledge technical support of Lawrence Quintana, Larry Rybarcyk, John Gomez, Fred Roybal, Mike Borden, Dan Fitzgerald, Richard Hutson, Joe Tiee, Charles R. Quick, Joey B. Donahue, John O'Donnell, Olin B. van Dyck, Richard Werbeck, Robert Macek and the many others at Los Alamos National Laboratory that helped in the completion of this work. Mark Gulley and Phil Keating have contributed much to this work. We would also like to thank Jim Hontas and William Miller of the University of New Mexico for helping us with many of these experiments. We also acknowledge Joseph Macek, Tom Bergeman, Péter Kálmán, János Bergou, and Joachim Burgdörfer for helpful theoretical discussions. This work was supported by the U.S. Department of Energy, in part by the Division of Chemical Sciences, Office of Basic Energy Sciences, Office of Energy Research.

REFERENCES

1. The production of neutral beam particles with energies in excess of 150 keV by neutralization of positive-ion beams by (stationary) electron capture is probably impractical. See J. H. Fink, W. L. Barr, and G. W. Hamilton, "A Study of Efficient High-Power, High-Energy Neutral Beams for the Reference Mirror Reactor", UCRL-52173, Lawrence Livermore Laboratory, Nov. 11, 1976. For a review of the electron capture problem at high energies see R. Shakeshaft and L. Spruch,"Mechanisms for charge transfer (or for any light particle) at asymptomatically high impact velocities", Rev. Mod. Phys. **51**, 369 (1979).
2. L. A. Pinnaduwage and L. G. Christophorou, J. Appl. Phys. **76**, 46 (1994).
3. A birdseye look into the future of accelerators can be found in E. J. N. Wilson, "Accelerators for the Twenty-First Century- A Review", CERN 90-05, 29 May 1990.
4. The idea of using H^- in accelerators was first introduced by Luis Alvarez, Rev. Sci Instrum. 22, 705 (1951). Not only did Alvarez suggest the stripping H^- in a Van de Graaff and accelerating the protons back down to ground potential, as is done in modern tandem machines, but he also realized the usefulness of negative ions for injection into circular orbits via foil stripping.
5. L. Rosen, Prog. Theor. Phys. Supplement 85, 168-189 (1985). See also Los Alamos National Lab Report **LA-UR-85-3437**, Nov. 1985.
6. Robert Macek and Arch Thiessen, private communication ,1995.
7. Some theoretical work is currently underway to compute the lifetime against dissociation in electric fields (J. M. Peak, private comm., 1995) and possible measurements are being proposed (H. A. Theissen, private comm., 1995).
8. H. C. Bryant, P. A. Lovoi, and G. G. Ohlsen, Phys. Rev. Lett. **27**, 1628-1631 (1971).
9. D. W. MacArthur, K. B. Butterfield, D. A. Clark, J. B. Donahue, P. A. M. Gram, H. C. Bryant, C. J. Harvey, W. W. Smith, G. Comtet. Phys. Rev. Lett., **56**, 282 (1986). See also "A Test of the Lorentz Transformation for Near Luminal Velocity Using an Atomic Interferometer," Jànos Bergou and Howard Bryant. Contributed paper SASP 1990 Obertraun Austria March 18-24, 1990. Published in SASP '90 (Symposium on Atomic and Surface Physics) T. D. Märk and F. Howorka, Eds. pp. 93-97.
10. H. P. Robertson, Rev. Mod. Phys **21**, 378 (1949).
11. A very thorough, and understandable, discussion on relativistic kinematics, including the relativistic Doppler shift, is given in "Classical Electrodynamics", 2nd edition, J.D. Jackson, Chapter 11, Wiley, New York (1975).
12. A good discussion of luminosity and its limitations with relativistic beams is given by C. Pellegrini, "Colliding-Beam Accelerators", Ann Rev Nuc Sci 22, 1- 24 (1972). Attention is drawn especially to the remarkable eq. 16 in this reference.
13. T.W.B Kibble, Phys. Rev. **150**, 1060 (1966).
14. V. Z. Slonim, Phys. Rev. A. **45**, 8034 (1992).
15. F.V. Hartemann, S.N. Fochs, G.P. Le Sage, N.C. Luhmann, Jr., J.G. Woodworth, M.D. Perry, Y.J. Chen, and A.K. Kerman, Phys. Rev. E. **51**, 4833

(1995).
16. Q. Su, J.H. Eberly, and J. Javanainen, Phys. Rev. Lett. **64**, 862 (1990).
17. R. Grobe and J.H. Eberly, Phys. Rev. Lett. **68**, 2905 (1992).
18. J.H. Eberly and K.C. Kulander, Science **262**, 1229 (1993).
19. E.P. MacKerrow and H.C. Bryant, "H^- Photodectachment Models using Realistic Laser Beams", to be submitted to Phys. Rev. A, June 1996.
20. C. Blondel and C. Delsart, NIMB **79**, 156-158 (1993), also Laser Physics **3**,699 (1993).
21. U. Fano and C.H. Greene, "Unfavored angular distributions", Phys. Rev. A. **22**, 1760 (1980).
22. A. Hershcovitch, Nucl. Instrum. Methods **A243**, 271-277 (1986).
23. L.Lipsky and M.J. Conneely, Phys. Rev. A**14**, 2193 (1976).
24. K.R. Lykke, K.K. Murray, and W.C. Lineberger, Phys. Rev. A **43**, 6104 (1991).
25. B. Gervais, C. Reinhold, and J. Burgdoerfer, Phys. Rev. A., to be published.
26. A.H. Mohagheghi, H.C. Bryant, P.G. Harris, R.A. Reeder, H. Sharifain, C.Y. Tang, H. Tootoonchi, C.R. Quick, S. Cohen, W.W. Smith, and J.E. Stewart, Phys. Rev. A **43**, 1345 (1991).
27. A.H. Mohagheghi, PhD Dissertation, University of New Mexico, Published as LANL Report No. **LA-11925-T**, 1990.
28. L.A. Wright, M.R. Franz, T.C. Genoni, Phys. Rev. A **32**, 1215 (1985).
29. G.H. Gillespie, R.S Jando, D.L. Moores, in Electronic and Atomic Collisions, Abstracts of Contributed Papers, XIII ICPEAC, Berlin 1983, Ed. J. Eichler et al. p. 402 (1983). See also G.H. Gillespie, Nucl. Instr. Meth. Phys Res. **B2**, 231 (1984) and **B10/11**, 22 (1985).
30. H. Friedrich, "Theoretical Atomic Physics", Springer-Verlag, Berlin, p. 27 (1990).
31. "Measurement of H^-, H^0, and H^+ cross sections produced by foil-stripping of 800 MeV H- ions", M. S. Gulley, P. B. Keating, H. C. Bryant, E. P. MacKerrow, W. A. Miller, D. C. Rislove, Stanley Cohen, J. B. Donahue, D. H. Fitzgerald, S. C. Frankle, David J. Funk, R. L. Hutson, R. J. Macek, M. A. Plum, N. G. Stanciu, O. B. van Dyck, C. A. Wilkinson, C. W. Planner, Phys. Rev. A **53**, 3201 (1996).
32. J.A. Stone and T.J Morgan, Phys. Rev. A **31**, 3612 (1985).
33. H.C. Bryant, M.S. Gulley, P.B. Keating, E.P. MacKerrow, W.A. Miller, D.C. Rislove, S.Cohen, J.B. Donahue, D.H. Fitzgerald, S.C. Frankle, D.J. Funk, R.L. Hutson, R.J. Macek, M.A. Plum, N.G. Stanciu, O.B. van Dyck, C.A. Wilkinson, C.W. Planner, Symposium on Atomic, Cluster, and Surface Physics (SACSP), Hintermoos/Maria Alm, Austria, March 20-26, 1994, Contributions pp. 124, 126.
34. P.B. Keating et al., to be published in Phys. Rev. A.
35. P.B. Keating, M.S. Gulley, H.C. Bryant, E.P. MacKerrow, W.A. Miller, D.C. Rislove, S.Cohen, J.B. Donahue, D.H. Fitzgerald, D.J. Funk, S.C. Frankle, R.L. Hutson, R.J. Macek, M.A. Plum, N.G. Stanciu, O.B. van Dyck, and C.A. Wilkinson, Phys. Rev. A **52** Dec. (1995)

36. A.J. Jason, D.W. Hudgings, and O.B. van Dyck, IEEE Trans. on Nuc. Sci., **NS-28**, 2704 (1981).
37. G.M. Stinson, W.C. Olsen, W.J. McDonald, P. Ford, D. Axen, and E.W. Blackmore, Nucl. Instr. Meth. **74**, 333 (1969). Data reported in this paper were included in the fit of Jason et al. (1981).
38. L.R. Scherk, Can. J. Phys. **57**, 558 (1979).
39. R.L. Hutson, Los Alamos National Laboratory Internal Report, **PSR-91-001**, Jan. 4, 1990.
40. U. Schramm, J. Berger, M. Grieser, D. Habs, E. Jaeschke, G. Kilgus, D. Schwalm, A. Wolf, R. Neumann, R. Schuch, Phys. Rev. Lett. **67**, 22 (1991)
41. F. B. Yousif. P. Van der Donk, Z. Kucherovsky, J. Reis, E. Brannen, J. B. A. Mitchell, T. J. Morgan, Phys. Rev. Lett. **67**, 26 (1991).
42. Andrew Watson, "Physicists Produce First Antiatom", Science **271**,147 (1996).
43. G. Baur, G. Boero, S. Brauksiepe, A. Bruzzo, W. Eyrich, R. Geyer, D. Grzonka, J. Hauffe, K. Kilian, M. LoVetere, M. Macri, M. Moosburger, R. Nellen, W. Oelert, S. Passaggio, A. Pozzo, K. Roehrich, K. Sachs, G. Schepers, T. Sefzick, R.S. Simon, R. Stratmann, F. Stinzing, M. Wolke, "Production of Antihydrogen", Phys. Lett. B, to appear.
44. Charles T. Munger, Stanley J. Brodsky, and Ivan Schmidt, "Production of relativistic antihydrogen atoms by pair production with positron capture and measurement of the Lamb Shift", Hyperfine Interactions **76**, 175 (1993). See also Charles T. Munger, Stanley J. Brodsky, and Ivan Schmidt,"Production of relativistic antihydrogen atoms by pair production with positron capture", Phys. Rev. D **49**3228 (1994).
45. J. Poirier, "A New Candidate for a Long-lived Neutral Particle at VHE/UHE from Stellar Sources?", 23rd International Cosmic Ray Conference, Calgary, July 19-30, 1993. Volume 4 Contributed Papers, pp 633-640.
46. Laser techniques for collimating and focusing neutral atomic beams with resonance radiation are discussed by V. S. Letokhov, Comments At. Mol. Phys. **19**, 119 (1987)
47. H. C. Bryant, B. D. Dieterle, J. B. Donahue, H. Sharifian, H. Tootoonch, D. M. Wolfe, P. A. M. Gram and M. A. Yates-Williams, Phys. Rev. Lett. **38**, 228 (1977).
48. "Effects of strong electric fields on resonant structures in H- photodetachment," H. C. Bryant, D. A. Clark, K. B. Butterfield, C. A. Frost, H. Sharifian, H. Tootoonchi, J. B. Donahue, P. A. M. Gram, M. E. Hamm, R. W. Hamm, J. C. Pratt, M. A. Yates, W. W. Smith. Phys. Rev. A **27**, 2889-2912 (1983).
49. D. W. MacArthur, K. B. Butterfield, D. A. Clark, J. B. Donahue, P. A. M. Gram, H. C. Bryant, C. J. Harvey, W. W. Smith, G. Comtet, Phys Rev. A **32**, 1921 (1985)
50. J. Callaway, Phys. Lett. **68A**,315 (1978).
51. J.T. Broad and W. P. Reinhardt, Phys. Rev. A **14**, 2159 (1976).
52. A. Stintz, Xin Miao Zhao, C.E.M. Strauss, W.B. Ingalls, G.A. Kyrala, D.J. Funk, H. C. Bryant, Phys. Rev. Lett.**75**, 2924 (1995)

53. D. B. Holtkamp and C. R. Quick, Nucl.Instrum. Meth.**A287**,348-352 (1990).
54. E.P. MacKerrow, O.B. van Dyck, report to Los Alamos Workshop on Prompt Radiation Safety, 1993.
55. E.P. MacKerrow, PhD Dissertation, University of New Mexico Department of Physics and Astronomy, 1995.
56. C.J. Johnstone, private communication, 1994.
57. S.L. Kramer, V. Stipp, C. Krieger, and J. Madsen, "High Efficiency Beam Splitting for H^- Accelerators," 1985 Particle Accelerator Conference, Vancouver, BC, Canada, May 13-16, 1985. (ANL-HEP-CP-85-31)
58. R. K. Sander, F. Archeleta, W. Harker, R. Martinez, C. R. Quick, R. Reeder, D. J. Taylor, F. Brumwell, C. Hummer, Rev. Sci. Instrum. **62**, 1893 (1991).
59. David J. Pegg, "Improvements in Neutral Particle Beam Sensing", unpublished report, **AFWL-TR-85-59**, February 1986, Air Force Weapons Laboratory, Air Force Systems Command, Kirland AFB, NM 87117-6008
60. G. Rohringer, "Particle Beam Diagnostics by Resonant Scattering", unpublished report, CR-1-783, DRC 78-35362 (1977), General Research Corp., Huntsville, Al.
61. W.H. McCulla, "Wire Shadow Beam Sensing Development at LANL", Presented to 1989 Neutral Particle Beam Technology Symposium and Scientific Interchange, Monterey, CA, July 17-21, 1989. J. A. Sullivan, W. H. McCulla, and J. D. Schneider, "Use of Wire Shadows for Accelerator Beam Diagnostics," Presented at the APS Spring Meeting, Baltimore, MD, April 18-21, 1988, and at the 1988 IEEE Particle Accelerator Conference, Capri, Italy, October 19-26, 1988.
62. Report of the NPB Beam Sensing and Control Panel, David C. Straw, Chairman, **WJSA-89-A-006**, May 17-18, 1989, W.J. Schafer Associates, Inc., Albuquerque NM 87106.
63. "Laser Diagnostics for H^- Beam Momentum and Momentum Spread," E. P. MacKerrow, S. Cohen, J. B. Donahue, D. R. Swenson, H. C. Bryant. Beam Instrumentation Workshop, Oct. 20-23 1993, Picacho Plaza Hotel, Santa Fe, NM. Also see Book of Abstracts, LANL report **LA-UR-93-3600**. Full paper published in AIP Conference Proceedings 319, Beam Instrumentation Workshop, Santa Fe, NM, October 1993, Robert E. Shafer, Editor (AIP, New York, 1994) 226-235.
    See also: "Non-invasive Diagnostics for H^- Ion Beams Using Photodetachment by a Focused Laser Beam," D. R. Swenson, E. P. MacKerrow, H. C. Bryant. Beam Instrumentation Workshop, Oct. 20-23 1993. Picacho Plaza Hotel, Santa Fe, NM. Also see Book of Abstracts, LANL report LA-UR-93-3600. (Full paper published in AIP Conference Proceedings, **319**, Op. Cit. pp. 343-352.)
64. "Measurement and Reduction of Momentum Spread in the LAMPF LINAC Beam," P. G. Harris, H. C.Bryant, A. H. Mohagheghi, R. A. Reeder, C. Y. Tang, J. B. Donahue, C. R. Quick, D. C. Rislove, H. Sharifian, H. Tootoonchi, T. C. Altman. Nucl. Instrum. Methods A **292**, 254-258 (1990).

CHAPTER 7

Neutral Beam–Atom Collisions Resulting in Electron Cusps

Dénes Berényi*

After giving a short survey of the cusp phenomenon in the electron spectrum emitted in ion–atom collisions, in general and in the case of neutral projectiles, different aspects of the pertaining experiments are treated, namely the production of the neutral beam, the realization of the beam–target interaction zone, and techniques for the analysis of electrons and ions from the collision, including coincidence arrangements and the data analysis. Finally, experimental results on the cusp for neutral projectiles, obtained by the above techniques, are discussed, and the future prospects of the field are summarized.

7.1 INTRODUCTION

7.1.1 On the cusp phenomenon in general

It has been known for more than two decades that there is a cusp-like formation (see Fig. 7.1) in the spectrum of electrons emitted in the forward direction at an energy corresponding to the projectile velocity in ion–atom collisions. The process is described by a projectile-centered continuum wave function. In the expression for the cross section of forward-ejected electrons, the difference of the velocity of the electron and the outgoing ion is in the denominator. Thus if the velocity of the electron is equal to that of the ion, a cusp emerges in the spectrum (see an older detailed survey in Ref. [1], some more recent surveys in Ref. [2–4], and some earlier papers: Ref. [5–8]).

As is well known, the cusp is produced mainly by two processes, namely electron capture to the continuum (ECC) and the electron loss to the continuum (ELC; this occurs only if the projectile has accompanying electron(s)).

*) Institute of Nuclear Research of the Hung. Acad. Sci. (ATOMKI), Debrecen, Hungary

There are, however, also other two-(multi-) electron mechanisms, such as e.g. transfer ionization (TI) or double electron loss in the continuum (DELC) (cf. e.g. Ref. [9–12]).

During recent years a number of interesting new results have been published on the cusp phenomenon. A detailed survey of all of these is not the intention of this chapter; only some results will be mentioned here.

The relative contribution of the different mechanisms (ECC, ELC, TI) to the production of the cusp as a function of the impact energy was studied by Kövér *et al.* [9] and Závodszky *et al.* [11]. Various studies were carried out on the shape of the cusp [12–16]. Jagutski *et al.* investigated the impact-parameter-dependence of the phenomenon [17]. A number of papers were published on the cusp in the case of solid targets (some of them e.g. in Ref. [18–20,43]).

Without any doubt one of the most interesting recent results in this field is the discovery of an ECC cusp with neutral projectiles [23], which is the subject of this paper. However, it should be emphasized that after more than two decades of research, having a lot of interesting results (including the recent interesting publications) and many different theoretical approaches as (cf. e.g. the paper of Jakubassa-Amundsen [2] and the references within), there is no consistent theoretical interpretation yet that describes and explains the measured data on the cusp (shape, cross section, width etc.), even though the phenomenon is interesting from different fundamental points of view (three body problem, screening, short and long range forces etc.).

7.1.2 Short history of cusp studies by neutral projectiles

The first cusp measurement in which neutral projectiles were used is that of Meckbach *et al.* [21] published in 1984. Here the double-differential cross sections of electrons emitted in H^0(105 keV) $\rightarrow$ He collisions were studied in a three-dimensional measurement. In the same year the cusp ("convoy electrons") was also observed for H^0 impact on C foils [22]. To follow the next important steps sketchily: undoubtedly, one of the most surprising results—as was mentioned above—was the discovery of an ECC cusp in the case of the neutral He^0 projectile in 1989 [23]. It was shown that the intensity of the cusp increased in the case of excited metastable neutral He projectiles [24].

Further details of the work on the cusp in general will not be given here. However, results on the cusp for neutral projectiles will be treated a bit more thoroughly in Section 7.9.

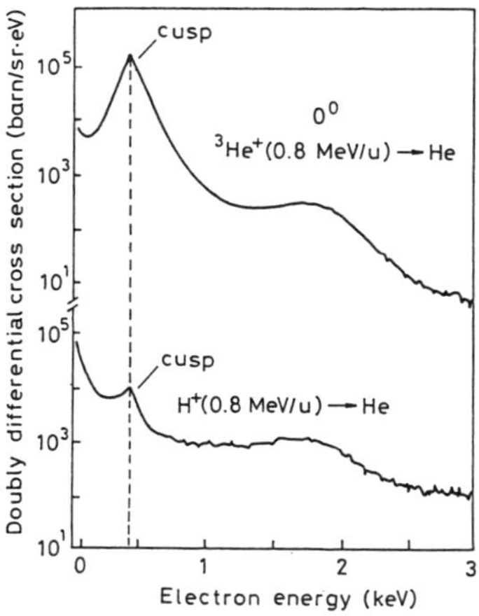

FIGURE 7.1 Typical spectra of electrons of continuous energy distribution with the "cusp" in the case of H^+, $^3He^+ \to He$ collisions (on the basis of Fig. 1 in Ref. [14]).

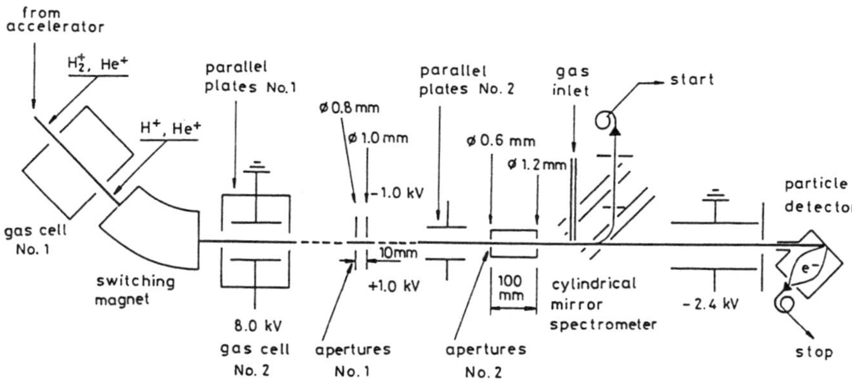

FIGURE 7.2 Arrangement in an experiment to study the cusp phenomenon at neutral projectile impact (based on Fig. 1 in Ref. [25]).

7.2 PRODUCTION OF THE BEAM OF NEUTRAL PROJECTILES

7.2.1 Neutral beam production in accelerators

In all experiments with neutral beams, small charged-particle accelerators were used. These include a Cockcroft–Walton generator [21], a pelletron

[22,27,29], and Van de Graaff accelerators of lower energy, namely of 1.5 [23], 2.5 [25,31], 7 MeV [26]. In these accelerators were produced primarily H^+ or He^+ ions in the energy range 0.1–3 MeV. Typical diameters of the momentum-analyzed ion beam were, e.g., 0.8 mm with 0.3° angular spread (Ref. [28]), and 0.5 mm and less than 0.1° angular spread (Ref. [34]).

The neutralization of the charged particle beam (H^+, He^+) was carried out by two different methods. One was the neutralization by electron capture in a gas cell (filled, e.g., with He) which was pumped differentially [21,23–25]. In the other case a neutralizer foil (carbon) was used with a foil-width of typically $\sim$10 μg/cm^2 [22,27–29]. The first method was usually used in the region of several hundred keV/u or below, the second at somewhat higher impact energies ($\sim$1 MeV/u and above). According to Yamazaki and Oda [27] a H^0 beam of 100–200 pA/u was obtained at 1–10 μA/u beam intensity for H^+ and H^{2+}.

In some cases, however, H^0 was produced by using a primary H_2^+ beam. These molecular ions were dissociated in a gas cell and the H^0 component was electrostatically selected [25,26]. In Ref. [27] a stripper foil was used to produce H^0 out of the H_2^+ beam.

To obtain a 200 keV H^0 beam in an accelerator having a minimum operating voltage of 250 keV, Trabold *et al.* [25] had the 400 keV H_2^+ beam dissociated in a gas cell. Then the H^+ beam, having half the energy of the original molecular ion beam, passed through another gas cell to produce a H^0 atom from of H^+ by electron capture in the gas. In both cells the pressure was roughly 10^{-5} mbar, more than an order of magnitude higher than in the beamline.

7.2.2 Cleaning and forming the neutral beam

The charged particles were separated electrostatically by a pair of deflector plates [21]. In the case of H^0 the deflector field eliminated the metastable component, too, by means of the Stark effect [22,27]. In the paper of Kroneberger *et al.* [26] it is mentioned that the charged-particle content of the H^0 beam is about 1%. As they say, it is probably produced by the entrance opening of the target chamber.

The details of cleaning and forming the neutral beams are reported in two papers, by Kuzel *et al.* [24] and Trabold *et al.* [25]. The general arrangement, with some actual values for sizes, distances, voltages etc., is shown in Fig. 7.2. There are some differences in the arrangement for H^0 and He^0. For electrostatic separation and quenching in the case of H^0, two sets of parallel-plate deflectors are in the line $\sim$1.5 m apart. The voltage of the first and the second sets was 8 kV and $\pm$1 kV, respectively. The applied voltage on the plates was enough to produce a H^0 beam consisting of ground-state atoms. As can be seen in Fig. 7.2, to define the neutral beam in this case two pairs of apertures were applied. The actual size of the neutral beam—achieved by using also a micrometrically movable slit system—was finally 1 mm^2, with an angular divergence $< 0.1^\circ$ and an intensity of less than 100 counts per second. The diameter of the neutral beam in the target region was less than 0.7 mm (Ref. [37]).

The above arrangement was not sufficient to achieve a neutral He^0 beam with ground-state atoms. In this case another gas cell (a quenching cell) was inserted between the apertures No. 1 and parallel plates No. 2. In this second cell the pressure was varied in order to modify the long-lived metastable content of the beam [24]. Here, however, there was no need for cell No. 1 in Fig. 7.2, and the two cells concerned were filled with air.

Because of the higher cross section for electron loss in the case of the metastable atoms (σ_L^*), the fraction (f) of metastable atoms in the beam can be changed by varying the pressure in the quenching cell. According to Ref. [24] the ratio of the metastable component to the total intensity of the beam (f) is as follows.

$$f = \frac{N^*}{N^0 + N^*} = \frac{N_i^*}{N_i^0} e^{-\mu(\sigma^* - \sigma_L^0)} \left[1 + \frac{N_i^*}{N_i^0} e^{-\mu(\sigma_L^* - \sigma_L^0)} \right]^{-1} \tag{7.1}$$

Here, N_i^* and N^* are the number of metastable atoms in the beam before and after the quenching cell, and N_i^0 and N^0 are the number of ground state atoms. Furthermore, σ_L^0 is the cross section for electron loss in the case of ground-state He^0 atoms and μ is the thickness of the quenching gas. Without the quenching cell the metastable component in the beam is between 25–35% [24,25,32] while using the quenching cell 0 percent was attainable.

7.3 THE INTERACTION ZONE

7.3.1 Gaseous targets

As a gaseous target a gas jet of He or Ar atoms was used (effusion gas or atomic beam target) [21,23–25]. It is described in detail in the paper of Trabold *et al.* [25]. According to this paper the diameter of the bore of the needle which produced the jet was 1 mm (at Meckbach *et al.* [21] 0.25 mm) and it was mounted on a long tube (diameter 2.5 mm, length 250 mm). The distance between the end of the needle and the edge of the projectile beam was 0.5 mm. The intensity of the gas jet was automatically controlled by a pressure gauge of high sensitivity which was connected to a gas valve. The relative target density can be determined or monitored by detecting Rutherford backscattering (e.g. in Ref. [31]). To ensure the condition of single collision, the linearity of the yield as a function of the pressure was generally checked [9,23].

7.3.2 Foil targets

In this case thin carbon foils were used as targets (1.5–60 $\mu g/cm^2$ [22], 20 $\mu g/cm^2$ [26,28]). They were located in the focus of the electron analyzer.

In Ref. [28] a surface barrier detector at 10° with respect to the beam direction was used to monitor the thickness of the carbon foil by elastically scattered particles. For comparison a calibrated foil was used. According to experience the foil-thickness increased by 5 to 10 percent during 5 to 10 min runs because of carbon buildup [26].

7.4 ELECTRON ANALYZERS

In the study of the cusp phenomenon good resolution of the electron spectrometer was not claimed in general due to the nature of the process itself (in particular, the broad energy distribution of the electrons in the cusp). The large acceptance angle of the equipment is important, however, especially in coincidence measurements.

In the cusp studies with neutral beams partly a double-stage cylindrical mirror electron spectrometer [23,31,33], partly a special system of cylindrical mirror electron spectrometer was used [25,34] (cf. Ref. [38] on cylindrical mirror analyzers in general). These spectrometers will be reported later in more detail.

In other cases the measurements were carried out using a conventional cylindrical mirror spectrometer [21,35], a magnetic sector electron analyzer [28] and a 30° parallel-plate electrostatic spectrometer [22,27,29]. In the first case the relative resolution was 0.08% and the half-angle of the acceptance cone, $\theta = 0.5°$. The magnetic spectrometer had a resolution of $(0.7\pm1)\%$ with an effective solid (acceptance) angle $(1.0 \pm 0.3) \times 10^{-5}$ sr. For the parallel-plate analyzer these figures were 1.4%, and 0.37° and 2.4° for the direction parallel and perpendicular to the rotation plane of the analyzer.

7.4.1 Cylindrical mirror electron analyzer by which the spectrum of emitted electrons can be measured from −180° to +180° relative to the beam direction [24,25,30,34,37,39]

The cylindrical mirror spectrometer can be rotated around a vertical axis (this axis crosses the beam at the point where the beam and the spectrometer axis cross each other). The angle between the spectrometer axis and the beam direction was 42.3°. The energy resolution was 3.14% and the angular acceptance 1.67° in the actual measurements (Ref. [25]) but they can be changed in a rather broad range (from 0.25 to 2.5°). The equipment is shown schematically in Fig. 7.3.

7.4.2 Distorted-field double-pass cylindrical mirror electron spectrometer [9,23,31,36]

A slit on the inner cylinder in the analyzer can be turned to set the emission angle of electrons to be measured. The resolution and acceptance angle values can be varied here in rather wide ranges. Actual values were 0.5% and 3.5° [23], 0.5% and 1.1° [31], 0.4% and 1.9° [33]. The sketch of the analyzer can be seen in Fig. 7.4.

7.4.3 Detection of electrons and efficiency calibration

The electrons at the focus of the analyzer were detected by a channeltron electron multiplier in all published measurements (see the references above). The channeltron pulses were counted in a multiscaler with standard techniques [27,29]. A voltage (+100 V in Ref. [34]) was maintained on the funnel of the channeltron to set the electron detection at a suitable range of the efficiency curve. In front of the channeltron, however, grid(s) was (were) placed for reducing the stray electrons (e.g. in Ref. [25,34]). Finally, the spectrometer was controlled by a microprocessor based electronic system [36].

The efficiency calibration of the analyzer (including that of the channeltron) was carried out by using the Rudd *et al.* values [40,41] for the cross section of the emitted electrons of the continuous energy distribution in H^+ – He or H^+ – Ar collisions [28,30,37]. The normalization carried out at different angles was extrapolated to 0° with the help of the data of Rudd *et al.* (as in the case of Ref. [42]). (In papers by Rudd *et al.* there are no data for 0°; the smallest angle they measured was 10°.)

7.5 ION AND ATOM DETECTION AND SEPARATION

7.5.1 Singles measurements

To detect neutral projectiles such as H^0 and He^0 which traversed the target, a thin carbon foil [39] (~20 $\mu g/cm^2$) was used to strip the neutral atoms. The ions were then detected in a biased Faraday cup. Knowing the stripping cross sections [44], the intensity of the beam of neutral projectiles can be determined. Singles measurements [21,24,25] were carried out by presetting a value of beam charge in the Faraday cup (see e.g. in Ref. [35]).

7.5.2 Coincidence measurements

In coincidence measurements another solution was realized. Here, first the outgoing projectiles were selected by an electric field, and a particle detector was moved to the different locations to detect the various ions (H^0, H^+; He^0,

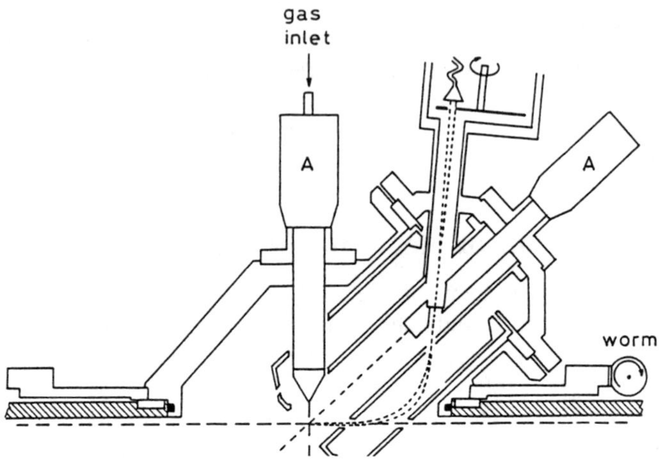

FIGURE 7.3 Schematic diagram of a cylindrical mirror electron spectrometer by which the electron energy distribution can be measured over the full angular range from $-180°$ to $+180°$. Reprinted from Bernardi, G., Suarez, S., Focke, P. and Meckbach, W., "Double differential electron distributions induced by 170 keV H^+ to He, H^0 to He and H^+-solid-foil interactions" [34], with kind permission of Elsevier Science - NL, Sara Burgerhartstraat 25, 1055 KV Amsterdam, The Netherlands.

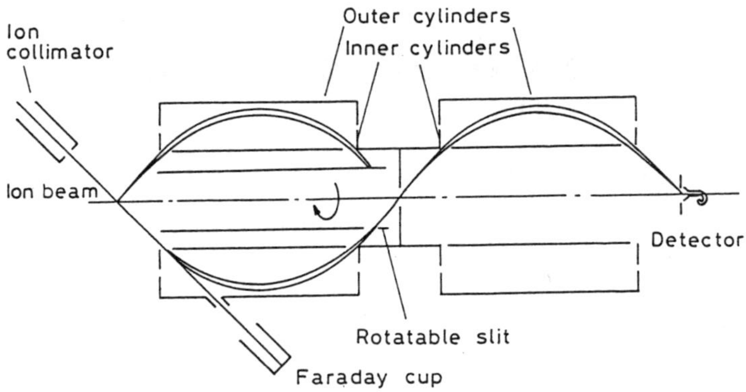

FIGURE 7.4 Sketch of a distorted-field double-pass cylindrical mirror electrostatic electron spectrometer Reprinted from Kover, A., Szabo, Gy., Berenyi, D., Gulyas, L., Cserny, I., Groeneveld, K.O., Hofmann, D., Koschar, P., and Brukhard, M., "Forward electron ejection in H^+-He, He^+-He and He^{2+}-He collisions" [36], with kind permission of IOP Publishing Limited.

He^+, He^{2+}) [23–25,37–39]. The detector used was made according to Rinn [45]. Here the ions to be detected enter an Al cylinder through a hole. Then the secondary electrons produced inside the cylinder are detected by a channel electron multiplier on one end of the cylinder [9]. Using this detection method there was no need for a stripping foil.

The detection efficiency in the coincidence studies was determined as follows: the sum of cusp electron intensities in the coincidence spectra was compared to its intensity in the singles spectrum [9,23]. In Ref. [9], e.g. a 58±7% efficiency was determined in this way. In an other work this value was 50±3% [24]. In Ref. [25], however, the normalization was carried out on the basis of the detected number of ions (atoms) in the different channels (corresponding to the different charge states) [24,25]. In this latter case, however, the relative efficiency of the detector must be checked. This was done by means of an electric pulse generator [25,39] from which statistically distributed signals were obtained. Both the number of the pulses and that of the outgoing projectiles were detected within the same time-window. By comparing the two spectra the relative efficiency was monitored during the measurements. For further detail see Ref. [25]. For absolute normalization even in these cases the former method (comparison of cusp intensities) was used.

7.6 COINCIDENCE ARRANGEMENTS

In recent measurements the cusp electron spectra have been studied in coincidence with the outgoing projectiles. Thus, it became possible to study the cusp stemming from different mechanisms (ECC, ELC) in the same process. The first such measurement was published by Kövér *et al.* [9] in 1989, and the first with neutral projectiles by Sarkadi *et al.* [23] in the same year. Later, however, coincidence measurements were carried out not only at 0° where the cusp can be observed [24,25,33] but at other angles where the cusp is not visible in the spectrum of electrons [37,39].

In all coincidence measurements practically the same equipment and procedures were used [9,23–25,33]. First the signals from the electron and particle detectors were processed by standard electronics and computer techniques. The time window in Ref. [24], e.g., was 400 nsec, and the time resolution was 15 nsec.

Random background coincidences were detected by a proper setting of the window in the time spectrum, and these were subtracted from the electron spectrum [25]. In Ref. [23], however, the coincidence counts were also corrected for loss because of the high count rate in the particle channels.

7.7 CORRECTIONS, ELIMINATION OF DISTURBING FACTORS AND UNCERTAINTIES

7.7.1 Some further corrections not mentioned in the previous sections

A number of different corrections have been already described in the previous sections: e.g. the correction for overall energy-dependent efficiency in Section 7.4.3 (including the unknown target intensity, the efficiency of the electron detector with the transmission function of the spectrometer) and that for chance coincidences, just above. Some other corrections were also applied. These are mentioned in the following paragraphs.

If we measure the electron spectrum without a target, the background due to residual gas is determined. In Ref. [9] it was below one percent. In this way, the dark current of the channeltron is also checked. A detailed description of the measurements of the residual gas contribution and the corresponding values can be found in Ref. [25].

7.7.2 Elimination of some disturbing factors

One of the disturbing factors to be eliminated is the magnetic field from the environment (Earth!) in the spectrometer region. In Ref. [34], e.g. a large, conveniently oriented single coil was used for that purpose. In this way the value of the field was reduced to less than 7 mG. In another case the field was reduced to 3% of its original value by a set of coils [25]. Another possibility to reduce this field is to apply soft magnetic material around the spectrometer region [9,36].

Another experimental problem to be solved is setting the spectrometer to 0° electron observation angle, relative to the direction of the projectile. In Ref. [25] and [31] this was done by setting different angles around the supposed 0° in both directions and looking for the maximum intensity of the cusp. The accuracy of this set was ±0.25° in Ref. [25].

The resolution correction is generally included in the overall efficiency correction (cf. Section 7.4.3) (cf. Ref. [23]).

7.7.3 Uncertainties

In the measurements concerned the absolute uncertainties are rather high [9,24,25,33,34,37,39]. The main source of uncertainty is the overall-efficiency calibration (see in Section 3.4.3), but there are contributions, sometimes substantial ones, from the statistical errors (these are especially high in some coincidence measurements), from the instability of the neutral projectile beam and that of the gas jet target, from background subtraction, and from the percentage of the metastable component in the case of the He^0 projectile beam.

The actual value of the absolute uncertainties was in the interval between 20 and 50%, mainly around 30–40%, but even higher values were found. The relative uncertainties were roughly half of the above values. In addition, in contrast to the really high values of uncertainties (even the relative ones), conclusions can still be drawn because the cross sections themselves varied dramatically [25].

7.8 DATA ANALYSIS

On the procedure of the analysis of the data there is not much written in the papers reporting on cusp studies using neutral projectile beams. The probable reason is that relative comparisons are performed among different cusp spectra in the majority of cases. The different corrections treated in the previous sections (as e.g. the overall-efficiency calibration in Section 7.4.3) can be regarded as the analysis of the measured data, too.

There are, however, some cases where further analysis of the data was done. One of them is the three-dimensional study

of the cusp phenomenon (i.e. the measurement of the electron yield as a function of both energy and detection angle around 0°) [21,31,34], where the so-called contour lines taken at different levels of the cusp are used to investigate the cusp [21,31]. In one of these papers [31] a separate subsection is devoted to the evaluation procedure (cf. also in Ref. [46]). Here according to Ref. [35] the shape of the cusp was characterized by the coefficients in the series expansion of the differential cross section. To determine these B_{ij} coefficients from the measured spectra, the spectrometer transmission function was separated according to the velocity and the angle

$$S(v_e, \Omega) = F(v_e)G(\theta)H(\phi) \tag{7.2}$$

where θ is the electron ejection angle relative to the direction of the projectile and ϕ is the corresponding azimuth angle. Approximating $F(v_e)$ by a Gaussian function and giving $G(\theta)$ and $H(\phi)$ on the basis of the actual geometry, B_{ij} can be determined from the formulae

$$Q(v_e, \theta) = \sum_{n,j=0} B_{nj}(v_p)U_{nj}(v_e, \theta) \tag{7.3}$$

and

$$U_{nj}(v_e, \theta) = \int (v'_e)^{n-1} P_j(\cos\theta')S(v_e, \Omega)dv_e \tag{7.4}$$

(v_e is the electron and v_p the projectile velocity, v'_e is the same for the electron in the projectile frame, P_j is the Legendre polynomial) by fitting (including the numerical integration of the formula of series expansion) the measured and calculated three-dimensional electron distribution by means of a least-squares fitting routine (see details in Ref. [31] and [46]).

The determination of the metastable-to-ground-state ratio in the case of the He^0 beam also should be mentioned here. In this case the singly differential cross sections were obtained as a function of the quenching gas pressure (see Section 3.22 and Fig. 7.2) by integrating the cusp yield in singles and coincidence measurements. By applying a fitting procedure (see the details in Refs. 24 and 47) not only the fraction of the metastable atoms in the beams but also the metastable-to-ground-state ratio for the different cusp production mechanisms could be determined.

7.9 RESULTS ON THE CUSP IN THE SPECTRUM OF ELECTRONS EMITTED IN NEUTRAL PROJECTILE–ATOM COLLISIONS

First the results obtained using a non-coincidence arrangement, then those obtained in coincidence measurements, will be shown. This is a classification according to instrumentation. However, there is a phenomenon-oriented classification behind it, because the coincidence arrangement makes it possible to differentiate among the different mechanisms contributing to the cusp in the same process.

7.9.1 Non-coincidence measurements

Here the studies with gas targets and those with foils will be treated in two different subsections.

Gas targets

One of the main characteristics of the cusp is its shape. Because the cusp in fact is a three-dimensional phenomenon (yield as a function of electron energy and emission angle around 0° relative to the projectile direction; see Fig. 7.5) one can study the cusp shape by means of contour lines taken at different levels of the cusp.

There are two studies with neutral projectiles carried out by the above method. One of them is the first cusp study using neutral projectiles by Meckbach *et al.* [21] and the other was published by Gulyás *et al.* [31]. In both cases H^0 projectiles were used. In the latter one, however, the results were also obtained with those for He^+ and He^{2+} projectiles.

The results for the cusp-shape in the case of H^0(105 keV) $\rightarrow$ He are rather of qualitative character. However, it can be stated that the distribution of electrons is not spherically symmetric. In addition, it was shown that the observed asymmetry cannot be explained by the calculations of Day and Briggs

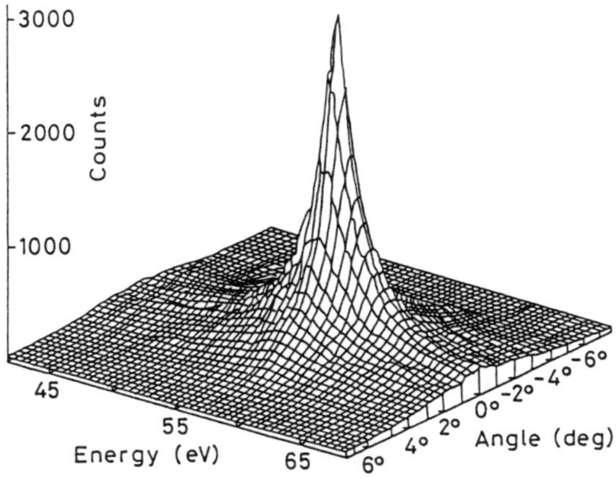

FIGURE 7.5 Three-dimensional cusp in the case of H^0(105 keV) → He (Ref. [21]).

based on a first Born-perturbation treatment [48,49]. It was supposed that higher terms should be included in the series expansion of the cross section.

In Ref. [31] the three-dimensional shape of the cusp was studied by means of the contour lines in the case of H^0(0.2 MeV/u) → Ne, Ar, Kr collisions. Here the experimental patterns were compared with the corresponding theoretical ones. The calculations were carried out in first-order Born approximation, taking into consideration the modifications by Hartley and Walters [50,51]. The patterns of the experimental and theoretical contour lines, however, showed no agreement at all (see e.g. Fig. 7.6). The disagreements were qualitatively explained by an ECC contribution of the ECC mechanism even in the case of the H^0 projectile.

The work of Kuzel *et al.* [30] should also be mentioned here. In this case, the goal of the whole study was not the investigation of the cusp. The electron loss phenomenon was studied (the differential cross section and other parameters) as a function of the angle of observation (including 0°) for H^0 projectile (0.5 MeV) and Kr target.

Foils

In all the studies published up to now, carbon foils were used as targets [22,28,43]. The target was bombarded in each case by H^0 projectiles (in the same studies charged particles were also used as projectiles) in the energy range from 0.9 to 3 MeV. The measurements were carried out at different angles (from 0° to 30°, using foils of different thickness (1.5–60 mg/cm^2). Experiments show that the shape of the cusp (here called "convoy electrons") depends on the target thickness (in the case of intermediate thickness from about 3 to 7 mg/cm^2; it consists of two components: a sharp cusp and a

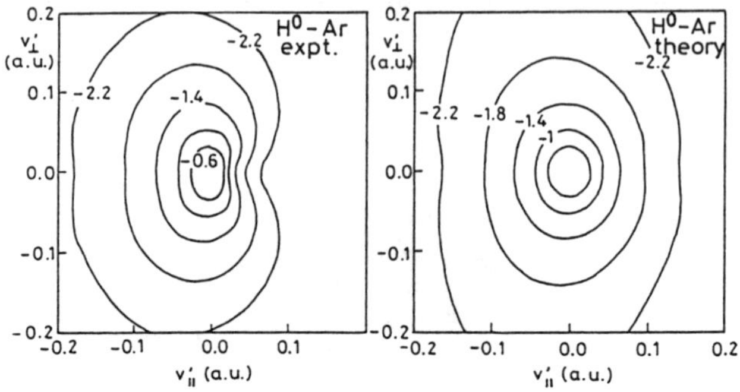

FIGURE 7.6 Deconvoluted experimental and the corresponding theoretical contour lines of the three-dimensional cusp for H^0(0.2 MeV/u) $\rightarrow$ Ar (Ref. [31]). The numbers at the contour lines indicate the logarithmic ratio of the given and the top level. On the x-axis the parallel and on the y-axis the perpendicular component of the projectile velocity relative to the direction of the beam is indicated.

broad shoulder towards lower velocities) and that there is a definite difference in the cusp shape depending on the projectile charge (H^0 or H^+). Fig. 7.7 shows the yield of convoy electrons as a function of foil thickness for H^0 and H^+. The difference was no longer observable when the ion charge came into equilibrium.

Barrachina *et al.* [52] was able to explain theoretically the observed shapes based only on the ELC (electron loss in the continuum) mechanism. The broader distribution of the convoy electron distribution stems from the direct electron loss process (at H^0) where these electrons are subjected to elastic and inelastic collisions. The sharp peak (cusp) in the convoy electron distribution is caused by the ELC process in the downstream surface of the foil. Furthermore it does not depend on the foil thickness because we are in the charge equilibrium region. Here indirect electron capture into the continuum prevails, namely first H^0 is produced out of H^+ by electron capture.

7.9.2 Coincidence measurements

By measuring the electron cusp spectrum in coincidence with the outgoing projectiles, the components of the cusp from the different mechanisms can be studied. The first such study with He^+ projectiles was carried out by Kövér *et al.* [9]. The surprising result, however, was obtained by Sarkadi *et al.* [23] when they found an ECC cusp, i.e. a cusp in the case of He^0(300 keV) $\rightarrow$ He, Ar collisions, while measuring the electrons in the cusp region in coincidence with the outgoing He^0 projectile. This process seems to be a capture of an electron into continuum states of the neutral He^0. Here the Coulomb field of

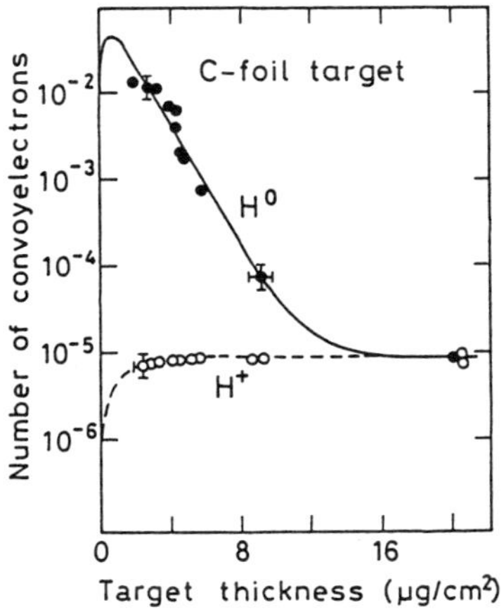

FIGURE 7.7 Convoy electron yield at 3 MeV H^0 and H^+ impact on C foil.

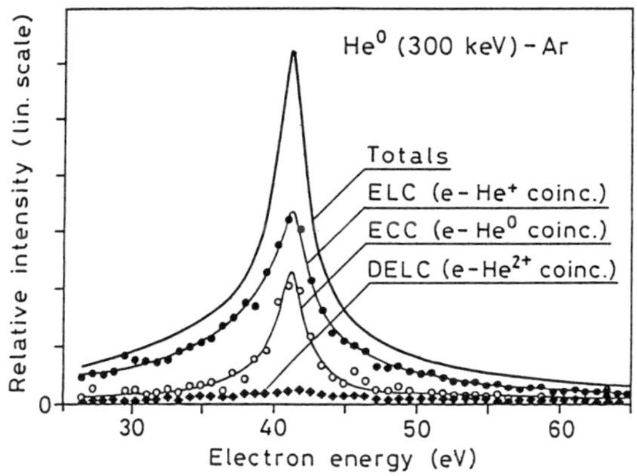

FIGURE 7.8 Cusps produced by different mechanisms for neutral He impact (from Ref. [23]).

the outgoing projectile is completely screened, so only interactions of short range are possible. The results of Sarkadi *et al.*, however, were confirmed by Trabold *et al.* [25] with 800 keV He^0. There is some disagreement between the two measurements on the width of the cusp.

In Fig. 7.8 the totals (non-coincidence spectrum) and the different components are shown for Ar targets. Here the shape of the ECC cusps is approximately symmetric and their width is narrower than that of ELC and ECC for He^+. For He targets the intensity of the ECC component exceeds ELC.

To explain the above surprising results a number of theoretical works were soon published. Jakubassa-Amundsen treated the issue in the framework of the impulse approximation [53] but the results of her calculations are in contradiction with the experimental data on the width of the cusp (they predict broader width for neutral than for charged projectiles). Barrachina [54] performed calculations in the frame of an *ab initio* theoretical model for the process supposing a low-energy virtual state of the electron–projectile system. According to the calculation a narrow cusp is expected. To check the validity of this theoretical approach, further measurements are desirable. Szótér supposes in his short comment [55], that the role of intermediate, non-metastable, autodetaching, excited negative ion states of the residual target ion in the capture of an electron into the continuum of the neutral projectile [56]. In this way an approximate qualitative agreement with the experiment could be attained on the cusp-shape. Finally, the work of Salin (unpublished work [57]) should be mentioned here, in which he supposed a mechanism with exchange contribution to target ionization: projectile electron ejection and simultaneous target electron capture by the projectile. Such an approach, however, seems to give too low a yield.

Similar experiments with H^0 were also carried out, and the results obtained are similar [13,14,25]. Trying to interpret these results on cusp shape, the difficulties in the interpretation of the experimental findings were summarized in Ref. [16].

In the case of He^0 projectiles, some further studies were also carried out. First of all should be mentioned the measurements of the relative contribution of the different mechanisms to cusp production as a function of the projectile impact energy in the case of He^0 – Ar collisions using the well-known coincidence arrangement [33,58]. A similar measurement for He^0 projectiles was carried out for He^+ by Kövér *et al.* [9]. As can be seen in Fig. 7.9 the ELC mechanism prevails over nearly the whole investigated region, but decreases monotonically with decreasing impact energy in the same region as the DELC. The ECC mechanism, however, is present at every impact energy here, and it increases in the direction of lower energies. The crossing point of the ECC and ELC curves is at a lower energy in this case than in the case of $He^+ \rightarrow$Ar collisions (cf. Fig. 4 in Ref. [9]).

The absolute double-differential cross section (DDCS) for He^0 and He^+ impact on Ar was compared on the ECC component in the region from 18.75 to 250 keV [59]. The results show that the DDCS values are only 20–40% higher for He^+ than those for He^0 (40% at 250 keV).

An experiment for the study of the impact-parameter dependence of the ECC cusp for He^0(0.1 MeV/u) $\rightarrow$ Ar collision should be also mentioned here [60]. The customary experimental arrangement used in these measurements was supplemented by a position-sensitive detector by which the scattering angle of He^0 could be measured (see further details for the instrumentation in Ref. [17] and [61]). Contrary to the difficulties in the evaluation of the conversion of scattering angles and recoil energies into impact parameters in the case of neutral collision partners [62,63], it could be stated that the impact-parameter dependence of the electron emission in the cusp region and in the binary encounter region have a similar course and is nearly constant. They show the same production mechanism.

As was already mentioned, the He^0 beam can contain a certain excited metastable component. The role of this beam component in cusp production has been investigated [24,64,65]. Here the cross sections for ECC, ELC and DELC processes were determined as a function of the pressure of the quenching gas (see Section 3.2.2). Concerning the above processes the following metastable-to-ground-state cross section ratios were obtained: 3.15, 10.2, 1.45. This means that the ECC cross section is an order of magnitude higher for He^0 in the metastable state than in the ground state. Also the difference in cusp shape was shown between He^0 and He^{0*} impact [66]. These findings indicate the role of Barrachina interpretation (see above) of the presence of ECC cusps with neutral projectile impact.

7.10 FUTURE PROSPECTS

If it is true that the whole cusp phenomenon is not completely interpreted theoretically (cf. Section 3.1), it is even more so for cusps produced by neutral projectiles. For example there are no theoretical curves to compare with experimentally determined relative contributions of the different mechanisms in Fig. 7.9.

Similarly many further measurements are missing and needed here to clarify the processes that produce the cusp at neutral projectile impact. It would be important to measure with pure metastable and pure ground-state He^0 beams [24]. Such studies are in progress now [67].

There are no measurements in this field besides the H^0 and He^0 projectiles. Investigations here with a C^0 beam are badly needed (there is a study with C^{n+}, including C^0, but only at 15° and 20°, where the cusp practically disappears [68]). A similar study for H^0 impact, as is shown in Fig. 7.9 for He^0, will be also very useful because H^0 is the simplest neutral projectile and so—supposedly—the theoretical interpretation of the findings would be less difficult.

It seems quite clear that the problems of this field have not been solved yet, and the field itself has not been concluded.

Finally, it should be mentioned that, as for the results of ion–atom collisions in general, those for neutral projectiles are perhaps even more interesting from the point of view of applications in other fields (e.g. astronomy, biology), in particular in the realization of controlled thermonuclear processes.

ACKNOWLEDGMENTS

The author is indebted to the following copyright owners for permission to reproduce certain figures: North Holland Publ. Co., Amsterdam (Figs. 7.3

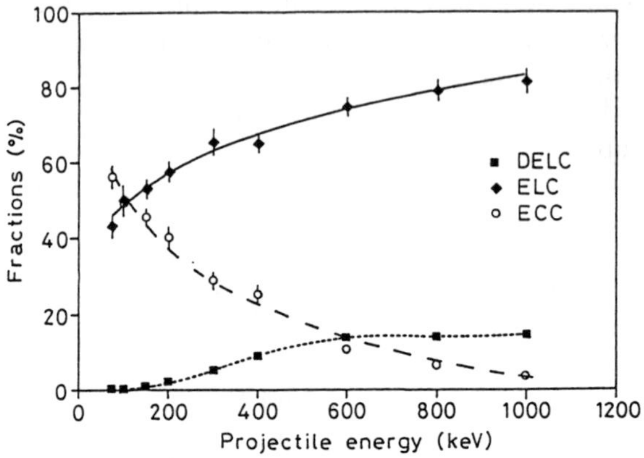

FIGURE 7.9 Relative distribution of the different mechanisms with neutral He projectiles as a function of the impact energy (from Ref. [33]).

and 7.7), Institute of Physics, Bristol, UK (Fig. 7.4); The American Physical Society, New York, USA (Figs. 7.5, 7.6, 7.8) and to the authors of the corresponding papers.

REFERENCES

1. Breining, M., Elston, S.B., Huldt, S., Liljeby, L., Vane, C.R., Berry, S.D., Glass, G.A., M. Schauer M., I.A. Sellin I.A., G.D. Alton G.D., Datz, S., Overbury, S., Lambert, R. and Suter M., *Phys. Rev.* **A25**, 3015 (1982).
2. Jakubassa-Amundsen, D.H., *Nucl. Instr. Meth.* **B86**, 82 (1994).
3. Hsu, C.C. and Yu, V.C., *Nucl. Instr. Meth.* **B86**, 90 (1994).
4. Garibotti, C.R. and Barrachina, R.O., *Nucl. Instr. Meth.* **B86**, 96 (1994).
5. Lucas, M.W. and Steckelmacher, W., in *High-Energy Ion–Atom Collisions*, ed. by Berényi, D. and Hock, G. (Springer Verlag, Berlin, 1988) p. 229.
6. Menendez, M.G., Huetz, A., Duncan, M.M., in *High-Energy Ion–Atom Collisions*, ed. by Berényi, D. and Hock, G. (Springer Verlag, Berlin, 1991) p. 24.
7. DuBois, R.D., in *High-Energy Ion–Atom Collisions*, ed. by Berényi, D. and Hock, G. (Springer Verlag, Berlin, 1991) p. 34.
8. Salin, A., *Nucl. Instr. Meth.* **B86**, 1 (1994).
9. Kövér, Á., Sarkadi, L., Pálinkás, J., Berényi, D., Szabó, Gy., Vajnai, T., Heil, O., Groeneveld, K.O., Gibbons, J. and Sellin, I.A., *J. Phys. B: At. Mol. Opt. Phys.* **22**, 1595 (1989).
10. Tanis, J.A., Schneider, D., Chantrenna, S., Prior, M.H., Hermann, R., Hutton, R. and G. Schiwietz, G., *Phys. Rev.* **A42**, 5776 (1990).
11. Závodszky, P.A., Sarkadi, L., Tanis, J.A., Berényi, D., Pálinkás, J., Plano, V.L., Gulyás, L., Takács, E. and Tóth, L., *Nucl. Instr. Meth.* **B79**, 67 (1993).

12. Plano, V.L., Haar, R.R., Tanis, J.A., Pálinkás, J., Sarkadi, L., Závodszky, P.A., Berényi, D., Khemliche, H., Prior, M.H., and Schneider, O., *Nucl. Instr. Meth.* **B86**, 181 (1994).
13. Kövér, Á., Sarkadi, L., Pálinkás, J., Gulyás, L., Szabó, Gy., Vajnai, T., Berényi, D., Heil, O., Groeneveld, K.O., Gibbons, J. and Sellin, I.A., *Nucl. Instr. Meth.* **B42**, 463 (1989).
14. Berényi, D., Sarkadi, L., Gulyás, L., Kövér, Á., Szabó, Gy. and Pálinkás, J., *Acta Phys. Hung.* **70**, 381 (1991).
15. Gulyás, L., Sarkadi, L., Pálinkás, J., Kövér, Á., Vajnai, T., Szabó, Gy., Végh, J., Berényi, D. and Elston, S.B., *Phys. Rev.* **A45**, 4535 (1992).
16. Berényi, D., Sarkadi, L., Kövér, Á. and Pálinkás, J., *Acta Phys. Hung.* **74**, 405 (1994).
17. Jagutzki, O., Koch, R., Skulartz, A., Kelbiech, C. and Schmidt-Böcking, H., *J. Phys. B: At. Mol. Phys.* **24**, 993 (1991).
18. Burgdörfer, J., in *High Energy Ion–Atom Collisions*, ed. by Berényi, D. and Hock, G. (Springer Verlag, Berlin, 1991) p. 193.
19. Kimura, K., Gibbons, J.P., Elston, S.B., Biedermann, C., DeSerio, R., Keller, N., Levin, J.C., Breinig, M., Burgdörfer, J. and Sellin, I.A., *Phys. Rev. Lett.* **66**, 25 (1991).
20. Kroneberger, K., Sigaud, G.M., Rothard, H., Heil, O., Albert, A., Maier, R., Schlösser, D., Schosnig, M., Trabold, H. and Groeneveld, K.O., *Nucl. Instr. Meth.* **B67**, 109 (1992).
21. Meckbach, W., Vidal, R., Focke, P., Nemirovsky, I.B. and Gonzales Lepera, E., *Phys. Rev. Lett.* **52**, 621 (1984).
22. Yamazaki, Y. and Oda, N., *Phys. Rev. Lett.* **52**, 29 (1984).
23. Sarkadi, L., Pálinkás, J., Kövér, Á., Berényi, D. and Vajnai, T., *Phys. Rev. Lett.* **62**, 527 (1989).
24. Kuzel, M., Sarkadi, L., Pálinkás, J., Závodszky, P.A., Maier, R., Berényi, D. and Groeneveld, K.O., *Phys. Rev.* **A48**, R1745 (1993).
25. Trabold, H., Sigaud, G.M., Jakubassa-Amundsen, D.H., Kuzel, M., Heil, O. and Groeneveld, K.O., *Phys. Rev.* **A46**, 1270 (1992).
26. Kroneberger, K., Clouvas, A., Schlösser, G., Koschar, P., Kemmler, J., Rothard, H., Biedermann, C., Heil, O., Brukhard, M. and K.O. Groeneveld, K.O., *Nucl. Instr. Meth.* **B29**, 621 (1988).
27. Yamazaki, Y. and Oda, N., *Nucl. Instr. Meth.* **194**, 323 (1982).
28. Koschar, P., Clouvas, A., Heil, O., Brukhard, M., Kemmler, J. and Groeneveld, K.O., *Nucl. Instr. Meth.* **B24/25**, 153 (1987).
29. Oda, N., Nishimura, F., Yamazaki, Y. and Tsurubushi, S., *Nucl. Instr. Meth.* **170**, 571 (1980).
30. Kuzel, M., Maier, R., Heil, O., Jakubassa-Amundsen, D.H., Lucas, M.W. and Groeneveld, K.O., *Phys. Rev. Lett.* **71**, 2879 (1993).
31. Gulyás, L., Szabó, Gy., Kövér, Á., Berényi, D., Heil, O. and Groeneveld, K.O., *Phys. Rev.* **A39**, 4414 (1989).
32. Horsdal-Pedersen, E., Heinemeier, J., Larsen, L. and Mikkelsen, J.V., *J. Phys.* **B13**, 1167 (1980).

33. Sarkadi, L., Závodszky, P.A., Kuzel, M., Maier, R., Pálinkás, J., Gulyás, L., Takács, E., Tóth, L., Berényi, D. and Groeneveld, K.O., *ATOMKI Annual Report 1992*. Debrecen, 1993. p. 47.
34. Bernardi, G., Suarez, S., Focke, P. and Meckbach, W., *Nucl. Instr. Meth.* **B33**, 321 (1988).
35. Meckbach, W., Nemirovsky, I.B. and Garibotti, C.R., *Phys. Rev.* **A24**, 1793 (1981).
36. Kövér, Á., Szabó, Gy., Berényi, D., Gulyás, L., Cserny, I., Groeneveld, K.O., Hofmann, D., Koschar, P. and Brukhard, M., *J. Phys. B: At. Mol. Phys.* **19**, 1187 (1986).
37. Kuzel, M., Heil, O., Maier, R., Lucas, M.W., Jakubassa-Amundsen, D.H., Farmery, B.W. and Groeneveld, K.O., *J. Phys. B: At. Mol. Opt. Phys.* **25**, 1839 (1992).
38. Risley, J.S., *Rev. Sci. Instr.* **43**, 95 (1972).
39. Heil, O., DuBois, R.D., Maier, R., Kuzel, M. and Groeneveld, K.O., *Z. Phys.* **D21**, 235 (1991).
40. Rudd, M.E., Toburen, L.H. and Stolterfoht, N., *At. Data Nucl. Data Tables* **18**, 413 (1976).
41. Rudd, M.E., Toburen, L.H. and Stolterfoht, N., *At. Data Nucl. Data Tables* **23**, 405 (1991).
42. Stolterfoht, N., Schneider, D. and Ziem, P., *Phys. Rev.* **A10**, 81 (1974).
43. Kemmler, J., Koschar, P., Heil, O., Biedermann, C., Rothard, H., Kroneberger, K., Lencinas, S., Groeneveld, K.O. and Sellin, I.A., *Nucl. Instr. Meth.* **B33**, 281 (1988).
44. Betz, H.D., *Rev. Mod. Phys.* **44**, 465 (1972).
45. Rinn, K., Müller, A., Eichenauer, H. and Salzborn, E., *Rev. Sci. Instr.* **53**, 829 (1982).
46. Gulyás, L., Kövér, Á., Szabó, Gy., Vajnai, T. and Berényi, D., in *High-Energy Ion–Atom Collisions*, ed. by Berényi, D. and Hock, G. (Springer Verlag, Berlin, 1991) p. 56.
47. Horsdal-Pedersen, E., Heinemeier, J., Larsen, L. and Mikkelsen, J.V., *J. Phys.* **B13**, 1167 (1980).
48. Day, M.H., *J. Phys.* **B13**, L65 (1980).
49. Briggs, J.S. and Day, M.H., *J. Phys.* **B13**, 4797 (1980).
50. Hartley, H.M. and Walters, H.R.J., *J. Phys.* **B20**, 1983 (1987).
51. Hartley, H.M. and Walters, H.R.J., *J. Phys.* **B20**, 3811 (1987).
52. Barrachina, R.O., Goni, A.R., Focke, P. and Meckbach, W., *Nucl. Instr. Meth.* **B33**, 330 (1988).
53. Jakubassa-Amundsen, D.H., *J. Phys. B: At. Mol. Opt. Phys.* **22**, 3989 (1989).
54. Barrachina, R.O., *J. Phys. B: At. Mol. Opt. Phys.* **23**, 2321 (1990).
55. Szótér, L., *Phys. Rev. Lett.* **64**, 2835 (1990).
56. Kunikeev, Sh.D. and Senashenko, V.S., *Zh. Eksp. i Teor. Fiz.* **102**, 826 (1992).
57. Salin, A., *Cusp electrons associated with fast neutral outgoing projectiles*, unpublished manuscript.
58. Kuzel, M., Maier, R., Sarkadi, L., Závodszky, P.A., Pálinkás, J., Berényi, D.

and Groeneveld, K.O., *IKF Jahresbericht 1992*. J.W. Goethe Univ., Frankfurt/M, 1993. p. 23.

59. Závodszky, P.A., Sarkadi, L., Kuzel, M., Pálinkás, J., Gulyás, L., Takács, E., Tóth, L., Berényi, D. and Groeneveld, K.O., Abstracts. *12th Int. Conf. on the Appl. of Accelerators in Res. and Industry*, Nov. 2-5, 1992. Denton, 1992. p. 152.
60. Kuzel, M., Maier, R., Sarkadi, L., Trabold, H., Jagutzki, O., Berényi, D., Schmidt-Böcking, H., Groeneveld, K.O. and Szabó, Gy., Abstracts. *12th Int. Conf. on the Appl. of Accelerators in Res. and Industry*, Nov. 2-5, 1992. Denton, 1992. p. 37.
61. Jagutzki, O., *Diplomarbeit*, IKF D418 (1989).
62. Trabold, H., Meier, R., Sarkadi, L., Berényi, D., Jagutzki, O., Kuzel, M., Kroneberger, K., Schosnig, M. and Groeneveld, K.O., *IKF Jahresbericht 1991*. J.W. Goethe Univ., Frankfurt/M, 1992. p. 28.
63. Maier, R., Sarkadi, L., Trabold, H., Kuzel, M., Jagutzki, O., Berényi, D., Schmidt-Böcking, H. and Groeneveld, K.O., Abstracts, *GSI:ESR-Workshop*, Sept. 14-17, 1992.
64. Kuzel, M., Sarkadi, L., Pálinkás, J., Závodszky, P.A., Maier, R., Berényi, D. and Groeneveld, K.O., *Abstracts of Contr. Papers, 18th ICPEAC*, Aarhus, July 21-27, 1993. p. 620.
65. Kuzel, M., Maier, R., Sarkadi, L., Závodszky, P.A., Pálinkás, J., Berényi, D. and Groeneveld, K.O., *IKF Jahresbericht 1992*. J.W. Goethe Univ., Frankfurt/M, 1993. p. 24.
66. Kuzel, M., Vikor, L., Závodszky, P.A., Sarkadi, L., Maier, R., Berényi, D. and Groeneveld, K.O., *ATOMKI Annual Report 1994*. Debrecen, 1995. p. 39.
67. Sarkadi, L., *et al.*, private communication, 1995.
68. DuBois, R.D., *Phys. Rev.* **A50**, 364 (1994).

CHAPTER 8

Bremsstrahlung and K X-Ray Production by Electron Bombardment of Thin-film and Gas Targets

C.A. Quarles*

In this chapter, current efforts to improve our understanding of both characteristic and continuum X-ray production are discussed. New experimental techniques designed to obtain more precise data for the single-photon emission spectrum with lower and better understood background are presented. Progress in the development of a detailed physical model for X-ray production based on the best theory currently available is described. Improvements in the comparison of experiment and theory are described. Finally, progress in the development of new coincidence experiments to provide a more detailed study of X-ray production is reviewed, and several areas are suggested for future work in this area.

8.1 INTRODUCTION

The study of bremsstrahlung and characteristic X-ray production is in its centennial year. Even after 100 years there is still much to learn. There are several recent reviews of various aspects of this field available, including: Seltzer and Berger (1985) [1]; Pratt and Feng (1985) [2]; Amusia and Pratt(1992) [3]; Hippler and Schneider (1994) [4]; and Pratt, *et al.* (1995) [5]. The purpose of this chapter is to describe some current efforts to test the details of our understanding of X-ray production. We will describe some of our recent and generally preliminary efforts to improve the comparison of theory and experiment. These efforts center on two areas: first, development

*) Texas Christian University.

of new experimental techniques to obtain more precise data on X-ray production with lower and better understood background; and second, development of a physical model based on the best theory to describe the experimental data in as much detail as possible. We will also describe initial results of our program of reanalysis of some earlier published data in comparison with the newly developed model. Finally we will indicate some of the major problems remaining and suggest some new areas for research in X-ray production that may be developed over the next few years as the field enters its second century.

The instrumentation for state-of-the-art measurement of thin-target characteristic X-ray and bremsstrahlung cross sections in the energy range of a few to a few hundred keV consists of solid state detectors for energy dispersive X-ray detection, special collimators for the electron beam and emitted photons to minimize backgrounds, and data analysis software that includes what is known about the bremsstrahlung theory, physical backgrounds, and detector response.

There are two reasons why an accurate description of X rays from thin targets is needed. First, thin-target bremsstrahlung is an important example of the fundamental electromagnetic interaction of electrons with atoms, molecules and ions. Second, it is required for the development of a more realistic model for the thick-target processes observed in nature.

8.1.1 Thin-target and thick-target bremsstrahlung

In this chapter we need to distinguish between thin-target and thick-target bremsstrahlung. Thin-target bremsstrahlung is the radiation of a photon in the single collision of an electron (or any charged particle) with an atom, ion or molecule. As a single interaction, it can be properly described by a cross section. In an experiment with a beam of electrons on a target, the condition for thin-target bremsstahlung is that the probability for energy loss by the beam as it traverses the target is very small and the probability of multiple interactions by the electron beam, including the dominant process of elastic scattering, is small. Thin target conditions can be realized with gas targets at low pressure, and with thin solid films. Whether a target is "thin" depends on the electron beam energy.

Typical X-ray tubes and most naturally occurring bremsstrahlung spectra do not satisfy the thin-target condition. Thick-target bremsstrahlung (TTB) is a multiple interaction process in which detailed information about electron energy loss, ionization, scattering and the effect of secondary electrons produced within the target is needed for a complete description. TTB is more properly described by a yield rather than a cross section. The spectrum depends on beam energy, target atomic number, angle of incidence of the electron and angle of emission of the X rays. The complexity of modeling this process has led to models which derive predicted spectra within the parameter space from interpolations. In some cases, experience is sufficient for the task and

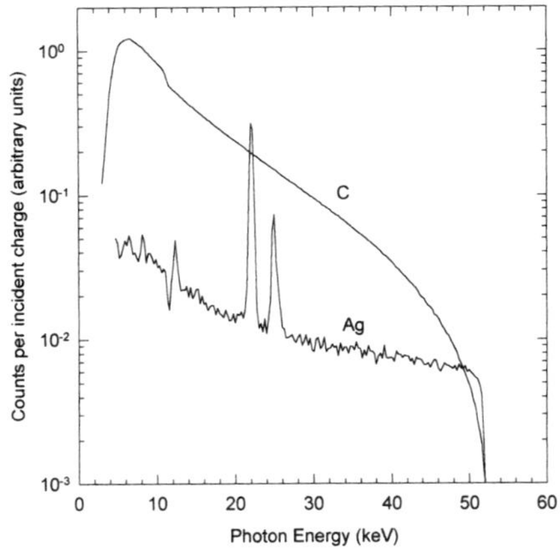

FIGURE 8.1 Plot of the counts (number of X rays) per incident charge versus photon energy at 90° for 52 keV electrons. Data is shown for photon transmission through a carbon (C) target thick enough to stop the incident electrons; and for a thin (50 $\mu g/cm^2$) silver (Ag) target to illustrate the difference between thin- and thick-target continuum spectra. The solid angle subtended at the target by the detector is the same for both spectra.

a detailed physical model, while it may be interesting, may not be required. Besides the X-ray tube, TTB spectra are observed in diverse natural and laboratory environments. TTB is produced in the atmosphere of a planet both by a spectrum of electrons from the solar wind and by secondary electrons produced from the bombardment of the atmosphere by heavy particles. TTB from the secondary electrons is a major component of the photon spectrum observed in PIXE experiments. TTB produced in the interaction of electrons with ions in a plasma provides a diagnostic technique for analysis of plasmas.

For some perspective, Figure 1 shows a thin-target spectrum for 50 keV electrons on a 50 $\mu g/cm^2$ Ag target and a TTB spectrum from 50 keV electrons on carbon. What is plotted is the radiated energy (or k times the number of photons per incident electron) versus photon energy k. The solid angle subtended at the detector is the same in both cases. The thin-target yield changes slowly with photon energy and scales as Z^2/β^2 where Z is the target atomic number, and β is the incident electron velocity in units of c. The thin target yield is finite at the kinematic endpoint, k_{max}. By contrast, the TTB yield rises from zero at k_{max} as $k_{max} - k$, and has significant target self-absorption at low k which depends on whether the spectrum is from reflection or transmission from the target. The TTB yield scales as Z instead of Z^2, due to the energy loss of the electrons within the target as the thick-target spectrum is produced. So, it is very easy for a TTB background to dominate a thin-target spectrum. In every thin-target experiment there is always some TTB. However, the thick-target effect can be reduced by arranging it to come from a much lower Z than the thin target, such as using an Al- or carbon-lined scattering chamber. Near k_{max} the thin target yield from a high Z target will almost always be observable even in the presence of a significant TTB background.

8.1.2 The present status of the theories

Bremsstrahlung from the atomic-field

The theory of bremsstrahlung from the electron as it scatters in the atomic field has been developed by Pratt, his students and co-workers. (Tseng and Pratt, 1971; and references in review articles listed above) [6] Their work provides the most accurate description of the bremsstrahlung process that is available. The results span a wide parameter space: from 2 to 92 in atomic number, from 1 to 500 keV in electron energy, from 0 to k_{max} in photon energy, and from 0° to 180° in photon angle. The predictions of this theory have been tabulated (Kissel, et al., 1983) [7]. Several versions of the interpolation program (Basic and Fortran) to give the cross section for any value of incident electron energy, target atomic number, photon energy and angle have been available since 1983.

The general agreement of bremsstrahlung theory and experiment is very good. It is fair to say that there is no significant disagreement between theory and experiment in the photon energy and angular distribution for any bremsstrahlung parameter set measured to date. Of course there are isolated discrepancies, but where these occur, there are some experimental details which are more likely than a breakdown of the theory to be the source of the discrepancy. The range of electron energy is from a few keV to several hundred keV. Targets from molecular hydrogen to uranium have been studied. Most measurements have been for a photon emission angle of 90°. The precision of absolute cross section measurements has been of the order of 20 to 25 percent. Comparisons between spectra, and ratios of data which reduce experimental errors, have determined the photon energy dependence to a precision of a few percent or better in some cases. Discrepancies for photon emission at 0° have recently been resolved and the new results for this case will be discussed below.

The role of experiment in this field is to test the theory at a sufficient number of points in the parameter space so that the theory can be used with confidence for parameters that have not been experimentally studied. The present status of this testing process will be summarized in this chapter.

Polarization Bremsstrahlung

Within the past several years, it has been recognized that in addition to atomic-field bremsstrahlung there is a second process, polarization bremsstrahlung (PolB), which in some cases can contribute significantly to the X-ray spectrum. (Tystovich,1992; Amusia, 1993; Korol,1995.) [8–10] Whereas atomic-field bremsstrahlung is the radiation by the projectile electron as it scatters in the field of the target atom, PolB is the radiation by the target atom as it is polarized by the projectile.

PolB has been widely discussed theoretically. However, no calculations are yet available for the cases of the photon energy and angular distribution for the photon and electron energies of interest in this chapter. It has been argued that the contribution of PolB to the total radiation spectrum observed is small when the radiated photon energy is much larger than the highest binding energy of electrons in the target atom. PolB is expected to contribute mainly in the region of characteristic X ray, or of radiation from transitions in the atom. (Korol, 1992) [11] At this point, no evidence for PolB contribution to the photon energy spectrum has been reported for electron bombardment in the energy range of interest in this chapter. The search for polarization effects is one of the main interests in experiments in progress.

Some calculation of PolB for specific energies and targets have been done. For example for energies above the 4d levels in Xe or La which have a large polarizability and thus produce an enhancement in the photon spectrum at photon energies of a few hundred eV. Enhancements in the photon spectra have been reported in La (Zimkina et al, 1981,1984) [12,13] and in Xe (Verkhovtseva et al, 1990) [14], but the measurements are difficult to interpret and have not yet been repeated.

Perhaps the most convincing evidence of PolB effects has been reported in proton bombardment of thick targets. In this case the atomic- field bremsstrahlung from the proton is negligible. Most of the observed radiation comes from a combination of PolB and atomic-field bremsstrahlung from secondary electrons knocked out in the target. It has been shown that PolB competes with and in some cases dominates bremsstrahlung from secondary electrons (Ishii and Morita,1984, 1985; Ishii, 1995) [15–17].

Characteristic X rays

Characteristic X-ray production, which in this chapter will refer mainly to K X rays, is typically assumed to arise from the ionization or excitation of the inner shell of the target atom, followed by the emission of an X ray. The X-ray emission is usually assumed to be incoherent with the production of the inner-shell vacancy. That is, it is assumed to occur on a time scale long compared to the time of formation of the vacancy or the time of transit of the electron. The X-ray intensity is then the product of the inner-shell ionization cross section and the fluorescence yield. Since the fluorescence yield can be measured in other interactions which produce an inner-shell vacancy (photoionization, or electron capture radioactive decay), the cross section for inner-shell ionization can be deduced from the experiment.

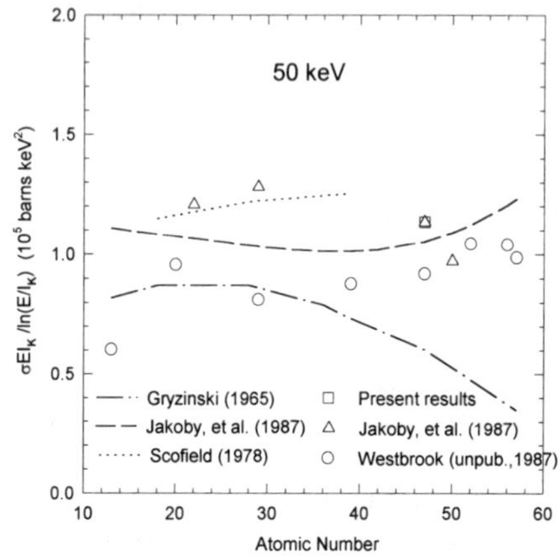

FIGURE 8.2 Plot of the scaled K-shell ionization cross sections versus atomic number for 50 keV electrons. A variety of theoretical calculations and data from various experiments are shown to illustrate the scatter in the data and theory for this case.

Total inner-shell, mainly K shell, ionization cross sections have recently been measured to a precision of about 2 to 10 %. There have been more measurements by different groups over the years than for bremsstrahlung. The earlier work (at the 15 to 25 % level) has been reviewed by Powell (1976) [18]. It is interesting that the energy dependence of the cross sections from near threshold to several times the ionization energy can be described by the simple Bethe theory. This theory is often used as a reliable model to fit the data. Even the variation of the so-called Bethe parameters with atomic number is typically less than about about 15 %.

There have been a variety of theoretical approximations, beyond the simple Bethe model, that have been used to describe the inner-shell ionization process. The predictions of the various theories often do not agree very well with each other or with experiment. This is illustrated in Figure 2 for 50 keV electrons, where the disagreement is highlighted by plotting the scaled total cross section, $\sigma \mathrm{E} I_K \,/\ln(\mathrm{E}/I_K)$, versus atomic number, where E is the incident electron energy and I_K is the ionization potential of the K shell. The disagreements are typically at the level of 25 % or so, within the two standard deviation level of most of the earlier experiments, but in worse agreement with recent experiments with better precision. The scatter among the experiments, among the theories, and between experiment and theory indicate that inner-shell ionization by electron bombardment is not as well understood as bremsstrahlung.

Since the bremsstrahlung theory is much better known than characteristic X-ray cross sections, and since bremsstrahlung always accompanies characteristic X rays, it is very attractive to determine the characteristic X-ray cross section more precisely by normalizing to the bremsstrahlung theory. This approach has been used to deduce K shell ionization cross sections, and promises to provide more precise measurements of the inner shell cross sections(Westbrook and Quarles, 1987) [19]. It is hoped that more precise experiments will, in turn, motivate increased theoretical effort to calculate the inner-shell ionization cross section in relation to the bremsstrahlung cross section. One caution, however, is that this normalization approach can lead to ionization cross sections that are systematically low unless careful consideration is given to the continuum backgrounds to the bremsstrahlung spectrum.

8.2 CURRENT EXPERIMENTAL TECHNIQUES IN PHOTON ENERGY SPECTRUM MEASUREMENTS FROM THIN TARGETS

8.2.1 The basic measurements: the photon energy and angular distribution

The advent of solid state energy dispersive detectors has facilitated the direct measurement of $d^2\sigma/d\Omega dk$, the cross section doubly differential with respect to the photon energy k and the solid angle Ω. Since the 60's, data on bremsstrahlung and characteristic x- rays have been taken almost exclusively with germanium or Si(Li) detectors . The detailed interpretation of the continuum energy spectrum is thus intimately tied up with the determination of the efficiency and response of the solid state detectors. A measurement of the cross section is made by positioning a well-collimated detector at a chosen photon emission angle with to respect to the electron beam and measuring the photon counting rate versus photon energy with a detector-preamp-amplifier-multichannel-analyzer electronics system. A schematic of the setup at TCU is shown in Figure 3.

The electron accelerator shown is a 300 kV Cockcroft-Walton accelerator. Experiments with gas cell targets have been done with a 150 kV accelerator and, at electron energies below 15 keV with an electron gun mounted directly onto or inside a target chamber. In our accelerators, a commercial TV tube electron gun is used. This gun is inexpensive, easy to replace and uses an indirectly heated oxide cathode. Directly heated tungsten filament guns, such as used in an electron microscope, have also been used when a higher beam current is desired or when a poor background gas pressure is expected. While a typical setup has one detector that is moved from angle to angle, our current setup employs three germanium detectors that take data simultaneously. This allows a more precise relative angular distribution to be measured.

Absolute cross sections in barns/sr/keV are determined from a measurements of counts by the following formula:

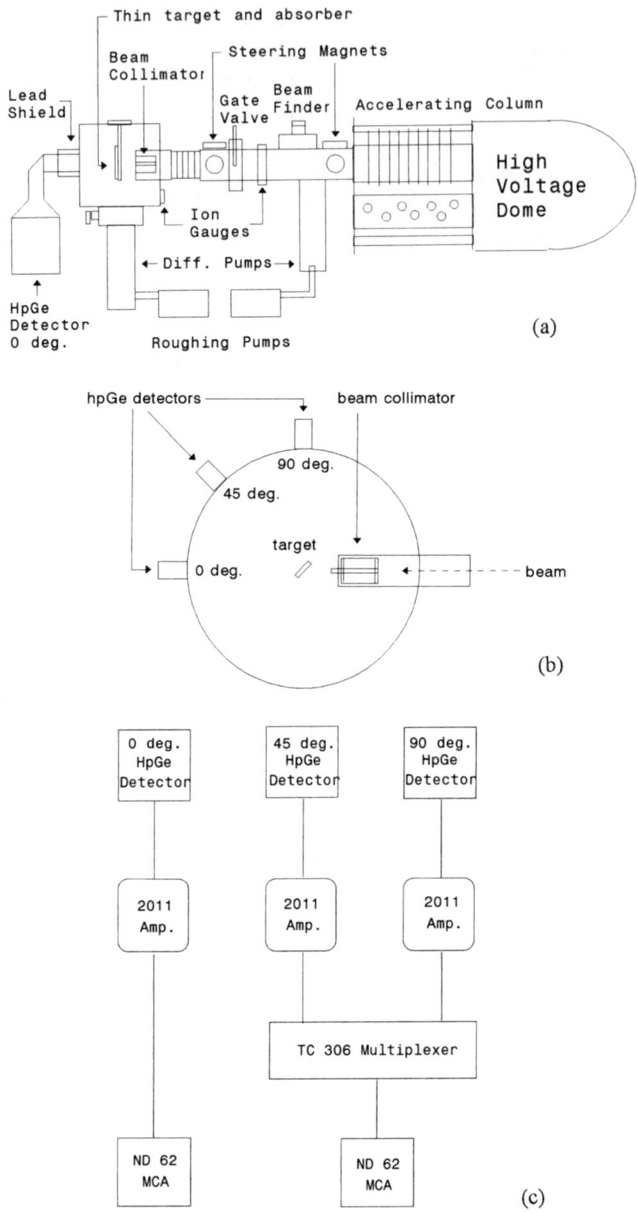

FIGURE 8.3 (a) Experimental setup currently in use at TCU employing three germanium detectors, showing Cockcroft-Walton electron accelerator, beam line, scattering chamber, and zero degree germanium detector. (b) Top view of the scattering chamber showing the position of the three germanium detectors. (c) Block diagram of the electronics used for measurement of the photon energy and angular distribution.

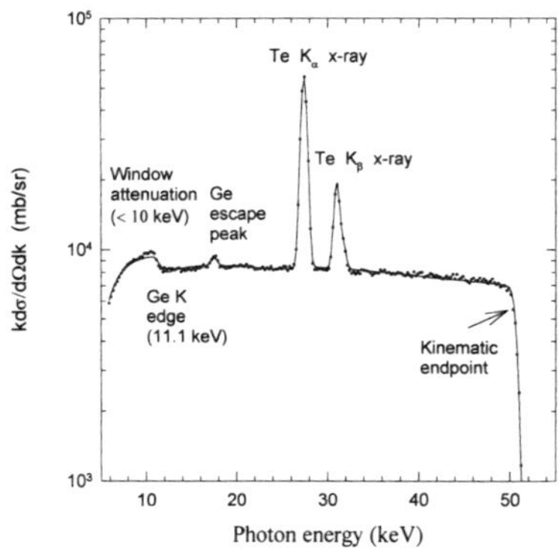

FIGURE 8.4 Thin-target spectrum of radiated photon energy, $kd\sigma/d\Omega dk$, versus photon energy k for 50 keV electrons on a Te target at 90°. This figure illustrates the general features observed in a typical measurement of the photon energy and angular distribution from a thin target under low background conditions. Indicated in the figure is the kinematic endpoint, characteristic X rays from the Te target, the germanium K X-ray escape peaks, the germanium K absorption edge, and the low energy attenuation by the detector window(s). The circles are data and the line is a fit to the data from the model discussed in the text.

$$\frac{d\sigma}{d\Omega dk} = \frac{N(k)}{N_0 \tau \epsilon(k) \Delta\Omega \Delta k} \tag{8.1}$$

where $N(k)$ is the number of counts at photon energy k, N_0 is the number of incident electrons, τ is the target thickness in atoms/cm^2, $\epsilon(k)$ is the energy dependent detector efficiency, Δk is the width of the energy channel, and $\Delta\Omega$ is the solid angle subtended at the target by the detector. In the case of characteristic K X rays, the total cross section is usually determined from the formula:

$$\sigma_K = \frac{4\pi N_k}{N_0 \tau \epsilon_K \omega_K \Delta\Omega} \tag{8.2}$$

where N_k is the number of K X rays in the spectrum summed over the K_α and K_β lines, and ω_K is the fluorescence yield. There is no energy width because the K X rays are assumed to be mono-energetic. The factor of 4π occurs because the X rays are assumed to be isotropic. If either of these assumptions is relaxed, we would return to Eq. (8.1) and determine the appropriate differential cross section. A typical photon energy spectrum is shown in Figure 4 where the major features common to all such spectra can be identified. What is plotted is $kd\sigma/d\Omega dk$ versus photon energy k. The sharp kinematic endpoint is shown at the electron beam energy of 50 keV. The theory predicts a finite cross section at the endpoint. Experimentally, the endpoint is spread by photon detector resolution and by variation in the electron beam energy. The counts above the endpoint are due to natural background, or in some cases may arise from pileup in the detector and electronics. The continuum bremsstrahlung is smooth and increases approximately as 1/k toward lower photon energy. The characteristic K X rays of the target (Te in this case) are seen as peaks. In this case, the K_α and K_β regions are resolved, but the K_{α_1} and K_{α_2} peaks and the structure in the K_β region are unresolved. The detector resolution of a few hundred eV is typical of Ge solid state detectors. The escape peaks from the X rays are also visible in this case at energies lower than the K X rays by the germanium K X-ray energy of 9.89 keV. The germanium K absorption edge is observable at 11.1 keV as a jump in the continuum. Below about 10 keV the count rate decreases exponentially due to photon absorption in the combination of vacuum and detector windows. In this case a mylar window isolated the vacuum from the detector and there was a Be window on the detector, as well as dead layer and front surface conducting region typical of the germanium detector. The germanium absorption edge is evidence of the loss of photo-peak detection efficiency as the photon energy approaches the K edge of the detector. Detailed measurement of the detector efficiency is primarily concerned with describing this effect. At the higher photon energy region, the detector efficiency is determined by the thickness of the detector crystal. Si(Li) detectors may be used for lower electron bombarding energy, but their efficiency falls off toward higher photon energy much more rapidly than germanium detectors.

8.2.2 Target thickness and target uniformity

Determination of target thickness is typically the largest source of error in a measurement with a thin-film target. Thin-film target thickness is characterized by the mass per unit area in units of μg/cm^2 since direct weighing of a measured area is still the most accurate way of determining the target thickness. Thin films that are either self-supporting or deposited on thin carbon backings are commercially available. Typical thicknesses in the neighborhood of $50\mu g/cm^2$ are obtainable with an accuracy of 20%. Target thickness error as low as 5% has been obtained in our lab (and others) by careful evaporation in well-controlled geometry coupled with weighing of samples with an electronic micro balance by directly depositing targets on a micro balance weight pan. The accuracy depends on the assumption that the actual target used is identical to the target weighed. Direct weighing of a target after use is also possible in principle for self- supporting targets. This has not generally been done because of the difficuulty of recovering a carefully measured area of the then-film target from the target frame.

The development of new techniques to improve the measurement of thin film thickness is very desirable. We are currently working on a technique using X-ray fluorescence in which targets of a particular material on mylar backing with a thickness of $\pm 5\%$ are used as standards for comparison with an unknown target in an identical geometry. It is expected that an accuracy approaching ±5% can be achieved with such a technique. Any technique that depends on calibration with known standards such as electron backscattering, Rutherford scattering, or alpha particle energy loss might be developed and improve the accuracy of commercially available films.

But, there is a significant problem remaining with thin-film targets even if the areal weight can be accurately determined. Thin films, whether self-supporting or on carbon backing, are stretched over a target frame and are wrinkled. The degree of wrinkling varies with target material and as shown in Ambrose *et al.*(1987) [20] can lead to substantial additional thickness error since the incident beam is not uniformly striking the target at a fixed angle. For targets with thicknesses of a few $\mu g/cm^2$ non-uniformity is a also problem. This too depends on the material. For very thin targets, clustering or island formation may lead to regions that are much thicker then the average weight per unit area would suggest.

The problems inherent with thin-film targets encourage the use of low pressure (a few mTorr) gas targets. Some measurements of photon spectra have been made with continuous flow gas jet targets, but these do not permit an absolute measurement to be made. Windowless gas cells can be used for absolute measurement in a differentially pumped configuration to allow the electron beam to interact with a well-defined volume of gas at known pressure. But so far this technique has been exploited in only a few cases. For absolute measurements, if the target "thickness" only depends on the cell pressure, it is necessary to control the flow, the cell temperature, and monitor the integrated pressure over the run. Because of the volume of interaction of the beam with the cell, it is more difficult to determine the solid angle subtended by the detector than for the thin film. However, the error involved in a calculation of the solid angle may be smaller than that due to non-uniformity and wrinkling of the thin film. Gas targets are not limited to the rare gases. There are many molecular gases of interest and there are a wide variety of organometallic gases in which a single heavy atom is bound to ethane or methane radicals. The effect on the bremsstrahlung spectrum of the hydrocarbons can be easily included in the theoretical model, and contributes only a small correction because of the Z^2 dependence of the thin-target bremsstrahlung.

8.2.3 Detectors: energy calibration, efficiency and response

The spectrum shown in Figure 4 and described above indicates most of the problems associated with the detector. The efficiency and response of germanium and Si(Li) detectors has been widely considered (see, for example, texts by Debertin, 1988, and Knoll, 1989) [21,22]. Energy calibration is done with radioactive sources that produce X rays or gamma rays of well-known energy. When the system is calibrated, the kinematic endpoint provides an accurate measure of the incident electron beam energy. If there are X rays

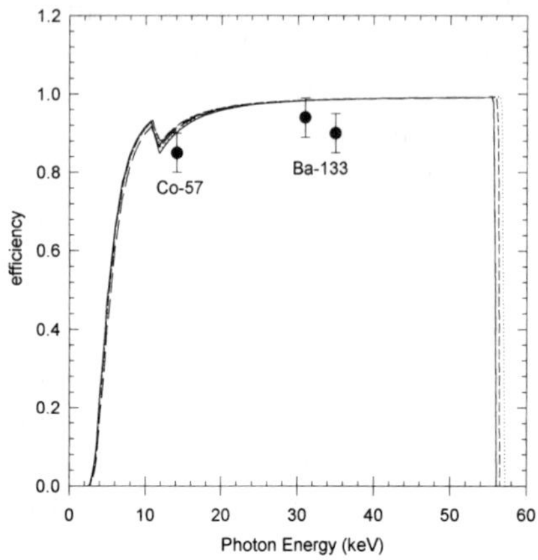

FIGURE 8.5 Typical photopeak efficiency curve for a germanium detector. The curves are from bremsstahlung spectra from 56 keV electrons on targets of various atomic numbers ranging from aluminum to uranium. The solid points are from calibrated radioactive sources with uncertainty at the 5 % level. The absolute scale of the bremsstrahlung-deduced efficiency curve could be normalized to the source data, however, that has not been done in this case. Several other radioactive sources could also be used to produce a few additional calibration points in this energy range. The advantage of the bremstrahlung spectrum approach is that it provides a smoothe, continuous measure of the detector efficiency.

from the target, then these and the germanium absorption edge often provide enough information for *in situ* energy calibration for each spectrum.

Calibrated radioactive sources are used to determine the detector solid angle if the target is a thin film, and may be used to measure the absolute photopeak efficiency at several X-ray or gamma ray energies. Typical radioactive sources that are useful for energy calibration and for efficiency measurements include Fe-55, Co-57, Cd-109, Ba-133, and Am-241. These sources are commercially available as point sources in calibrations at the 2 to 5 % level traceable to NBS standards. Of course, the solid angle and efficiency of a detector depend critically on the detector collimation. These factors must be determined *in situ* for a particular detector, collimator and window system being used.

A typical photopeak efficiency curve for a well-collimated Ge detector in the 50 keV energy range is shown in Figure 5. The lines are determinations of the energy dependence of the efficiency from the thin-target bremsstrahlung produced from several targets ranging from Al to U. The thin-target bremsstrahlung spectrum measured in a low background experiment has an application as a very good measurement of the detector efficiency as a function of energy, especially in regions for which radioactive sources do not provide adequate calibration points. This approach has been used now by several experimenters (Altman, *et al.*, 1987; Weathers, *et al.*, 1991) [23,24].

Either the bremsstrahlung theory can be taken as the normalization or the efficiency curve can be normalized to a selected gamma ray from a calibrated radioactive source. Several such calibration points from Co-57 and Ba-133 are also shown in Figure 5. At higher energies not shown in the figure, the efficiency falls off due to the reduction in photoionization cross section with increasing photon energy. At lower energies, the efficiency is mainly determined by the windows between the detector crystal and the source. An escape peak such as seen in Figure 4 due to escape of the characteristic K x- ray from the detector is often evident in the photon spectrum accompanying any X-ray or gamma ray peaks. For a continuum, of course, no specific escape peaks are seen. But the escape effect still plays an important role in the photopeak efficiency of the detector, and is responsible for the characterisitc dip in the photopeak efficiency curve at energies above the Ge (or Si) K absorption edge.

Little correction is usually required for the response of a germanium detector at the energies considered here, that is from a few keV to about 100 keV. It is typical to model the detector response by a Gaussian peak at the incident photon energy (or photopeak) with a small low energy shoulder or tail due to both Compton effects and incomplete charge collection in the detector.

In order to deduce the correct energy dependence for a continuum spectrum it is generally important to deconvolute the effect of detector response. This is the effect of a tail on the Gaussian photopeak that can contribute to a build up of counts at lower photon energy. However, as we discuss below, deconvolution is not necessary for the cases we have studied so far except in the special regions of the germainum K absorption edge and kinematic endpoint of the bremsstrahlung spectrum. This means that the tail of the Gaussian photopeak detector response is small enough in this energy range to make any distortion of the continuum spectrum negligible. However, at higher incident electron energy (say, > 150 keV) and for photon energies where the photopeak efficiency is much lower relative to the Compton effect, it may be important to consider deconvolution in fitting the continuum spectrum over a broad photon energy range.

Finally, it should be pointed out that the detector response to a monoenergetic photon cannot easily be determined by observing X-ray or gamma ray lines from a radioactive source. The effects of the source thickness and backscattering in the source are often much larger than the actual response of the detector. We have examined this effect for a particular germanium detector by comparing the X-ray spectrum from a thin target with that from a corresponding source, for example, silver K X rays from a Cd-109 source and K X rays from a thin silver target.

8.2.4 Solid angle and charge collection

Determination of detector solid angle is straightforward with thin-film targets and a well-collimated beam and detector. A simple geometric computation is usually sufficiently accurate. The solid angle (in product with the efficiency) can be measured with a calibrated radioactive source in the position of the target so that the geometry of the experiment is duplicated as close as possible. For this purpose, a source with an X ray or gamma ray with an energy in a region known to have nearly 100% efficiency for the detector in question is convenient.

For gas cells, the determination of the solid angle from the extended volume of the typical cell is somewhat more difficult. The most direct way to calculate the solid angle is to use a Monte Carlo model of the geometry including as much as is known about the beam profile, cell geometry and detector collimation. Such Monte Carlo calculations are not difficult or time consuming with desk top computers available today.

Charge collection is done by collecting the electron beam in a Faraday cup. Some care must be taken to correct for elastic scatter by the target out of the Faraday cup and to minimize electron loss from backscatter (which can be 10 to 15 % from a flat surface). Thus for an electron beam, a carbon-lined Faraday cup with a high length-to-diameter ratio is desirable. For higher electron energy, it is also necessary to have adequate lead shielding around the carbon cup to minimize charge loss due to photoionization by high energy photons as they escape from the Faraday cup. Commercial current integrators coupled with good Faraday cup design can provide accuracy in charge collection of a few percent or better.

8.2.5 Background processes important in interpreting photon spectra

It is impossible to measure a bremsstrahlung spectrum that is completely free of background. The target-in–target-out measurement of background that works for most scattering experiments does not suffice for bremsstrahlung measurements. The simplest background that is not corrected by this standard technique arises from electrons that elastically scatter in the target into the detector collimator or window and produce TTB background that is seen by the detector. Removing the target will remove the background, but not measure it!

TTB is ubiquitous. It arises from anything that electrons can scatter into. For example, the detector can "see" through the target to a vacuum chamber wall into which electrons can elastically scatter in the target. In order to adequately describe the thin-target spectrum, we must correct for the TTB background. A self-consistent treatment of the thin-target spectrum and the TTB contribution must be made. In early experiments, the background was estimated by putting a photon absorber between the detector and target or by magnetically deflecting the scattered electrons from the detector. While such techniques were adequate for earlier experiments that did not achieve today's precision, and while deflection can in some cases reduce (but never eliminate) the background, the preferred approach now is to accept the background, design the experiment to minimize it as much as possible and then include a model for the background in the analysis of the data.

Minimization of some backgrounds

First, consider the background process where the incident electron elastically scatters in the target into the detector or vacuum window. The scattered electron will then produce TTB in the window. A simple approximate model for this background can be worked out. Here the TTB from the window is transmitted through the window to the detector and so exhibits the same absorption characteristics as the window produces on the thin-target bremsstrahlung from the target. The number of photons per incident electron per sr per keV is given approximately by the simple Kramers-Kulenkampf-Dyson formula (Storm, 1972) [25]

$$n(k) = \frac{2.76 \times 10^{-6} Z_W (k_{max} - k)}{4\pi k} \Delta\Omega_d \Delta k \tag{8.3}$$

where Z_W is the atomic number of the window, and $\Delta\Omega_d$ is the solid angle subtended at the window by the detector.

The electron spectrum n_e on the window is given by the elastic scattering formula:

$$n_e = N_0 \tau 1.98 \times 10^{-26} Z^2 \frac{1-\beta^2}{\beta^4} \frac{1}{sin^4\theta/2} \Delta\Omega_W \tag{8.4}$$

where N_0 is the number of incident electrons, τ is the target thickness in atoms/cm^2, and $\Delta\Omega_W$ is the solid angle subtended at the target by the window.

The TTB from the window is the product of equation 3 and 4. To discuss the size of this background let us add it to the approximate thin-target bremsstrahlung produced in the target to get the total number of photons $N(k)$ detected at energy k:

$$N(k) = N_0 \tau Z^2 [0.435 \times 10^{-32} Z_W \frac{k_{max} - k}{k} \frac{1-\beta^2}{\beta^4} \frac{1}{sin^4\theta/2} \Delta\Omega_W \Delta\Omega_d + 0.5 \times 10^{-27} \frac{1}{k\beta^2} \Delta\Omega] \Delta k \tag{8.5}$$

where the first term in the brackets is the TTB and the second term is the thin-target bremsstrahlung. Factoring:

$$N(k) = N_0 \tau 10^{-27} \frac{Z^2}{k\beta^2} \Delta\Omega \Delta k [0.435 \times 10^{-5} Z_W (k_{max} - k) \times \frac{1-\beta^2}{\beta^2} \frac{1}{sin^4\theta/2} \frac{\Delta\Omega_W \Delta\Omega_d}{\Delta\Omega} + 0.5] \tag{8.6}$$

Comparison of the two terms in the brackets shows the conditions for TTB to be significant and suggests ways to minimize this effect. Notice that the TTB background is independent of the target atomic number, but depends linearly on the atomic number of the window. Usually windows are Be or carbon or Mylar and so have a low Z. The background for $\theta = 45°$ is about 10 times that for $\theta = 90°$. Most thin-target experiments have been performed at $\theta = 90°$. The TTB is smallest near the kinematic endpoint. As k decreases, TTB becomes increasingly important. The maximum contribution for non-relativistic electron energy is essentially independent of electron energy, or in other words, the effect of TTB increases as k/k_{max} decreases.

Finally, the magnitude of TTB can be affected significantly by the geometry as indicated by the ratio of solid angles. The worst case is to have the window very close to the detector crystal. Here, $\Delta\Omega_W \sim \Delta\Omega$ and $\Delta\Omega_d$ could approach 2π sr. A better arrangement is to locate the window about half way between the target and detector. This would be an electron absorber or perhaps a vacuum window, not the typical Be detector window. Then, $\Delta\Omega_W \sim 1/4\Delta\Omega$ so the ratio becomes $\frac{1}{4}\Delta\Omega_d$, which can be $\sim 10^{-4}$ sr with good collimation.

A second type of TTB background comes from electrons that elastically scatter in the target into the vacuum chamber wall. They produce TTB in the wall that can be seen by the detector. In this case the simple KKD approximation for TTB is not enough. Here the TTB spectrum is affected by the self-absorption in the wall, which will be very different from that in the detector window considered above. As discussed below, the effect can be modeled by adding a broad Gaussian peak to the TTB spectrum that goes to zero at the endpoint but produces a signature "hump" in the continuum that is easily recognized with experience. In this case, the scattering angle is the supplement of the photon emission angle. Hence, the effect increases significantly as the photon angle is increased beyond 90° since this corresponds to forward electron scattering into the wall. This wall background can be reduced by good collimation and distant chamber walls with as low Z as possible so that $\Delta\Omega_W < \Delta\Omega$ and $\Delta\Omega_d < \Delta\Omega$

TTB can also arise from sources other than electrons elastically scattered in the target: for example, electrons scattered in the beam collimators, or photons produced in the Faraday cup that backscatter into the detector. In this case the TTB would not depend on the target atomic number or thickness. This type of background, however, would tend to subtract out in a target-in–target-out measurement.

Collimator design to reduce background

The principle of good collimator design for electron experiments is simple: first, make sure the electrons hit only low Z materials so the TTB is minimized; and second, back up these regions with adequate photon absorbers. Practically, a lead-filled carbon collimator works well as an electron beam collimator and as a photon detector collimator if electrons are going to be scattered from the target into the detector collimator. The carbon wall of the collimator should be as thin as possible consistent with stopping all electrons of a given energy. The lead must be thick enough to adequately absorb the maximum energy photons that are produced. Thus, the dimensions of collimators differ greatly as the electron beam energy is changed. Adequate collimation and shielding for electron energies less than 25 keV is simple to achieve with thin collimators. Shielding and collimation become increasingly difficult as the energy increases. This is the current limitation to achieving adequate signal-to-noise for experiments with gas targets above 25 keV. Even with thin-film targets that are as much as 10^4 times thicker than gas targets, with a corresponding signal- to-noise advantage, the collimator and shielding requirements have limited most recent experiments to energies less than about 150 keV. Earlier experiments at energies of 500 keV and above typically used well-shielded NaI detectors and were not concerned with the precision of the photon energy dependence or detailed comparison with theory that we are interested in today.

8.2.6 Experimental errors

The error in the cross section measurement depends on the errors in each of the factors in Eq. 1 or 2. For thin-film targets, the largest error in usually in the target thickness with 20 % typical and 5 % achievable. The errors in charge collection (N_0), solid angle, and efficiency can be of the order of a few percent. For gas targets, the target thickness depends mainly on gas pressure, which can be better known, but then the solid angle typically represents a larger uncertainty. Efficiency and detector response errors depend on the detector and not on the target type.

8.3 DEVELOPMENT OF A PHYSICAL MODEL FOR THE PHOTON ENERGY SPECTRUM

8.3.1 The Monte Carlo approach and the fitting procedure

We have developed a Monte Carlo program for the multi-parameter nonlinear least squares fit of the photon energy and angular distribution spectrum to a physical model. The physical model describes the photon energy spectrum observed in the detector from photons originating in the target from bremsstrahlung and inner-shell ionization, from TTB backgrounds, and includes the detector energy calibration, response and efficiency. The model parameters are constrained by approximate knowledge of their values, which provide good initial values for the Monte Carlo fit. The best fit is found subject to a normalization of the area of the model to the integrated number of counts over a selected energy range. Thus, the data is normalized to the bremsstrahlung theory, which does not depend on any free parameters. The normalization factor is of course the product of the incident charge, the target thickness, the absolute solid angle, and the absolute efficiency. If these factors were measured with precision, the agreement of the normalization factor with the product of these factors would measure the agreement of the data with the theoretical bremsstrahlung cross section. In most of the fits discussed here, the absolute factors have not yet been measured with high precision, and the results are normalized to the bremsstahlung theory instead of to the product of these factors. Thus, the inner-shell ionization cross sections, for example, are determined as absolute cross sections normalized to the bremsstrahlung theory.

8.3.2 Atomic-field bremsstrahlung theory

For computation of the bremsstrahlung theory, we are using a True Basic interpolation program (KQP), which uses an ASCII file benchmark data set (Kissel, 1983) [7]. A FORTRAN version of the interpolation program is also available. The current generation Monte Carlo fitting program is written in Visual Basic. This version runs under Windows on a PC and computes the bremsstrahlung theory from an executable version of the KQP interpolation program. There are no free parameters in the bremsstrahlung theory. The theory can be evaluated for targets consisting of up to three elements. If the target is a molecule, such as TbF_3, the relative weights are used and stoichiometry is assumed. For targets deposited on carbon backings, the relative weight of the carbon and target elements are used based on the weight per unit area of the target and backing. Thus the contribution of the backing can be included in the bremsstrahlung theory. Alternatively, a data run consisting of the backing alone can be subtracted from the target-plus-backing run. Then the model would include only the target elements. Both approaches have been used in fitting with essentially equivalent results. At present the element weights in the target are not parameters; but, considering the uncertainty in the target-backing thickness, it may be useful at times to make the relative weights of target components constrained parameters in the fit.

Figure 6 illustrates the photon energy dependence of the theoretical atomic-field bremsstrahlung cross section at 50 keV for several atomic numbers (Figure 6a), and also the dependence of the photon energy spectrum for different radiated photon angles for one target atomic number (Figure 6b). This figure will provide a perspective on the shape to the theoretical spectrum and how it depends on the parameters of photon emission angle and target atomic number.

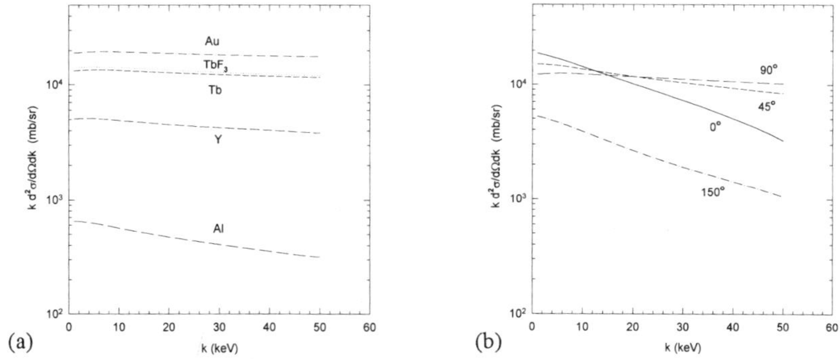

FIGURE 8.6 The theoretical atomic-field bremsstrahlung cross section plotted as $\mathrm{kd}^2\sigma/\mathrm{d}\Omega\mathrm{dk}$ versus photon energy k for 50 keV incident electron energy for (a) various targets at the photon emission angle of 90°. (b) various photon emission angles for silver.

8.3.3 Energy calibration and electron beam energy

While the energy calibration of the detector and electronics is measured by use of radioactive sources, the precise values of the energy per channel and the zero of energy are fit parameters. If there are X-ray peaks in the data or if the germanium absorption edge is well resolved, then the energy calibration parameters that best fit the data can be determined for each run. There is variation in these parameters for the best fit from run to run of the order of a few tenths of a percent. This is due to normal drifts in the stability of the detector and electronics over a run of several hours or over a period of days from run to run. In cases when both the K and L shell X rays are observed in the data, it is necessary to include a small second order term in the energy calibration. A small non-linearity is commonly observed when accurate energy calibration of detector-MCA systems is performed.

When the energy per channel and zero have been determined, then the electron beam energy, k_{max}, and the energy width of the end point can be fitted. The beam energy for a run can be determined from the fit to within about $\pm$ 50 eV. The resolution of the endpoint is a combination of detector resolution and variation in the beam energy from the electron accelerator, and typically determined to about 0.5%.

8.3.4 Modeling the characteristic X rays

The characteristic X-ray peaks occur at well-known energies, which are fixed in the program. The theoretical relative intensities of the X rays are taken as starting parameters. We have found good agreement with the K_α to K_β ratio and with K_{α_1} to K_{α_2} for targets studied. To obtain the best fit to the X-ray regions, it is usually necessary to allow the relative intensities in the

K region to vary within narrow constraints. We model the X-ray peaks by Gaussian peaks with the centers fixed at the known X-ray energies. The only parameter is the energy width, or resolution, which is usually taken to be a constant. The relative X-ray intensities and the overall characteristic X-ray intensity relative to the bremsstrahlung theory are parameters. Thus when a best fit is obtained, we obtain the product of inner-shell ionization cross section and fluorescence yield normalized to bremsstrahlung theory.

In some cases when both L and K X rays are observed in the data, it is necessary to take the variation of energy resolution with photon energy into account. A linear variation is adequate. In fitting the peaks, other peak shapes have been tried, such as Breit-Wigner and Lorentzian. The best fit is found with a simple Gaussian shape with (in some cases) a small low-photon-energy tail due to Compton effect. We were especially interested in the Breit-Wigner peak shape. It had been thought that polarization bremsstrahlung could lead to a resonant contribution in the X-ray region, which would interfere with the characteristic X rays that arise from ionization and lead to a Breit-Wigner peak added to the Gaussian. However, the data examined to date do not support such an addition.

8.3.5 Modeling the thick-target bremsstrahlung backgrounds

The TTB background is modeled as described in the equations 3-5. The overall intensity of the TTB is a parameter. In addition, for some runs, an adequate model of the TTB requires the addition of a broad peak or "hump" in the background, which is constrained to go to zero at k_{max}. The TTB "hump" is described by a modified Gaussian peak described by three parameters: intensity, center and width. Contributions of TTB background are illustrated in the fits shown to the thin target and gas target data discussed below.

8.3.6 Modeling detector efficiency, detector windows, and K X-ray escape

The fitting program calculates the photon efficiency from an interpolation of tabulated photon attenuation coefficients (Storm and Israel Tables, 1970, available as an ASCII file) [26]. The photon energy dependence of the detector efficiency is modeled by the following function:

$$\epsilon(k) = (1 - exp(-\mu(k)t_d)) \times exp(-\sum_i^n \mu_i(k)t_i) \times ESCAPE(k) \qquad (8.7)$$

where μ is the total absorption coefficient of the detector, t_d is the detector thickness, $\mu_i(k)$ is the absorption coefficient of the i th window, and t_i is the ith window thickness. The windows include a vacuum window, detector window, gold conducting surface on the detector(if any) and any deadlayer. The efficiency is determined by the detector thickness at high photon energy and by the windows at low photon energy.

The probability of escape of a detector K X ray depends on the K-shell photoelectric absorption cross section, $\sigma_{ph}(k)$. The proportionality constant, A, which is a fit parameter, depends on the detector geometry and collimation. So the correction to the photopeak efficiency is given in Eq. 7 by the factor: $ESCAPE(k) = (1 - A\sigma_{ph}(k))$. The photoelectric absorption cross sections for germanium or silicon are interpolated from the ASCII file of the Storm and Israel Tables. Escape reduces the photopeak efficiency at energy k due to loss of the detector K X rays, but there is also an added contribution to the counts at k from escape due to photons incident at energy k plus detector K X ray. This effect is usually neglected in discussing the efficiency from monoenergetic photons, unless the escape peak from a higher energy gamma happens to overlap a lower energy gamma, but is important in the case of a continuum. This additional additive effect of escape is included in the fitting program. For Ge, the escape correction has little effect on efficiency for k greater than about 30 to 35 keV, but reduces the efficiency as k approaches the absorption edge, as can be seen in Figure 5. There is no escape correction for k less than the K absorption edge, but there is an apparent increase in efficiency below the edge due to loss from incident photons with energy above the edge. This one-parameter model provides both for the observed energy dependence of the reduction in efficiency above the K absorption edge, and for the intensity of the escape peak associated with characteristic X-ray peaks. The effect of detector resolution and response on the absorption edge is discussed below.

8.3.7 Modeling the detector response: Compton tails, deconvolution and the kinematic endpoint

For photon energies of interest in this chapter (a few keV to 100 keV) and for well-collimated detectors, detector response from a low energy tail to the photopeak is a very small effect. However, since we are concerned with determining the energy dependence of a continuum spectrum we have considered deconvolution of the detector response. Early versions of the fitting program incorporated detector response deconvolution as an option. Deconvolution of the Gaussian peak detector response is important at the kinematic endpoint and at the detector K absorption edge because these are theoretical points of discontinuity, which are broadened by the Gaussian detector resolution and also, in the case of the endpoint, by electron energy loss and energy resolution of the electron beam. For other regions of the spectrum, the concern is that a significant low-energy tail in the detector response function could lead to an enhancement in the number of counts at lower photon energy. Deconvolution would eliminate this effect on the continuum spectrum. However, when a reasonable low-energy tail is assumed for the detectors used, the effect of deconvolution on the continuum spectrum is found to be negligible. As mentioned above, one cannot deduce the response tail directly from radioactive source spectra. Any tail observed with sources is typically due to source thickness and Compton backscatter in the source itself. The smallness of the actual detector response tail can be verified in the fit obtained to the characteristic X-ray peaks in an actual electron beam on target run. Here a small approximately- exponentially-decreasing Compton tail can be observed as a small shape distortion on the low photon energy side of the Gaussian, but we see no significant or extended low-energy tail that would influence the fit to the continuum or require deconvolution. So in the present fitting program we have not deconvoluted the response, only added a small exponentially decreasing tail, described by two parameters, to each X-ray Gaussian peak. In fact, in most fits described here the tail is too small to be included.

At present, the model approximates the end point region by matching the upper half of a Gaussian to the bremsstrahlung theory. This is usually adequate, but does lead to a worse fit to the continuum within a few keV of the endpoint. Similarly, the germanium K absorption edge is smoothed out in the experiment by the detector Gaussian resolution. The "poorer" fit between the model and the data at these two regions can be seen in the Figures. Both of these effects can be treated very well by a deconvolution of the data with a Gaussian detector response. However, since we do not need a deconvolution over the whole spectrum, we have not run the deconvolution for most cases since it is very time consuming. In the future, we plan to modify the fitting program to deconvolute only over the narrow regions around the absorption edge and the endpoint.

8.3.8 Variation in energy calibration of background run

In the fitting of the data from the difference experiment that we will describe below, it was found necessary to allow the energy calibration of the absorber run to vary slightly from that of the target-plus-absorber run. This had the effect of removing a small apparent background "hump", which was really created in the difference data by the difference between two spectra when a small drift in the energy calibration had occurred.

8.3.9 Improvements to the model

In addition to the refinement to the present model of deconvolution of the spectrum near the endpoint and germanium absorption edge, the model could be improved by a better model for TTB background and by incorporating predictions for polarization bremsstrahlung when they become available.

We can compute TTB directly in the linear slowing-down approximation from the tabulated thin-target cross sections using tabulated electron energy loss from Berger and Seltzer's tables (Berger, 1964, available as an ASCII file) [27]. This calculation has been carried out for the case of TTB transmission spectra for 70 keV electrons for photons emitted at 45° and 90° and agrees well with experiment (Ambrose, *et al.* 1991) [28]. This project has not yet been developed further to provide TTB for different geometries or for reflection spectra. It is believed that such a more-detailed TTB model could be an improvement over the present model in describing the TTB background, especially in the case where it arises from electrons scattered into the target chamber wall.

It is hoped that detailed calculation of polarization bremsstrahlung for energies accessible to solid state detectors will be available soon so that it could be added to the present model. We understand that it may soon be possible to have calculations for electron energies up to 25 keV for the rare gases (Korol, private communication) [29]. Even if the contribution is small as expected, say only a few percent enhancement over normal bremsstrahlung on the photon energy or angular distribution, the experimental techniques are sensitive enough to detect the polarization bremsstrahlung effect. The only exception would be if the effect is simply identical to normal bremsstrahlung in photon energy and angular distribution so that only the absolute scale of the bremsstrahlung is enhanced by a few percent, well below the prospects for absolute measurements in the foreseeable future.

8.4 DISCUSSION OF TYPICAL RESULTS: COMPARISON OF MODEL AND DATA

For all the data analyzed to date with the fitting program described above, the data is well described by the model with a reasonable choice of the parameters. That is, the energy dependence of the photon spectrum agrees well with the bremsstrahlung theory when effects due detector efficiency and small TTB backgrounds are included. One may ask: would a disagreement with theory ever be found using this procedure? A disagreement with theory would show up as a case for which a good fit to the model cannot be obtained with a reasonable and consistent choice for the fit parameters. But, one is required to look very carefully at all the factors such as TTB backgrounds before claiming a disagreement with theory. Thus, while some variation in energy calibration, detector response, detector efficiency and escape effects—and in some cases in the TTB background—is expected from run to run, consistency in the parameters is demanded. To date in many runs, we have not found any significant discrepancies between experiment and theory that we have not been able to attribute to some small background effects. The fit is sensitive enough to show quickly whether an incorrect target element or angle has been used to compute the theory.

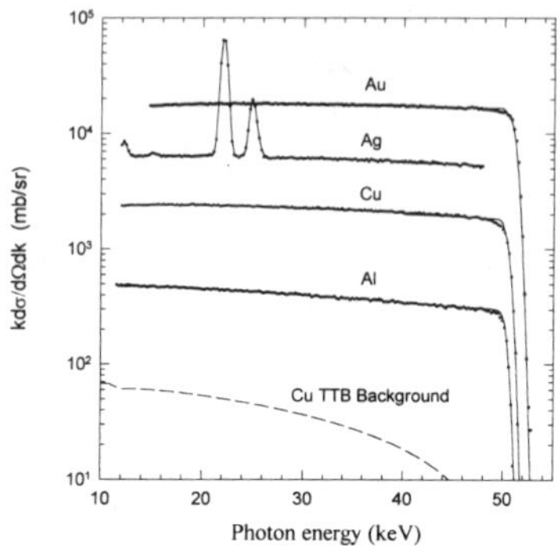

FIGURE 8.7 Plot of $kd\sigma/d\Omega dk$ versus photon energy k at 90° for 50 keV electrons on Al, Cu, Ag and Au. The circles are the data and the solid line is a fit of the model discussed in the text. the dashed line is the TTB background for the Cu target to illustrate the level of TTB background typical of a low background experiment at 90°. A similar level of TTB background is found for each target shown.

8.4.1 Thin-film target experiments

Data for electron energies of 50 to 140 keV and for photon emission of 30°, 45°, 60°, 90°, and 135° for thin film targets of Al, Cu, Ag, and Au were published by Quarles and Heroy (1981) [30]. The results for the shape function (the scaled doubly differential cross section normalized to the theoretical photon energy spectrum) was in general agreement with the theory (Tseng et al., 1979) [31].

Data for 90° emission from 50 to 100 keV electrons on thin-film targets with Z ranging from 6 to 92 were published by Ambrose et al (1987) [20]. The data were absolute measurements of bremsstrahlung with errors of about 25 %, dominated by target thickness and target wrinkling errors. The photon energy dependence of the spectrum was determined to about 3 % by considering the ratio of data for two targets compared with the ratio of the theory. In this case, the major errors such as target thickness were scale errors but were not photon-energy-dependent errors. Ambrose *et al.* (1987) did not consider the characteristic x- rays in their analysis. A few cases of the Ambrose et al. (1987) data have now been fitted with the present model and very good results are obtained. It is possible to determine the level of TTB background in the runs and to measure the ionization cross section from some of the runs. Time has not permitted fitting the entire data set available; however, it is planned to reanalyze a substantial amount the data with the latest fit program since it is a very good data set with low background, which would be difficult to repeat.

Typical spectra for 90° are shown in Figure 7 for 50 keV electrons on Al, Cu, Ag and Au targets. The targets are each about 50 μg/cm^2 on 15 μ

g/cm^2 carbon films (except for Al, which was self-supporting). In all cases there is a TTB background at the level of about 3%, which is shown for Cu. This background is probably due to scattering in the chamber wall. Although the wall was carbon-lined, the chamber was small so there was a significant solid angle for the TTB. The vacuum window was about halfway between the target and the detector, which minimizes the TTB from the window. In the photon energy region above the characteristic peaks (above about 25 keV for the spectra shown), only the TTB, detector efficiency and escape parameters affect the fit. The data are shown fit only to about the Ge absorption edge, and characteristic X rays are omitted from the fit except for Ag. When the characteristic K X-ray peaks for Ag are included, a good fit is obtained with the $K_{\alpha 1}$ to $K_{\alpha 2}$ and K_α to K_β ratios in good agreement with theory.

It is difficult to see how to improve these spectra much, although a larger target chamber would further reduce the TTB from scattering into the wall and perhaps also reduce any scattering from the Faraday cup. Data for 125 and 150 keV show an increasing TTB background as would be expected for the small target chamber and amount of shielding employed.

Thin-film experiments with electron beam energy below about 50 keV have not been reported. For lower electron energy plural or multiple scattering effects begin to become important for the 50 μg/cm^2 targets. In general, no experiment has been done in the transition region between thin- and thick-target, the region where the electron loses aa substantial amount of energy but does not come to rest in the target.

8.4.2 Designing the next generation of thin-film experiments

Where do thin-film experiments go from here? What is needed is to select a case for careful study in which an accurate absolute measurement can be made in the 50 to 100 keV energy range. A photon emission angle of 90° is the optimum choice for minimized TTB background. With careful attention to design of the geometry and shielding from photons scattered from the Faraday cup, it has been shown (Ambrose *et al.*, 1987) [20] that it is possible to do an experiment where the accuracy is limited by the target thickness error. A higher energy would permit a thicker target but requires much more shielding to reduce the TTB background. Thus there may be a trade off in error due to target thickness and TTB background. The background problems have limited published results to 100 keV or less although the energy may be extended to 150 keV with analysis of existing, unpublished data with the present model. For electron energies from 50 to 150 keV, targets of 50 μg/cm^2 are thin enough. We are presently trying to develop some techniques to improve the determination of target thickness for a few selected targets, which could then be used to measure the absolute cross section for several benchmark cases, such as Ag and Au at 50 and 100 keV. The rest of the data could then be normalized to these experimental benchmarks. At present, a goal of a 5 % error in the absolute bremsstrahlung cross section for a few selected cases seems achievable. While target wrinkling may still be a limiting factor for self-supporting films, this may be alleviated by depositing the thin film on a thick carbon absorber in an experiment arrangement discussed below.

8.4.3 Thick-absorber difference experiments

A new approach was introduced recently (Ambrose *et al.*, 1991) [28] to measure the bremsstrahlung spectrum produced at 0° and to reduce the background at other forward angles. The idea is to place an electron absorber of low Z directly behind the target. This will produce a large thick-target spectrum added to the thin-target spectrum from the target. However, if the presence of the thin target is assumed not to change significantly the electron spectrum incident on the absorber, then a target-in target-out measurement is possible. The thin-target spectrum can be determined by scaling the target-out run by the ratio of charge collected and subtracting it from the target-in run. In the fit to the data, the charge scale factor is multiplied by a ratio parameter, which is near one.

This approach seems to be the only practical way to study the spectrum at 0°. In this case the detector looks directly down the electron beam line and even very small drifts in the beam can greatly change the photon spectrum seen. In the first use of this technique, a discrepancy between theory and experiment was observed at 0°. However, that discrepancy has now been determined to arise from the run-to-run variation in the photon spectrum produced within the beam collimator. Also, it was found that drift of the energy calibration from run to run can lead to a small apparent background "hump" in the thin-target difference spectrum. This "hump" was removed when the energy calibration of the absorber run was varied a few percent. Now, there is no significant discrepancy between the theory and data at 0°.

The present setup in the TCU lab (Figure 3) allows the spectrum at three angles to be measured simultaneously by three germanium detectors. Preliminary measurements have been made for a few targets at 50 and 100 keV. Results for 45°, 90° and 0° are shown in Figures 8 to 10 for 50 keV electrons on Ag. Here we can see the effect of angle on the continuum spectrum and the constancy of K X-ray production with angle. In fitting this data, no TTB background is found. The charge collection ratio and the energy calibration of both target-in and target-out runs are fit parameters. The effect of allowing the energy calibration to vary is illustrated by comparing Figure 10 and Figure 11.

In Figure 11, the apparent background or "hump" is seen, but it disappears when the energy calibration of the absorber run is varied. The variation in this case is only a few tenths of a percent in the energy per channel and a few percent in the zero. This variation is well within what is observed in

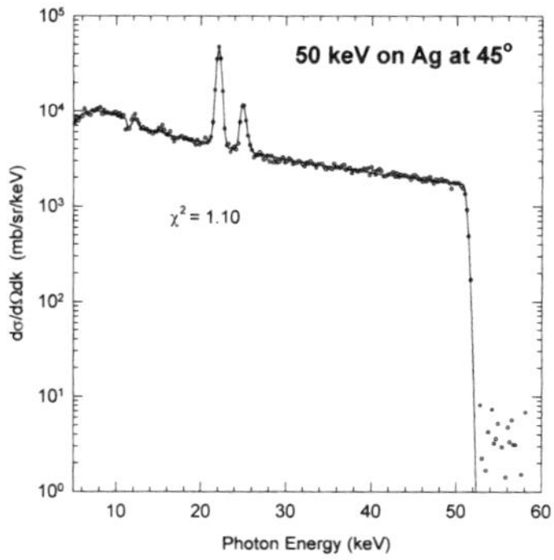

FIGURE 8.8 Plot of $d\sigma/d\Omega dk$ versus photon energy k at 45° for 50 keV electrons on Ag. The circles are the data and the solid line is a fit of the model discussed in the text. The value of chi square for this fit is shown in the figure.

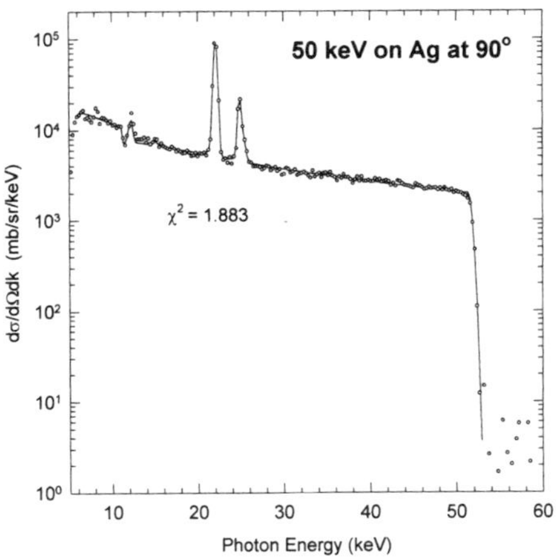

FIGURE 8.9 Plot of $d\sigma/d\Omega dk$ versus photon energy k at 90° for 50 keV electrons on Ag. The circles are the data and the solid line is a fit of the model discussed in the text. The value of chi square for this fit is shown in the figure.

run-to-run drift with our setup.

The charge collection ratio parameter is not expected to be exactly one due to a small difference in electron backscattering from the absorber when the target is in place. This effect can be measured by observing the change in charge collected by the absorber when the target is moved in and out, by normalizing to some other fixed quantity, such as the charge collected on the interior of the electron beam collimator. Generally we have found that the charge collection ratios vary by less than a percent from angle to angle within the same run. But there have been some exceptions to this and we still are

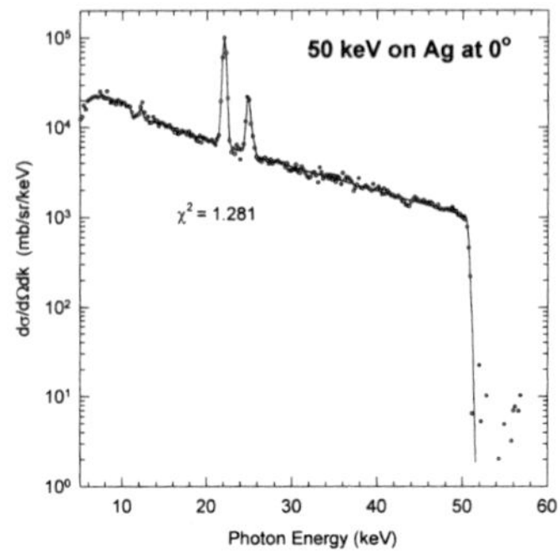

FIGURE 8.10 Plot of $d\sigma/d\Omega dk$ versus photon energy k at 0° for 50 keV electrons on Ag. The circles are the data and the solid line is a fit of the model discussed in the text. The value of chi square for this fit is shown in the figure.

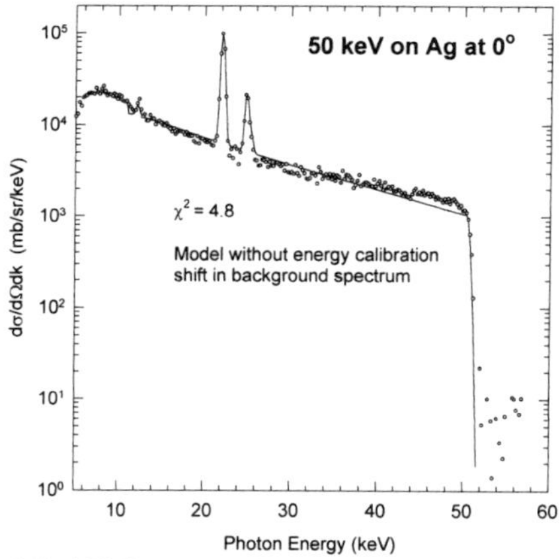

FIGURE 8.11 Plot of $d\sigma/d\Omega dk$ versus photon energy k at 0° for 50 keV electrons on Ag showing the "hump" in the spectrum when the energy calibration in background run is not varied, compare with Figure 10.

studying the consistency of the ratio. At this time we believe that a small systematic angular difference in the charge collection ratio may be expected due to the angular dependence of the thick absorber spectrum with the target in place.

For both the bremsstrahlung and the inner-shell ionization cross section measurement made in the target-plus-absorber configuration, a correction for electron backscattering in the absorber is required. The thick-target electron backscattered spectrum is essentially continuous and constant with electron energy. The integrated backscattered spectrum from carbon is about 10 % of the incident beam (August *et al.*, 1989) [32]. This effect has been included in the correction for the K-shell ionization cross section. It has not yet been included in the bremsstrahlung continuum, but it is small at 0°, since the bremsstrahlung from the backscattered spectrum would be produced at 180°.

The limitation of this absorber difference technique is largely statistical. Adequate statistics in the difference of the absorber spectra demand increasing run time as the atomic number of the thin film is reduced or as the electron beam energy is increased. Because of the high count rate from the thick absorber, the beam intensity can not be increased without unacceptable pile-up corrections. At present we have not tried to obtain good statistics data for electron energies above 50 keV or for Z ¡ 47. Futhermore, at angles other than 0°, the correction in the model for electron backscattering from the absorber into the target may prove to be less accurate than the correction for TTB in the detector window.

8.4.4 Gas target experiments

Gas target data, until very recently, has been limited to electron energies of 15 keV or less. As such, the gas data and the thin-film data do not overlap but compliment each other.

There is the data of Hippler *et al.* (1981) at 10 keV for He, Ne, N_2, Kr, Xe and UF_6; [33] of Semaan and Quarles(1981, 1982) for 6 to 10 keV on the Ne, Ar, Kr, Xe and dimethyl-Hg; [34,35] and the data of Estep and Quarles(1987) for 5, 10 and 15 keV on H_2, He, Ne and N_2. [36] The agreement for Ne and N_2 of Hippler et al. and Estep and Quarles is good, while there is some difference in the slope of the photon energy dependence for He. Whereas there is some disagreement with theory for the UF_6 data, the Hg data agrees well with theory. All of the gas target data is relative cross sections or measurement of the photon energy dependence. In addition there is data on hydrocarbon gases and on sulfur-bearing molecular gases. (Quarles and Estep, 1985 and 1986) [37] [38] All the gas target data to date has been taken with a Si(Li) detector.

One of our recent efforts has been to extend measurements to higher bombarding energy and to use a differentially pumped gas cell to permit absolute cross section measurements. Only preliminary results from this work are available as yet.

Typical spectra for 25 keV are shown in Figures 12 to 14. As can be seen, the TTB background ranges from 10% to 40%, about 10× higher than that in the thin-film data. Characteristic K X rays from Fe and Cr are seen in the spectra. This background comes from photons produced by electrons scattered into the stainless steel target chamber, which can be seen through the detector collimators. Results for K-shell ionization for 25 keV for Ar and Kr are in good agreement with the trend of earlier measurements and with the theory of Jakoby *et al.* (1987), [39] even with the large TTB background.

Figure 15 shows a spectrum for 52 keV electrons on Ar. Here the TTB background is higher than that for 25 keV and has a pronounced "hump" in the higher k region that is typical of background that has penetrated the shielding.

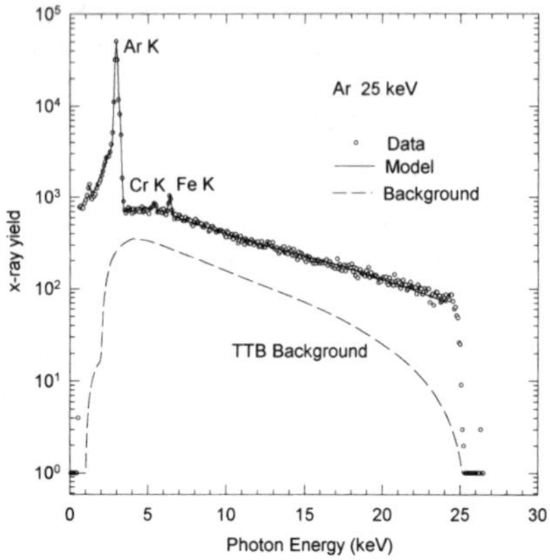

FIGURE 8.12 Plot of X-ray yield versus photon energy at 90° for 25 keV electrons on Ar. Background characteristic K X rays from Cr and Fe and the TTB background shown as the dashed line are due to electron scattering into the stainless steel target chamber.

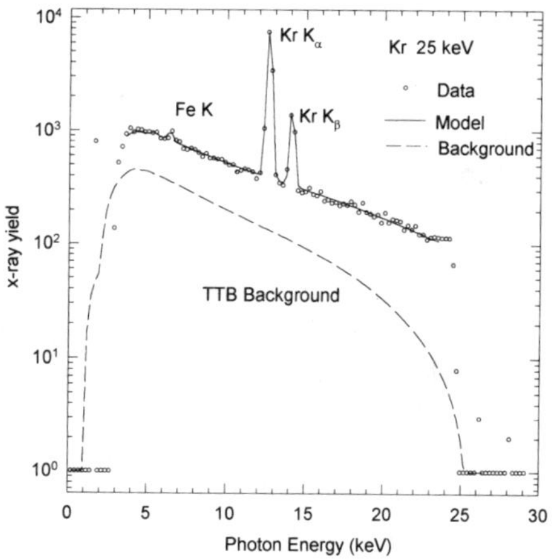

FIGURE 8.13 Plot of X-ray yield versus photon energy at 90° for 25 keV electrons on Kr. Background characteristic K X rays from Fe and the TTB background shown as the dashed line are due toelectron scattering into the stainless steel target chamber.

Characteristic X rays are also seen from Fe due to electrons scattered into the stainless steel scattering chamber and from Pb from incomplete absorption of photons produced in Pb shielding. Even with this "worst case" of TTB background, it is still possible to obtain a good fit and determine a reasonable value for the Ar K shell ionization cross section. However, this case is shown only to illustrate the extent of the TTB background problem for gas targets with higher electron energy. We require data with the background reduced by

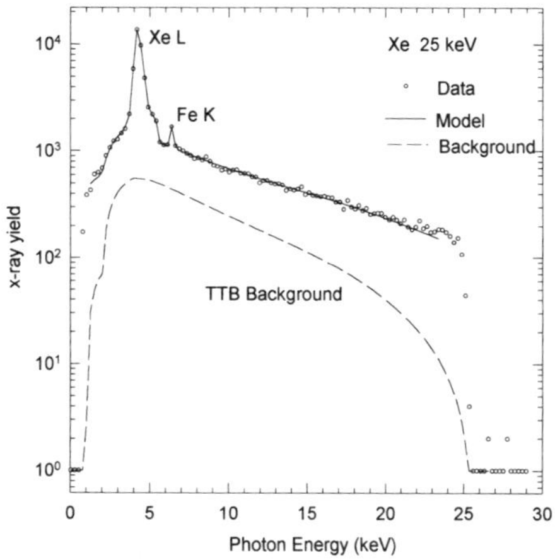

FIGURE 8.14 Plot of X-ray yield versus photon energy at 90° for 25 keV electrons on Xe. Background characteristic K X rays from Fe and the TTB background shown as the dashed line are due to electron scattering into the stainless steel target chamber.

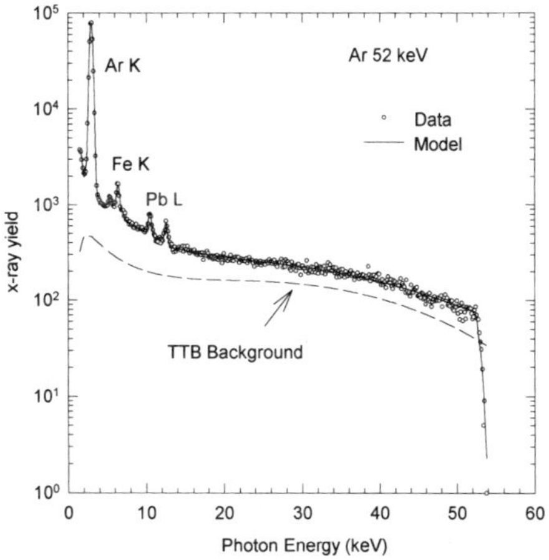

FIGURE 8.15 Plot of X-ray yield versus photon energy at 90° for 52 keV electrons on Ar, illustrating the larger TTB background at higher electron bombarding energy. The background characteristic X rays for Fe are from the target chamber and the X rays from Pb are from the lead collimater, which was not carbon lined.

about an order of magnitude before we can be as confident of the agreement of the data with the bremsstrahlung theory and of the deduction of the inner-shell ionization cross section as we are with the thin-film target data.

The data are not yet on an absolute scale since the solid angle for the cell geometry has not yet been determined absolutely. Efforts to go to 50 to 100 keV have met with problems of excessive background from electrons scattered into the scattering chamber, the detector collimator and photons Compton scattered from the Faraday cup. At these higher energies, TTB background approaches 90 % and dominates the thin target spectrum.

8.4.5 Designing the next generation of gas target experiments

Signal-to-noise ratio for gas targets is a problem as the beam energy is increased above 25 keV. Adequate shielding for a thin-film target at 100 keV is insufficient when the target is 10^4 times thinner. In principle, it is only necessary to increase the shielding to achieve adequate photon absorption. But our experimental design has met with several other problems related to electrons scattering in the target gas toward the detector. Part of this problem has arisen from the desire to have a very thin detector window to permit the measurement of characteristic X rays of a few keV with good efficiency, such as the Ar K X rays, for example. With a thin detector window, it is mandatory to deflect the scattered electrons from the window with very good efficiency to eliminate penetration of the window by the higher energy electrons. Magnetic deflection of the scattered electrons can reduce or eliminate the background from electron penetration of the detector window. However, unless care is taken to insure that the deflected electrons stop in a region well shielded from the detector, the deflection can introduce a significant TTB background. Our current detector collimator design has a collimator followed by a permanent magnet to deflect the electrons before the final detector collimator. The electrons stop in the region between the collimators, but this has not yet been adequate to reduce the photon background sufficiently for beam energies above 25 keV. It is planned to replace the permanent magnet with an electromagnet, so the deflection can be controlled and tuned for different beam energies. Also, with the electromagnet, the deflected electrons can be dumped in a second Faraday cup that can be well shielded from the detector. It is clear that the deflection system, while it will permit using thin window or windowless detectors, will produce a TTB bremsstrahlung background that is more complex than that described above. Here the background will arise from electrons stopped in the walls of the collimator before the deflection magnet. It is expected that the TTB background will have a significant "hump" due to incomplete photon absorption in the collimator. It is believed that the background can be identified by comparing runs with and without an electron absorber inserted before the first collimator.

8.4.6 K X-ray cross section measurements

The characteristic X-ray peaks can be seen in several of the Figures; both K and L X-ray spectra are seen. Determination of the K shell ionization cross section by normalizing to the atomic-field bremsstrahlung cross section is straightforward, and the fit program determines the normalized X-ray yield. Only the fluorescence yield is needed to determine the cross section from the fitted X-ray yield for the K shell. In case of the L shell, it is necessary to fit the peaks in detail to determine the ionization of the individual shells. Coster-Kronig transition corrections must also be made.

We have used the present model to determine K-shell ionization cross sections at 25 keV for both Ar and Kr, albeit with somewhat high TTB backgrounds. While we have an extensive data set from Ambrose *et al.*(1987) that include characteristic X rays, we have not yet carried out the analysis of that data with the present model except for Ag. In the thick-absorber difference experiments, we have also studied in detail only Ag at 50 keV. The results are promising, however. We find isotropy in the K X-ray rate from 0°, 45° and 90° within a relative error of 3%. We find a result of 59.8 ± 0.6 b for the cross section. This is the most precise result so far reported for this cross section. This result is in very good agreement with the recent results of Jakoby, et al (1987) and with their theoretical model. It is clear that similar results could be obtained for targets with other atomic numbers, within the limitations of the thick absorber technique discussed above. Also, in the fit, the $K_{\alpha 1}$ to $K_{\alpha 2}$ and K_{α} to K_{β} ratios in good agreement with theory.

Based on preliminary studies and analysis of the data of Ambrose et al (1987) [20] at 50, 75 and 100 keV for a range of Z's, we believe that the preliminary K-shell ionization cross section data reported by Westbrook and Quarles (1987) [19] at 100 keV is systematically low by about 10 to 20 % due to the neglect of TTB background in the normalization to atomic-field bremsstrahlung theory. Some of the data in the work of Westbrook and Quarles was taken over a limited photon energy range with a Si(Li) detector to provide X-ray spectra with good energy resolution. So, in this case it is very difficult to determine the extent of any small TTB background. While data for some elements were taken with both germanium and Si(Li) detectors, and good agreement was found, we now recognize that there is some TTB in the Ge detector data where the full bremsstrahlung spectrum is seen. Not including even a small TTB background can systematically yield a low value for the ionization cross section. It is expected that reanalysis of that subset of the data that includes the entire photon energy range of the bremsstrahlung spectrum, with the present model, which includes TTB background, will provide accurate K-shell ionization cross sections.

8.5 BEYOND THE PHOTON ENERGY SPECTRUM: COINCIDENCE EXPERIMENTS

While the focus of this chapter is on the photon energy and angular distribution, the so called doubly differential cross section, there is renewed interest in the triply differential cross section: the measurement of the emitted photon and the scattered electron in coincidence. This interest is stimulated by the new computer code developed by Tong, Shaffer and Pratt (Pratt, et al, 1995) [5]. Coincidence experiments in the field of bremsstrahlung date from the first experiment by Nakel in 1967 and include a limited number of measurements (Nakel, 1967; Faulk and Quarles, 1974; Aehlig, *et al.*, 1977; Komma and Nakel, 1982) [40–43]. This early era of experimental work was compared with the Born approximation and with an approximation (Elwert and Haug, 1969) [44], which was valid only for low atomic number. The experiments were generally not in very good agreement with the theory. So, the availability of new accurate theoretical predictions may be expected to stimulate a new round of experimental studies of the triply differential cross section for bremsstrahlung. Nakel and his students have pursued increasingly sophisticated coincidence experiments over the years using polarized electrons. Their work has been reviewed recently (Nakel, 1994) [45].

The coincidence between a K X ray and one scattered (or ejected) electron was first observed by Quarles and Faulk (1973) [46]. This process has been treated as a doubly differential cross section, $d^2\sigma/d\Omega_e dE$. That is, the emission of the X ray is assumed to be isotropic and the coincidence with the X ray serves to tag the event as coming from inner-shell ionization. If X-ray isotropy is not assumed, then the measured cross section is $d^3\sigma/d\Omega_e d\Omega_\gamma dE$. Over the years there have been several other experimental studies of this process (Sandner, 1990; Genz, 1990; Komma and Nakel, 1982) [47,48,43]. One common feature of the experimental results to date is that a higher than predicted coincidence rate is observed when the electron has nearly the maximum energy permitted. This apparent enhancement when one electron has high energy (and the other electron has low energy) may be an effect of polarization bremsstrahlung.

8.5.1 The proposed experimental setup for electron-photon coincidence experiments

The development of an electron-photon coincidence setup for a new generation of triply differential cross sections measurements has recently been described (Liu, 1993; Ambrose, 1995) [49,50]. The block diagram of the electronics is shown in Figure 16. The software for data acquisition and analysis has been written in Visual Basic (Ambrose, 1995). The electronics of the setup limits the number of electron and/or photon detectors to 12. However, the requirements of large solid angle and adequate detector shielding (not to mention cost of detectors) will limit the setup to four or five detectors. In this case we plan to have two detectors for electrons and two or three detectors for photons. This will permit the study of two or three photon emission geometries at the same time for each electron scattering angle. Thus, we can look for anisotropies in characteristic X-ray emission, for example. In some cases where the electrons share nearly equally the available energy we can look for

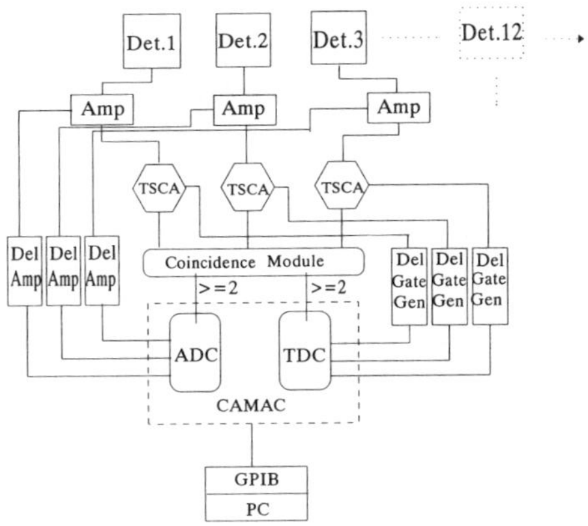

FIGURE 8.16 Block diagram of the electronics for a multi-particle (electron or photon) coincidence experiment set-up at TCU. Reprinted from Vinod Ambrose, C. A. Quarles, "An accelerator based multi-coincidence electron - photon spectrometer for the measurement of bremsstrahlung and inner-shell ionization cross sections," *Nucl. Instrum. Meth.* **B99** (1995) 170, with kind permission of Elsevier Science - NL, Sara Burgerhartstraat 25, 1055 KV Amsterdam, The Netherlands.

both electrons in the final state and examine triple coincidences such as (e, γ2e) processes or double bremsstrahlung such as(e, 2γ) or even (e, $e2\gamma$). The very small coincidence rate for these triple coincidence processes will no doubt preclude their measurement simultaneously with the double coincidence processes, but experimental conditions can be selected to emphasize these less common processes.

The main difference between the proposed experiments and earlier ones is the use of an energy-dispersive Si(Li) detector for the detection of the electron and measurement of its energy. Previous experiments have employed magnetic spectrometers to select the electron. The primary purpose of the spectrometers has been to eliminate the elastically scattered electrons and reduce the electron detector counting rate and thus the accidental coincidence rate. The magnet designed by Nakel has been especially effective in that it has a broad energy acceptance for energies below the elastic peak. In the proposed setup the accidental coincidence rate, while larger than the case with a magnetic spectrometer, is still acceptable. The coincident energy spectrum of the real coincidences can be determined by subtracting a background spectrum. This is illustrated in Figure 17 for one geometry for 70 keV electrons on a Cu target. The real coincidences are seen as a horizontal band at the Cu K X-ray energy, corresponding to the inner-shell ionization process, and as a diagonal band corresponding to the bremsstrahlung process. It is expected that the larger solid angle of the present set up will permit data acquisition with good statis-

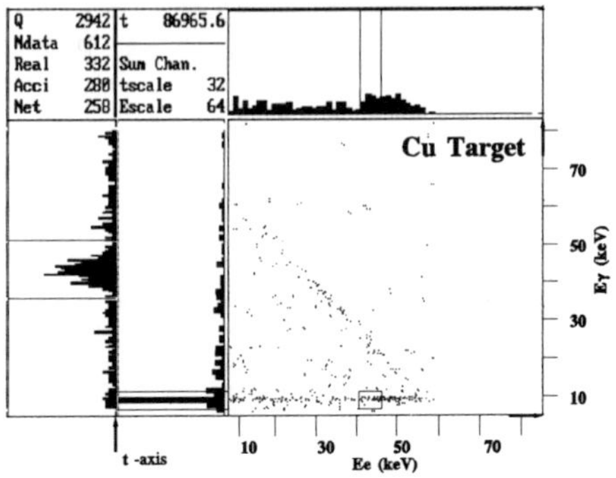

FIGURE 8.17 An electron-photon energy coincidence spectrum for 70 keV electrons on a Cu target. The K-shell ionization process shows up as a horizontal band centered on the Cu K X-ray energy of 8.9 keV, and the bremsstrahlung process shows up as a diagonal band. The sum spectra from the counts in each detector are shown along the left side and top. On the far left, the time spectrum is shown with the time axis indicated. The time scale is from -200 to +200 ns. The run shown had 612 events in a run time of 86965.6 sec. The small box shown selects a region of the coincidence spectrum for calculation of the coincidence yield shown in Figure 18. Reprinted from Jingai Liu and C. A. Quarles, "An accelerator based electron - photon coincidence spectrometer for the study of inelastic electron - atom collisions," *Nucl. Instrum. Meth.* **B79** (1995) 825, with kind permission of Elsevier Science - NL, Sara Burgerhartstraat 25, 1055 KV Amsterdam, The Netherlands.

tics in shorter run times than previous experiments, and allow the exploration of a wider range of measurement geometries.

Preliminary results with two detectors are illustrated in Figure 18. These results are an analysis of a coincidence spectrum such as shown in Fig. 17 for 70 keV electrons on Cu. Here we have plotted the coincidence yield versus electron and photon energy rather than the cross section since the response of the Si(Li) detector to electrons is yet to be determined and modeled in detail. Preliminary measurements with monoenergetic electrons indicate that the Si(Li) detector response effect is small for high energy electrons, but becomes increasingly important as the electron energy is decreased. The absorption in the detector window and the detector response will limit the lower energy at which reliable measurements can be made to above about 20 keV. So, it is proposed to select an initial electron energy of 75 - 100 keV. This represents a compromise between large solid angle, adequate detector shielding to eliminate coincident cross-talk between the detectors, and a sufficiently broad range of measurable scattered electron energy. Initial targets will be Cu and Ag because these have K X rays that permit the study of the K-shell ionization differential cross section as well as bremsstrahlung, and Au or UF_4 to investigate the bremsstrahlung from a high Z target. The selection of

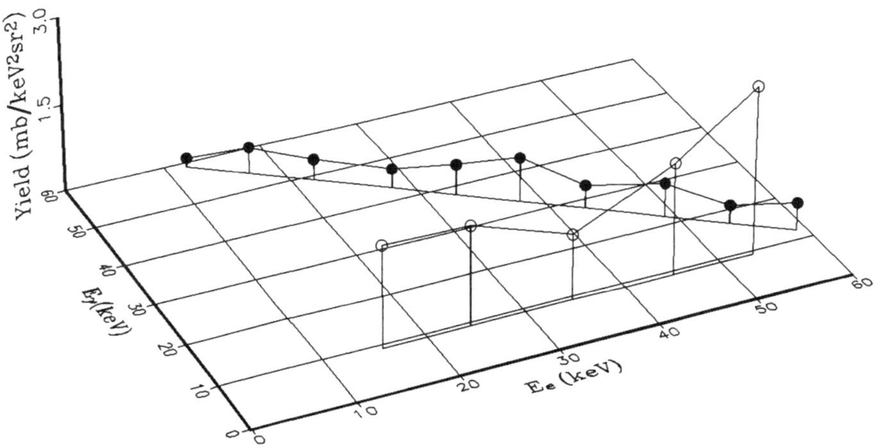

FIGURE 8.18 Preliminary results for the coincidence yield versus photon and electron energy for 70 keV electrons on Cu.. The yield is normalized to the number of incident electrons and to the target thickness but does not include correction for the efficiency or response of the electron detector. The open circles correspond to the differential cross section for K-shell ionization ($E_\gamma =$ 8.9 keV), and the solid circles are from the triply differential cross section for bremsstrahlung seen as a diagonal ($E_\gamma + E_e =$ 70 keV). Reprinted from Jingai Liu and C. A. Quarles, "An accelerator based electron - photon coincidence spectrometer for the study of inelastic electron - atom collisions," *Nucl. Instrum. Meth.* **B79** (1995) 825, with kind permission of Elsevier Science - NL, Sara Burgerhartstraat 25, 1055 KV Amsterdam, The Netherlands.

experimental conditions will be coordinated with the theorists (Shaffer, private communication) [51] so that the cross section cases measured can also be calculated.

8.6 CONCLUSIONS

There are still many areas for further study in X-ray production by electron bombardment. The following are some of the areas in which progress may reasonably be expected over the next several years. Most but not all of these areas have been alluded to already in this chapter. No doubt, there are still some surprises in store.

It is important to have more precise absolute cross sections for the photon energy and angular distribution. A goal of 5% for a few select "benchmark" cases seems within reach. The major obstacle in thin-film targets at higher energy is the precision of the determination of target thickness. New techniques for improved target-thickness determination are needed. Use of thin films directly deposited on carbon absorbers may overcome the target wrinkling problem. For gas targets, reduction of background remains a major concern as the electron energy is increased.

Measurements of bremsstrahlung at angles other than 90° are needed to broaden the range of parameters tested. While recent measurements at 0° are helpful, data for a wider range of angles, including angles greater than 90°, also remain important. By simultaneously measuring the spectrum at different angles with multiple detectors, it should be possible to realize a precision on the relative angular distribution of the order of a few percent, similar to that presently obtained for the photon energy dependence. This precision, or perhaps even better, is needed if effects due to polarization bremsstrahlung are to be observed.

The experiments up to now on the photon energy and angular distribution have been inclusive. That is they have not distinguished the final state of the target. Coincidence experiments in which the bremsstrahlung in coincidence with a recoil ion is observed would allow investigating processes such as radiative ionization.

Only a few bremsstrahlung measurements have been made on molecules. For higher energy electrons, the bremsstrahlung from the molecule is the incoherent sum of that from the constituent atoms; that is, the bremsstrahlung is additive. However, when the electron energy is a few keV and the photon energy is less than a keV, interference effects in the photon spectrum have been predicted, but have not yet been observed.

More precise studies of both K- and L-shell ionization cross sections are needed. Experiments that systematically examine both the energy and atomic number dependence of the cross section are necessary if the apparent discrepancies between theory and experiment are to be elucidated and resolved.

A new generation of measurements of the bremsstrahlung triply differential cross section is now timely with the availability of new calculations at the level of accuracy that before has only been possible for the photon energy and angular distribution. With the electron-photon coincidence measurement, the parameter space is further increased. The role of experiment continues to be to provide a test of a selection of well-chosen points in the vast parameter space. Selection of parameters for study need to be made with both experimental and theoretical (calculational) limitations in mind so that there can be maximal overlap between theory and experiment. The experimental setup for the triply differential cross section can also provide new data on the doubly differential cross section for inner-shell ionization, and perhaps can be tuned for new data on such processes as double bremsstrahlung (e,2γ), and processes with three particles in the final state such as (e,2eγ) and (e,e2γ).

Bremsstrahlung spectra have been used to determine the photon energy dependence of the efficiency of X-ray detectors. This technique is complimentary to the usual radioactive sources or X-ray fluorescence techniques, and depends on a theory that is independent of those that the other techniques depend upon. Further work is needed to establish this technique and develop it into a useful and widely available alternative. So far, only the thin-target spectrum has been used, but use of the thick-target spectrum needs to be investigated.

Development of the thick-target model spectra and additional experimental studies of thick-target spectra are needed for a range of electron energies, target atomic numbers and for different geometries.

The transition from thin- to thick-target has not yet been investigated, and there may be a number of cases where this is of more than academic interest, such as in gases at moderate to high pressure, or in electron collision with clusters.

ACKNOWLEDGMENTS

The author acknowledges with much appreciation the support for research provided by the Welch Foundation, the Chevron Petroleum Technology Co., the W. A. "Tex" Moncrief Chair, and the TCU Research Fund. Special thanks to Vinod Ambrose for his work on the Monte Carlo program and his assistance in preparing this chapter. Thanks to the students and postdocs who have carried out much of the work discussed here: Joseph Altman, Rajkumar Ambrose, Vinod Ambrose, Lee Estep, Darren Kahler, Halim Lehtihet, Tammy Lehtihet, Jingai(Annie) Liu, Mars Semaan, Greg Westbrook, and Jason Zhao. Many thanks for the help of Michael Murdock and Jim Williams in the TCU machine shop.

REFERENCES

1. S. M. Seltzer and Berger, M. J., *Nucl. Inst. Meth. Phys. Res.* **B12** 95 (1985).
2. Pratt, R. H. and I.- J. Feng, in *Atomic Inner Shell Physics*, ed. B. Crasemann New York: Plenum Press, 1985, p. 533.
3. Amusia, M. Y. and R. H. Pratt, *Commun. At. Mol. Phys.* **28** 247 (1992).
4. Hippler, R. and H. Schneider, *Nucl. Inst. Meth. Phys. Res.* **B87** 267 (1994).
5. Pratt, R. H., C. David Shaffer, N. b. Avdonina, Xiao-Min Tong and Viorica Florescu, *Nucl. Inst. Meth Phys. Res.* **B99** 156 (1995).
6. Tseng, H. K. and R. H. Pratt, *Phys. Rev.* **A3** 3 (1971).
7. Kissel, L., C. A. Quarles, and R. H. Pratt, *Atom. Data Nucl. Data Tables* **28** 381 (1983).
8. Tystovich, V. N. and I. M. Oiringel, editors, it Polarization Bremsstrahlung of Particles and Atoms, New York: Plenum Press, 1992.
9. Amusia, M. Y. and A. V. Korol, *Nucl. Inst. Meth Phys. Res.* **B79** 146 (1993).

10. Korol, A. V., A. G. Lyalin, A. S. Shulakov, and A. V. Solov'yov, *J. Phys. B: At. Mal. Opt.Phys.* **28** L155 (1995).
11. Korol, A. V., *J. Phys.* **B25** L341 (1992).
12. Zimkina, T.M., A. S. Shulakov, and A. P. Brajko, *Fiz. Tverd. Tela.* **23** 2006 (1981).
13. Zimkina, T.M., A. S. Shulakov, A. P. Brajko, A. P. Stepanov and V. A. Fomichev, *Fiz. Tverd. Tela.* **26** 1981 (1984).
14. Verkhovtseva, E. T., E. V. Gnatchenko, B. A. Zon, A. A. Nekipelov, and A. A. Tkachenko, *Zh. Eksp. Teor. Fiz.* **98** 797 (1990).
15. Ishii, K. and S. Morita, *Phys. Rev. A* **30** 2278 (1984);
16. Ishii, K. and S. Morita,*Phys. Rev. A* **31** 1168 (1985).
17. Ishii, K., *Nucl. Inst. Meth Phys. Res.* **B99** 163 (1995).
18. Powell, C. J., *Rev. Mod. Phys.* **48** 33 (1976).
19. Westbrook, G. L. and C. A. Quarles, *Nucl. Inst. Meth. Phys. Res.* **B24/25** 168 (1987).
20. Ambrose, R, J. C. Altman, and C. A. Quarles, *Phys. Rev. A* **35** 529 (1987).
21. Debertin, K. and R. G. Helmer, *Gamma- and X-ray Spectrometry with Semiconductor Detectors*, Amsterdam:North Holland, 1988.
22. Knoll, Glenn F., *Radiation Detection and Measurement 2nd ed.*, New York: John Wiley & Sons, 1989.
23. Altman, J. C., R. Ambrose, C. A. Quarles, and G. Westbrook, *Nucl. Inst. Meth Phys. Res.* **B24/25** (1987) 1028.
24. Weathers, D. L., J. L . Duggan, M. R. McNeir, Y. C. Yu, F. D. McDaniel, C. A. Quarles, H. Lehtihet, and D. Kahler, *Nucl. Inst. Meth. Phys. Res.* **B56/57** 964 (1991).
25. Storm, E., *Phys. Rev.* **A5** 2328 (1972).
26. Storm, E. and H. I. Israel, *Nucl. Data Tables* **7** 565 (1970).
27. Berger, M. J. and S. M. Seltzer, *NASA* **SP3012** (1964).
28. Ambrose, R., D. L. Kahler, H. E. Lehtihet, and C. A. Quarles, *Nucl. Inst. Meth. Phys. Res.* **B56/57** 327 (1991).
29. Korol, A. V., *private communication.*
30. Quarles, C. A. and D. B. Heroy, *Phys. Rev. A* **24** 48 (1981).
31. Tseng, H. K., R. H. Pratt, and C. M. Lee, *Phys. Rev. A* **19** 187 (1979).
32. August, H. J., and J. Wernisch, *Phys. Status Solidi* **A114** 629 (1989).
33. Hippler, R., K. Saeed, I. McGregor, and H. Kleinpoppen, *Phys. Rev. Lett.* **46** 1622 (1981).
34. Semaan, M and C. A. Quarles, *Phys. Rev.* **A 24** 2280 (1981).
35. Semaan, M and C. A. Quarles, *Phys. Rev A* **26** 3152 (1982).
36. Estep, Lee and C. A. Quarles, *Physica* **145C** 369 (1987).
37. Quarles, C. A. and Lee Estep, *Phys. Letters* **110A** 387 (1985).
38. Quarles, C. A. and Lee Estep, *Phys. Letters* **114A** 9 (1986);
39. Jakoby, C., H. Genz, and A. richter, *Jour. de Physique* **48** C9-487 (1987).
40. Nakel, W., *Phys. Lett.* **A25** 569 (1967).
41. Faulk, J. D. and C. A. Quarles, *Phys. Rev.* **A9** 732 (1974).
42. Aehlig, A., L. Metsger and M. Scheer, *Z. Phys. A* **281** 205 (1977).

43. Komma, M. and W. Nakel, *J. Phys. B* **15** 1433 (1982).
44. Elwert, G. and E. Haug, *Phys. Rev.* **183** 90 (1969).
45. Nakel, W. , *Phys. Lett. C* (1994).
46. Quarles, C. A. and J. D. Faulk, *Phys. Rev. Lett.* **31** 859 (1973).
47. Sandner, W. and C. E. Theodosiou, *Phys. Rev.* **A42** 5208 (1990).
48. Genz, H., *X-ray and Inner Shell Processes*, eds. T. A. Carlson, M. O. Krause and S . T. Manson, New York: AIP Press, 1990 p. 375.
49. Liu, Jingai and C. A. Quarles, *Nucl. Inst. Meth Phys. Res.* **B79** 825 (1993).
50. Ambrose, V. and C. A. Quarles, *Nucl. Inst. Meth Phys. Res.* **B99** 170 (1995).
51. Shaffer, David, *private communication.*

CHAPTER 9

Parametric X Rays

Alexander V. Shchagin* and Xavier K. Maruyama†

The parametric X-ray radiation (PXR) is a new type of monochromatic, polarized, directed, tunable radiation arising in the vicinity of Bragg directions of a crystal when relativistic particles are passing through it. For the last ten years PXR properties have been intensively investigated experimentally by researchers in the former SU, the USA, and in other countries. In this review the nature and main properties of PXR as a new source of a bright monochromatic X-ray beam are considered: energy, width, smooth tuning of the PXR spectral line; fine structure and absolute differential yields of PXR in the vicinity of and at angular distances from Bragg directions; angular spread of the PXR beam; the influence of incident electron energy and of the density effect on the PXR properties; linear polarization of PXR; background in PXR spectra. Experimental setups for linacs and the results of measurements are discussed. Experimental data are compared to theoretical calculations at PXR energies between 5 and 400 keV for incident electron energies ranging from 15 to 1200 MeV. Possible applications of PXR as a source of a bright, tunable X-ray beam are discussed.

9.1 INTRODUCTION

The radiation generated by a charged particle moving through a medium with a periodically varying permittivity was first predicted by Fainberg and Khizhnyak [1] in 1957. Later, the radiation of charged particles moving through a crystal in the X-ray band was investigated by Ter-Mikaelian [2]. The properties of this radiation were theoretically considered in papers [3–12], where several variants of theoretical descriptions of parametric X-ray radiation (PXR) were proposed. After experimental discovery of PXR in 1985 in Tomsk

*) corresponding author. Kharkov Institute of Physics and Technology
Kharkov 310108, Ukraine. email: shchagin@av.kharkov.ua
†) Department of Physics, Naval Postgraduate School
Monterey, CA 93940, USA

[13] experimental studies of PXR properties were undertaken in the former Soviet Union in the late 80's, described in refs. [14–16] (now Russia), [17] (now Armenia), [18–21,38] (now Ukraine). In the 90's, investigations in this field were begun in the USA [22–24], Japan [26], Canada [24], and Germany [27].

9.1.1 The nature of PXR

One of the basic results of the electrodynamic theory is that a charged particle, moving rectilinearly and uniformly, does not emit any radiation. Therefore, the surprising discovery in the 30's of coherent radiation by a charged particle moving rectilinearly and uniformly through a homogeneous medium was commended with the Nobel Prize in 1958. This radiation, usually named as Cherenkov or Vavilov-Cherenkov radiation [28–30], is possible at a certain angle to the particle velocity vector $\vec{V}$ only if the particle velocity $\left|\vec{V}\right|$ exceeds the phase velocity of the electromagnetic wave in the medium $\frac{c}{\sqrt{\varepsilon}}$, where $\varepsilon = \varepsilon(\omega)$ is the medium permittivity, and c is the light velocity. In the 40's, Ginzburg and Frank predicted transition radiation to arise as the charged particle crosses the boundary of two media with different permittivities [31]. But the Cherenkov radiation and the transition radiation were not the last unexpected radiation effects for a particle moving rectilinearly and uniformly. The next radiation effect of surprise was predicted theoretically by Fainberg and Khizhnyak in ref. [1]. They discovered that a charged particle moving in the medium will emit coherent radiation even at $\left|\vec{V}\right| < \frac{c}{\sqrt{\varepsilon}}$, provided that the parameters of medium permittivity vary periodically. In ref. [1] they called this radiation "parametric Cherenkov radiation".

The well-known Huyghens construction for the Cherenkov radiation is shown in fig. 9.1. The charged particle is moving through the homogeneous medium with velocity $\left|\vec{V}\right| > \frac{c}{\sqrt{\varepsilon}}$. The particle-excited spherical secondary waves from continuously distributed points P1 - P3 lag behind the particle located at point P4. Therefore, coherent Cherenkov radiation is generated without any phase shift, due to the interference of secondary waves, on condition that

$$\cos\theta = \frac{c}{\sqrt{\varepsilon}V}, \tag{9.1}$$

where θ is the angle between the vector $\vec{V}$ and the unit vector $\vec{\Omega}$ for the radiation direction. Condition 9.1 can be written in vector form as

$$c - \sqrt{\varepsilon}\vec{V}\cdot\vec{\Omega} = 0 \tag{9.2}$$

In fig. 9.2 the Huyghens construction is shown for the parametric Cherenkov radiation. Here the charged particle is moving through the medium with a

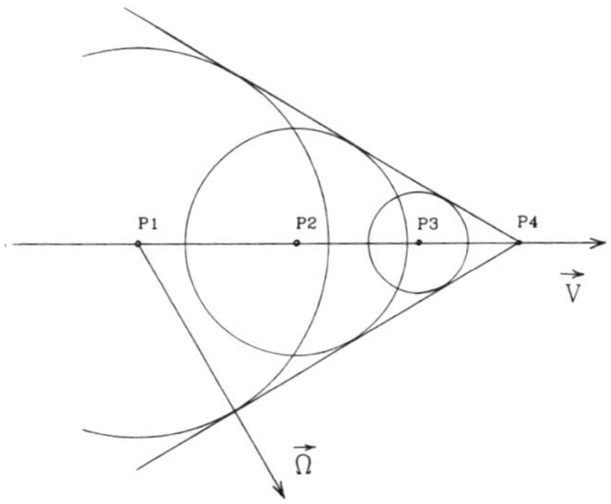

FIGURE 9.1 The Huyghens construction for Cherenkov radiation. The charged particle is moving through a homogeneous medium with velocity $\left|\vec{V}\right| > \frac{c}{\sqrt{\varepsilon}}$. The spherical waves of secondary radiation, excited by the particle in the medium, are emitted from continuously distributed points $P1 - P3$. The direction of coherent radiation $\vec{\Omega}$ may be found from equations 9.1, 9.2.

periodically varying permittivity at a velocity $\left|\vec{V}\right|$ lower than the phase velocity of the electromagnetic wave in the medium $\frac{c}{\sqrt{\varepsilon_0}}$, where $\varepsilon_0 = \varepsilon_0(\omega)$ is the constant part of the medium permittivity. The medium periodicity goes along the reciprocal lattice vector $\vec{g}$. Note that $|\vec{g}| = \frac{2\pi}{d}$, where d is the period in direct space. The bold lines in fig. 9.2 show the periodic structure and may be considered as a family of crystallographic planes, if the periodic medium is a single crystal. In contrast to the Cherenkov radiation, the particle-excited spherical waves of secondary radiation come from the periodically-sited points P1 - P3. The corresponding spherical waves of secondary radiation are shown by circles C1-C3. They run ahead of the particle sited at point C4, because $\left|\vec{V}\right| < \frac{c}{\sqrt{\varepsilon_0}}$. But for the arbitrarily chosen observation direction $\vec{\Omega}$ coherent monochromatic radiation may occur due to the interference of secondary waves at certain phase shifts, if the spacing between equal phase surfaces C1-C3 is equal to a whole number n_λ of wavelengths λ

$$\frac{c}{\sqrt{\varepsilon_0}} \triangle t - \vec{V} \cdot \vec{\Omega} \triangle t = n_\lambda \lambda, \tag{9.3}$$

where $\triangle t = \frac{2\pi}{\left|\vec{g} \cdot \vec{V}\right|}$ is the secondary radiation flash periodicity at points P1-P3. In fig. 9.2 the coherent radiation waves in the arbitrarily chosen direction $\vec{\Omega}$ are shown for $n_\lambda = 1$. Using the formula $\omega = \frac{c}{\sqrt{\varepsilon_0}} \frac{2\pi}{\lambda}$ and eq. 9.3, we may find the frequency ω_{CR} and energy E_{CR} of parametric Cherenkov radiation

$$E_{CR} = \hbar\omega_{CR} = n_\lambda \frac{c\hbar \left|\vec{g} \cdot \vec{V}\right|}{c - \sqrt{\varepsilon_0}\vec{V} \cdot \vec{\Omega}}, \tag{9.4}$$

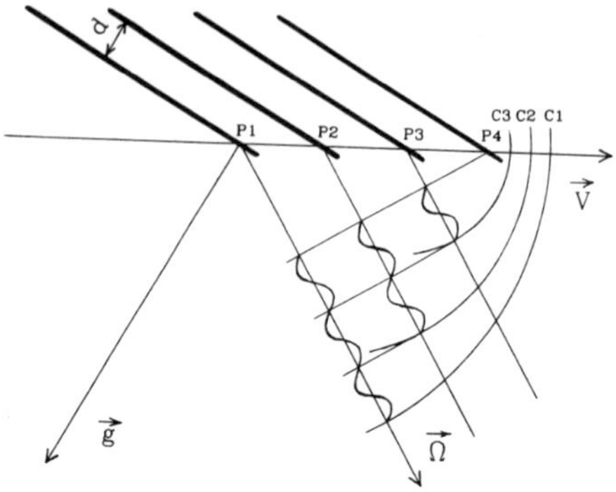

FIGURE 9.2 The Huyghens construction for parametric Cherenkov radiation and for parametric X-ray radiation from a crystal. The charged particle is moving through a periodical medium with velocity $\left|\overrightarrow{V}\right| < \frac{c}{\sqrt{\varepsilon_0}}$. The bold lines show the permittivity periodicity $\varepsilon = \varepsilon(x)$ with period d. The vector $\overrightarrow{g}$ is the reciprocal lattice vector. The spherical waves of secondary radiation $C1 - C3$, excited by the particle in the medium, are emitted from periodically distributed points $P1 - P3$. For any observation direction $\overrightarrow{\Omega}$ the frequency and the energy of coherent radiation is given by equations 9.4, 9.5. The formation of coherent radiation is shown for the case where the distance between neighboring spheres of equal phases is equal to one radiation wavelength.

In ref. [1] the corresponding equation was obtained for a layered medium at $\overrightarrow{V} \parallel \overrightarrow{g}$. For a crystal medium the formula of type 9.4 was derived in ref. [2].

In particular, in the case of homogeneous medium $n_\lambda = 0$ or $d \to \infty$, therefore the numerator in 9.4 is equal to 0 and radiation is possible only if the denominator is 0. So, only common Cherenkov radiation (see 9.2) can take place in a homogeneous medium. In the following formulae for a crystal medium n_λ will be included in $\overrightarrow{g}$, because the radiation from a crystal is not possible for every n_λ due to peculiarities of the crystal structure.

In the X-ray band at $\hbar\omega \gg E_K$, where E_K is the K-shell ionization energy of medium atoms, the permittivity is usually less than unity, $\varepsilon_0 = 1 - \left(\frac{\omega_p}{\omega}\right)^2$. Therefore, the phase velocity $\frac{c}{\sqrt{\varepsilon_0}} > |\overrightarrow{V}|$, and the usual Cherenkov radiation is impossible, whereas the parametric Cherenkov radiation can occur for any crystal orientation and any observation angle. Also the nature of parametric Cherenkov radiation completely differs from channeling radiation [32], which occurs in the forward direction only from a channeling particle.

The parametric Cherenkov radiation is a nice effect and since 1957 it has been considered theoretically by numerous authors from different points of view. As a result, about ten different names for this effect may be found in the literature. We will use the term "parametric X-ray radiation" (PXR), which has been accepted in most experimental publications for the last ten years. More detailed discussions about the nature of parametric Cherenkov

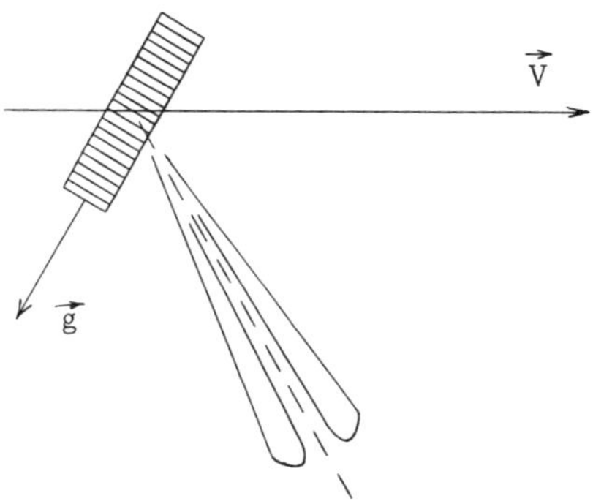

FIGURE 9.3 PXR generation in the Laue geometry. The electron beam passes through a crystal slab with velocity $\vec{V}$. The vector $\vec{g}$ is the reciprocal lattice vector for crystallographic planes shown by hatching on the target body. The dashed line shows the Bragg direction for initial direction $\vec{V}$ relative to the crystallographic planes. The cone of the PXR beam with the gap in the middle of the cone is generated around the Bragg direction.

radiation in X-ray band and about the nomenclature can be found in refs. [11,12].

9.1.2 PXR reflection

Ter-Mikaelian was the first to theoretically investigate, in ref. [2], PXR properties in the X-ray band from a relativistic charged particle moving through a single crystal (he called it "resonance radiation"). From his theory it may be seen that the PXR angular distribution has a sharp conical maximum (reflection) with the gap in the middle, found around the Bragg direction. The angular size of the PXR reflection is about or greater than $\frac{1}{\gamma}$, where γ is the relativistic factor of the incident particle.

By analogy with the ordinary diffraction of X rays in a crystal, there are two schemes for PXR generation: one is the Laue geometry shown in fig. 9.3, and the other is the Bragg geometry shown in fig. 9.4. In both cases, the beam of charged relativistic particles with velocity $\vec{V}$ passes through a crystal-radiator, and a narrow PXR beam (reflection) of conical shape is generated around the Bragg direction (dashed lines) with respect to the crystallographic planes. Most of the experiments described below were carried out in the Laue geometry.

In this chapter we shall describe mainly the recent results and investigations of PXR differential properties. Differential measurements are usually performed with a small-size X-ray spectrometric detector to provide angular resolution much better than the fine structure of the PXR reflection shown in figs. 9.3 and 9.4. To establish the validity of a theory, the results of spec-

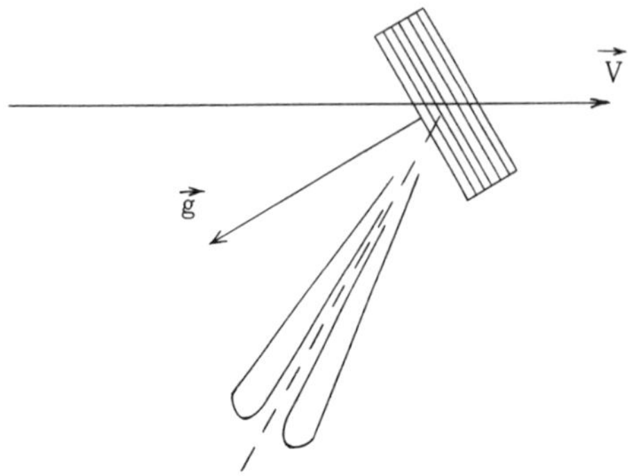

FIGURE 9.4 PXR generation in the Bragg geometry. In contrast to fig. 9.3, the crystallographic planes lie parallel to the slab surface. The notation is the same as in fig. 9.3.

trometric differential measurements are directly compared to calculations by differential theoretical formulae.

9.2 PXR FROM A SINGLE CRYSTAL

9.2.1 Experimental

The first semi-differential spectrometric measurements of PXR were performed in Tomsk [14,15] using an X-ray detector with a slit collimator. These measurements at electron beam energies of 500 and 900 MeV may be considered as differential only for one dimension and as integral for the second dimension.

The first completely differential measurements were performed in Kharkov [19,33,34] with the use of a circular collimator placed in front of the X-ray detector. To ensure differential measurements, the collimator size allowed to pass to the detector an angular part that is very small relative to the total PXR reflection (see figs. 9.3, 9.4). Below we shall describe these experiments in more detail.

The experimental measurements were performed at the exit of the linac LUE-40 at the Karkhov Institute of Physics and Technology at electron beam energies $E_e = 15.7 - 25.7$ MeV. In the first of these investigations [19] the PXR differential yield in arbitrary units was obtained by normalizing the PXR quanta number to the quanta number of the characteristic X-ray K-radiation of target atoms. After setting the experimental procedure of absolute measurements, simultaneous experiments were carried out to investigate the absolute Si K-ionization cross section [36], the PXR absolute differential yield in the vicinity of the Bragg direction (BD) [33,37], and the PXR absolute differential yield at a significant angular distance from BDs [34].

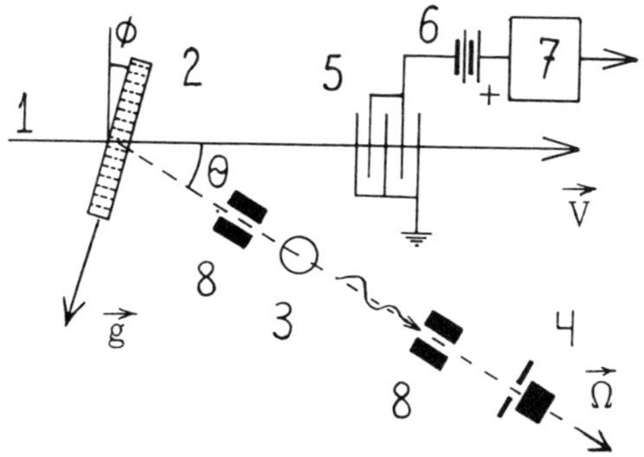

FIGURE 9.5 The experimental setup for incident electron energies 15.7 – 25.7 MeV. Electron beam 1 is shown by vector $\vec{V}$. Vector $\vec{g} = \langle 111 \rangle$ is the reciprocal lattice vector for the crystallographic planes (111) shown by hatching on Si single crystal target 2. Target 2 is placed in a vacuum 3-axis goniometer and the angle ϕ is the angle of crystal rotation. The Bragg direction is coincident with the observation direction $\vec{\Omega}$ if $\phi = \frac{\theta}{2}$. The Si target thickness is 17 μm. The Si(Li) X-ray detector with lead collimator 4 is at a distance of 4520 mm from the target at a fixed angle $\theta = 305.9$ $mrad$. The observation direction is shown by the unit vector $\vec{\Omega}$. Cleaning magnet 3 and collimators 8 are to reduce the background in X-ray spectra. Secondary emission monitor 5 and charge-to-code convertor 7 with battery 6 are for beam charge monitoring. Reprinted from A. V. Shchagin, V.I. Pristupa, N. A. Khizhnyak, "K-shell ionization cross section of Si atoms by relativistic electrons," *Nucl. Instrum. Meth.* **B84** (1994) 9, with kind permission of Elsevier Science - NL, Sara Burgerhartstraat 25, 1055 KV Amsterdam, The Netherlands.

The experimental layout is shown in fig. 9.5, taken from ref. [36]. The beam of accelerated electrons *1* passes in vacuum through a target *2* and a secondary emission monitor *5*. A 17 μm thick silicon single crystal plate was used as a target *2*. The crystallographic plane $(\bar{2}20)$ was parallel to the plate surface. The target was placed in a vacuum 3-axis goniometer and was prealigned so that the $\langle \bar{2}20 \rangle$ axis would be parallel to the velocity vector $\vec{V}$ of the beam electrons, and the crystallographic plane (111) would be perpendicular to the detection plane and parallel to the vector $\vec{V}$. The measurements of orientational dependences in figs. 9.8,9.9, 9.11,9.12, 9.13,and 9.14 were performed as a function of the target rotation angle ϕ relative to the axis perpendicular to the detection plane in the direction of the detector. At $\phi = 0$ the crystallographic plane (111) is parallel to the electron beam and the reciprocal lattice vector $\vec{g} = \langle 111 \rangle$ is perpendicular to the electron beam.

The current pulse frequency of the accelerator was 50 Hz, and the beam pulse duration was 4 μs. To avoid X-ray spectrum distortions due to the pile-up effect, the mean beam current value was chosen to be such that the counting rate of spectrometer should be below 5 Hz. Typically, it was a few nA. The

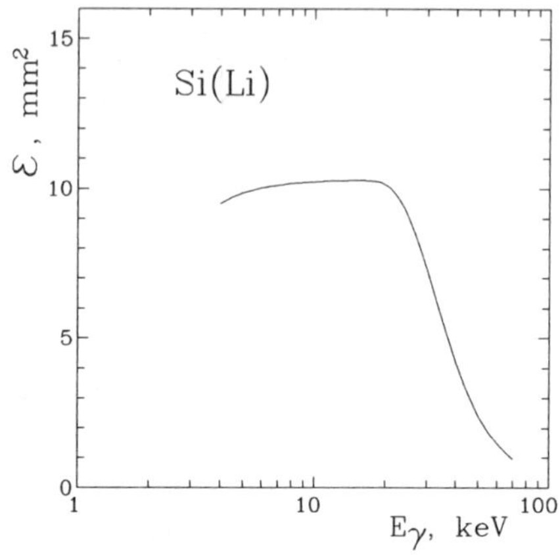

FIGURE 9.6 Calculated effective area of the Si(Li) X-ray detector with the lead collimator as a function of incident photon energy E_γ. The diameter of the lead collimator is 3.7 *mm*.

charge of the beam that has passed through the target during the exposure time was measured by secondary-emission monitor *5* and by charge-to-code converter *7* with battery *6*. The conversion factor of the secondary-emission monitor was measured using the Faraday cup of the linac.

The spectra were measured using a Si(Li) X-ray detector *4* with liquid nitrogen cooling. The detector was placed at a distance of 4520 *mm* from the target, at angle $\theta = 305.9$ *mrad* to the beam axis, and was connected to the goniometer chamber by a vacuum tube. The unit vector $\vec{\Omega}$ shows the observation direction. The cleaning magnet *3* was placed in the middle of the vacuum tube to protect the detector from electrons scattered in the target. Collimators *8* were used to reduce the background in X-ray spectra. A lead collimator with a circular hole was installed in front of detector *4*. The diameter of the detector aperture with this collimator was 3.7 *mm*. The diameter of electron beam spot on the target was about 7 *mm*. So, the angular resolution of detecting system of about 2 *mrad* was much less than the PXR reflection angular size $\frac{1}{\gamma}$, which is about 20 *mrad* at an incident electron energy $E_e = 25$ MeV.

The energy resolution of the X-ray spectrometer was found to be 286 eV at $E_\gamma = 6.4$ keV. The detector efficiency as a function of X-ray quanta energy was measured through the use of a set of intensity-calibrated X-ray radioactive sources. The calculations of the detector effective area for the Si(Li) detector of about 4 *mm* thickness are shown in fig. 9.6. The photon energy band 4-21 keV is best suited to absolute measurements of the PXR yield (see figs. 9.7, 9.8, 9.9) due to the maximal and almost constant value of effective area. At higher photon energies, the detector efficiency decreases due to increasing detector transparency for incident photons. At lower photon energies the detector efficiency is reduced due to photon flux attenuation in the Be window, gold and dead layers of the detector.

The experimental procedure for PXR investigations has been described in detail in ref. [21]. The absolute measurements technique has been described in detail in ref. [36]. The absolute measurements error of quanta number in figs. 9.11, 9.12, and 9.16 did not exceed 15% of the measured value [33].

Similar layout and procedure were used in another experiment [20] performed in Kharkov on the accelerator LUE-2 for PXR measurements at $E_e = 1200$ MeV. In that experiment, PXR was detected from the crystallographic plane (220) of a Si target, 50 μm thick, with a Ge(Li) detector placed at angle $\theta = 17.88\ mrad$.

9.2.2 PXR coherence, spectral peak energies and widths

Experimental studies of PXR coherence properties were first performed in refs. [19,20]. The typical X-ray spectrum measured in refs. [19,34] is shown in fig. 9.7. The 1.8 keV peak is due to the characteristic X-ray radiation of Si target atoms produced after K-shell ionization by the incident electrons [36]. The 12.1 keV PXR peak has high intensity because it is observed close to the BD for $\vec{g} = \langle 111 \rangle$ [19,33]. The weak 4.0 and 16.1 keV PXR peaks are observed at a significant angular distance from the corresponding BDs [34]. Note the low background in the measured spectra.

In figs. 9.8 and 9.9 from ref. [33] the peak energies (except the characteristic K-radiation) in the measured X-ray spectra versus the target orientation are shown by full circles. Solid lines show the calculations of coherent radiation energy according to the formula [2,19]

$$E_{CR} = \hbar\omega_{CR} = \frac{c\hbar \left| \vec{g} \cdot \vec{V} \right|}{c - \sqrt{\varepsilon_0}\vec{V} \cdot \vec{\Omega}}, \tag{9.5}$$

where $\hbar$ is the Planck constant divided by 2π, ε_0 is the constant part of the medium permittivity, $\vec{g}$ is the reciprocal lattice vector, $\vec{\Omega}$ is the unit vector in the direction of the detector, and c is the light velocity. For instance, $c\hbar|\vec{g}| = 3.954$ keV for $\vec{g} = \langle 111 \rangle$ in Si at room temperature. The solid lines E, D, C show the calculations by formula 9.5 for the rows of reciprocal lattice vectors $\vec{g}_{1i}$=...$\langle \bar{1}\bar{1}3 \rangle$,$\langle 11\bar{1} \rangle$...; $\vec{g}_{2i}$=...$\langle \bar{1}\bar{1}5 \rangle$, $\langle 111 \rangle$, $\langle 33\bar{3} \rangle$...; $\vec{g}_{3i}$=...$\langle 004 \rangle$,$\langle 220 \rangle$..., respectively. From this figure one can see that the experimental data for all peaks at any target orientation are in equally excellent agreement with the coherent radiation energies calculated by formula 9.5. Note that the radiation associated with the reciprocal lattice vector $\vec{g} = \langle 111 \rangle$ from the row $\vec{g_{2i}}$ is detected to have a much higher intensity than the contribution of other vectors from the row $\vec{g_{2i}}$, because the center of PXR reflection

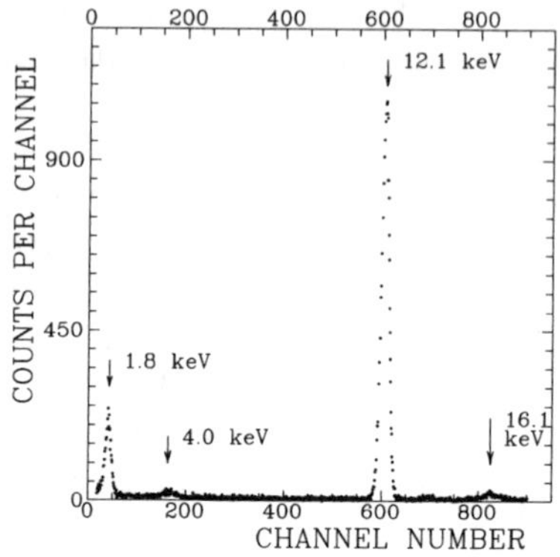

FIGURE 9.7 From ref. [34]. Experimental X-ray spectrum at $\phi = 142.9\ mrad$ and at $E_e = 25.0$ MeV without background subtraction. The experimental conditions are shown in fig. 9.5. The 1.8 keV spectral peak is due to characteristic K-radiation of Si target atoms excited by relativistic electrons. The 4.0 and 16.1 keV peaks are due to PXR at significant angular distances from the corresponding Bragg directions. The 12.1 keV peak is mainly due to PXR in the vicinity of Bragg direction for $\vec{g} = \langle 111 \rangle$. The 4.0, 12.1, 16.1 keV peaks are associated with the rows of reciprocal lattice vectors $\vec{g_{1i}}$, $\vec{g_{2i}}$, $\vec{g_{3i}}$, respectively. Reprinted from A. V. Shchagin, V.I. Pristupa, N. A. Khizhnyak, "Parametric X-ray radiation from relativistic electrons in a crystal in the vicinity of and at angular distance from a Bragg direction," *Nucl. Instrum. Meth.* **B99** (1995) 277, with kind permission of Elsevier Science - NL, Sara Burgerhartstraat 25, 1055 KV Amsterdam, The Netherlands.

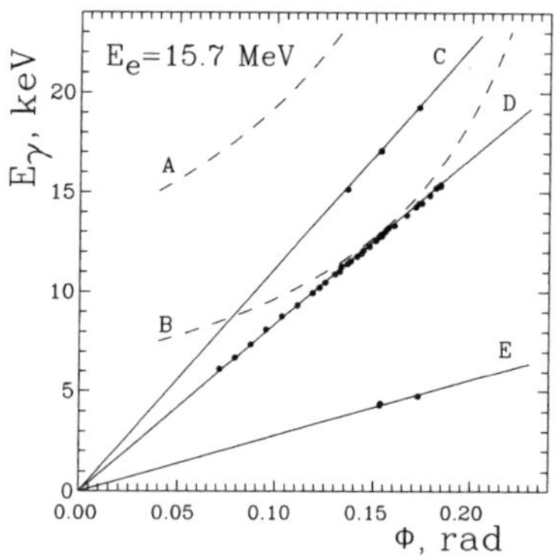

FIGURE 9.8 From ref. [33]. The PXR spectral peak energies in radiation spectra measured versus the crystal rotation angle ϕ (full circles) at $E_e = 15.7$ MeV. The solid lines E, D, C show the coherent radiation energies calculated by formula 9.5 for the rows of reciprocal lattice vectors $\vec{g_{1i}}$, $\vec{g_{2i}}$ and $\vec{g_{3i}}$, respectively. The dashed lines B, A show the Bragg energies calculated by formula 9.6 in the direction of detection for $\vec{g} = \langle 111 \rangle$ and $\vec{g} = \langle 220 \rangle$, respectively. The experimental conditions are shown in fig. 9.5.

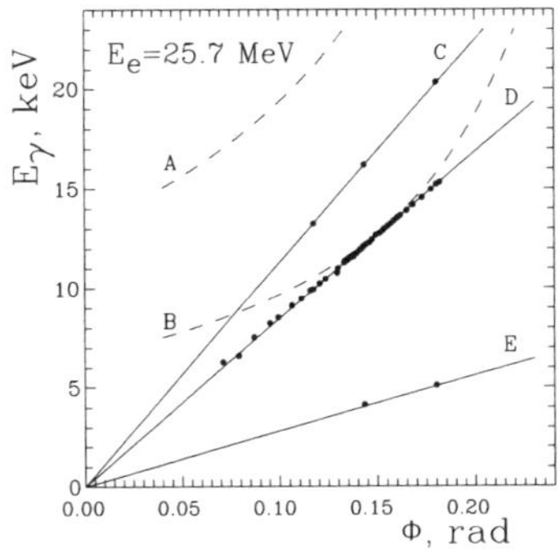

FIGURE 9.9 From ref. [33]. The same as in fig. 9.8 at $E_e = 25.7$ MeV.

from the plane (111) at $\phi \approx \theta/2$ is directed close to the detector direction $\overrightarrow{\Omega}$, whereas all the rest reflections from the vectors of rows $\overrightarrow{g}_{1i}$, $\overrightarrow{g}_{2i}$, $\overrightarrow{g}_{3i}$ are directed at large angles to the detector.

The peaks in the spectra associated with rows $\overrightarrow{g_{1i}}$ and $\overrightarrow{g_{3i}}$ are caused by the sum of radiations from the reciprocal lattice vectors which compose these rows. From formula 9.5 it is evident that all vectors of each row give identical radiation energies in the geometry of our experiment. The line in which the vectors of each row come to an end is perpendicular to $\overrightarrow{V}$ and the scalar products in the numerator of formula 9.5 have identical values. So, each reciprocal lattice vector from the row gives the same E_{CR}. By analogy with a similar known effect for coherent bremsstrahlung [2], this effect was called a "row effect" for PXR in ref. [19].

The agreement between the experimental data for all the peaks observed and the calculations by formula 9.5 confirms the coherence of radiation in the vicinity of and at a significant angular distance from the centers of PXR reflections (or, in other words, from corresponding Bragg directions). The same results were obtained at $E_e = 1200$ MeV for PXR quanta energies up to 400 keV [20].

Note that in the experiments [19,33,34] the registration of weak coherent-radiation peaks at a significant angular distance from the centers of PXR reflections was possible due to high spectral and angular resolutions of the detecting system, and also to the low background in the spectra.

In 1986-1988, at scientific meetings in the former SU, there were several discussions about the nature of X-ray radiation observed in the vicinity of the Bragg direction. The alternative point of view was that this monochromatic radiation might be due to an ordinary diffraction in the crystal target of usual bremsstrahlung from the target and/or from the accelerator. To clear up this uncertainty, in an oral paper [38] presented by one of the authors of this chapter and in ref. [19] the experimental energies were compared to

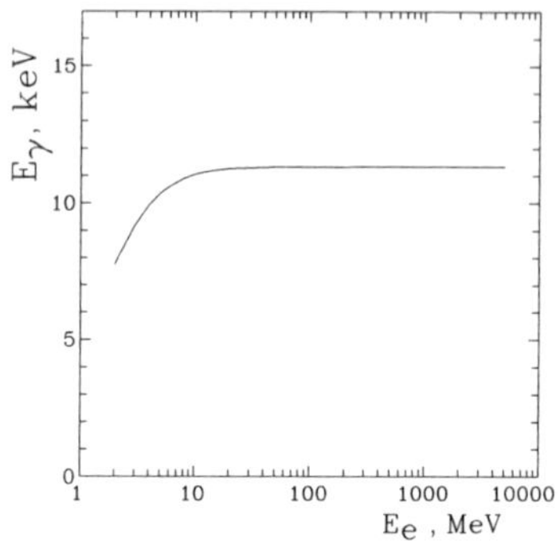

FIGURE 9.10 The calculations of PXR peak energy for $\vec{g} = \langle 111 \rangle$ in the vicinity of Bragg direction at $\phi = 133.3$ *mrad* as function of incident electron energy E_e. The experimental conditions are shown in fig. 9.5.

calculations of Bragg energy for the detector direction using formula 9.6. The calculations of energies E_B determined by the Bragg law in a crystal for the detector direction $\vec{\Omega}$ are presented by dashed lines in figs. 9.8 and 9.9.

$$E_B = \frac{c\hbar \left| \vec{g} \right|^2}{2\sqrt{\varepsilon_0} \left| \vec{\Omega} \cdot \vec{g} \right|} \tag{9.6}$$

The calculation by formula 9.6 for the reciprocal lattice vector $\vec{g} = \langle 111 \rangle$ (line B) is close to the experimental data and to the calculation by formula 9.5 only at $\phi \approx \theta/2$, when the center of the PXR reflection from the (111) plane is in the vicinity of the detector direction, and the divergence increases with disorientation. The calculation by formula 9.6 for the reciprocal lattice vector $\vec{g} = \langle 220 \rangle$ (line A) has given the lowest energy in comparison with energy values for all other vectors of the rows $\vec{g_{1i}}$ and $\vec{g_{3i}}$, but even this energy much differs from the experimental data. E_B is the energy of quanta that may be diffracted by the crystal from the bremsstrahlung radiation flux to the direction of the detector. The influence of bremsstrahlung diffraction is insignificant because a separate spectral peak with energy E_B was not found in measured spectra [19].

Thus the PXR energy peaks calculated by the Bragg formula 9.6 has given values close to the experimental values only in the direction close to that of the PXR reflection center. In the general case, the energies of PXR peaks are described by formula 9.5, which takes into account the coherent nature of PXR. It is of interest to note that the coherent radiation energy of the particle moving in periodical media, described by formula 9.5, does not depend on a particular interaction mechanism. This formula is true both for the coherent bremsstrahlung radiation arising due to the interaction of particles with the nuclear crystal subsystem [2], and the PXR arising due to the interaction of a particle with the electron crystal subsystem. For example, in ref. [39] were performed differential measurements of coherent bremsstrahlung spectra at $\theta < \frac{1}{\gamma}$ and successful comparison of spectral peaks energies to calculations by a formula similar to 9.5. To determine the mechanism of radiation, it is necessary to measure the absolute differential yield of the coherent radiation observed.

The width at half maximum of the PXR spectral peak for a detector of small angular size at $\phi \approx \theta/2$ may be obtained easily from formula 9.5

$$\Delta E_{CR} \approx E_{CR} \frac{\Delta\theta}{tg(\theta/2)}, \tag{9.7}$$

where $\Delta\theta$ is the polar angular resolution of the experiment. The $\Delta\theta$ value should include the detector size and the beam spot size on the target, and also the electron beam divergence in the target (including electron multiple scattering). Formula 9.7 was first confirmed experimentally in ref. [19]. The coherent radiation energy E_{CR} is a weak function of incident electron energy (see formula 9.5). To demonstrate this property, fig. 9.10 shows the PXR energy calculated by formula 9.5 as a function of electron energy at a fixed geometry. The PXR energy is independent of E_e at $V \to c$.

The experimental and theoretical results shown in figs. 9.8 and 9.9 and in refs. [19,20,33] are an excellent demonstration of possibilities for precise smooth tuning of PXR frequency in a wide X-ray band (5-400 keV) using simple target rotation.

9.2.3 The PXR absolute differential yield

The fine structure of PXR reflection was first investigated with high angular and energy resolution in arbitrary units in ref. [19] and in absolute units in ref. [33].

The PXR in the vicinity of a Bragg direction

In figs. 9.11 and 9.12, taken from ref. [33], the measured number of PXR quanta per electron and per unit solid angle in the vicinity of the Bragg direction for $\vec{g} = \langle 111 \rangle$ at $E_e = 15.7$ and 25.7 MeV is shown by full circles. The statistical errors of measurements are shown by vertical bars. The absolute measurement error did not exceed $\pm 15\%$. The calculations of the absolute measurement error are similar to those described in ref. [36].

The angular resolution of measurements was much better than the angular size of the PXR fine structure. The multiple scattering of incident electrons in a thin target was insignificant. Therefore, the experimental data were compared directly to the calculated PXR differential yield. The calculations

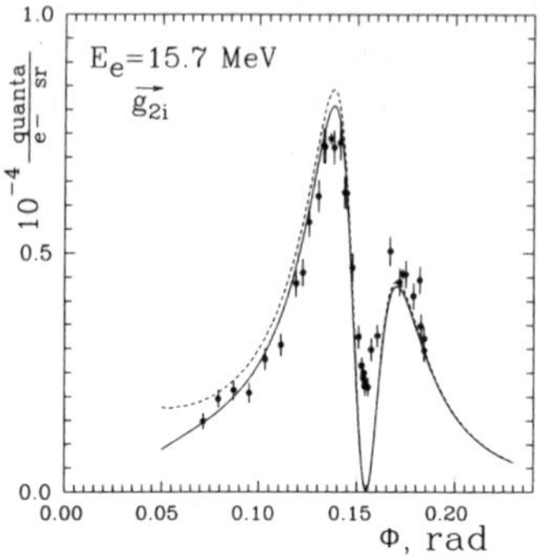

FIGURE 9.11 From ref. [33]. The absolute number of PXR quanta for the spectral peak associated with the row $\overrightarrow{g_{2i}}$ measured at $E_e = 15.7$ MeV versus the crystal rotation angle ϕ (full circles with statistical errors). The measurements were performed close to the Bragg direction for $\vec{g} = \langle 111 \rangle$ from row $\overrightarrow{g_{2i}}$. The Bragg direction for $\vec{g} = \langle 111 \rangle$ is directed to the detector at $\phi = \frac{\theta}{2} \approx 153$ *mrad.* The solid (dashed) line represents the calculations by formula 9.8 for $\overrightarrow{g_{2i}}$ with (without) taking into account the PXR attenuation in the target. The experimental conditions are shown in fig. 9.5.

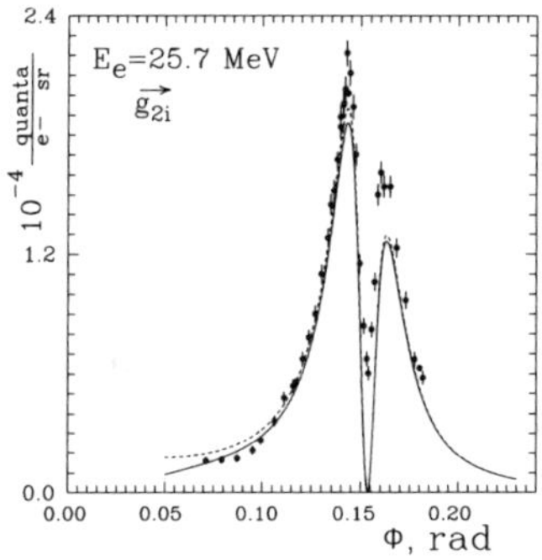

FIGURE 9.12 From ref. [33]. The same as in fig. 9.11 at $E_e = 25.7$ MeV.

were carried out by the formula for the number of quanta, derived in ref. [19] from the kinematical theory [2]

$$Y_{\vec{g}} = \frac{dN}{nd\Omega} = \frac{e^2 \omega L \left|\chi_{\vec{g}}(\omega)\right|^2}{2\pi\hbar\varepsilon_0^3 V \left(\frac{c}{\sqrt{\varepsilon_0}} - \vec{V}\cdot\vec{\Omega}\right)} \cdot$$

$$\left| \frac{\left(\omega\sqrt{\varepsilon_0}/c\right) \vec{\Omega} \times \left[\left(\omega\varepsilon_0/c^2\right)\vec{V} + \vec{g}\right]}{\left[\left(\omega\sqrt{\varepsilon_0}/c\right)\overrightarrow{\Omega_\perp} - \overrightarrow{g_\perp}\right]^2 + \left(\frac{\omega}{V}\right)^2 \left[(1/\gamma)^2 + \left(\frac{V}{c}\right)^2 (1-\varepsilon_0)\right]} \right|^2, \tag{9.8}$$

where dN is the number of quanta with frequency $\omega = \omega_{CR}$ (see formula 9.5) emitted in the solid angle $d\Omega$ as n particles with the charge e and relativistic factor γ pass through a crystal of a thickness L; $\chi_{\vec{g}}(\omega)$ is the Fourier component of the variable part of permittivity; $\vec{\Omega}_{\perp}$, $\vec{g}_{\perp}$ are the components of $\vec{\Omega}$, $\vec{g}$ perpendicular to $\vec{V}$. At $\hbar\omega_{CR} \gg E_K$ the constant part of the medium permittivity is $\varepsilon_0 = 1 - \left(\frac{\omega_P}{\omega}\right)^2$, where $\omega_P = \sqrt{\frac{4\pi e^2 \rho_e}{m_e}}$ is the plasma frequency [2]. For instance, $\hbar\omega_P = 31.1$ eV in Si single crystal. Later, a similar formula was derived in ref. [9] from a semi-quantum expression. The results of calculations of the PXR absolute differential yield for the radiation associated with the reciprocal lattice vector $\vec{g} = \langle 111 \rangle$ are shown by the lines in figs. 9.11 and 9.12. The contribution from other vectors of the row $\vec{g}_{2i}$ to the peak is much less than 1%, as was calculated in ref. [33].

The dashed (solid) line is the calculation without (with) taking into account the PXR attenuation in the target. For the calculation without attenuation the factor $L = \frac{T}{|\vec{t} \cdot \vec{v}|}$ was used. With the attenuation taken into account, the factor L was used in the form

$$L = T_e \left| \frac{\vec{t} \cdot \vec{\Omega}}{\vec{t} \cdot \vec{v}} \right| \left[1 - \exp\left(-\frac{T}{T_e \left| \vec{t} \cdot \vec{\Omega} \right|} \right) \right] \tag{9.9}$$

where $T_e = T_e(\omega)$ is the e-fold attenuation length of radiation with frequency ω; T is the target plate thickness; $\vec{t}$ is the unit vector perpendicular to the target plane; $\vec{v} = \frac{\vec{V}}{V}$. Formula 9.9 is applicable to the arbitrary target and detector orientations relative to the beam for Bragg and Laue geometries [10].

It is necessary to note good agreement of the measured absolute differential yield of PXR in the vicinity of Bragg direction at $E_e \ll E_{crit}$ with that calculated by formula 9.8, derived from the kinematical theory [2]. For the first time this result was presented in ref. [33]. Recently, it has been confirmed in ref. [27] at $E_e < 10$ MeV.

PXR at a significant angular distance from the Bragg direction

Figs. 9.13 and 9.14 show the results of measurements and calculations from ref. [34] of the absolute differential yield of coherent radiation for the rows of reciprocal lattice vectors $\vec{g}_{1i}$ and $\vec{g}_{3i}$ at $E_e = 25.7$ MeV. The number of quanta for the rows is calculated as a sum of PXR radiations for the closest reciprocal lattice vectors which compose these rows; in other words, from closest-from-remote PXR reflections

$$\begin{aligned} Y_{\vec{g}_1} &= Y_{\vec{g}=\langle\bar{1}\bar{1}3\rangle} + Y_{\vec{g}=\langle 11\bar{1}\rangle} \\ Y_{\vec{g}_3} &= Y_{\vec{g}=\langle 004\rangle} + Y_{\vec{g}=\langle 220\rangle} \end{aligned} \tag{9.10}$$

The calculations [34] show that values of the contributions from other vectors of the infinite rows $\vec{g}_{1i}$ and $\vec{g}_{3i}$ to the yield are much less. The items in

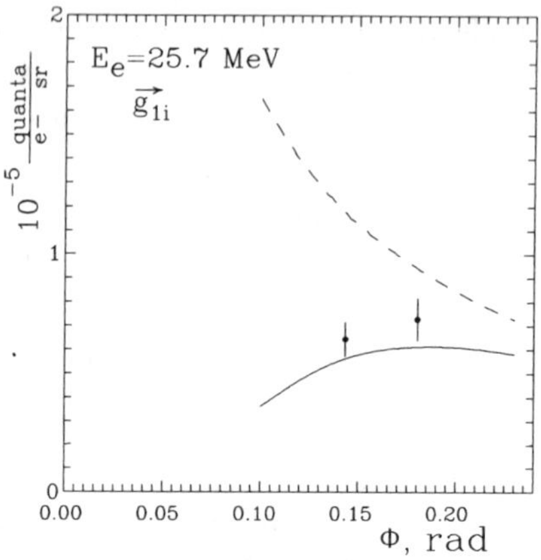

FIGURE 9.13 From ref. [34]. The absolute number of PXR quanta measured at $E_e = 25.7$ MeV for the spectral peak associated with the row of reciprocal lattice vectors $\overrightarrow{g_{1i}}$ versus the crystal rotation angle ϕ (full circles with statistical errors). The dashed (solid) line represents the calculation without (with) taking into account the PXR attenuation in the target using formula 9.8. The experimental conditions are shown in fig. 9.5. Reprinted from A. V. Shchagin, V.I. Pristupa, N. A. Khizhnyak, "Parametric X-ray radiation from relativistic electrons in a crystal in the vicinity of and at angular distance from a Bragg direction," *Nucl. Instrum. Meth.* **B99** (1995) 277, with kind permission of Elsevier Science - NL, Sara Burgerhartstraat 25, 1055 KV Amsterdam, The Netherlands.

each sum 9.10 have comparable values. Similar results for the rows $\overrightarrow{g_{1i}}$ and $\overrightarrow{g_{3i}}$ at $E_e = 15.7$ MeV are presented in ref. [34]. The experimental and calculated data for the row effect are in good agreement despite the fact that the angular distances from the detector to the centers of corresponding PXR remote reflections are about $10\frac{1}{\gamma}$ [34]. To our knowledge, the PXR at significant angular distance from Bragg directions and row effect influence was not studied in independent investigations yet.

It should be noted that there is good agreement between the measured absolute differential yield and the one calculated with due regard to attenuation both in the vicinity (see also ref. [19]) of and at a significant angular distance from BDs. This agreement indicates that in both cases the main contribution to the radiation observed comes from the interaction of incident particles with the electron subsystem of the crystal or, in other words, the PXR, and the influence of coherent bremsstrahlung is insignificant [35]. So, formulae 9.8, 9.9, and 9.10 may be used to estimate the PXR differential yields for forthcoming PXR applications.

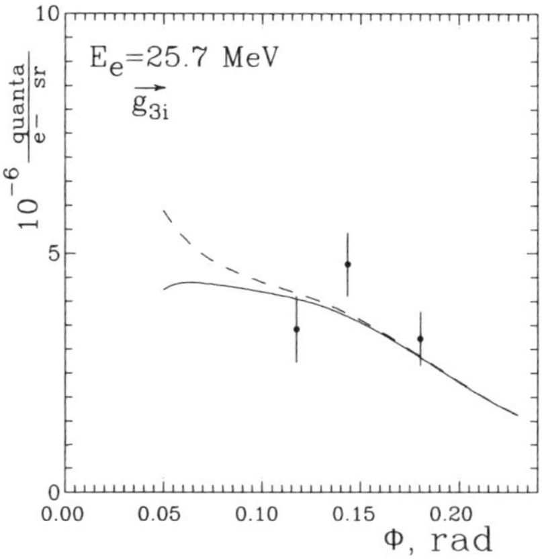

FIGURE 9.14 From ref. [34]. The same as in fig. 9.13 at E_e = 25.7 MeV for the spectral peak associated with the row of reciprocal lattice vectors $\vec{g_{3i}}$. Reprinted from A. V. Shchagin, V.I. Pristupa, N. A. Khizhnyak, "Parametric X-ray radiation from relativistic electrons in a crystal in the vicinity of and at angular distance from a Bragg direction," *Nucl. Instrum. Meth.* **B99** (1995) 277, with kind permission of Elsevier Science - NL, Sara Burgerhartstraat 25, 1055 KV Amsterdam, The Netherlands.

9.2.4 Background in PXR spectra

The background in PXR spectra may exert an appreciable influence on the efficiency of the application where monochromatic radiation is required. The background conditions in PXR spectra were discussed in ref. [34]. where it was shown that the spectral density of incoherent background in PXR spectra may be about or less than 0.01 part of the spectral density in the PXR peak in the vicinity of the Bragg direction (see fig. 9.7). Slight spectral peaks from remote PXR reflections may be considered as a coherent background, and their energies and intensities may be calculated using formulae 9.5, 9.8, and 9.10. Higher-order PXR spectral peaks may be considered as a coherent background, too, if the monochromatic first-order PXR spectral line is necessary for applications. To avoid a second-order peak, the crystallographic plane (111) in diamond-like crystals should be used, because the structure factor for the (222) plane is zero. Yields for third- or higher-order radiation peaks are comparable to or less than 0.01 times the first one at $E_e \ll E_{crit}$. The characteristic X-ray radiation from the target material (see fig. 9.7) may be absorbed by a thin film. So the resulting spectral peak/background ratio in the vicinity of BD may be about or greater than 100 even at a moderate incident electron energy.

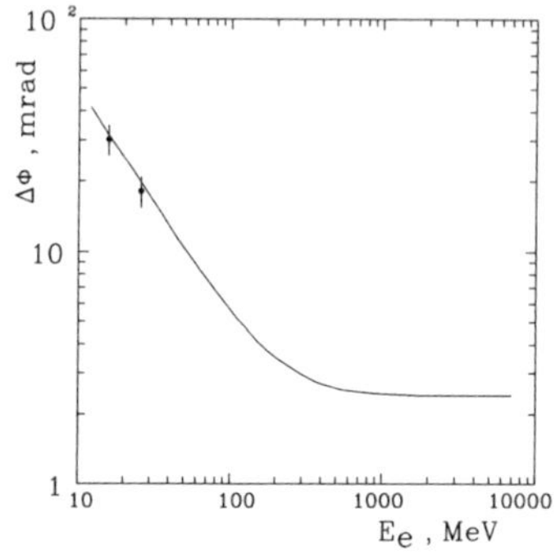

FIGURE 9.15 From ref. [33]. The angular distance between left and right maxima in figs. 9.11 and 9.12 versus the incident electron energy. Full circles are the experimental data at E_e = 15.7 MeV and E_e = 25.7 MeV. The calculations shown by the solid line were performed using equation 9.8 and a simple formula 9.11. The differences between both calculations are about or less than 1%. Therefore, the results of both calculations are shown by one line. The experimental conditions are shown in fig. 9.5.

9.2.5 The energy dependences of PXR properties

From formula 9.5 and fig. 9.10 one can see that the quantum energy in PXR peaks is practically independent of the incident electron energy at $\left|\vec{V}\right| \to c$. However, in ref. [33] a strong dependence of the angular distribution and the PXR yield on the incident electron energy was observed. Let us consider these dependences, because the proper choice of incident electron beam energy is important for real applications.

Angular size of PXR reflection

From formula 9.8 it follows that the PXR reflection has a conical shape around the BD and an angular size $1/\gamma$ at $E_e \ll E_{crit}$. Figs. 9.11 and 9.12 show the fine structure of the reflection on scanning the reflection direction by crystal rotation at a fixed detector. Note that the PXR distribution has its minimum near the exact Bragg geometry for $\phi = \theta/2$.

In fig. 9.15, taken from ref. [33], numerical calculations using equation 9.8 for the angular distance between the right and left maxima of the PXR yield (see figs. 9.11 and 9.12) are shown. The same line shows the calculations using a simple formula [33].

$$\Delta\phi = \sqrt{\frac{1}{\gamma^2} + \left(\frac{\omega_p}{\omega_R}\right)^2} \tag{9.11}$$

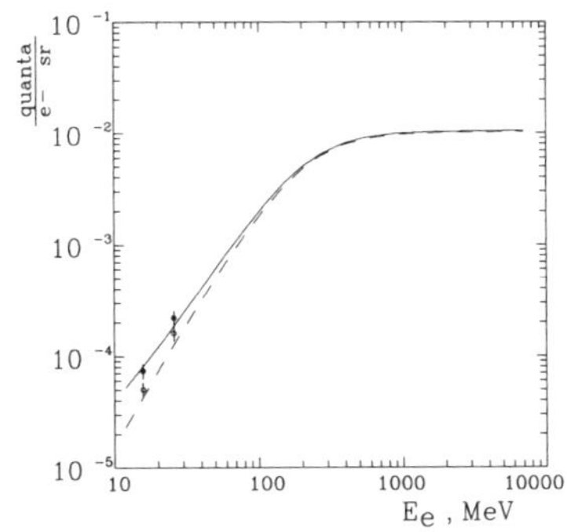

FIGURE 9.16 From ref. [33]. The differential yield of PXR quanta in the left (full circles and solid line) and right (open circles and dashed line) maxima from figs. 9.11 and 9.12 versus the incident electron energy. The calculations by equation 9.8 are shown by lines. The circles represent the experimental data at $E_e = 15.7$ MeV and at $E_e = 25.7$ MeV with absolute measurements error bars. In the calculations for each incident electron energy value the angle ϕ of target rotation was chosen to provide the maximum yield for the left and right maxima from figs. 9.11 and 9.12. The experimental conditions are shown in fig. 9.5.

The difference between both calculations are comparable to or less than 1%. It is seen from fig. 9.15 and formula 9.11 that the angular size of the PXR reflection $\Delta\phi$ decreases with an increasing incident electron energy as $\Delta\phi \approx \frac{1}{\gamma}$ up to the critical energy $E_{crit} = \frac{\hbar\omega_R}{\hbar\omega_p} mc^2$, where $\hbar\omega_R$ is the energy of PXR quanta in the vicinity of the reflection center. In the case considered, we have $\hbar\omega_R = 12.9$ keV and $E_{crit} = 212$ MeV. At $E_e \gg E_{crit}$ the angular size of the reflection does not depend on incident electron energies and $\Delta\phi = \frac{\omega_p}{\omega_R} = 2.4$ *mrad*. In fig. 9.15, the experimental results [33] are shown by full circles. It is seen that the experimental data on the angular size of PXR reflection at $E_e \ll E_{crit}$ are in good agreement with the two mentioned calculations. So, formula 9.11 is appropriate for calculating the PXR reflection angular size for practical applications. Note that at a fixed angle ϕ (fixed target) the angular size $\Delta\phi$ is equal to the angle between the PXR cone axis and the maximum of PXR differential yield (see figs. 9.3 and 9.4).

Square-law growth of PXR differential yield

Fig. 9.16, taken from ref. [33], shows the calculation by formula 9.8 of the PXR differential yield in the left (solid line) and right (dashed line) maxima of the orientational dependences of the PXR yield (see figs. 9.11 and 9.12). It is seen from fig. 9.16 that up to the critical energy E_{crit} the PXR differential yield at the maxima increases with increasing incident electron energy nearly as the square of the incident electron energy. At $E_e \gg E_{crit}$, the PXR differential yield does not depend on E_e. In fig. 9.16, the results of the measurements [33] are shown by full circles for the left maximum of figs. 9.11 and 9.12 and by open circles for the right maximum of the same figures. One can see that the experimental results [33] for the PXR differential yield at the maxima for $E_e \ll E_{crit}$ are in good agreement with the calculation using equation 9.8.

The square-law growth of the PXR differential yield was predicted theoretically in ref. [5]. This square-law behavior at $E_e \ll E_{crit}$ was shown experimentally for the first time in ref. [33]. This result was recently confirmed in ref. [27] at electron energy below 10 MeV.

Thus, the above-discussed experimental investigations of spectral and angular differential characteristics of PXR as a function of incident electron energy at $E_e \ll E_{crit}$ fully confirm the validity of the kinematical theory [2].

9.2.6 The influence of the density effect

In ref. [33] it was first shown that the density effect influence is the physical reason why the linear decrease in the angular size of PXR reflection (fig. 9.15) and the square-law growth of the PXR differential yield at maxima (fig. 9.16) are stopped as the incident electron energy reaches $E_e \approx E_{crit} = \frac{\hbar\omega_R}{\hbar\omega_p} mc^2$. The density effect for charged particles moving through condensed media is a well-known effect in high-energy physics [2]. For example, the density effect leads to saturation of ionization losses and to suppression of bremsstrahlung for a low photon energy. It is due to the influence of the condensed medium polarization in the field of the incident particle [2].

PXR may be considered as a result of diffraction of the virtual photons associated with the relativistic charged particle from crystallographic planes of the crystal through which the particle moves. The spectral-angular distribution of virtual photons of the relativistic charged particle moving through a medium with the average permittivity $\varepsilon_0 = 1 - (\omega_p/\omega)^2$ for the photon energy $\hbar\omega \ll E_e$ at $\gamma \gg 1$ may be found, for example, in [8,33]

$$\frac{dN}{d\omega d\Omega} \sim \frac{\varepsilon_o}{\omega} \left[\frac{\theta_1}{\theta_1^2 + 1/\gamma^2 + (\omega_p/\omega)^2} \right]^2 , \qquad (9.12)$$

where θ_1 is the angle between the particle velocity vector $\overrightarrow{V}$ and the virtual photon wave vector $\overrightarrow{k}$, and ω_p is the plasma frequency of medium. Only the field components of the incident particle with a frequency of about ω_R take part in the formation of the PXR reflection with the quantum energy in the reflection center $\hbar\omega_R$. From the analysis of expression 9.12 it follows [33] that the angular and energy dependences of the PXR reflection are similar to those of the distribution of virtual photons with frequency $\omega = \omega_R$. Let us consider this analogy in more detail for $E_e \ll E_{crit}$ and $E_e \gg E_{crit}$.

In the energy region $E_e \ll E_{crit}$ that corresponds to $1/\gamma^2 \gg (\omega_p/\omega_R)^2$, the presence of medium has no influence on the distribution 9.12: the maxima in the angular distribution are located at angles $\theta_{1\,\max} = 1/\gamma$, and the number of quanta at the maxima increases as $\frac{dN}{d\omega d\Omega} \sim \gamma^2$. This corresponds to the above-shown calculations of PXR properties by equation 9.8 confirmed by experimental results [33]. Note that $\Delta\phi$ is the angular distance between PXR maxima in the orientational dependences (see figs. 9.11 and 9.12 and formula 9.11). At a fixed crystal, $\Delta\phi$ corresponds to $\theta_{1\,\max}$, that is, the angular distance between the PXR reflection cone axis and the PXR maximum yield. The influence of the medium at $E_e \ll E_{crit}$ is negligible, and the distribution of virtual photons 9.12 is similar to that for a particle moving through vacuum.

However, at $E_e \gg E_{crit}$, which corresponds to $1/\gamma^2 \ll (\omega_p/\omega_R)^2$, as seen from formula 9.12, the influence of the medium polarization on the particle field leads to independence of the components with frequency ω_R of distribution 9.12 from the incident electron energy. For PXR, this means the transition from decreasing angular size of the reflection to stabilization at $E_e \gg E_{crit}$ ($\Delta\phi = \theta_{1\,\max} \to \omega_p/\omega_R$, see formulae 9.11 and 9.12), and the transition from the square-law growth of quanta amount at maxima to saturation at $E_e \gg E_{crit}$ (see fig. 9.16 and formula 9.12).

So, the stabilization at $E_e > E_{crit}$ of the number of quanta, of the angular distribution and other properties of the PXR reflection takes place due to the influence of the density effect of the medium on the field of the incident particle moving through this medium. Unfortunately, the action of the density effect leads to cessation of square growth of the PXR differential yield as a function of the incident electron energy. Therefore, it makes no sense to use an electron beam energy higher than $E_e = E_{crit}$ in order to obtain a high PXR flux. The stabilization of the PXR yield and the angular distribution at $E_e > E_{crit}$ have been observed experimentally in refs. [14,26]. In ref. [33] those results were first explained as a manifestation of the density effect influence on PXR properties.

The density effect influence on PXR properties is taken into account (without discussion of physical reason) in formulae in refs. [2,7,19] and is not taken into account in formulae in refs. [5,27].

The angular size of the PXR reflection $\theta_{1\,\max}$ is much greater than the angular band $|\chi_{\vec{g}}(\omega)|$, appropriate for the dynamic effects in a crystal [8,9], at any γ

$$\theta_{1\,\max} = \Delta\phi = \sqrt{\frac{1}{\gamma^2} + \left(\frac{\omega_p}{\omega_R}\right)^2} \gg \left(\frac{\omega_p}{\omega_R}\right)^2 > |\chi_{\vec{g}}(\omega)| \qquad (9.13)$$

So the density effect influence leads to the impossibility of dynamical PXR [3,4] for a significant angular part of the PXR reflection, even at $E_e \to \infty$ [33].

9.2.7 Linear polarization of PXR

As discussed above, the PXR reflection usually has a conical shape. It may be seen from the theory [2] that the linear polarization directions are oriented

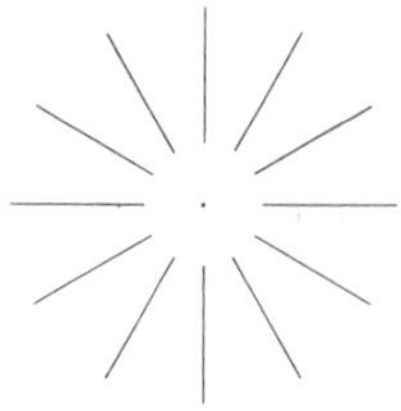

FIGURE 9.17 Linear polarization directions in the PXR reflection are shown by radial lines on the plane perpendicular to the Bragg direction (see figs. 9.3 and 9.4).

radially around the reflection center. Fig. 9.17 shows the plane perpendicular to the Bragg direction (dashed lines in figs. 9.3 and 9.4). The linear polarization directions are shown by radial lines around the reflection center in fig. 9.17. If the detector is of the same size or greater than the PXR reflection, the average polarization should generally be close to zero. However, by collimating the corresponding differential part of the PXR reflection, the X-ray beam with an arbitrary direction of linear polarization may be chosen for applications. This was confirmed experimentally in ref. [16]. In particular, if the reciprocal lattice vector $\vec{g}$ is in the detection plane, the linear polarization direction is in the detection plane too (parallel polarization). Recently, the image of PXR angular distribution for radiation polarized in the direction perpendicular to the plane where the vectors $\vec{V}$ and $\vec{g}$ lie (perpendicular polarization), has been observed in ref. [24] first using the coordinate-sensitive detector.

This image, taken from ref. [24], is shown in fig. 9.18. Due to peculiarities of diffraction in the crystal at $\theta = \frac{\pi}{2}$, the PXR radiation with a parallel polarization is suppressed and only the radiation with a perpendicular polarization is possible. This may be seen, for example, from calculations shown in figs. 9.19 and 9.20 and the corresponding discussion. Therefore, fig. 9.18 shows two identical X-ray beams instead of a usual conical distribution with an angular size of about or more than $\frac{1}{\gamma}$. The image is due to the sum of PXR reflections for $\vec{g} = \langle 022 \rangle$ and $\vec{g} = \langle 044 \rangle$. For both reflections $E_e > E_{crit}$. The experimental angular distance between these beams (13 $mrad$) is somewhat greater than $2\frac{1}{\gamma} = 11.3$ $mrad$ due to the above discussed density effect influence. The angular size of each beam is much less than $\frac{1}{\gamma}$. Both beams in fig. 9.18 are polarized in the vertical direction [24].

9.2.8 Discussion and calculations of PXR properties from a single crystal

Thus, the above-discussed experimental investigations of spectral, angular and polarization characteristics of PXR fully confirm the validity of the

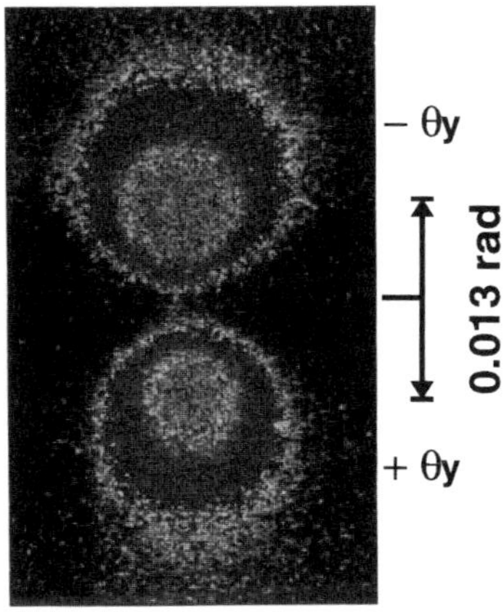

FIGURE 9.18 From ref. [24]. The image of the PXR angular distribution measured by a position-sensitive X-ray detector in the Laue geometry at $E_e = 230$ MeV. The Si single crystal target thickness was 20 μm. The Si slab surface was parallel to the (100) plane, which is perpendicular to the (022) plane. The coordinate-sensitive detector was placed at $\theta = \frac{\pi}{2}$, the angle of crystal rotation $\phi = \frac{\pi}{4}$. The image is due to the sum of PXR reflections for $\vec{g} = \langle 022 \rangle$ and $\vec{g} = \langle 044 \rangle$, because the PXR differential yields for both reflections have comparable values [24]. The plane, where the vectors $\vec{g}$ and $\vec{V}$ are lying, cuts the image along the horizontal line in the middle of the image. The two X-ray beams with perpendicular (vertical in the figure) linear polarizations are clearly seen.

kinematical theory [2] and the formulae used above to calculate PXR from a thin single crystal. Recently, investigation [27] of the differential PXR yield has provided an independent confirmation for a part of the above-mentioned results at $E_e < 10$ MeV. Unfortunately, reliable measurements of the PXR absolute differential yield at energies of several hundreds keV and at $E_e > E_{crit}$ are lacking. Yet, Ter-Mikaelian's theory [2] seems reliable enough, and the formulae under consideration may be used with confidence in calculations for real applications. Probably, the influence of coherent bremsstrahlung [25] may be used for enhancement of radiation yield in some cases.

To demonstrate the possibilities of the PXR as a source of bright polarized X-ray beams, figs. 9.19 and 9.20 show calculations of the PXR differential yield for parallel and perpendicular polarizations under typical experimental conditions. In fig. 9.21 calculations of the corresponding PXR quanta energies are given. The calculations in figs. 9.19-9.21 as functions of the observation angle θ relative to the incident electron beam were performed by one of the authors (A. V. Shchagin) using equations 9.5, 9.8. For each fixed observation angle, the crystal alignments were chosen to provide the maximum yield of PXR polarized in the corresponding direction. The consideration of figs. 9.19 to 9.21 may be helpful in choosing the geometry, the crystal type and thick-

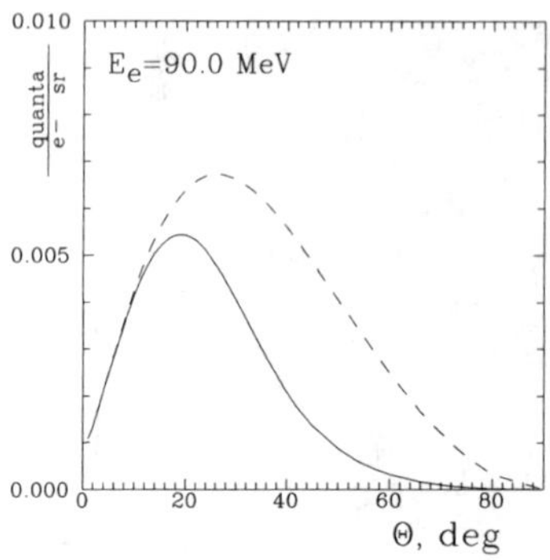

FIGURE 9.19 Calculations of the PXR differential yield in the Laue geometry for $\vec{g} = \langle 111 \rangle$ as a function of observation angle θ. The thickness of a Si slab is 60 μm, $\vec{g} = \langle 111 \rangle$ is parallel to the slab surface, the incident electron beam energy $E_e = 90$ MeV. For each observation angle θ, the crystal slab alignments chosen so as to provide the maximum PXR differential yield, polarized in the plane where the vectors $\vec{V}$ and $\vec{g}$ are lying. The solid (dashed) line shows the results of calculations with (without) taking into account the attenuation of PXR in the crystal.

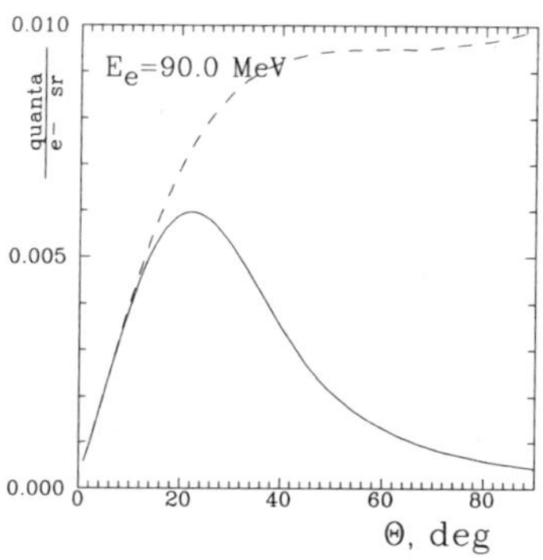

FIGURE 9.20 The same as in fig. 9.19 for the linear polarization direction perpendicular to the plane where the vectors $\vec{V}$ and $\vec{g}$ are lying. For each observation angle θ, the alignment of the crystal slab is chosen so as to provide the maximum PXR differential yield, polarized in the direction perpendicular to the plane where the vectors $\vec{V}$ and $\vec{g}$ are lying.

ness for each specific application, and also in estimating the PXR differential yield and the PXR spectral line energy. Note that the influence of the PXR attenuation in the crystal is significant at low PXR energies.

Let us consider the behavior of the polarization and of the angular distribution of the PXR reflection from a thin single crystal without regard for attenuation. At $\theta < 20°$ the differential yields are nearly the same for both

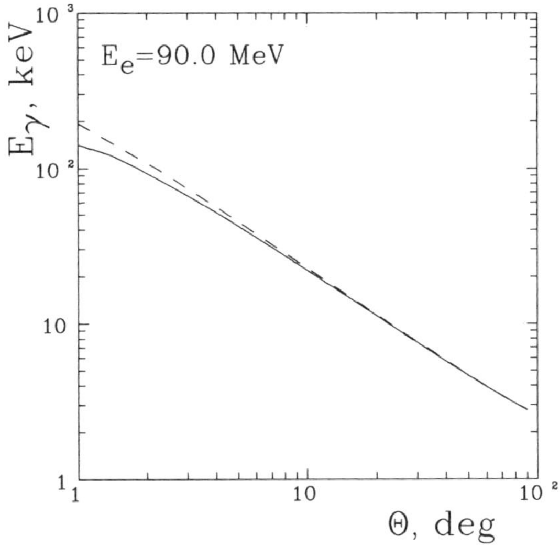

FIGURE 9.21 Energy of PXR quanta as a function of the observation angle θ, the conditions being the same as in figs. 9.19 and 9.20. The solid (dashed) line shows the results of calculations for the parallel (perpendicular) linear polarization direction.

polarization directions, and the PXR reflection shape is close to a perfect conical form. At $\theta > 25°$ the PXR yield for the parallel polarization begins to fall, while the PXR yield for the perpendicular polarization grows continuously. The PXR reflection has the form of a distorted cone. At $\theta = 90°$, the PXR yield is maximum for the perpendicular polarization and is zero for the parallel polarization, similarly to the usual diffraction of real X rays in a crystal. So, at $\theta = 90°$ the PXR reflection transforms to two X-ray beams of equal energy with distinct perpendicular linear polarization (see fig. 9.18, taken from ref. [24]).

The obvious 3-dimension figures for calculated PXR angular distribution may be found in refs. [7,10].

9.3 PXR FROM A MOSAIC CRYSTAL

Theoretical investigation of PXR properties from a mosaic crystal was performed in ref. [6]. The first successful observation and investigation of PXR from a mosaic crystal were made in ref. [22]. Also in this paper the PXR was first observed in the Bragg geometry. In fig. 9.22, taken from ref. [22], the experimental PXR spectrum from mosaic graphite is shown. It was indicated in [22] that the anomalously high PXR yield for high-order PXR spectral peaks cannot be described by the existing theory. The origin of this unexpected disagreement was not found in ref. [22]. So, in contrast to a single crystal case,

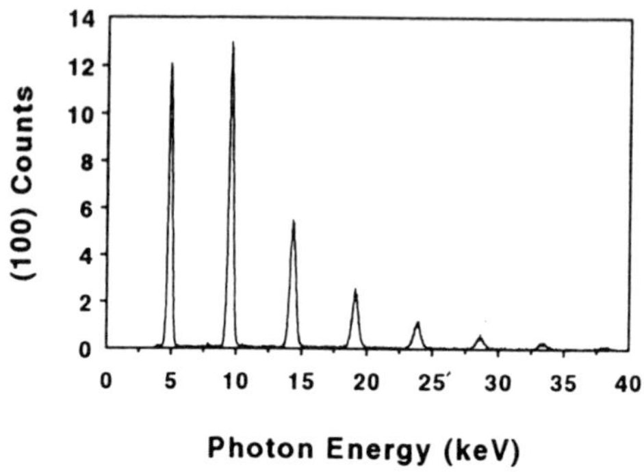

FIGURE 9.22 From ref. [22]. Experimental PXR spectrum measured in the Bragg geometry from the (002) plane of a mosaic graphite crystal at $\theta = \frac{\pi}{4}$ by the Si(Li) detector for the angle of crystal rotation $\phi = \frac{\pi}{8}$. The plane (002) is parallel to the target slab surface. The target thickness is 1.39 mm. The mosaic spread is 0.4°. The incident electron energy $E_e = 90$ MeV. The first-order PXR energy is 4.88 keV. Several high-order PXR spectral peaks have unexpected high yields [22].

the main properties of PXR from a mosaic crystal need further theoretical and experimental investigations.

9.4 POSSIBILITIES FOR PXR APPLICATIONS

In recent decades numerous applications of a monochromatic X-ray beam have been developed. Now the most popular source of an X-ray beam is synchrotron radiation. The number of storage rings for generating X-ray beams is increasing in the world every year.

It is evident that PXR may be used as a new source of monochromatic polarized X-ray beams enabling the tuning of direction, polarization and frequency of radiation. The angular divergence of a PXR beam is close to $\frac{1}{\gamma}$ at a moderate electron beam energy, the 5-400 keV frequency band observed in the current investigations may be extended. For generating a monochromatic X-ray beam, a linac with a crystal radiator appears to be simpler, easier, more compact and cheaper than a storage ring with a complicated monochromator. For example, PXR seems more convenient for X-ray beam generation in outer space, because a linac, unlike a storage ring may be placed in a single space vehicle. Ref. [23] showed the advantages of PXR for generating hard X rays in comparison with synchrotron radiation even under stationary conditions. PXR may be used in different fields as an alternative to synchrotron radiation sources. Real crystals have sets of numerous crystallographic planes. As a result, a set of numerous X-ray beams in different directions propagate from a crystal situated in the electron beam. Many of those X-ray beams may be utilized simultaneously much as the synchrotron radiations from different parts of a storage ring are utilized.

The properties of the PXR beam depend on the charge, energy, divergence of incident particles and on the properties of a crystal. Therefore, PXR may be used for measurements of beam [40] and crystal [24,41] properties, for crystal alignment in particle beam [42].

A collimated, tunable PXR monochromatic polarized beam may be applied for precision calibration of spectral, polarization, orientational characteristics when creating new generations of X-ray and gamma-ray telescopes, optics, detectors and other X-ray equipment [34].

Scanning of PXR quanta energy around the K or L edges of certain elements may be applied in medical diagnostics [23], to measurement of contamination, to the control of nuclear materials circulation in the world [34], to the search for and measurement of minerals, and to molecular holography, among other applications. PXR quanta energy scanning is realizable via simple mechanical scanning of crystal orientation (see figs. 9.8 and 9.9 and also obvious fig. 2 in ref. [19]).

Recently in ref. [43] heavy-element tomography has been developed using a tunable X-ray beam source, based on high-order ordinary diffraction of ordinary bremsstrahlung. PXR as a source of monochromatic, tunable X-ray beam seems very convenient for this application. Also, PXR has good prospects as a source of short, powerful pulses of polarized X-ray beam, excited in a crystal by a bunched electron beam [44].

ACKNOWLEDGMENTS

One of the authors (A.V. Shchagin) is grateful to N.A. Khizhnyak for discussion about the nature of PXR; to R.B. Fiorito and D.W. Rule for discussion about the data in figs. 9.18 and 9.22 and for permission to reproduce the figs. from refs. [22,24] in this review; and also to L. F. Prokopchuk for help in preparing this paper. This paper was made possible in part due to Grant # K4Q100 from the Joint Fund of the Government of Ukraine and International Science Foundation and due to the Grant from AIP.

REFERENCES

1. Fainberg Ya. B. and Khizhnyak N.A., *Zh. Ehksp. Teor. Fiz.* **32,** 883-895 (1957).
2. Ter-Mikaelian M.L., *High-Energy Electromagnetic Processes in Condensed Media* (Wiley-Interscience, New York, 1972).
3. Baryshevskii V.G., *Channeling, Radiation and Reactions in Crystals at High Energies* (BGU publ., Minsk, 1982) [in Russian].
4. Garibian G.M. and Yang C., *X-ray Transition Radiation* (Acad. Sciences of Arm. SSR, Yerevan, 1983) [in Russian].

5. Dialetis D., *Phys. Rev. A* **17,** 1113-1122 (1978).
6. Afanas'ev A.M. and Aginyan M.A., *Zh. Ehksp. Teor. Fiz.* **74**, 570-579 (1978).
7. Feranchuk I.D. and Ivashin A.V., *J. Physique* **46,** 1981-1986 (1985).
8. Bazylev V.A. and Zhevago N.K, *Radiation of fast particles in matter and external fields* (Nauka, Moscow, 1987) [in Russian].
9. Nitta H., *Phys. Lett. A* **158,** 270-274 (1991).
10. Caticha A., *Phys. Rev. B* **45,** 9541-9550 (1992).
11. Garibian G.M. and Yang C., *Nucl. Instr. and Meth. A* **248,** 29-30 (1986).
12. Buts V.A. and Khizhnyak N.A., 78-89; Fainberg Ya.B. and Khizhnyak N.A., 278-291; in *Plasma Electronics* (Editor Kurilko V.I., Naukova Dumka, Kiev, 1989) [in Russian].
13. Vorobiev S.A., Kalinin B.N., Pak S. and Potylitsin A.P., *Zh. Ehksp. Teor. Fiz., Pis'ma* **41,** 3-6 (1985).
14. Adishchev Yu.N., Didenko A.N., Mun V.V., Pleshkov G.A., Potylitsin A.P., Tomchakov V.K., Uglov S.R. and Vorobiev S.A., *Nucl. Instr. and Meth. B* **21,** 49-55 (1987).
15. Adishchev Yu.N., Babadzhanov R.D., Vorobyev S.A., Kalinin B.N., Mun V.V., Pak S., Pleshkov G.A., Potylitsin A.P. and Uglov S.R., *Zh. Ehksp. Teor. Fiz.* **93,** 1943-1950 (1987).
16. Adishchev Yu.N., Verzilov V.A., Potylitsin A.P., Uglov S.R. and Vorobiev S.A., *Nucl. Instr. and Meth. B* **44,** 130-136 (1989).
17. Avakyan R.O., Avetisyan A.Eh., Adishchev Yu.N., Garibian G.M., Danagulyan S.S., Kizogyan O.S., Potylitsin A.P.,Taroyan S.P., Ehlbakyan G.M. and Yang C., *Zh. Ehksp. Teor. Fiz. Pis'ma* **45**, 313-316 (1987).
18. Adeishvili D.I., Blazhevich S.V., Boldyshev V.F., Bochek G.L., Vit'ko V.I., Morokhovskij V.L. and Shramenko B.I., *Dokl. Akad. Nauk SSSR* **298,** 844-846 (1988).
19. Shchagin A.V., Pristupa V.I. and Khizhnyak N.A., *Phys. Lett. A* **148,** 485-488 (1990).
20. Morokhovskii V.L. and Shchagin A.V., Zh. Tekh. Fiz. **60,** no. 5, 147-150 (1990); *In English: Sov. Phys. Tech. Phys.* **35,** 623-624 (1990).
21. Adeishvili D.I., Blazhevich S.V., Bochek G.L., Kulibaba V.I., Lapko V.P., Morokhovskii V.L., Fursov G.L. and Shchagin A.V., Prib. Tekhn. Ehksp. **3,** 50-52 (1989); for correction see Prib. Tekhn. Ehksp. **6,** 4 (1989).
22. Fiorito R.B., Rule D.W., Maruyama X.K., DiNova K.L., Evertson S.J., Osborne M.J., Snyder D., Rietdyk H., Piestrup M.A. and Ho A.H., *Phys. Rev. Lett.* **71,** 704-707 (1993).
23. Fiorito R.B., Rule D.W., Piestrup M.A., Qiang Li, Ho A.H. and Maruyama X.K., *Nucl. Instr. and Meth. B* **79,** 758-761 (1993).
24. Fiorito R.B., Rule D.W., Piestrup M.A., Maruyama X.K., Silzer R.M., Skopik D.M. and Shchagin A.V., *Phys. Rev. E* **51,** Part 1, R2759-R2762 (1995).
25. Kleiner V.I., Nasonov N.N. and Safronov A.G., *Phys. Stat. Sol (b)* **181**, 223 (1994)
26. Asano S., Endo I., Harada M., Ishii S., Kobayashi T., Nagata T., Muto M., Yoshida K. and Nitta H., *Phys. Rev. Lett.* **70,** 3247-3250 (1993).

27. Freudenberger J., Gavrikov V.B., Galemann M., Genz H., Groening L., Morokhovskii V.L., Morokhovskii V.V., Nething U., Richter A., Sellschop J.P.F. and Shul'ga N.F., *Phys. Rev. Lett.* **74**, 2487-2490 (1995).
28. Jelly J.V., *Cherenkov Radiation and its Applications* (Lnd.: Pergamon Press, 1958).
29. Zrelov V.H., *Vavilov-Cherenkov Radiation and its Application in High Energy Physics* (Atomizdat, Moscow, 1968) [in Russian].
30. Frank I.M., *Vavilov-Cherenkov Radiation. Theoretical Aspects.* (Nauka, Moscow, 1988) [in Russian].
31. Ginzburg V.L. and Frank I.M., *Zh. Ehksp. Teor. Fiz.* **16,** 15-28 (1946).
32. Andersen J.U., Bonderup E. and Pantell R.H. , *Ann. Rev. Nucl. Part. Sci.* **33**, 453-504 (1983).
33. Shchagin A.V., Pristupa V.I. and Khizhnyak N.A., *Absolute differential yield of parametric X-Ray radiation*, Preprint KFTI 93-32, Kharkov, 1993; Proceedings of International Symposium on Radiation of Relativistic Electrons in Periodical Structures, September 6-10, 1993, Tomsk, Russia, p.62-75. See also reviews [37].
34. Shchagin A.V., Pristupa V.I. and Khizhnyak N.A., Nucl. Instr. and Meth. B **99**, 277-280 (1995).
35. Blazhevich S. V., Bochek G. L., Gavrikov V. B., Kulibaba V. I., Maslov N. I., Nasonov N. N., Pirogov V. N., Safronov A. G. and Torgovkin A. V., *Phys. Lett.* **A195**, 210 (1994).
36. Shchagin A.V., Pristupa V.I. and Khizhnyak N.A., *Nucl. Instr. and Meth. B* **84,** 9-13 (1994).
37. Shchagin A.V. and Khizhnyak N.A., *Nucl. Instr. and Meth. B.* **119** 115-122 (1996); Shchagin A. V., *The Future of Accelerator Physics, The Tamura Symposium Proceedings*, Austin, TX, November 1994, AIP Press, pp. 359-377.
38. Kas'an S.V., Morokhovskii V.L., Pristupa V.I. and Shchagin A.V., Proceedings of XVIII All-Union Meeting for Physics of Interaction of Charged Particles with Crystals, Moscow, May 1988 (MGU, Moscow, 1989), pp. 77-79 [in Russian].
39. Watson J.E. and Koehler *J., Phys. Rev. B* **25,** 3079-3090 (1982).
40. Kalinin B.N., Potylitsin A.P., Verzilov V.A., Vnukov I.E., Endo I., Harada M., Kobayashi T., Kuwamoto T., Yoshida K., Muto M., and Nitta H., *Nucl. Instr. and Meth. A* **350,** 601-604 (1994).
41. Amosov K.Yu., Kalinin B.N., Potylitsin A.P., Sarychev V.P., Uglov S.R. Verzilov V.A., Vorobiev S.A., Endo I. and Kobayashi T., Phys. Rev. E **47,** 2207-2209 (1993).
42. Morokhovskii V. L., Shchagin A. V., *Prib. Techn. Ehksp.* **3**, 36–38 (1989); in English: *Instruments and Experimental Technique* **32**, 529-532 (1989).
43. Bertschy M., Grittin M., Jolie J., Mondelaers W. and Warr N., *Nucl. Instr. and Meth.* **B103**, 330-338 (1995); Bertschy M., Jolie J. and Mondelaers W., *Appl. Phys.* **A62**, 437-443 (1996).
44. Shchagin A. V., Parsa Z., to be published.

CHAPTER 10

Hard X-ray Microbeam Production and Application

P. Chevallier* † and P. Dhez*‡

Progress in microanalysis is evidently linked to the production of microbeams able to locally excite a sample. Electron microscopy is certainly the most widespread method, but the sample must be under vacuum and the technique suffers from rather poor detection limits. X rays of a few keV are very attractive for the realization of a photon microprobe allowing the sample to be kept in air. The present article summarizes progress achieved and forseen in the production of intense hard X-ray microbeams on the recently developed high brillance synchrotron radiation sources and gives some examples of applications.

10.1 INTRODUCTION

There is a growing need for trace elemental analysis at levels less than one part per million by weight in nearly every scientific field. This first step of bulk level trace elemental analysis is rapidly becoming insufficient and techniques capable of mapping the distribution of each element at the highest lateral resolution are desired.

For qualitative and quantitative analysis, X-ray fluorescence (XRF) is often used as a "finger printing" technique for characterizing the bulk elemental composition of a material. Mapping the lateral distribution of elemental composition has been performed using X-ray emission microprobes. The great majority of these microprobes uses electrons as projectiles; a few others employ high energy protons as a means of excitation. Only very recently, photon microprobes have become available, installed at various Synchrotron Radiation facilities.

---

*) LURE, University Paris XI, Bat. 209D, 91 405 Orsay Cedex, France.
†) LPAN University Paris VI, 75 2352 Paris Cedex, France.
‡) LSAI University Paris-Sud, Bat 350, 91 405 Orsay Cedex, France.

As we shall see, the photon is a very interesting projectile, and should undoubtedly lead to the best microprobe if enough photons could be concentrated into approximately μm^2 spots. Due to the very small aperture of all hard X-ray optics this cannot be obtained with the best conventional X-ray tubes but is now possible with Synchrotron Radiation (SR), the characteristics of which have allowed the development of new X-ray optical devices. Thus, efficient photon microprobes can only be installed on modern SR facilities and it seems that they will always remain a limited analytical tool that will never supplant the wide use of other kinds of microprobes.

Lateral resolution at the micron scale is now reached with a wide variety of modern techniques. SIMS (Secondary Ion Mass Spectroscopy) [1] offers this resolution with excellent depth resolution as well, but is destructive and quantification is difficult in many cases [2]. LAMMA (Laser Microprobe Mass Analysis) [3] is a very useful technique but it suffers from the unreproducible character of the laser induced ionisation and difficulties are often encountered in interpreting the mass spectra [4]. EPMA (Electron Probe Micro Analysis) [5] presents extremely high lateral resolution. Unfortunately, due to the spread of the electron beam below the surface of the sample, the X-ray fluorescence spectrum arises from a sphere of about 1 μm radius or larger. Moreover, this technique is limited to the determination of major constituents because of the high background generated by the incident and secondary electrons as they slow down through the sample. Micro-PIXE offers high sensitivity (down to the ppm range) and can reach the micron scale lateral resolution [6] but with severe damage to sensitive samples. Micro-PIXE, however, can be easily coupled with Nuclear Reaction Analysis (NRA) for the determination of light elements, and with Rutherford Back Scattering (RBS) for depth profile determination, which is of great interest in the characterisation of many samples. Finally, all these techniques require the sample to be in vacuum which may be impossible with some very volatile, too large, or living samples.

Sub-micron spatial resolution is now achievable in the hard X-ray range thanks to the advent of new X-ray optics in conjunction with SR. Thus micro-Synchrotron Radiation X-Ray Fluorescence (micro-SRXRF) has appeared in recent years as an interesting quantitative micro analytical technique with high sensitivity that can be used on any kind of samples in air, except for the lightest elements. Moreover, associated with other techniques such as Extended X-ray Absorption Fine Structure (EXAFS), X-ray Absorption Near Edge Structure (XANES), and diffraction, very useful chemical and structural information will be obtainable.

In this paper we shall give the emphasis to hard X-ray microbeam production ($E > 1$ keV), explaining the various approaches to reach such a goal. Then examples of applications, namely in X-ray fluorescence, will be presented.

10.2 MICROBEAM PRODUCTION

For most of the analytical methods using X rays, beam optimization is specific to each kind of sample and needs to be done by considering all together optic, source, and detector. In addition, beam size alone is not a sufficient criterion; a microbeam with not enough photons to perform measurements is useless. To receive a useful signal from the sample under study, the first requirement is to focus a sufficient number of photons within a convenient energy band, containing a minimum of scattering and contamination by harmonics, and having a correct polarization. These simultaneous conditions are necessary for obtaining a signal with a good counting rate and a high signal-to-noise ratio.

The real goal is to get a sufficient number of photons, inside a convenient energy band pass, with a minimum of scattering and harmonics contamination, and a correct polarization, to obtain a signal with good statistical counting rate and high signal over noise ratio.

It has to be mentioned that, the sample usually being set at 45° from the incident beam, due to the hard X-ray penetration, the lateral resolution will be spoiled in the horizontal direction with a thick, low-absorbing matrix. If one really wishes to attain a 1 μm resolution, and if the penetration depth is larger, then the sample thickness should not be more than 1 μm.

We will not develop here all the aspects of these preliminary remarks. In this chapter we will only comment on a few aspects, especially those about source size and divergence of X rays emitted by storage rings or associated devices. Although some experiments are using wavelength-dispersive spectrometers to analyze the emitted fluorescence, we will not consider this optical part but only the excitation part.

Before we review the various possibilities for different sources and energy ranges, let us point out that the optical qualities of X-ray beams for fluorescence microprobe are less demanding than those for the diffraction and scattering techniques. That is mainly because parallelism and spectral purity are not so stringent.

Most of the beams are shaped by combining several optical systems. The more frequently used schemes will be mentioned when describing one of the elements for the first time.

Finally, it is important to point out that the sample scanning system is a very important part of a microprobe, although it is not discussed here. With charged particles the sample is kept fixed and the beam is scanned over a given area to get elemental mapping. This is not yet possible with photons, although it is under study [7,8]. The sample as then to be scanned and this introduces a smearing of the map due to the unavoidable backlash of any mechanical system.

10.2.1 X-ray sources

As we shall see, a general characteristic of any hard X-ray optic is its small aperture (typically a few mrad). Then, in order to obtain a very intense beam (we shall see that the required flux for a good X-ray microprobe should be of the order of 10^{10} $ph/s/\mu m^2$) and a small focal spot (μm scale or less) the most severe criterion of the X-ray source is its brilliance ($ph/s/\mu m^2/mrd^2/0.1\%BW$).

In this sense the beam delivered by third generation SR rings is the only one that can be used efficiently to develop a real X-ray microprobe, as its brilliance is at least 10 orders of magnitude higher than that of the best available X-ray tube.

When a light charged particle (e^- or e^+), moving at a relativistic velocity, undergoes a change in the direction of its velocity vector, it loses part of its energy as electromagnetic radiation. For example, when an electron (or positron) enters the magnetic field of a bending magnet, it radiates energy (Synchrotron Radiation).

Most generally, a SR machine is a storage ring where high energy electrons (or positrons) are injected and kept on their orbit through RF feeding to compensate the energy lost by radiation at every turn. Depending on the machine, the particle energy is in the 1-10 GeV range and the ring diameter goes from 4 to more than 100 meters. The particles, grouped in one (or many) bunch, travel in the ring under vacuum (10^{-10} torr typically) and are deviated by dipole bending magnets. A beam line can be adapted, tangent to the ring, in each bending magnet, and each time the bunch passes in front of such a line, a flash of SR is sent to the experimental areas installed along the line. The duration of this flash is directly linked to the length of the bunch. When working in the X-ray regime the beam lines are usually isolated from the storage ring by a thin beryllium window allowing the experiments to work in primary vacuum when possible.

Due to relativistic contraction, the light is emitted in a cone of very small aperture (typically $\theta = 1/\gamma$ where γ is the ratio of the particle total energy to its rest mass energy), centered on the velocity vector of the emitting particle—that is, tangent to its orbit. Thus, in the horizontal plane the SR appears as a layer situated in the orbital plane and with very low vertical divergence (a fraction of a mrad.).

This light presents unique properties that make SR an ideal source for nearly any experiment where interaction of electromagnetic radiation with matter is involved, except in the laser energy range. Some of the main characteristics are:

- a very intense and continuous energy spectrum that starts in the far infrared and extend more or less in the X-ray regime, depending on the particle energy in the machine and its radius of curvature.
- a very low angular divergence, typically of the order of the rocking curve

of perfect crystals, that renders monochromatisation of the beam very efficient. The excitation energy can then be continuously tuned in a large domain, simply by rotating the crystals of the monochromator. This low divergence is also perfectly adapted for more sophisticated optics that could hardly be used on more conventional sources, as we shall see in the following;

- exceptionally high brilliance for the third generation machines;
- a high degree of horizontal polarisation when looking in the orbital plane;
- pulsed time structure to the emitted light.

Many efforts had to be made to overcome the high heat load on the first optical elements or slits set in the beam line, and severe shielding of the experimental area must be ensured, a consequence of which is that the whole experiment must be confined in a closed hutch and then be entirely remote controlled.

In the last generation machines the straight sections between two dipole magnets are widely used to install two kinds of insertion devices:

- wigglers that shift the energy spectrum in the hard X-ray regime.
- undulators, where the whole energy spectrum is contracted to a nearly monochromatic line and its harmonics, the energy of which can be tuned by changing the gap between the magnets. These sources offer exceptionally high brilliance.

More details about SR and its applications can be found in the literature [9–12].

10.2.2 Pinholes and collimators

Synchrotron sources (magnet, wiggler or undulator) are known to be so brilliant that a simple collimator, correctly set in the orbital plane, allows to select the more appropriated part of the source for intensity and polarisation.

In some experiments [13] a pinhole is the only optical part to shape the beam at the convenient size. Selecting the central part of the synchrotron beam is convenient to increase the polarization and to get high flux in a white beam. Keeping a usable photon flux with presently existing sources limits the spatial resolution to about 5 to 10 microns. This minimum size is very dependent on the source brilliance but can be decreased by using convenient focusing optics.

Alternatively, a set of pinholes having different diameters can be inserted to reach the desired spatial resolution and to adjust the photon flux. Another possibility is to use two perpendicular slits to define the selected area. Slits built with two parallel cylindrical rods are known to give low scattering and are especially interesting at high energy.

Additional absorbing filters and plane mirror or Bragg reflectors are often used to better define the excitation energy range. This is interesting for fluorescence analysis as it improves the signal-to-background ration, and absolutely necessary for XANES or EXAFS studies that require high monochromaticity.

As indicated later, pinhole or collimators are also frequently combined with focusing mirrors or crystals to decrease the beam size. An efficient pinhole capable of stopping abruptly the hardest X ray is not easy to find, especially if one wishes to limit scattering. Because scattering and fluorescence emitted by the optical elements are significant limiting factors for high spatial resolution, it is also interesting to set a collimator as close as possible of the front of the sample. The size of this last collimator is adjusted so as to just remove the scattering of the prepared beam without touching the intense beam.

10.2.3 Focusing optics

Focusing has the advantages of increasing the collected photon flux, decreasing the size of the spot impinging the sample and restricting the spectrum initially emitted by the source. High spatial resolution studies being time consuming, an ideal system should give the possibility to vary the spot size, as a zoom does. That would allow a first fluorescence study with low resolution in view to select the most interesting area for further high resolution measurements.

For a given source, a good parameter to compare the efficiency of different optical schemes is the theoretical compression factor, which includes the angular aperture and the surface of the source collected by the optic. Due to the optics defects the real obtained gain can be less than the calculated values and dependent on the desired spot size. Collimating the source itself to decrease the focused spot size is useful to study its ultimate size, even if no real low concentration sample could be studied with it. Wires, grids, thick microelectronic masks drawn on membranes, transmitting gratings, and knife edge are the most common high-contrast objects, used to measure the effective spatial resolution, either in transmission or in fluorescence.

We need to mention that scattering around the focus spot could be a limiting factor for resolution. In defining the spot size different criteria could be used and so the FWHM alone is not a sufficient parameter. Round and high contrasted spots are scarcely obtained, so the beam profile plotted on a semilogarithmic scale, in two perpendicular directions, is a better criterion because it include not only scattering but also astigmatism and overlapping orders for diffracting focusing optics such as Fresnel or Bragg-Fresnel lenses.

Three main systems are used on X-ray microfocusing beam lines:

- glancing incidence mirrors based on the total reflection phenomenon is the first one.

- crystals or, more recently, interferential mirrors especially developed for X rays are very popular.
- transmission or reflecting gratings are the most recent.

Outside all practical manufacturing aspects, the theoretical limitations of these X-ray focusing optics have been extensively studied [14–16]. It is interesting to notice that in recent years, thanks to progress in grinding and polishing, a few figured optics have been approaching the predicted theoretical limits. Each one being abundantly described in specialized articles or books, [17–22], we will only mention their main features and present limitation, especially for energies over 1 keV. Starting from the simplest we will finally indicate a few new systems that are potentially interesting but not yet completely evaluated, like capillary optics.

Single- and double-focusing glancing incidence total reflection mirrors

According to Snell's law, total reflection of X-ray photons occurs when its glancing angle on a flat surface becomes smaller than a critical value, θ_c. This critical angle is given by $\cos\theta_c = n$ where n is the index of refraction. In the X-ray regime this index is always smaller but very nearly equal to unity so that the critical angles are extremely low. For low atomic weights mirrors, the critical angle can be calculated from $\theta_c = \frac{20.3\sqrt{\rho}}{E}$, where θ_c is in mrad, E in keV and ρ in $g \cdot cm^{-3}$: For lead glass ($\rho = 5.2\ g \cdot cm^{-3}$) θ_c is on the order of 0.2° for energies around 10 keV.

Total reflection mirrors are low energy band pass the cut-off energy off which can be adjusted through the proper choice of the glancing incidence angle and the coating material. Quite reliable X-ray indices [23], and formula including scattering effect, give the possibility to evaluate the reflected spectra for a given material and incidence angle. Glancing incidence of about 10 mrad are necessary around 1 keV and 1 mrad on the 10 keV range. Such low angles limit the optical aperture and so the collected photon flux. In addition, glancing incidence focusing optics suffer from astigmatism, spherical, and chromatic aberrations.

Because X-ray scattering from surfaces is an important limitation of sharp focusing, one usually selects a blank of material able to be highly polished. If the reflectivity of this material is not convenient, another material is coated over the blank. Glass can be used for low flux. On high brilliance sources, heat load that will distort the surface of the mirrors has to be taken into account. Super polished silicon mirrors, cooled at liquid nitrogen temperature, have been successfully tested, and more recently, special cooper alloys machined with cooling channel close to the surface have been used [24].

Although a single toroidal or elliptical mirror can in principle provide good focusing, only recently have a few ground mirrors reached the theoretical per-

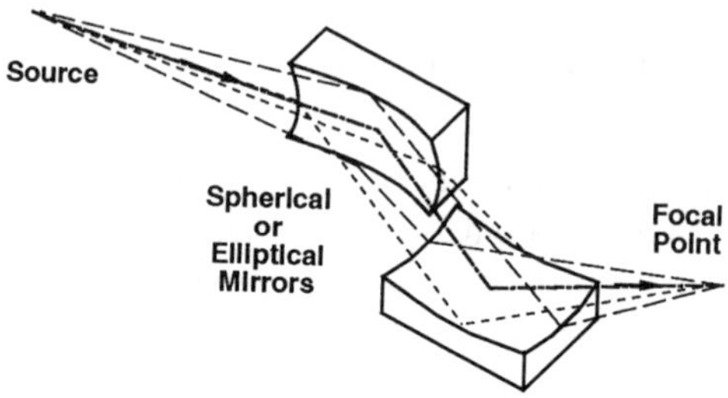

FIGURE 10.1 Kirkpatrick-Baez focusing optics uses two spherical, ellipsoidal or cylindrical mirrors, mounted with crossed curvatures as indicated. The common focus of both mirrors is obtained by the right choice of the radius of each one. Total or Bragg reflection and ground or dynamically bent mirrors can also be used.

formance that allows focusing to less than one micron. For a large radius of curvature, dynamically bent flat mirrors have been developed and mechanical benders have been carefully explored [25]. New possibilities of so-called "bimorphe" mirrors, continuously paved with flat piezo pieces glued at the rear and shaped electrodes, have been recently tested but their length and radius of curvature are still quite limited [26]. More advanced adaptative X-ray optics, with local piezo actuators, similar to those used for astrophysics mirrors, are under development [27]. Because the incidence angle is small (typically 1 mrad), the mirrors length can reach one meter for high energy X rays; in such a case segmented mirrors are an alternative, but they are costly, and time consuming to tune.

Spherical, double reflecting mirrors are the oldest, and remain the favourite, solution on existing beam lines for X-ray focusing. Figure 10.1 shows a scheme of the most widely used system named Kirpatrik-Baez (KB) mirrors [28]. This system consist of two spherical or cylindrical mirrors mounted so as to cross each other's curvatures in order to cancel each other's astigmatism. On-line angular and between-mirror-distance adjustments are necessary to reach the best focusing. The length of each mirror can be different to match the divergence in vertical and horizontal directions and, if necessary, the second mirror can be shortened to give more space close to the sample. Spherical surfaces are usual on such systems because low roughness and precise radius of curvature are easier to manufacture. Recently, elliptical and ellipsoidal mirrors, having a shape close enough to the theoretical one, with no faceting and very low roughness, have demonstrated the possibility to decrease the astigmatism [29]. Such surfaces are difficult to grind but, for large radius, flat mirrors can be dynamically bent to nearly the proper shape with the advantage of on-line fine radius adjustment. This was recently demonstrated on a third generation synchrotron source with graded period multilayer mirrors [30].

Wolter type focusing systems [31], built with two confocal ellipsoid and hyperboloid mirrors, have also been built, but they are much more complex to grind and expensive than KB systems. Such a device was tested at the Photon Factory [32], but is no longer used.

Total reflection KB and Wolter system can be set directly on a synchrotron source or behind a Bragg monochromator having the desired band pass. When placed before the focusing element, double crystal and multilayer monochromators offer the opportunity to keep a fixed monochromatized exit beam, so that the energy may be changed continuously without affecting either the size or the absolute position of the focused spot. This is not the case of focusing optics built directly with Bragg or Fresnel reflectors.

Crystals and multilayer focusing optics

From perfect to mosaic crystals and multilayers it is possible to select a convenient Bragg reflector having the desired band pass for a given experiment. It can be shaped so as to simultaneously focus the incident beam.

A flat, asymmetrically cut crystal is the simplest system to increase the photon density but the gain is quite low compared to focusing.

Conical focusing large band pass system have also been prepared by shaping highly oriented mosaic pyrolytic graphite crystals [33]. Unfortunately the mosaic spread spoils the focal spot, and its size is typically in the 500 μm to 1 mm range, depending on the focal distance and the selected energy. Combined with a collimator, it has been successfully used by Sparks et al to build the very first hard X-ray microprobe (100 μm) installed on a SR machine [34].

For a better but still low-focusing solution, double cylindrical crystal monochromators have been developed. In such a system the two crystals are parallel and their curvatures are crossed as indicated in figure 10.2. Special anticlasting techniques are applied by machining deep groves on the back of the crystal to avoid bending in both direction (saddle effect) [19].

For line focusing, the most usual system is the flat, thin, triangular crystal. In this system the base of the triangle is kept fixed and the top is pushed by a linear actuator to adjust the radius of curvature to get the best focal line. The sides of the triangle can be specially ground so as to improve real cylindrical curvature [35]; nevertheless the line width stays in the few hundreds of μm range. As the incidence angle changes along the height of the triangle, the corresponding Bragg-reflected wavelengths are focused together on a line. Thus we have a broad band pass monochromator where the photons coming out of the focused line are angularly dispersed in accordance with their wavelength. Then a position-sensitive detector, set at the right distance, can register the whole diffracted energy range. Figure 10.3 is a scheme showing how such a bent triangular crystal can be combined with another reflector to get spot focusing like in the KB system, but still in the tenth of a μm diameter range. The second reflector can be of any type; for example a multilayer has been

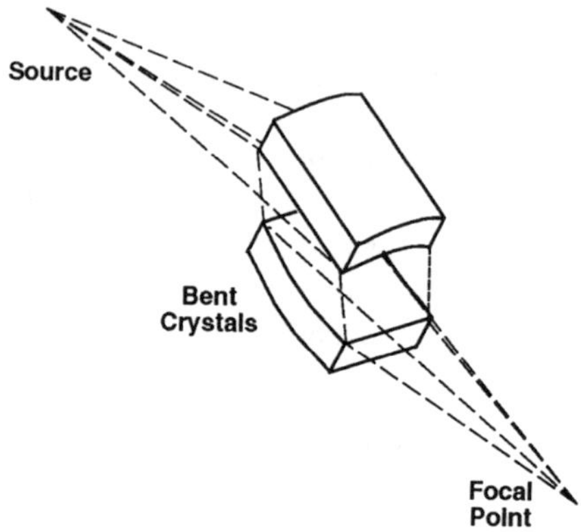

FIGURE 10.2 Cylindrically bent crystals are low focusing optics adapted to the double crystal geometry mounting, frequently used on synchrotron sources. The radius of curvature needs to be adjusted when scanning the wavelengths. To obtain a real cylindrical curvature, and avoid saddle shapes, the rear of the crystal is especially machined with deep groves, not shown here.

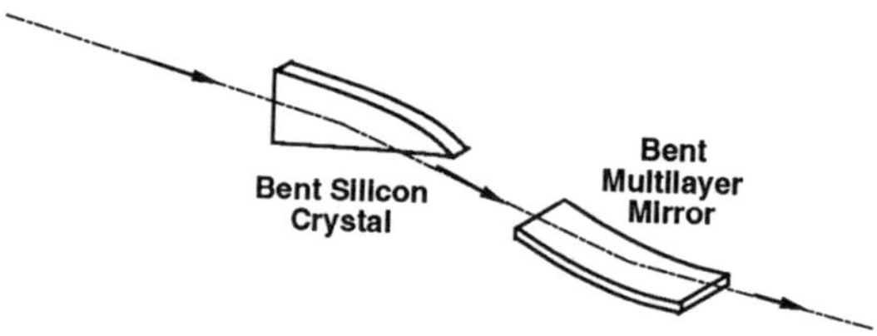

FIGURE 10.3 Example of KB focusing optics with two dynamically bent cylindrical Bragg reflectors combining a crystal and a multilayer. In this case, the thin silicon crystal has a triangular shape which provides the cylindrical curvature by a simple pushing rod at the apex of the triangle. The multilayer is coated on a super polished flat piece of glass next bent by a four point bender.

used by Fourme *et al.* at LURE to remove the third order of a Si[111] first triangular crystal [36].

Horizontal and vertical focusing can also be obtained, but for a specific energy, using a single ellipsoidally bent crystal. This was first used at Daresbury [37]. For high resolution the focal spot (typically $10 \times 20\ \mu m^2$) has to be further collimated.

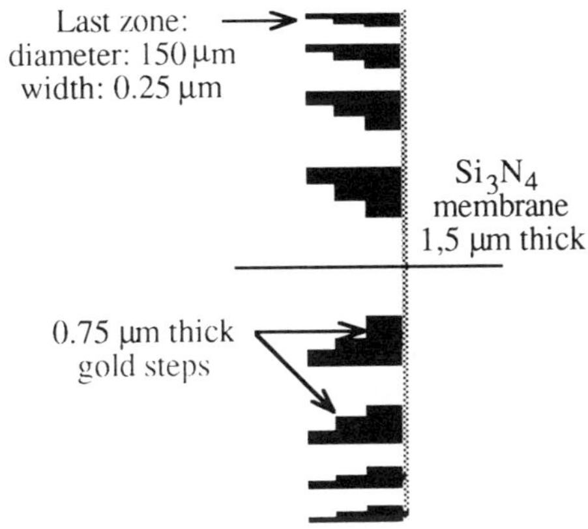

FIGURE 10.4 The use of Fresnel lenses could be pushed up to several keV if thick enough zones could be prepared so as to give a good contrast. Thickness, diameter and width of the rings indicated on this figure have been calculated for 8 keV. The stair like profile of the rings, designed to increase the efficiency of the lens, may be obtained with adaptated microtechnologies.

Multilayers are artificial interferential mirrors prepared by alternate deposition of two materials choosen for their high optical index contrast at the considered wavelength on a flat (or curved) super-polished rigid substratum. According to the present state of the art in coating methods, the smallest period giving noticeable reflectivity cannot be lower than 2 *nm*. That currently limits normal incidence multilayer focusing optics to energies lower than 0.3 keV. Schwarzchild type microscopes, built with two confocal spherical mirrors coated with multilayers, have reached 0.3 microns resolution, but only below 300 eV [38]. Normal incidence is not usable for energies higher than 300 eV, so astigmatism spoils the focusing properties of these optics for lower incidence angle. Typically, when working around 10 keV, the incidence angle on now-available multilayers is on the order of 1 degree. This is quite a small angle, but nevertheless almost ten times larger than the corresponding total reflection angle for the same energy range on any conventional mirror, and so leads to mirror lengths ten time smaller. For this reason, above 1 keV, KB systems coated by multilayer have been early tested and are still actively developed [30,29].

Focusing grating

Fresnel transmission lenses, with successive opaque and transparent circular zones, were first applied in the visible range and are the most well known kind of focusing gratings. The necessary continuous decrease of the grating period from the central to the external zone (the radius of the nth zone is proportional to $\sqrt{n}$) leads to a maximum lens diameter depending on the technology used. Because the ultimate spot size is set by the smallest grating period, electron beam technology is preferred for high resolution zone plates. On synchrotron sources, focal spots about 0.2 μm in diameter, which corresponds to the predicted theoretical limit, have been obtained in the so-called water window (0.4 - 1 keV). In this energy range, soft X-ray microscopes are built with a condenser lens followed by an objective lens [39].

Unfortunately the efficiency of such a transmitting Fresnel lens decrease at higher energy because no material presents the right absorption and phase contrast between the successive zones. Special technologies are under development to built thick enough zone plates, even with adapted groves profile to increase the efficiency. Figure 10.4 is an example of lens profile for 8 keV photons, prepared by the LIthography Galvanoformung Abformung (LIGA) technology [40]. So far no device has been built around such high energy X-ray transmission Fresnel lens, so practical limitations for spatial resolution and maximum energy are not yet known.

The quite small diameter of transmitting Fresnel lens (typically a few 100 μm) matches well the progresses of the high brightness synchrotron source beam size, but thermal stability could become a limitation for such focusing devices when placed as the first optical element in the beam line.

The basic principle and advantage of high resolution diffractive focusing optics, demonstrated on transmitting Fresnel zones, can be kept on reflecting focusing gratings. This was first suggested by Aristov in 1986 [41], and holds for normal and glancing incidence, and even for Bragg reflection or total reflection. In the X-ray range, depending on the desired band pass, the Bragg reflection can be obtained with crystals or multilayers. Combining the Fresnel grating diffraction and the Bragg monochromatisation, this new kind of focusing is usually named Bragg-Fresnel Lens (BFL). This new optical solution avoids the absorbing material problem of transmission grating Fresnel lenses and seems also better adapted to thermal stability. As for transmitting Fresnel lenses, the ultimate spatial resolution is dependent on the size of the smallest groove, thus electron beam technology is used to pattern the gratings. Ion etching is next applied to mill the gratings. Adjusting the groove depth in principle allows preparation of more efficient phase-contrast lenses and reduced zero-order contamination. For phase gratings, another tested solution is to coat the Bragg reflector with a well-calculated thickness of material giving the right dephasing by transmission and to stop the etching at the bottom of the coated layer [42].

A few microprobes have been developed around different kinds of BFL and focus spot evaluated. They are described in section 10.3.2.

Two linear Bragg-Fresnel gratings, set in the KB geometry, can produce μm scale beam spots. This was demonstrated at LURE (Orsay, France) [43], where a 100 μm diameter pinhole placed in the SR direct beam and acting as

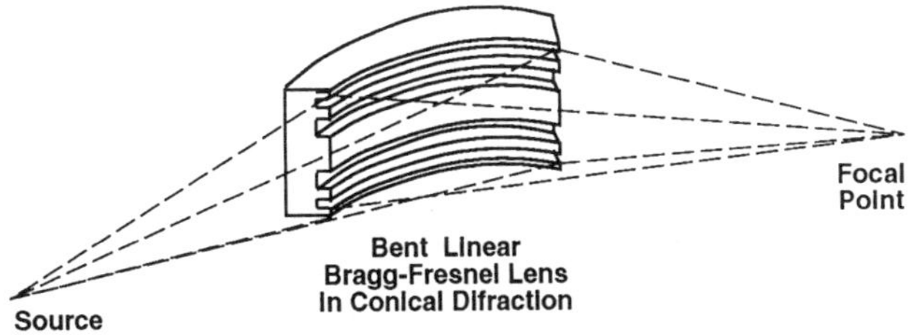

FIGURE 10.5 Example of a Bragg-Fresnel focusing optic. Focusing in one direction is obtained by a linear grating ion etched on a crystal and used in the conical mode. Bending the crystal allows to adjust the focusing in the other direction.

the photon source was imaged on a 2 × 3 μm^2 focal spot.

Double focusing can also be achieved with a single linear BFL used in the conical geometry when properly bending the substratum as shown figure 10.5. This was first shown by Snigirev et al [44] who have obtained very fine focusing for up to 60 keV X rays on ion-etched crystals.

Such double focusing can be obtained with a flat Bragg reflector with curved grooves as early demonstrated on Si[333] crystal at 8 keV for normal incidence as shown in figure 10.6 [45]. Under normal incidence the corresponding groove shapes are circular zones identical to a transmitting grating but need back scattering reflection which is not very convenient for a microprobe arrangement. In fact, in the 10 keV energy range, very narrowly spaced diffracting planes must be used (Ge[777], for example) the reflectivity of which are of course very low. With this kind of arrangement the lateral resolution will be improved because of reduced astigmatism, but at the cost of the intensity. For example the direct beam of a very parallel electron beam undulator source, called a "low-β" undulator (130 x 90 μm^2) of the third generation European Synchrotron Radiation Facility (ESRF, Grenoble, France) was imaged on a 0.7 μm focal spot (FWHM) with an intensity of 10^8 ph/s [46].

For glancing incidence, circles become ellipses. Figure 10.7 presents, together with a linear BFL, the scheme of two other types of developed focusing systems. The first one (figure 7b) , called on-axis BFL, is made by drawing the complete elliptical grating (note that the ellipses are not exactly confocal so as to keep the phase of the reflected beam). The drawback of such on-axis lenses, as for Fresnel transmitting lens, is that all the successive grating orders are centred along the direction of the reflected beam. The resulting overlapping between zero and first order will be discussed later with experimental results.

Because Fresnel lenses are like holograms, any part of the lens pattern can be selected to reproduce the image. This principle is illustrated in the third scheme (figure 7c) with the so called off-axis BFL, which enables to spatially separate the various orders [47,48]. It was shown that this does not result

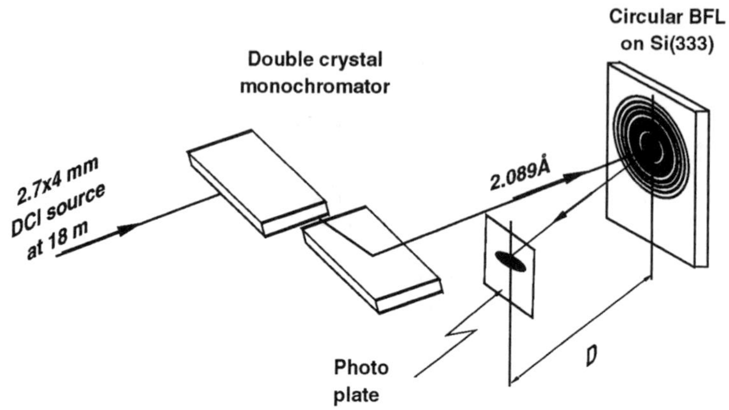

FIGURE 10.6 Two dimensional focusing of hard X rays by a phase circular Bregg-Fresnel lens in the case of Bragg backscattering.

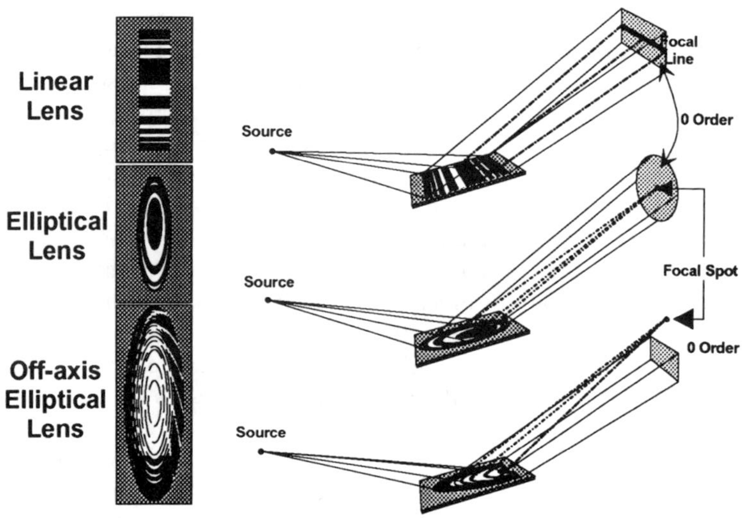

FIGURE 10.7 Linear Bragg-Fresnel grating ensures one dimension focusing whereas Bragg-Fresnel gratings with curved grooves provide 2D focusing with a single element. They are equivalent to reflecting holograms. On each figure the reflected zero order is represented by a shaded area. With centred systems (a and b), called on axis, the focused spot is at the centre of the zero order, so the spot is surrounded by undesirable photons that lower the contrast. A second design (c) called off-axis, is to prepare a lens by using only one part of the hologram in such a way to spatially separate the focus spot and the zero order.

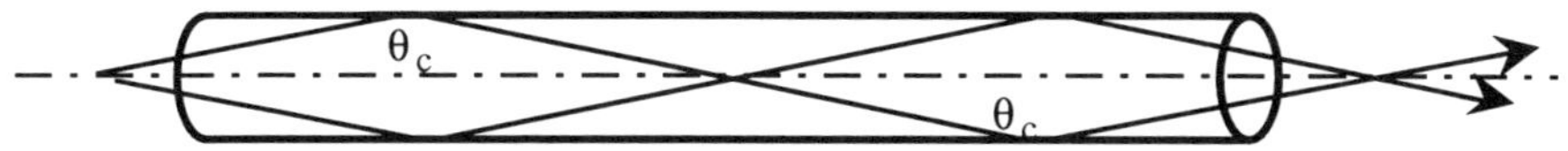

FIGURE 10.8 Principle of a cylindrical glass capillary used under glancing incidence to reflect and guide X rays. This is a collecting rather than a focusing element.

in loss of intensity because one can choose a part of the complete pattern near the apogee of the ellipses where technology doesn't limit the drawing of successive lines, but it does cause a slight reduction in lateral resolution due to enhanced astigmatism [49].

BFL X-ray optics have been developed only during the last few years and have demonstrated their ability to obtain micron spot size when used on the last generation of high brilliance synchrotron sources. They are now a good alternative to KB systems for efficient focusing below 1 μm.

Capillary optics

Total reflection inside a glass tube of small diameter acts like an optical fibre for visible light but such a guide can also be used for X rays. Due to total reflection inside the capillary wall, the beam can be squeezed inside this waveguide with a resulting gain in intensity compared to a simple collimator (see figure 10.8). This was shown as early as 1931 [50].

Strictly speaking, this is not a focusing optic but a collecting optic that allows the transportation of intense flux. In other words it by-passes the loss of intensity (the $1/d^2$ law) encountered when a sample is set far from the source of a conventional X-ray tube. The intensity loss is small as long as the capillary is kept straight enough to fulfil the total reflection condition. Typically these especially prepared capillaries are made of glass or of borosilicate. The external diameter is about 1 cm and the capillary diameter is in the range of 100 μm, but sometimes as small as 4 μm, for a length of 10 cm. They were first proposed for use in a table-top X-ray microprobe about 10 years ago [51].

Present technology allows the manufacture of polycapillary tubes, bundles of 500 to 3000 parrallel capillaries. Their internal diameter is in the range of 4 to 17 μm and the external diameter of the bundle goes from 350 to 500 μm. The typical length is around 20 cm but can reach 80 cm.

Kumakov [52] was apparently the first to assemble bundles of polycapillaries with appropriate curvature for concentrating wide X-ray sources on submillimeter spots. The expected flux should be very high since a large radius of curvature is apparently able to keep noticeable intensity even in the 10 keV range. The resulting spot size is very dependent on the bending precision of

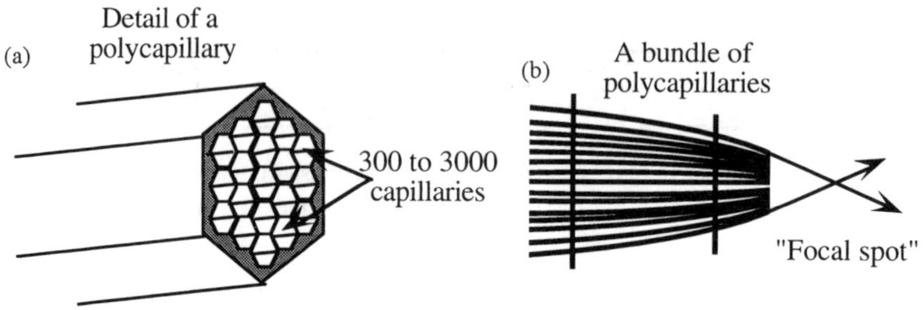

FIGURE 10.9 (a) Principle of a polycapillary tube. It consists of 300 to 3000 capillaries of 4 to 17 μm diameter each. Figure b shows how such polycapillaries can be bundled together to produce intense X-ray microbeams.

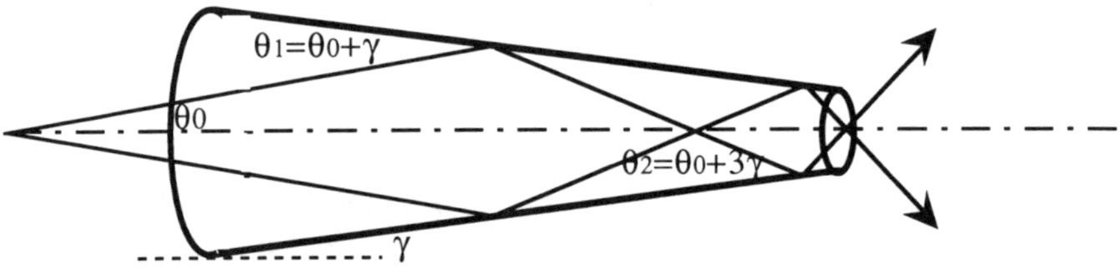

FIGURE 10.10 Principle of a tapered glass capillary that squeezes X rays on a small spot. For minimum beam size the sample must be placed directly on the output of the capillary

each individual guided beam and on the distance from the end of the capillary. An unavoidable consequence of this basic idea is that the beam is quite divergent as schematically indicated figure 10.9, so the sample has to be kept very close to the focal spot.

A single capillary tube with a conical shape can be used to collect and focus hard X rays, as was demonstrated by Stern *et al.* [53]. In such a capillary the glancing incidence angle of each photon will increase by twice the cone angle at each reflection, as can be seen from figure 10.10. As soon as this glancing angle gets larger than the critical angle the reflectivity drops almost to zero and the photon is lost.

Using this focusing device, extremely small spots can be produced. For example, a spot of white X rays of 90 $\pm$ 5 nm and a resolution of 50 nm has been obtained by Bilderback *et al.* at CHESS with enough intensity to perform microfluorescence imaging and Laue diffraction [54].

A disadvantage of such tapered capillaries is that the exciting beam is quite divergent, so the optimum resolution can only be reached if the sample is set very close to the end of the tube. This divergence is mainly set by the cone opening angle, and the focal depth can be evaluated to lie between 20 to 100 times the tip diameter of the capillary. This might turn out to be very uncomfortable when studying samples in special environments (high pressure, furnace, etc.).

Another particularity of these optics is that they act as low-pass energy filters because of the total reflection phenomenon mentioned earlier. For a white incident beam the output spectrum is continuous in energy up to a sharp cutoff, the energy of which depends strongly on the material of the wall of the capillary. This is of a great interest, for example, for Laue diffraction.

X-ray capillary optics are now commercially proposed. Specific methods to improve their fabrication and reliability in performance are actively explored by several groups in the main synchrotron centres [55]. Namely, tapered fibres and single-reflection fibres are under study and could become good candidates in specific cases. Thanks to the activities focused on this new kind of X-ray optics it will be soon possible to evaluate more precisely its potential and the niches where it could be applied.

10.3 X-RAY MICROPROBES

10.3.1 Comparison of micro-SRxrf with EPMA and micro-PIXE

Many comparisons between these different microanalysis systems, all of which use the induced X-ray fluorescence, have been published. Recent, exhaustive reviews have recently been published by Janssens *et al.* [56,57].

Minimum detectable Limit

The variety of excitation beams now available (photons, electrons, heavy charged particles ranging from proton to uranium) and of detectors (energy or wavelength dispersive) gives rise to many different combinations for X-ray fluorescence analysis. As long as we are only concerned with the localization and identification of a particular element at the lowest concentration possible and with the best lateral resolution, physical circumstances should help to decide which is the best arrangement.

The minimum detectable limits (MDL) can be calculated by comparing the fluorescence intensity (S) with the square root of the background (B) under the fluorescence peak. In other words, for a given sample, the optimum combination of excitation beam and detector is the one that maximises the ratio $S/\sqrt{B}$.

We shall now examine the different parameters that can enhance this ratio.

Intensity The number of incident projectiles should be very high since MDLs only improve with the square root of the intensity. As a result the

TABLE 10.1 Energy dissipation (in keV) in different 1 μm thick targets by various projectiles for the production of one characteristic X ray.

Projectile	Photon		Electron		Proton	
Energy (keV)	5	20	20	50	3000	5000
$Al(K_\alpha)$	5		50		170	
$Cu(K_\alpha)$		14	1900	830	31000	10000
$Au(L_\alpha)$		13	1700	910	71000	21000

choice of the beam enters into consideration and the ability of the sample to withstand the high intensity without damage must also be considered.

Charged particles would seem more favourable than photons because they can be rather easily focused on small samples through magnetic lenses. Technically, intensities of the order of a μA can be obtained on small areas, meaning up to 10^{12} to 10^{13} projectiles per second (not at the micron scale of course). We pointed out that conventional X-ray sources cannot compete, but because of its intensity and very low angular divergence, Synchrotron Radiation becomes an interesting tool for producing photon microbeams.

When considering the sample's ability to support such high flux, SR becomes really very interesting. In Table 1 we report the calculated energy (in keV) left in various 1 μm thick targets when one characteristic X-ray line is produced by different projectiles of different energies. The interest of monochromatic photons is obvious and can be easily understood. The photon which has not interacted passes through the sample without leaving any energy whereas a charged particle loses its energy during its slowing down in a quite short range whether or not it has created a useful inner-shell vacancy.

Fluorescence cross sections The X-ray-induced fluorescence cross section must be large. Photoionisation cross sections are generally much larger than inner-shell ionisation cross sections for charged particles of usual energy for fluorescence analysis. Moreover, for a given energy, the photoionisation cross section scales with Z^4 whereas the inner-shell ionisation probability by charged particles drops very rapidly, nearly as Z^{-4}. This behaviour is most interesting in trace element analysis.

Background The background under the considered fluorescence peak should be as low as possible. In order to see whether this background can be reduced we must understand its origin. For this discussion we shall refer to calculations from Jaklevic [58].

For electron excitation most of the count rate contributes to a continuous background and in this case MDLs are in the 100 to 1000 ppm range. This background is due to the bremsstrahlung accompanying the slowing down of the incident electron as well as the electrons ejected during collisions. It cannot be avoided but in principle may be minimized when using lower energy electrons (although ionization cross sections will decrease also) and samples with low-Z matrix.

In this sense proton excitation is much more interesting and the X-ray excitation!proton vs. photon increase in sensitivity, usually in the ppm range, is mainly due to a drastic reduction in background shape and overall intensity. Bremsstrahlung due to the slowing down of the proton is negligible. In addition the contribution of the ejected electrons is quite low and limited to the lower energy part of the spectrum because these electrons are usually ejected with very small kinetic energy.

Under excitation by X rays of continuous energies, the continuous background is due to the scattering (Rayleigh and Compton) of the incident beam from the sample, and the MDLs are in the range 1-10 ppm.

The situation changes when using monochromatic photons. In this case, the scattered events are reduced to a sharp peak having an energy higher than that of the fluorescence lines to be observed. The background, under these conditions, is only due to the low energy tail of the total absorption peak observed with Si(Li) detectors. With modern state-of-the-art Si(Li) detectors this leads to a reduction of the background by more than four orders of magnitude . Of course the signal is reduced also, but not as much as might be feared. This results in much lower MDL's.

Required photon flux

If one wishes to record the fluorescence spectra with a lateral resolution of 1 μm in a μm thick sample and at the ppm level, within an acquisition time of one second, then one needs about 10^{10} ph/s compared to 10^{12}-10^{14} protons to do so.

All these numbers are extremely high considering that these projectiles are intended to pass through a 1 μm^2 area. Although such intense fluxes are not impossible to get with photons and newly developed X-ray optics installed on high-brilliance undulator sources, it seems completely impossible to reach them with charged particles. Furthermore, it is very well known that the sample cannot sustain such high doses. For example, 2 pA μm^{-2} seems to be a limit for biological samples.

10.3.2 X-ray fluorescence microprobes

Horowitz and Howell [59] first demonstrated the feasibility of X-ray microscopy in the soft X-ray range, but the pioneer work in the hard X-ray regime for X-ray microanalysis is that of Sparks, *et al.* in the late seventies [34,60,20].

In this chapter we shall describe the various kinds of X-ray microprobes now in operation, and present their main characteristics. Quite a few general reviews concerning the application of these microprobes to trace element analysis and other techniques have already been published [61–64].

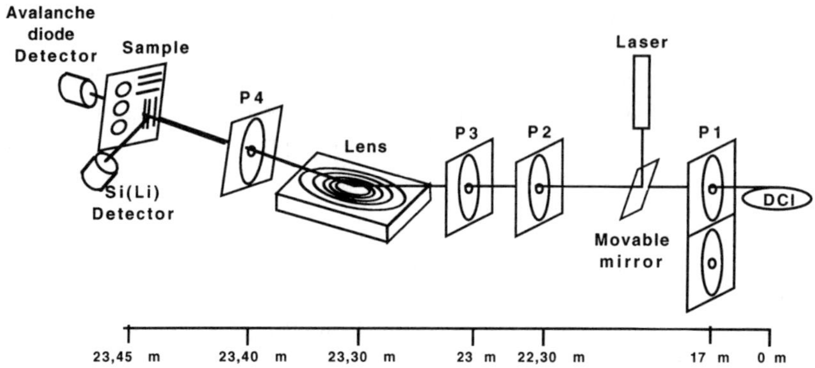

FIGURE 10.11 Scheme of the experimental set up of the LURE-IMT photon microprobe. Pinhole P1 acts as the photon source. P2 and P3 are collimators to reduce scattering and P4 a 20 μm pinhole to cut most of the zero order diffracted beam.

The LURE-IMT microprobe

Description The present design of the X-ray fluorescence photon microprobe uses a single, flat, elliptical Bragg-Fresnel Multilayer Lens (BFML) made in IMT RAS (Chernogolovska, Moscow district, Russia). A schematic of the experimental set up is shown in figure 10.11. It has been installed on line D15 of the DCI storage ring (Orsay, France) since July, 1992 [65].

Because of the large size of the electron beam in this first generation machine, optional entrance pinholes (50 or 100 μm diameter) were added 6.3 m upstream from the lens to simulate a smaller apparent X-ray source at the cost of lower flux. After the beryllium window, which isolates the X-ray beamline under vacuum from the rest of the experiment which is under air, a second pinhole (200 μm diameter) is installed to eliminate most of the scattered photons.

The elliptical BFML position is adjusted in the direct beam by a five-axis multi-stage goniometer (y, z, θ_x, θ_y, θ_z) that was built by "Radicon Ltd" in St Petersburg (Russia). To further restrict the scattered beam, and cut down most of the undiffracted zero-order, a 20 μm diameter pinhole is positioned 10 cm after the lens on a two axis stage (y, z).

The sample to be analysed is positioned in the image plane within an accuracy of 0.1 μm and adjusted at the focal distance within 1 μm steps by another three-axis Micro-Controle remote-controlled multi-stage. This stage can be set perpendicularly to the beam for transmission tests or at 45° for fluorescence mapping.

The positioning of the lens is the most critical tuning parameter to obtain the best focal spot and maximum intensity. To take full advantage of the high degree of horizontal polarization, the small axis of the lens should be set in the orbital plane. The incident angle should fit the Bragg angle to better than 0.05° for maximum intensity, and, to reduce aberrations, the long axis should be aligned with the beam direction at better than 0.05°. A very rough positioning is first made using a laser that simulates the X-ray beam path. In a second step we use an X-ray CCD camera built at LURE with 10 μm resolution, which nevertheless allows a fairly good alignment. This takes no more than half an hour. The final positioning is achieved using high resolution photographic films (X-ray or holographic films) and observing, under microscope, the shape of the registered focal spot. This step may take up to two hours, sometimes more, but the whole procedure is neither that difficult nor that time consuming.

Finally the X-ray fluorescence spectrum is recorded with a 13 mm^2 area Si(Li) detector of 150 eV energy resolution for 5.9 keV photons. An aluminum collimator is installed on the detector axis for shielding and to limit the acceptance angle. Scattering in the detector is further reduced by placing it in the plane of the synchrotron ring perpendicular to the incoming beam. Pulses from the pre-amplifier are sent to a Tennelec 244 amplifier and then processed through a Nucleus Multichannel Analyzer card set in an IBM PC computer. When working in transmission mode a Si avalanche diode detector of 1.5 mm^2 area, operating at room temperature, is used. In fluorescence mode and with thin samples this detector is also useful for monitoring the transmitted beam.

The Bragg-Fresnel Multilayer Lens (BFML) In this new kind of X-ray optical system, a multilayer acts as the Bragg reflector and ensures monochromatization of the white incident X-ray beam. The Fresnel zone structure, made by electron beam lithography and ion etching techniques on the multilayer substratum, provides diffraction focusing of the reflected X-ray beam. In principle, BFML keeps the advantage of the Fresnel zone plates, which offer high spatial resolution when used in transmission. Moreover, when used in the reflection mode the grating in the multilayer presents high thermal stability, as demonstrated widely through the use of multilayer X-ray mirrors. Such a high mechanical stability is of primary importance considering the photon flux it will undergo with the advance of ultra-bright synchrotron sources.

The first test of BFML with synchrotron radiation was made in the soft X-ray range at Siberia-2 SR facility in Moscow in 1990 by Aristov *et al.* [66] and in the hard X-ray range at LURE in 1991 [43,67]. This first kind of photon microprobe used an arrangement of two linear BFML mounted on two independent goniometers in the KB focusing scheme. Alignment of the microprobe was rather time-consuming but the results were very encouraging since a 100 μm diameter X-ray source could be imaged on a 2 x 3 μm^2 focal spot.

By that time theoretical considerations led Aristov and Erko [68] to the conclusion that BFML, with elliptical zone structures, would provide two-dimensional focusing. Such lenses were constructed at the IMT RAS and

TABLE 10.2 Typical lens parameters (description of the lens used on our setup).

Wavelength (Energy)	Dimensions (s × l) axis	Multilayer period (nm)	Minimum zone width (s and l) axis	Focal length (cm)
0.1 nm (12.4 keV)	59 μm × 2.48 mm	3.25	250 nm 10.7 μm	15
0.154 nm (8 keV)	360 μm × 14 mm	2.98	300 nm 6 μm	40
0.09 nm (13.8 keV)	210 μm × 14 mm	2.98	300 nm 10.4 μm	40

tested in LURE in 1992 [69]. It is this kind of lens that is now operational on our photon microprobe.

The lenses now used are prepared as phase lenses, the optimum profile depth corresponding to the maximum first order diffraction efficiency, calculated with the formula for phase conical multilayer gratings [68]. The ion etching of the lens was done on 60 bilayers with 3.2 nm period tungsten/carbon multilayer mirror coated by magnetron sputtering on a thick, highly polished silicon substratum. An ALCATEL SCM-651 coating system, designed with the help of B. Vidal (LOE, Marseille, France), was used for the multilayer preparation. The main characteristics of the lens used here are given in Table 2.

The diffraction aberration characteristics of this kind of optics can be calculated using aberration formulas developed by Aristov and Erko [68]. It appears that, because the present technology allows to draw a large number of Fresnel zones, the contribution of these aberrations can be neglected. The theoretical resolution limit is then mainly fixed by the minimum zone width of the lens and in practice it is the magnification of the lens that will contribute to the focal spot size in our installation.

A drawback of this type of lens is that the focal spot is surrounded by a halo due to undiffracted zero-order photons. We have measured that there are about as many photons in this halo as in the focal spot. Its contribution can be reduced by positioning a slit after the lens, as far from the lens as possible. Unfortunately this causes difficulties in setting the detector and the microscope near the sample.

The off-axis Bragg-Fresnel multilayer lens is a very interesting solution to this problem, recently proposed and tested by Mirone et al. [47]. In this case the focal spot and the zero order undiffracted beam are separated in space and so can be very easily stopped with an absorber set before the sample.

Tests and results The LURE microprobe was tested in different ways and on two different machines: DCI (LURE, Orsay, France) and ESRF (European Synchrotron Radiation Facility, Grenoble, France). We investigated well-known objects to test both the spatial resolution and the sensitivity of the microprobe.

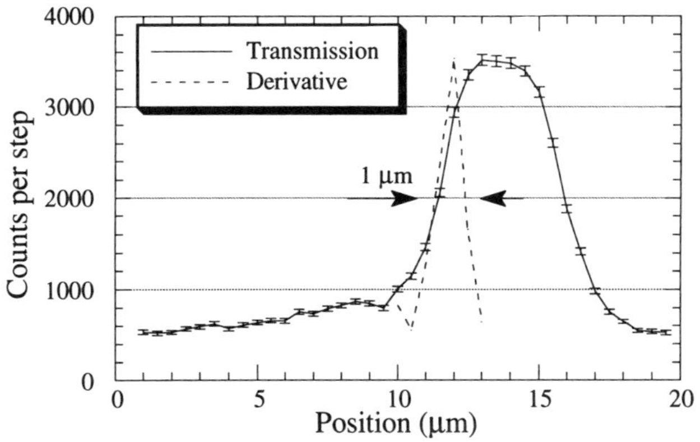

FIGURE 10.12 Line scan across a 5 μm large gold strip. The derivative shows a 1 μm beam size.

Spatial Resolution At the 1 μm level, spatial resolution starts to be difficult to measure. We used high resolution photographic plates detectors, and knife edge objects either in transmission or in fluorescence mode.

For the transmission mode we used a specially prepared gold mask, with 5 μm large and 7 μm thick strips, supported by a SixCx membrane. Figure 12 shows the variation of the absorbed X-ray flux versus position when the strip is moved across the focused beam. Derivation of the transmitted flux is also shown. A resolution of 1 μm was obtained with good contrast [70].

Photon Flux and Lens Efficiency By measuring the X-ray fluorescence emitted from a thick metallic foil set at the focal spot, one can calculate the number of incident photons. At DCI, the total intensity in the spot was measured to be 7×10^5 photons per second in the case of a 100 μm input pinhole and a stored current of 260 mA in the machine. This leads to a flux of 6.7×10^4 photons per second per μm^2 and per 100 mA in the band pass of the multilayer for 12.4 keV photons.

At the ESRF, the microprobe was installed on beam line 9 (Troika) which is a high-brilliance undulator X-ray source, well-adapted to our kind of optics. We measured 2×10^9 photons per second, which means a flux of 1.6×10^8 photons per second per μm^2 per 100 mA in the band pass of the undulator [71]. These results are presented in Table 3.

The overall efficiency was measured by comparing the number of photons striking the lens (in the band pass of the multilayer) to that passing in the focal spot for 12.4 keV photons. Thus defined we found an efficiency of 16%, a value in quite good agreement with the theoretical prediction of 26% for a classical amplitude-phase grating. This indicates the phase character of the obtained diffraction [70].

TABLE 10.3 Performance of the LURE-IMT Microprobe.

Machine X-ray source	DCI - D15 bending magnet	ESRF- ID10 undulator
Input energy spectrum	white	11.8 keV 4th harmonic
Output energy	12.4 keV	11.8 keV
FWHM of the focal spot (microns^2)	2x2 (100 μm pinhole) 1x1 (50 μm pinhole)	2.5 (V) x 5 (H)
Photon flux (phot $s^{-1}\mu m^{-2}(100mA)^{-1}$)	6.7×10^4	1.6×10^8

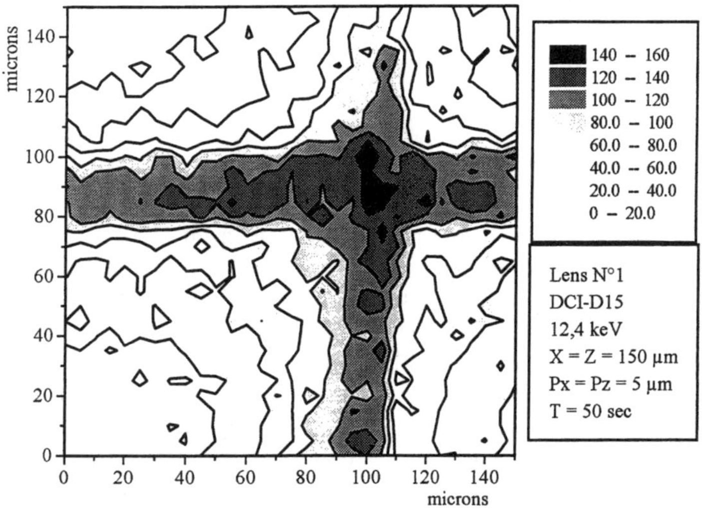

FIGURE 10.13 Scan of a nickel cross made of two perpendicular 20 μm diameter wires embedded in a resin.

Examples

Fluorescence test of a nickel cross The object is a cross made of two perpendicular 20 μm diameter nickel wires embedded in a resin. Nickel can be easily excited with a 12.5 keV photon energy. Figure 10.13 shows a two dimensional map of a 150 x 150 μm area around the cross. Such an object is also interesting for demonstrating the shadowing effect of the emitted X ray which appears for the vertical wire.

Minimum Detectable Limits: comparison with micro-PIXE Synthetic glasses of pure oxide were analyzed at the ESRF and also at the Pierre Süe PIXE microprobe (Laboratoire Pierre Süe, LPS, Saclay, France) by M. Mosbah *et al.* [72] These glasses contained known trace concentrations

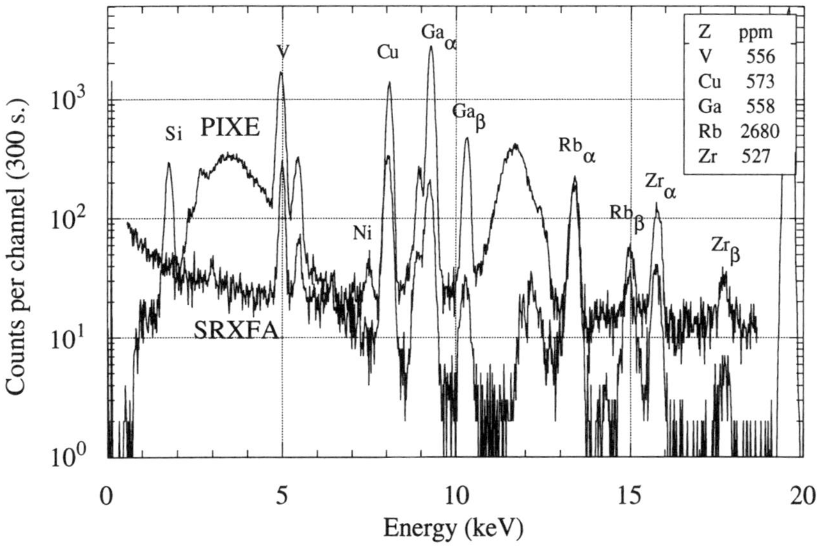

FIGURE 10.14 Comparison of the X-ray fluorescence spectra from the same sample obtained with the LURE-IMT photon microprobe and the P.Süe (Saclay, France) micro-PIXE installation. Acquisition time and solid angle are the same for both spectra but the micro-PIXE data are obtained under vacuum whereas with the photon-microprobe there was an air path of 6 cm.

of several elements between vanadium and niobium. The X-ray fluorescence spectra of one of theses glasses observed with both microprobes are shown in figure 10.14.

For micro-SRXRF, the excitation energy of the undulator was set at 11.8 keV to reduce the intensity and avoid saturation of the detector. Higher order diffracted photons (energy = 23.6 keV) could still excite characteristic lines of elements heavier than Rb with interesting efficiency. The air path between the sample and the detector (6 cm, solid angle 3.6 millisteradian) and the rather high excitation energy makes it unfavourable for the observation of light elements. The PIXE spectrum was obtained under 3 mC of 2.5 MeV proton excitation in a vacuum chamber, and the solid angle was 13 millisteradian. To avoid saturation of the detector with the low energy part of the bremsstrahlung spectrum it is necessary to place an aluminium filter between the sample and the detector, which absorbs X rays from low atomic mass elements.

Up to now, SRXRF and PIXE probes have easily achieved MDLs of 1 to 10 ppm depending on the material under study and optimal conditions [73,74]. From the spectra such as the one presented in figure 10.14, one can compute the MDLs and compare the sensitivity of both kinds of microprobes under usual working conditions (see figure 10.15).

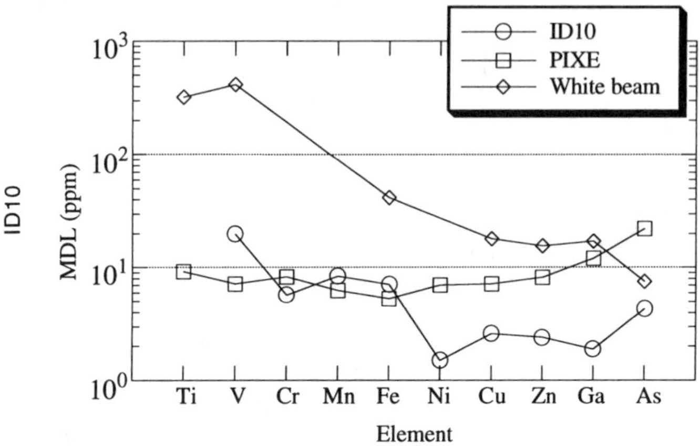

FIGURE 10.15 Minimum detectable limits calculated from spectra like that of figure 14. Results are shown for the LURE-IMT photon microprobe installed at the ESRF (ID10) and under white beam excitation (ESRF bending magnet) and are compared with those obtained with the P. Süe micro-PIXE installation.

This comparison highlights a better sensitivity for elements from V to Fe with PIXE and from Ni to As with SRXRF, as might be predicted from the behaviour of the ionization cross sections. In fact, as this was our first experiment, we know that the MDLs reached at the ESRF can be pushed downward by a factor of ten by optimization of the incident energy at the maximum intensity coming from the undulator (a factor of five was measured), better shielding of the detector (perhaps a factor of two) and reduction of the sample-to-detector distance (at least a factor of five). At the same time better MDLs for heavy elements could be obtained with micro PIXE with higher energy protons, but that would probably not change drastically the advantage of photon excitation for this kind of element.

Up to now, the LURE-IMT microprobe has mainly been used for mineralogical studies such as micrometeorites [75], glass inclusions in volcanic rocks [75,72] and liquid inclusions in coesite-bearing rocks [76].

The Photon Factory X-ray microprobe

Description This X-ray microbeam system is installed on beam line 4A (bending magnet) of the Photon Factory (Tsukuba, Japan). It is based on a Kirkpatrick-Baez optic in combination with a double crystal (or multilayer) monochromator [77].

A plan view of the experimental arrangement is shown in figure 10.16. Slits 1, located 9 m from the source, are used to adjust the beam size whereas

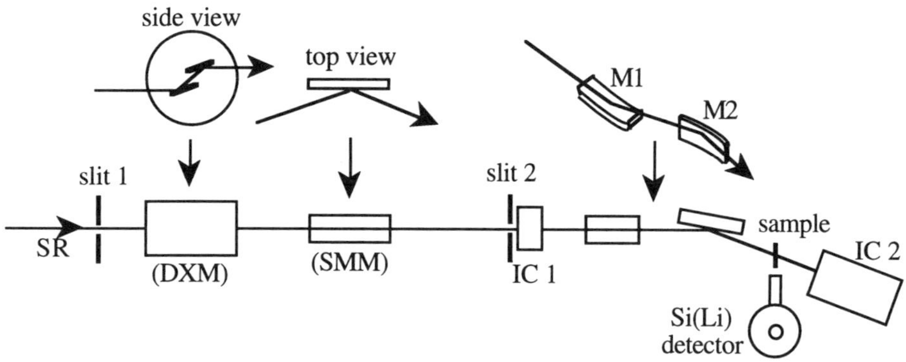

FIGURE 10.16 Plan view of the Photon Factory X-ray microprobe. Two elliptical mirrors (M1 and M2) are used in the KB scheme to focus monochromatic X rays provided either by a fixed exit double crystal monochromator (DXM) or a single multilayer (SMM). Slits 1 and 2 helps to define the final beam size. IC1 and IC2 are two ionisation chambers used to monitor the X-ray flux and to perform EXAFS or XANES experiments in the transmission mode.

TABLE 10.4 Photon Factory X-ray microprobe mirrors parameters [77].

	First mirror (M1)	Second mirror (M2)
Focusing	Vertical	Horizontal
Magnification	80:1	200:1
Long axis (m)	6.88	6.88
Short axis (mm)	8.06	5.13
f_1[a] (m)	13.598	13.700
f_2[b] (mm)	170	68.5

[a] f_1 is the source–mirror distance
[b] f_2 is the mirror–focal point distance

the aim of slits 2 is to reduce scattered radiation. In order to reduce spherical aberrations [78,79] two elliptical platinum-coated fused quartz mirrors (Canon Inc.) have replaced the two spherical ones first used. The parameters of these mirrors are given in Table 4.

Depending on the entrance slits size and the and monochromator, different spot sizes and fluxes are available. Some typical values are presented in Table 5.

Although the beam size obtained with this kind of optics is a little larger than that reported for BFML, and although the alignment time is longer, this scheme offers many interesting advantages. Mainly, in combination with the fixed exit double crystal monochromator, it offers the unique opportunity

TABLE 10.5 Photon Factory X-ray microprobe mirrors parameters [77].

Monochromator	Beam size $h \times v$ (μm^2)	Flux $ph/s/300mA$	X-ray energy (keV)
Double	3.0×4.2	6.5×10^6	10.5
crystal	6.1×6.1	1.2×10^8	7.1
Multilayer	5.5×6.1	8.0×10^8	12.0
	6.7×5.6	1.6×10^10	6.9

of tuning the X-ray energy continuously, whereas BF optics are drawn for a specific energy.

Very recent development of this microprobe, tested on beam line 16X of the Photon Factory (multipole wiggler) using higher magnification mirrors, led to a 0.7×0.9 μm^2 minimum spot size with a photon flux of 2×10^8 ph/s/300 mA for 9 keV X-rays and a multilayer monochromator [29].

An example of application This microprobe has been used in a large number of studies of various materials, mainly biological and geological samples.

Very spectacular applications can be found in the literature, such as the elemental distribution of S, Ca, Cu and Zn in 10 μm thick slices of human hair [77]. A 150 x 150 μm^2 map, recorded with 3 μm scan steps, was obtained with only 10 s acquisition time per pixel. Keeping in mind that the nominal concentration of copper is only 10 ppm, this gives a good idea of how powerful such X-ray microprobes can be for the study of trace element distribution.

The Daresbury X-ray microprobe

The Daresbury (UK) microprobe was installed on a bending magnet of the Synchrotron Radiation Source in 1989 [80,81]. It presents two original features compared to the facilities described above.

First, the focusing device is a concave ellipsoidally bent Si[111] crystal. The sagital radius is 100 mm and the meridian radius 5740 mm. It is fixed on a four-axis stage (two translations and two rotations) 80 meters from the source and focuses 15 keV photons (0.3 keV band width) 38 cm away on a 10×20 μm^2 (FWHM) spot. The flux density is about 10^4 $ph/s/mA/\mu m^2$ [80].

Second, the sample is placed in a vacuum chamber to reduce scattering and the argon peak. This is the only X-ray microprobe to offer this opportunity, which is most interesting for the study of light elements.

It seems that this microprobe was entirely devoted to microfluorescence analysis. Some of the main topics that were studied were human hair [82], hamster tooth germs [83,37], and rat brain sections to investigate the role of zinc [84]. On this installation it was also shown [84] that the ratio of the area of the fluorescence peak to that of the Rayleigh + Compton peaks could be used to extract reliable concentrations, just as is commonly done in conventional XRF [85]. This important fact will greatly assist quantitative analysis.

The Berkeley Microprobe

The Berkeley (USA) group has for a long time been concerned with microbeam production [86] and has proceeded to many measurements on various SR machines. The microprobe has been very recently installed on beam line 10.3 (bending magnet) of the Advanced Light Source (ALS). It uses two multilayer-coated spherical mirrors in the KB scheme to produce a beam spot in the micrometer range [30].

The first experiments are focused on the study of defects and cracks of ceramic materials, the distribution of doping elements in semiconductors, geophysical inclusions in rocks, pollution, and archeology.

At present the elemental sensitivity is at the fentogram level for K to Zn and the acquisition time is about 60 seconds.

The Brookhaven Microprobe

The Brookhaven (USA) microprobe undoubtedly was one of the first X-ray microprobes in use. It has been installed on beam line X26A (a bending magnet) of the National Synchrotron Light source (NSLS) at Brookhaven since 1985 [13,87,88]. It runs to a large extent as a collimated microprobe with a beam size of 8 μm, directly on the white beam, and only 10 m from the electron beam. This is the simplest way of producing a microbeams but it has been very successful, as is attested to by the number of publications. A focusing mirror and/or a high-resolution monochromator can be installed for special experiments.

A very important part of the beam time has been devoted to micro-fluorescence analysis; this was especially so before 1990. This activity is still of major interest, and quite recently a wavelength-dispersive analyser was installed, which seems a unique case [89].

Since the beginning, this microprobe has been used in a great variety of scientific fields. In chemistry and materials science chemical speciation [90], catalyst analysis [91], and soil and sediments study [92] are a few examples of subjects. In geology, the study of pyroxene (compared with results obtained with electron microprobe [93]), elemental mapping of fluid inclusions in various rocks [94], and analysis of rare earth elements [74] have opened new fields of investigation. A more general survey of the contribution of X-ray microprobes in this field was published by Sutton et al [95]. In astrophysics, micrometeorite and particularly chondritic particles have been extensively studied for their trace element concentration [96], and new populations of meteorites discovered [97]. Applications in biomedical sciences are numerous [98]; we can mention protein crystallography [99], metabolism and deposition of lead in bones [100], cartilage growth [101], and the correlation between the concentration of various elements and arthritic disease [102]. Even microbeam therapy has been emphasized [103]. Finally, environmental studies are rapidly developing [104] due to the sensitivity of the method and its high lateral resolution, well-adapted to investigating microsamples like fly ash particles [105,106].

This microbeam has also been used for pioneer investigations in the development of micro-EXAFS, XANES, and micro-diffraction, as will be seen later.

The Brookhaven group is also far in advance concerning X-ray microtomography [107]. For these experiment they use a mechanical collimator to produce a 10 x 10 μm^2 X-ray beam in which the sample is scanned and rotated. From the set of X-ray fluorescence spectra 3D images can be reconstructed. This has been used in various studies, such as the determination of polymerization particle morphology [108], catalysts [109], thermal spray coating imperfections measurement [110], and porous media [111].

Laboratory microprobes

We said earlier in this chapter that X-ray microprobes could only be installed on modern SR machines. This is true when ultimate performance is required, such as elemental mapping at the ppm range with 1 second per pixel acquisition time, or for chemical mapping of these elements, as we shall see in the next section.

In fact the advent of microcapillaries has rendered possible the realization of quite efficient laboratory X-ray microprobes. We must remember that these optics where first developed on conventional X-ray tubes and that Hirsch and Kellar showed radiographic images of X-ray microbeams with a 35 μm capillary as early as 1951 [112].

Laboratory microprobes where first built in Sweden [113] and USA [114]. During the last ten years many such instruments have been constructed [115–117]. They all use a single microcapillary, the entrance of which is set in the X-ray tube, as near to the anode as is technologically possible.

The lateral resolution is typically in the 100-200 μm range when using a cylindrical capillary, but can drop down to 20 μm with ultra fine focus X-ray tubes or with tapered capillaries [118]. They have provided interesting results in various fields [119], such as material science [120] and biology [118], and this is only the beginning.

10.3.3 Related techniques of investigation

Elemental mapping at the μm scale and at the ppm level is very valuable in many research fields. By now more and more information is required and, for example, one of the great successes of micro-PIXE is to give a depth profile of the elemental concentration when used in conjunction with micro-RBS or Resonant Nuclear Reactions.

X-ray microbeams also allow very interesting investigations of the sample and, in principle, chemical and structural information can be obtained.

Micro-diffraction

One of the most staightforward applications of X-ray microbeams is micro-diffraction. There is a real demand in this field to get structural informations on samples that might be difficult to crystalize, for which only micron-scale crystals can be produced (some proteins, for example), or to study small inclusions in mineral or metallurgical samples, the size of which is often in the μm range also.

Thus micro-WAXS (micro-Wide Angle X-ray Scattering) and micro-SAXS (micro-Small Angle X-ray Scattering) are under development and have started to provide useful information, as recently demonstrated at the ESRF [121,122].

An advantage of many of the X-ray optics here described is their rather long focal length—the focal length of the BFML we now use is of 40 cm. Thus enough space is left around the sample to introduce special environment apparatus like cryostats, furnaces, or high pressure cells. We can predict that these hard X-ray microbeams will be ideal for the study of matter under extreme high pressure constraint, as the anvils are only a few microns in diameter.

Microcapillary-made microbeams offer a unique possibility to perform Laue diffraction because these optics do not act as a monochromator but rather as a low-pass energy filter. The only drawback seems to be the strong divergence of the outcoming beam (typically 3 mrad) which obliges the sample to be placed no more than a few microns from the end of the capillary, and this will spoil the resolution on the recording plate. Very spectacular and encouraging preliminary results were obtained by Bilderback et al [54] who recorded the diffraction pattern of a 5×10^{-3} μm^3 single crystal in a 10 min exposure with a 0.36 μm beam diameter. Another example of the interest of microcapillaries was shown by Singer et al [123] who gave evidence of the hydrolysis of a stable intermediate in catalysis by trypsin using dynamic Laue diffraction studies.

A final interesting conclusion drawn from all these preliminary micro diffraction experiments is that they allow a much higher signal-to-noise ratio than can be obtained with slits intersecting a much larger incident beam.

Micro-EXAFS

EXAFS is one of the most widespread applications of SR. This acronym is attached to the oscillations observed after the absorption edge when measuring the X-ray absorption coefficient of a compound. These oscillations extend from the edge up to a few hundred of eV over the edge and they are a fingerprint of the nature, number, and distance of the surrounding atoms. Thus very interesting chemical information can be extracted from an EXAFS spectrum.

Micro-EXAFS has only very recently appeared and the first experiment seems to be that of Hayakawa et al in 1991 [124] performed at the 20 μm scale on minerals, using an ellipsoidal mirror as focusing device. For this kind of experiment a fixed-exit double crystal monochromator of high energy resolution is needed (typically $\Delta E/E = 10^{-4}$) and is usually placed before the focusing optic.

More recently the EXAFS spectrum of iron in magnetite was reported by Iida *et al.* [125] with a 6.1 × 6.1 μm^2 beam size. The complete spectrum was obtained in the transmission mode in about 10 minutes, which proves that such a microprobe can be applied to practical samples. For low concentrations ($<$ 0.1% typically) or thick samples the spectrum can only be recorded in the fluorescence mode and at the cost of a longer acquisition time.

Such experiments are also currently carried out at NSLS in Brookhaven with a focal spot in the 50-100 μm range.

These first investigations prompted many other ideas, some of which are now under development, namely the opportunity to perform, at the same time, X-ray diffraction and EXAFS under ultra high pressure [126].

Micro-XANES

XANES represents the fine structure near the absorption edge of a compound. In a simplified model, the energy of the edge is related to the oxidation state of the studied element, and its shape is mainly determined by the kind of ligands that are attached to it. Just like with EXAFS, valuable chemical information is extracted from the XANES spectra. An interesting point with this spectroscopy is that the XANES signal only extend a few tens of eV across the absorption edge, as compared to about 1 keV for EXAFS. Thus the characteristics of the focal spot (size, absolute position on the sample, intensity) are much more likely to be kept during the scanning of the necessary energy range. Acquisition time will be shorter by about an order of magnitude).

For example, Iida *et al.* [125], obtained a chemical state map of iron . The sample comprised two natural iron oxides, hematite and magnetite. The K_α fluorescent distribution of iron, as well as the transmitted intensity, were recorded in order to build a 160 μm square map with 4 μm steps and five (or nine) second collection times at two different energies: 7.122 keV (where the XANES structure starts to appear) and 7.4 keV (well over the K iron edge).

Since the absorption curve of each chemical state shows a unique energy dependence, we have a set of two equations that allows to extract the concentration of each of the two iron species in the sample.

$$F1 = C_A I_1^A + C_B I_1^B \tag{10.1}$$

$$F2 = C_A I_2^A + C_B I_2^B \tag{10.2}$$

Here the subscript 1 and 2 stands for the two energies and A and B for the two species. F is the recorded fluorescence intensity, C the concentration and I the normalized fluorescence intensity of each chemical state, measured from the complete XANES spectra of both species. This needs to be done only once. If the sample is thin enough, the same procedure can be applied in the transmission mode.

Micro XANES is an active area of the scientific program of the Brookhaven microprobe. It has mainly been used for the study of micrometeorites—for example the chromium valency in individual lunar olivine grains was determined [127,128]—but also in material science [129] and soil study [130].

Micro-depth profiling

It is well-known that for a given energy X rays striking a flat surface with an incident angle smaller than the critical angle θ_c will undergo total reflection. In fact, under this condition the transmitted wave, the evanescent wave, travels only a few nanometers in the sample. This very short range, and its dependence on the incident angle, keeping $\theta_i < \theta_c$, has been used for quite a long time to extract depth concentration profiles of impurities lying just under the surface of the sample [131,132].

In these experiments, the exiting beam arrives under glancing incidence and a detector records the variation of the the X-ray fluorescence intensity in a direction perpendicular to the surface while the incident angle is changed. From these data the depth profile can be calculated. In such a configuration the incident angle is so small that the exciting beam illuminates the whole surface of the sample. The depth information is therefore an average of the depth profile over the entire sample.

Recently, Noma and Iida [133], have proposed a very clever application of the X-ray microprobe to reconstruct depth profile maps. The optical reciprocity theorem states that the same information can be obtained when looking at the fluorescence signal at grazing exit as when the exciting beam arrives at the sample under normal incidence.

The experimental set up is shown figure 10.17. In this case the incident beam is a microbeam delivered from the X-ray microprobe and the fluorescence spectrum is recorded under grazing exit angle. By keeping the exciting beam fixed and recording the fluorescence spectrum while tilting the sample, a depth profile at this point of incidence is obtained. A second mode of investigation is to keep the exit angle fixed, which means we are exploring a given depth in the sample, while the incident microbeam is scanned over the sample. A distribution map at a given depth can then be obtained.

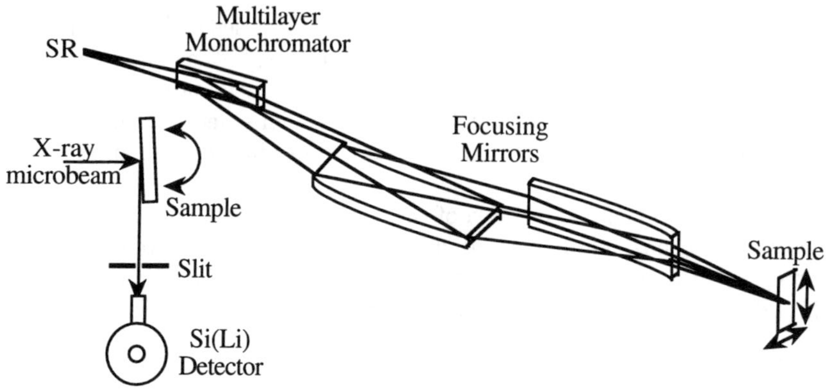

FIGURE 10.17 Layout of the experimental set up for micro X-ray depth profiling with the Photon Factory photon microprobe.

10.4 CONCLUSION

With the advent of SR and of new X-ray optics, photon X-ray fluorescence microprobes have been actively developed. Although this analytical tool will always remain of limited access, its performance for trace element mapping should be far ahead of what can be reached under electron excitation or even with micro PIXE installations.

Very interesting results where obtained at LURE DCI (Orsay, France), especially to demonstrate new possibilities and to test new optics, but due to the low brilliance of the DCI source it is yet impossible to use our microprobe for mapping. In fact such photon microprobes are perfectly adapted to the undulator sources of the latest generation synchrotron storage rings like ESRF or APS.

The high lateral resolution now reached with photon microprobes will certainly soon be better than 1 micron under usual working conditions. In fact, due to the noticeable penetration length of X rays, it would be interesting to measure how the spatial resolution will increase when the spot size becomes much smaller than the penetration length. In addition, the small amount of energy deposited in the sample even at maximum flux, the high sensitivity for heavy elements, and the fact that the whole set-up can operate under atmospheric conditions makes it especially interesting for life science studies.

ACKNOWLEDGMENTS

We are indebted to F. Rousseaux (L2M, CNRS France) for producing the gold mask used as a test object, to M. Maurette and C. Engrand

(CSNSM,Orsay, France) for the meteorite samples and assistance during the experiment, to M. Mosbah (LPS / CNRS / CEA, Saclay, France) for her help and analysis of glass standards in micro PIXE and micro SRXRF, to M. Brunel (Laboratoire de Cristallographie, Grenoble, France) for useful discussion of results and help, to B. Vidal (LOE, Marseille, France) for providing perfect quality multilayers .

The staff of LURE (in particular F. Legrand, M. Idir, G. Soullié A. Mirone and P. Populus) and ESRF (and especially A. and I. Snigirev, A. Freund and A. Grübel) are fairly thanked for excellent operating conditions and their help before, during and after the experiment.

We are also very grateful to Drs. Iida (Photon Factory), Jones (Brookhaven) and Van Langevelde (Amsterdam) for sending us their most relevant results in this field.

The present work is supported by Ultimatech and PICS contracts from the CNRS (France), and we are grateful to Prof. Aristov for the support from IMT-RAS (Russia).

REFERENCES

1. Castaing R., and Slodzian G., *J. Microscopy,* **1**, 395 (1962)
2. Michiels F., Van Hoolst W., van Espen P., and Adams F., *J. Am. Soc. Mass. Spectrom.* **1**, 37 (1990)
3. Denoyer E., Van Grieken R., Adams F., and Natush D., *Anal. Chem.* **26A**, 54 (1982)
4. Adams F., and Mauney T., *Adv. Mass Spectrom.* **10**, 507 (1986)
5. Goldstein S.I., Newbury D.E., Echlin P., Joy D.C., Fiori C., and Lifshin E., *Scanning Electron Microscopy and X-ray Microanalysis*, New York: Plenum Press, 1984, ch. 1, pp. 5-7.
6. Vis.R.D., *Fres. Z. Anal. Chem.* **337**, 622 (1990)
7. Roshchupkin D.V., Brunel M., de Bergevin F., and Erko A.I., *N.I.M.* **B72**, 471 (1992)
8. Tucoulou R., Schelokov A., Roshchupkin D.V., Brunel M., Ortéga L., and Chevallier P., *Optics Comm.* **118**, 175 (1995)
9. Winick H., and Doniach S., *Synchrotron Radiation Research*, New York: Plenum Press, 1980.
10. Koch E.E., *Handbook on Synchrotron Radiation, Vol. 1*, Amsterdam: North-Holland, 1983.
11. Marr G.V., *Handbook on Synchrotron Radiation, Vol. 2*, Amsterdam: North-Holland, 1987.
12. Brown G.S., and Moncton D.E., *Handbook on Synchrotron Radiation, Vol. 3*, Amsterdam: North-Holland, 1991.
13. Gordon B. M. , and Jones K. W., *N.I.M.* **B10/11**, 293 (1985).
14. Aristov V.V., Basov Yu., Redkin S.V., Snigirev A.A., and Yunkin V.A., *N.I.M.* **A261**, 72 (1987).
15. Aristov V.V., Erko A.I., and Martinov V.V., *Rev. Phys. Appl.* **23**, 1623 (1988).

16. Erko A.I., *J. de Phys. IV* **C9**, 245 (1994)
17. Deslattes R.D., *N.I.M.* **172**, 201 (1980)
18. Deslattes R.D., *N.I.M.* **177**, 147 (1980)
19. Sparks C.J., Ice G.E., Wong J., and Batterman B.W., *N.I.M.* **194**, 73 (1982)
20. Ice G.E. and Sparks C.J., *N.I.M.* **222**, 121 (1984)
21. Dhez P.,*X-ray Microscopy III*, Michette A., Morrison G., and Buckley C. Eds, Heidelberg: Springer-Verlag, 1992, pp 107-113
22. Michette A.G., *Optical systems for soft X rays* New York: Plenum Press, 1986.
23. Henke B.L., Gulikson F.H., and Davis J.C., *Atomic Data and Nuclear Data Tables* **54**, 181-342 (1993)
24. Marot G., Freund A., Rossat M., Joksch H., Ziegler E., Berman L.E., Chapman D., Hastings J.B., Iarocci M., van der Laske G., and Lazarz N., *Rev. Sci.Instrum.* bf 63, 477 (1992)
25. Underwood J., *Space Science Instr.* bf 3, 259 (1977)
26. Susini J., Labergerie D;, and Zhang L., *Rev. Sci. Instrum.* **66**, 2229 (1995)
27. Dezoret D.,*Thesis: University Paris-sud* 1995
28. Kirkpatrick P., and Baez J., *J. Opt. Soc. Am.* **38**, 766 (1948)
29. Iida A., and Hirano K., *N.I.M.* in press.
30. Underwood J.H., Thompson A.C., Kortright J.B., and Chapman K.C., *N.I.M.*in press
31. Wolter H., *Ann. Phys. 6th Ser.* **10**, 286 (1952)
32. Goshi Y., Aoki S., Iida A., Yamaji H., and Sahuri K., *Adv. X-ray Anal.* **31**, 495 (1988)
33. Grigirieva I.G., Antonov A.A., and Baryshev V.B., *Synchrotron Radiatiion News* **3**, 15 (1990)
34. Sparks C.J.Jr., Raman S., Yakel H.L., Gentry R.V., and Krause M.O., *Phys.Rev.Lett.* **38**, 205 (1977)
35. Tolentino H., Baudelet F., Dartyge E., Fontaine A., Lena A., and Tourillon G., *N.I.M.* **A289**, 307 (1990)
36. Fourme R., Dhez P., Benoit J.P., Dubuisson J.M., and Frouin J.,*Rev. Scient. Instrum.* i**63**, 982 (1992)
37. Van Langevelde F., Bowen D.K., Tros G.H.J., Vis R.D., Huizing A., and De Boer D.K.G., *N.I.M.* **A292**, 719 (1990)
38. Zhengquan T., MacDowell A.A., La Fontaine B., Bjorkholm J.E., Tennant D., Taylor D., Himel M., Freeman R.R., Waskiewicz W.K., Spector S., Ray Chaudhuri A.K., Stulen R.H., Ng W., and Cerrina F., *Rev. Scient. Instrum.* **66**, 2241 (1995).
39. Niemann B., Schneider G., Guttmann D., Rudolph D., and Schmahl G., *X-ray Microscopy IV*Aristov V.V., and Erko A.I. Eds , Chernogolovka: Institute of Microelectronics Technology, 1994, pp 66-75
40. Di Fabrizio E., *Rev. Sci. Instrum* in press
41. Aristov V.V., Gaponov S.V., Erko A.I., and Salaschenko N.N., *Optics News* **12**, 128 (1986)
42. Basov Yu., Roshchupkin D.V., Schelokov I.A., and Yaskin A.E., *X-ray Mi-*

*croscopy IV*Aristov V.V., and Erko A.I. Eds , Chernogolovka: Institute of Microelectronics Technology, 1994, pp 593-600
43. Erko A.I., Khzmalian E., Panchenko L., Redkin S., Zinenko V., Chevallier P., Dhez P., Khan-Malek C., Freund A.,.and Vidal B., *X-ray Microscopy III*, Michette A., Morrison G., and Buckley C. Eds, Heidelberg: Springer-Verlag, 1992, pp 217-219
44. Snigirev A., *Rev. Sci. Instrum.* **66**, 2053 (1995)
45. Basov Yu., Pravdivtseva T.L., Snigirev A., Belakhovsky M., Dhez P., and Freund A., *N.I.M.* **A308**, 363 (1991)
46. Snigirev A., Snigireva I., Engström P., Lequien S., Suvorov A., Hartman Ya., Chevallier P., Idir M., Legrand F., Soullié G., and Engrand C., *Rev. Sci. Instrum.* **66**, 1461 (1995)
47. Mirone A., Idir M., Dhez P., Soullié G., and Erko A.I., *Optics Com.* **111**, 191 (1994)
48. Idir M., Mirone A., Soullié G., Dhez P., Guerin Ph., and Ladan F.,*Optics Comm.* **119**, 633 (1995)
49. Soullie G., *Thesis: University Paris-Sud* 1996
50. Jentzsch F., and Kellar J.N., *Z. Techn. Phys.* **12**, 185 (1931)
51. Rindby A., *N.I.M.* **A249**, 536 (1986)
52. Kumakov F.F., *Phys. Rep.* **191**, 289 (1990)
53. Stern E.A., Kalman Z., Lewis A., and Lieberman K., *Appl. Opt.* **27**, 5135 (1988)
54. Bilderback D.H., Hoffman S.A., and Thiel D.J., *Science* **263**, 201 (1994)
55. Engström P., Larsson S., Rindby A., Buttkewitz A., Garbe S., Gaul G., Knöchel A., and Lechtenberg F., *N.I.M.* **A302**, 547 (1991)
56. Janssens K.H., Van Langevelde F., Adams F.C., Vis R.D., Sutton S.R., Rivers M.L., and Jones K.W., *Adv. in X-ray Anal.* **35**, 1265 (1992).
57. Janssens K., Adams F., Rivers M.L., and Jones K.W., *Scanning Microscopy Supplement* in press.
58. Jaklevic.M.J., *Proceedings of the Energy Research and Development Administration: X and g Ray Symposium. Conf 760539* Ann Arbor, Michigan, May 1976.
59. Horowitz P., and Howell.J.A., *Science* **178**, 608 (1972)
60. Sparks C.J. Jr., *Synchrotron Radiation Research* Winick H, and Doniach S., Eds., New York: Plenum Press, 1980, ch 14, pp 459-512.
61. Iida A. and Gohshi Y., *Handbook on Synchrotron Radiation, Vol 4* Ebashi S., Koch M., and Rubinstein E., Eds, Elsevier Science Publishers, 1991, ch 9, pp 307-348.
62. Babyshev V., Kulipanov G., and Skrinsky A., *Handbook on Synchrotron Radiation, Vol 3* Brown G.S., and Moncton D.E., Eds, Amsterdam: North-Holland, 1991, ch 16, pp 639-688.
63. Jones K.W., *Handbook on X-Ray Spectrometry: Methods and Techniques* Van Grieken R., and Markowicz A., Eds., New York: Marcel Dekker Inc, 1992, ch 8, pp 411-452
64. Chevallier P., Dhez P., Erko A., Firsov A., Legrand F., and Populus P., N.I.M.

in press
65. Chevallier P., Dhez P., Legrand F., Erko A., Agafonov Tu., Panchenko L.A., and Yakshin A., JTMT Vol 14(3) in press
66. Aristov V.V., Erko A.I., Firsov A.A., Gaponov S.V., Salaschenko N.N., Zabelin A.V., Rybakov B.V., and Stepanov E.P., *Proceedings of the 2d European Conference in X-ray Synchrotron Radiation Research* Balena A., Bernieri E.,and Mobilio S., Eds, Bologna : SIF,1995, Vol 25, 275
67. Dhez P., Erko A.I., Kahzmalian Z., Vidal B., and Zinenko V., *Appl. Optics* **31**, 1 (1992).
68. Aristov V.V., and Erko A.I., *X-ray Optics*1991 Moscow:Nauka.
69. Chevallier P., Dhez P., Legrand F., Erko A.I., and Vidal B., *Inst. Phys. Conf. Ser.* **130**, 617 (1992)
70. Erko A.I., Agafonov Yu., Panchenko L.A., Yuakshin A., Chevallier P., Dhez P., and Legrand F., *Optics Com.* **106**, 146 (1994)
71. Chevallier P., Dhez P., Legrand F., Idir M., Soullié G., Mirone A., Erko A.I., Snigirev A., Snigireva I., Suvorov A., Freund A., Engström P., Als Nielsen J., and Grübel. A., *N.I.M.* **A354**, 584 (1995)
72. Mosbah M., Clocchiatti R., Michaud V., Piccot D., Chevallier P., Legrand F., Als Nilsen G., and Grübel.G., *N.I.M.* **B104**, 481 (1995)
73. Czamanske G.K., Sisson T.W., Campbell J.L., and Teesdale W.J., *American Mineralogist* **78**, 893 (1993)
74. Chen J.R., Chao E.C.T., Back J.M., Minkin J.A., Rivers M.L., Sutton S., Cygan G.L., Grossman J.N., and Reed M.J., *N.I.M.* **B75**, 576 (1993)
75. Legrand F., *Thesis: University of Paris-sud*, 1995
76. Philippot P., Chevallier P., Chopin C. and Dubessy J. *Contrib. Mineral. Petrol.* **121**, 29 (1995)
77. Iida A., and Noma T., *N.I.M.* **B82**, 129 (1993).
78. Suzuki Y., Uchida F., and Hirai Y., *Jpn. J. Appl. Phys.* **28**, L1660 (1989)
79. Suzuki Y., Uchida F., *Rev. Sci. Instrum.* **63**, 578 (1990)
80. Van Langevelde, F. Tros G.H.J.,.Bowen D.K., and Vis R.D., *N.I.M.* **B49**, 544 (1990).
81. Vis R.D., and van Langevelde F.,.*N.I.M.* **B54**, 417 (1991).
82. Van Langevelde F., Tros G.H.J., Vis R.D.,.*JTMT* **7**, 242 (1990).
83. Tros G.H.J., Van Langevelde F., and Vis R.D., *N.I.M.* **B50**, 343 (1990).
84. Van Langevelde F., and Vis R.D., *Anal.Chem.* **63**, 2253 (1991).
85. Kubo H., and Smythe W.R., *Anal.Chem.* **51**, 1194 (1979).
86. Underwood J.H., Thompson A.C., Wu Y., and Giauque R.D., *N.I.M.* **A226**, 296 (1988)
87. Chen J.R., Chao E.C.T., Minkin J.A., Back J.M., Jones K.W., Rivers M.L., and Sutton S.R., *N.I.M.* **B49**, 533 (1990).
88. Sutton S.R., Rivers M.L., Bajt S., and Jones K.W., *N.I.M.* **B75**, 553 (1993).
89. Rivers M.L., Thorn K.S., Sutton S., Jones K.W., and Bajt S., *Lunar Planet. Sci.* **24**, 1203 (1993)
90. Gordon B.M., and Jones K.W., *Biological Trace Element Research* Subramanian K.S., Iyengar G., and Okamoto K., Eds., Washington: American

Chemical Society, 1991, Ch 23, pp 290-305.
91. Jones K.W., Spanne P., Webb S.W., Conner W.C., Beyerlein R.P., Reagan W.J., and Dautzenberg F.M., *N.I.M.* **56/57**, 427 (1991)
92. Bertsch P., Hunter D., Sutton S.R., Bajt S., and Rivers M.L., *Environ. Sci. Tech.* **28 (5)**, 980 (1994)
93. Treiman A.H., and Sutton S.R., *Geochimica et Cosmochimica Acta* **56**, 4059 (1992)
94. Vanko D., Sutton S.R., Rivers M.L., and Bodnar R., *Chem. Geology* **109**, 125 (1993)
95. Sutton S.R., Rivers M.L., Bajt S., Jones K.W., and Smith J.V., *N.I.M.* **A347**, 412 (1994)
96. Flynn G.J., Sutton S.R., and Bajt S., *Lunar Planet.Sci.* **24**, 495 (1993).
97. Flynn G.J., and Sutton S.R., *Lunar Planet. Sci.* **21**, 541 (1991)
98. Jones K.W., Gordon B.M., Schilovsky G., Spanne P., Xue Dejun , Bockman R.S. and Saubermann A.J., *Microbeam Analysis* Ingram P., and Michael J.R. Eds, San Francisco: San Francisco Press Inc., 1990, pp 401-404
99. Getzoff E.D., Jones K.W., McRee D., Moffat K., Rivers M.L., Schildkamp W., Singer P.T., Spanne P., Sweet R.M., Teng T.Y., and Westbrook E.M., *N.I.M.* **B79**, 249 (1993)
100. Jones K.W., Bockman R.S., and Bronner F., *Neuro Toxicology* **13**, 835 (1992)
101. Vittur F., Tuniz C., Paoletti S., Rizzo R., and Jones K.W., *Biochem. Biophys Res. Commun.* **188**, 1010 (1992)
102. Rizzo R., Grandolfo M., Godeas C., Jones K.W., and Vittur F., *J. of Experimental Zoology* **273**, 82 (1995)
103. Slatkin D.N., Spanne P., Dilmaniam F.A., and Sandborg M., *Medical Physics* **19 (6)**, 1395 (1992)
104. Tuniz C., Zanini F., and Jones K.W., *.N.I.M.* **B56/57**, 877 (1991).
105. Török Sz., Faigel G., Jones K.W., Rivers M.L., and Sutton S.R.,*X-ray Spectrometry* **23**, 3 (1994)
106. Török Sz., Faigel G., Osan J., Jones K.W., Rivers M.L., Sutton S.R., and Bajt S., *Adv. in X-ray Anal.* **37**, 711 (1994)
107. Spanne P., and Jones K.W., *Synchrotron Radiation News* **7 (6)**, 6 (1994)
108. Jones K.W., Spanne P., Lindquist W.B., Conner W.C., and Ferrero M., *N.I.M.* **B68**, 105 (1992)
109. Ferrero M., Sommer R., Spanne P., Jones K.W., and Conner W.C., *J. Polymer Science: Part A: Polymer Chemistry* **31**, 2507 (1993)
110. Spanne P., Jones K.W., Herman H., and Riggs W.L., *J. Thermal Spray Tech.* **2 (2)**, 121 (1993)
111. Spanne P., Thovert J.F., Jacquin C.J., Lindquist W.B., Jones K.W., and Adler P.M., *Phys. Rev. Lett.* **73**, 2001 (1994)
112. Hirsch P.B., and Kellar J.N., *The Proc. of the Phys. Soc., Section B* **64 (5)** (1951)
113. Engström P., Larsson S., Rindby A., and Stocklassa B., *N.I.M.* **B36**, 222 (1989)
114. Carpenter D.A., Taylor M.A., and Holcombe C.E., *Adv. in X-ray Anal.* **32**,

115 (1989)

115. Kobayashi Y., Fukumoto N., and Kurahashi M., *Measurement Science and Technology* **2**, 183 (1991)
116. Pella P.A., and Fang L., *Adv. in X-ray Anal.* **35**, 1063 (1992)
117. Shimomura S., and Nakazawa H., *Adv. in X-ray Anal.* **35**, 1289 (1992)
118. Shakir N., Engström P., Larsson S., and Rindby A., *N.I.M.* **B52**, 194 (1990)
119. Larsson S., *Thesis: University of Goteborg* 1991
120. Carpenter D.A., Taylor M.A., and Lawson R.L., *JTMT* **12 (2)**, 141 (1995)
121. Engström P.,Fielder S. and Riekel C.*Rev.Sci.Instrum* **66**, 1348 (1995).
122. Riekel C., and Engström P., *N.I.M.* **B97**, 224 (1995).
123. Singer P.T., Carty R.P., Berman L.E., Schlitchting I., Stock A., Smalas A., Cai Z., Mangel WF., Jones K.W., and Sweet R.M., *Phil. Trans. R. Soc. London* **A340**, 285 (1992)
124. Hayakawa S., Gohshi Y., Iida A. Aoki S., and Sato K.,*Rev. Sci. Instrum.* **62**, 2545 (1991).
125. Iida A., Noma T., Hayakawa S., Takahashi M.,and Gohshi Y, *Jpn. J. Appl. Phys.* **32 Suppl 32-2**, 160 (1993)
126. Itie J.P., Polian A., and San Miguel A., *Jpn. J. Appl. Phys.* **32 Suppl 32-2**, 711 (1993)
127. Sutton S.R., Jones K.W., Gordon B., Rivers M.L., and Smith J.V., *Lunar and Planet. Sci.* **22**, 1411 (1991)
128. Sutton S.R., Jones K.W., Gordon B., Rivers M.L., Bajt S., and Smith J.V., *Geochim. et Cosochim. Acta* **57**, 461 (1993)
129. Bajt S., Sutton S.R., and Delaney J., *Geochimica et Cosmochimica Acta* in press.
130. Schulze P., and Sutton S.R., *J. Soil Sci. Soc. of America* in press
131. Parratt L. G., *Phys. Rev.* **95**, 359 (1954).
132. Brunel M., and de Bergevin F., *Acta Cryst.* **A42**, 299 (1986)
133. Noma T., and Iida A., *Rev. Sci. Instrum.* **65**, 837 (1994)

CHAPTER 11

Study of the Interaction of Highly Charged Ions with Surfaces by X-ray Spectroscopy of Hollow Atoms

G. Giardino and J.P. Briand*

We present in this chapter some experiments and specific techniques concerning the interaction of slow, highly charged ions with surfaces. This chapter deals mainly with the study of the X rays emitted by "hollow atoms" formed above, at, or below various surfaces (metals, semiconductors and insulators). It describes the "atomic clock" property of hollow atoms and its use for the study of the interaction at the subangstrom and subfemtosecond level. Instrumental techniques (X-ray spectroscopy, ion deceleration, etc.) and some illustrations of their uses will be presented.

11.1 INTRODUCTION

The interaction of slow, highly charged ions with surfaces has been studied only recently. The very first experiments in this field (1988-1990) dealt mainly with the study of the ejected electrons [1–3] and the dynamics of the incoming ions above the surface [4–6]. The sputtering of surface atoms and the study of the characteristic in-flight emission of the ions began in 1990 [7–10]. This chapter aims to study the physics and methods based on the analysis of the characteristic X rays and Auger electrons emitted in flight by the ions during their approach or penetration of a surface.

*) Laboratoire de Physique Atomique et Nucléaire-Institut du Radium, Université Pierre et Marie Curie, 4 Place Jussieu, 75252 Paris Cedex 05, France (ERS CNRS No112)

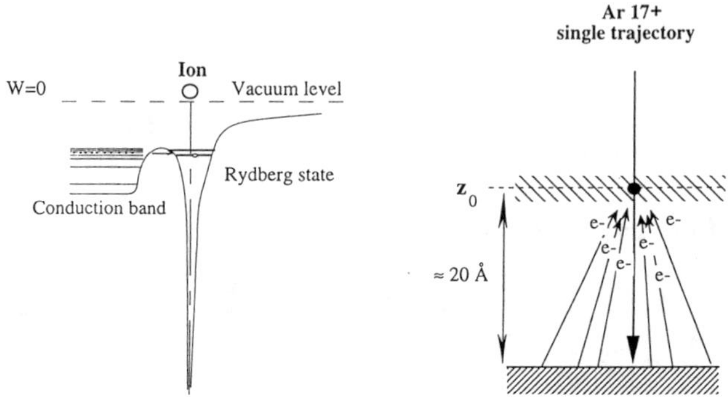

FIGURE 11.1 Electron transfer from a metal target to an ion.

11.2 HOLLOW ATOMS

A highly charged ion approaching a neutral atom may capture one or more electrons from the atom. For energetic reasons—the binding energy of the electron after its capture must be roughly equal to that of electron of the target atom (*resonant capture*)—this capture populates high Rydberg states of the ion. It is thus a very selective process of capture. The capture by the same ion of a second electron is usually more difficult, owing to the increase of its ionization potential, and occurs at closer distances and smaller geometric cross sections. The single-capture process usually dominates. Since a metal is an infinite reservoir of electrons having the same binding energy (Fig. 11.1), one can then expect that the ion is instantaneously neutralized by electrons located in its Rydberg states at relatively large distances from the surface.

This process is difficult to identify because the inverse process of capture—resonant ionization—may limit ion neutralization. Moreover, the very high electric fields appearing during the interaction may hinder the capture processes (*field ionization*: the outermost shells raise up into the continuum), and the existence of such highly inverted levels may also be questionable for fundamental reasons of quantum mechanics [11]. Fortunately the number of electrons extracted is very large compared to the nuclear charge of the ion, and the "flying" electrons dynamically screen the ion (*space charge effect*), allowing capture in lower n levels where the existence of hollow atoms is more likely.

After being (supposedly) captured in high Rydberg states these electrons experience a long series of Auger transitions [12]. The Auger transition being much more probable between close levels, a large number of steps are required to fill the innermost shells. The hollow atom is then reionized, but also re-fed all along a very complex series of transitions. These transitions between high-n levels having very low energies lie in the low-energy electron continuum, due to the ejection of free electrons, and these transitions are not resolvable. One

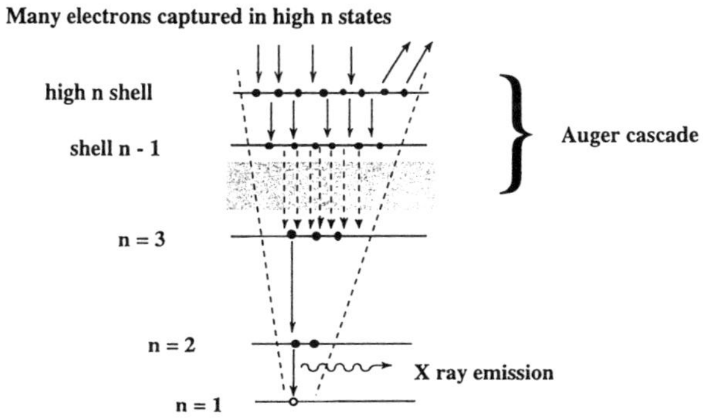

FIGURE 11.2 Schematic view of the decay of a hollow atom.

only observes the last steps of the cascade, i.e., the Auger or X-ray transitions between the K, L, M, N shells, the energy of which is much higher (Fig. 11.2).

It is the purpose of this chapter to review the experimental techniques that are used to study these transitions, as well as the most salient and recent results obtained in the study of ion-surface interactions.

These highly excited species take quite a long time to decay to the ground state, since there are many steps to reach the innermost shells by Auger transitions, and at the usual kinetic energies of the ions (10 keV/q at the exit of the most conventional ECR or EBIS ion sources) they reach the surface before having the opportunity to decay above it. At the surface the ions are peeled off [11] because the outermost electrons are weakly bound and cannot survive the collision. So most of the time it is extremely difficult to observe the rare events taking place outside the surface. In order to increase the time needed by the ion to reach the surface one must either decelerate the ion in electrostatic devices or study the interaction at very grazing incidence. We will review decelerating techniques in the last section of this chapter.

Below or at the surface the ion, after being reionized, must be neutralized, and another kind of hollow atom is formed [9] that populates lower excited states of the ions (e.g., M and N shells for Ar ions) at much shorter distances. These excited states then decay below the surface, within the first few monolayers, and these states were seen early in the history of these studies. By studying the X-ray spectra observed at high resolution on a metal surface we observed in 1990 [9] clear signatures of the formation of exotic species we named *hollow atoms*, where the M and N shells of the ion are closed and the L and K shells remain empty, or quasi-empty. There are thus different kinds of hollow atoms: those formed at, or below the surface we shall name *surface hollow atoms*, and those more complicated species, formed and decaying outside the surface, that shall be called *Rydberg hollow atoms*. The latter, whose existence now seems to be clearly established by X-ray spectroscopy, will be

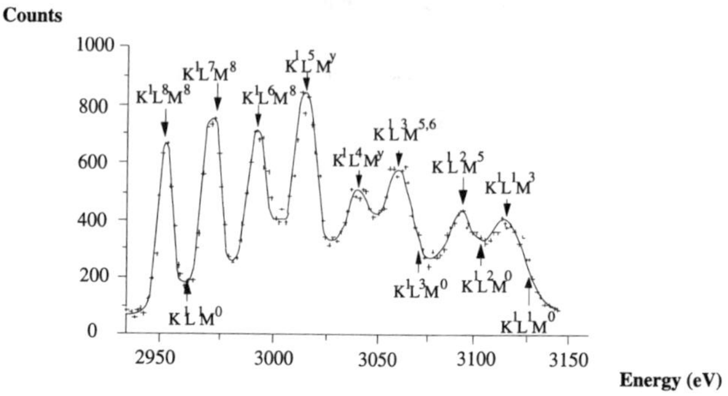

FIGURE 11.3 Kα L^x array of X-ray satellites for Ar observed at high resolution [9].

discussed extensively in the next sections of the chapter.

11.2.1 The atomic clock property of the hollow atoms

Hollow atoms of both kinds (surface hollow atoms and Rydberg hollow atoms) decay to their ground state through a long series or Auger transitions, finally filling the innermost shells. One will only observe the last steps of this filling, in particular the filling of the eight holes present in the L shell in both kinds of hollow atoms. As recently demonstrated [12,13], for Ar ions the filling of the L shell proceeds through eight LMM (LMN...) Auger transitions. The transition probabilities for each of these transitions, now well known theoretically [13,14] as well as experimentally [13], are of the order of a few 10^{-16} s^{-1}.

As soon as some electrons are present in the L shell the remnant K hole may be filled either by emission of a Kα line or by a KLL Auger transition. This transition takes place in the presence of a certain number x of L electrons. In Fig. 11.3 we present the Kα spectrum observed when any number statistically present. This spectrum, obtained at high resolution as discussed in the next section, exhibits eight well separated satellite lines [9] corresponding to the eight occupation states of the L shell. There is thus a competition, at the last steps of the filling of the inner holes, between the filling of the L shell mainly through LMM Auger transition and the filling of the K hole(s) shell, mainly by KLL Auger emission or Kα transitions.

If—as it is the case for Rydberg hollow atoms—the filling of the L shell is slow compared to that of the K shell (many steps are necessary to reach the L shell), as soon as one (or two) electron(s) has arrived in the L shell, the Kα line may be emitted (Fig. 11.4).

The KL^x array of satellites is then strongly peaked at KL^1 and KL^2, as shown in Fig. 11.5.

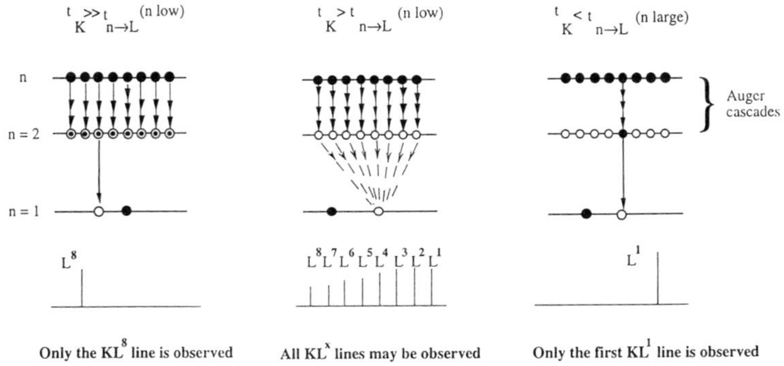

FIGURE 11.4 The clock property of the hollow atoms.

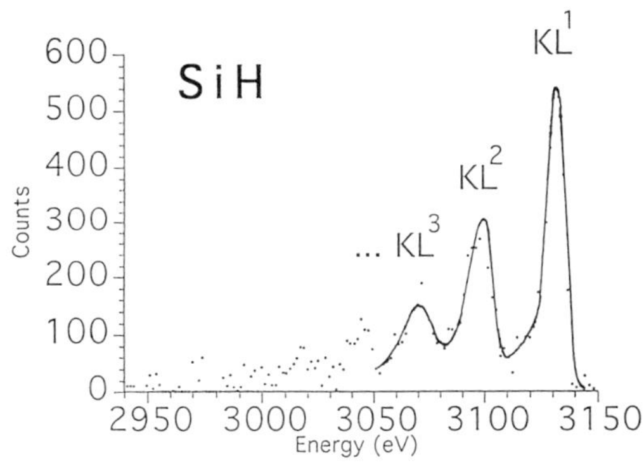

FIGURE 11.5 KL^x array of satellites emitted by Rydberg hollow atoms (Ar^{17+} ions at 1 eV/q at normal incidence on SiH). Reprinted with permission from *Phys. Rev. Lett.* **77** (1996) 1452.

If, by contrast, the filling of the L shell is very fast (which happens with surface hollow atoms, the electrons of which come directly from the M (or N) shell), the Kα line may be emitted in presence of any number x of L spectator electrons, as shown in Fig. 11.3.

The relative intensity of the KL^x line provides information on the timing of the interaction: the filling of the L shell plays the role of a sandglass, and the K hole (emission of a Kα line) that of a flash illuminating the actual configuration.

11.2.2 A time cascade approach

A quantitative description of this L-shell filling is given using the general equations of the radioactive-series growth and decay, the Bateman equations.

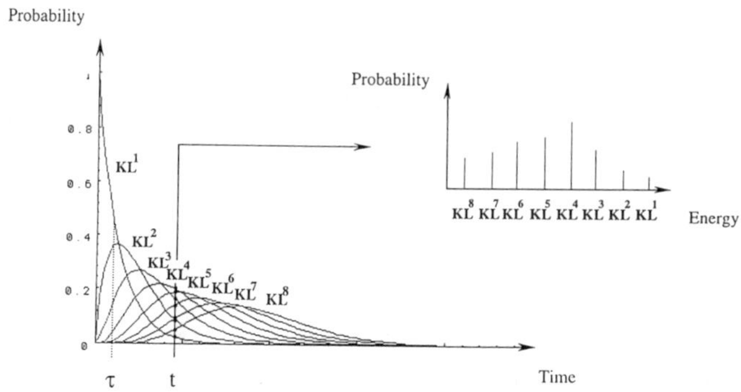

FIGURE 11.6 Probability of finding the hollow atom in a KL^x state ($l_1 = l_2 = ...$) and the corresponding KL^x spectrum observed at t.

Knowing the half-lifetimes of different radioactive nuclei, it is easy to deduce the quantity of each species present at a given time. A similar situation exists in the case of the Auger cascade filling the L shell of the hollow atom. It is possible to calculate the relative probability of finding a hollow atom in the KL^x configuration (x electrons in the L shell during the X-ray L→K transition—see the next section) at a given time t measured in units of the decay lifetime, as illustrated in Fig. 11.6.

The shape of the KL^x configuration and the relative intensities of the line obtained in these cases can be indentified from the $K\alpha$ L^x array of X-ray satellites for Ar observed at high resolution (Fig. 11.4).

11.2.3 Snapshots of the L shell filling with two K holes

The best illustration of this stepwise filling of the L shell is obtained for Ar^{18+} when two K holes are present (a fully stripped ion). In such a situation two $K\alpha$ transitions are emitted, one after the other (the so-called hypersatellite and satellite cascade) [15]. During the time between the emission of these two lines, the L shell is continuously filled and one expects to observe for the second transition a KL^x array of satellite lines having a larger number of L electrons than the first one. This has been observed experimentally, and is schematically presented in Fig. 11.7.

In this case, two L→K transitions of very different energy take place while Auger cascade occurs (Fig. 11.7a). The high resolution spectrum (Fig. 4) can be recorded in each case giving the main number x_1 of electrons in the L shell (Fig. 11.7b). The number of L electron changes because of that transition, but also and mainly because of the continuous filling of the L shell by Auger cascade. So the second L→K transition occurs with a number x_2 of electrons in the L shell: $L^{x_2} \rightarrow K^{-1}$ transition, centered on the sixth high resolution line

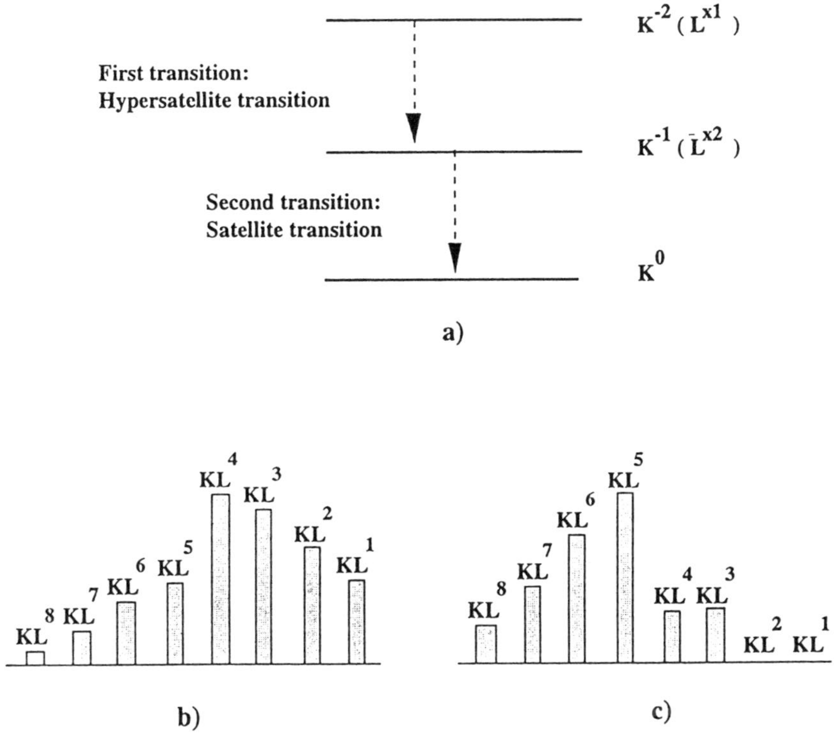

FIGURE 11.7 Hypersatellite and satellite transitions. The number of L electrons changes between the two transitions.

($x_2 \approx 6$) (Fig. 11.7c) without any KL^1 or KL^2 line (no possibility to have one or two electrons in the L shell) [16]. These measurements, which have been done for different ions [13], have provided numbers in good agreement with calibrated partial lifetimes for the LMM Auger transition [13,14].

The exact energy of all these $K\alpha L^x$ lines, as well as the more recently observed $K\beta L^x$ and $K\gamma L^x$ transitions, gives us the exact configuration of the ion at the time of the filling of the K hole.

These eight KL^x satellites play the role of time markers, the lines of higher energy (KL^1, KL^2) being emitted before those of lower energy ($KL^{8,7,\cdots}$). These time markers may be converted into a mean decay path $x = vt$. At velocities of the order of a few 10^4 ms^{-1} (a few eV/q), an L hole filling takes place along a few tenths of an angstrom, i.e., well below the atomic size. At much higher velocity (10^6 ms^{-1}), the filling of the L shell vacancies extends over a few atomic rows. This atomic clock has the exact period needed to study the above, at, or below surface interaction of highly charged ions.

11.3 X-RAY SPECTROSCOPY OF HOLLOW ATOMS

The formation of hollow atoms has been studied on different targets (metals, insulators and semiconductors) and in different experimental situations (normal incidence and grazing incidence, different kinetic energies). The shape of the spectrum changes under the various conditions. The X rays emitted in flight by the ions may be observed either at low resolution and high transmission (Si(Li) detectors), or at high resolution and low transmission (crystal spectrometers). Both techniques have been used. The X-ray spectra emitted by Ar^{17+} ions impinging on metal surfaces were first observed by Donets with a Si(Li) detector [8]. The ions were produced at low intensity ($\approx 10^6$ ions s^{-1}) by the very first EBIS source [11]. These X rays have been further analyzed by Briand et al. [9] with a crystal spectrometer. This high resolution spectroscopy allowed the first observation of the hollow atoms. These experiments have been carried out with the more intense beams produced by the Minimafios ECR source of Grenoble (10^9 ions s^{-1}). The X-ray spectra have been studied, for the high Z ions, from Fe up to U with the (low intensity) beam, typically 10^3–10^4 ions s^{-1}, produced by the EBIT or Super-EBIT source of the Lawrence Livermore National Laboratory, with a Si(Li) detector. More recently, the X-ray spectra of the light ions C, N, and O have also been observed at the Lawrence Berkeley National Laboratory with high resolution, low energy Si(Li) detectors.

As discussed below, the study of X rays seems to be the most powerful technique for studying the time interaction with surfaces (the atomic clock effect). The main advantages of the X-ray spectroscopy relative to Auger electron spectroscoppy can be summarized as follows.

The energy of the X rays are not affected by the penetration of the ions below a surface. One can then study the reactions below the surface, at the surface, or outside the surface. In Fig. 11.8 we present the KLL Auger spectra of N^{6+} observed at a relatively high ion energy, i.e., when most of the observable decay takes place below the surface, and at low energy and at grazing incidence (Fig. 11.10), i.e., when a substantial fraction of the decay occurs outside the surface [17].

In the first case one observes a very broad line which reflects the straggling of the Auger electrons below the very first monolayers of the target. A small bump superimposed on this peak was attributed to the very small fraction of the ions decaying outside the surface. This KLL Auger spectrum is compared to the X-ray spectra (Fig. 11.9) recently observed in the same experimental conditions with a Si(Li) detector of relatively good resolution (≈ 120 eV fwhm at 500 eV).

The energy of this ray, which is found to be exactly equal to that of the neutral atom, demonstrates that the L shell is closed at the time of the filling of the K hole. Then the complex structures observed in Auger spectroscopy

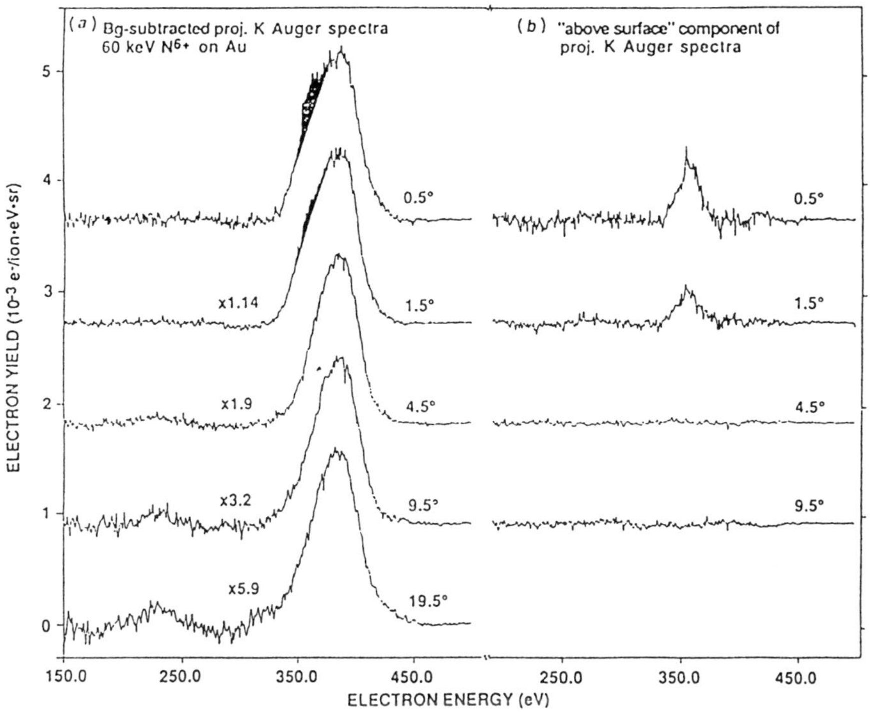

FIGURE 11.8 KLL Auger spectra of N^{6+} on gold surfaces at various incident angles. In black, at 0.5°: the outside decay signal. (b) above surface component [17]

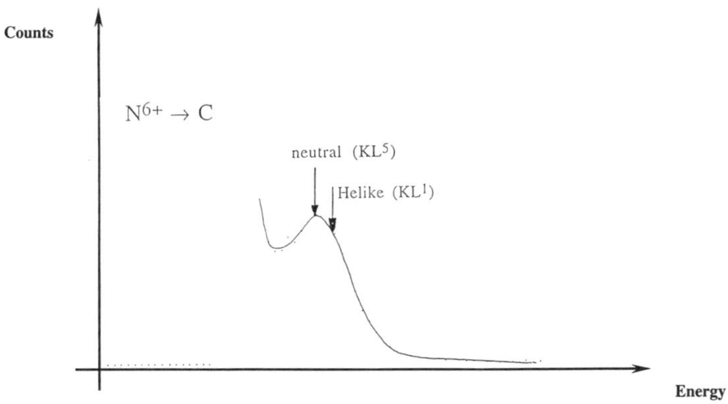

FIGURE 11.9 Kα-ray spectrum of N^{6+} on C surfaces (below-surface interaction). To be published, S. Bardin, Ph.D.; see also ICPEAK 1995)

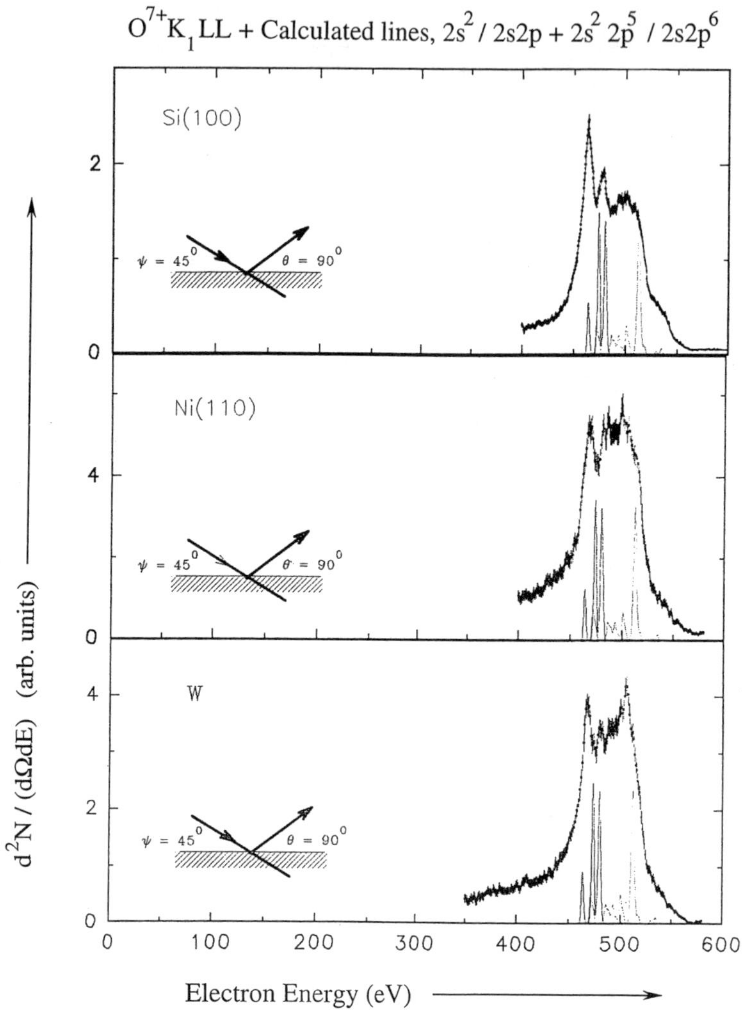

FIGURE 11.10 O^{7+} high resolution KLL Auger spectra (sum of above and below surface decays). Courtesy of R. Morgenstern.

definitely come from the straggling of these electrons in the very first atomic monolayers of the surface and not from a complex array of KL^x satellites, as observed in crystal spectrometer.

At much lower ion energies the KLL Auger spectra exhibit some resolved structures [10]. Some of these lines were attributed to the outside decay, others to the below surface decay. Unfortunately the fine structure of each of the KL^x components extends over a much larger energy range than the mean energy

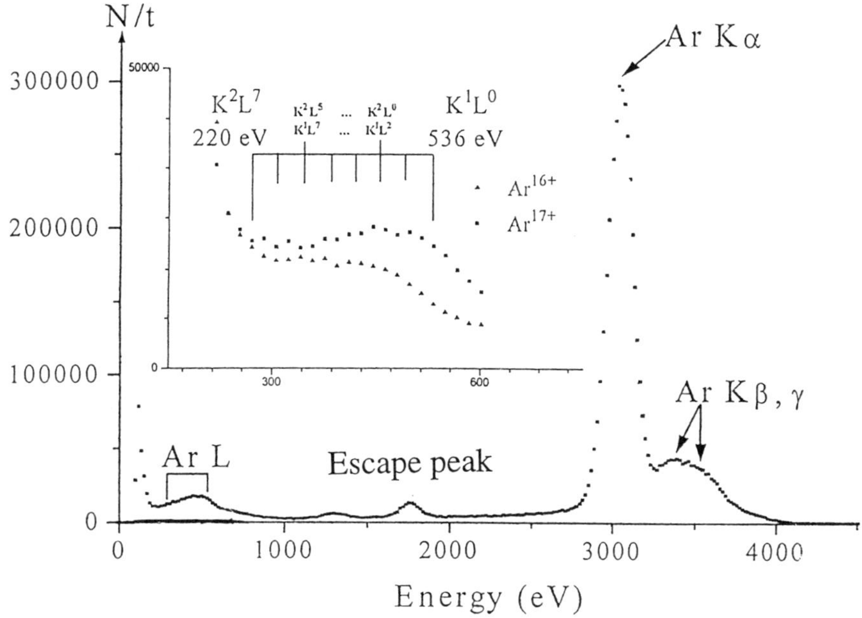

FIGURE 11.11 $K_{\alpha,\beta,\gamma}$ and $L_{\alpha,\beta}$ Ar spectrum observed with a Si(Li) detector at low resolution during collision of Ar^{17+} ions at 10 keV/q on Si targets. Reprinted with kind permission from *Phys. Rev. A* **54** (1996) 4136.

difference, and one can only clearly identify the lowest energy component of KL^2 (above surface decay: KL^1 is not observable in Auger spectroscopy), and the highest energy component of $KL^{6,7}$ (below surface decay).

The energy of the electrons is very sensitive to the electric fields present near the surface. As discussed below, the study of Auger emission by decelerated ions must be done in an equipotential energy region excluding any direct deceleration in biasing the surface, as discussed in the next paragraph.

The energy of the electron is very sensitive to the Doppler effect. At the energy of the ions extracted from the ECR sources the Doppler shifts are quite large. This may be considered either as an experimental difficulty, or used, as first done by de Zwart [6], to study the direction of the emitted ions.

In Fig. 11.11 we present the K X-ray spectrum of Ar^{17+} impinging on metal targets at 10 keV/q kinetic energy and normal incidence, observed with a Si(Li) detector. As described above, we can see by comparing this spectrum to that observed at high resolution (Fig. 11.3) that in measuring the mean

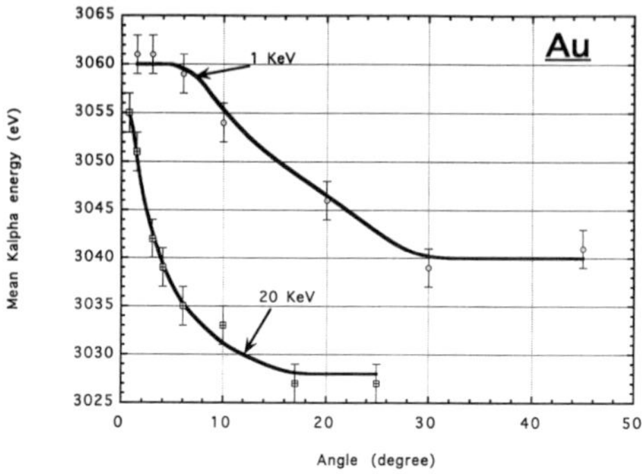

FIGURE 11.12 Variation of mean Kα energy of Ar^{17+} ions impinging on gold surfaces, as a function of the incident angle.

energy of the Kα line one can extract the mean number of L electrons present at the time of the K-hole decay. One can also deduce from this spectrum the relative intensity of the Kα and Kβ lines, a number that is impossible to measure in a crystal X-ray spectroscopy, where the two X rays must be recorded in different experiments owing to the relatively small energy range which can be recorded at one time. The Si(Li) detector has been used very often when the intensity of the beam was too low to allow the use of a crystal spectrometer, or when a great number of spectra have to be recorded in a small time (a typical X-ray spectrum, like those presented in Fig. 11.3, needs at minimum half an hour of exposure). Moreover, the absolute intensity of the Kα line may be related to the mean fluorescence yield (each KL^x line has a specific fluorescence yield ranging from 0.1–1).

The Si(Li) detector has been successfully used in these three cases:

The study of the outside decay of the ions at grazing incidence [12,18,19]. This kind of experiment is usually very difficult to carry out with a crystal spectrometer, owing to the small geometric interaction of the ion beam with the target at very grazing angle. The observable outside decay clearly increases (the mean energy of the Kα line, i.e., of the intensity of the KL^1 and KL^2 satellites, increases), as the normal velocity decreases, i.e., when the incident angle decreases.

We present fig. 11.12 as an illustration of the use of the Si(Li) detector. Here, the asymptotic limit of the mean value of L electrons present at the time of the K X-ray emission clearly shows the effect of the image acceleration of the ions near a surface (the internal dynamics of the decay of the hollow atoms does not change with the "normal" velocity of the ion when this energy becomes lower than that due to the image acceleration).

The study of the X-ray rate while decelerating the ions down to zero by biasing the target. In this case, as also discussed below, many data

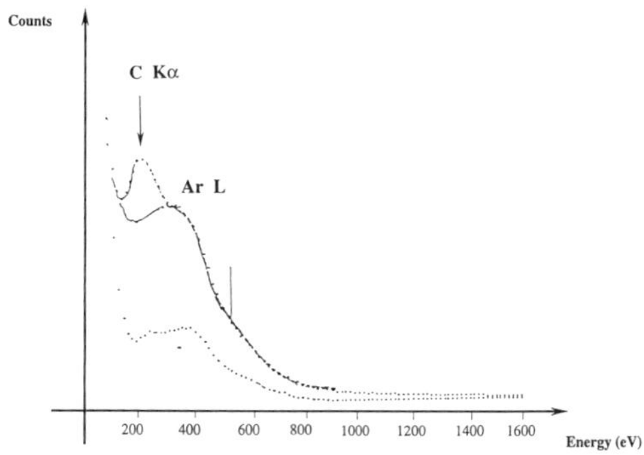

FIGURE 11.13 Ar L X ray and C K X ray in $Ar^{16+} \rightarrow C$ collision.

points have to be recorded over a large range of decelerating voltages. The exact energy of the ion depending on the fluctuation of the plasma potential in an ECR source, these data points have to be taken during a time of stability of the plasma which should not be too large. In this kind of experiment one can extract two pieces of information: the cut-off voltage, which depends on the target (and plasma potential) and but not on the exact energy of the line, and the mean fluorescence yield (the measurement of which needs a high stability of the beam intensity) (Fig. 11.26).

Comparison of the K and L lines As shown in Fig. 11.11, one can study with a Si(Li) detector simultaneously both the K and L lines, and also the low energy X rays emitted by the cathode. For instance, it has been possible to demonstrate, by comparing the relative intensities of the K and L lines of Ar (eight L transitions are emitted for one K transition), that for Ar the L shell is mainly filled by Auger transitions and not by side feeding [16], as it is, partially, for light atoms.

Another application of the Si(Li) detector is presented in Fig. 11.13 illustrating the collision of Ar^{16+} ions on C and Si targets. It clearly appears, by the observation of target C K lines, that some of the electrons captured by the Ar ion come from the K shell of the surface, a result which elucidates collision mechanism below a surface [17].

Si(Li) detectors have been used extensively to study high-Z hollow atoms [18,19] (Fig. 11.14), using the much higher-Z, low-intensity ions produced at the Livermore EBIT (10^3–10^4 ions s^{-1} up to Th^{80+}),.

These detectors have also been used recently [20] to discover inner Auger excitation in uranium ions (ionization of an innermost shell of an ion, the M shell of which was initially closed via a spontaneous dielectronic excitation).

Obviously the high-resolution X-ray spectroscopy (crystal spectroscopy) provides a larger set of data than those obtained with Si(Li). It allows, as discussed above, the exploitation of the atomic clock property of the hollow

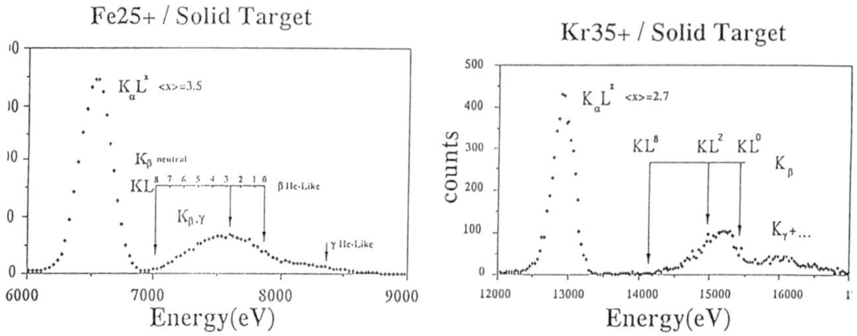

FIGURE 11.14 Fe and Kr X-ray spectra in Fe^{25+} and Kr^{35+} collisions at 7 keV/q on solid targets at normal incidence [13]. Reprinted from J.P. Briand, B. d'Etat, D. Schneider, M. Clark, V. Decaux, "On the mechanism of formation of hollow atoms below a surface", *Nucl. Instrum. Meth.* **B87** (1994) 138, with kind permission of Elsevier Science - NL, Sara Burgerhartstraat 25, 1055 KV Amsterdam, The Netherlands.

atoms and the study of the time-dependence of the interaction. This kind of experiment cannot, however, be carried out with low intensity beams (¡10^9 ions s^{-1}), or with low Z elements with small fluorescence yields. They have only been done with powerful ECR sources.

These most powerful ECR sources cannot provide fully stripped ions beyond Z=18 (Ar) and do not provide more than a few 10^9 or 10^{10} particles per second of Ar^{17+} ions. This reduces greatly the number of available techniques of crystal X-ray spectroscopy to those having the largest transmission. For instance, one can exclude flat crystal spectrometer devices, the transmission of which does not exceed 10^{-9}. For Ar, if one takes into account (i) the ion fluorescence yield, (ii) the transmission of berylium windows (at 3 keV one must work in vacuum), (iii) the efficiency of the detectors at such an energy using the special geometry of an ion-surface collision, the overall efficiency of a curved crystal spectrometer lies between 10^{-8} and 10^{-9}. A curved (R≈1m) Ge ⟨111⟩ crystal spectrometer was once used [12] to study the interaction of Ar^{17+} ions at grazing incidence. In this case we were dealing with an extended X-ray source due to the small area of interaction of the ion beam with the target (this interaction took place, in this experiment, over 10 cm—a huge surface for collecting all emitted X rays). To date, most experiments have been carried out mainly with a flat mosaic crystal spectrometer, the transmission of which is larger by one order of magnitude than that of a curved crystal.

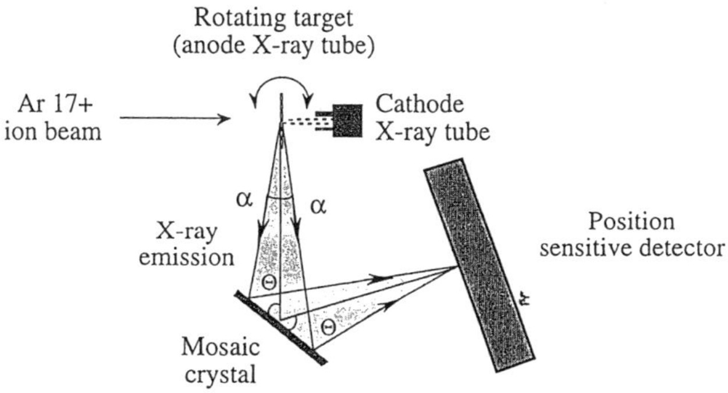

FIGURE 11.15 High energy resolution device: mosaic crystal and position-sensitive detector.

11.3.1 High resolution devices

We can separate all the lines composing the Kα and Kβ by using the reflection on a mosaic crystal according to Bragg's law. For each photon of energy transition KL^x, we can tune a (θ, 2θ) spectrometer, and detect the position of each photon using a localization detector (Fig. 11.15).

The principle of this spectrometer is to exploit the focusing property of a flat mosaic crystal, which allows rather large horizontal acceptance angles. One can find on the market some pyrolytic graphite crystals, the random angle of which ranges from 0.8° up to 2°. These numbers, which have to be compared to the transmission of flat, pure monocrystalline crystal that lies in the range of few arc·second, clearly show the excellent transmission of such devices. The reflectivity of the crystal (15-40 %) allows an easy detection of X rays in the range of 2-20 keV. The best overall transmission rates of these spectrometers are of the order of 10^{-7} (including all experimental constraints). One must however realize that a Kα spectrum of Ar extends over eight lines and, in the best conditions, a spectrum having reasonable statistics cannot be obtained in less than one hour with an Ar^{17+} beam of the order of few nanoamps (electric). The study of Ar^{18+} ions, the intensity of which is 10 to 80 times smaller than that of Ar^{17+}, makes such experiments, with extremely interesting fully stripped ions, very difficult.

The counterpoint to the good transmission of these devices is that they need an extremely good mechanical precision because the resolution of flat crystal spectrometers is mainly defined by the width of the X-ray source. At 3 keV and a distance of 17 cm, one eV extends over 80 mm. One must work at looking at the X rays emitted at 90° with respect to the ion beam with a take-off angle a as small as possible (Fig. 11.15). The emittance of the ECR sources being quite poor (100 π mm mrad), the size of the available beams never exceeds 4×8 mm which, at take-off angles of 3°, leads to an apparent X-ray source of 400 μm, i.e., 5 eV. It is then necessary to get a better resolution (2-3 eV) to use very small take-off angles which also require an extremely accurate positioning—under UHV—of the target with respect to the crystal. In these detectors the detection of the X rays has mainly been done with Xenon filled position-sensitive proportional counters of the backgammon type. The metallic anode (Ag, Pd, Sn) of an X-ray tube is selected to frame energetically the energy spectra of Kα and Kβ lines.

11.3.2 Other high-resolution energy devices

At lower X-ray energy (1 keV), Bonnet *et al.* [21] used a flat pseudocrystal spectrometer associated with a Soller slit and stepping detection. This device allowed the observation of some of the KL^x lines of Ne^{9+} ions. The resolution and transmission in this difficult energy range are however very poor.

Recently, an attempt has been made at Cornell University to detect the very low energy X rays (280-600 eV) emitted by C, N, O ions using flat multilayers.

Another exciting, but difficult, device, the bolometer, has recently been developed and used to study the X rays emitted during ion-surface interactions [22]. This bolometer is a micro-calorimeter working at extremely low temperature, which, instead of detecting the charges induced by an X ray in a solid, measures the heat excess due to the slowing down of the photoelectrons. The resolution of such a detector, which in the best cases can reach 6 eV (fwhm), is then close to that of a crystal spectrometer, but it is much more efficient. We present in Fig. 11.16 the first spectrum obtained recently at Livermore with such a detector. This spectrum clearly illustrates the efficiency and future capability of this kind of detector.

11.4 X-RAY STUDY OF FULLY DECELERATED HIGHLY CHARGED IONS ON SURFACES

To date, the main interest in the field of highly charged ion-surface interactions is the study, at very low energy, of ions impinging metal, insulator or semiconductor surfaces. Most of the ion sources (ECR and EBIS) provide ions of few keV/q up to 20 keV/q kinetic energy (minimum voltages needed to extract the ions either from an ECR plasma, or an EBIS source).

The ions must thus be at least partly decelerated at the exit of the source through, e.g., an Einzel lens (Figs 17 and 18). Obviously, such deceleration from typically 10-20 keV/q down to, e.g., 1 keV/q leads to an important loss of intensity, angular divergence, or beam size.

A full deceleration of the ions can be performed for X-ray studies, but not for charged particle studies, without losing any ions simply by biasing the target at the source potential. All experiments dealing with the study of charged particles (Auger electrons emitted in flight by the ion, scattering, specular

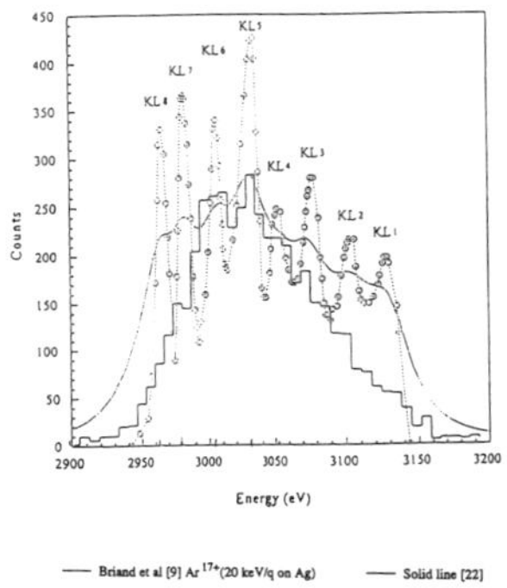

FIGURE 11.16 Kα X-ray spectrum observed with a bolometer in Ar^{17+} →Be collisions, at 7 kev/q and normal incidence. The scaled line is the experimental spectrum obtained at Livermore, the solid line is the fit of the scaled line, and the dotted line is our experimental spectrum. Reprinted from M. LeGros, E. Silver, D. Schneider, J.M. Donald, S. Bardin, R. Schuch, N. Madden, and J. Beeman, *Nucl. Instr. and Meth. A* **357** (1995) 113, with kind permission of Elsevier Science - NL, Sara Burgerhartstraat 25, 1055 KV Amsterdam, The Netherlands.

reflection of incoming ions, or sputtering of surface ions) were usually carried out by polarizing the whole device at the high voltage of the source.

11.4.1 Weak decelerating setup

Beam line

In Fig. 11.17 we present an experimental setup developed at the AIM facility in Grenoble with a 10 GHz ECR ion source.

The ions extracted from the source at 20kV (1 kV ECR potential, -19 kV extraction nozzle) are decelerated down to 1kV in a double Einzel lens structure [23], and fully decelerated on the target in the experimental zone.

We show (Fig. 11.18) a schematic view of the potential bias along the beam line. The intensity of the, e.g., Ar^{17+} beam of this ECR source was of the order of 10^7 ions per second before the double Einzel lens and 10^6 after the lens. The beam at the focal point has a diameter of $\approx$ 2mm and $\approx 2°$ angular divergence (Fig. 11.21).

Focusing pipe and ion backscattering

The last deceleration step from1 keV/q down to zero obtained by biasing the target at 1kV (total deceleration) may lead to an unacceptable angular divergence of the beam and the backscattering of the ions for certain trajec-

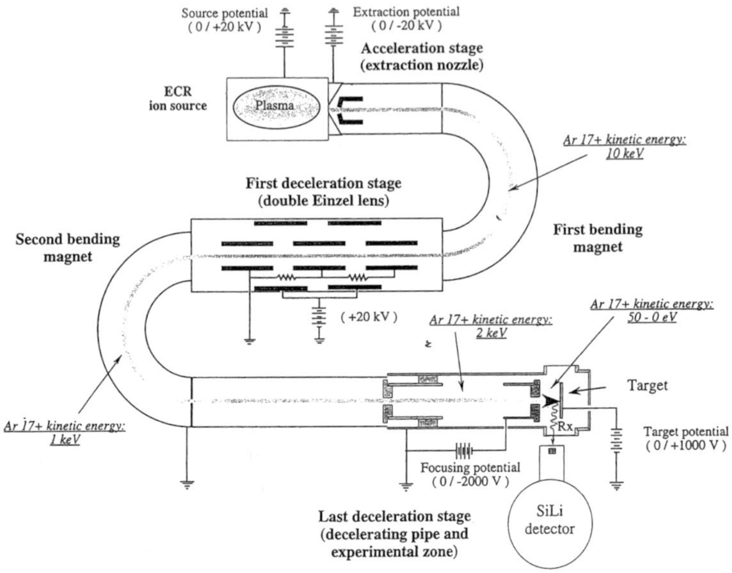

FIGURE 11.17 Ar^{17+} ion beam line from ECR ion source to experimental zone. Low resolution measurement with Si(Li) detector.

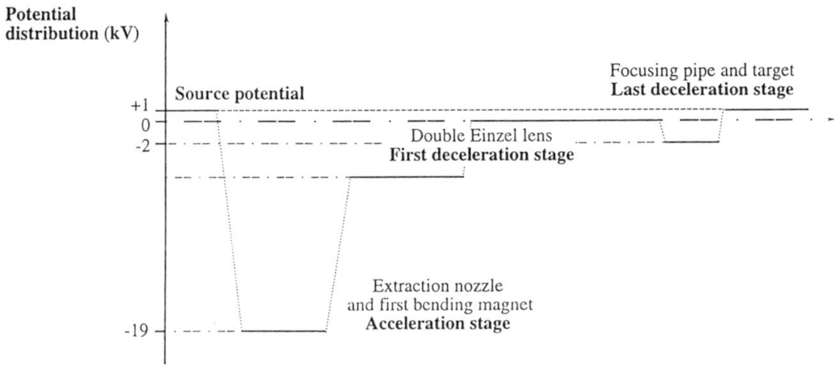

FIGURE 11.18 Potential biasing along the beam line.

tories to the wall, which is at ground (Fig. 19a). This reaccelerates the ions and leads to the emission at the walls of X rays coming from energetic ions.

This difficulty has been overcome by using a focusing lens in front of the target. As seen in Fig. 11.19b, the beam is focused onto the target (which does not really change the normal approach velocity; obviously this would not have been the case at grazing incidence). The design of the focusing pipe leads to a thin spot of ions on the target in any case, i.e., with or without biasing the target. When the target bias is tuned at $(1+\epsilon)$ keV, one observes a backward

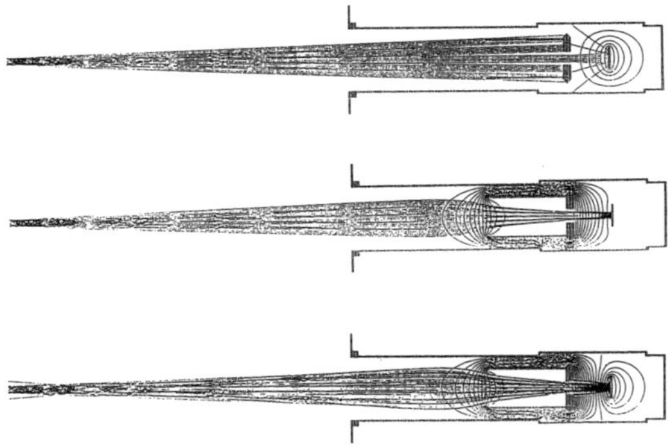

FIGURE 11.19 Ar^{17+} trajectories in the last focusing pipe (experimental setup). Calculated 1 keV Ar^{17+} ions, for beam pipe (a) at ground, for two values of ϵ: (b) 0 V, and (c) +10 V.

motion of the Fig. 19c).

It is thus possible to have a perfect control of the backscattering of the ions. By moving the Si(Li) detector to the left part of the Be window of the experimental setup, one hides the ion interactions on the exit diaphragm of the pipe (formation of hollow atoms on metal at 3 keV and Bremsstrahlung spectrum), and obtains only the spectra of ions interacting with the target.

As shown in Fig. 11.21 an optimum compromise is reached when the focusing pipe is biased at a voltage two times larger than the acceleration potential of the ions at the exit of the source.

11.4.2 Strong beam deceleration with single gap setup

Similar experiments of decelerated ion beams have been carried out by using a single gap strong deceleration of the beam (ion beam of 12 keV kinetic energy at the beamline entrance down to 0 eV on the target). These experiments were done after a charge state separation by a magnet dipole at the immediate exit of the 14 GHz ECR Caprice ion source in Grenoble (12kV extraction potential on the nozzle) (Fig. 11.20). As shown above, the potential biasing the focusing lens is $\approx$ -25 kV while the target is biased at 12 kV for a total deceleration.

The polarization bias of the focusing lens then allows one to obtain a thin spot on the target (2 mm diameter) and a full transmission of the ion beam. A detailed schematic view of the focusing pipe with power supplies and low resolution Si(Li) detector is given Fig. 11.21 for a 1 keV Ar^{17+} ion beam.

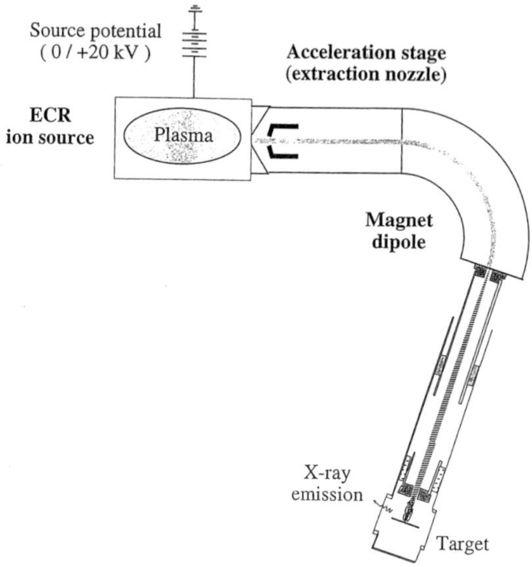

FIGURE 11.20 Strong decelerating device in one gap without any loss of ions.

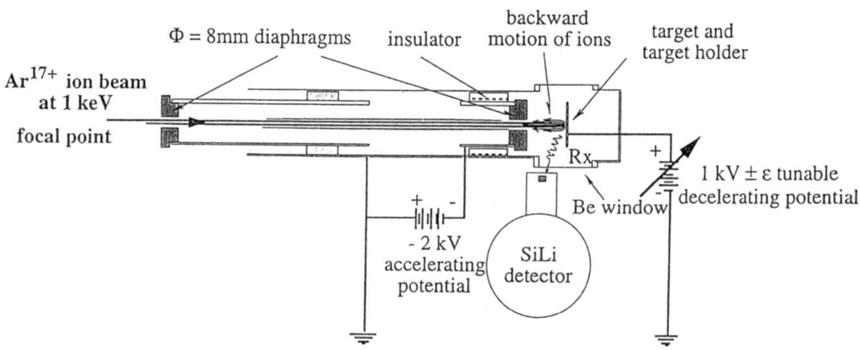

FIGURE 11.21 Last deceleration (1 keV→0 eV) pipe. Low energy resolution measurement with a Si(Li) detector.

Beam intensity and electric fields in the pipe

This device is the only way to use the 5 nA Ar^{17+} ion beam delivered by the 14 GHz ECR source with a high resolution crystal spectrometer. This structure uses large decelerating potentials in a single gap ($\Delta V \approx 35$ kV, i.e., 1.4 MV m^{-1} from the pipe to the target region).

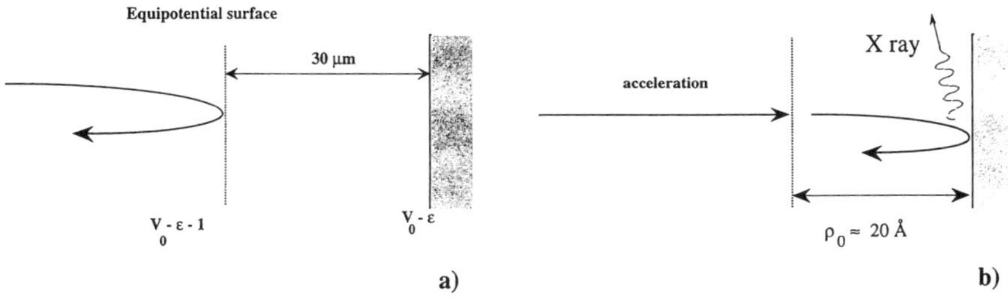

FIGURE 11.22 (a) V_0-ε-1 equipotential position in front of the target; (b) Minimum approach distance ρ_0 for electronic capture.

In such a device, all the elements are close to the disruptive potential breakdown, and one must be careful of noise coming from electron Bremssthralung on the biased elements.

Parallel plate capacitor configuration

The design of this setup leads to a potential distribution similar to a parallel plate capacitor between the pipe and the target. Close to the target, in the decelerating region, equipotentials can be seen as planes and are parallel to the surface of the target. In this design, the potential gradient is of the order of 30 μm/V, and potential bias on the target can be tuned accurately within 1 Volt without any flatness modification of the potential lines (Fig. 11.22a).

The target bias can be tuned around $V_0 \pm \epsilon$ (V_0 is the source potential), in such a way that the exact region in which the ions are backscattered is accurately defined.

Types of interaction and the respective spectrum configurations

Using these focusing properties and the possibility of controlling the ion's distance of approach from the target, as described above, we studied the whole ion kinetic energy range from 10 keV down to 0 eV (Fig. 11.23). The high-resolution spectra show drastic variations corresponding to above-, at-, and below-surface interactions.

At high velocities, the spectrum is mainly centered on the $KL^{5,6}$ lines corresponding to below-target surface interaction. In this case, the ion is quickly re-fed by electrons during the Auger cascade and the emission of the Kα line occurs at any time during the L shell filling (see Fig. 11.2).

At lower energy, the ions partly decay outside the surface, and the KL^x distribution starts peaking up at low x of KL^x satellites, as shown Fig. 23. At the lowest energies, one observes pure outside decay (no refeeding).

Image effect and potential barrier

Experiments at very low energy have been carried out at high resolution on various target surfaces (metals, semiconductors, and insulators). Above the surfaces, at a distance far above the minimum distance ρ_0 at which they can capture some electrons from the target, the ions experience an acceleration towards the surface due to their own *image*, the collective response of the electron gas in the metal. An ion of, e.g., zero kinetic energy will then drop irremediably onto the surface, and it is thus not possible to study the interaction of this ion with the surface at energy close to zero. In our experimental decelerating device, where the external applied electric field ($\approx 10^6$ V m^{-1}) is much weaker than that due to the ion's image close to the surface ($\approx 10^9$ V m^{-1}), the ion is reflected by the potential barrier at quite a large distance (1000 Å) by tuning the decelerating potential (Fig. 11.22a) (variation in position of the potential barrier), or it drops irremediably into the potential well (Fig. 11.24).

Recently we demonstrated that with insulator or semiconductor targets, this image acceleration is overbalanced by the local distribution on the surface of positive charges, vacancies created by removing electrons. In this case, the ion is backscattered after passing a certain time below ρ_0, i.e., capturing and losing a certain number of electrons. Therefore the ion never touches the surface, and one may observe a pure outside decay of Rydberg hollow atoms and its characteristic KL^1 X-ray spectrum (Fig. 11.25a) (above-surface interaction), whereas with metal targets, the spectrum peaks on the KL^2 satellites—fast feeding of the M shell during "at surface" interactions followed by extremely fast LMM Auger transitions, providing additional L-shell filling (Fig. 11.25b).

11.4.3 Kinetic energy resolution

As shown above, the decrease of the X-ray rate, signalling a penetration of the ion below ρ_0, is obtained in changing the deceleration potential by a few Volts. The precision of the penetration below ρ_0 depends experimentally on the kinetic energy resolution ΔE of the ion while interacting with the target. The X-ray rate also changes with the fluorescence yield depending on the number of electrons in the L and M shells, and, so, depends on the kind of target. We present Fig. 11.26), a typical plot with a metal target, where the counting rate is measured as a function of the decelerating bias, observed with ions coming from an ECR source.

In order to obtain good precision in measuring cut-off voltages, it is necessary to reduce the dispersion energy ΔE of the ions.

Energy resolution in ECR sources

To date, all the low-energy experiments presented here have been carried out with ECR ion sources. In such sources, the extracted ions are formed

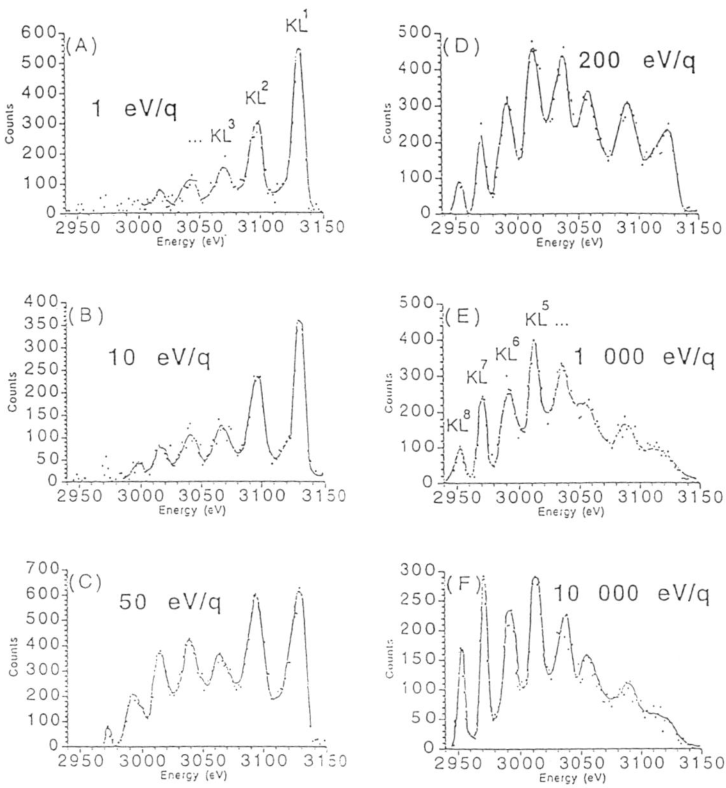

FIGURE 11.23 High resolution spectra with Ar^{17+} for various ion kinetic energies. Reprinted with kind permission from *Phys. Rev. A* **54** (1996) 4136.

in a magnetically confined plasma (bottle and n-polar magnetic coils). The extracted ions are those escaping the plasma with different kinetic energies, accelerated by the plasma potential and the biased extraction nozzle.

Experiments show a fluctuating ion kinetic energy of around 5 eV, which depends on (i) the stability of the plasma: argon and oxygen pressure mixture, (ii) the stability of the injected RF heating, (iii) the changes of the RF reflection according to the plasma configuration and injected gas pressure, (iv) the non-homogeneous ionic distribution in the plasma, etc. One observed these instabilities by the amplitude fluctuations of the beam intensity (up to 10%) and also the frequency of these instabilities (from 0.5 s up to some 10 s). In such situations, the ion energy dispersion increases. Moreover, because of the diameter of the plasma chamber ($\approx$ 15 cm) and the relatively high diameter of the extraction nozzle ($\approx$ 10 mm), these sources deliver beams of high angular aperture (some d° aperture for some mm beam diameter), leading to an emit-

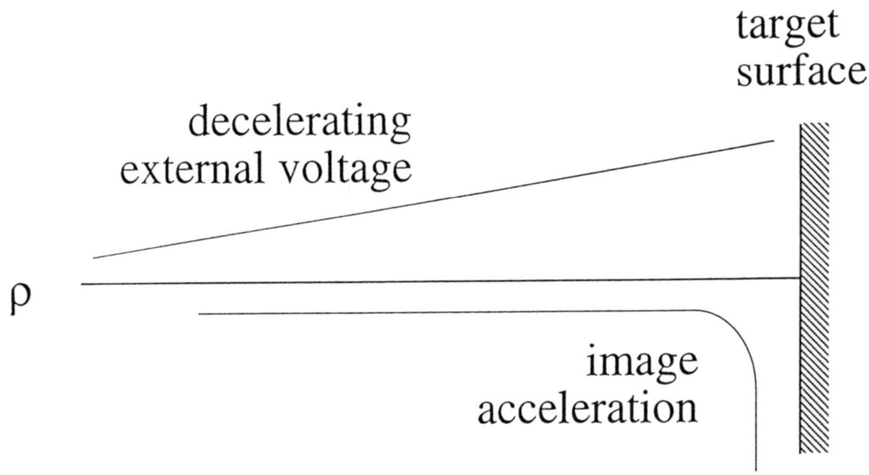

FIGURE 11.24 Linear decelerating voltage and potential well induced by the image acceleration.

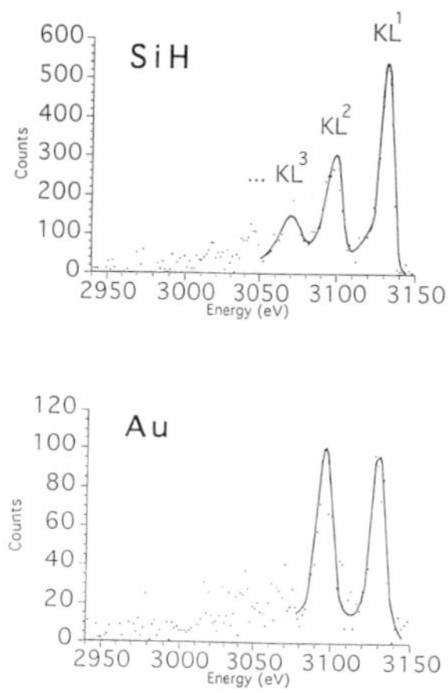

FIGURE 11.25 Fully decelerated Ar^{17+} spectra on SiH and Au targets. Reprinted with kind permission from *Phys. Rev. Lett.* **77** (1996) 1452.

tance of the order of 100 π mm mrad at the focal point before the focusing pipe.

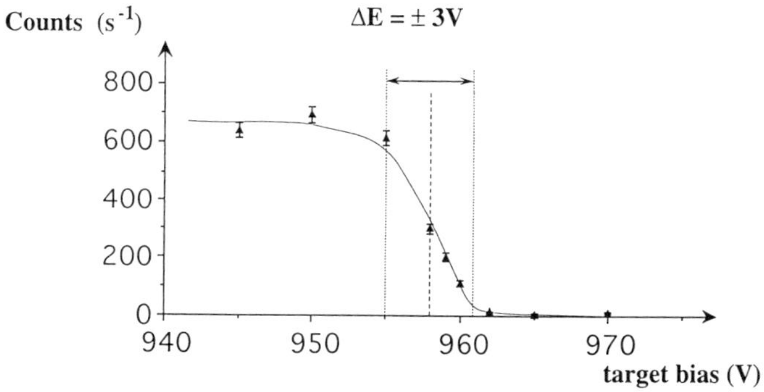

FIGURE 11.26 X-ray rate as a function of the decelerating voltage.

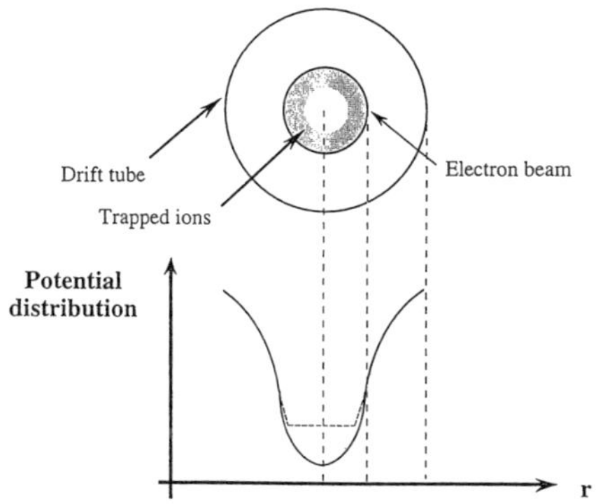

FIGURE 11.27 Radial potential well transformation by electron beam neutralization.

Energy resolution in EBIS sources

One can expect better results with EBIS sources. In this kind of source the potential well depth in the electron beam is proportional to the beam intensity, and is typically from a few Volts up to 100 V for the usual sources (radial trapping) (Fig. 11.27). The potential of the longitudinal barrier (longitudinal confinement) gives the exit ion energy after the source, with a very high stability and with an energy dispersion of the order of the potential well in the electron beam.

A way to reduce the potential well and, so, the ΔE dispersion of the extracted ions would be to reduce the electron beam intensity, but this weakens the ion beam. The increase of the electron beam increases the well depth, but also degrades the inherent beam qualities of the electron beam heating.

Recent work concerning electronic space charge neutralization (ion beam cooling) [24] has shown that electron beam neutralization leads to lower transversal electronic temperatures without changing the ionic longitudinal motion (1%). This neutralization is obtained by filling the electron well potential with ions (more than 50% neutralization). The electron beam is not transversly fully neutralized (neutralization mainly occurs in the center of the electronic beam) because of the transverse ion energies, which leads to a radial leak at the periphery of the electronic beam. Such a configuration induces a diminution and a flattening of the potential well in the center of the electron beam (dashed line Fig. 11.27). The action of the magnetic field used for electron confinement, and such a centered neutralization, induce low energy dispersion of the ions in the axial region.

REFERENCES

1. M. Delaunay, R. Geller, J. de Bernardi, P. Ludwig, and P. Sortais, *Surface Science* **197**, L273-280 (1988).
2. M. Delaunay, M. Fehringer, R. Geller, D. Hitz, P. Varga, and H. Winter, *Phys. Rev. B* **35**, (1987).
3. M. Fehringer, M. Delaunay, R. Geller, P. Varga, and H. Winter, *Nucl. Instr. and Methods B* **23**, 245-247 (1987).
4. S.T. Zwart, T. Fried, U. Jellen, A.L. Boers, A.G. Drentje, *J. Phys. B* **18**, L623 (1985).
5. S.T. de Zwart, T. Fried, D.O. Boerma, R. Hockstra, A.G. Drentje, and A.L. Boers, *Surf. Sci. Lett.* **177**, L939 (1986).
6. De Zwart S.T., Thesis, University of Groningen (1987), and *Nucl. Inst. and Methods B* **23**, 239 (1987).
7. F. Meyer, C.C. Havener, S.H. Overbury, K.J. Reed, K.J. Snowdon, and D.M. Zehner, *J. Phys. (Paris), Colloq* **50**, C1-263 (1989).
8. E.D. Donets, *Nucl. Instr. and Methods B* **9**, 522 (1985).
9. J.P. Briand, L. de Billy, P. Charles, S. Essaba, P. Briand, R. Geller, J.P. Desclaux, S. Bliman, and C. Ristori, *Phys. Rev. Lett.* **65**, 159 (1990).
10. S. Schippers, J. Limburg, J. Das, R. Hoekstra, and R. Morgenstern, *Phys. Rev. A* **50**, 540 (1994).
11. J.P. Briand, in *X-ray and Inner-Shell Processes (X90)*. Knoxville, Tennessee, 1990, edited by Thomas A. Carlson, Manfred O. Krause, Steven T. Manson, AIP, Conf. Proc 215. AIP, New York, (1990), p513.
12. B. d'Etat, J.P. Briand, G. Ban, L. de Billy, J.P. Desclaux, and P. Briand, *Phys. Rev. A* **48**, 1098 (1993).
13. J.P. Briand, B. d'Etat-Ban, D. Schneider, M.A. Briere, V. Decaux, J.W. McDonald, and S. Bardin, *Phys. Rev. A* **53**, 1 (1996).
14. N. Vaeck and J. Hansen, *At. MolPhys.* **28**, 3523 (1995).
15. J.P. Briand, P. Chevallier, M. Tavernier, and J.P. Rozet, *Phys. Lett.* **27**, 777 (1971).
16. J.P. Briand, L. de Billy, P. Charles, S. Essabaa, P. Briand, R. Geller, J.P. Desclaux, S. Bliman, and C. Ristori, *Phys. Rev. A* **43**, 565 (1991).
17. F. Meyer, S.H. Overbury, C.C. Havener, P.A. Zeijlmans van Emmichoven, D.M. Zehner, *Phys. Rev. A* **44**, 723 (1991).

18. M. Schulz, C.L. Cocke, S. Hagmann, M. Stockli, and H. Schmidt-Bocking, *Phys. Rev. A* **4**, 1653 (1991).
19. J.J. Bonnet, A. Fleury, M. Bonnefoy, M. Chassevent, T. Lamy, A. Brenac, A. Simionovici, H.J. Andra, and S. Andriamonje, *Z. Phys. D* **21**, S343 (1991).
20. R. Schuch, D. Schneider, D.A. Knaff, D. de Witt, J.M. Donald, M.H. Chen, M.W. Clark, and R.E. Marrs, *Phys. Lett.* **70**, 1073 (1993).
21. J.J. Bonnet, A. Fleury, M. Bonnefoy, M. Chassevent, T. Lamy, A. Brenac, A.Simionovici, H.J. Andra, and S. Andriamonje, *Z. Phys. D.* **2**, 343 (1991).
22. M. LeGros, E. Silver, D. Schneider, J.M. Donald, S. Bardin, R. Schuch, N. Madden, and J. Beeman, *Nucl. Instr. and Meth. A* **357**, 110 (1995).
23. T. Lamy, G. Lamboley, D. Hitz, and H.J. Andra, *Rev. Sc. Instr.* **61** (1), 336 (1990).
24. F. Varenne, Thesis. Université Blaise Pascal, Clermont-Ferrand (1996).

CHAPTER 12

Auger Electron Spectroscopy of Target Atoms

B. Sulik* and N. Stolterfoht†

An introduction into the Auger process is presented including basic concepts, notation, classification and a brief history of target Auger spectroscopy studies in ion-atom collisions. Experimental methods and apparatus relevant for the field of target Auger spectroscopy are reviewed. Emphasis is given to the electron spectrometers including concrete realizations. Selected experimental studies are surveyed to illustrate the potentials of the experimental methods and provide insight into the trends in the field. The treated topics are multiple ionization of target atoms, angular distribution of the Auger emission, and post-collision interaction studies.

12.1 INTRODUCTION

The Auger process is a spontaneous decay mode of highly excited atoms or ions by electron emission. It competes with the well-known radiative decay. Since the observation of the process by Pierre Auger [1,2] in 1925 and the subsequent theoretical formulation by Wentzel [3] in 1927, the Auger effect has received increasing attention in many fields including atomic physics, nuclear physics, solid state physics, surface physics, and chemistry.

The transition resulting in Auger electron emission is caused by the electron-electron interaction. Generally, the Auger process involves two active electrons interacting with each other in the field of the atomic nucleus screened by the other (passive) electrons. The initial state is discrete and its energy lies above the first ionization threshold. The final state implies the ejected electron and the residual ion. When the Auger transition takes place in the isolated

*) Institute of Nuclear Research of the Hungarian Academy of Sciences (ATOMKI), P.O. Box 51, H-4001 Debrecen, Hungary.

†) Hahn-Meitner Institut Berlin GmbH, Bereich Festkörperphysik, Glienickerstr. 100, D-14109 Berlin, Germany.

atom, it depends only on the atomic properties. Therefore the study of the Auger effect yields direct information about the structure and the dynamics of the atom. For non-isolated atoms, the Auger process is rather sensitive to the surrounding electronic structure, which makes it widely applicable in chemistry and surface analysis.

To prepare the initial state one needs to create an inner-shell or core vacancy in the investigated atom or ion by means of photon, electron or ion/atom impact. Then absolute or relative Auger production cross sections can be determined from the measured line intensities. From these data cross sections for populating particular initial states can be derived, providing detailed information about vacancy production mechanisms like collisional excitation, ionization or electron capture.

It should be realized that the Auger process is relatively slow in comparison to time scales characteristic of the atom. For instance, the Auger process takes place during several hundreds of revolutions of orbiting outer-shell electrons. Thus it follows from the uncertainty principle that the natural spread of the Auger electron energy is small in comparison with the characteristic atomic energies. On the other hand, in comparison with the radiative transitions in outer shells, the Auger process is very fast and, hence, the corresponding width of the Auger lines is comparably large.

Moreover, the lifetime of the Auger or autoionization states is usually much longer than the atomic collision times. This is generally true for incident protons or electrons. For energetic ions, let us consider a collision where the projectile velocity is equal to the mean orbital velocity of the inner-shell electron to be ionized. Then it is reasonable to take the effective collision time equal to the time of one orbital revolution of the inner-shell electron. This yields an Auger lifetime several hundred times longer than the collision time. When reducing the projectile velocity by a factor of 10, the Auger lifetime still remains several tens of times longer. Therefore, a separate treatment of the vacancy production process and the Auger process is a reasonable approximation.

In the present chapter, high resolution Auger electron spectroscopy of target atoms is considered for energetic ion-atom collisions. A brief history and a short analysis of the specific features of target Auger spectroscopy is presented on an introductory level. Emphasis is given to the experimental method and apparatus. Finally, a few experimental studies have been selected to illustrate the potentials of the methods and give an impression about the tendencies in the field.

The separate treatment of target spectroscopy and projectile spectroscopy in this book reflects the historical fact that they have developed rather separately from the early 1970's. Projectile Auger spectroscopy and the related instrumental aspects are detailed in the forthcoming two chapters of the present book.

For a wider overview of the field of target Auger spectroscopy, the reader is referred to the comprehensive articles that review the work about the Auger effect. The theory of autoionization and the Auger effect has been discussed in detail by Burke [4], Burhop and Asaad [5] Chattarji [6], McGuire [7], Åberg and Howat [8], and Stolterfoht [9] . Atomic structure calculations related to Auger line energies and transition probabilities are widely referred e.g., in the review of Stolterfoht [9] and in a recent paper of Kabachnik *et al.* [10].

Surveys about experimental studies of ion-induced Auger electrons have been given by Ogurtsov [11], Niehaus [12], Rudd and Macek [13] Sevier [14], Mehlhorn [15–17], Matthews [18], Berényi [19,20] and Stolterfoht [9,21,22]. Some of them [16,20–22] cover both target and projectile Auger spectroscopy. Specific experimental studies of electron spectra have been discussed by Stolterfoht [23–25], Berényi [19], Mann *et al.* [26] and Rudd *et al.* [27]. Related work concerning excitation mechanisms in ion-atom collisions have been reviewed, e.g., by Garcia *et al.* [28], Kessel and Fastrup [29], Meyerhof and Taulbjerg [30], Briggs and Macek [31], and McGuire [32].

Target Auger studies by electron and photon impact have been surveyed e.g., by Siegbahn *et al.* [33], Krause [34], Carlson [35], and Mehlhorn [15–17]. Application for chemistry and surface analysis is detailed in the works of Siegbahn *et al.* [36] and Riviere [37]. Electron emission from slow atomic collisions has been overviewed by Niehaus [38].

In the following, we restrict our considerations to the high-resolution Auger spectroscopy of free atoms, i.e., gas or vapour targets, at the beamlines of heavy particle accelerators. The impact energy region is larger than a few tenth of keV/u. Throughout this chapter atomic units ($m_e = e = \hbar = 1$) are used unless otherwise stated.

12.1.1 Brief history

In the early 1960's, electrostatic electron spectrometers became widely used in studies of the Auger effect and autoionization at gaseous or vapour targets by electron and photon impact (see e.g., refs. [39–42]). The high resolution abilities of the spectrometers made possible a detailed analysis of spectral structures.

The first use of energetic ions to obtain spectroscopic information about Auger and autoionization states started in 1964. Rudd [43] measured the first high resolution spectra of He autoionization electrons produced by H^+ and H impact. The use of incident ions has advantages compared to photon and electron impact, since they may excite states which are otherwise not accessible. However, the use of ions has also disadvantages. Ion impact produces line broadening which may inhibit high-resolution studies. The decaying target atom may be influenced by the slowly emerging charged projectile producing the post-collision effect. It was first observed by Barker and Berry [44] in early experiments of ion-induced autoionization electrons. Moreover, line broadening is produced by kinematic effects [9,45]. In particular, heavy ions transfer large recoil momenta so that the emitter moves in different directions during the decay [46]. For example, the lines in the Ne K Auger spectrum cannot be resolved when they are produced in Ne + Ne collisions at energies of a few hundred keV [47].

For the target atom, broadening effects can be reduced substantially when the impact energy is increased. In 1973 Ne-Auger spectra were measured with high resolution by Matthews *et al.* [48] using oxygen projectiles of about 30 MeV. The spectra confirmed previous observations in the field of X-ray spectroscopy which have indicated considerable outer-shell ionization in addition to the inner-shell vacancy production [49]. In fact, with very heavy projectiles, the target atom can be stripped to a few-electron system as shown by Stolterfoht *et al.* [50] for 200-MeV Xe^{31+} projectiles.

The possibility of removing several electrons in a single collision and reducing the kinematic broadening effects opened the field of high-resolution Auger spectroscopy of few electron systems. Several groups performed studies of target Auger electron produced by energetic heavy-ion impact (see e.g., Matthews *et al.* [48], Schneider *et al.* [51], Mann and Folkmann [52], Woods *et al.* [53], Kádár *et al.* [54], and Matsuo *et al.* [55]).

By varying the charge state of heavy projectiles in fast collisions, different degrees of multiple ionization of the same target species can be achieved. From the set of high-resolution spectra, a large portion of satellite lines associated with different initial vacancy states can be identified. With this method, 2s and 2p subshell ionization probabilities or full subshell vacancy population distributions associated with K-shell ionization of Ne were determined by Kádár *et al.* [56,57] and Sulik *et al.* [58] in 5.5 MeV/u H^+, N^{2+}, N^{7+}, Ne^{3+}, Ne^{10+}, Ar^{6+}, Ar^{16+} + Ne collisions. This kind of experimental study is valuable for understanding the mechanism and the statistical aspects of multiple vacancy production in heavy ion-atom collisions.

Finer details of the excitation or multiple ionization dynamics can be experimentally studied by measuring the angular distribution of the emitted Auger electrons. The observed anisotropy of the Auger lines provides information about the non-statistical population of the magnetic electron or vacancy substates, i.e., the alignment of the charge cloud, in the collision. The anisotropy of the Li $1s2s2p\ ^2P$ lines excited by 50-400 keV H^+ and He^+ projectiles was studied and interpreted by Ziem *et al.* [59]. The first evidences of nonstatistical population and anisotropic satellite distribution in multiply ionized neon were provided by Matthews *et al.* [60] and Stolterfoht [23]. The first experimental determination of the anisotropy parameters for satellite Auger lines in the spectrum of neon, doubly ionized by 5.5 MeV/u heavy ions, was reported by Ricz *et al.* [61,62]. These data triggered a long discussion in atomic structure theory [10]. By coincidence methods, Sarkadi *et al.* [63] and Gutenkunst *et al.* [64] separated the electron capture and ionization contributions to the vacancy production on the L_3 subshell of argon and magnesium by proton bombardment. A strong alignment with steep impact velocity dependence has been found for the capture process in both cases, while the integral alignment was nearly zero. The observed impact velocity dependence has been reproduced only qualitatively by capture theories.

Detailed information about the post-collision interaction (PCI) of the outgoing Auger electrons with other collision fragments can be extracted from high-resolution Auger spectra by analyzing the shape of the individual lines. Under ion impact, PCI is a rather complicated effect with strong impact velocity and angular dependence. Systematic studies have been performed on this topic by Arcuni [65], Takács *et al.* [66], Sarkadi *et al.* [68], and Ricz [67].

It is worth mentioning here that both PCI and angular distributions are widely investigated by electron impact (see e.g., refs. [69–71]). Auger studies on atomic and molecular species at synchrotrons have also been reported recently (e.g., refs. [72,73]). Systematic work on target species by merging the potentials of the accelerator, electron-impact and synchrotron facilities together, could be a future trend in atomic physics.

12.2 THE AUGER EFFECT

In a simple picture, the Auger effect may be regarded as electron transitions between atomic orbitals shown in Fig. (12.1). The example refers to a vacancy in the shell A being filled by an electron from the shell B. The excess energy is given to an electron in the shell C so that it is transferred to a continuum state of well-defined energies.

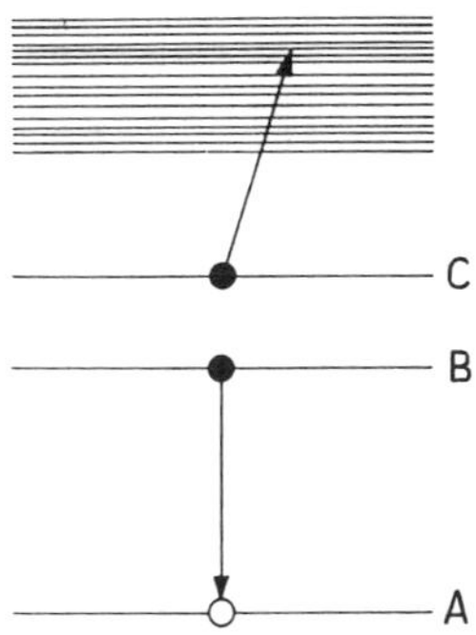

FIGURE 12.1 Schematic diagram illustrating the Auger effect

The spontaneous transition causing the emission of an electron is also called autoionization. In principle, there is no difference between the Auger effect and autoionization. However, for historical reasons it is common use to distinguish between different categories of initial states. The Auger effect refers to atoms ionized in an inner shell whereas autoionization is attributed to atoms doubly excited in the valence shell. However, it should be emphasized that this distinction loses significance when inner and outer shells are simultaneously excited or ionized. In the following, the notation Auger effect is used to cover both effects.

An individual Auger transition is defined by specifying fully the initial and final states involved. In multi-electron systems it is convenient to consider vacancies rather than electrons [74]. For example, an individual Auger transition is denoted by K-$L_1L_{23}(^1P)$, where the LS-coupling scheme is applied to fully specify the final state. Also, the transition K-$L_1L_{23}(^3P)$ may be considered. Each transition gives rise to a line in the corresponding Auger spectrum as shown in Fig. (12.2). The example refers to the K-Auger spectrum of Ne excited by 4.2-MeV H^+ impact [75].

All individual Auger transitions possible for given (sub)shells are summarized to (sub)groups using the notation K-L_1L_{23} or K-LL for example. Similarly, groups are summarized to series where only the shell of the initial state is specified, e.g. K-series or L-series. The present notation may also be applied to multi-vacancy states often created by heavy particle impact. Suppose that n L-vacancies are initially produced in addition to the K-vacancy, one may write KL_n-L_{n+2}. Of course, also subshells as well as terms may be specified.

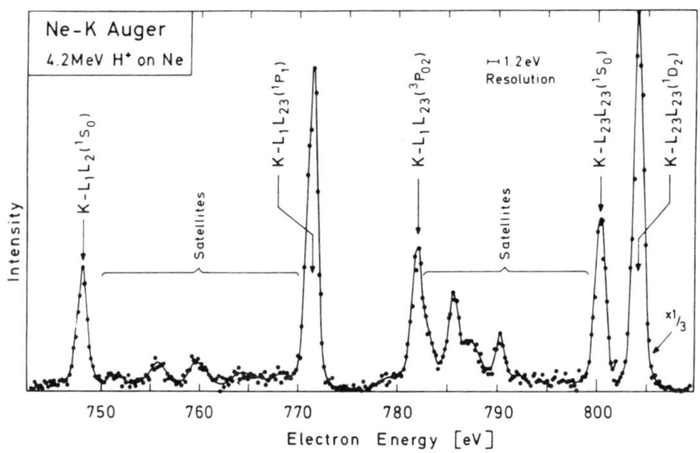

FIGURE 12.2 Neon-K Auger spectrum produced by 4.2-Mev/u H^+ impact. Reprinted from N. Stolterfoht, H. Gabler and U. Leithauser, "High-resolution Ne Auger spectrum produced in 4.2 MeV H^+ - Ne collisions." [75], with kind permission of Elsevier Science - NL, Sara Burgerhartstraat 25, 1055 KV Amsterdam, The Netherlands.

An Auger line which refers to the transition from a single vacancy state to double vacancy states is denoted a normal line or *diagram line*. Other lines are called *satellites*. For instance, the lines attributed to initial multi-vacancy states are satellites. Furthermore, lines due to double vacancies initially in the inner shell are called *hypersatellites*. The distinction between normal lines and satellite lines is very useful for incident electrons, photons and light ions, which produce primarily initial single vacancy states. However, for heavy ions which produce primarily multi-vacancy states, this distinction loses significance. Very often, heavy ions produce Auger spectra which consist exclusively of satellite lines.

Auger transitions imply definite selection rules referring to the final state of the *total* system. The initial state and the final state are specified by the total angular momenta J_i and J_f and their magnetic quantum number M_i and M_f, respectively. In general, $\Delta J = 0$ and $\Delta M = 0$ for the Coulomb, spin-orbit and spin-spin interactions. Only the hyperfine interaction can change them by ± 1. Parity is always conserved, $\Delta \Pi = 0$. The total orbital momentum L and the total spin S are conserved only for the Coulomb interaction.

Further notations are commonly used to classify the Auger effect. A transition is called a Coster- Kronig (CK) transition [76] when the initial vacancy and one final vacancy occupy subshells within one shell. If the initial vacancy and both final vacancies are produced in the same principal shell, the transition is called Super-Coster-Kronig (SCK) transition. It is only mentioned here that higher-order Auger processes are also possible where three or more electrons are affected [12,42,77–81].

The Auger process is characterized by the transition energy which is defined as the difference between the energies of the initial state and the final ionic state. The transition energy is equal to the energy of the ejected electron. Furthermore, the Auger process is characterized by the transition rate P_{if} which is equal to the number of Auger transitions per unit time. The transition rate corresponds by way of the uncertainty principle to the partial Auger width Γ_{if}^a, which has the dimension of energy. Actually, the quantities P_{if} and Γ_{if}^a are equal in magnitude in atomic units. It should be pointed out that the transition rate as well as the transition energy are defined for isolated atoms and, hence, they are typical atomic structure quantities.

If there are several Auger decay channels for a given initial state, the corresponding partial Auger widths add to the Auger width

$$\Gamma_i^a = \sum_f \Gamma_{if}^a \tag{12.1}$$

of the initial state. Analogously, the Coster-Kronig width Γ_i^{ck} and the radiative width Γ_i^x are obtained. These widths contribute to the total with

$$\Gamma_i = \Gamma_i^a + \Gamma_i^{ck} + \Gamma_i^x \tag{12.2}$$

of the initial state. Then it follows for the natural line width

$$\Gamma_i^n = \Gamma_i + \Gamma_{f'} \tag{12.3}$$

of an Auger line, where $\Gamma_{f'}$ is the total width of the final states which may further decay. The difference between the partial width Γ_{if}^a and the natural line width Γ_i^n should be noted. The partial width Γ_{if}^a is uniquely specified by the initial and the final state, whereas Γ_i^n depends on transitions other than from i to f and/or on radiative transitions.

The branching ratio into a specific Auger channel is the partial Auger yield

$$a_{if} = \Gamma_{if}^a / \Gamma_i \tag{12.4}$$

The decay in all Auger channels defines the Auger yield

$$a_i = \Gamma_i^a / \Gamma_i \tag{12.5}$$

Similarly, the Coster-Kronig yield f_i and the fluorescence yield ω_i are obtained. From eq. (12.2) it is seen that

$$a_i + f_i + \omega_i = 1.$$

Information about the widths and yields are obtained from the related line spectra. A transition from an initial state to a final state gives rise to a specific line whose intensity I_{if} is a measure for the corresponding excitation probability and branching ratio.

It is important to realize that the Auger line intensity is not proportional to the transition rate P_{if}, but it is proportional to the Auger yield a_{if}. Often, a_{if} is close to unity, i.e.,, when the decay to one final state f dominates. In this case, the intensity I_{if} probes primarily the excitation probability Q_i. Information about the transition rate P_{if} may then be obtained from the analysis of the natural width of the Auger line.

Neither the theory of ion-atom collisions nor that of the Auger effect will be discussed here. Comprehensive works have been reported in Section 12.1. From an experimental point of view, however, it is important to provide a general picture about the energy and angular distribution of the electrons emitted in the Auger process, i.e., about the differential Auger production cross sections.

For fast projectiles, the collision and the Auger decay can be treated as successive, independent processes. Then the single differential Auger emission cross section for a particular Auger transition $i \to f$ (i.e., a particular Auger line) is written in the factorized form [8]

$$\frac{d\sigma}{dE} = \frac{\Gamma_{if}^a / 2\pi}{(E - E_a)^2 + (\Gamma_i^n / 2)^2} \, \sigma_i \, . \tag{12.6}$$

Here E_a is the nominal energy of the Auger transition $i \to f$, while σ_i is the production cross section of the initial state i for the collision. Eq. (12.6) provides a symmetric Lorenzian line-shape characterized by the natural line-width Γ_i^n.

A similar factorization can be achieved in the description of the angular distribution of the Auger electrons in non-coincidence measurements [16]

$$I(\Theta) = \frac{I_0}{4\pi}\left(1 + \sum_{k=2}^{k_{\max}} \alpha_k A_{k0} P_k(\cos\Theta)\right), \tag{12.7}$$

where the parameter α_k depends only on the Auger process, while the alignment parameter A_{k0} is determined exclusively by the relative populations of the magnetic substates in the collision process. The total Auger intensity I_0 is an integral over the solid angle. The $P_k(\cos\Theta)$ are Legendre polynomials and Θ is the polar angle relative to the beam direction. The index k runs over the even numbers. In simple systems, $k_{\max}$ is a small number, e.g., for single electron or single vacancy initial states on the L_{23} subshell, $k_{\max} = 2$.

It is noted that replacing I_0 by $d\sigma/dE$ (from Eq. (12.6)), Eq. (12.7) directly yields the double differential Auger electron emission cross section, provided σ_i represents the sum of the magnetic substate population cross sections.

The above formulas show general symmetries in the Auger electron emission, i.e., a symmetry in the line-shape and a forward-backward symmetry in the angular distribution. This fact originates from the separate treatment of the collision process and the Auger process. Any asymmetry observed in the experimental data indicates that this separation is not justified for the system studied.

12.3 EXPERIMENTAL METHOD AND APPARATUS

There are only a few studies covering the majority of the instrumental and methodological details relevant for the field of accelerator based target Auger spectroscopy (e.g., refs. [13,18,25]). Many related aspects, however, have been carefully analyzed in different papers (e.g., refs. [9,14,16,27,57,82–85]. In the present work, an attempt is made to *mention* all the important topics providing references, and summarize the basic knowledge for planning experiments and collecting data. Some instrumental developments are also reported in the present Section.

12.3.1 General considerations

Fig.(12.3) shows the schematic diagram of an experimental setup and the geometry of the collision region, typical for electron-spectroscopic studies at ion-beams impinging on a gas target. The arrangement is common for most non-coincidence experiments where secondary electrons are measured. At this stage, we try to minimize the applied approximations. An arbitrary density distribution of the target atoms $n(\mathbf{r})$ is considered. Similarly, the beam flux $n_b(\mathbf{r}_\perp)$ is allowed to vary arbitrarily in two-dimensions. Background counts as well as the scattering of the emitted electrons on the target gas are not considered here, since they can be treated separately.

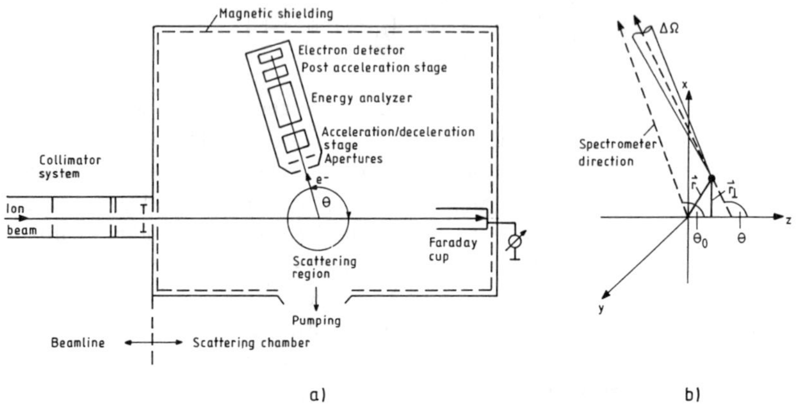

FIGURE 12.3 Typical experimental setup used for electron measurements in ion - gas atom collisions (a), and the geometry of the collision region (b).

The number of electrons counted at a nominal energy E_0 (set by the spectrometer voltages) and a nominal polar angle Θ_0 is as follows.

$$N_e(E_0, \Theta_0) = \int_{V,E} dV\, dE\, \frac{d^2\sigma[E, \Theta(\mathbf{r})]}{d\Omega\, dE}\, n(\mathbf{r})\, n_b(\mathbf{r}_\perp)\, \Delta\Omega(\mathbf{r})\, t(E, E_0,\mathbf{r})\, \eta(E) \quad (12.8)$$

Here E is the energy of the emitted electron, Θ is the polar angle of emission relative to the beam direction (z-axis) and $d^2\sigma/d\Omega dE$ is the double-differential cross section (DDCS) for electron emission in the collision process. The solid angle $\Delta\Omega$ is defined by the first slit (or object slit, see below in Fig. (12.5)) of the spectrometer. For monoenergetic electrons emitted with kinetic energy E from point $\mathbf{r}$, the inner part of the spectrometer is fully characterized by the dimensionless transmission function $t(E, E_0,\mathbf{r})$. In general, the efficiency of the electron detector $\eta(E)$ is also energy dependent.

The above picture may be used as a guide to the planning of a specific experiment. The quantities in Eq. (12.8) should be accurately measurable. Usually, reasonable approximations are made. The maximum value of $\Delta\Omega$ should be small compared to the angular variation of DDCS. Similarly, the energy region ΔE where the transmission t is nonzero should be narrow compared to the energy variations of DDCS. Then, it is justified to put the DDCS term in front of the integral. The detector efficiency is a slowly (if at all) varying function of E and can be considered as constant within the ΔE region.

In most experiments, the beam is well collimated into a small spot-size. Accordingly, the collision volume is small, and one does not need to know the distribution of the beam flux n_b, only the total flux N which is measured by a Faraday cup and current integrator. If the target is a gas-cell, the target density n is approximately constant and one has a homogeneous line target.

If the distance of the first slit (object slit, see Fig. 12.5 below) of the spectrometer from the collision region is large compared to the beam diameter and the object slit size, the transmission function of the spectrometer t can be factorized to separate $\mathbf{r}$-dependent and E-dependent terms. Consequently, the spectrometer is well characterized by an effective product of the solid angle and the target length $(l\Delta\Omega)_{eff}$ (depending on Θ), a maximum transmission T, and an effective energy window $\Delta E(E)$ [13,25].

Finally, the right side of Eq. (12.8) is obtained as a product and one gets the well known formula for the DDCS (see e.g., [27])

$$\frac{d^2\sigma(E,\Theta)}{d\Omega\, dE} = \frac{N_e}{N\, n\, T\, (l\Delta\Omega)_{eff}\, \Delta E\, \eta}, \tag{12.9}$$

which can be used in most cases. It is more advanced, however, to start from Eq. (12.8), and check the conditions of all the simplifying steps when planning an experimental setup. For non-homogeneous gas jet targets (see below) e.g., it is more appropriate to determine an effective quantity $(nl\Delta\Omega)_{eff}$ than to use the product $n(l\Delta\Omega)_{eff}$.

In general, the most difficult task is to measure absolute DDCS with reasonable accuracy. Instrumental and methodological aspects have recently been analyzed in detail by Rudd *et al.* [27]. In target Auger spectroscopy, the most important quantities to study are the line energies, relative line intensities, and line shapes. Therefore, in the following, we concentrate on relative measurements. Since the majority of the problems are common for absolute and relative measurements, some items are only mentioned or briefly treated here. For more details, the reader is referred to the recent review [27].

Particular aspects are characteristic of Auger spectroscopy. In order to achieve reliable spectroscopic information, e.g., the kinetic energy of the Auger electrons has to be measured with high absolute accuracy. Hence, the calibration of the energy scale as well as accuracy, and long-term stability are essential problems. Therefore, emphasis is given to these topics in the following discussion.

12.3.2 Gas targets and vacuum conditions

For typical gas target measurements, the scattering chamber is operated under high vacuum conditions yielding a base pressure of typically less then 10^{-6} mbar (without external gas inlet). Electron spectroscopy studies are generally performed under single collision conditions. Gas jet or gas cell targets have both been used in past.

The gas jet is usually a simple tube between 0.2-1 mm diameter operated at distances of a few mm from the ion beam. A gas jet is typically oriented perpendicular to the scattering plane defined by the beam and the spectrometer. A more directed gas beam can be achieved by using a bundle of narrower tubes. A common solution is to use a small piece of a microchannel plate, where the diameter of the individual tubes is a few tenth of microns. The advantage of a more directed gas beam is that the target is more localized. Furthermore, a better ratio of the target-region pressure to the base pressure can be achieved by the same pumping power. With such gas jets, pressures of a few 10^{-3} Torr are typically obtained in the target region. Gas jets are simple and compact, leaving space around the collision region. However, they must be properly aligned even for relative measurements. The main disadvantage of them is that neither the value of the average target pressure nor the inhomogeneous density distribution of the target can readily be measured.

To obtain higher target pressures, differentially pumped gas cells may be used. Gas cell targets are better suited for absolute cross section measurements than gas jets, since the target density is constant in the collision region. Moreover, the target pressure may be directly measured. However, for electron DDCS measurements the cell should be prepared with many holes allowing the electrons to enter the spectrometer at different angles. It may be complicated to construct an adequate arrangement for differential pumping. Another problem is the scattering of both the ion beam and emitted electrons in the target gas within the cell. The latter causes not only a loss in the counting due to the electron absorption by the targer gas but also a scattering into other angular channels, distorting the measured angular distributions [13]. Furthermore, already small scattering on the walls or slits of the gas cell may easily produce a large number of stray electrons.

The gas pressure in the scattering center must be limited so that the electron detectors are not affected. Moreover, care must be taken to establish single-collision conditions. For instance, to preserve the incident charge state the projectile should not undergo charge-changing collisions. The probability W that a particle traveling through a gas region of length l and pressure P suffers a collision is [25]

$$W = \sigma \, l \, P \, c_v$$

where σ is the relevant total cross section and $c_v = 2.4 \; 10^{16}$ 1/cm^{3}mbar is the constant converting the pressure of mbar into number of particles per cm^3. With a path length of 0.3 cm and a pressure of 4 10^{-3} mbar, typical for a gas jet, it follows that W is negligibly small ($\leq 10^{-2}$) if the cross section does not exceed the values of 10^{-15} cm^2 [25].

When the gas pressure is sufficiently low to fulfill the single collision condition, a large beam current can, in principle, produce space charging by ionizing a significant portion of the target gas along its trajectory [27]. This effect is more pronounced for heavy ion impact. Space charge yields line shifting and broadening, which makes it especially unpleasant in Auger spectroscopy. Unavoidable beam fluctuations may further increase broadening. To eliminate the effect, the beam current must be kept at a sufficiently low value, where the line positions are stable.

12.3.3 Electron spectrometers

For the energy analysis of electrons various types of electrostatic spectrometers are used. Generally, the theory of the electrostatic spectrometers is relatively complex so that only general aspects are discussed here. For more details the reader is referred to the articles by Rudd [13], Roy and Carette [86], Dahl [87], Wannberg *et al.* [88], Afanasjev and Javor [89], Granneman and Van der Wiel [90], Leckey [91], and Roy and Tremblay [92].

Fig. 12.4 shows different types of energy dispersive electrostatic spectrometers utilized in ion-atom collision experiments. Each spectrometer consists essentially of two plates producing a well defined electric field. The notations of the spectrometers refer primarily to the shape of the plates. There are the parallel plate spectrometer [93,94], the cylindrical plate spectrometer [95], the toroidal spectrometer [96–99], and its two simpler and widely used versions, namely the cylindrical mirror analyzer (CMA) [100,101], and the concentric hemispherical analyzer (CHA) [33] or its sector version, the spherical plate spectrometer [102]. It should be noted here that our aim is to provide brief information about the spectrometers most often used in ion-atom collision experiments. Different names for both the spectrometers and their parts can be found in the literature. Many authors prefer to use e.g., the word 'electrode' instead of 'plate' or 'plain mirror' instead of 'parallel plate'. For a more systematic classification, the reader is referred to Refs. [89,91].

It is rather common to make a distinction between spectrometer and analyzer in the literature. Usually, the analyzer is considered as the energy dispersive element of the spectrometer, which latter contains all the elements from the first aperture to the detector including accelerator/decelerator lenses. In this sense, Fig. (12.4) shows analyzers. The distinction, however, is not always unique. In the following we talk about spectrometers, but no acceleration/deceleration lenses are considered in the formulas. However, when the term 'analyzer' is included in a commonly used abreviation, e.g., CMA or CHA, both terms are used with the same meaning.

The deflection angle α_D is a characteristic parameter of any spectrometer. In Fig.(12.5) slit geometries and corresponding electron trajectories are shown for a parallel plate spectrometer with a deflection angle $\alpha_D = 60°$. (It is common to refer to this spectrometer by its entrance angle $\alpha_0 = 30°$). It is noted, however, that the following discussion deals with general aspects so that it applies to other spectrometers, too.

Electrons from the source enter the electric field through the entrance slit. After appropriate deflection the electrons leave the field by passing the exit slit. The slit denoted 'object' defines the beam of electrons before passing the entrance slit. Another defining slit denoted 'image' is located at the image of the object. The object and image may coincide with the entrance and exit slit, respectively. Let us denote the widths of the four slits by w_o, w_{ent}, w_{ex}, and w_i. The corresponding heights of the slits are denoted by h_o, h_{ent}, etc. On a plane perpendicular to the electron trajectories the projected effective width

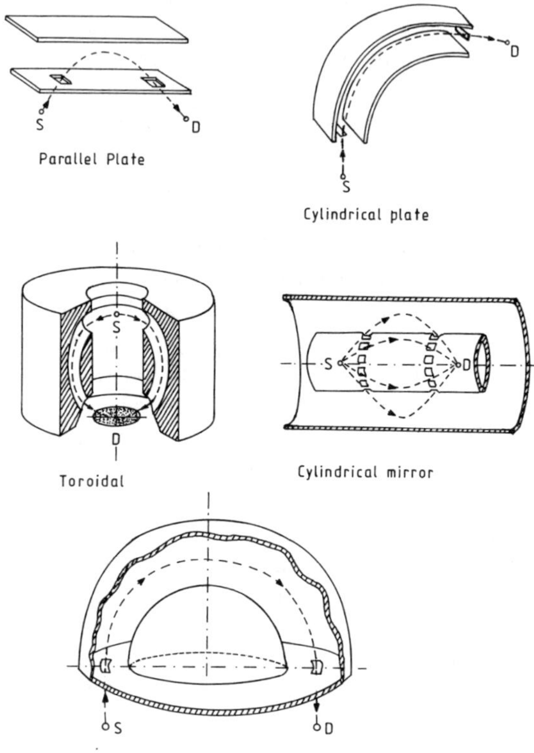

FIGURE 12.4 Different types of electrostatic electron spectrometers.

of the slits is smaller. It is useful to give an independent notation s_o, s_{ent}, etc., to the projected slit widths. In the specific case of $\alpha_D = 60°$, $s_x = \sin(30°)w_x$, i.e., $s_x = w_x/2$.

In the following, the angle measured in the plane of the electron trajectory (i.e., in the graphical plane of Fig. 12.5), α will be referred to as polar angle. The angle measured in the direction perpendicular to the plane of the electron trajectory, β will be referred to as azimuthal angle. To avoid conflicts, the notation of the polar and azimuthal angles relevant for the electron spectrometers (α, β) have been chosen to be different from those relevant for the collision system (Θ, Φ).

Let L be the distance between the crossing points of the electron trajectories with the object and image slits. For electrons originating from a point source the object slit determines the polar acceptance angle $\Delta\alpha_o = s_o/d_{so}$ where d_{so} is the source - object slit distance. One may define a different polar angle $\Delta\alpha_s$ from the opposite direction. The maximum value of $\Delta\alpha_s$ is the angular range

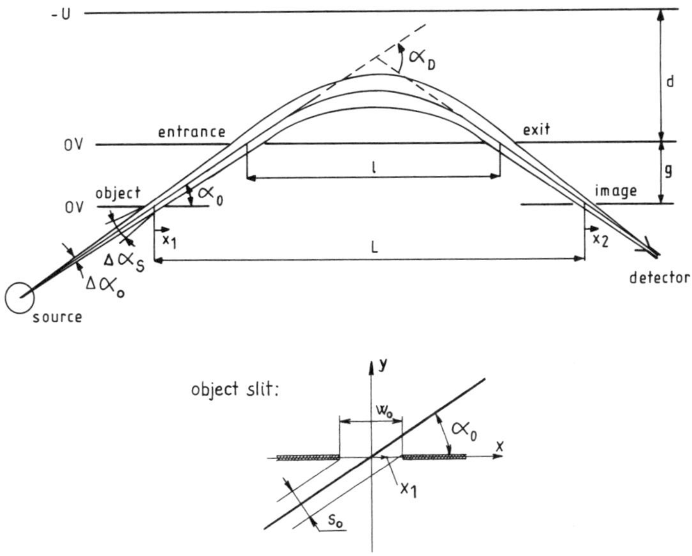

FIGURE 12.5 Geometry of a parallel plate spectrometer ($\alpha_D = 60°$).

of all the possible trajectories passing through the spectrometer, considered at the object slit. Hence, this maximum $\Delta\alpha_s$ determines the maximum source size in the polar direction. Electrons emitted outside this angular region cannot be counted. If, however, the source size is smaller, $\Delta\alpha_s$ is determined by the source size.

The maximum possible divergence of the electron beam with respect to the polar entrance angle α_0 is $\Delta\alpha = \Delta\alpha_o/2 + \Delta\alpha_s/2$. Let the analogous azimuthal quantities for the electrons (in the direction perpendicular to the electron trajectory plane) be $\Delta\beta_o$, and $\Delta\beta_s$, and the maximum azimuthal divergence $\Delta\beta$.

Now, it is possible to introduce some characteristic quantities of the spectrometer. The solid angle is obtained as $\Delta\Omega = \Delta\alpha_o\Delta\beta_o$. It is determined exclusively by the position of the (point) source and the geometry of the object slit of the spectrometer. Let us recall from Eq. (12.8) that, at a given position of a point source, the spectrometer between the object slit and the image slit is characterized by its transmission t. The product of the spectrometer transmission and the solid angle $\Delta\omega(\mathbf{r}) = \Delta\Omega(\mathbf{r})\ t(\mathbf{r})$ characteristic to a point source in an arbitrary position $\mathbf{r}$ can be considered as an effective solid angle.

The effective solid angle $\Delta\omega(\mathbf{r}_0)$ characteristic of a point source in its *central* position is often referred to as *brightness*. Varying the position of the point source in the vinicity of its central place, $\Delta\omega$ becomes smaller. The accepted area of the source can be estimated as $(d_{so}\Delta\beta_s + h_o)(d_{so}\Delta\alpha_s + w_o)$. Averaging over this area, a mean effective solid angle $\overline{\Delta\omega}$ can be determined ($\overline{\Delta\omega} < \Delta\omega(\mathbf{r}_0)$). For extended sources, the measure of the efficiency of the spectrometer is the *luminosity* which is equal to the product of the mean effective solid angle and the accepted area of the source.

In ion-atom collision experiments, the source area is the product of the length l and the diameter of the beam seen by the spectrometer. In most cases, the beam diameter does not appear explicitly in Eq. (12.9), but it is implied in the beam current $J = N/t$. The beam current usually increases when the beam diameter increases. Hence, the luminosity of the spectrometer enters into Eq.(12.9) through the quantities $(l\Delta\Omega)_{eff}$ and N. This shows that for a given resolution the spectrometer with large luminosity is favored, as a high count rate is desirable.

Both the solid angle and the accepted source area are determined by the focusing power of the spectrometer. The most important focusing parameters and the energy resolution of a spectrometer are determined by the function $x_2(x_1, \alpha, \beta, E)$ which gives the position of the electron trajectory at the image slit, as a function of its position and direction at the object slit and the kinetic energy of the electron. The explicit expression of x_2 is called the basic equation of the analyzer (see e.g., [13]). The Taylor expansion of x_2 up to first order in energy and up to the first nonvanishing angular terms has a unique form for electrostatic spectrometers [13,82]

$$x_2 = M\,x_1 + D\,\frac{E - E_0}{E_0} + P\,(\alpha - \alpha_0)^n + A\,(\beta - \beta_0)^m, \tag{12.10}$$

where $m, n \geq 2$. The quantities with index '0' are the nominal values belonging to the central electron trajectory ($x_1 = x_2 = 0$). The nominal energy of the spectrometer E_0 is determined by the electrode potential U and geometric factors. The geometric ratio $c = U/E_0$ is denoted as *spectrometer constant.*

The expansion parameters in Eq. (12.10) have special meanings. For spectrometers without acceleration or deceleration between the object and image slits, the *magnification* is $M = \pm 1$. The increment of $|x_2|$ resulting from the variation of the electron energy E characterizes the *dispersion* of the spectrometer, D. It has a dimension of length. For the parallel plate spectrometer in Fig. (12.5), $D = l$. The dispersion should be large to obtain small ΔE, i.e. good resolution for a given image slit size w_i.

The angular difference $|\alpha - \alpha_0|$ attributed to the increment of $|x_2|$ characterizes the focusing power of the spectrometer in the direction of the polar angle. Let us recall here that $|\alpha - \alpha_0| \leq \Delta\alpha$. The focusing power is significant when the increment of $|x_2|$ is small for large values of $\Delta\alpha$. Since $\Delta\alpha \ll 1$, good focusing can be obtained when the low power terms cancel in the Taylor expansion. With $n = 2$ in Eq. (12.10), only the first term is cancelled and the spectrometer is said to focus in first order.

The derivatives in the Taylor expansion depend on the geometric parameters of the spectrometer. In general it follows that the number of derivatives which may be cancelled is equal to the number of adjustable geometric parameters. In the example given in Fig. (12.5) there are two such parameters, i.e. the deflection angle α_D and the distance between the planes of the object and entrance slit g. With a suitable choice ($\alpha_D = 60°$, $g = \sqrt{3}/6\, l$) it is possible to cancel the first two terms, i.e. the spectrometer focuses in second order ($n = 3$). In spectrometers where the object and the image slit coincide with the entrance and the exit slit, respectively, the distance g is fixed to zero. In this case only the parameter α_D is adjustable. Therefore, only one term may be cancelled and, consequently, the spectrometer focuses in first order only.

The polar aberration term P in Eq. (12.10) shows how much the first nonvanishing term in $\Delta\alpha$ shifts the position of the trajectory at the image slit. Hence, the (polar) focusing power is characterized by n and P.

In addition to the focusing with respect to the polar angle the spectrometer may have the capability to focus in azimuthal direction. Accordingly, these spectrometers are denoted as doubly focusing instruments. The toroidal, the cylindrical mirror spectrometer and the concentric hemispherical analyzer focus in the azimuthal direction in any order. This follows from the fact that the plates of these spectrometers are curved with respect to the azimuthal angle. Likewise, the spectrometers whose plates are flat with respect to that angle do not focus in the azimuthal direction. They are denoted as singly focusing instruments. In Eq. (12.10) the azimuthal aberration is accounted by A and m. It is noted here that for spectrometers without azimuthal focusing $m = 2$, while for those focusing in any order in the azimuthal direction, $A = 0$.

By varying the kinetic energy of the emitted electrons while E_0 is kept constant and counting the detected electrons, one gets the transmission function characteristic of the spectrometer. It usually has a triangular or trapezoidal form distorted by the angular aberrations. The energy resolution of the spectrometer ΔE is defined as the FWHM of the transmission function. The accurate determination of ΔE is rather complicated. Instead, the base width of the transmission function ΔE_B is calculated [89]

$$\frac{\Delta E_B}{E_0} = \frac{w_o + w_i}{D} + c_P(\Delta\alpha)^n + c_A(\Delta\beta)^m, \tag{12.11}$$

where the constants c_P and c_A are the polar and azimuthal aberration coefficients, respectively. It is usually a good approximation to take the energy resolution equal to the half of the base width, $\Delta E = \Delta E_B/2$ [89]. The important message of Eq. (12.11) is that the quantity which is determined by the geometry of the spectrometer is not the absolute but the relative energy resolution. This is generally true for energy dispersive electrostatic spectrometers.

There are some other important operational requirements of most analyzers, namely the fringe field plates. For reasons of clarity, they are not shown in Figs. (12.4) and (12.5). Obviously, it is not practical to construct analyzers that are much larger in size than the distance between the slits. However, the electrons passing trough the analyzer must see only uniform fields to avoid severe aberrations. Therefore, the field of the analyzer must be terminated at each end to avoid field distortion in the aperture region. The termination can be realized by a series of electrodes (e.g. narrow plates for the parallel plate, ring shaped electrodes for the cylindrical mirror, etc.) whose potentials are set by dividing resistors to ensure a smooth transition between the analyzer electrode potentials. Another solution for the termination is the use of ceramic surfaces coated with resistor films.

However, fringe fields are produced by the slits too. The larger the ratio of the slit size to the distance of the analyzer electrodes, more strongly the electrons are deflected by fringe fields around the slits. When characterizing a spectrometer, it is important to quantitatively know this effect. Fringe field distortions of this kind can be significantly reduced by covering the slits with high transmission metal grids. Of course, fringe fields and electron scattering on the grids and the edges of the slits decrease the maximum value of the transmission.

Spectrometers for target Auger spectroscopy

Various aspects have to be considered when a choice is made between the various types of spectrometers. Favorable features of the spectrometers are high resolution, efficiency, and simplicity of design. High resolution is required when individual Auger lines are to be measured. Natural line width and separation of adjacent lines are of the order of 0.1 eV to 1 eV. Accordingly, resolutions $\Delta E/E$ of 10^{-4} to 10^{-3} are required using, e.g., 1000 eV electrons. In principle such resolutions may be realized by reduction of the spectrometer slits or by deceleration of the electrons. The latter method takes advantage of the fact that $\Delta E/E$ is usually constant so that the reduction of E decreases ΔE. The deceleration method is advantageous, since the loss in the spectrometer efficiency is usually lower than that originating from the reduction of the slit sizes. Furthermore, the deceleration method allows variation in ΔE during the measurements.

The simplicity of the spectrometer design is related to disturbance effects involved in the electron measurements. Significant experimental problems are produced by stray electrons reaching the detector and by spurious electric and magnetic fields. For instance, low energy electrons are significantly affected by the earth magnetic field. For high-resolution measurements it is required to reduce the earth magnetic field by a factor of about 100. This can be done by Helmholz compensation coils [13] or by Mu metal shielding vessel (see Fig.12.3) placed inside the scattering chambers [13,103].

At a given target size and desired luminosity, the resolution of an electrostatic spectrometer can be improved by increasing its size. Spurious electric fields and slit-edge effects can be significantly reduced in this way, too. Unfortunately, longer electron trajectories make the arrangement more sensitive to remaining magnetic fields. A better magnetic shielding in a larger chamber, however, increases the cost of the apparatus significantly.

Similarly, spurious electric fields must be avoided. They are generally produced by electrons collected at insulating surfaces (e.g. contaminated by finger prints). Such fields are critical near the target region as they may produce energy shifts of the electrons of some eV. Spectrometers, which can be placed outside the target region, combined with gas-jet targets minimize such effects. Different temperature-dependent contact potentials of the materials used in the scattering chamber and the spectrometer shift the line positions. The best solution to avoid these effects is to coat all the metal surfaces with a slide of gold or graphite.

The most serious problem for electron spectroscopy in ion-atom collisions is stray electrons producing background signals. Electrons are similar to diffusing gas atoms in finding their way through tiny holes (Actually, they are worse since they tend to multiply—see the electron multiplier). Openings are always present in order that the spectrometer can be evacuated. Hence, considerable effort has to be made to prevent spurious electrons from reaching the detector. Experience shows that it is easier to control the spurious effects when instruments of simpler design are used.

A significant source of stray electrons is the backplate scattering. All the electrons from the target with energies larger then the actual pass energy of the spectrometer hit the negative voltage plate (see Fig. (12.5)). They are not only scattered but multiplied. The electric field of the analyzer accelerates these electrons in the direction of the base plate. Many of them may arrive at the detector producing background counts. To avoid this problem, one may use a grid instead of a solid backplate. The backplate can be coated by materials of low secondary emission (e.g., graphite). Collimators and a repulsive electric field can be used in front of the detector. The best solution, however, is the use of a double pass or tandem spectrometer (see e.g. Fig. (12.7)). In this case, the electron beam is analyzed by two stages of the spectrometer consecutively. Stray electrons produced by backplate scattering in the first analyzer will enter but not pass the second stage, since their energies and angles are different from those of the signal electrons. Tandem spectrometers have other advantages, too, relevant in high resolution spectroscopy. They provide more degrees of freedom to find the solution for high luminosity and high resolution at a specific target geometry.

Let us consider now the spectrometers most often used in high resolution Auger studies at the beamlines of accelerators.

Parallel plate The single focusing parallel plate spectrometer has been widely used in many fields in atomic collision physics including target Auger spectroscopy (see e.g., [23,24,51,75,103,104]). Its dispersion is sufficiently high to achieve good energy resolution. Focusing in first order as well as second order has been realized for this spectrometer. The first order focusing type (with $\alpha_D = 90°$) has an object slit located at the entrance slit [93]. The parallel plate spectrometer that focuses in second order has separate object and entrance slits and $\alpha_D = 60°$ [94]. The luminosity of the parallel plate spectrometer is small for point-like sources but rather good for extended sources. Fringe field distortions are relatively small. Source and detector positions well outside the spectrometer are advantageous. Compact tandem versions are also widely used (see e.g. [105]). Deceleration or acceleration lenses can be conveniently fitted to the spectrometer. The main advantages are the simplicity in construction, compact arrangement, and small size. This makes it easily applicable in angular distribution measurements.

In order to illustrate the resolving power of a parallel plate spectrometer ($\alpha_D = 90°$) used with deceleration, a few diagram lines of the Ar L-Auger spectrum are shown in Fig. 12.6 [106]. The data were taken by electron impact on a gas-jet target. The Auger electrons were decelerated down to 3 eV pass energy. The half width of the spectrometer function was measured to be 90 ± 15 meV.

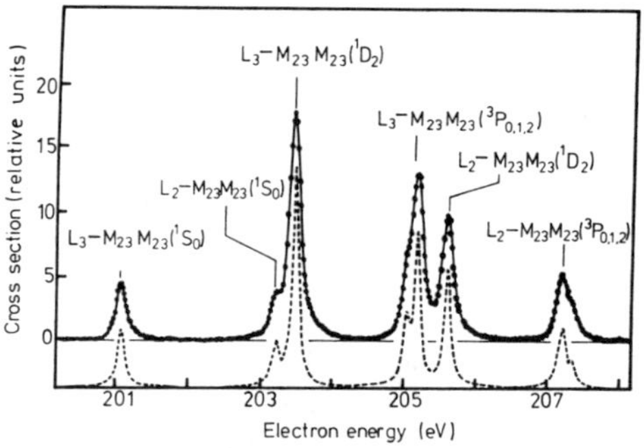

FIGURE 12.6 Part of the Ar L-MM Auger spectrum excited by 2.5 keV electrons. Full curve: result of computer fit; broken curve: deconvoluted spectrum. Reprinted from D. Ridder, J. Dieringer and N. Stolterfoht, "Fine-structure splitting of the L_2 $^3P_{0,1,2}$ Auger lines in argon" [106], with kind permission of IOP Publishing Limited.

From Fig. 12.6, it is clearly seen that the finite energy resolution broadens the natural line widths only slightly. The fine structure splitting of the multiplets can be studied quantitatively. One should note, however, that it was necessary to use an electron beam current of 1 mA to achieve good statistics.

Cylindrical plate The cylindrical plate spectrometer is a single focusing instrument with a first order focus at $\alpha_D = 127°$. It has been used in atomic collision experiments since the early sixties (see the review of Rudd and Macek [13] and references therein). Its main characteristics are similar to the parallel plate but the construction is a bit more complicated, the size is larger, the luminosity is slightly smaller, and fringe field distortions are stronger.

Concentric hemispherical analyzer (CHA) Studies performed by means of CHA and its sector versions have been reported by many authors in studies concerning target Auger spectroscopy (see e.g., refs. [26,48,52,107]). With $\alpha_D = 180°$, it is doubly focusing with first-order focus in the polar direction, and any-order focus in the azimuthal direction. In the sector versions of CHA ($\alpha_D < 180°$,), second order focus can also be obtained. For extended sources, it is possible to achieve high resolution and high luminosity simultaneously. The construction is rather complicated. Commercial versions are also available. Source and detector positions are outside the analyzer, therefore it is well suited to the use of acceleration or deceleration lenses and for angular distribution measurements.

Cylindrical mirror analyzer (CMA) The solid angle of the spectrometer increases considerably when a larger azimuthal angular range is accepted. This is realized by the cylindrical mirror analyzer [108]. Moreover, CMA has a remarkably large polar acceptance angle, too. Among the frequently used spectrometers CMA has the largest value of brightness at a given resolution. Design and construction is relatively simple. A special feature of CMA is that it has several focal conditions. It can focus e.g., from point to point and from point to ring in both first and second order. In the most often used point-to-point double focusing arrangement, $\alpha_D = 84°$. Because of its large solid angle the spectrometer is widely used in the field of electron spectroscopy for chemical analysis (ESCA) [37].

The application of CMA in collision experiments is not simple. The target region is localized inside the inner cylinder. It is difficult to use high density gas targets there since the diffusing gas may not readily be pumped away. Preacceleration or deceleration is difficult to perform. If the ion beam is directed along the axis of the spectrometer, providing the largest solid angle, the electron observation angle relative to the incident beam is fixed to $\alpha_0 = 42°$ and cannot be varied. Part of the difficulties were removed by Toburen [109] by using a CMA constructed from half cylinders and locating it relatively far away from the target region. Woods *et al.* [53] rotated the axis of a CMA by 132° with respect to the ion beam axis so that electrons ejected from 90° to 174° in the laboratory frame were analyzed. In this way they achieved a high luminosity at backward angles. This is especially advantageous when the total Auger production is to be measured, since the continuous electron background is relatively small in this angular range.

New approaches were made by Varga *et al.* [110] and Kövér *et al.* [111,112]. Varga *et al.* [110] constructed a distorted field version of CMA. By calculating the electron trajectories in different systems, they found that a box-type termination of the cylinders (see Fig. (12.7a) provides good focusing properties and energy resolution with object and image focal points outside the box. In the application, Kövér *et al.* [111,112] directed the ion beam through the target region of the distorted-field tandem CMA on such a way, that the ion beam could pass one of the entrance slits for the electrons. This arrangement was similar to that of Woods *et al.* [53] but the aim was different. Evidently, it was necessary to make a hole on the outer cylinder of the CMA to allow the beam to enter the Faraday cup. In such a way the different electron trajectories in the CMA belong to different polar emission angles relative to the beam direction. Hence, within a limited region an angular distribution measurement can be performed *by rotating only the electron detector* which is mounted on an angle defining cylinder. Typically, gas-jet targets have been used in the experiments. The schematic view of the spectrometer and its possible positions relative to the beam are shown in Fig (12.7).

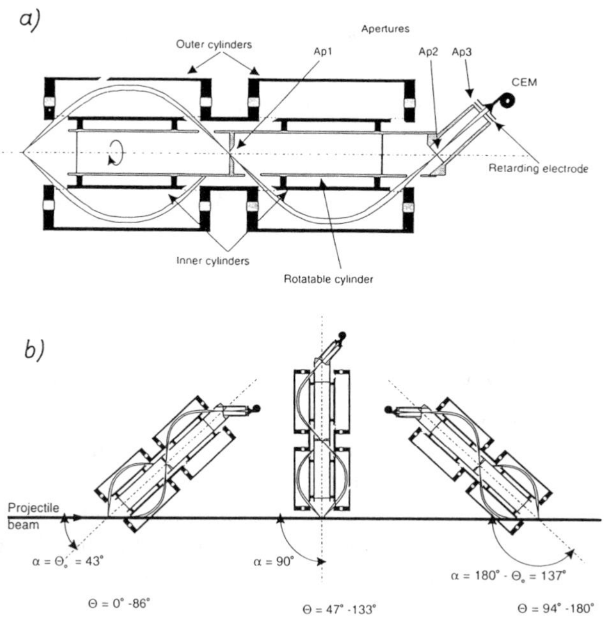

FIGURE 12.7 Schematic view of a distorted field tandem CMA spectrometer (a) and its possible positions for angular distribution measurements (b). Reprinted from Á. Kövér, D. Varga, I. Cserny, E. Szmola, Gy. Mórik, L. Gulyás and K. Tokési, "A distorted field cylindrical mirror electron analyzer, 2. Performances and application for studying ion - atom collisions" [112], with kind permission of Elsevier Science - NL, Sara Burgerhartstraat 25, 1055 KV Amsterdam, The Netherlands.

The arrangement in Fig. (12.7) has many advantages. It is possible, e.g., to measure angular distributions in fine steps, even around 0° and 180°. There are no changes in the target region when the observation angle is changed. Large solid angle and high resolution of the original CMA has been retained. The main disadvantage is that there is no simple way to apply a pre-deceleration lens. Hence, the energy resolution is limited by the (variable) slit sizes of the spectrometer.

Also, when performing angular distribution measurement using the spectrometer shown in Fig (12.7), for varying the observation angle one still needs to change the detector position. An entirely different approach, measuring the total angular distribution simultaneously, was made by Varga *et al.* [54,57,113,114]. They constructed a multidetector instrument which accepts the full range of the polar angle Θ relative to the beam direction. The spectrometer is divided into several angular sectors, each of which is equipped with a channeltron detector allowing for the individual registration of the electrons. Hence, electrons may be measured simultaneously under 13 polar angles relative to the beam direction including 0° and 180°. The schematic diagram and a perspective view of their spectrometer is displayed in Fig (12.8).

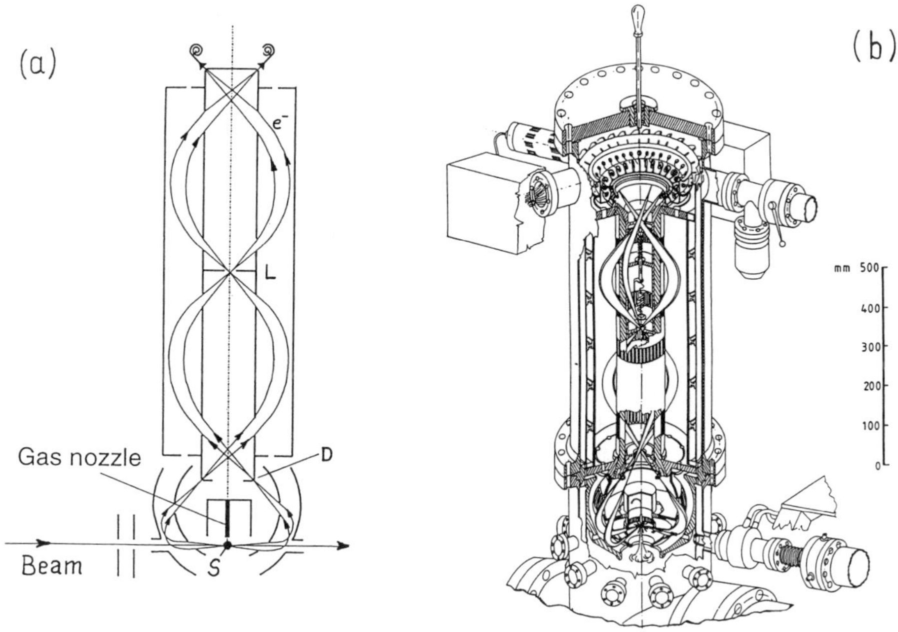

FIGURE 12.8 Schematic cross section (a) and persective view (b) of the spherical + double cylindrical mirror spectrometer for simultaneous angular distribution measurements. Part (b) is reprinted from D. Varga, I. Kádár, S. Ricz, J. Végh, Á. Kövér, B. Sulik and D. Berényi, "A spherical mirror – double cylindrical mirror electron spectrometer for simultaneous energy and angular distribution measurements" [114], with kind permission of Elsevier Science - NL, Sara Burgerhartstraat 25, 1055 KV Amsterdam, The Netherlands.

The spectrometer consists of a spherical mirror and a double-pass CMA. The spherical mirror collects all electrons ejected from the source S in the azimuthal plane and focuses them into a ring D at the entrance of the first CMA. The double-pass CMA focuses ring to ring. On the image ring, 13 channeltron detectors are located in 15° angular steps from 0° to 180°. Between the two CMA stages there is a focal point at L. The spherical mirror has large angular focusing power and a fairly poor (10%) energy resolution. Its role is only to project the source to the focal ring D with proper angular conditions for the CMAs. Resolution and solid angle are determined by the double pass CMA. The source - detector distance is 795 mm. The carefully collimated ion beam with a diameter of 2.5 mm crosses the spherical mirror. Its deflection by the electric field is negligible for ion impact energies larger than 100 keV. A gas-jet target is located above the source point S and directed downwards. The energy resolution can be improved by a spherical ring deceleration lens located around the target. It is constructed with finite-size holes at all the angular channels. No grids are used in the target region. CMA slit sizes can be changed from outside the vacuum chamber. In this way, the relative energy resolution can be varied between 0.05 % and 0.6 % without deceleration. At 0.6 % relative resolution, the solid angle is ≈ 0.15 % per angular channel. We would like to mention here that the spectrum shown in Fig. 12.6 has also been measured by means of the above triple-pass spectrometer with the same instrumental resolution. However, in contrast to the 1 mA, in that case the electron beam current needed to arrive at good statistics in the single angular channels within a few hours was below 1 μA. This difference shows that a large CMA has excellent luminosity.

A few experimental results obtained with the help of the above two spectrometers are presented in the forthcoming sections.

It is recalled here that both CMA and CHA are special cases of the toroidal spectrometer (see Fig. 12.4) [96]. For toroidal spectrometers other than CMA or CHA, however, both theory and design is rather complicated. First applications in e-2e measurements were reported only recently [97–99].

Other spectrometers

There are instruments using methods other than electrostatic field deflection for the energy analysis of the electrons. For instance, instead of electric fields one may apply magnetic fields to deflect the electrons. The magnetic spectrometers are often chosen to analyze high-energy electrons such as those produced by nuclear beta decay [115]. In atomic collision experiments, magnetic spectrometers are utilized when electrons above about 20 keV are to be measured.

Moreover, there are non-dispersive instruments applying the retarding field method (see e.g., ref. [116]). Generally, these spectrometers have huge solid angles. However, the measured curve has to be differentiated to achieve the corresponding electron spectrum. The differentiation creates significant statistical errors so that the advantage of the large solid angle is partially cancelled. A method to achieve dispersion with this spectrometer is the focusing of the near-zero energy electrons which barely have passed the retarding field threshold onto the detector [116]. Unfortunately, in this case it is difficult to avoid problems produced by background electrons.

A further dispersive instrument is the time-of-flight spectrometer [117]. The time-of-flight method, however, requires pulsed incident beams or electron-ion coincidence measurements. The spectrometer itself is relatively simple. It consists essentially of a drift tube. The instrumental effort has to go into the electronic part, which must be able to handle pulses in the subnanosecond region. Since electrons are relatively fast, only energies below about 200 eV have been treated by the time-of-flight method [117]. Pre-deceleration may extend this energy range. The solid angle of the time-of-flight spectrometer is much larger than the electrostatic analyzers. This makes it well suited for coincidence measurements, where high resolution electrostatic analyzers provide low count rates.

12.3.4 Electron detectors

After passing the spectrometer, the energy-analyzed electrons are detected by electron multipliers. Three main type of multipliers are most often used for detecting electrons in the few eV - few keV kinetic energy range.

The first is the traditional, high sensitivity multiplier constructed from discrete dynodes. It provides a high efficiency ($\eta = 0.8 - 0.98$) which is rather stable in a wide energy range. The accepted count rate is a few hundreds of kHz, rather large. An advantage of it is the large sensitive area, while the related disadvantage is its large size. Unfortunately, this detector must be kept continuously in high vacuum to avoid surface damages and efficiency degradation.

In general, the channel electron multiplier (channeltron) has a lower efficiency ($\eta = 0.3 - 0.8$) and a limited count rate (80-100 kHz). Depending on the manufacturer and the technology, their efficiency may vary with the electron energy in different ways. It is usually constant or only slightly changing above 100 eV. It can be kept at open air for long times without significant degradation but not without the change of efficiency. The small size makes it attractive in many applications.

The main deficiency of the channeltrons is not always recognized. It is the dependence of the efficiency on the place of the electron impact within the sensitive area. The less sensitive part is the center, where the electrons fall deep into the tube and the multiplication factor drops by a factor of 2-8. It may happen that a sharp focusing into the center of the detector decreases the count rate by a factor of five [118].

Channel plate (microchannel plate) is also widely used for electron detection. Its efficiency is usually more stable and less energy dependent than that of channeltrons. However, they are not advised to keep in open air for a longer time. Channel plates are able to accept as high (or sometimes higher) count rates as the discrete-dynode multipliers. Their main drawback is the relatively large dark count rate. In some energy-dispersive spectrometers, channel plates are used as position-sensitive detectors located in the focal plane of the spectrometer, in order to take a wider region of the energy spectrum in one step. They are also used for the simultaneous angular distribution measurement in toroidal spectrometers. Since they have extremely good timing properties, channel plates are the most advanced detectors in time of flight spectrometers and in any coincidence measurements.

We would like to emphasize that a signal of a detector can only be considered as a count if it passes the discriminator. All kinds of multipliers are very sensitive on the proper setting of the high voltage. The optimization of the signal-to-noise ratio and the right setting of the discrimination level are also essential to utilize the potential efficiency of the detector. If all these conditions are fulfilled, the determination of the detector efficiency is still a hard task. Different methods (independent current measurement, reference proportional counter, etc.) are referred by Rudd *et al.* [27]. Since the efficiency of multipliers of any art is usually not very stable in time, only absolute calibration is advised for absolute measurements.

12.3.5 Data acquisition and spectrometer control

Electrostatic electron spectrometers, as well as acceleration or deceleration lenses, need well-calibrated, low-noise, accurate high voltage power supplies. In high-resolution Auger studies, the necessary overall accuracy is typically better then 10^{-4}. The setting of the high voltage power supplies is usually controlled by a computer via digital/analog converters. The spectra are taken successively, collecting the same amount of charge by the beam-Faraday cup for every step. Electron signals provided by the discriminator(s) are also counted by the computer. The spectra are usually taken in several consecutive runs to ensure the possibility for reproducibility and reliability checks to be performed later.

The above setup and procedure is generally well known. The reliability test of the measured spectra is usually performed by visual comparison of the different scans. If, however, the cross section of the studied process, i.e., the number of counts in one scan, is small, less approximate methods should be applied. Such mathematical procedure for post-measurement reliably test has been developed and used by Végh *et al.* [85]. Finally, we would like to call attention to the fact that slow power supplies for the varied spectrometer voltages and slow current integrators cannot be used in electron spectroscopy. For obvious reasons, these electronic elements have to be much faster than the possible minimum time interval for taking one energy point of the spectrum.

It is noted here that although accurate power supplies are essential in high-resolution studies, the best way to calibrate the energy scale of the full experimental setup is to measure reference Auger lines with accurately known energies.

12.3.6 Collision kinematics

After the collision, both projectile and recoil ions are moving. Since the Auger process is slow compared to the collision time, the Auger electron always originates from a moving emitter. Such kinematic effects may destroy the available energy and angular resolution. A detailed analysis of the kinematic effects has been performed Rudd and Macek [13]. More recent work is treated by Stolterfoht [9] (see also the references therein) for both target and projectile Auger spectroscopy. Here we recall only the results most relevant to target Auger spectroscopy.

Let us assume that the Rutherford formula is a good approximation for treating the energy transfer between the nuclei of the two colliding atoms. Then, for not too small impact energies, the energy transfer to the target nucleus *at a fixed impact parameter* is inversely proportional to the projectile impact energy, $T_2 = const/T_1$ [25]. Sometimes, this relation is quoted as *kinematic paradoxon* [9]. Most of the atomic collision mechanisms can be associated with a certain impact parameter region. For K-shell ionization, e.g., the average impact parameter is of the order of the K-shell radius (and it may increase slightly with increasing impact velocities).

Kinetic energy transfer to the target atom yields line broadening. As a rough estimation in energetic ion-atom collisions, the velocity distribution of the recoil target atoms can be considered as being isotropic in a plane perpendicular to the beam direction. Accordingly, line broadening originating from the Doppler effect shows a maximum at 90° observation angle and a minimum at 0° and 180°. Heavy ion impact at relatively low energies may completely destroy the resolution and wash out any structures from the Auger spectrum.

Fortunately, due to the kinematic paradoxon, Doppler broadening can be significantly reduced by increasing the impact energy of the heavy ions. The effect is illustrated in Fig.(12.9), comparing Ne K-Auger spectra produced by 500 keV Ne^+ [103] and 30 MeV O^{5+} [119] bombardment. Although the instrumental resolutions are nearly equal for the two spectra, the 30 MeV O^{5+} spectrum exhibits line structures that are totally washed out in the 500 keV Ne^+ spectrum. This shows that high-energy accelerators are very useful for the measurement of electrons from recoil target atoms. Indeed, high resolution Auger spectroscopy studies have been performed using either light ions or a few MeV/u heavy ions.

For heavy ions of a few MeV/u or higher energy range, however, nuclear reaction channels may also be open. Among other problems, nuclear reactions can significantly contribute to the background of the atomic physics measurement. Neutrons and high-energy γ photons can easily provide stray electrons inside or in front of the electron detectors. For strongly collimated beams the majority of the nuclear background signal originates from the first beam-collimator aperture. Often, a negative correlation exists between the first-slit current and the Faraday current. Obviously, such a negative correlation may

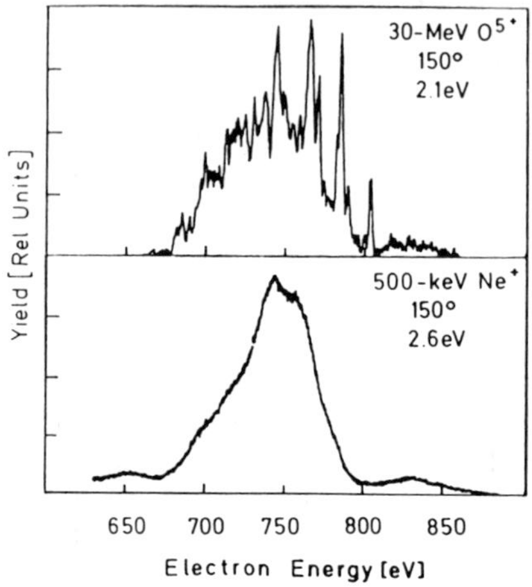

FIGURE 12.9 Ne K-Auger spectra excited by 30 MeV O^{5+} impact [119] and 500 keV Ne^{+} [103] impact. The observation angle is 150° and the energy resolution (FWHM) is indicated in the figure. Reprinted with permission from *Phys. Rev. A* **12** (1975) 1313.

yield much larger errors in the spectrum than a more intensive background component which is proportional to the collected charge. Therefore the first collimator aperture must be located as far away as possible from the scattering chamber. At pulsed beam accelerators, coincidence methods can also be utilized to reduce the nuclear component of the background [85].

12.4 EXPERIMENTAL STUDIES

In this section, a few experimental studies are selected to illustrate the potentials of the methods and provide insight into the tendencies in the field. Three topics are considered.

12.4.1 Multiple ionization

The possibility to remove several electrons in a single collision and to reduce the kinematic broadening effects opened the field of high-resolution Auger spectroscopy of systems with high degree of multiple ionization. The effect of heavier projectile impact is demonstrated in Fig (12.10) [50] where various Ne K-Auger spectra are compared [50,75,119,120]. It is well known in the field of Auger spectroscopy that the mean energy of the satellite lines decreases as the number of L-shell vacancies increases. Therefore, from the spectra in Fig 12.10 it is concluded that the degree of outer-shell vacancy production increases strongly with increasing nuclear charge of the projectile. Moreover, it is seen that for the lightest and heaviest projectiles the spectrum consists of only a limited number of lines. These lines belong to one or two vacancy states.

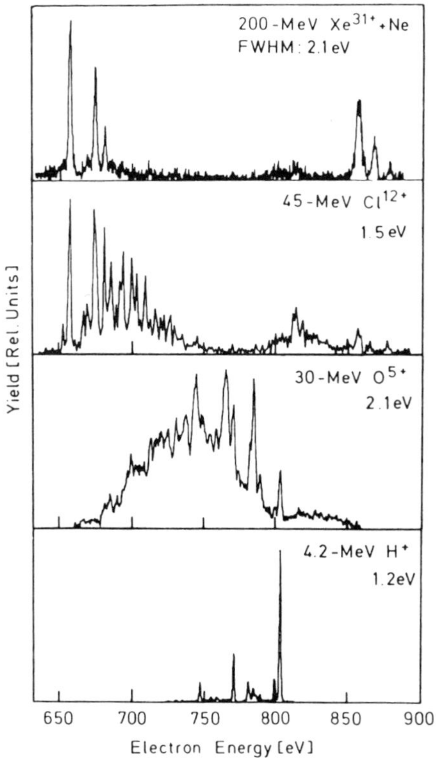

FIGURE 12.10 Comparison of Ne K-Auger spectra produced by 4.2 MeV H^+ [75], 30 MeV O^{5+} [119], 45 MeV Cl^{12+} [120] and 200 MeV Xe^{31+} [50] impact. Reprinted from N. Stolterfoht, D. Schneider, R. Mann and F. Folkmann, "Auger emission for highly ionized neon produced in 200 MeV Xe^{31+} + Ne collisions" [50], with kind permission of IOP Publishing Limited.

Several experimental studies have been devoted to determining the degree of multiple ionization of the Ne L-shell by performing high-resolution Auger electron measurements (see e.g. refs. [9,26,51,54,56,58]). In those experiments, a large variety of heavy ions was used. The analysis was based on the independent particle model (IPM) for multiple ionization. Within the framework of IPM, the vacancy population distribution P_n on the L-shell is binomial [121,122]

$$P_n = \binom{8}{n} p_L^n \,(1 - p_L)^{8-n},$$

where the parameter p_L is the average one-electron ionization probability for the L-shell associated with K-shell ionization. Analogously, a double binomial distribution with two parameters can be used, assuming that the ionization probabilities are different for the 2s and 2p subshells. Within this framework, the determination of two or three members of the distribution is sufficient to derive a p_L value (or p_{2s} , p_{2p} values). Only a few lines in the Auger spectrum, belonging to two or three initial vacancy states, must be identified. Then, with the help of calculated branching ratios (see Eq. (12.4)), the relative populations of two or three initial vacancy states and the corresponding p_L value can be extracted. With this method, a linear connection between the centroid energy of the Ne K-LL Auger group and the average number of L-shell vacancies was established and verified experimentally [51,54].

Obviously, the above method is strongly model-dependent. The reliability of the derived probabilities depends on the applicability of IPM for the given collision system and the quality of the branching ratio calculations. A model-independent experimental determination of the total $2s - 2p$ vacancy distribution can be performed by identifying the majority of the lines in the Auger spectrum. For the neon target, many satellite Auger lines have been observed and identified by Matthews *et al.* [48] in H^+, He^+, and 30 MeV/u O^{5+} + Ne collisions.

As is shown in Fig. (12.10), different degrees of multiple ionization of the same target species can be achieved by varying the charge state of heavy projectiles in fast collisions. If the resolution is sufficiently high, and the degree of multiple ionization is varied in fine steps, a large portion of satellite lines associated with different initial vacancy states can be identified from the interrelated spectra. Such a systematic study was performed by Kádár *et al.* [56,57]. In those experiments, neon gas target was bombarded by 5.5 MeV/u H^+, N^{2+}, N^{7+}, Ne^{3+}, Ne^{10+}, Ar^{6+}, and Ar^{16+} ions at the beamlines of proton and heavy ion cyclotrons. The spectra were collected by the three-stage multidetector spectrometer displayed in Fig. (12.8) with deceleration at a constant 175 eV pass energy. The resolution was 1.05 eV for the 804.5 eV K-$L_{23}L_{32}$ diagram line of neon. Characteristic energy regions of the K-Auger spectra measured in Ar^{16+} -Ne and Ne^{3+} -Ne collisions are shown in Fig. (12.11).

In the line identification procedure calculated Auger transition energies and intensity ratios, experimentally determined target and projectile Auger spectroscopy data for F-, O-, Li-, and Be-like states, and optical spectroscopy data were used as references. Line-groups that changed their intensities together, when going from one projectile to another, were selected and analyzed separately before starting the next iteration of fitting and identification. As a result of this long procedure, 104 lines were identified from the observed 152 [57]. The intensity of the identified lines gives 80-85% of that of the total K-LL spectrum. All the non-identified lines were ordered to a given number of the L-shell vacancies. In such a way, full 2s-2p subshell vacancy distributions have been determined for Ne^{3+}, Ne^{10+},and Ar^{6+} projectiles. The distributions are

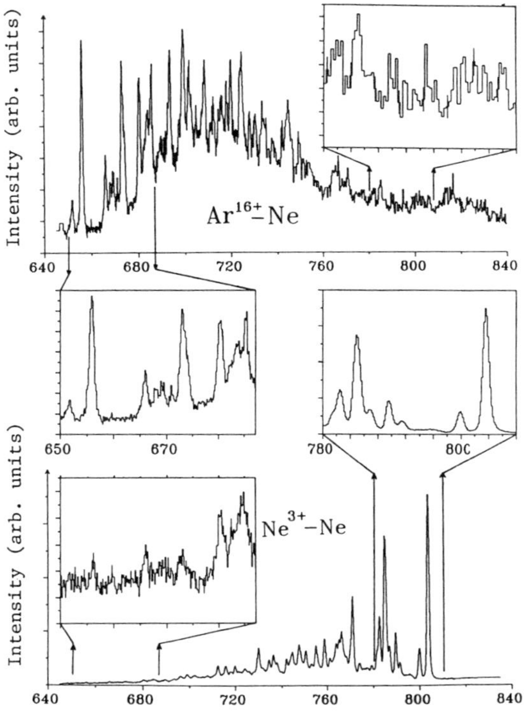

FIGURE 12.11 Selected characteristic energy regions of the K-Auger spectra measured in 5.5 MeV/u Ar^{16+} -Ne and Ne^{3+} -Ne collisions. From [57].

presented in (Fig (12.12)). A two-parameter binomial distribution (shadowed bars) has been fitted to the experimental data (black bars). The typical value of the relative statistical error of the experimental data is 5 % (not shown here).

A close inspection of Fig. 12.12 shows that the fitting is rather good for $Ne^{10+}$and Ar^{6+} but out of the statistical error limits for Ne^{3+}. The applicability of the independent particle model has not been strongly confirmed but no systematic deviations have been found from its predictions. The empty bars display the results of an independent calculation, performed by using a non-perturbative geometric binary encounter model of Sulik *et al.* [58,123]. In general the agreement between experiment and the independent theory is not much worst than between the experimental data and the fitted distributions.

For other projectiles listed above, only parts of the spectra were measured with good statistics. However, they provide sufficient information to extract subshell ionization probabilities within the IPM framework [58]. The data are displayed in Fig. 12.13, as a function of a scaling variable emerging from the

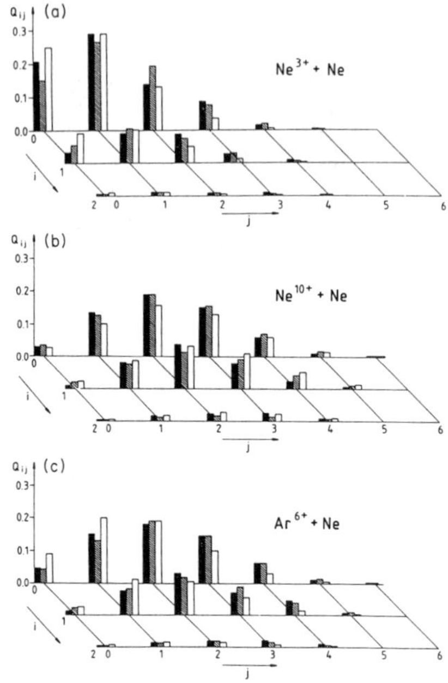

FIGURE 12.12 Experimentally determined vacancy configuration distributions (black bars) compared to the the double binomial distribution fitted to the experimental data (shadowed bars) and to the theory [58] (empty bars). Here i and j are the number of $2s$ and $2p$ vacancies, respectively. From [57].

theory [58,123]. In the present case, the scaling variable is proportional to the effective charge of the projectile. It is clearly seen that both the scaling properties and the absolute values are well-reproduced by the binary encounter model.

12.4.2 Angular distribution studies

Nonstatistical population of the magnetic substates in ion-atom collisions may happen by excitation to a $j > 1/2$ subshell or by creating a vacancy in it [16]. The latter can be performed by ionization, excitation or electron capture. For studying the collision mechanism and the fine details of the atomic/ionic structure, it is useful to start with the simplest systems. Considering single excitation followed by autoionization, the simplest target is Li. It has uniquely s electrons in the ground state $1s^2 2s\ ^2S$. Moreover, a *single* $1s - 2p$ excitation provides excited states that can decay by Auger process.

The angular distribution of the $1s2s2p\ ^2P$ Auger lines of Li excited by H^+ and He^+ impact was studied by Ziem *et al.* [59]. Spectra are taken at different

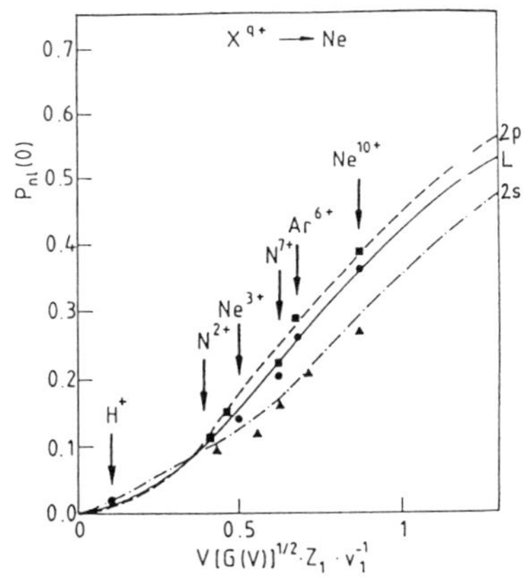

FIGURE 12.13 Experimental and theoretical ionization probabilities for Ne L-subshells produced by the impact of 5.5 MeV/u ions. Reprinted from B. Sulik, I. Kádár, S. Ricz, D. Varga, J. Végh, G. Hock and D. Berényi, "A simple theoretical approach to multiple ionization and its application for 5.1 and 5.5 MeV/u X^{9+} - Ne collisions" [58], with kind permission of Elsevier Science - NL, Sara Burgerhartstraat 25, 1055 KV Amsterdam, The Netherlands.

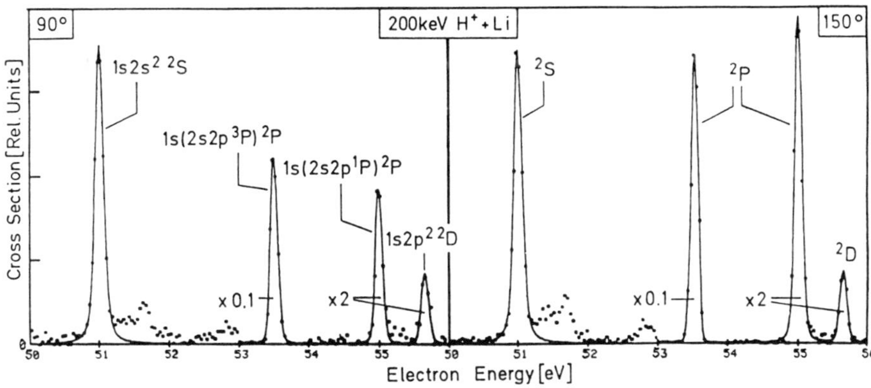

FIGURE 12.14 Lithium autoionization spectra measured at different angles by 200 keV H+ impact. The full curves result from least-square fit of spectrometer functions to the data. Reprinted with permission from [59].

polar angles relative to the beam direction as displayed in Fig. 12.14.

The $1s2s^2\ ^2$S line is isotropic. Therefore it can be used for normalization. Then, the two members of the $1s2s2p\ ^2$P doublet show definite anisotropy. Moreover, their angular dependence is different. For these lines, Eq. (12.7) becomes very simple. There is only one term in the sum ($k = 2$), and $\alpha_2 = 1$

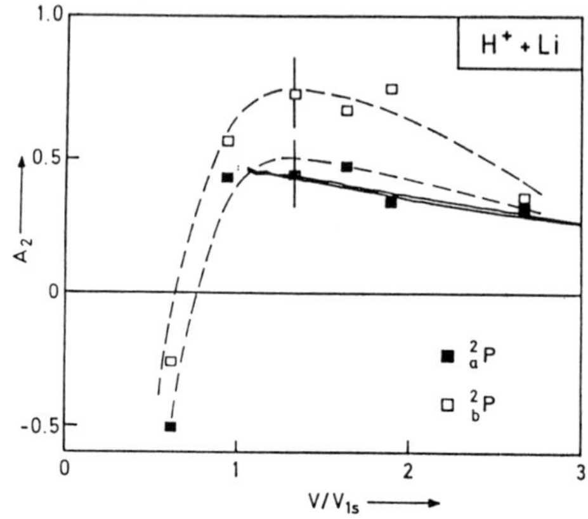

FIGURE 12.15 Alignment parameter A_{20} as a function of the projectile velocity v in units of v_{1s}. Data points are given for the $^2\mathrm{P}_a$ (full symbols) and $^2\mathrm{P}_b$ (open symbols) terms. The full lines are PWBA results multiplied by 0.3. The dashed curves are plotted to guide the eye. Reprinted with permission from [59].

(see e.g., ref. [16]).

$$I(\Theta) = \frac{I_0}{4\pi}\left(1 + A_{20}P_2(\cos\Theta)\right) \tag{12.12}$$

Hence, the alignment parameter A_{20} can be derived directly from the relative intensities. The expression for the alignment parameter in the LS coupling scheme is [16]

$$A_{20} = 2\frac{\sigma(^2P,0) - \sigma(^2P,1)}{\sigma(^2P,0) + 2\sigma(^2P,1)}\,. \tag{12.13}$$

In principle the magnetic substate excitation cross sections σ should be calculated separately for the two members of the doublet, i.e., for $1s(2s2p\ ^3\mathrm{P})\ ^2\mathrm{P}_a$ and $1s(2s2p\ ^1\mathrm{P})\ ^2\mathrm{P}_b$. However, within the LS coupling scheme, the calculated alignment is equal for the two members of the doublet. Ziem *et al.* performed plane wave Born (PWBA) calculations to determine the theoretical alignment parameter. Experiment and theory are compared in Fig. 12.15.

Although PWBA reproduced the tendency of the experimental data at high impact velocities, it failed to give account the absolute values. Moreover, the observed difference between the alignment parameters for the members of the doublet could not be explained within the given theoretical framework. Since double excitation was found to be non-negligible, probably higher-order processes are responsible for the discrepancies. A later experimental study performed at much higher impact velocities (95 MeV/u Ar^{18+} on Li) [124] provided much lower double excitation rates and a common alignment parameter for the two doublet lines in agreement with PWBA calculations.

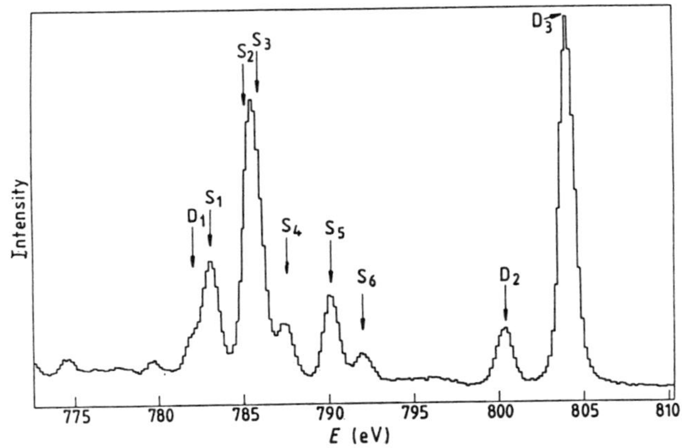

FIGURE 12.16 Ne K-LL Auger lines in the 782-805 eV energy region (5.5 MeV/u Ne^{3+} - Ne collision, 165° obsevation angle). The line denoted by S_4 belongs to the $1s2s^22p^5\ ^1P$ -$1s^22s^22p^3\ ^2P$ transition, while S_5 belongs to that of $1s2s^22p^5\ ^1P$ -$1s^22s^22p^3\ ^2D$. Reprinted from S. Ricz, I. Kádár, V.A. Shchegolev, D. Varga, J. Végh, D. Berényi, G. Hock, and B. Sulik, "Identification and angular distribution of the KL – $LL_{2,3}L_{2,3}$ satellites in the Ne K Auger spectra From the 5.5 MeV u^{-1} Ne^{3+} – Ne collision process" [61], with kind permission of IOP Publishing.

The first evidence of nonstatistical population and anisotropic satellite distribution in multiple ionized Li-like neon were provided by Matthews *et al.* [60] and Stolterfoht [23]. The 7-fold ionization of Ne is a complex collision. Without isotropic reference lines in the spectra, anisotropy parameters have not been determined.

Concerning nonstatistical vacancy production, the simplest case is to create one vacancy on a closed L-shell. The lowest Z element with closed L-shell is neon. For Ne, however, one gets an Auger transition at a single L-shell vacancy only if the K-shell is ionized. Hence, an anisotropic angular distribution of Auger lines is expected in the case of double ionization with one vacancy on the K-shell and another on the L_3 subshell. In a systematic study, Ricz *et al.* [61,62] measured the angular distribution of the K-LL satellite lines of neon doubly ionized by 5.5 MeV/u H^+, N^{2+}, Ne^{3+}, Ne^{10+} and Ar^{6+} ion impact investigating the projectile charge dependence of the alignment at a constant impact velocity. The projectile velocity dependence of the same process was studied by Takács *et al.* [125] in 0.7-2.1 MeV/u H^+ + Ne collisions. In both cases, the measurements were performed by means of the triple-pass multichannel spectrometer shown in Fig. 12.8. A typical spectrum containing the oxigen-like satellite region is shown in Fig. 12.16. As an example, the angular distribution of the lines denoted by S4 and S5 are plotted in Fig. 12.17. The solid curves are the fitted distributions.

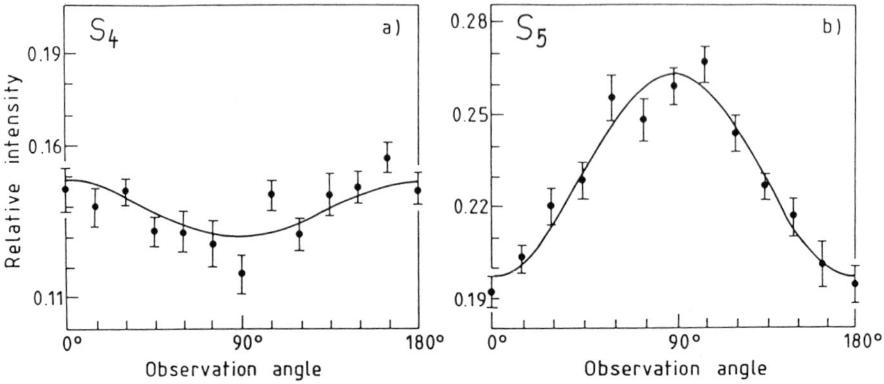

FIGURE 12.17 Angular distribution of the $1s2s^2 2p^5\ ^1P$ -$1s^2 2s^2 2p^3\ ^2P$ (S_4) and the $1s2s^2 2p^5\ ^1P$ -$1s^2 2s^2 2p^3\ ^2D$ (S_5) Ne Auger transitions. (5.5 MeV/u Ne^{3+} - Ne collision) The solid curve is the fitted angular distribution. Reprinted from S. Ricz, I. Kádár, V.A. Shchegolev, D. Varga, J. Végh, D. Berényi, G. Hock, and B. Sulik, "Identification and angular distribution of the KL – $LL_{2,3}L_{2,3}$ satellites in the Ne K Auger spectra From the 5.5 MeV u^{-1} Ne^{3+} – Ne collision process" [61], with kind permission of IOP Publishing.

An interesting point here is that different Auger lines have different α_k values. They can have even opposite sign. If two lines originate from the same initial state, i.e., the alignment parameter is strictly common for the two transitions, then the ratio of the two α_k values is a *measured* quantity. Calculated α_k values are very sensitive on the theoretical methods if more than one partial wave contributes to the transition [10,16]. Therefore, the measured angular distributions provide important information for atomic structure and Auger theories independently of the particular collision process. A detailed analysis of coherence and correlation in the Ne K-LL Auger decay, based on the above experimental data [61,62,125], has recently been performed by Kabachnik *et al.* [10].

Like for the excitation of Li, first order theories within the LS coupling scheme provide one common alignment parameter A_{20} for the L_3 subshell in the ionization case, too. Hence, one may expect that by means of realistic α_k values *all* the measured Auger lines provide the same alignment parameter. For most of the investigated lines, Ricz *et al.* [61,62] confirmed this within the experimental error. It supports the conclusion that the LS coupling scheme is applicable for K+L vacancy states of neon.

Another important feature of the K+L-shell double ionization is that the characteristic impact-parameter region is of the order of the K-shell radius and small compared to that of the L-shell. Therefore, L-shell ionization is governed by small impact parameter collisions. Hence, in Eq. (12.13) one

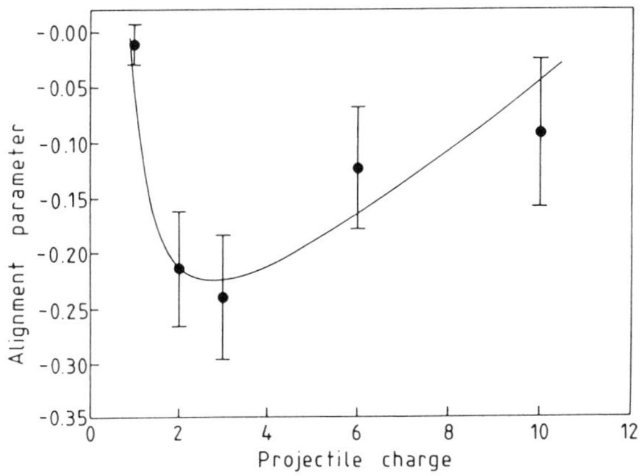

FIGURE 12.18 Alignment parameter versus projectile ion charge for the $1s2s^22p^5\ ^{1,3}P$ state of Ne in 5.5 MeV/u H^+, N^{2+}, Ne^{3+}, Ar^{6+} and Ne^{10+} - Ne collisions. The solid line is drawn to guide the eye. Reprinted from S. Ricz, J. Végh, I. Kádár, B. Sulik, D. Varga and D. Berényi, "Fast multicharged ion-induced satellite Auger electron angular distributions" [62], with kind permission of Elsevier Science - NL, Sara Burgerhartstraat 25, 1055 KV Amsterdam, The Netherlands..

may replace cross sections by ionization probabilities. The method can be considered as a differential measurement providing more information about fine details of the collision mechanism than total cross section studies do.

The results of the systematic study of Ricz *et al.* [61,62] are summarized in Fig. 12.18. A strong dependence of the alignment on the projectile charge has been found. First order semiclassical calculations predict a common value of the alignment parameter, namely -0.3 for all projectiles (see also Fig. 12.19) [125]. This is in fairly good agreement with the N^{2+} and Ne^{3+} data. For H^+ projectile, the contribution of the entirely isotropic shake-off mechanism is not negligible. When taking it into account, A_{20} is reduced to -0.07 for H^+ [125]. It is still significantly larger than the experimental value. At higher charge states, higher-order effects could be responsible for the decreasing absolute value of the alignment parameter.

The experimental point for H^+ in Fig. 12.18 is the same as that displayed by a full symbol in Fig. 12.19, where the projectile velocity dependence is displayed in proton-neon collision [125]. In Fig. 12.19 it is clearly seen that using realistic atomic wave functions and taking into account the contribution of the isotropic shake-off ionization mechanism, a good agreement has been achieved between theory and experiment. The only exception is the 5.5 MeV point where theory provides much stronger alignment than the experiment.

For heavier targets, a vacancy in the closed L shell can decay by L-MM, L-MN, etc. Auger transitions. Hence the alignment of *single ionized* atoms

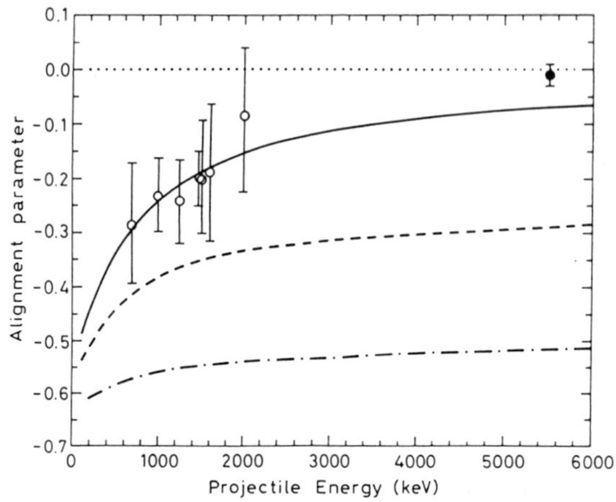

FIGURE 12.19 The projectile energy dependence of the alignment parameter for the $1s2s^22p^5\ ^{1,3}P$ state of Ne in H^+- Ne collisions. Dot-dashed line, the results of the calculations within the framework of the straight-line semiclassical approximation with hydrogen-like wave functions; dashed line, the same calculation with Hartree-Slater wave functions; solid line, Hartree-Slater results combined with the shake-off contribution. From ref. [125].

can also be studied by Auger spectroscopy. It is noted here that for heavier elements, this single L-vacancy alignment is usually studied by X-ray spectroscopy. Since here the total cross sections enter into Eq. (12.13), the alignment parameter provides information about average quantities. In the high impact velocity region, convenient for high resolution Auger studies, the anisotropy is usually small. If, however, different vacancy production mechanisms are distinguished, the partial alignment parameters can be very large. This is demonstrated in Fig. 12.20 [63,64].

The alignment of the L_3 subshell induced by electron capture for proton impact has been studied by Sarkadi *et al* [63] for Ar target, and by Gutenkunst *et al* [64] for Mg target in wide impact energy regions. Both experiments were based on the measurement of the angular distribution of the L-MM Auger electrons (diagram lines, see Fig. 12.6) in coincidence with the outgoing neutral H atoms.

In both cases, strong and velocity dependent alignment has been found for the capture process. The total alignment measured without coincidence is nearly zero in the investigated impact energy region, in good agreement with PWBA-based effective alignment calculations [63]. It is clearly seen that neither of the displayed capture theories agree with the experimental data quantitatively. Only the Oppenheimer-Brinkmann-Kramers (OBK) approximation reproduces the tendencies for both collision systems and provides a qualitative agreement with the Mg data. Fig. 12.20 clearly demonstrates how

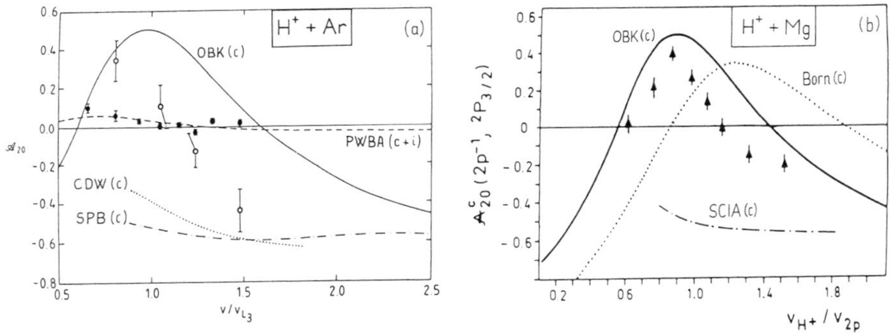

FIGURE 12.20 L3-subshell alignment parameters for proton bombardement as a function of the relative collision velocity: **a)** for Ar target [63], open circles: results obtained by the coincidence detection of the Auger electrons with the outgoing H° atoms; full circles: results of non-coincidence measurements; lines: theoretical calculations. **b)** for Mg target [64], symbols: corrected coincidence data for the electron capture process; lines: theoretical calculations. Part **a** is reprinted from L. Sarkadi, T. Vajnai, J. Pálinkás, Á Kövér, J. Végh and T. Mukoyama, "L_3-subshell alignment of Ar following charge-changing collisions with protons" [63], with kind permission from IOP Publishing Limited. Part **b** is reprinted from A. Gutenkunst, S. Zuccati and W. Mehlhorn, "L_3 – subshell alignment of magnesium after electron capture by protons" [64], with kind permission from IOP Publishing Limited.

important it is to perform differential experiments to really test theories.

12.4.3 Post-collision interaction

In a general sense, post collision interaction (PCI) is a time-dependent Coulomb interaction between the collision fragments produced in an atomic collision. Concerning the Auger process, the manifestations of PCI are line shift, line-shape distortion and line-intensity variation. Since the first observation of the phenomenon by Barker and Berry [44] in slow ion-atom collisions, it has been studied extensively both experimentally and theoretically (see e.g., the reviews [67,126] and references therein). PCI can be significant for the majority of atomic collision systems, like photon, electron, or ion impact. Detailed information about the post-collision interaction can be extracted from high-resolution Auger spectra by analyzing the shape of the individual lines.

Post collision interaction has two important aspects. The basic research aspect is that the study of the PCI effect provides valuable data relevant to three and many-body collision theories. The practical aspect is connected to the precise determination of the position and the intensity of the Auger lines.

Accurately known PCI line-shapes can improve the evaluation procedure and help to extract reliable spectroscopic information from the measured spectra.

For an outgoing heavy charged projectile, the PCI-distorted Auger line-shape can be well treated within the framework of a semiclassical theory [127–129]. A fully quantum mechanical continuum distorted wave (CDW) treatment of the PCI effect in ion-atom collisions has been developed by Barrachina and Macek [130]. For fast collisions, the two theories predict very similar dependence of the Auger line-shape on the impact energy and the Auger electron emission angle in good agreement with experiment. It is noted here that the CDW theory predicts not only line shape but also intensity variation due to the focusing Coulomb attraction exerted by the projectile on the Auger electron ejected into the forward direction. Conclusive experimental evidence and theoretical explanation of the Coulomb focusing effect in slow He+ - He collisions was presented by Swenson *et al* [131]. Experimental indications of intensity variations in the region where the velocity of the Auger electron $\mathbf{v}_a$ and that of the outgoing projectile $\mathbf{v}_p$ are nearly equal were recently reported for Ne target [67].

If the projectile-target separation is large at the moment of the Auger decay and the projectile-electron relative velocity $\mathbf{v}_p - \mathbf{v}_a$ is not too small, the semiclassical theory leads to an approximate analytical expression for the line-shape

$$P(E) = \frac{\Gamma^a_{if}/2\pi}{(E-E_a)^2 + (\Gamma^n/2)^2} \frac{\pi\xi}{\sinh(\pi\xi)} \exp\left(2\xi \arctan\frac{2(E_a - E)}{\Gamma^n}\right). \quad (12.14)$$

Here the Lorenzian line-shape of Eq. (12.6) is distorted by a factor containing the asymmetry parameter ξ. The connection between the asymmetry parameter and the line shift is linear

$$\Delta E = \Gamma^n \xi/2. \quad (12.15)$$

The asymmetry parameter is expressed by the Auger electron velocity and the projectile velocity

$$\xi = \frac{Z_p}{v_p}\left(\frac{v_p}{|\mathbf{v}_p - \mathbf{v}_a|} - 1\right). \quad (12.16)$$

It is clearly seen that the predicted asymmetry depends on the emission angle of the Auger electron. Moreover, there is a singularity at $\mathbf{v}_p = \mathbf{v}_a$, i.e., where the outgoing projectile and the Auger electron leave the collision region in the same direction, with the same velocity. At the singularity, the energy of the Auger electron is equal to the so called cusp electron energy ($E_{cusp} = v_p^2/2$). The above simple formulas are not valid near the singularity, but they correctly show many characteristic features of the PCI effect.

Under ion impact, PCI is a rather complicated effect with strong impact velocity and angular dependence. The number of outgoing charged particles

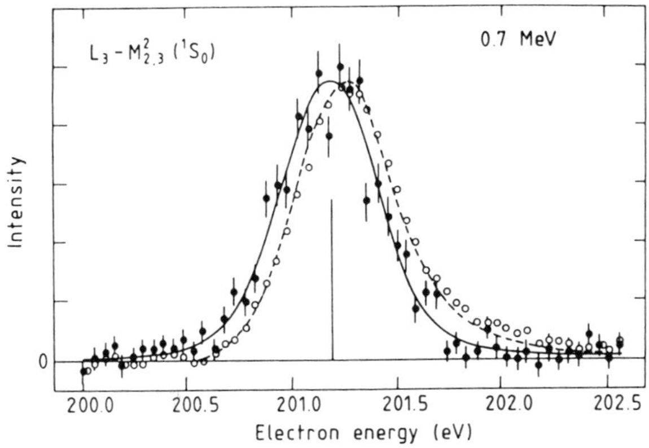

FIGURE 12.21 Electron spectra of the L_3-$M_{2,3}M_{2,3}$ $^1S_0$ Auger line of argon excited by 0.7 MeV protons and observed at 0°. The two spectra have been obtained in the same experiment by detecting the Auger electrons in coincidence with the outgoing H^0 atoms (full symbols) and without the coincidence condition (open symbols). The amplitudes of the peaks are normalized to each other after background substraction. The curves through the experimental points are results of computer fitting based on the lineshape given by Eq. (12.14). Reprinted from L. Sarkadi, T. Vajnai, J. Végh and Á Kövér, "A new method for measurement of post-collision interaction effects for p on Ne collisions" [68], with kind permission of IOP Publishing Limited.

can be large. It was shown by Ricz *et al.* [132] that at high impact velocities where the PCI effect between the outgoing projectile and the Auger electron is negligible, directly ionized electrons still provide a significant PCI distortion of the Auger lines. However, there is a unique situation, namely the electron capture from an inner shell of the target by a singly charged projectile. In this case the Auger electron is the only outgoing charged particle. Accordingly, no post-collision effect is expected. The undistorted Auger line shape and transition energy can be directly measured and compared to those distorted by the PCI effect. The coincidence study of Sarkadi *et al.* [68] clearly demonstrates the effect of PCI for 0.7 MeV proton-argon collision around 0° as shown in Fig. 12.21. The Auger line measured in coincidence with the outgoing neutral projectiles provides a symmetric shape. In contrast, the line measured without coincidence condition, i.e., which belongs predominantly to ionization, is shifted to higher energies and has an asymmetric shape.

It is noted here that there exists an alternative model-independent method to determine the natural line shape and PCI distortions. It is based on the fact that the line shape of a long-living metastable Auger transition is practically not affected by PCI. The outgoing charge fragments are typically far away when the Auger emission occurs. Since the natural width of the metastable states is much smaller than that of the other initial states, these

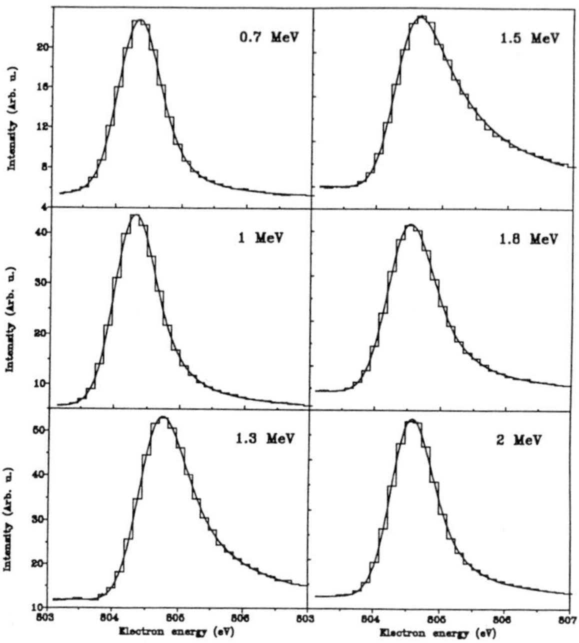

FIGURE 12.22 The Ne K-$L_{2,3}L_{2,3}$ $^1D_2$ Auger line excited by proton bombardment at different impact energies at 0° observation angle. The curves through the experimental points are results of computer fitting with use of the distribution given by Eq. (12.14). Reprinted from E. Takács, L. Sarkadi, S. Ricz, B. Sulik and L. Tóth, "Measurement of post-collision interaction effects for p on Ne collisions" [66], with kind permission of IOP Publishing Limited.

lines can be used for the experimental determination of the spectrometer function [106,132]. Hence, the shape and asymmetry of the other lines in the spectrum can be reliably determined during the fitting procedure.

An experimental study of the PCI effect around the singularity predicted by Eq. (12.16) was performed by Takács *et al.* [66]. The distortion of the Ne K-$L_{2,3}L_{2,3}$ $^1D_2$ diagram Auger line (E_a = 804.5 eV) has been studied by proton bombardment in the energy range of 0.7-2.0 MeV at different polar emission angles by means of the triple pass spectrometer. Since the cross section of the electron capture to continuum (ECC) process was fairly small for the considered collision system, background substraction could easily be done even in the zero degree spectra. The proton energy belonging to the cusp energy of 804.5 eV is 1477 keV. The spectra collected at 0° observation angle are shown in Fig. 12.22 after background substraction. The resonance-like behavior of line-shift and asymmetry around the expected singularity at 1477 keV is clearly demonstrated. It is also seen that the semiclassical theory

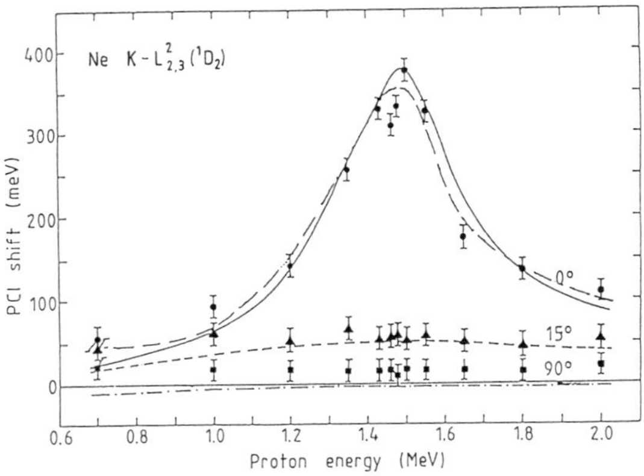

FIGURE 12.23 The dependence of the PCI shift of the K-$L_{2,3}L_{2,3}$ $^1D_2$ Auger line of neon on the impact proton energy at 0°, 15°, and 90° observation angles. The experimental data in the figure have been obtained by fitting the lines with Eq. (12.14) and deriving the line-shift from the asymmetry parameter by the help of Eq. 12.15. The full curve represents the theory of van der Straten and Morgenstern [128]. The broken curve represents the results obtained using the theory of Barrachina and Macek [130]. An integration over the finite solid angle of the spectrometer is included in the theoretical curves. Reproduced from E. Takács, L. Sarkadi, S. Ricz, B. Sulik and L. Tóth, "Measurement of post-collision interaction effects for p on Ne collisions" [66], with kind permission of IOP Publishing Limited.

(Eq.(12.14)) provides an excellent fit (solid line) of the experimental data (step curves).

The results are summarized in Fig. 12.23, showing the dependence of the PCI shift of the K-$L_{2,3}L_{2,3}$ $^1D_2$ Auger line of neon on the proton impact energy at 0°, 15°, and 90° observation angles. It is clearly seen that the resonant like behavior is quantitatively reproduced by both semiclassical and CDW theories at 0° and 15° observation angles. Of course, an integration over the finite solid angle of the spectrometer is included in the theoretical curves. The disagreement at 90° can be identified as PCI effect originating from the directly ionized electrons (see ref. [132]).

The level of our understanding related to the PCI effect in fast ion-atom collisions seems to be rather satisfactory concerning the adequate line-shapes for spectroscopic applications. Questions remain open, however, e.g., in connection with the observed intensity variations at high impact velocities [67] which cannot be reproduced by the present theories.

ACKNOWLEDGMENTS

The authors would like to thank T. Zouros, D. Varga, L. Sarkadi, Á. Kövér, J. Végh and S. Ricz for their critical reading of the manuscript and their valuable comments. This work was supported by the Hungarian National Science Foundation (OTKA, Contract No. T 016636) and the Hungarian-German Intergovernmental ST Cooperation Programme (Contract No. B.129)

REFERENCES

1. P. Auger, Compt. Rend. **177**, 169 (1923); **180**, 65 (1925).
2. P. Auger, J. Phys. Radium **6**, 205 (1925).
3. G. Wentzel, Z. Phys. **43**, 524 (1927).
4. P.G. Burke, in *Advances in Atomic and Molecular Physics*, Vol. 4, eds. D.R. Bates and I. Estermann (Academic Press, New York, 1968) p. 173.
5. E.H.S. Burhop and W.N. Asaad, in *Advances in Atomic and Molecular Physics*, Vol. 8 (Academic Press, New York, 1972) p. 164.
6. D. Chattarji, *The Theory of Auger Transitions* (Academic Press, New York, 1976).
7. E.J. McGuire, in *Atomic lnner-Shell Processes*, Vol. I, ed. B. Crasemann (Academic Press, New York, 1975) p. 293.
8. T. Åberg and G. Howat, in *Handbuch der Physik*, Vol. 31, eds. S. Flgge and W. Mehlhorn (Springer Verlag, Berlin, 1982) p. 469.
9. N. Stolterfoht, Physics Reports, **146**, 317 (1987).
10. N. M. Kabachnik, J. Tulkki, H. Aksela and S. Ricz, Phys. Rev. A **49**, 4653 (1994).
11. G.N. Ogurtsov, Rev. Mod. Phys. **44**, 1 (1972).
12. A. Niehaus, in *Atomic Inner-Shell Physics*, ed. B. Crasemann (Plenum Press, New York, 1985) p. 377.
13. M.E. Rudd and J. Macek, in *Case Studies in Atomic Physics*, Vol. 3, eds. E.W. McDaniel and M.R.C. McDowell (North-Holland, Amsterdam, 1973) p. 47.
14. K.D. Sevier, *Low Energy Electron Spectrometry* (J. Wiley Interscience, New York, 1972).
15. W. Mehlhorn, *Electron Spectrometry of Auger and Autoionizing States: Experiment and Theory*, Lecture Notes, University of Aarhus (1978) unpublished.
16. W. Mehlhorn, in *Atomic Inner-Shell Physics*, ed. B. Crasemann (Plenum Press, New York, 1985) p. 119.
17. W. Mehlhorn, in *X-ray and Inner Shell Processes* (AIP Conf. Proc. 215), eds. T.A. Carlson, M.O. Krause and S.T. Manson (AIP, New York, 1990) p. 465.
18. D.L. Matthews, in *Methods of Experimental Physics*, Vol. 17, ed. P. Richard (Academic Press, New York, 1980) p. 433.
19. D. Berényi, in *Advances in Electronics and Electron Physics*, Vol. 56 (Academic Press, New York, 1981) p. 411.
20. D. Berényi, in *X-Ray Spectroscopy in Atomic and Solid State Physics* NATO-ASI Series Vol. B-187, eds. J. Gomes Ferreira and M.T. Ramos (Plenum Press, New York, 1988) p. 25.
21. N. Stolterfoht, J. El. Spectrosc. Relat. Phenom. **67**, 309 (1993).

22. N. Stolterfoht, Physica Scripta **T51**, 36 (1994).
23. N. Stolterfoht, in *Scientific and Industrial Applications of Small Accelerators*, IEEE Trans. Nucl. Sci. 76CH1175-8 NTS (1976) p. 311.
24. N. Stolterfoht, *Topics in Current Physics*, Vol. 5: *Structure and Collisions of lons and Atoms*, ed. I.A. Sellin (Springer Verlag, Berlin, 1978) p. 155.
25. N. Stolterfoht, in *Fundamental Processes in Energetic Atomic Collisions*, eds. H.0. Lutz, J.S. Briggs and H. Kleinpoppen (Plenum Press, New York, 1983) p. 295.
26. R. Mann, F. Beyer and F. Folkmann, in *XII lntern. Conf. on the Physics of Electronic and Atomic Coliisions, lnvited Papers*, ed. S. Datz (North-Holland, Amsterdam, 1982) p. 683.
27. M.E. Rudd, Y.-K. Kim, D.H. Madison and T.J. Gay, Rev. Mod. Phys. **64**, 441 (1992).
28. J.D. Garcia, R.J. Fortner and T.M. Kavanagh, Rev. Mod. Phys. **45** 111 (1973).
29. Q.C. Kessel and B. Fastrup, in *Case Studies in Atomic Physics*, Vol. 3, eds. E.W. McDaniel and M.C. McDowell (North-Holland, Amsterdam, 1973) p. 137.
30. W.E. Meyerhof and K. Taulbjerg, Ann. Rev. Nucl. Sci. **27**, 279 (1977).
31. J.S. Briggs and J.H. Macek, in *Advances in Atomic, Molecular and Optical Physics*, Vol. 28, eds. D. Bates and B. Bederson (Academic Press, New York, 1990) p. 1.
32. J.H. McGuire, in *Advances in Atomic, Molecular and Optical Physics*, Vol. 29, eds. D. Bates and B. Bederson (Academic Press, New York, 1992) p. 217.
33. K. Siegbahn, C. Nordling, G. Johannson, J. Hedman, P.F. Heden. K. Hamrin, U. Gelius. T. Bergmark, L.O. Werme, R. Manne and Y. Baer, *ESCA Applied to Free Molecules* (North-Holland, Amsterdam, 1968).
34. M.O. Krause, in *Atomic lnner Shell Processes*, Vol. II, ed. B. Crasemann (Academic Press, New York, 1975) p. 33.
35. T.A. Carlson, *Photoelectron and Auger Spectroscopy* (Plenum Publishing Corporation, New York, 1975).
36. K. Siegbahn et al., *ESCA-Atomic, Molecular, and Solid-State Structure by Means of Electron Spectroscopy*, Nova Acta Reg. Soc. Sci. Uppsallienis, Ser. IV (1967) p. 20.
37. J.C. Riviere, *Surface Analytical Techniques* (Clarendon Press, Oxford, 1990).
38. A. Niehaus, Physics Reports **186**, 149 (1990).
39. W. Mehlhorn, Phys. Lett. **21**, 155 (1966).
40. W. Mehlhorn, Z. Physik **160**, 247 (1960).
41. H.W. Berry, Phys. Rev. **121** 1714 (1961).
42. T.A. Carlson and M.O. Krause, Phys. Rev. Lett. **14**, 390 (1965).
43. M.E. Rudd, Phys. Rev. Lett. **13**, 503 (1964); **15**, 580 (1965).
44. B.R. Barker and H.W. Berry, Phys. Rev. **151**, 14 (1966).
45. M.E. Rudd, T. Jorgensen Jr. and D.J. Volz, Phys. Rev. Lett. **16**, 929 (1966); and Phys. Rev. **151**, 28 (1966).
46. Yu.S. Gordeev and G.N. Ogurtsov, Zh. Eksp. Teor. Fiz. **60**, 2051 (1971) [Sov. Phys.-JETP **33** 1105 (1971)].

47. R.K. Cacack, Q.C. Kessel and M.E. Rudd, Phys. Rev. A **2**, 1327 (1970).
48. D.L. Matthews, B.M. Johnson, J.J. Mackey and C.F. Moore, Phys. Rev. Lett. **31**, 1331 (1973).
49. D. Burch, P. Richard and R.L. Blake, Phys. Rev. Lett. **26**, 26 (1971).
50. N. Stolterfoht, D. Schneider, R. Mann and F. Folkmann, J. Phys. B **10**, L281 (1977).
51. D. Schneider, M. Prost, R. DuBois and N. Stolterfoht, Phys. Rev. A **25**, 3102 (1982).
52. R. Mann and F. Folkmann, J. Physique **40**, CI-236 (1979).
53. C. Woods, R.L. Kauffman, K. Jamison, N. Stolterfoht and P. Richard, Phys. Rev. A **12**, 1393 (1975).
54. I. Kádár, S. Ricz, V.A. Shchegolev, B. Sulik, D. Varga, J. Végh, D. Berényi and G. Hock, J. Phys. B **18**, 275 (1985).
55. T. Matsuo, J. Urakawa, A. Yagishita, Y. Awaya, T. Kambara, M. Kase and H. Kumagai, J. Phys. B **16** L239 (1983).
56. I. Kádár, S. Ricz, V.A. Shchegolev, D. Varga, J. Végh, D. Berényi, G. Hock, and B. Sulik, Phys. Lett. A **115**, 439 (1986).
57. I. Kádár, S. Ricz, J. Végh, B. Sulik, D. Varga and D. Berényi, Phys. Rev. A **41**, 3518 (1990).
58. B. Sulik, I. Kádár, S. Ricz, D. Varga, J. Végh, G. Hock and D. Berényi, Nucl. Instr. Meth. B **28**, 509 (1987).
59. P. Ziem, W.H.E. Schwarz, D. Ridder and R. Schilling, Japanese Journal of Applied Physics, **17**, Supplement 17-2, 367 (1978).
60. D.L. Matthews, R.J. Fortner, D. Schneider and C.F. Moore, Phys. Rev. A **14** 1561 (1976).
61. S. Ricz, I. Kádár, V.A. Shchegolev, D. Varga, J. Végh, D. Berényi, G. Hock, and B. Sulik, J. Phys. B **19**, L411 (1986).
62. S. Ricz, J. Végh, I. Kádár, B. Sulik, D. Varga and D. Berényi, Nucl. Instr. Meth. B **61**, 411 (1991).
63. L. Sarkadi, T. Vajnai, J. Pálinkás, Á Kövér, J. Végh and T. Mukoyama, J. Phys. B **23**, 3643 (1990).
64. A. Gutenkunst, S. Zuccati and W. Mehlhorn, J. Phys. B **27**, 533 (1994).
65. P.W. Arcuni, Phys. Rev. A **33**, 105 (1986).
66. E. Takács, L. Sarkadi, S. Ricz, B. Sulik and L. Tóth, J. Phys. B **24**, L381 (1991).
67. S. Ricz, Nucl. Intr. Meth. B **86**, 119 (1994).
68. L. Sarkadi, T. Vajnai, J. Végh and Á Kövér, J. Phys. B **24**, L381 (1991).
69. H. Ishii, Y. Iketaki, T. Watanabe, T. Takayanagi, K. Wakiya and H. Suzuki, Phys. Rev. A **43**, 134 (1991).
70. Birgit Lohmann, J. Phys. B **24**, L249 (1991).
71. Q. Zengh, A.K. Edwards, R.M. Wood and M.A. Mangan, Phys. Rev. A **52**, 3940 (1995).
72. C.D. Caldwell and M.O. Krause, Phys. Rev. A **47**, R759 (1993).
73. D.H. Lee, B.M. Johnson, N.A. Guardala, J.L. Price, G.A. Glass, M.F. Stumborg and K.W. Jones, Nucl. Instr. Meth. B **79**, 261 (1993).

74. H.A. Bethe and R. Jackiw, *Intermediate Quantum Mechanics* (W.A. Benjamin Inc., New York, 1968) p. 135.
75. N. Stolterfoht, H. Gabler and U. Leithauser, Phys. Lett. A **45**, 351 (1973).
76. D. Coster and R.L. Kronig, Physica **2**, 13 (1935).
77. T.A. Carlson and M.O. Krause, Phys. Rev. Lett. **17**, 1079 (1965).
78. M.O. Krause, T.A. Carlson and W.E. Moddeman, J. de Phys. (Paris) **32** C4-139 (1971).
79. V.V. Afrosimov, Yu.S. Gordeev, A.N. Zinoviev, D.H. Rasulov and A.P. Shergin, *IX Intern. Conf. on the Physics of Electronic and Atomic Collisions, Abstracts*, eds. J.S. Risley and R. Geballe (University Press, Seattle, 1975) p. 1068.
80. T. Åberg and J. Utriainen, Phys. Rev. Lett. **22**, 1346 (1969).
81. P. Richard, J. Oltjen, K.A. Jamison, R.L. Kaufmann, C.W. Woods and J.M. Hall, Phys. Lett. **54** A, 169 (1975).
82. C.E. Kuyatt, in *Methods of Experimental Physics*, Vol. 7A, eds. B. Bederson and W.L. Fite (Academic Press, New York, 1968) p. 1.
83. I. Kádár, D. Berényi, S. Ricz, B. Sulik, D. Varga and J. Végh, in *Spectroscopy and Collisions of Few-Electron Ions*, eds. M. Ivascu, V. Florescu and V. Zoran (World Scientific, Singapoore, 1989) p. 434.
84. J.C. Nickel, P.W. Zetner, G. Shen and S. Trajmar, J. Phys. E: Sci. Instrum. **22**, 730 (1989).
85. J. Végh, I. Kádár, S. Ricz, B. Sulik, D. Varga and G. Székely, Nucl. Instr. Meth. A **281**, 605 (1989).
86. D. Roy and J.-D. Carette, Canadian J. Phys. **49**, 2138 (1971).
87. P. Dahl, *Introduction to Electron and Ion Optics*, (Academic Press, New York, 1975).
88. B. Wannberg, U. Gelius and K. Siegbahn, J. Phys. E **7**, 149 (1974).
89. V.P. Afanasev and S.J. Javor, Zh. Tech. Fiz. **45** 1137 (1975).
90. E.H.A. Granneman and M.J. Van der Wiel, in *Handbook of Synchrotron Radiation*, Vol. 1, ed. E.E. Koch (North Holland, Amsterdam, 1983) pp. 369-462.
91. R.C.G. Leckey, J. El. Spectr. Rel. Phenom. **43**, 183 (1987).
92. D. Roy and D. Tremblay, Rep. Prog. Phys. **53**, 1621 (1990).
93. G.A. Harrower, Rev. Sci. Instr. **26**, 850 (1955).
94. T.S. Green and G.A. Proca, Rev. Sci. Instr. **41**, 1409 (1970).
95. A.L. Hughes and V.J. Rojanski, Phys. Rev. **34**, 284 (1929).
96. F. Tofoletto, R.C.G. Leckey and J.D. Ridley, Nucl. Instr. Meth. B **12**, 282 (1985).
97. G. Richmond, G. Bagley and T. Reddish, in *XVIII. Int. Conf. on the Physics of Electronic and Atomic Collisions, Abstracts*, eds. T. Andersen, B. Fastrup, F. Folkmann and H. Knudsen (Aarhus, Denmark, 1993) p. 800.
98. R.W. Van Boeyen and J.F. Williams, in *XIX. Int. Conf. on the Physics of Electronic and Atomic Collisions, Abstracts*, eds. J.B.A. Mitchell, J.W. McConkey and C.E. Brion (Whistler, Canada, 1995) p. 504.
99. K. Yamazaki, A. Danjo, M. Yoshino, H. Tanuma and N. Kobayashi, in *XIX. Int. Conf. cn the Physics of Electronic and Atomic Collisions, Abstracts*, eds.

J.B.A. Mitchell, J.W. McConkey and C.E. Brion (Whistler, Canada, 1995) p. 809.

100. H. Hafner, J.A. Simpson and C.E. Kuyatt, Rev. Sci. Instr. **39**, 33 (1968).
101. J.S. Risley, Rev. Sci. Instr. **43**, 95 (1972).
102. E.M. Purcell, Phys. Rev. **54**, 818 (1938).
103. N. Stolterfoht, D. Schneider, D. Burch, B. Aagaard, E. Bøving and B. Fastrup, Phys. Rev. A **12** 1313 (1975).
104. N. Solterfoht, D. Schneider, D. Burch, H. Wiemann and J.S. Risley, Phys. Rev. Lett. **33**, 59 (1974).
105. A. ltoh, T. Schneider, G. Schiwietz, Z. Roller, H. Platten, G. Nolte, D. Schneider and N. Stolterfoht, J. Phys. B **16**, 3965 (1983).
106. D. Ridder, J. Dieringer and N. Stolterfoht, J. Phys. B **9**, L137 (1976).
107. D.L. Matthews, B.M. Johnson, J.J. Mackey, L.E. Smith, W. Hodge and G.F. Moore, Phys. Rev. A **10**, 1177 (1974).
108. E. Blauth, Z. Phys. **147**, 228 (1957).
109. L.H. Toburen, Phys. Rev. A **3**, 216 (1971).
110. D. Varga, Á. Kövér, L. Kövér and L. Redler, Nucl. Instr. Meth. A **238**, 393 (1985).
111. Á. Kövér, Gy. Szabó, D. Berényi, L. Gulyás, I. Cserny, K-O. Groeneveld, D. Hofmann and M. Burkhard, Phys. Rev. A **34**, 2751 (1986).
112. Á. Kövér, D. Varga, I. Cserny, E. Szmola, Gy. Mórik, L. Gulyás and K. Tokési, Nucl. Instr. Meth. A **373**, 51 (1996).
113. D. Varga, J. Végh, Á. Kövér, S. Ricz and A. Domonyi, Atomki Reports, **23**, 80 (1981).
114. D. Varga, I. Kádár, S. Ricz, J. Végh, Á. Kövér, B. Sulik and D. Berényi, Nucl. Instr. Meth. A **313**, 163 (1992).
115. A. Fahlmann, S. Hangstrom, K. Hamrin, R. Nordberg, C. Nordling and K. Siegbahn, Ark. Fys. **31**, 479 (1966).
116. D.A. Hutchital and J.D. Rigden, *Electron Spectroscopy*, (North Holland, Amsterdam, 1972) p. 79.
117. D.J. Lynch, L.H. Toburen and W.E. Wilson, J. Chem. Phys. **64**, 2616 (1976).
118. Gy. Vikor and S. Ricz, private communication.
119. D. Burch, N. Stolterfoht, D. Schneider, H. Wiemann and J.S. Risley, Nuclear Physics Annual Report 1974 (Univereity of Washington, Seattle, 1975) unpublished.
120. D. Schneider, C.F. Moore and B.M. Johnson, J. Phys. B **9**, L153 (1976).
121. D.F. Burch, in *Proc. lntern. Conf. on lnner-Shell Ionization Phenomena*, eds. R.W. Fink et al., Conf.-720 404 (USAEC, Oak Ridge, 1973).
122. J.M. Hansteen and O.P. Mosebeck, Phys. Rev. Lett. **29**, 1361 (1972).
123. B. Sulik, G. Hock and D. Berényi, J Phys. B **17**, 3239 (1984).
124. N. Stolterfoht, X. Husson, D. Lecler, R. Köhrbrück, B. Skogvall, S. Andriamonje and J.P. Grandin, in *XVII. Int. Conf. on the Physics of Electronic and Atomic Collisions, Abstracts*, eds. I.E. McCharty, W.R. McGillivray and M.C. Standage (Griffith University, Brisbane, Australia, 1991) p. 393.
125. E. Takács, S. Ricz, J. Végh, I. Kádár, J. Pálinkás, B. Sulik, L. Tóth, D.

Berényi and N.M. Kabachnik, Phys. Rev. A **50**, 1197 (1994).
126. A. Russek and W. Mehlhorn, J. Phys. B **19**, 911 (1986).
127. M.Yu. Kuchiev and S.A. Sheinerman, Sov. Phys.-JETP **63**, 986 (1986); Sov. Phys.-USP **32**, 596 (1989).
128. P. van der Straten and R. Morgenstern, J. Phys. B **19**, 1361 (1986).
129. P. van der Straten, R. Morgenstern and A. Niehaus, Z. Phys. D **8**, 35 (1988).
130. R.O. Barrachina and J.H. Macek, J. Phys. B **22**, 2151 (1989).
131. J.K. Swenson, C.C. Havener, N. Stolterfoht, K. Sommer and F.W. Meyer, Phys. Rev. Lett. **63**, 35 (1989).
132. S. Ricz, I. Kádár and J Végh, Nucl. Instr. Meth. B **40/41**, 77 (1989).

CHAPTER 13

Zero-Degree Auger Electron Spectroscopy of Projectile Ions

Theo J.M. Zouros* and Do-Hyung Lee†

The technique of zero-degree Auger projectile spectroscopy (ZAPS) is reviewed with particular emphasis on *high resolution* studies in energetic ion-atom collisions. The basic kinematic principles are discussed, aspects of the experimental apparatus including spectrometer and target cell are described in detail and selected topics of data analysis such as absolute normalization of cross sections to the Binary Encounter peak, and cross section determination of long-lived states are treated. Practical applications such as the precise determination of the projectile energy and the metastable beam fraction are also described. Finally, illustrations are presented from the use of this technique to explore basic collision mechanisms in double-capture, resonant transfer-excitation, electron-electron excitation and ionization, meV Coster-Kronig production and autodetachment of negative ions.

13.1 INTRODUCTION

In the past two decades considerable attention has been paid to the study of ion-atom collisions using highly charged ion projectiles. By limiting the number of electrons on the projectile and by using simple targets such as He or H_2, considerable simplification of the collision system is attained. Thus, the study of such fundamental atomic collision processes as excitation, ionization and capture, becomes considerably more tractable from both the experimental and the theoretical points of view. More recently, interest has focused on high

*) Department of Physics, University of Crete P.O. Box 2208, 71003 Heraklion and Institute of Electronic Structure and Laser, P.O. Box 1527, 71110 Heraklion, Crete, GREECE.
e-mail:tzouros@physics.uch.gr
†) Tennessee, Knoxville, TN 37996 and Oak Ridge National Laboratory, Physics Division, Bldg. 5500, MS-6377, Oak Ridge, TN 37831, USA.
e-mail:dhlee@utkux.utk.edu

resolution studies of projectile ions, which can provide state-selective information with important practical bearing on controlled thermonuclear fusion, astrophysics, and the possible creation of vuv and x-ray lasers [1,2]. Existing theories that have successfully predicted total cross-sections for these processes can now be tested to the next order of sophistication and accuracy by comparing with *state-selective differential* cross section measurements. Furthermore, few-electron systems are expected to provide one of the simplest testing grounds for studying the many-particle problem at large, and in particular, the role played by electron-electron interactions. [3–11]

In general, the lighter the projectile, the fewer the important channels available for ionization, capture or excitation of the projectile. Light projectiles have small fluorescence yields but much larger Auger electron yields, making the use of Auger spectroscopy particularly attractive for investigating low-Z projectile autoionizing states formed in the collision.

Projectile electron spectroscopy has been utilized for many years [12,13], since it enjoys a basic advantage over target electron spectroscopy, in that a large variety of highly charged ions have become increasingly available at collision energies and charge states readily controllable by the experimenter. However, kinematic line broadening effects [14,15], which become particularly large at high collision energies, have limited the usefulness of this technique, to rather low resolution ($> 1\%$) investigations [13]. These broadening effects were somewhat reduced at forward or backward observation angles [16] and studies of He- and Li-like projectile states at intermediate collision energies (~ 50 kev/u) were reported [17,18] at an observation angle of 6.4°. However, substantial reduction of kinematic broadening can be achieved at an observation angle of 0° or 180° with respect to the beam direction. This entails special instrumental provisions at these angles since the beam would have to pass through the spectrometer entrance slit without obstruction and still maintain good resolution. Today, the prevailing method uses a tandem spectrometer in which two analyzers are joined together in series. The first analyser really acts as a deflector, since fairly large slits have to be used to allow for the unobstructed passage of the ion beam, with a further aperture on the back side for the beam to exit. The second analyser with much narrower slits provides the high energy resolution. Resolution is further improved by decelerating electrons prior to energy analysis. Typical 0° spectra are shown in Fig. 13.1.

The first tandem spectrometers reported measurements at 180° for low keV/u collisions [19–21] and at 0° for fast (1-5 MeV/u) [22,23] and intermediate (15-150 keV/u) [24–26] collision energies. It was soon clear that high resolution measurements of unprecedented resolution ($\sim 0.1\%$ or better) were indeed possible at these angles and could provide state-selective information about projectile ions for a great variety of different collision systems and processes. [22,23,27,26,28] Thus, for example, zero-degree studies of basic collision processes such as transfer-excitation [24,29], double electron capture

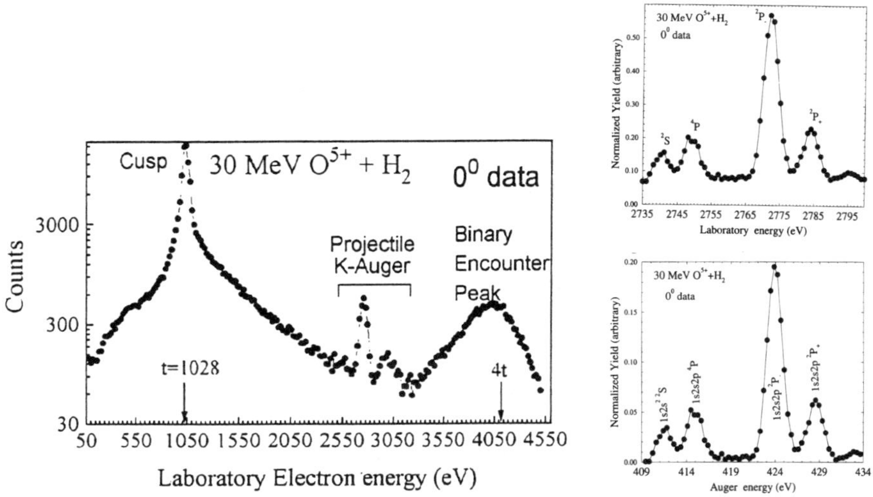

FIGURE 13.1 Typical electron spectra taken at 0° observation angle for 30 MeV O^{5+} collisions with H_2 using a tandem electron spectrometer. Left: The electron energy range from 50-4500 eV is displayed in low resolution showing the basic features of such a spectrum: (a) the Cusp peak at electron energy $t = \frac{1}{2}mV_p^2$, (b) the Oxygen (projectile) K-Auger peaks, (c) the Binary Encounter electron (BEe) peak centered near an energy of $4t$, resulting from target ionization. Right: High resolution spectrum of just the largest projectile K-Auger peak (left) obtained by decelerating the electrons prior to energy analysis to an energy $\varepsilon_{pass} = 200$ eV. Four major lines are revealed that can be associated with excitation. Top: laboratory frame. Bottom: projectile rest frame after background subtraction.

[30–32,25,33,34], resonance-transfer excitation [35–39], e-e excitation [40–42] and e-e ionization [43,8] yielded new state-selective cross section information about collision mechanisms. Studies of atomic structure [44–47], Rydberg states produced in gases or foils [48,27,49,50], low energy Coster-Kronig transitions [51,52], autodetachment of negative ions [53–57], as well as a variety of other collision phenomena [58–62] involving high resolution spectroscopy of charged ions were also very successful. Zero-degree projectile electron spectroscopy studies have been particularly effective at electrostatic accelerators, where the collision energy can be changed easily, thus providing means to study the energy dependence of the collision mechanisms.

Today, the technique of zero-degree Auger projectile spectroscopy (ZAPS) is being used by an increasing number of accelerator facilities in the US, Europe, South America and Japan to investigate properties of highly charged ions at collision energies ranging from a few keV to hundreds of MeV. There have already been a number of reviews and progress reports [48,27,63,26,64–66,7,28,67,68] on high resolution and zero-degree Auger projectile spectroscopy. Here we limit our presentation to details of the technique that are not so well known. In particular, after briefly discussing some of the basic principles of zero-degree projectile electron spectroscopy, we take a more detailed look at various aspects of the experimental set-up. We go on to the determination of *absolute* differential cross sections for prompt and long-lived autoionizing states by normalizing to the Binary Encounter peak. Practical applications such as the determination of the projectile beam energy and the metastable beam fraction are also discussed. Finally, examples from the recent literature are provided, demonstrating both the methodology and the physics that can be studied by applying ZAPS to the investigation of few electron projectiles in collisions with H_2 and He targets.

13.2 PRINCIPLES OF PROJECTILE AUGER ELECTRON SPECTROSCOPY

The basic goal of ZAPS is to provide high resolution electron spectra from which both atomic structure and cross section information can be obtained. To date, ZAPS has been particularly effective in providing state-resolved K-Auger spectra for collisions of low-Z ions ($2 \leq Z \leq 10$) with H_2 and He. To obtain high quality spectra ZAPS takes advantage of the kinematic properties of electrons ejected from moving emitters, as well as the inner-shell vacancy production mechanism of "ion surgery" applicable in fast collisions with light targets. These two aspects are described next.

13.2.1 Kinematics of electrons from moving emitters

Kinematic effects can severely influence the shape, width and energy of the projectile Auger lines. The analysis of such effects [69,14,70,15,71,72,27] is in general complicated and model dependent since the projectile Auger electron is emitted from the *scattered* projectile. However, in energetic collisions of a few MeV/u or larger, of particular interest here, projectile ions are mostly scattered through very small angles (~mrads) with minimum energy loss and negligible effect on the trajectory of the emitted electrons. We may thus assume for simplicity that the projectile ion scattering angle is zero and use simple velocity vector addition, to directly determine the projectile-to-laboratory frame transformations, particularly in the case of zero-degree observation, where the best resolution is possible.

Using the velocity addition diagrams shown in Fig. 13.2, the electron kinetic energy $\varepsilon' = \frac{1}{2}m\mathrm{v}'^2$ in the projectile rest frame is related to the corresponding laboratory kinetic energy $\varepsilon = \frac{1}{2}m\mathrm{v}^2$ by:

$$\varepsilon' = \varepsilon + t - 2\sqrt{\varepsilon t}\cos\theta \tag{13.1a}$$

or equivalently,

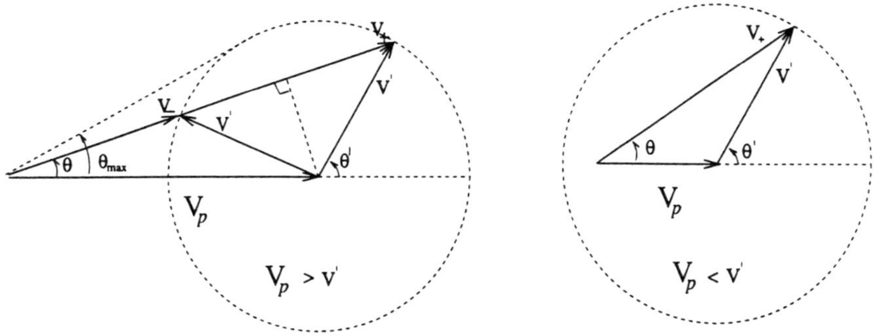

FIGURE 13.2 Projectile electron kinematics. The electron velocity in the laboratory frame $\mathbf{v} = \mathbf{V}_p + \mathbf{v}'$ where $\mathbf{V}_p$ is the projectile velocity and $\mathbf{v}'$ is the electron velocity in the projectile rest frame. The electron ejection angles are θ and θ' in the two frames, respectively. For $\mathrm{V}_p > \mathrm{v}'$ two solutions, $\mathrm{v}_{\pm}$, are possible with $\theta \leq \theta_{max}$. Reprinted from "High resolution studies of two-electron processes observed in energetic ion-atom collisions via zero-degree Auger spectroscopy", T. J. M. Zouros, D. Schneider, and N. Stolterfoht, *Nuc. Instr. Meth.* **B31** (1988) 349, with kind permission from Elsevier Science - NL, Sara Burgerhartstraat 25, 1055 KV Amsterdam, The Netherlands.

$$\varepsilon = \varepsilon' + t + 2\sqrt{\varepsilon' t}\cos\theta' \tag{13.1b}$$

where θ and θ' are the ejection angles of the electron in the laboratory and projectile rest frame, respectively. The quantity t is equal to the energy of an electron moving with the velocity of the projectile, $t = \frac{1}{2}mV_p^2 = \frac{m}{M_p}E_p$, where E_p and M_p are the projectile energy and mass, respectively. It is convenient to introduce the universal dimensionless parameter:

$$\zeta \equiv \sqrt{\frac{t}{\varepsilon'}} \tag{13.2}$$

For fast emitters ($V_p > v'$ or equivalently $\zeta > 1$), as shown in Fig. 13.2, there can be two possible solutions ($\pm$) obtained by solving Eq 13.1a for ε as a function of θ:

$$\varepsilon_{\pm}(\theta) = \varepsilon' \left(\zeta\cos\theta \pm \sqrt{1-\zeta^2\sin^2\theta}\right)^2 \qquad \left(\zeta > 1, 0 \leq \theta \leq \theta_{max} = \sin^{-1}\tfrac{1}{\zeta}\right) \tag{13.3a}$$

$$\varepsilon(\theta) = \varepsilon' \left(\zeta\cos\theta + \sqrt{1-\zeta^2\sin^2\theta}\right)^2 \qquad \left(\zeta \leq 1, 0^\circ \leq \theta \leq 180^\circ\right) \tag{13.3b}$$

Clearly, $\zeta^2\sin^2\theta \leq 1$, results in a lower limit on the electron energies accessible to our spectrometer. However, for $\theta = 0^\circ$, the whole energy range is accessible and we have from Eq. 13.3:

$$\varepsilon_+ = (\sqrt{\varepsilon'} + \sqrt{t})^2 = \varepsilon' (1+\zeta)^2 \qquad (\text{all } \zeta,\ \theta = 0^\circ) \tag{13.4a}$$

$$\varepsilon_- = (\sqrt{\varepsilon'} - \sqrt{t})^2 = \varepsilon' (1-\zeta)^2 \qquad (\zeta > 1,\ \theta = 0^\circ) \tag{13.4b}$$

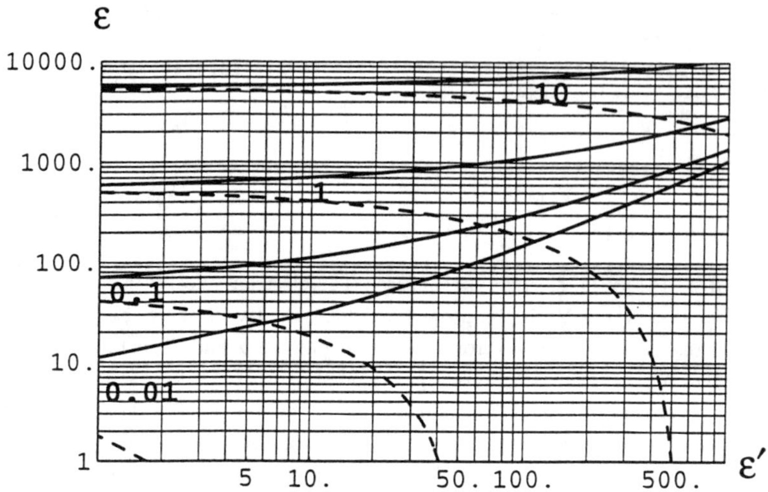

FIGURE 13.3 Relation between rest frame energies ε' (eV) and the laboratory frame ε (eV) at $\theta = 0°$ for $E_p = 0.01, 0.1, 1, 10$ MeV/u. Continuous lines correspond to the (+) solutions, while dashed lines to the (−) solutions of Eq. 13.4.

where the (+) sign refers to the higher and the (−) sign to the lower laboratory electron energy corresponding to rest frame ejection angles $\theta' = 0°$ or $180°$, respectively, in the case of fast emitters. Thus, a doubling of the Auger line is observed in the laboratory frame [27]. In Fig. 13.3 the relation between the rest frame energies ε' and the laboratory frame energy ε is given for four different collision energies and $\theta = 0°$.

Kinematic effects arising from the above relations affecting the detected electron yields and line shapes are discussed next.

Line broadening

Clearly, as indicated in Eqs. 13.1, both the finite acceptance angle $\Delta\theta$ and the uncertainty in the projectile energy, Δt, can give rise to an uncertainty $\Delta\varepsilon$ in the detected laboratory electron energy ε, independent of the specific energy resolution of the spectrometer utilized. This inherent energy uncertainty results in the overall broadening of the Auger lines.

In general, for non-zero ($\theta \neq 0°$) observation angles the dominant effect arises from the finiteness of $\Delta\theta$. However, this so called kinematic line broadening, ΔB_θ, is much reduced when electron observations are performed at $\theta = 0°$. In this case, and depending on the energy resolution of the accelerator utilized, the line broadening ΔB_t arising from the finite uncertainty Δt, can

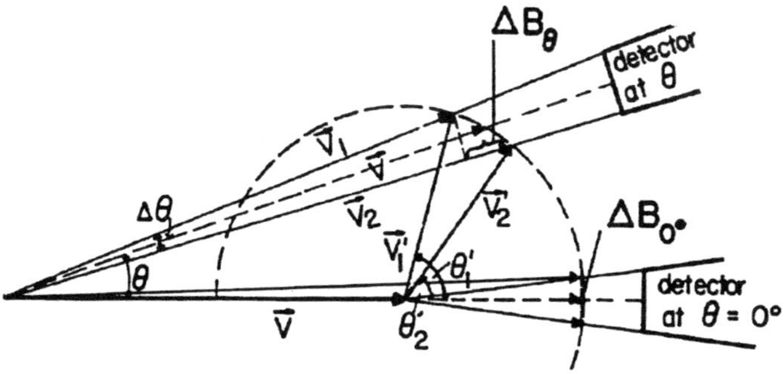

FIGURE 13.4 Geometric demonstration of kinematic line broadening for spectrometer with acceptance angle $\Delta\theta$ placed at the laboratory observation angle θ. Two electrons having identical speeds $v_1' = v_2'$ in the projectile frame but ejected at different angles θ_1' and θ_2' will be detected in the lab within $\Delta\theta$ with different speeds $v_1 \neq v_2$, resulting in a kinematic energy broadening, $\Delta B_\theta \sim |\frac{1}{2}mv_1^2 - \frac{1}{2}mv_2^2|$. At $\theta = 0°$ (or 180°), ΔB_θ is minimized.

become as significant as $\Delta B_{0°}$. In the next two sections we discuss these two broadening effects in more detail.

a. Kinematic line broadening

Electrons detected at non-zero laboratory angles ($\theta \neq 0°$), as demonstrated in Fig. 13.4, can originate from a range of different angles θ' in the projectile rest frame due to the finite acceptance angle $\Delta\theta$ of the electron spectrometer. Thus, two electrons having the *same* energy ε' and *different* ejection angles θ_1' and θ_2', in the projectile rest frame, detected within an acceptance angle $\Delta\theta$, will correspond to *different* laboratory velocities $\vec{v}_1$ and $\vec{v}_2$ with corresponding energies $\varepsilon_1 \neq \varepsilon_2$, as indicated in Fig. 13.4. A net energy broadening ΔB_θ results, with $\Delta B_\theta \sim |\varepsilon_1 - \varepsilon_2|$. However, at $\theta = 0°$ and 180°, as seen by a careful inspection of Fig. 13.4, ΔB_θ is substantially reduced. Furthermore, it is clear that ΔB_θ increases with collision energy.

The kinematic broadening may be computed exactly:

$$\Delta B_\theta^{exact} = \begin{cases} |\varepsilon(\theta + \Delta\theta/2) - \varepsilon(\theta - \Delta\theta/2)| & \text{for} \quad \theta \in (\frac{\Delta\theta}{2}, \pi - \frac{\Delta\theta}{2}) \\ |\varepsilon(\theta + \Delta\theta/2) - \varepsilon(0°)| & \text{for} \quad \theta \in [0, \frac{\Delta\theta}{2}] \; or \; [\pi - \frac{\Delta\theta}{2}, \pi] \end{cases} \tag{13.5}$$

Further insight can be gained by expanding ΔB_θ^{exact} in a Taylor series in powers of $\Delta\theta$:

$$\Delta B_\theta^{exact} \approx \left| \sum_n \frac{\partial^n \varepsilon(\theta)}{\partial \theta^n} \frac{(\Delta\theta)^n}{n!} \right| = \left| \sum_n \Delta B_\theta^{(n)} \right| \qquad (13.6)$$

By differentiating Eq. 13.1a with respect to θ and using Eq. 13.3, the expansion coefficients can be obtained in terms of ε' and θ. Thus, to first order in $\Delta\theta$ we have:

$$\Delta B_\theta^{(1)} = -2\Delta\theta \sin\theta \, \frac{\sqrt{\varepsilon t}}{1 - \sqrt{\frac{t}{\varepsilon}} \cos\theta} \qquad (13.7a)$$

$$= \mp 2\Delta\theta \sin\theta \, \varepsilon' \, \zeta \frac{(\zeta \cos\theta \pm \sqrt{1-\zeta^2 \sin^2\theta})^2}{\sqrt{1-\zeta^2\sin^2\theta}} \qquad (13.7b)$$

which is seen to depend on $\sin\theta$ and therefore $\Delta B_\theta^{(1)} = 0$ at $\theta = 0°$ or $180°$. For $\theta = 0°$ or $180°$ we thus need to calculate the second order broadening contribution $\Delta B^{(2)}$, which is now dominant. For $\theta = 0°$ (remembering to set the effective angular range to $\Delta\theta/2$), we use the above procedure to obtain: [27]

$$\Delta B_{0°}^{(2)} = (\frac{\Delta\theta}{2})^2 \frac{\varepsilon}{|1 - \sqrt{\frac{\varepsilon}{t}}|} \qquad (13.8a)$$

$$= (\frac{\Delta\theta}{2})^2 \, \varepsilon' \, \zeta (1 \pm \zeta)^2 \qquad (13.8b)$$

Typical values of $\Delta B_{0°}^{(2)} \sim 0.5 - 10$ eV for Auger energies $\varepsilon' \sim 100 - 1000$ eV and collision energies, $E_p \sim 1-10$ MeV/u. A rough estimate for $5° < \theta < 170°$ shows that $\Delta B_{0°}^{(2)}/\Delta B_\theta^{(1)} \sim \Delta\theta$. For typical detector acceptance angles $\sim 1°$ ($\Delta\theta \sim 0.02$ radians), $\Delta B_{0°}^{(2)} \sim 100 - 400$ times smaller than $\Delta B_\theta^{(1)}$. Thus, substantial reduction of the kinematic broadening occurs at $\theta = 0°$ where the first-order term vanishes. In Fig. 13.5 (top), the ratio of $R = \Delta B_\theta^{exact}/\Delta B_{0°}^{exact}$, is plotted as a function of ζ and θ. The inaccessible region marked by the white space results from the limitation $\theta \leq \theta_{max}$ of Eq. 13.3a. Large values of R indicate regions of large kinematic broadening. A ridge of maximum broadening is observed around $\theta = 50°$ which spreads to larger angles for decreasing values of ζ. It is reminded that cylindrical mirror analyzers are normally used at 42.3° in target spectroscopy. [13] Clearly this would be a bad choice for high resolution projectile electron spectroscopy. At backward angles near $\theta \sim 180°$, a region of relatively low kinematic broadening is seen to exist, as indicated in Fig. 13.5 (top), which remains largely unexplored by high resolution projectile spectroscopy.

Computing the relative broadening, $R_{\Delta B_\theta^{(2)}}$, we obtain the simple result: [73]

$$R_{\Delta B_\theta^{(2)}} \equiv \frac{\Delta B_\theta^{(2)}}{\varepsilon} = (\frac{\Delta\theta}{2})^2 \, \zeta \qquad (\theta = 0° \text{ or } 180°) \qquad (13.9)$$

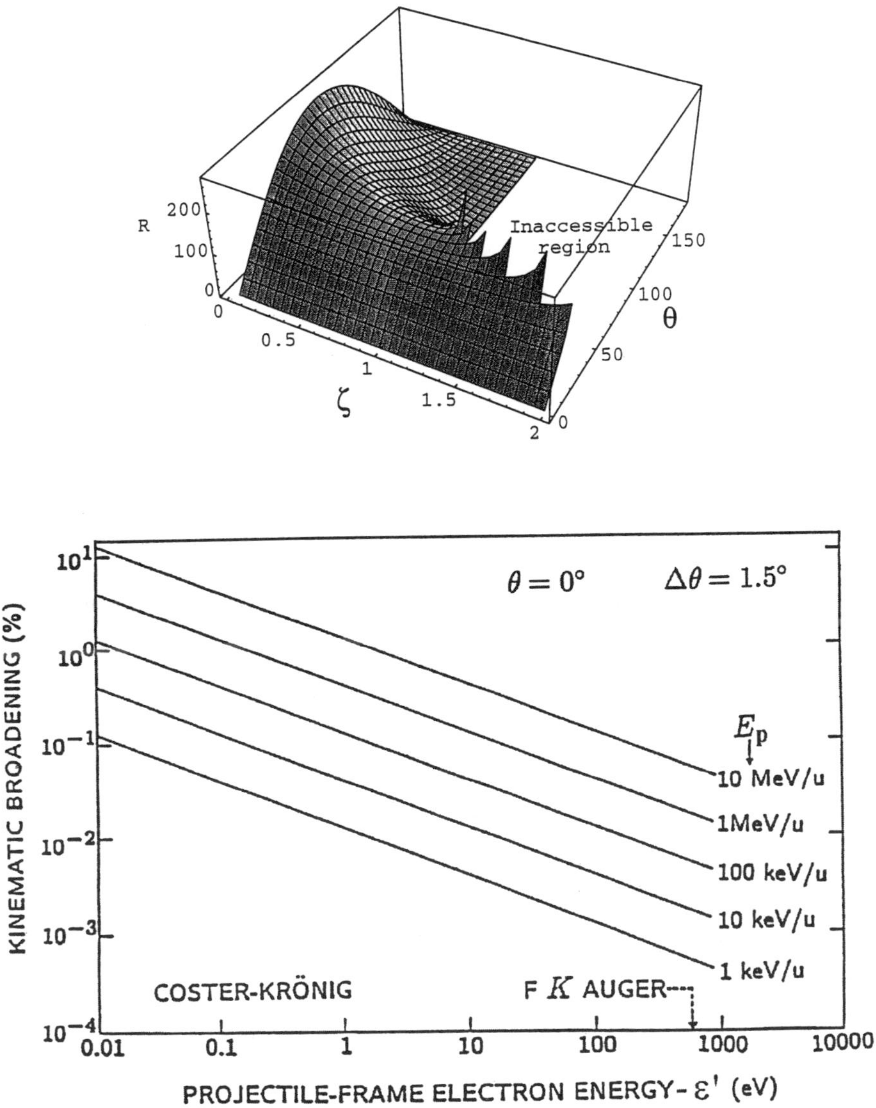

FIGURE 13.5 Top: $R = \Delta B_{\theta}^{exact}/\Delta B_{0°}^{exact}$ (see Eq. 13.5) plotted as a function of $\zeta = \sqrt{t/\varepsilon'}$ and θ (degrees) for an acceptance angle $\Delta\theta = 1.5°$. Bottom: Relative kinematic broadening, $\Delta B_{0°}^{(2)}/\varepsilon$ (see Eq. 13.9) plotted as a function of Auger energy for different projectile energies E_p (adapted from Ref. [73]).

$R_{\Delta B^{(2)}_{0°}}$ is plotted in Fig. 13.5 as a function of the Auger energy ε' for a few characteristic values of the projectile energy. Clearly, ZAPS applications where Auger energies are large and the collision slow, such as in heavier-Z projectiles increasingly available at storage rings, EBIS and ECR sources [30–32,25,33,34] can be expected to have the best resolution.

It is interesting to compute $\Delta B^{(2)}_{\theta}$ for $\theta = 180°$. The accessible energy range is now limited though, since $\zeta < 1$. The following relations hold:

$$\varepsilon = (\sqrt{\varepsilon'} - \sqrt{t})^2 \qquad (\zeta \leq 1,\, \theta = 180°) \tag{13.10a}$$

$$\Delta B^{(2)}_{180°} = (\frac{\Delta\theta}{2})^2 \frac{\varepsilon}{1 + \sqrt{\frac{\varepsilon}{t}}} \tag{13.10b}$$

$$= (\frac{\Delta\theta}{2})^2 \varepsilon' \zeta (1 - \zeta)^2 \tag{13.10c}$$

Comparing the 2nd order broadenings at $\theta = 0°$ and 180° for $\zeta < 1$ we have:

$$\frac{\Delta B^{(2)}_{180°}}{\Delta B^{(2)}_{0°}} = \left(\frac{1 - \zeta}{1 + \zeta}\right)^2 \tag{13.11}$$

indicating that $\Delta B^{(2)}_{180°}$ is always smaller than $\Delta B^{(2)}_{0°}$. Furthermore, since electrons are kinematically shifted to lower energies at this observation angle, smaller deceleration factors can be used entailing smaller losses in transmission. Thus, in principle, good quality measurements should be possible at 180°, with even higher resolution than at 0°. A clear drawback is the energy limitation of $\zeta < 1$ (part of the inaccessible region of Fig. 13.5) making this method impractical for fast collisions ($t > 1000$ eV) unless higher-Z projectiles with larger Auger energies are studied. The few measurements reported at 180°, to date, have indeed been performed at very low collision energies (few keV/u) with good results [19–21].

b. Line broadening due to energy spread of the emitter

The emitter energy-spread broadening ΔB_t due to the variation in t, the reduced projectile energy, can be computed to first order from $\Delta B^{(1)}_t = \Delta t \partial\varepsilon/\partial t$ [27], using Eq. 13.1a. For $\theta = 0°$ observation we obtain:

$$\Delta B^{(1)}_t = \left(\frac{\Delta t}{t}\right) \varepsilon' \zeta \,|1 \pm \zeta| \qquad (\theta = 0°) \tag{13.12}$$

Comparing this to the 2nd order kinematic broadening $\Delta B^{(2)}_{0°}$ given by Eq. 13.8b we obtain:

$$\frac{\Delta B^{(1)}_t}{\Delta B^{(2)}_{0°}} = \frac{\left(\frac{\Delta t}{t}\right)}{(\frac{\Delta\theta}{2})^2} \frac{1}{(\zeta \pm 1)} \qquad (\theta = 0°) \tag{13.13}$$

TABLE 13.1 Calculated line broadenings for 5-33 MeV collisions of F^{6+} ions. Corresponding spectra appear in Fig. 13.25 (left). The K-State zero-degree tandem spectrometer parameters given in Table 13.2 have been used in this calculation together with a beam energy spread $\Delta t/t = 7 \times 10^{-4}$, Auger energy $\varepsilon'_A = 567.8$ eV and width $\Gamma' = 80.5$ meV where $\Gamma = |1+\zeta|\Gamma'$ (see Eq. 13.16). The pass energy ε_{pass} used and the instrumental energy width $\Delta\varepsilon_{ins} = R_{ins}\varepsilon_{pass}$ are also listed. The limiting acceptance angle, $\Delta\theta_{lim}$ (see Eq. 13.15) is seen to be almost double the spectrometer acceptance angle $\Delta\theta = 1.17°$ suggesting that better optimization of spectrometer parameters is possible for the study of this particular Auger line. The computed limiting pass energy ε_{pass}^{lim} (see Eq. 13.19) indicates that resolution can also in principle be improved.

E_p (MeV)	t (eV)	ζ $\sqrt{\frac{t}{\varepsilon'_A}}$	ε_+ (eV)	$\Delta B_{0°}^{(2)}$ (meV)	$\Delta B_t^{(1)}$ (meV)	Γ (meV)	Γ_{tot} (meV)	$\Delta\varepsilon_{ins}$ (meV)	ε_{pass} (eV)	$\Delta\theta_{lim}$ (°)	ε_{pass}^{lim} (eV)
4.75	137	0.492	1263	64.6	291	119	191	2756	100	2.5	7
14.25	411	0.851	1946	172	626	147	357	5513	200	2.2	13
16.63	480	0.919	2092	200	701	153	395	4135	150	2.2	14
19	549	0.983	2233	228	775	158	434	5513	200	2.2	16
26.125	754	1.153	2631	315	986	171	545	5513	200	2.1	20
33.25	960	1.300	3004	406	1189	183	654	5513	200	2.0	24

For tandem accelerators, $\Delta t/t \sim 3-9(\times 10^{-4})$ and is seen to be of the same order as $(\Delta\theta)^2$ for $\Delta\theta \sim 1° - 1.5°$. Hence, ΔB_t may be larger than $\Delta B_{0°}^{(2)}$ for typical ZAPS applications ($\zeta = 1.1 - 2$ for the (+) solution). In Table 13.1 an example of the calculated broadenings is shown.

We can combine the two broadenings and the stretched natural line width Γ (see Eq. 13.16 below) in quadrature [27] to obtain an estimate of the *total* laboratory line width (FWHM), Γ_{tot}:

$$\begin{aligned}\Gamma_{tot} &= \sqrt{(\Delta B_t/2)^2 + (\Delta B_{0°}/2)^2 + \Gamma^2} \\ &\approx \sqrt{(\Delta B_t^{(1)}/2)^2 + (\Delta B_{0°}^{(2)}/2)^2 + \Gamma^2} \qquad (13.14)\end{aligned}$$

assuming ΔB_t and $\Delta B_{0°}$ to relate to the *base* width of the Auger line profile. We note that a general analysis should also include effects due to azimuthal broadening, the variation in the projectile scattering angle and other effects which have been ignored in this analysis since they are usually small in ZAPS applications. More details can be found in Refs. [69,14,70,15,71,72,27] and references therein.

Clearly to obtain experimental resolutions of the order of the natural line width Γ the sum of the main broadenings considered here has to be minimized. The experimental resolution and the kinematic broadening, can be decreased, within limits, by decreasing the dimensions of the analyser apertures, increasing the spectrometer size and/or decreasing the pass energy ε_{pass}. However, as mentioned above, the usual limiting factor is the beam energy spread. On tandem accelerators terminal ripple (typically a few kV per charge state), affects the highest projectile charge states the most. For example, a 35 MeV F^{8+} beam at the K-State tandem has a typical $\Delta t/t \sim 7 \times 10^{-4}$. LINACs using time bunching will be affected even more, typical values being about double those for tandems. Storage rings using electron cooling have smaller values, with $\Delta t/t \sim 0.3 - 1.5 \times 10^{-4}$ depending on beam current and energy. On these machines, effective beam currents are large ($\sim \mu$A), making them particularly attractive for high resolution measurements. To date ZAPS has not been used with stored ions.

Clearly, if $\Delta B_{0^\circ}^{(2)} \sim \Delta B_t^{(1)}$ both broadenings must be reduced simultaneously if better resolution measurements are to be possible. Thus, setting $\Delta B_{0^\circ}^{(2)} = \Delta B_t^{(1)}$ a useful lower limit for the spectrometer acceptance angle may be established:

$$\Delta\theta \geq \Delta\theta_{lim} \equiv \frac{2}{|1 \pm \zeta|}\sqrt{\frac{\Delta t}{t}} \tag{13.15}$$

Line shifting

The Auger line energy ε' is shifted by the additional two terms on the right hand side of Eq. 13.1b to a higher energy ε in the lab frame. For fast projectiles ($\zeta >> 1$) the shift follows primarily from the second term, t. Furthermore, for $\zeta >> 1$, as already discussed, Auger lines can only be observed within a narrow range of forward angles.

The effect of line shifting may be favorably applied to measurements where the Auger energy ε' is very small. By kinematically shifting the Auger lines to higher energies, basic instrumental problems such as contact potentials or inadequate shielding of the earth's magnetic field may be overcome with little effort. In fact, using this technique, electrons from the Auger decay of high Rydberg states [48,27,51,52] or shape-resonances of negative ions [53–57] formed in ion-atom collisions have been measured with high resolution. Measurements of autoionization energies in the meV range, otherwise impossible to access, have been reported. In Fig. 13.6, spectra with two autoionizing lines [28,52] having rest energies of 60 meV and 100 meV are shown. Other examples are shown in Fig. 13.27 when discussing studies of negative ions.

For very fast projectiles the electron energies of the (+) solution can become too large for the typical spectrometer range (maximum energies $\sim 5 - 6$ keV). In this case, the lower energy solution (−) may be used [27].

Line stretching

The width of the Auger lines is changed according to the third term on the right hand side of Eq. 13.1b. For zero-degree measurements, we have: [27]

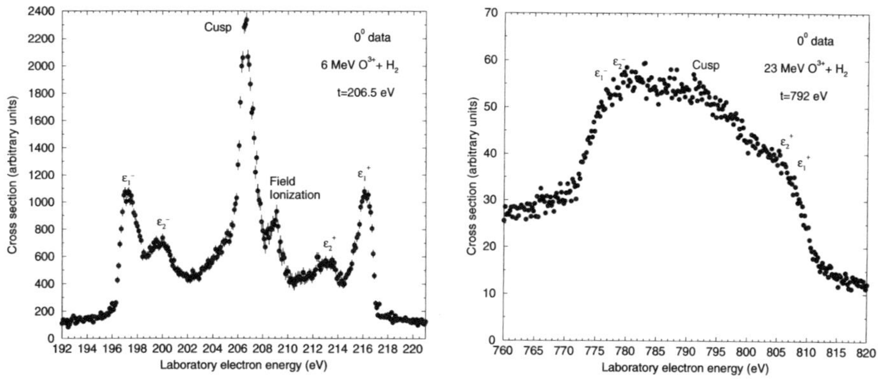

FIGURE 13.6 Zero-degree observation of meV electrons from autoionizing states produced by excitation in O^{3+} collision with H_2. [52,74] The two lines at $\varepsilon_{1,2}^{\pm}$ on either side of the Cusp have rest frame energies $\varepsilon_1' \sim 100$ meV and $\varepsilon_2' \sim 60$ meV, respectively. For both spectra $\varepsilon_{pass} = 10$ eV giving an instrumental resolution $\Delta\varepsilon_{ins} = 280$ meV. Left: 6 MeV collisions, $\Delta B_{0^\circ}^{(2)} \sim 1.01$ eV with $\Delta\theta = 1.17^\circ$ and $\Delta B_t^{(1)} \sim 0.3$ eV. The small peak just to the right of the Cusp is due to field ionization of projectile Rydberg electrons [75–77] inside the first stage of the tandem spectrometer. Right: Same spectrum but for 23 MeV collisions, $\Delta B_{0^\circ}^{(2)} \sim 7.45$ eV and $\Delta B_t^{(1)} \sim 1.1$ eV. Here the kinematic broadening, which increases with projectile velocity, completely washes out the peak structure. Other reasons for the strange looking spectrum are under investigation. A beam energy spread of $\frac{\Delta t}{t} = 7 \times 10^{-4}$ was assumed in the calculation of $\Delta B_t^{(1)}$.

$$\Delta\varepsilon = |1 \pm \zeta| \, \Delta\varepsilon' \qquad (\theta = 0^\circ) \tag{13.16}$$

resulting in stretching of the line shape for the (+) solution and compression for the (−) solution. The effect of line stretching may be favorably applied to measurements requiring extremely high resolution, such as the measurements of natural line widths. Light atoms have Auger line widths $\Gamma_a' \sim 0.1$ eV. Hence, it is useful to stretch the line kinematically so that its width becomes comparable to the spectrometer resolution. In Fig. 13.7 Auger lines resulting from intermediate energy He^+ + He collisions [24,27] are shown in the lab frame and also after transformation to the projectile frame.

The ^3P line width is seen to be 0.10 eV in the projectile rest frame. Since this line is known to have a much narrower natural line width (0.010 eV see Ref. [78]) the measured width is clearly due to the much larger instrumental resolution. The stretching factor $(1+\zeta)$ is 2.3 and helps appreciably to determine the natural width of the ^1S line which amounts to 0.14 eV and 0.32 eV in the projectile and laboratory frames, respectively, after the instrumental resolution has been unfolded.

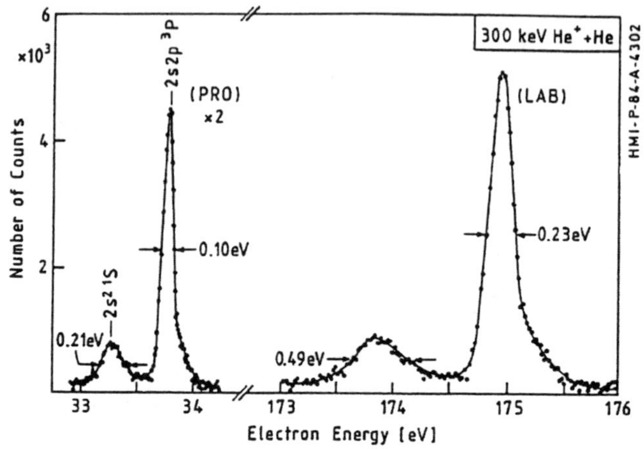

FIGURE 13.7 Kinematic line stretching and enhancement. Projectile autoionization lines (right) in 300 keV collisions of $^3He^+$ + He are kinematically stretched-out and enhanced when observed in the laboratory frame (LAB). The same spectrum transformed to the projectile rest frame (PRO) is shown on the left. Note the factor of 2 used to multiply the rest frame electron yield. This spectrum was obtained by decelerating electrons to a pass energy, $\varepsilon_{pass} = 7$ eV. The spectrometer had an acceptance angle $\Delta\theta \sim 2.1°$ giving a kinematic broadening of $\Delta B^{(2)}_{0°} = 0.076$ eV. Reprinted from "High resolution studies of two-electron processes observed in energetic ion-atom collisions via zero-degree Auger spectroscopy", T. J. M. Zouros, D. Schneider, and N. Stolterfoht, *Nuc. Instr. Meth.* **B31** (1988) 349, with kind permission from Elsevier Science - NL, Sara Burgerhartstraat 25, 1055 KV Amsterdam, The Netherlands.

It should be emphasized that stretching and broadening refer to qualitatively different effects. The first one originates from the transformation properties implied in Eqs. 13.1, while the second arises primarily due to the finite acceptance angle of the electron spectrometer, as shown in Eq. 13.8. Accordingly, the stretching effect leaves intrinsic structures of the Auger spectrum unchanged, whereas the broadening effect tends to wash them out.

Line enhancement

Another consequence of projectile-electron kinematics is that the laboratory electron yield is enhanced with respect to the rest frame yield by the stretching factor. This was already evident in the example of Fig. 13.7. The frame transformation for the double differential cross sections is given by: [27]

$$\frac{d^2\sigma}{d\varepsilon d\Omega} = |1 \pm \zeta| \frac{d^2\sigma}{d\varepsilon' d\Omega'} \qquad (\theta = 0°) \qquad (13.17)$$

resulting in enhancement of the laboratory lines for the (+) solution and diminishment for the (−) solution.

Angular compression

The well known "beaming" effect observed for X rays emitted from relativistic emitters, is also observed for electrons ejected from fast projectile ions. Electrons emitted *isotropically* in the rest frame will be ejected into a narrow beam in the forward direction in the laboratory frame. The transformation of the angular range can be calculated from the angle transformation obtained directly from Fig. 13.2:

$$\sin\theta = \sqrt{\frac{\varepsilon'}{\varepsilon}}\,\sin\theta' = \frac{\sin\theta'}{\sqrt{1+2\zeta\cos\theta'+\zeta^2}} \tag{13.18}$$

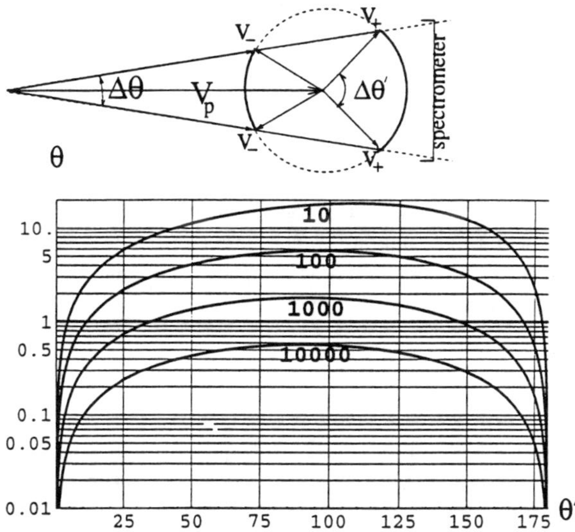

FIGURE 13.8 Angular compression in very fast emitters. Top: The angular range $\Delta\theta'$ in the forward and backward directions in the projectile rest frame (thick dark arcs) is compressed into the spectrometer acceptance angle $\Delta\theta$. Bottom: Relation between θ (degrees) and θ' (degrees) for $\zeta^2 : 10, 10^2, 10^3, 10^4$ (see Eq. 13.18). For $\zeta^2 = 10^4$, the whole rest frame angular range $\theta' = 0° - 180°$ is seen to be compressed into the laboratory range of $\theta = 0° - 0.6°$ completely covered by the typical spectrometer acceptance angle $\Delta\theta \sim 1.2°$. The range $0° \le \theta' \le 90°$ gives rise to the (+) solutions, while the range $90° \le \theta' \le 180°$ gives rise to the (-) solutions.

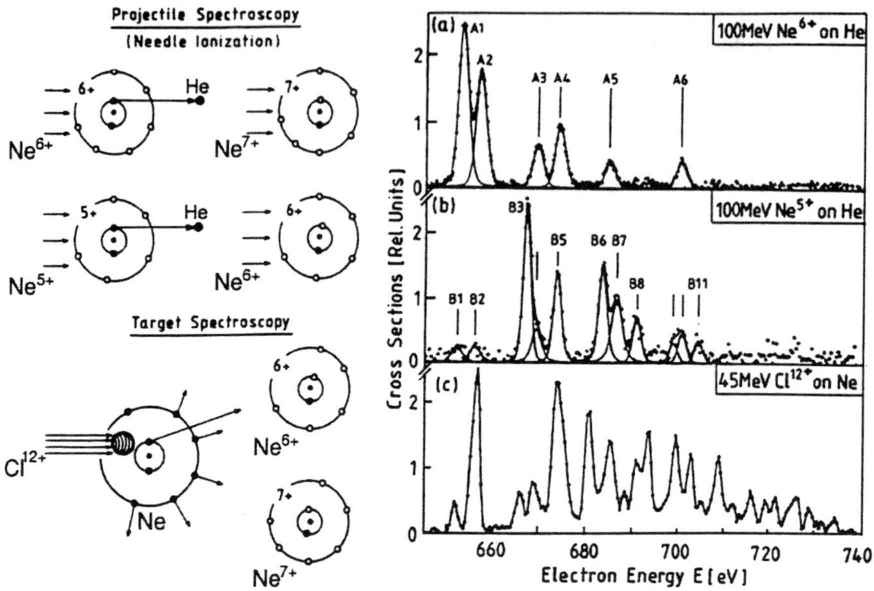

FIGURE 13.9 Demonstration of needle ionization: When a projectile collides with light target atoms, an inner vacancy is produced while its outer-shell structure is preserved with only a single final projectile charge-state emerging. Simple spectra result, as shown in the Ne K-Auger spectra on the right. The observed Auger lines in (a)-(b) are characteristic of Li- and Be-like ions resulting from the removal of a single 1*s* electron from the impinging configuration. In the opposite situation (c), encountered from an example in target spectroscopy, a *heavy* projectile collides with a heavy target, resulting in many different target states which give rise to complicated Auger spectra [79] with severe line blending (figures adapted from Ref. [27]).

For large values of ζ^2 this can result in severe compression of the rest frame angular distribution as demonstrated in Fig. 13.8. In such cases, e.g. for $\zeta \sim 10^2$, angular information will be washed out due to the resulting averaging over the large angular range in the rest frame [52]. Thus, angular compression and kinematic broadening are seen to be closely linked. In the example shown in Fig. 13.6, for 6 MeV O^{3+} collisions, $\zeta^2 \sim 2000$ and thus the rest frame range $|\theta'| \leq 22°$ gets compressed into the acceptance angle $\Delta\theta = 1°$ of the spectrometer at $\theta = 0°$. For 23 MeV collisions beaming is more extreme with $|\theta'| \leq 50°$ compressed into $\Delta\theta = 1°$.

We finally note that when $\varepsilon < \varepsilon'$, as for example in the case of backward observation angles for slow emitters (see Eq. 13.3b), Eq. 13.18 shows that an angular expansion occurs instead of compression.

13.2.2 Ion surgery

It is well known from target spectroscopy that light projectiles such as photons, electrons or protons have the ability to selectively excite or ionize a single $1s$ target electron without disturbing the outer-shell electrons [80,81]. This effect has also been used advantageously in high resolution *projectile* K-x-ray spectroscopy, [82–85] using light targets such as H_2 or He. The situation is shown schematically in Fig. 13.9 (left) and has been referred to as "ion surgery" [63] and in the case of ionization (excitation) as "needle" ionization (excitation) [22,23].

Ion surgery conditions prevail in fast collisions [44,63], where the probability for electron capture and double-excitation processes is small. Then projectile K-Auger lines result predominantly from the $1s \longrightarrow n\ell$ excitation or $1s$ ionization of the incoming ion. The line identification is greatly facilitated due to the simple structure of the spectra as shown in the example of Fig. 13.9 (right) (a)-(b). Much more complicated spectra can arise in the opposite situation, shown in Fig. 13.9 (c), where many different target charge states are produced resulting in severe line blending [79]. These target spectra result from the ionization of a target by a *heavy* projectile. Ion surgery has played an important role in almost all ZAPS measurements to date. It is specifically invoked in section 13.5.1 and used in the determination of the fraction of metastable ions in a beam of Be-like ions.

13.3 EXPERIMENTAL SET-UP

Both target and projectile electron spectroscopy use very similar apparatus. Thus, the beam line, scattering chamber, electronics and data acquisition are common in both cases. Here we shall only present details that are particular to ZAPS.

The experimental set-up for ZAPS used at the HMI-Berlin [22] is shown in Fig. 13.10. Most other setups have been modeled after this one. In Fig. 13.11 the KSU setup is shown [76].

13.3.1 Gas target

A differentially pumped target cell is used to give a well defined interaction length. Excitation and decay of the projectile takes place primarily within this cell. The gas cell should be operated at pressures sufficiently low for single collision conditions to prevail for *every* observed line of interest [25]. Typical gas cells used are about 5-10 cm long and run at pressures of a few mTorr. A capacitive manometer can be used for accurate absolute pressure readings. Differential pumping of the cell is usually performed by the main chamber pump. The chamber pressure with gas cell running should be better than $1-2\times10^{-5}$ Torr to ensure a low electron background and stable operation of

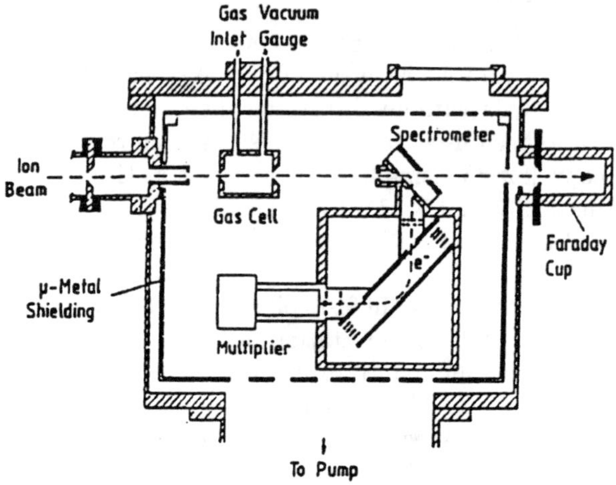

FIGURE 13.10 ZAPS set-up used at the Hahn-Meitner Institute in Berlin. Reprinted from "High resolution studies of two-electron processes observed in energetic ion-atom collisions via zero-degree Auger spectroscopy", T. J. M. Zouros, D. Schneider, and N. Stolterfoht, *Nuc. Instr. Meth.* **B31** (1988) 349, with kind permission of Elsevier Science - NL, Sara Burgerhartstraat 25, 1055 KV Amsterdam, The Netherlands. [22,23]

the electron detector. Typical cell apertures have diameters of $\sim 2-3$ mm. At KSU, the target cell, seen in the bottom of Fig. 13.11, is *doubly*-differentially pumped, using an extra 450 lt/s turbo-molecular pump directly on the cell. This allows for excellent running conditions with usual chamber pressures in the $1-5\times10^{-6}$ Torr range with target pressures in the $1-10\times10^{-3}$ Torr range [76,73].

It is useful to have a target cell that can by electrically floated. Then by placing a voltage on the cell, electrons produced inside the cell can be separated from electrons produced between the target cell and the spectrometer. Metastable states can thus be identified as demonstrated in Fig. 13.12.

13.3.2 The tandem electron spectrometer

ZAPS measurements to date have predominantly used the tandem spectrometer design first built by the Hahn-Meitner Institute (HMI) atomic physics group in Berlin in 1983 [22–24] and shown in Fig. 13.10. Variations of the HMI design have been used in fast collision studies at KSU [77,76] and WMU [86] and in slow-intermediate collisions at ANL [29]. Other designs have also been reported, such as a tandem composed of two spherical analyzers for 180° observation [19], a large *single*-stage hemispherical spectrometer at GSI-Darmstadt [31], a small spherical-sector spectrometer at ORNL for use with negative

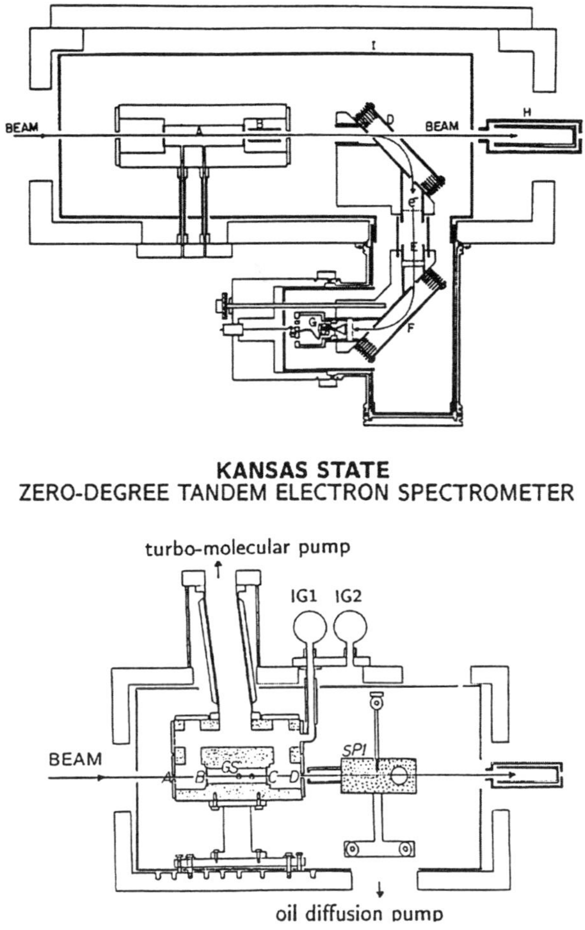

FIGURE 13.11 Kansas State University (KSU) ZAPS set-up. Top: (A) target gas cell, (B) deflector, (D) first stage used as a deflector, (E) focusing lens and retarding grids, (F) second stage used for high resolution analysis, (G) channeltron with electron suppression grid, (H) Faraday cup, (I) μ-metal shielding. Deflector (B) could be independently biased to deflect electrons when needed.
Bottom: Side view of above apparatus showing details of the double differentially pumped gas cell. This cell is pumped by a specially dedicated turbo-molecular pump directly attached to the gas cell assembly (GS) which could be electrically floated. Collimators A and D with apertures of 1.6 and 4.0 mm in diameter could also be biased independently and were used to monitor the beam. The target gas itself is kept between collimators B and C each having apertures of 2.4 mm in diameter (from Ref. [73]).

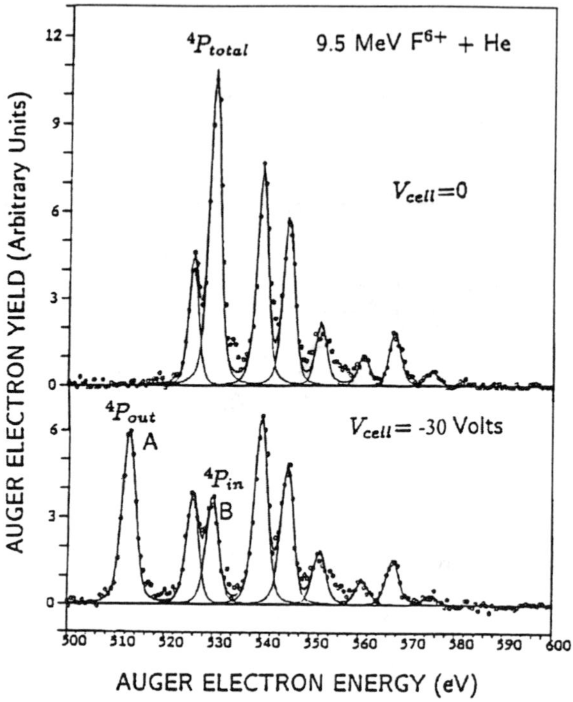

FIGURE 13.12 Identification of the metastable $1s2s2p\,^4P$ Auger line by biasing the target cell. Metastable states upon biasing are seen to give rise to two different lines originating from electrons produced inside (peak B) and outside (peak A) the target gas cell. The gas cell (GS) in Fig. 13.11 was biased with -30 V with respect to the spectrometer entrance slit held at ground potential. Cell collimators (A) and (D) are also held at ground potential. The electrons ejected inside the cell are thus shifted to higher energies by 30 eV in the laboratory frame. Both spectra have already been transformed back to the rest frame. In the bottom figure, the spectrum has been shifted back to its unbiased range before transforming it to the rest frame (from Ref. [73]).

ions [57] and a tandem composed of a parallel plate deflector mounted onto a hemispherical analyzer at RIKEN [33].

Principles of operation

The HMI tandem spectrometer design consists of two 45° parallel-plate analyzers connected together in series. The entrance analyzer acts as a deflector, sweeping all electrons of interest into the direction of the second analyzer where the high resolution analysis takes place. The deflector has a relatively large entrance slit to allow the ion beam to enter with the minimum of slit scatter-

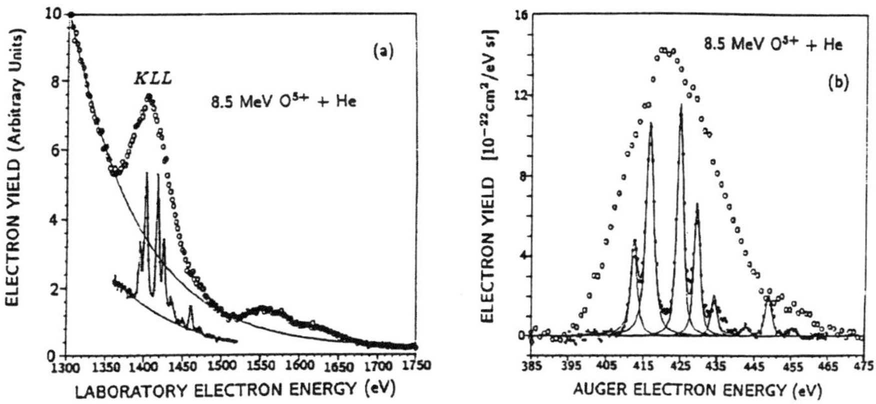

FIGURE 13.13 O^{5+}+He collisions. Circles: Low resolution K-Auger spectrum taken without deceleration. Continuous line spectrum: High resolution blow up of KLL lines taken with deceleration. The structureless continuous line is a background polynomial fit. (a) Laboratory frame, (b) Projectile rest frame after background subtraction (from Ref. [73]).

ing. At zero-degrees there are many forward-scattered background electrons. The basic function of the deflector is to reduce this background.

A well known method widely used to obtain high resolution spectra is to decelerate the electrons prior to energy analysis. The instrumental energy resolution $R_{ins} = \Delta\varepsilon_{ins}/\varepsilon$ is a constant determined by the fixed geometry of the spectrometer [87]. Thus, by lowering the electron energy ε, $\Delta\varepsilon_{ins}$ is also necessarily lowered. In the tandem spectrometer this is performed between the two stages. Following deceleration the electrons pass through the second analyzer with energy ε_{pass}. For the spectra of Fig. 13.7, taken with the HMI tandem spectrometer having a resolution $R_{ins} = 3\%$, a final $\Delta\varepsilon_{ins} \sim 0.210$ eV was attained by decelerating electrons from around 175 eV down to 7 eV. This resulted in an overall resolution of $R_o = 0.210/175 = 0.12\%$. For the spectra of Fig. 13.6, taken with the KSU tandem spectrometer having a resolution $R_{ins} = 2.8\%$, electrons were decelerated down to 10 eV. This resulted in a final $\Delta\varepsilon_{ins} = 0.280$ eV with a corresponding overall resolution $R_o = 0.280/216.5 = 0.129\%$.

Deceleration stage

Deceleration is produced by the two parallel high transmission grids shown in Figs. 13.10 and 13.11, just before the entrance to the second stage. These are usually separated by 2-3 mm.

The great improvement in resolution due to deceleration comes at a big cost in transmission. The electron transmission under deceleration can be shown to be roughly proportional to the inverse of the deceleration factor, F, defined as $F \equiv \varepsilon/\varepsilon_{pass}$. Thus, large deceleration factors are usually accompanied by big losses of count rate. For this reason high beam intensities are very important.

Clearly, if $\Delta\varepsilon_{ins} \sim \Gamma_{tot}$ (where Γ_{tot} has been defined in Eq. 13.14), no substantial gain in resolution may be attained by further decreasing the pass energy. Thus, a limiting pass energy ε_{pass}^{lim} may be established and since $\Delta\varepsilon_{ins} = R_{ins}\varepsilon_{pass}$ we obtain:

$$\varepsilon_{pass} \geq \varepsilon_{pass}^{lim} = \frac{\Gamma_{tot}}{R_{ins}} \tag{13.19}$$

Values of ε_{pass}^{lim} have already been given in the example of Table 13.1.

Improvements to the electron transmission of the KSU tandem were accomplished by using a cylindrical einzel lens biased with voltage V_L as shown in Fig. 13.14. In the HMI design no extra lens is used. However, some focusing has been noted for small values of the deceleration factor. This probably results from the inherent lens properties of such a two-cylinder design [88]. Design criteria for such a deceleration system are discussed in some detail in Ref. [13].

Further improvements in the overall transmission of the HMI tandem were recently noted [89] when the entrance and exit slits of the first stage were covered with high transmission grid to reduce fringing field distortion effects [90]. The improvement in transmission is clearly demonstrated in the simulation of Fig. 13.15 produced with charge particle trajectory program SIMION [89]. It is interesting to note that no adverse effects were observed, at least at low energy collisions where this was experimentally tested, even though the entrance grid is hit by the ion beam.

Voltage scheme

Three different voltages, biasing the spectrometer plates of the entrance and exit stages, need to be scanned to record a spectrum. In Fig. 13.14, a schematic diagram of the KSU tandem spectrometer and the analyzing voltages used are shown in detail. When no deceleration is used, $\varepsilon_{pass} = \varepsilon$, V_R is grounded and only V_1 and V_2 are scanned. When decel mode is used, all three voltages V_1, V_2 and V_R are scanned together in such a way as to maintain a fixed pass energy, ε_{pass}. The voltage scheme is given in Fig. 13.14 (left). The voltages were controlled by a computer via 12- or 16-bit digital-to-analog converters.

Spectrometer design

The 45° parallel plate analyzers were chosen for their simplicity [91,87,92,93,90]. However, this type of analyzer focuses electron trajectories in

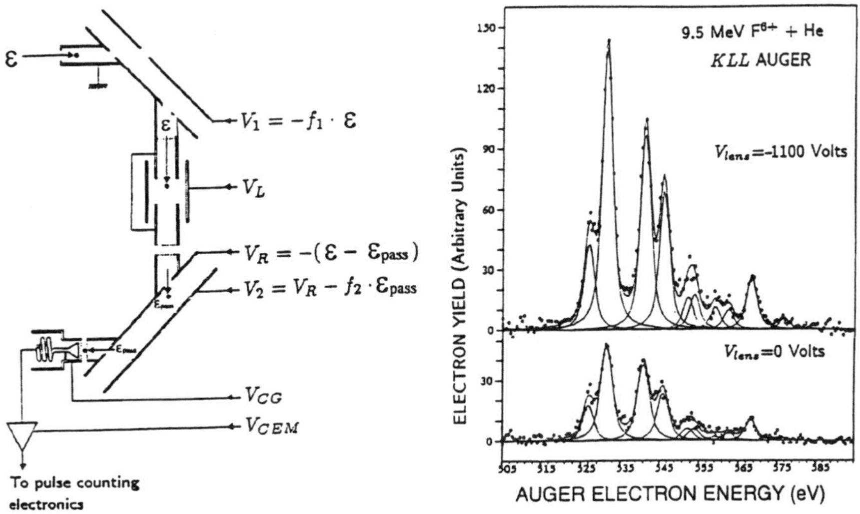

FIGURE 13.14 Left: Voltage scheme for KSU tandem electron spectrometer. Electrons enter the first stage with energy ε, are deflected by 90°, are decelerated to an energy ε_{pass} and analyzed by the second stage. On exit they are accelerated back to their original energy and detected by the channeltron. High transmission grids are placed just after the exit slit of both first and second stages to reduce field fringing. An extra grid in front of the channeltron allows for a small fixed negative voltage bias to suppress low energy electron background. The spectrometer constants f_1 and f_2 for each stage, are indicated. V_1, V_2 and V_R were set by a computer program and stepped every time the number of ions collected in the Faraday cup reached a predetermined limit. In deceleration mode, ε_{pass} is fixed thus maintaining a constant transmission through the second stage. In non-deceleration mode, $\varepsilon_{pass} = \varepsilon$ and $V_R = 0$, resulting in a transmission that depends on the energy ε. The einzel lens voltage V_L was set empirically for maximum transmission and left constant under deceleration operation. Its effect is shown in the spectra on the right. Right: Einzel lens performance under deceleration operation. Here, $\varepsilon_{pass} = 150$ eV and the mean laboratory energy, $\bar{\varepsilon} = 1600$ eV. Thus, the overall deceleration factor is $F \sim 10.7$. The lens is seen to increase the yield by a factor of 2.5 (figures adapted from Ref. [73]).

only one direction and only to first order [87,92]. Higher transmission can be attained by hemispherical analyzers [94–96,90,92] which are double-focusing, but have yet to be used extensively in zero-degree studies. We shall not go into the details of spectrometer design and their focusing properties but refer the reader to the many existing reviews and articles [87,94–96,92,90,97]. Both HMI and KSU tandem spectrometers were made out of brass for HV operation, with an aluminum version of the HMI spectrometer fabricated for UHV operation. High quality insulators with low outgasing rates such as Vespel (polyimide) or Delrin are used in the construction [90]. Guard electrodes

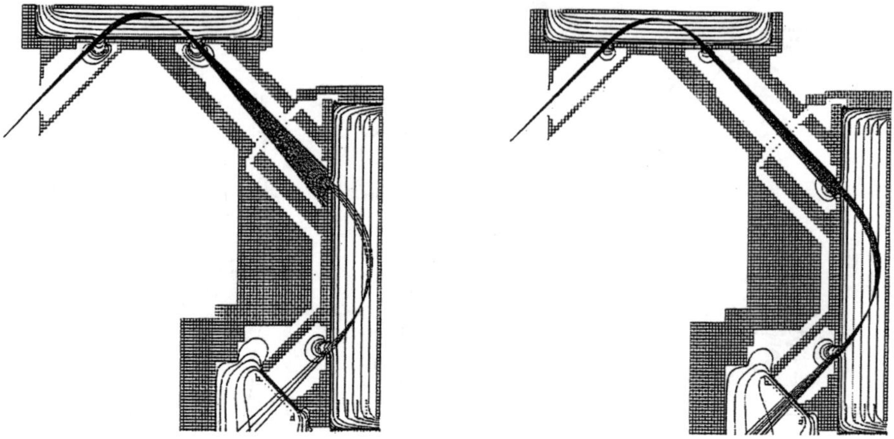

FIGURE 13.15 Effect of fringing fields around the slits of the first stage of the HMI tandem spectrometer. No deceleration is used. The electron trajectories are shown leaving from the source at the upper left, traversing the two 45° parallel plate analyzers and finally detected on the bottom right of each figure. Left: Strong fringing field–no grid is used. Right: Weak fringing field produced by placing a high transmission grid on both slits. When the fringing field is reduced improved transmission results. The simulation was performed with program SIMION. The electron energy is 1 keV. The equipotential lines are clearly shown indicating the fringing fields [89].

were utilized particularly on the second stage to further compensate for field distortion caused by fringing.

In Table 13.2 typical values of various spectrometer parameters for both the KSU and HMI tandem designs are listed for comparison. The basic equations for computing these parameters are also given in the table. In calculating the acceptance angles it is assumed that the length of the electron trajectory between foci does not contribute. Thus, distances along the parabolic trajectories from entrance to exit slits in each analyzer are not included in the calculation of $\Delta\phi$, since in this direction there is first-order focusing, while they are included in the calculation of $\Delta\theta$ where there is no focusing. The length $1.148l_i$ is the length of the parabolic trajectory in stage i between slits separated by length l_i with $i = 1$ or 2.

As seen in Table 13.2, the biggest difference between the two spectrometers is the much larger separation between stages in the KSU design as well as the much larger gas cell (midpoint) distance to the spectrometer (y_c). Thus, the HMI design is seen to be very compact with a 6 times larger effective solid angle. The KSU spectrometer, however, with its reduced acceptance angle has a correspondingly smaller kinematic broadening and the longer electron flight paths help to further reduce background from stray electrons. Furthermore,

TABLE 13.2 Design parameters of the KSU [76] and HMI [22] tandem 45° parallel-plate spectrometers. All lengths are in mm. Lengths in the HMI set-up vary depending on whether intermediate (10-150 keV/u) or high (3-5 Mev/u) energy collisions are studied.

	KSU		HMI		
Parameter	stage 1	stage 2	stage 1	stage 2	Definition
slit separation (l)	69	69	35	50	center-to-center
entrance slit width (w^i)	3.3	2.5	4	1.5	
entrance slit length (t^i)	4.7	9.9	10	10	
exit slit width (w^o)	4.7	1.3	4	1.5	
exit slit length (t^o)	10	9.8	10	10	
field plate separation (d)	21.8	21.2	10.5	15	
spectrometer constant (f)	0.63	0.614	0.6	0.599	$f = \frac{2d}{l}$
Resolution: $R = \Delta\varepsilon/\varepsilon$					
Base (R_B)	11.6%	5.6%	23.0%	6.0%	$R_B = \frac{w^i+w^0}{l} + 2(\Delta\phi/2)^2 + \frac{1}{2}(\Delta\theta/2)^2$
FWHM (R_{ins})	5.8%	2.8%	11.5%	3.0%	$R_{ins} \equiv R_B/2$ see Ref. [92]
distance between stages (S)	120		54		exit slit 1–entrance slit 2
gas cell length (l_c)	100		50 (100)		
gas cell distance (y_c)	202		95 (120)		cell mid-point–entrance slit 1
Point-source					
Full acceptance angles (°):					Point-source at y_c
Focusing direction ($\Delta\phi$)	0.17		0.41 (0.35)		$\Delta\phi = \frac{w_2^0}{\sqrt{2}(y_c+S)}$
Non-focusing direction ($\Delta\theta$)	1.17		2.3 (2.1)		$\Delta\theta = \frac{t_2^0}{y_c+S+1.148(l_1+l_2)}$
Solid angle ($\Delta\Omega_0$ in μSr)	59		289 (225)		$\Delta\Omega_0(y_c) = \Delta\theta\Delta\phi$

the large gas cell pressures that could be used due to the double-differential pumping of the cell somewhat compensated for the loss in solid angle.

Electron detectors

The energy-analyzed electrons can be detected using various types of detectors. The HMI spectrometer has been used primarily with a high gain electron multiplier. This detector provides a rather energy-independent electron detection efficiency η, has a large detection area and can be run at high count rates ($\sim$ 500 kHz). The main drawback of this detector, however, is that it has to be kept constantly under vacuum to avoid serious efficiency degradation. The KSU spectrometer uses a channeltron detector. It also has a high efficiency $\eta \sim 0.5 - 0.8$, but has a smaller detection area and a smaller maximum count rate ($\sim$ 100 kHz). Channeltrons can remain in the open air without any problem when not in operation. Channel plates have also been used [98–100], enjoying most of the advantages of both multiplier and channeltron detectors. However, their higher dark current and greater fragility seems to have somewhat discouraged their use to date. In future more efficient devices, utilizing position sensitive detection, channel plates can be expected to play a dominant role.

The independent efficiency calibration of electron detectors is rather tedious and somewhat controversial. [101] In electron spectroscopy measurements most cross sections are normalized to known absolute target Auger electron standards. The authors have developed a different technique to measure the overall efficiency by *in situ* measurements of the Binary Encounter electron (BEe) peak [102] and normalization to well-accepted *calculated* BEe cross sections. This method is described in detail in the next section.

13.4 DATA ANALYSIS–SELECTED TOPICS

Here we discuss particular aspects of ZAPS data analysis related to absolute cross section determination of prompt and metastable states. These topics have not received much attention previously.

13.4.1 Absolute cross section determination: Binary Encounter peak method

The laboratory double-differential cross section (DDCS)) for electron production are given in general by: [8,103]

$$\frac{d^2\sigma}{d\varepsilon d\Omega} = \frac{N^e}{N_0\, n\, T(\varepsilon)\, l_c \Delta\Omega\, \Delta\varepsilon\, \eta(\varepsilon)} \tag{13.20}$$

where $N^e = N^e(E_p, \varepsilon, \theta)$ is the measured electron count at a particular collision energy E_p, electron energy ε and detection angle θ. n and l_c are the target gas density and the gas cell length. $\Delta\Omega$ is the effective solid angle. For prompt states, $\Delta\Omega$ is the just the point source solid angle $\Delta\Omega_0(y_c)$ for a source a distance y_c from the spectrometer entrance slit as given in Table 13.2. For long lived states, the electrons can be ejected at any point along the path after the collision in the target cell and a more elaborate calculation for $\Delta\Omega$ is required. This is presented in section 13.4.2. The transmission $T(\varepsilon)$ is the overall transmission of the electron spectrometer and $\eta(\varepsilon)$ is the intrinsic electron detector efficiency. The energy band-pass of the spectrometer, is $\Delta\varepsilon = R_{ins}\, \varepsilon$. In deceleration mode $\varepsilon = \varepsilon_{pass}$, (see section 13.3.2) and thus is constant. R_{ins}, the instrumental energy resolution of the high resolution stage of the spectrometer is also a constant. N_0 is the number of incident projectiles.

In the special case where the beam of incident particles has a strong metastable component which contributes to some or all of the observed Auger lines (as for example in the case of Be-like O^{4+} beams having almost 60% of the projectiles in the $1s^2 2s2p\, ^3P$ metastable state), Eq. 13.20 will not in general be correct [8]. However, Eq. 13.20 may still be used to define the normalized double differential yield (NDDY). The NDDY must then be multiplied by a correction factor to give the DDCS. These corrections are discussed in sections 13.4.2 and 13.5.1.

Assuming $N_0, n, l_c, \Delta\Omega_0(y_c), R_{ins}$ and ε are well known, $T(\varepsilon)$ and η are the most important parameters to determine. These can be obtained through careful experimentation or may be deduced by comparison to accepted cross section "standards". Even in the later case, these standards are only known for some particular energy or angular ranges and thus the need to extrapolate to a different range might introduce various new assumptions and errors.

We present a method [102,38] for determining the product $T(\varepsilon)\eta$ in ZAPS by normalizing electron DDCS to *calculated* Binary Encounter electron (BEe) DDCS produced in collisions of *bare ions* with H_2 or He targets [102]. This method is particularly well suited to zero-degree observation in energetic collisions and has the advantage that it provides a direct and accurate *in situ* absolute calibration (normalization) of the overall efficiency $T(\varepsilon)\eta$ over a large electron energy range $\varepsilon \sim 1 - 5$ keV.

Binary Encounter electrons [102] are target electrons ionized through direct, hard collisions with energetic projectiles, giving rise to a broad energy distribution centered approximately at a laboratory electron energy $\sim 4t\cos^2\theta$ [104,14]. The BEe DDCS are well described in the impulse approximation (IA) for $\theta = 0°$ observation [102]. As seen from the projectile frame, the target electron suffers a 180° Rutherford scattering off the ion with cross section $\left(\frac{d\sigma}{d\Omega'}\right)_R$ [102]. Thus, the following simple formula, obtained in the IA, gives the laboratory BEe DDCS (in a.u.), for collisions of *bare* ions with H_2 and He targets for $\theta = 0°$: [102]

$$\left(\frac{d^2\sigma}{d\varepsilon d\Omega}\right)_{BEe} = \left(\sqrt{\frac{t}{\varepsilon'}} + 1\right)\left(\frac{d\sigma}{d\Omega'}\right)_R \frac{J(p_z)}{\sqrt{2t} + p_z} \tag{13.21a}$$

where

$$\left(\frac{d\sigma}{d\Omega'}\right)_R = \frac{Z_p^2}{16\varepsilon'^2} \tag{13.21b}$$

$$p_z = \sqrt{2}(\sqrt{\varepsilon' + \varepsilon_I} - \sqrt{t}) \tag{13.21c}$$

$$\varepsilon = (\sqrt{t} + \sqrt{\varepsilon'})^2$$

Thus, $\left(\frac{d^2\sigma}{d\varepsilon d\Omega}\right)_{BEe}$ and the laboratory electron energy ε are given as functions of ε'. $J(p_z)$ is the Compton profile of the target electron, representing the probability of finding such an electron with momentum p_z along the z-axis taken along the collision direction. The Compton profile for H_2 or He is experimentally known and given analytically [105]. The ionization energies ε_I are 15.5 and 24.5 eV for H_2 and He, respectively. The BEe DDCS formula as given by Eqs. 13.21 is very convenient and simple to use and has been found to be in excellent agreement with measured DDCS in a variety of collision systems for bare ions with $Z_p = 1 - 9$ [102]. The validity of the IA description

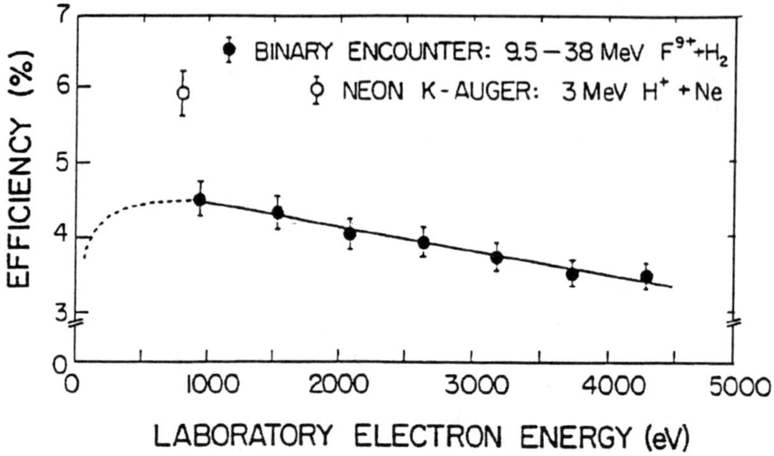

FIGURE 13.16 Overall absolute spectrometer efficiency $T(\varepsilon)\eta$ plotted as a function of the laboratory electron energy ε. The solid line was interpolated using the data points (solid circles) obtained by normalizing the $F^{9+} + H_2$ BEe yields to the IA calculation (see text). The dashed line was extrapolated using the results of Ref. [111]. The open circle is the efficiency measured using the known Ne target K-Auger cross section [112] at $\varepsilon \approx 800$ eV for 3 MeV H^+ + Ne collisions. The error bars are calculated from statistics alone. The Ne K-Auger data have an overall absolute uncertainty of 20% [112] (adapted from Ref. [73]).

of the BEe peak is also supported by more sophisticated calculations based on the CDW-EIS formulation, in which two-center effects due to the influence of both the projectile and target charges on the ejected electrons are included [106,107].

For partially stripped ions ($q \neq Z_p$) the BEe DDCS can still be calculated by the above simple formula [108] but the Rutherford cross section needs to be replaced by the cross section for elastic scattering in the field of a *non*-bare ion requiring a more elaborate calculation [109,110]. In this case, BEe DDCS are known to *increase* with decreasing charge-state q [108].

By comparing the IA calculation of Eq. 13.21 to *measured* BEe DDCS, $T(\varepsilon)\eta$ is directly determined for $\varepsilon \sim 3t - 5t$ around the BEe peak. By varying the collision energy (t), a much larger electron energy range can be covered. The overall efficiency $T(\varepsilon)\eta$ determined in this way is shown as a function of ε in Fig. 13.16 for the KSU spectrometer [102].

When high resolution spectra (from partially stripped ions) are obtained using deceleration in the tandem spectrometer, further loss of transmission occurs and $T(\varepsilon_{pass}, F)$ must be determined, where $F = \varepsilon/\varepsilon_{pass}$ is the deceleration factor. This can in principle be accomplished as discussed above by directly measuring bare BEe DDCS in *deceleration* mode. However, in

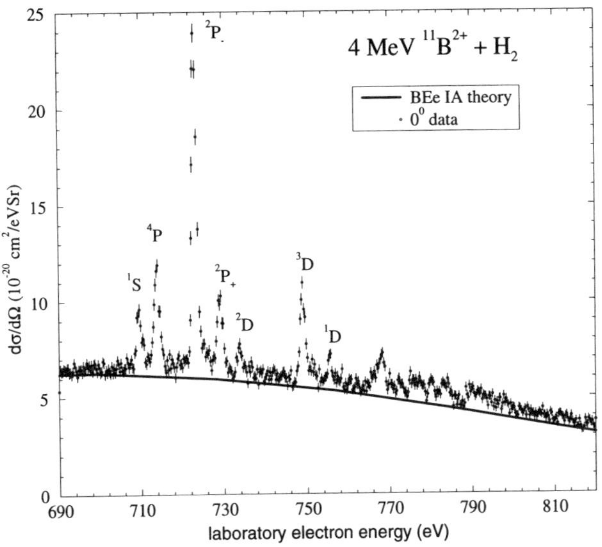

FIGURE 13.17 High resolution 0° data from 4 MeV $B^{2+} + H_2$ collisions [113]. The Auger DDCS have been converted to an absolute scale by *normalizing* to calculated BEe DDCS for $B^{2+} + H_2$ collisions, which have in turn been scaled from bare $B^{5+} + H_2$ BEe DDCS, as described in the text.

practice, since beam currents can be small for bare ions this method may be impractical. In this case, a slightly different procedure is followed:

(a) BEe DDCS $\left(\frac{d\sigma}{d\varepsilon d\Omega}\right)_{q=Z_p}$ for *bare* ions are measured in *low* resolution and compared to calculated IA BEe DDCS, thus determining the product $T(\varepsilon)\eta$ as already discussed [102].

(b) BEe DDCS $\left(\frac{d\sigma}{d\varepsilon d\Omega}\right)_q$ for *partially-stripped* ions of charge-state q, are measured in *low* resolution and converted to absolute DDCS using the value of $T(\varepsilon)$ determined in (a). The overall efficiency $T(\varepsilon)\eta$ should be the same as in (a) since it depends only on the electron energy which is the same for both cases (a) and (b) [108].

(c) High resolution DDCS, for the ion of charge-state q is now measured using deceleration. The measured spectrum includes Auger lines superimposed on the BEe peak determined in (b) as shown in Fig. 13.17. Thus, by normalizing the high resolution DDCS to the absolute BEe DDCS determined in (b), $T(\varepsilon_{pass}, F)$ is fixed and the final *absolute* DDCS may be obtained in the laboratory frame.

This method of absolute normalization is limited to fairly fast collisions ($E_p > 0.5$ MeV/u) for which the BEe peak is well separated from the Cusp

TABLE 13.3 Lifetimes τ_J (ns) of the $1s2s2p\,^4P$ state for light ions.

Ion	$J = \frac{1}{2}$	$J = \frac{3}{2}$	$J = \frac{5}{2}$	References
He^-	10.7 μs	12 ± 2 μs	350 ± 15 μs	[116,117]
C^{3+}	3.69	13.2	117	[118]
O^{5+}	0.91	3.34	27.67	[118]
F^{6+}	0.57	1.84	15.90	[118]
Ne^{7+}	0.40	1.6	10.4	[119]

peak. Furthermore, it relies on the availability of intense bare ion beams not always available at the projectile energy of interest. Typical minimum beam currents with light ions are $\sim 5 - 10$ nA for low resolution BEe studies and $\sim 50 - 100$ nA for high resolution work. For less intense beam currents more conventional normalization methods can be used with some care to obtain absolute DDCS for non-bare ions, including direct comparisons to theoretical BEe DDCS of *partially* stripped ions. Studies of BEe DDCS for few-electron projectiles are still under active investigation. [108–110,114,115] For collisions with H_2 and He, theory seems to be in good agreement with experiment for few-electron ions [109,110]. In general, more work is needed in establishing reliable absolute standards for DDCS at zero-degrees.

13.4.2 Single differential cross section determination of long-lived states

Upon integrating the DDCS given by Eq. 13.20 over the electron energy for each Auger line (e.g. area under the peak) one obtains the single differential cross section (SDCS). In the case of long-lived projectile Auger states one faces the additional difficulty that part of the measured yield N^e originates outside the target cell along the path between the cell and the spectrometer resulting in a very different effective solid angle than in the case of prompt Auger transitions. This has been demonstrated in Fig. 13.12. For such a metastable state, we may set the effective solid angle, $\Delta\Omega$ (see Eq. 13.20), equal to $G_\tau \Delta\Omega_0(y_c)$ and compute the correction factor G_τ.

The most commonly encountered metastable state in ZAPS is the $1s2s2p\,^4P$ state. Lifetimes τ_J depend on the total angular momentum component J and are given in Table 13.3 for a few light ions. For the Li-like oxygen and fluorine ^4P states in 0.5-2 MeV/u collisions the effect of the long lifetimes gave $G_\tau \sim 1.5 - 1.7$ showing this to be a substantial correction [73]. Typical decay curves are shown in Fig. 13.18 for 4-32 MeV collisions of O^{5+}. We note that for low-Z ions, the different J components of the Auger line cannot usually be resolved.

The correction factor G_τ is computed as follows. The number of projectiles excited to the metastable state J in collisions with a differential target length element dx is:

$$dN_0^J = N_0\, n\, \sigma_J\, dx \qquad (0 \leq x \leq l_c) \tag{13.22}$$

where σ_J is the *production* cross section for populating state J in the collision. Within the time $t = y/V_p$ and $t + dt$ later, $dN^J(y)$ metastable states decay:

$$dN^J(y) = dN_0^J\, \frac{e^{-y/V_p\tau_J}}{V_p\tau_J}\, dy \qquad (0 \leq y \leq L - x) \tag{13.23}$$

where $L = y_c + l_c/2$ is the distance from the cell entrance to the spectrometer entrance slit. Defining $\frac{dw_J}{d\varepsilon d\Omega}\xi_J$ as the double differential probability for electron emission into the direction (θ, ϕ) with energy ε and Auger yield ξ_J, we integrate all contributions along the beam trajectory in front of the spectrometer to obtain the total detected yield, N_J^e, as follows:

$$N_J^e = N_0\, n\, \sigma_J\, \frac{dw_J(\theta = 0°)}{d\varepsilon d\Omega}\xi_J\, l_c\, G_{\tau_J} \Delta\Omega_0(y_c)\, \Delta\varepsilon\, T(\varepsilon)\eta(\varepsilon) \tag{13.24}$$

with

$$G_{\tau_J} \equiv \left[\frac{1}{l_c \Delta\Omega_0(y_c)} \int_{x=0}^{l_c} dx \int_{y=0}^{L-x} dy\, \frac{e^{-y/V_p\tau_J}}{V_p\tau_J}\, \Delta\Omega_0(L - x - y)\right] \tag{13.25}$$

where we have assumed for simplicity that $\frac{dw_J}{d\varepsilon d\Omega}$ does not vary much along the integration path.

We note that $\sigma_J \frac{dw_J(\theta=0°)}{d\varepsilon d\Omega}\xi_J$ can be identified with the DDCS of Eq. 13.20 for the particular state component J. Adding all J contributions we obtain:

$$N^e = \sum_J N_J^e = N_0\, n\, l_c \Delta\Omega_0(y_c)\, \Delta\varepsilon\, T(\varepsilon)\eta\, \sigma \frac{dw}{d\varepsilon d\Omega} \sum_J a_J \xi_J G_{\tau_J} \tag{13.26}$$

In Eq. 13.26 we have assumed statistical production of the states J with $\sigma_J = a_J \sigma$, where σ is the total production cross section (for all J states) and a_J the statistical weight with $a_J = (2J+1)/\sum_J (2J+1)$. We have also assumed that w_J does not depend strongly on J and set $w_J = w$. If we use the J-averaged Auger yield $\xi \equiv \sum_J a_J \xi_J$ we obtain from Eq. 13.26 the averaged value for G_τ:

$$G_\tau = \frac{\sum_J (2J+1)\xi_J G_{\tau_J}}{\sum_J (2J+1)\xi_J} = \frac{\sum_J a_J \xi_J G_{\tau_J}}{\xi} \tag{13.27}$$

It is seen, that in the limit $V_p\tau_J \to 0$ (prompt states) and $l_c \to 0$ (point source) $G_\tau = 1$ as expected.

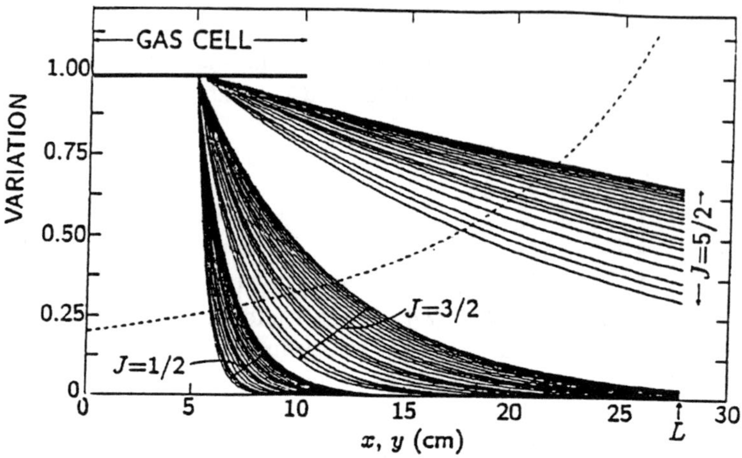

FIGURE 13.18 Decay curves, $e^{-y/V_P\tau_J}$ for the $O^{5+}(1s2s2p)^4P_J$ states (assumed to be produced at the center of the gas cell) for different projectile energies ranging from 4-32 MeV. A large part of the states are seen to decay outside the target cell volume, some even beyond the spectrometer. The dashed line shows the variation of solid angle $\Delta\Omega_0(y)$ as a function of y from $y = 0$ to $y = L$. The spectrometer entrance is at L (adapted from Ref. [73]).

13.5 PRACTICAL APPLICATIONS

13.5.1 Determination of metastable beam fractions in Be-like beams

The method of ZAPS has been used to determine the metastable beam fraction, F_m, of the long-lived $1s^22s2p\,^3P$ state in Be-like C^{2+}, O^{4+} [8] and Ne^{6+} [27] ions in collisions with H_2 and He. The basic premise underlying the determination of F_m is that the cross section for $1s$ ionization of the ground state $1s^22s^2\,^1S$ is the *same* as that for the metastable state $1s^22s2p\,^3P$. This premise, experimentally verified in high resolution K x-ray studies, has been used before in the determination of the metastable beam fraction of He-like ions [85]. The K-shell ionization cross section σ_K^g and σ_K^m of the ground and metastable states, respectively, are assumed to be equal, i.e. $\sigma_K^g = \sigma_K^m$, if the ionization process is unaffected by differences in the outer shell configurations. In this case, "needle" ionization conditions apply, as already discussed in section 13.2.2 for collisions with light targets such as H_2 and He.

Schematically, if T represents the target (He or H_2) and P the Be-like projectile of charge-state q, for $1s$ ionization of the ground state followed by Auger relaxation we have:

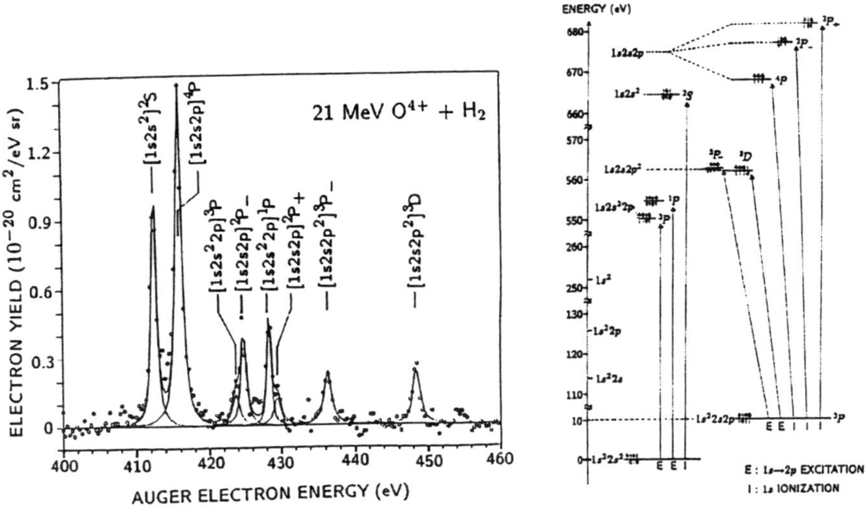

FIGURE 13.19 Normalized zero-degree Auger electron yield for 21 MeV O^{4+} collisions with H_2 after subtraction of background continuum electrons and transformation to the projectile frame. The $1s2s^2\,{}^2S$ line results from $1s$ ionization of the ground state. The ${}^4P$, ${}^2P_-$ and ${}^2P_+$ states result from $1s$ ionization of the metastable $1s2s2p^2\,{}^3P$ ion. All other lines are due to the decay of Be-like configurations produced by excitation. The transition scheme is given on the right (from Ref. [73]).

$$
\begin{aligned}
P^{q+}(1s^2 2s^2) + T(1s^2) &\longrightarrow P^{q+1}(1s2s^2\,{}^2S) + T(?) + e^- &&\text{(1s ionization)}\\
&\qquad\qquad \lfloor\!\longrightarrow P^{q+2}(1s^2) + e^- &&\text{(Auger)}
\end{aligned}
$$

In the case of $1s$ ionization of the metastable state, however, the metastable ion, upon losing a $1s$ electron, can autoionize giving rise to the three distinct lines in the electron spectra [44], as shown in Fig. 13.19:

$$
P^{q+}(1s^2 2s2p\,{}^3P) + T(1s^2) \longrightarrow \left\{
\begin{array}{l}
P^{q+1}[1s2s2p\,{}^4P] \\
P^{q+1}[1s(2s2p\,{}^3P)\,{}^2P_-] + T(?) + e^- \quad \text{(1s ionization)} \\
\underbrace{P^{q+1}[1s(2s2p\,{}^1P)\,{}^2P_+]}
\end{array}\right.
$$

$$
\lfloor\!\longrightarrow P^{q+2}(1s^2) + e^- \qquad \text{(Auger)}
$$

The cross sections σ_K^g and σ_K^m can be directly determined from the NDDY, $dY/d\Omega$, measured for each of the four states using Eq. (13.20):

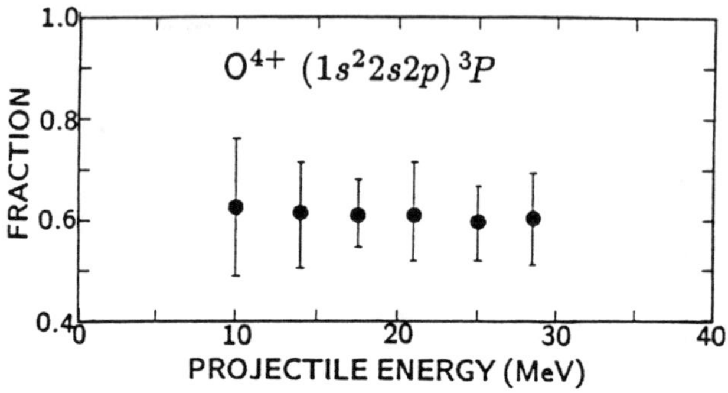

FIGURE 13.20 Metastable $(1s^2 2s2p\,^3P)$ beam fraction in energetic O^{4+} + He and H_2 collisions determined by ZAPS (adapted from Ref. [8]).

$$\sigma_K^g = \left(\frac{N_0}{N_g}\right) \frac{4\pi}{\xi_{2S}} \frac{dY(^2S)}{d\Omega} = \frac{4\pi}{F_g} \frac{1}{\xi_{2S}} \left[\frac{dY(^2S)}{d\Omega}\right] \tag{13.28}$$

$$\sigma_K^m = \left(\frac{N_0}{N_m}\right) \sum_X \frac{4\pi}{\xi_X} \frac{dY(X)}{d\Omega} = \frac{4\pi}{F_m} \left[\frac{1}{\xi_{^4P}} \frac{dY(^4P)}{d\Omega} + \frac{1}{\xi_{^2P_-}} \frac{dY(^2P_-)}{d\Omega} + \frac{1}{\xi_{^2P_+}} \frac{dY(^2P_+)}{d\Omega}\right] \tag{13.29}$$

where ξ_X is the Auger yield of state X. F_g and F_m are the beam fractions in the ground or metastable state, respectively. It is seen that $\frac{4\pi}{\xi_X}\frac{dY(X)}{d\Omega}$ are the *state production* yields, determined by integrating the NDDY given in Eq. 13.20 over electron energy (i.e. area under the peak for each state X) and emission angle, assuming isotropic emission.

Equating Eqs. (13.28) and (13.29) and using $F_g + F_m = 1$ we may solve for F_m to obtain:

$$F_m = \frac{\left[\frac{1}{G_\tau \xi_{4P}} \frac{dY(^4P)}{d\Omega} + \frac{1}{\xi_{2P_-}} \frac{dY(^2P_-)}{d\Omega} + \frac{1}{\xi_{2P_+}} \frac{dY(^2P_+)}{d\Omega}\right]}{\left[\frac{1}{\xi_{2S}} \frac{dY(^2S)}{d\Omega} + \frac{1}{G_\tau \xi_{4P}} \frac{dY(^4P)}{d\Omega} + \frac{1}{\xi_{2P_-}} \frac{dY(^2P_-)}{d\Omega} + \frac{1}{\xi_{2P_+}} \frac{dY(^2P_+)}{d\Omega}\right]} \tag{13.30}$$

The correction for the long lifetime of the $(1s2s2p)^4P$ state is seen to be explicitly included in G_τ [120]. More details, including comparisons to a different method based on the coefficients of fractional parentage of the states can be found in Ref. [8].

The experimentally determined fraction F_m, using Eq. (13.30), is shown in Fig. 13.20 for O^{4+} projectiles. Both He and H_2 targets gave very similar results and thus only the average for both targets is shown. F_m is seen to be fairly constant within the experimental error over the whole range of projectile energies, consistent with other similar determinations [27,85].

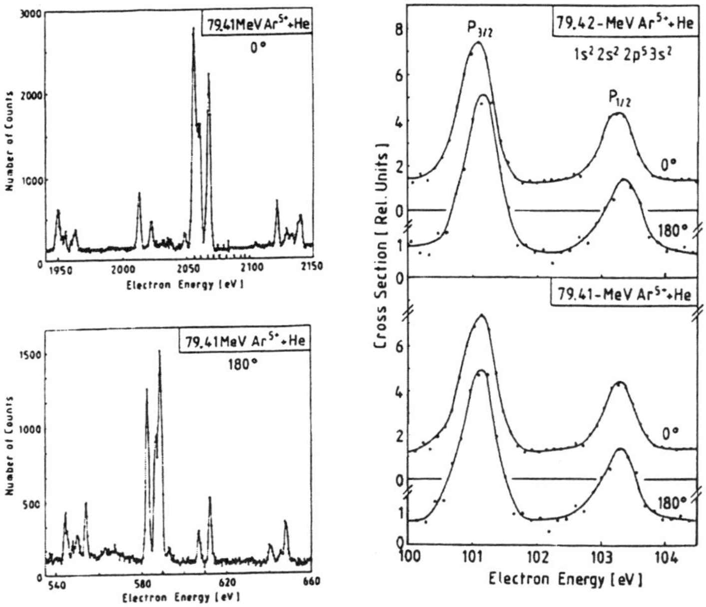

FIGURE 13.21 Ar-L-Auger spectra from collisions of Ar^{5+} with He. Left: Laboratory spectra correspond to rest frame ejection angles $\theta' = 0°$ (top) and 180° (bottom). Right: Projectile rest frame spectra. When a projectile energy of 79.42 MeV is chosen a slight mismatch between the rest frame lines is observed (top). However, when an energy of 79.41 MeV is used (bottom) good agreement is obtained. Reprinted from N. Stolterfoht, "High-resolution Auger spectroscopy in energetic ion atom collisions" [27], with permission.

13.5.2 Determination of the projectile beam energy–Accelerator energy calibration

The Auger line energies can be accurately determined only if the projectile energy is also accurately known. If t and ε are known, then the Auger energy ε' can be directly obtained via Eq. 13.1a. Measured laboratory electron energies have a relative uncertainty of $\sim 0.1 - 0.3$ eV. By taking a high resolution scan of the Cusp energy region a very good determination of t can be obtained. This method can been used to calibrate the energy of accelerators with the same or superior accuracy than possible with more conventional techniques [121,122].

In cases where both the forward (+) and backward (-) laboratory electron energy lines $\varepsilon_\pm$ can be observed (energy doubling), only possible for $t > \varepsilon'$ (see example in Fig. 13.6), a different method can be used to accurately determine both t and ε', via Eq. 13.4a. This is demonstrated in Fig. 13.21 where the projectile velocity is adjusted until the corresponding Auger lines coincide in the projectile frame. [27] This method yields high accuracies of $\sim 10^{-4}$.

In usual Auger spectra, small inaccuracies of 1-2 eV in the line energies can originate from contact potentials [123]. These are usually eliminated by making all components along the path of the electrons out of the same material. However, in zero-degree spectroscopy of fast projectiles, as a result of kinematic shifting of the lines to much higher energies, the role of contact potentials is much reduced [27].

Finally, spectra can also be calibrated by comparison to known target Auger lines [123]. Thus, for example, Ar-L_{23}-Auger lines in the 150-200 eV range [124], Ne-K-Auger lines in the 740-810 eV range [125,80] or Ar-KLL-Auger lines in the 2,500-2,700 eV range [126] have been used in the past.

13.6 STATE-SELECTIVE CROSS SECTION MEASUREMENTS

ZAPS has provided the field of ion-atom collisions with unique state-selective cross section information. A variety of measurements of single and double electron capture, transfer-excitation, excitation and ionization cross sections of few-electron projectile ions in collisions mostly with He and H_2 have been reported in recent literature. We present just a few of these as indicative of the quality of the information that can be obtained and the physical insights that can be gained.

13.6.1 Double electron capture

Double-electron-capture (DEC) studies can provide information on the role played by electron-electron interactions and correlation effects during the collision. With the advent of intense highly-charged ion sources such as EBIS and ECR (see the chapters on highly charged ion sources in this volume) and their expanding use in accelerator-based atomic physics, ZAPS has played an increasingly important role. To date, there have been many ZAPS measurements in the keV/u energy range where DEC is large. At higher collision energies less work has been reported.

150-500 keV $^3He^{2+}$ + He collisions

This was one of the first collision systems ever studied by ZAPS providing unique information on state-selective DEC in the 50-150 keV/u energy range in which very few measurements have been performed. In the collision process,

$$^3He^{++} + ^4He(1s^2) \rightarrow He^{**}(2lnl') + He^{++} \qquad \text{(2e}^- \text{ capture)}$$
$$\quad\quad\quad\quad\quad\quad\quad\quad \hookrightarrow He^{+}(1s) + e^{-} \qquad \text{(Auger)}$$

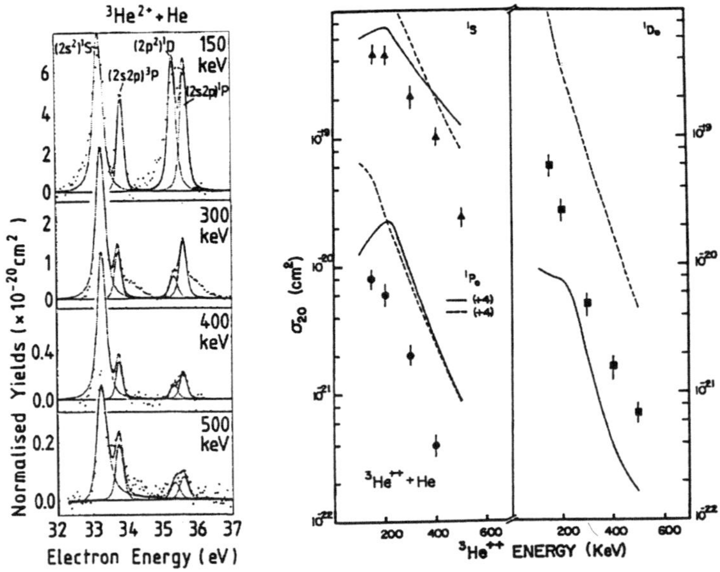

FIGURE 13.22 0° measurements of double electron capture for 150-500 keV $^3He^{2+}$+He collisions. Left: Electron spectra (projectile rest frame) for different projectile energies. Right: Projectile energy dependence of double-electron capture cross sections. Coupled channel AO+ calculations are also shown (lines). Dashed line: Both electrons captured with same probability. Continuous line: Each electron is captured with a different probability to account for possible relaxation of the target core after the capture of the first electron. Reprinted from "High resolution studies of two-electron processes observed in energetic ion-atom collisions via zero-degree Auger spectroscopy", T. J. M. Zouros, D. Schneider, and N. Stolterfoht, *Nuc. Instr. Meth.* **B31** (1988) 349, with kind permission from Elsevier Science - NL, Sara Burgerhartstraat 25, 1055 KV Amsterdam, The Netherlands.

two electrons are captured by the projectile. Doubly excited He**(2lnl′) (n $\geq$2) projectile states were observed to be formed. Absolute cross-sections were measured for the three lowest He autoionization states $(2s^2)^1S$, $(2p^2)^1D$ and $(2s2p)^1P$, which were well resolved in the overall electron spectra as demonstrated in Fig. 13.22 (left). Note that the ^3P state, also seen in these spectra, cannot be produced by DEC due to spin conservation. It is produced by double collisions [25].

In Fig. 13.22 (right), the cross-sections obtained are displayed as a function of projectile energy. Results are compared with an AO+ coupled channel calculation [127]. This is an independent-particle-model calculation which does not explicitly include the effects of electron-electron interactions. Furthermore, effects from excitation or ionization were not included. Thus, these calculations are only expected to be valid to within a factor of two [127].

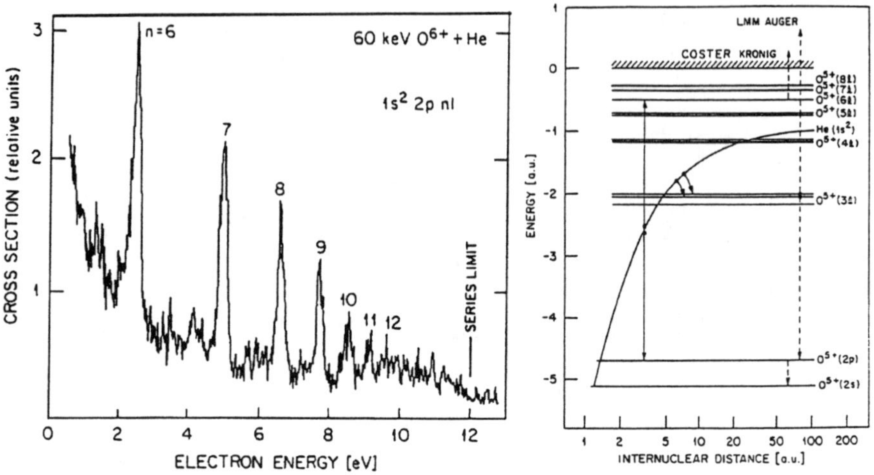

FIGURE 13.23 Double electron capture in 60 keV $O^{6+}(1s^2)$+He collisions. Left: 0° Coster-Kronig electron spectra (projectile rest frame). Right: Schematic diagram of orbital energies for the collision system. In the Coster-Kronig transitions, electrons are ejected from states with $n \geq 6$, due to the $2p \to 2s$ transition (from Ref. [30]).

60 keV O^{6+} + He collisions

This collision system studied by ZAPS provided some of the first evidence for correlation effects during DEC [30]. In this collision system, either a correlated one-step or uncorrelated two-step DEC can occur:

$$O^{6+}(1s^2) + {}^4He(1s^2) \to O^{4+}(1s^2 2pnl) + He^{++} \qquad \text{(correlated 2e}^-\text{-capture)}$$
$$\hookrightarrow O^{5+}(1s^2 2s) + e^- \qquad \text{(Coster-Kronig)}$$

$$O^{6+}(1s^2) + {}^4He(1s^2) \to O^{4+}(1s^2 3l3l') + He^{++} \qquad \text{(uncorrelated 2e}^-\text{-capture)}$$
$$\hookrightarrow O^{5+}(1s^2 2p) + e^- \qquad \text{(LMM Auger)}$$

In the first case, Coster-Kronig electrons result for $n \geq 6$ with lines in the 2-12 eV range in the projectile frame. In the second case, LMM Auger electrons result with lines in the 15-50 eV range in the projectile frame. In Fig. 13.23 (left) a spectrum of the Coster-Kronig lines interpreted as due to correlated DEC is shown. In Fig. 13.23 (right) a schematic diagram explains the two different DEC processes.

In slow collisions single electron capture is rather selective, mostly populating a single projectile orbital. Thus, two electrons captured in two independent events (uncorrelated DEC) mostly populate closely lying orbitals, leading to very nearly equivalent electron configurations. In Fig. 13.23, the $3l3l'$ configurations are seen to be populated by such a process. However, the production of non-equivalent electron configurations, as in the $2p6l$ configuration, leads to Coster-Kronig emission as seen in Fig. 13.23, thus providing a strong signature of correlated double capture. The correlation mechanism is seen to be auto-excitation, a process similar to autoionization except that the electron is excited to a Rydberg level rather than the continuum. [67]

13.6.2 Projectile-electron–target-electron interactions

In the past decade a variety of dielectronic processes, well known from electron-ion collisions, have also been identified experimentally in the interaction between inner-shell projectile-electrons and target-electrons during fast ion-atom collisions [11]. These electron–electron (e-e) interactions give rise to novel correlation mechanisms exposing the *dynamic* role of electrons in such collisions. We present a few selected results from ZAPS measurements of RTE, e-e Excitation and e-e Ionization. These three processes are explained schematically in Fig. 13.24.

Extensive studies of RTE [7,128], eeE and eeI [10] have shown that they can be understood within the same simple, intuitive framework based on the impulse approximation (IA) [73,11]. In the projectile frame and for fast enough collisions, the target-electron interacts with the projectile ion (nucleus+electrons) as a *free* particle with a momentum distribution broadened by its orbital motion. Thus, ion-atom e-e processes become linked to their electron impact counterparts [73,11]. In the projectile frame the IA gives the following simple result (in a.u.):

$$\sigma_{ee}(E_p) = \int \sigma_{eI}(\varepsilon')\, \frac{J(p_z)}{\sqrt{2t} + p_z}\, d\varepsilon' \tag{13.31}$$

This is very similar to the IA formula developed for the BEe production with p_z, $J(p_z)$, ε', and t defined as in Eqs. 13.21. The main difference is that the elastic-scattering (Rutherford) cross section has been replaced by the projectile electron impact cross section σ_{eI} of interest, and the BEe DDCS with the corresponding e-e ion-atom cross section $\sigma_{ee}(E_p)$. An integration over the momentum (energy) distribution of the impinging target electron is indicated. The connections between ion-atom processes and their electron impact counterparts are presented in Table 13.4.

Considering the simplicity of this approach surprisingly good agreement with experiments has been found. In particular, for RTE [128,7], e-e Excitation [64,42], e-e Ionization [129–131,8,9,132,133,10,11] and Binary Encounter electron production [102], the IA results are in good to excellent agreement with measured cross sections for a variety of ions in collision with H_2 and He targets.

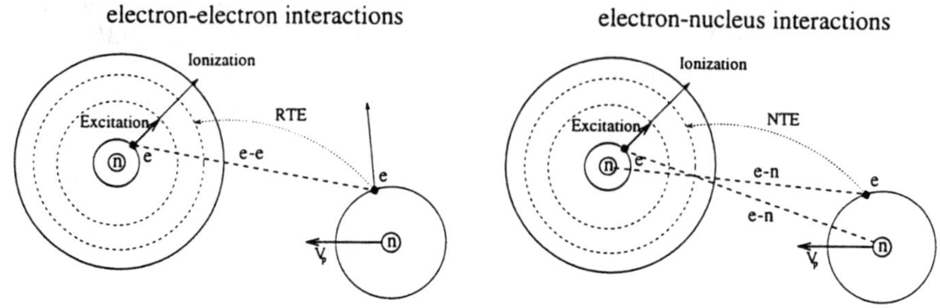

FIGURE 13.24 Projectile excitation can be mediated either through electron-electron (e-e) interactions (left) or electron-nucleus (e-n) interactions (right). Left: The target *electron* interacts with the projectile electron, either exciting it to a discrete (e-e Excitation) or continuum state (e-e Ionization). In the case where excitation is accompanied by simultaneous capture of the exciting target electron the process is known as Resonance Transfer Excitation (RTE). In e-e interactions the excitation of both electrons is correlated and the target is always excited or ionized. In the limit where the target electron can be considered to be *free*, these processes would correspond to the electron-ion scattering processes of electron impact Excitation, Ionization and Dielectronic Capture, respectively. Right: The target *nucleus* interacts with the projectile electron either exciting it to a discrete (e-n Excitation), or continuum state (e-n Ionization). In the case where excitation is accompanied by simultaneous capture of an electron the process is known as Non resonance Transfer Excitation (NTE). During e-n interactions, the excitation of the projectile is independent of the final state of the target electron, which usually remains in the ground state.

TABLE 13.4 Collision processes linked via the impulse approximation (IA)

Ion-Atom collision processes	$\Longleftrightarrow$ IA	**Electron-Ion** collision processes
Resonance Transfer Excitation (RTE)	$\leftrightarrow$	Dielectronic Capture (DC) or time-reversed Auger
e-e Excitation (eeE)	$\leftrightarrow$	electron Impact Excitation (eIE)
e-e Ionization (eeI)	$\leftrightarrow$	electron Impact Ionization (eII)
Binary Encounter electrons (BEe)	$\leftrightarrow$	Elastic electron Scattering

In general both e-e and e-n interactions are present simultaneously during the collision both giving rise to the same excited projectile state. Thus, signatures have to be discovered by which the e-e contributions can be identified over and above the usually stronger e-n contributions. We note that Eq. 13.31 gives the cross section *only* for the e-e interaction. The usually competing e-n interaction may be computed independently, within a Born treatment and just added to the e-e cross section. New theoretical approaches are becoming available in which both e-e and e-n interactions are included in a four-body theory and their contributions are added coherently, with the possibility of interference [134,135]. In other cases, the e-n "background" can be almost completely eliminated by a judicious choice of the experimental parameters [11]. In the following sections the different e-e signatures are demonstrated from recent results using ZAPS.

Resonance Transfer Excitation (RTE)

The *free* electron analogue of RTE is the time-reversed Auger effect, also known as dielectronic capture (DC). It is reminded that a free electron cannot be captured without losing energy. In dielectronic capture, this energy is lost to projectile excitation. Excitation to a particular projectile state requires exactly the right amount of energy and therefore this process is highly resonant. This can also been seen from the principle of detailed balance applied to the time-reversed Auger process. Thus, the cross section for dielectronic capture is well approximated by a delta-function in the electron collision energy ε' with $\sigma_{DC} \sim \delta(\varepsilon' - \varepsilon_A)$, ε_A being the Auger energy of the particular doubly-excited state of the projectile formed by dielectronic capture. Substituting σ_{DC} for σ_{eI} in Eq. 13.31 and performing the now trivial integration over the energy profile of the "beam" of target electrons we obtain (in a.u.):

$$\sigma_{RTE} \sim \frac{J(p_z)}{\sqrt{2t} + p_z} \tag{13.32a}$$

$$p_z = \sqrt{2}\,(\sqrt{\varepsilon_A + \varepsilon_I} - \sqrt{t}) \tag{13.32b}$$

Since $J(p_z)$ is maximum for $p_z = 0$ it is seen that at $t = (\varepsilon_A + \varepsilon_I)$, σ_{RTE} will be a maximum. Thus, RTE is seen to have a resonance at $E_p = \frac{M_p}{m}(\varepsilon_A + \varepsilon_I)$.

The doubly excited projectile state produced by RTE can decay back down to the ground state either by photon or Auger electron emission. Thus, the RTE resonance signature may be observed in the collision energy dependence of characteristic X rays or Auger electrons of the projectile. A resonance, at the predicted energy would verify RTE. The first observation of an RTE resonance in 1982 used K X rays in coincidence with projectile ions whose charge state was reduced by one due to electron transfer [136,137]. However, the X rays were measured in low resolution and thus more than one state could contribute. Just a few years later ZAPS provided the first *state-selective* measurements [23,35–38,138–140] of RTE. Some of these results are shown in Fig. 13.25. The RTE resonance for the Be-like fluorine ^3D state occurs around 20 MeV and is formed by $1s \to 2p$ excitation and capture to a $2p$ orbital. The Auger energy for this state is 567.8 eV, as seen from the Auger spectrum on the left. We note that by using H_2 as a target, e-n contributions due to NTE were minimized [37].

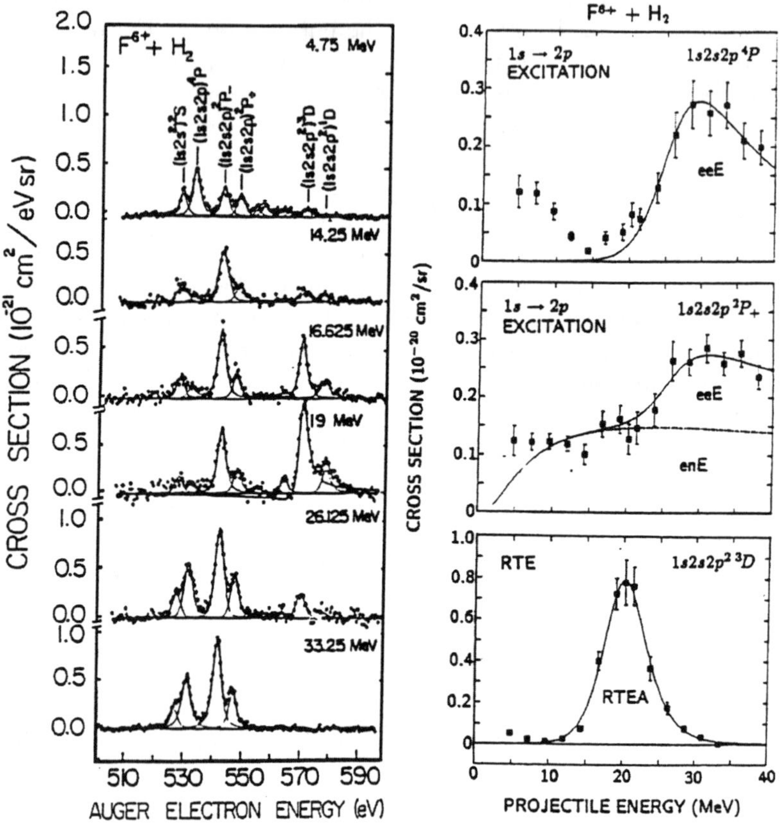

FIGURE 13.25 0° measurements of RTE and e-e Excitation in 5-33 MeV F^{6+}+H_2 collisions. Left: High resolution DDCS. Line assignment indicates the first four lines correspond to Li-like lines resulting from $1s \rightarrow 2s, 2p$ excitation. The higher energy lines correspond to Be-like lines formed by TE. Note the resonance behavior of the ^3D and ^1D lines as the collision energy is varied across the resonance energy at $\sim$ 20 MeV clearly identifying these lines to be produced by RTE. Also notice the behavior of the ^4P line which is first seen to decrease with collision energy and then after $\sim$ 19 MeV to start increasing again. This behavior indicates the onset of electron-electron Excitation, once the threshold for eeE is crossed. The projectile energy dependence of the single differential cross sections for the ^4P, $^2P_+$ and ^3D states (areas under the peaks) are plotted on the right. The $^2P_+$ state is formed by contributions from both enE and eeE. The ^4P can only be formed by eeE with spin exchange and by Transfer-Loss below threshold. Direct $1s \rightarrow 2p$ excitation is forbidden due to spin conservation. Furthermore, eeE contributions are zero below the excitation threshold. The ^3D state shows the characteristic resonance structure of RTE. Continuous lines are IA calculations, while the chain line is a Born calculation for the enE contribution (adapted from Ref. [73]).

electron-electron Excitation (eeE) and Ionization (eeI)

In the case of electron impact excitation or ionization of ions, there are no resonances as in DC to signal the onset of the e-e interaction. However, in both cases the electron must have enough energy to excite or ionize the ion. Thus, a threshold exists below which the cross section for electron impact is zero. This threshold should also be reflected in the ion-atom counterpart of these electron impact processes (see Table 13.4). In Fig. 13.25 (left) apart form the RTE lines already discussed the rest of the Auger lines are due Li-like configurations, which are produced by $1s \rightarrow 2\ell$ ($\ell = 0, 1$) excitation. In Fig. 13.25 (right) the differential cross sections measured at $\theta = 0°$ for $1s2s2p\,^2P_+$ and $1s2s2p\,^4P$ Auger production and their energy dependence are shown. $\Delta E_{1s\rightarrow 2p} \sim 723$ eV for F^{6+}, which is seen to correspond to a mean collision energy of about 25 MeV. While this would be a sharp threshold for electron impact excitation, in the case of e-e Excitation it is seen that the ion-atom threshold is smeared-out due to the much broader collision energy distribution of the target electron "beam".

It is interesting to note that in the example shown in Fig. 13.25, involving the production of the ^4P state in the excitation process $F^{6+}(1s^22s)\,^2S + H_2 \rightarrow F^{6+}(1s2s2p)\,^4P$, neither a direct e-e nor e-n excitation can give rise to the ^4P state since this would involve a spin flip of the projectile electron. It is well known that a pure Coulomb interaction cannot change the spin of the electron. However, if the active target and projectile electrons are *exchanged* it is possible to obtain a projectile ^4P state from an initial ^2S state. It is seen in Fig. 13.25(right top) that only e-e contributions are needed for agreement with the data above threshold, indicating the absence of e-n contributions, as expected. In the case of the production of the $(1s2s2p)^2P_+$ state both e-e and e-n can contribute and are thus added incoherently as indicated. It is reminded that the ^4P state is metastable and therefore the effective solid angle corrections due to the long lifetime, already discussed in section 13.4.2, must be properly applied to obtain the differential cross section shown in Fig. 13.25.

The case of e-e Ionization is shown in the energy dependence of K-shell ionization cross sections in Fig. 13.26. In the case of $1s$ ionization of C^{2+} the threshold for electron impact ionization is 340 eV which translates into a mean collision energy of ~ 7.5 MeV. For ionization, no sudden change in the energy dependence of the cross section is observed as is seen for e-e Excitation. The reason for this is simple: electron impact excitation involves discrete states and therefore has a very well defined excitation energy ΔE with a sharp threshold. Electron impact ionization involving transitions to a continuum of final states does not have a well defined excitation energy, thus leading to a broadened threshold [8,115].

We finally note that many new significant measurements of e-e Ionization have recently appeared, using techniques other than electron spectroscopy. These use mostly coincidence techniques between recoil target and scattered

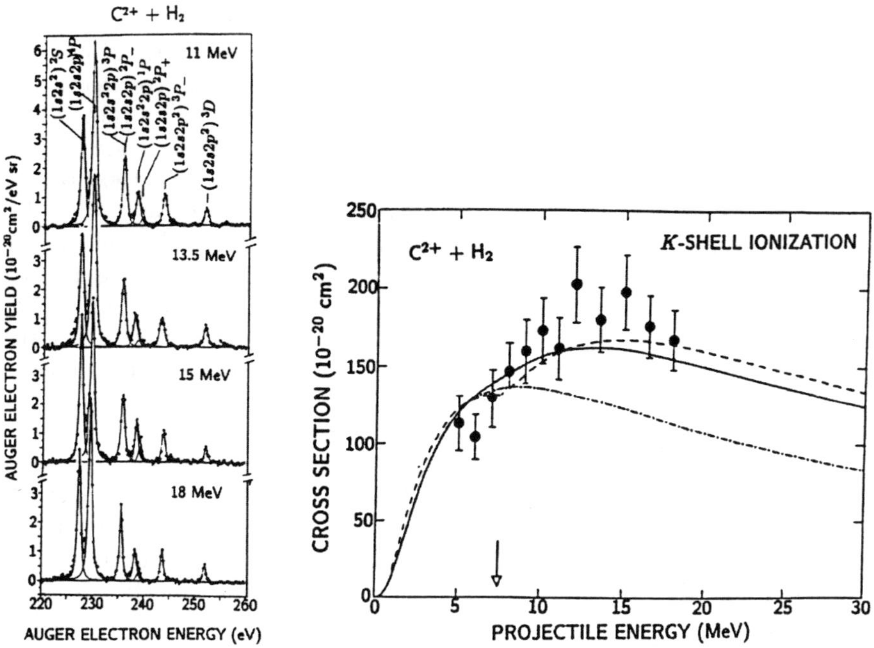

FIGURE 13.26 0° measurement of K-shell projectile ionization in 5-18 MeV C^{2+}+H_2 collisions. Left: High resolution normalized electron yields after subtraction of background continuum electrons and transformation to the projectile frame. Line assignment is similar to the case of O^{4+} shown in Fig. 13.19. The $1s2s^2\,{}^2S$ line results from $1s$ ionization of the ground state. The ${}^4P$, ${}^2P_-$ and ${}^2P_+$ states result from $1s$ ionization of the metastable $1s^22s2p\,{}^3P$ ion. All other lines are due to the decay of Be-like configurations produced by excitation. Right: Measured K-shell projectile ionization cross section, σ_K. Continuous line: PWBA-IA calculation of $\sigma_K = \sigma_{enI} + \sigma_{eeI}$. Dashed line: PWBA screening-antiscreening calculations of σ_K. Dash-Dotted line: PWBA calculation of σ_K considering only $e-n$ interaction. Arrow indicates projectile energy corresponding to the $1s$ ionization threshold for electron impact of the same velocity (adapted from Ref. [8]).

projectiles to reduce or isolate the contribution of competing e-n interactions [129,131,9,141,132].

13.6.3 Shape resonances of negative ions

The collisional detachment of an electron from a negative ion is a fundamental process in atomic physics. This process is particularly interesting when it involves a *resonance*; a transiently-formed state that decays via autodetachment [142]. This resonance lies energetically above its parent state such as the ground state of Li or the excited state of He($1s2p\,^3P$), as shown in the inserts of Fig. 13.27. Shape resonances have been studied in low energy electron–atom scattering [143–145] and, to a lesser extent, by photodetachment [146,147]. Recent observations of shape resonances in negative ion collisions with gas targets include the $H^-(2s2p\,^1P^o)$ [53–55], the $He^-(1s2p^2\,^4P)$ [56], and the $Li^-(2s2p\,^3P)$ and tentatively the $B^-(2p^2\,^1D)$ states [57].

Fig. 13.27 shows electron energy spectra obtained with ZAPS in collisions of 100 keV Li^- and He^- ions with a He gas target. A single-stage spherical-sector spectrometer with mean radius of 35 mm and external pre-retardation was used to achieve high resolution. A channeltron was used to detect the electrons. The spectrometer had an acceptance angle $\Delta\theta \sim 2^\circ - 6^\circ$ depending on where the electrons were detached in the target gas cell. The cell had a length of 37 mm and was placed about 40 mm from the spectrometer entrance slit, which was 2 mm in diameter. Typical measured negative ion currents were 100 pA. Also shown in the inserts are partial energy level diagrams for the Li/Li^- and He/He^- systems. A signature of the Li^- shape resonance is exhibited on both sides of the Cusp peak due to kinematic line doubling (see discussion in section 13.2.1) and the resonance state is identified as the $2s2p\,^3P^o$ state. It is noted that excitation of this resonant state from the ground state of Li^- requires the occurence of an electron *spin exchange* in the collision [148,143], as in the case of the e-e Excitation of the $1s2s2p\,^4P$ state discussed earlier. In the case of He^- the resonance structure dominates and the Cusp peak is not distinguishable, apparently due to a large shape resonance cross section and a very small ratio of resonance energy to the Cusp energy t. Shown in the figures as solid curves are least-square fits to the data points, which are essentially the predicted electron energy spectra. From the fits the energy and width of the resonance state are extracted.

The DDCS consist of contributions arising from the direct (non-resonant) and resonant detachment processes: [53,56,57]

$$\frac{d^2\sigma}{d\varepsilon' d\Omega'} = \left(\frac{d^2\sigma}{d\varepsilon' d\Omega'}\right)^{NR} + \frac{\alpha\epsilon' + \beta}{1 + \epsilon'^2}, \qquad (13.33)$$

where $(d^2\sigma/d\varepsilon' d\Omega')^{NR}$ is the direct (non-resonant) contribution which can be parameterized by using $\Sigma_l \Sigma_n a_{ln}(\varepsilon')P_l(\cos\theta')$ with a_{ln} being expressed in powers of ε' and P_l being Legendre polynomials with l being the angular momentum of the outgoing electron. The reduced energy variable is defined as $\epsilon' = 2(\varepsilon' - \varepsilon'_r)/\Gamma'$, where the primes refer to projectile frame parameters as used consistently through out this article. In principle, the resonant and non-resonant detachment channels may interfere. This possibility is included by using the Shore (or Fano) parameters α and β. The resonance is characterized in the usual manner by an energy ε'_r and a width Γ'.

The contribution from the resonances is shown in the bottom curves in the

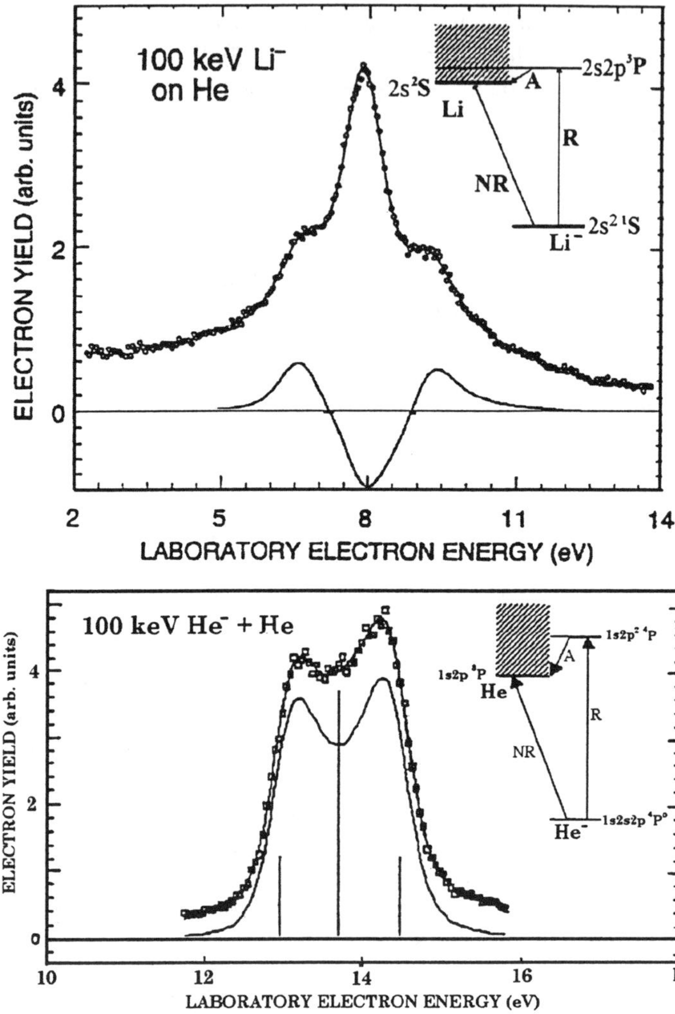

FIGURE 13.27 0° measurement of shape resonances using ZAPS in 100 keV Li^- and He^- + He collisions. The spectra indicate the central Cusp peak and the kinematically doubled resonance structures. The solid curve through the data points represent least square fits, while the bottom curves indicate the contribution of the shape resonances including interference. Inserts: Partial energy level diagrams for Li/Li^- and He/He^- systems. Resonant (R) and non-resonant (NR) detachment is shown. The shape resonance decays via autodetachment (A) leaving the residual atom in the ground state, in the case of Li and in the $1s2p\,^3P$ state for He. The resonance parameters extracted from the fits are for Li, $\varepsilon_r' = 50 \pm 6$ meV and $\Gamma' = 64 \pm 25$ meV [57] and for He, $\varepsilon_r' = 11$ meV and $\Gamma' = 8$ meV (adapted from Ref. [57]).

figure. It can be seen that there exists appreciable destructive interference, for the case of Li^- detachment, in the region of the cusp corresponding to the threshold region in the ion frame. This could possibly be due to the presence of a virtual state such as the theoretically predicted ^3S state [144]. The extracted values of the resonance parameters allow for the identification of the observed structure with the $2s2p\,^3P$ shape resonance state in Li^-, which until now [57] has defied experimental investigation even though it has been the subject of numerous calculations over the past few decades [143,144,149]. This is primarily due to the experimental difficulty of detecting the very low energy electrons ejected in the autodetaching decay and the fact that the state is not efficiently optically coupled to the ^1S ground state of Li^-, thereby prohibiting photodetachment studies. These problems however are overcome using ZAPS since the electron energies are now kinematically shifted into a more experimentally accessible energy region. For example, in the measurements at 100 keV the ratio of the energies of the detached electrons in the laboratory frame to those in the ion frame was more than two orders of magnitude. The measured energy and width of the $He^-(1s2p^2\,^4P)$ shape resonance [56] agreed well with the more accurate results obtained by Walter *et al.* [147] who employed photoabsorption to selectively excite this resonance state.

13.7 FUTURE PROSPECTS

Zero-degree Auger projectile spectroscopy can be relied to provide state-selective cross section information about ion-atom collision processes for years to come, particularly for low Z-ions ($Z \leq 10$) where comparable x-ray techniques are much more tedious. Past and present work has focused primarily on singles K-Auger spectra of few electron ions in collisions with H_2 and He targets, conditions which give rise to rather simple and well resolved lines.

Future applications could be extended to heavier projectiles, L-Auger studies and to ions with more than 4 electrons, as well as collisions utilizing heavier, crystalline or laser-excited targets. High resolution measurements in coincidence with the scattered projectile could provide information on the orientation [99,100] of the intermediate formed states and help identify collision processes under non-needle ionization conditions. Measurements of electrons in coincidences with recoils could provide more information about e-e interactions [115], particularly when combined with the technique of recoil ion momentum spectroscopy [150] (see also corresponding chapter in this volume).

In the last few years we have also witnessed the coming of age of ion storage rings and ion traps for use in atomic physics, providing new possibilities for studying both ion-atom and electron–ion collision processes with unprecedented beam energy resolution and intensity. ZAPS has yet to be used in these new collision environments. In particular, applied to the study of cooled ions in storage rings, ZAPS should in principle provide electron spectra of unprecedented quality given the much higher beam intensity and smaller energy spread of these beams. Applications to the study of ions excited by electron impact are also under way [151].

The viability of these new applications of ZAPS will necessarily require the use of much more efficient electron spectrometers utilizing position sensitive electron detectors [99,100] and possibly double focusing devices. Already such analyzers are under construction and are expected to provide the 2nd generation ZAPS devices.

ACKNOWLEDGMENTS

We are particularly indebted to our colleagues Pat Richard at Kansas State University and Nico Stolterfoht at HMI-Berlin. We would like to thank them for their unmitigated support and guidance through out our extensive collaboration. Much of the reported development of ZAPS was undertaken in the stimulating and well equiped environments of the HMI and Kansas State collision laboratories.

We would also like to thank Geraldo Sigaud of PUC-Rio for his critical reading of the manuscript, Gabor Toth and Martin Grether for communicating unpublished results and Nektarios Sartzetakis for his kind technical support with the electronic conversion of the figures.

T.J.M.Z. would like to acknowledge partial support by the Division of Chemical Sciences, Office of Basic Energy Science, Office of Energy Research, U.S. Department of Energy and NATO collaborative research grant CRG-910567.

REFERENCES

1. R. K. Janev and H. Winter, Phys. Rev. **117**, 265 (1985).
2. H. P. Summers and W. J. Dickson, in *Recombination of Atomic Ions*, NATO Advanced Study Institute Series B: Physics, edited by W. G. Graham, W. Fritsch, Y. Hahn, and J. Tanis (Plenum Publishing Corp, New York, 1992), Vol. 296, pp. 31–48.
3. R. L. Becker, A. L. Ford, and J. F. Reading, Phys. Rev. A **29**, 311 (1984).
4. N. Stolterfoht, in *Spectroscopy and Collisions of Few-Electron Ions*, edited by M. Ivascu, V. Florescu, and V. Zoran (World Scientific, Singapore, New Jersey, London, 1989), p. 342.
5. N. Stolterfoht, Phys. Scr. **42**, 192 (1990).
6. J. H. McGuire, Adv. At. Mol. & Opt. Phys. **29**, 217 (1992).
7. T. J. M. Zouros, in *Recombination of Atomic Ions*, NATO Advanced Study Institute Series B: Physics, edited by W. G. Graham, W. Fritsch, Y. Hahn, and J. Tanis (Plenum Publishing Corporation, New York, 1992), Vol. 296, pp. 271–300.
8. D. H. Lee *et al.*, Phys. Rev. A **46**, 1374 (1992).
9. E. C. Montenegro, W. E. Meyerhof, and J. H. McGuire, Adv. At. Mol. & Opt. Phys. **34**, 249 (1994).

10. T. J. M. Zouros, in *Applications of Particle and Laser Beams in Materials Technology*, NATO Advanced Study Institute Series E: Applied Sciences, edited by P. Misailides (Kluwer Academic Publishers, Netherlands, 1995), Vol. 283, pp. 37–52.
11. T. J. M. Zouros, Comments At. & Mol. Phys. **32**, 291 (1996).
12. I. A. Sellin, in *Topics in Current Physics*, edited by S. Bashkin (Springer, Heidelberg, 1976), Vol. 1, p. 265.
13. D. L. Matthews, in *Atomic Physics: Accelerators*, Vol. 17 of *Methods of Experimental Physics*, edited by P. Richard (Academic Press, New York, 1980), pp. 433–527.
14. M. E. Rudd and J. H. Macek, in *Case Studies in Atomic Physics*, edited by E. W. McDaniels and M. C. McDowell (North–Holland, Amsterdam, 1972), Vol. 3, pp. 47–136.
15. N. Stolterfoht *et al.*, Phys. Rev. A **12**, 1313 (1975).
16. R. Bruch *et al.*, Phys. Rev. A **19**, 587 (1979).
17. M. Rødbrø *et al.*, J. Phys. B **10**, L483 (1977).
18. M. Rødbrø, R. Bruch, and P. Bisgaard, J. Phys. B **12**, 2413 (1979).
19. R. Morgenstern, A. Niehaus, and U. Thielmann, Phys. Rev. Lett. **37**, 199 (1976).
20. R. Morgenstern, A. Niehaus, and U. Thielmann, J. Phys. B **10**, 1039 (1977).
21. R. Morgenstern, A. Niehaus, and G. Zimmermann, J. Phys. B **13**, 4811 (1980).
22. A. Itoh *et al.*, J. Phys. B **16**, 3965 (1983).
23. A. Itoh *et al.*, Phys. Rev. A **31**, 684 (1985).
24. A. Itoh *et al.*, J. Phys. B **18**, 4581 (1985).
25. T. J. M. Zouros, D. Schneider, A. Itoh, and N. Stolterfoht, Phys. Rev. A **35**, 1963 (1987).
26. T. J. M. Zouros, D. Schneider, and N. Stolterfoht, Nucl. Instrum. Methods Phys. Res. Sect. B **31**, 349 (1988).
27. N. Stolterfoht, Phys. Rep. **146**, 315 (1987).
28. N. Stolterfoht, J. Electr. Spectr. and Rel. Phenom. **67**, 309 (1994).
29. M. L. A. Raphaelian, H. G. Berry, and N. B. Mansour, Phys. Rev. A **43**, 4071 (1991).
30. N. Stolterfoht *et al.*, Phys. Rev. Lett. **57**, 74 (1987).
31. R. Mann and H. Schulte, Z. Phys. D **4**, 343 (1987).
32. R. Mann, Phys. Rev. A **35**, 4988 (1987).
33. H. A. Sakaue *et al.*, J. Phys. B **23**, L401 (1990).
34. H. A. Sakaue *et al.*, J. Phys. B **24**, 3787 (1991).
35. J. K. Swenson *et al.*, Phys. Rev. Lett. **57**, 3042 (1986).
36. T. J. M. Zouros, D. Schneider, and N. Stolterfoht, J. Phys. B **21**, L671 (1988).
37. T. J. M. Zouros *et al.*, Phys. Rev. A **40**, 6246 (1989).
38. T. J. M. Zouros, C. P. Bhalla, D. H. Lee, and P. Richard, Phys. Rev. A **42**, 678 (1990).
39. D. H. Lee *et al.*, Phys. Rev. A **44**, 1636 (1991).
40. T. J. M. Zouros, D. H. Lee, and P. Richard, Phys. Rev. Lett. **62**, 2261 (1989).
41. S. Ricz, B. Sulik, N. Stolterfoht, and I. Kádár, Phys. Rev. A **47**, 1930 (1993).

42. B. Sulik *et al.*, Phys. Rev. A **52**, 387 (1995).
43. T. J. M. Zouros, D. H. Lee, P. Richard, and J. M. Sanders, Nucl. Instrum. Methods Phys. Res. Sect. B **56/57**, 107 (1991).
44. R. Bruch *et al.*, Phys. Rev. A **35**, 4114 (1987).
45. P. Focke *et al.*, Phys. Rev. A **40**, 5633 (1989).
46. I. Kádár *et al.*, Phys. Rev. A **44**, 2900 (1991).
47. M. Sataka *et al.*, Phys. Rev. A **44**, 7290 (1991).
48. D. Schneider *et al.*, Phys. Rev. A **34**, 169 (1986).
49. W. D. Zeitz, R. Kowallik, and D. Schneider, Phys. Rev. A **39**, 43 (1989).
50. B. D. DePaola *et al.*, J. Phys. B **29**, 1247 (1996).
51. C. Liao *et al.*, Nucl. Instrum. Methods Phys. Res. Sect. B **98**, 324 (1995).
52. T. J. M. Zouros *et al.*, Nucl. Instrum. Methods Phys. Res. Sect. B **99**, 27 (1995).
53. L. H. Andersen, J. P. Bangsgaard, and J. Sørensen, Phys. Rev. Lett. **57**, 1558 (1986).
54. M. M. Duncan and M. G. Menendez, Phys. Rev. A **39**, 1534 (1989).
55. F. Penent, J. P. Grouard, J. L. Montmagnon, and R. I. Hall, J. Phys. B **25**, 2831 (1992).
56. P. A. Zavodsky, L. Sarkadi, L. Víkor, and J. Pálinkás, Phys. Rev. A **50**, 899 (1994).
57. D. H. Lee, W. D. Brandon, D. Hastorp, and D. J. Pegg, Phys. Rev. A **53**, R633 (1996).
58. Y. Yamazaki *et al.*, Phys. Rev. Lett. **57**, 992 (1986).
59. Y. Yamazaki *et al.*, Phys. Rev. Lett. **61**, 2913 (1988).
60. J. K. Swenson *et al.*, Phys. Rev. Lett. **63**, 35 (1989).
61. J. K. Swenson *et al.*, Phys. Rev. Lett. **66**, 417 (1991).
62. D. H. Lee *et al.*, Nucl. Instrum. Methods Phys. Res. Sect. B **56/57**, 99 (1991).
63. N. Stolterfoht *et al.*, Nucl. Instrum. Methods Phys. Res. Sect. B **24/25**, 168 (1987).
64. T. J. M. Zouros *et al.*, Nucl. Instrum. Methods Phys. Res. Sect. B **40/41**, 17 (1989).
65. P. Richard, in *X-Ray and Inner-Shell Processes*, edited by T. A. Carlson, M. O. Krause, and S. T. Manson (AIP, New York, 1990), p. 315.
66. B. D. DePaola, Nucl. Instrum. Methods Phys. Res. Sect. B **56/57**, 154 (1991).
67. N. Stolterfoht, Phys. Scr. **T51**, 39 (1994).
68. N. Stolterfoht, in *Two-Center Effects In Ion-Atom Collisions*, edited by T. J. Gay and A. F. Starace (American Institute of Physics Conference Series, New York, 1996), Vol. 362, p. 163.
69. Y. S. Gordeev and G. N. Ogurtsov, Sov. Phys. JETP **33**, 1105 (1971).
70. J. R. Risley, A. K. Edwards, and R. Geballe, Phys. Rev. A **9**, 1115 (1974).
71. A. Gleizes, P. Benoit-Cattin, A. Bordenave-Montesquieu, and A. Merchez, J. Phys. B **9**, 476 (1976).
72. P. Dahl, M. Rødbrø, B. Fastrup, and M. E. Rudd, J. Phys. B **9**, 1567 (1976).
73. D. H. Lee, Ph.D. dissertation, Kansas State University, 1990, (unpublished).
74. T. J. M. Zouros and M. Wilson, in *Europhysics Conference Abstracts*, edited

by R. C. Thompson (European Physical Society, Edinburgh, UK, 1995), Vol. 19A, p. 67.
75. D. Schneider *et al.*, Nucl. Instrum. Methods Phys. Res. Sect. B **24/25**, 173 (1987).
76. D. H. Lee *et al.*, Nucl. Instrum. Methods Phys. Res. Sect. B **40/41**, 1229 (1989).
77. B. D. DePaola *et al.*, Nucl. Instrum. Methods Phys. Res. Sect. B **40/41**, 187 (1989).
78. P. G. Burke and D. D. McVicar, Proc. Phys. Soc. **86**, 989 (1965).
79. D. Schneider, C. F. Moore, and B. M. Johnson, J. Phys. B **9**, L153 (1976).
80. N. Stolterfoht, H. Gabler, and U. Leithäuser, Phys. Rev. Lett. **45**, 351 (1973).
81. N. Stolterfoht, D. Schneider, R. Mann, and F. Folkmann, J. Phys. B **10**, L281 (1977).
82. H. Tawara, P. Richard, K. A. Jamison, and T. J. Gray, J. Phys. B **11**, L615 (1978).
83. H. Tawara *et al.*, Phys. Rev. A **20**, 2340 (1979).
84. H. Tawara *et al.*, Phys. Rev. A **19**, 1960 (1979).
85. M. Terasawa *et al.*, Phys. Rev. A **27**, 2868 (1983).
86. M. Zhu *et al.*, Nucl. Instrum. Methods Phys. Res. Sect. B **98**, 351 (1995).
87. K. D. Sevier, *Low Energy Electron Spectrometry* (Wiley, New York, 1972).
88. D. W. O. Heddle, *Electrostatic Lens Systems* (Adam Hilger, Bristol, 1991).
89. M. Grether, Ph.D. dissertation, TU-Berlin, 1995.
90. J. H. Moore, C. C. Davis, and M. A. Coplan, *Building Scientific Apparatus* (Addison-Wesley, London, 1983).
91. G. A. Harrower, Rev. Sci. Instrum. **26**, 850 (1955).
92. V. P. Afanas'ev and S. Y. Yavor, Sov. Phys. Tech. Phys. **20**, 715 (1976), [Translation of Zh. Tekh. Fiz. **45**, 1137-70 (1975)].
93. W. Steckelmacher and M. W. Lucas, J. Phys. E **12**, 961 (1979).
94. P. Allard and J.-D. Carette, Can. J. Phys. **49**, 2132 (1971).
95. Y. Delage and J.-D. Carette, Can. J. Phys. **49**, 2118 (1971).
96. D. Roy and J.-D. Carette, Can. J. Phys. **49**, 2138 (1971).
97. N. Stolterfoht, in *Fundamental Processes in Energetic Atomic Collisions*, edited by H. O. Lutz, J. S. Briggs, and H. Kleinpoppen (Plenum, New York, 1983), Vol. 103, pp. 295–318.
98. M. Mack *et al.*, Phys. Rev. A **39**, 3846 (1989).
99. H. Khemliche, M. H. Prior, and D. Schneider, Phys. Rev. Lett. **74**, 5013 (1995).
100. M. H. Prior and H. Khemliche, in *XIX International Conference on the Physics of Electronic and Atomic Collisions (Abstracts 19th ICPEAC)*, edited by L. J. Dubé, J. B. A. Mitchell, J. W. McConkey, and C. E. Brion (American Institute of Physics Conference Series, Woodbury, NY, 1995), Vol. 360, p. 557.
101. F. Bordoni, Nucl. Instrum. & Methods **97**, 405 (1971).
102. D. H. Lee *et al.*, Phys. Rev. A **41**, 4816 (1990).
103. B. Sulik and N. Stolterfoht, in *Accelerator-based atomic physics techniques and applications*, edited by S. M. Shafroth and J. C. Austin (American Institute

of Physics Conference Series, New York, 1996).
104. N. Stolterfoht *et al.*, Phys. Rev. Lett. **33**, 59 (1974).
105. J. S. Lee, J. Chem. Phys. **66**, 4906 (1977).
106. P. D. Fainstein, V. H. Ponce, and R. D. Rivarola, J. Phys. B **22**, 1207 (1989).
107. L. Gulyás, P. D. Fainstein, and A. Salin, J. Phys. B **28**, 245 (1995).
108. P. Richard *et al.*, J. Phys. B **23**, L213 (1990).
109. C. P. Bhalla and R. Shingal, J. Phys. B **24**, 3187 (1991).
110. D. R. Schultz and R. E. Olson, J. Phys. B **24**, 3409 (1991).
111. R. L. Arnoldy, P. O. Isaacson, D. F. Gats, and L. W. Choy, Rev. Sci. Instrum. **44**, 172 (1973).
112. C. W. Woods *et al.*, Phys. Rev. A **13**, 1358 (1976).
113. T. J. M. Zouros *et al.*, in *DAMOP*, edited by APS (APS, New York, 1994).
114. T. J. M. Zouros *et al.*, Phys. Rev. A **49**, 3155 (1994).
115. T. J. M. Zouros *et al.*, Phys. Rev. A **53**, 2272 (1996).
116. T. Brage and C. F. Fisher, Phys. Rev. A **44**, 72 (1991).
117. T. Andersen *et al.*, Phys. Rev. A **47**, 890 (1993).
118. M. H. Chen, B. Crasemann, and H. Mark, Phys. Rev. A **27**, 544 (1983).
119. S. Schumann, K. O. Groenveld, G. Nolte, and B. Fricke, Z. Phys. A **289**, 245 (1979).
120. We note that correction factor G_τ is equivalent to factor $\frac{1}{G_\Omega}$ introduced in Ref. [8].
121. R. Mann, S. Hagmann, and L. Weitzel, Nucl. Instrum. Methods Phys. Res. Sect. B **34**, 403 (1988).
122. H. I. Hidmi, P. Richard, and I. Ben-Itzhak, Nucl. Instrum. Methods Phys. Res. Sect. B **88**, 313 (1994).
123. W. Mehlhorn, Electron spectrometry of Auger and autoionizing states: Experiment and Theory, Univ. of Aarhus summer lecture notes (unpublished) (1978).
124. N. Stolterfoht, D. Schneider, and P. Ziem, Phys. Rev. A **10**, 81 (1974).
125. M. O. Krause *et al.*, Phys. Lett. **31A**, 81 (1970).
126. J. J. Mackey *et al.*, J. Phys. B **7**, L447 (1974).
127. A. Jain, C. D. Lin, and W. Fritsch, Phys. Rev. A **39**, 1741 (1989).
128. J. A. Tanis, in *Recombination of Atomic Ions*, NATO Advanced Study Institute Series B: Physics, edited by W. G. Graham, W. Fritsch, Y. Hahn, and J. Tanis (Plenum Publishing Corp, New York, 1992), Vol. 296, pp. 241–257.
129. H. P. Hülskötter, W. E. Meyerhof, E. Dillard, and N. Guardala, Phys. Rev. Lett. **63**, 1938 (1990).
130. M. B. Shah and H. B. Gilbody, J. Phys. B **24**, 977 (1991).
131. E. C. Montenegro, W. S. Melo, W. E. Meyerhof, and A. G. de Pinho, Phys. Rev. Lett. **69**, 3033 (1992).
132. W. Wu *et al.*, Phys. Rev. Lett. **72**, 3170 (1994).
133. R. Moshammer *et al.*, Phys. Rev. Lett. **73**, 3371 (1994).
134. R. Gayet, J. Hanssen, A. Martinez, and R. Rivarola, Comments At. & Mol. Phys. **30**, 231 (1994).
135. R. Gayet, J. Hanssen, and L. Jacqui, J. Phys. B **28**, 2193 (1995).

136. J. A. Tanis *et al.*, Phys. Rev. Lett. **49**, 1325 (1982).
137. J. A. Tanis *et al.*, Phys. Rev. Lett. **53**, 2551 (1984).
138. D. H. Lee *et al.*, in *12th Int. Conf. on Atomic Physics: Abstracts of Contributed Papers*, edited by W. Baylis, G. W. F. Drake, and J. W. McConkey (Univ. of Michigan, Ann Arbor, MI, 1990).
139. R. C. Parameswaran, C. P. Bhalla, B. P. Walch, and B. D. DePaola, Phys. Rev. A **43**, 5929 (1991).
140. R. Parameswaran *et al.*, Phys. Rev. A **47**, 3801 (1993).
141. R. Dörner *et al.*, Phys. Rev. Lett. **72**, 3166 (1994).
142. H. S. W. Massey, *Negative Ions* (Cambridge Univ. Press, New York, 1976).
143. D. W. Norcross, J. Phys. B **4**, 1458 (1971).
144. A.-L. Sinfailam and R. K. Nesbet, Phys. Rev. A **7**, 1987 (1973).
145. J. Yuan and Z. Zhang, J. Phys. B **22**, 2751 (1989).
146. H. C. Bryant *et al.*, Phys. Rev. Lett. **38**, 228 (1977).
147. C. W. Walter, J. A. Seifert, and J. R. Peterson, Phys. Rev. A **50**, 2257 (1994).
148. N. F. Mott and H. S. W. Massey, *The Theory of Atomic Collisions* (Oxford University Press, London, 1965).
149. I. I. Fabrikant, Opt. Spectros. (USSR) **53(2)**, 131 (1982).
150. J. Ullrich *et al.*, Comments At. & Mol. Phys. **30**, 285 (1994).
151. A. Frank, A. Müller, A. Liedtke, and H. Schnell, Annual report, Institut für Strahlenphysik Universität Stuttgart (1995).

CHAPTER 14

Molecular Imaging with Fast Ion Beams

Daniel Zajfman and Oded Heber *

This chapter presents the basic principles and recent developments in the field of molecular imaging techniques with fast ion beams. It is explained why imaging of molecular dissociation makes it possible to increase our understanding of dissociation processes as well as molecular structure, and enable full control over the parameters (such as internal excitation and kinetic energy release) that dominate the breaking up of molecules. The two most relevant geometries, in which imaging detectors have been used, are described, and their respective advantages and disadvantages are discussed. Examples of imaging for the study of various dissociation processes and molecular geometry are also presented.

14.1 INTRODUCTION

The use of imaging for the study of molecular dissociation processes has mushroomed in many laboratories during the last decades. It is a very powerful tool for the study of the internal mechanisms leading to molecular dissociation, and for the investigation of molecular structure as well as for the investigation of branching ratio of the dissociation products. These ratios are important, not only for the basic understanding of molecular dissociation, but also because they have practical aspects in many fields such as astrophysical chemistry [1], plasma physics [2], and others.

As an example, let us assume that a diatomic molecule AB dissociates in two fragments A+B, due to an external interaction. The first values of interest are probably the energy threshold for the dissociation and the dissociation cross section. However, these values have physical meaning only if the initial state of the molecule (electronic, vibrational and rotational states) and the

---

*) Department of Particle Physics, The Weizmann Institute of Science, Rehovot, 76100, Israel

final states of the fragments (electronic state only for atomic fragments) are known.

More specifically, for the comparison between the experimental results for the dissociation of AB into A+B and theoretical calculations to be valid, a detailed knowledge of the initial state of the molecule and final state of the products is required. Since in all processes of dissociation some kinetic energy is released (in the molecule center-of-mass frame of reference) to the dissociation products, it can be shown that, in many cases, the measurement of the asymptotic velocities of all fragments can yield unique information on the initial quantum state of AB as well as on the final states of A+B. The measurement of the fragment asymptotic velocities from a dissociating molecule is called molecular imaging.

Various molecular processes such as photodissociation [3], electron dissociation [4], atom-molecule collision dissociation [12], and molecular ion - solid interaction [6,7] have been studied using imaging techniques. In this Chapter, we will restrict ourselves to the technique of molecular imaging with fast ion beams. This includes the interaction of fast molecular ion beams with electrons and atoms and the interaction of fast atomic ions with molecules. We will concentrate mainly on experiments where the correlated velocities of all the fragments from a single dissociation are measured. This requires that the measurement be done one molecule at a time, so that fragments reaching the detector simultaneously originate from the same molecule. As will be shown below, this condition puts stringent requirements on the detector, but it enables the understanding of the full "reaction path" of a specific dissociation process.

In the next Section, a short description of molecular dissociation processes will be presented, with emphasis on the parts which are relevant to imaging technique. This will be followed by a section devoted to the basic principles behind molecular imaging with fast beams, as well as the general requirements for the detection of the final velocities of the dissociating fragments. The next part presents various detectors and electronics used in present imaging experiments, and the last section includes a few examples of molecular imaging with fast ion beams.

14.2 MOLECULAR DISSOCIATION PROCESSES

When a molecule dissociates as a result of a perturbation which is fast compared to the dissociation time, the process can in general be described in two steps. The first part is the perturbation itself, which depends on the type of interaction, the collision energy and the initial state of the molecule. In this step, the molecule is brought from an initial bound state to a dissociative state (see Fig. 14.1). The second step is the dissociation process, which depends mainly on the molecular properties (i.e. potential curves or surfaces), and has only minor dependence on the type of interaction (i.e. electron, atom or photon) which started the dissociation. Most often, the dissociation occurs along a dissociative state, although sometimes, the dissociation can be due to a tunelling process, in which case there is an intermediate state which is quasi-bound (i.e. a local minimum exists in the potential curve) [8]. In any case, when the dissociation process is complete, the released fragments have some kinetic energy, relative to the frame of reference of the initial molecule. The asymptotic velocities of the fragments contain important information on the initial state of the molecule, and/or on the final states of the fragments. This information reflects mainly the molecular properties, and is almost independent of the initial interaction, thus making it possible to track the 'reaction path', in terms of the potential curves that the molecule followed during the dissociation. Hence, the measurement of the final velocities of the fragments permits a detailed understanding of the dissociation process.

In the following, we will very briefly review the few types of molecular dissociation that are relevant to the imaging technique with fast ion beams. This is not intended to be a tutorial on molecular dissociation principles, but more an introduction that is biased toward what is needed to understand the rest of this Chapter. For simplicity, we will describe the relevant quantities based on the dissociation of a diatomic molecule.

14.2.1 Collision-induced dissociation: continuous energy release

Let us assume that a diatomic molecule AB in its ground ro-vibrational state can be described as an harmonic oscillator within the Born-Oppenheimer approximation (see Fig. 14.1). In this case, the nuclear wave function is a Gaussian and the probability P(r) of finding the two nuclei at a relative distance r is given by:

$$P(r) = A \exp -\frac{(r-r_0)^2}{\sigma^2} \tag{14.1}$$

where

$$\sigma = \sqrt{\frac{\hbar}{\mu\omega}}, \tag{14.2}$$

r_0 is the equilibrium bond length, μ is the reduced mass of the molecule, and ω is the harmonic oscillator zero-point frequency. Assume now that due to an external perturbation such as in a fast molecule-atom or molecule-electron collision, the molecule undergoes a transition from the bound state to a dissociative state E(r) (see Fig. 14.1).

If the perturbation is fast, the transition will be so called "vertical", and the molecule will start to dissociate along the dissociative state E(r). The relevant question is 'What are the final kinetic energy distributions of the fragments

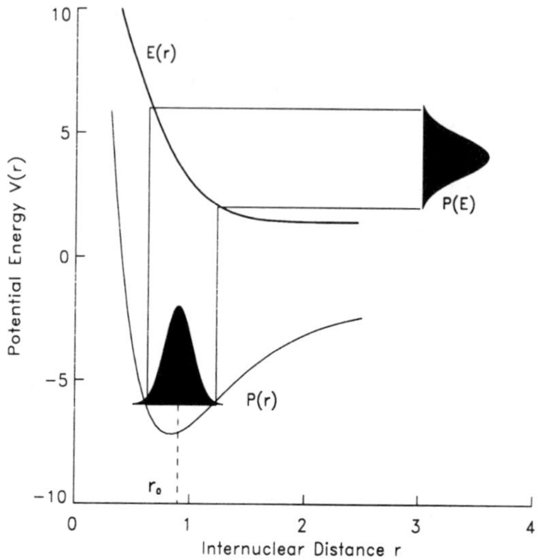

FIGURE 14.1 Schematic description of the transformation from P(r) to P(E) in a collision induced dissociation yielding to a continuous energy release (arbitrary units).

A and B after the dissociation is completed?'. The answer can be based on classical or quantum arguments, which give in this case the same solution. Classically, one needs to consider probability conservation: The probability to find the fragments A and B with a relative kinetic energy between E and E+dE is given by:

$$P(E)dE = P(r)dr \rightarrow P(E) = P(r)\left|\frac{dr}{dE}\right| \tag{14.3}$$

Graphically, the P(E) distribution can be represented using the projection (or reflection) of the initial probability distribution on the energy axis, through the dissociation curve (See Fig. 14.1).

Thus, if the function E(r) is known, one can easily calculate P(E). A simple case, which will also be studied as an example later, is when E(r) is the Coulomb potential. In this case, $E(r)=q_a q_b/r$, where q_a and q_b are the final charges of the fragments A and B, respectively. The distribution of kinetic energy P(E) can be deduced from Eq. 3:

$$P(E) = A\frac{q_a q_b}{E^2}\exp\left[-\frac{\frac{q_a q_b}{E} - r_0}{\sigma^2}\right] \tag{14.4}$$

If the relative kinetic energy distribution of the fragments P(E) can be measured, one can reconstruct the original vibrational wave function (squared)

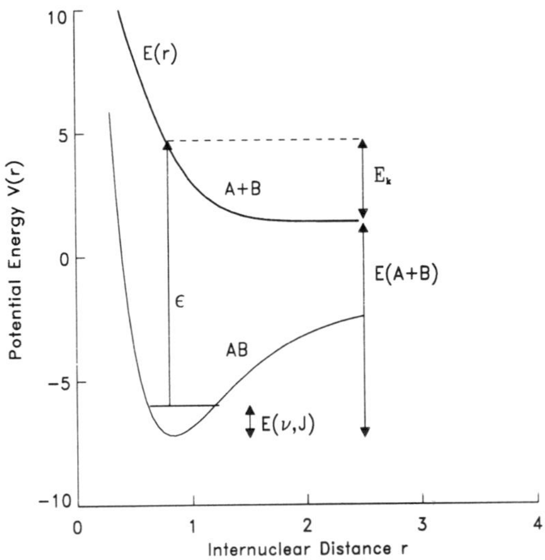

FIGURE 14.2 Schematic description of the energy release between the fragments A and B for collision induced dissociation yielding to a discrete energy release (arbitrary units).

of the molecule AB. Although more complicated, it is possible to show that such reconstruction is feasible even for polyatomic molecules [9]. On the other hand, if the initial molecular structure (and quantum state) is known, one can use the measurement of the kinetic energy release (or the asymptotic velocities) to map out the dissociative potential curve E(r).

The above description is just a simple example, as in most cases, many dissociative states can be populated and as such, the distribution P(E) is in fact due to a superposition of release kinetic energy distributions for each state, with the proper population weight.

14.2.2 Collision-induced dissociation: discrete energy release

A different type of collision induced dissociation is, for example when a molecular ion AB^+ captures a free electron, or is photodissociated, and the bound molecule undergoes a transition to a dissociative state as in the previous case. Here again, the molecule will start to dissociate along a certain (dissociative) potential curve E(r), and some kinetic energy will be released to the fragments A and B. However, in this case, the kinetic energy release in the dissociation is simply based on energy conservation (see Fig. 14.2).

Let us assume that the molecule AB is originally in a given vibrational and rotational state with quantum numbers ν and J, respectively. Then if the molecule undergoes a transition from the bound state to a dissociative state E(r) after transfer of energy ϵ, the molecule will start to dissociate along the state E(r). The final relative kinetic energy release E_k of the fragments A and B is given by:

$$E_k = E(\nu, J) + \epsilon - E(A+B) \tag{14.5}$$

where E(ν,J) is the initial energy of the molecule AB in the vibrational state ν and rotational state J, relative to the minimum of the potential well of the initial electronic state of AB, and E(A+B) is the internal (electronic) energy of the fragments A and B (depending on their final quantum states) relative to the same point (see Fig. 14.2). E(A+B) is also the asymptotic limit of E(r) for $r \rightarrow \infty$.

Thus, in this case, for a single initial state and a single final state, only a well-defined kinetic energy is released, and not a distribution as in the previous section. Hence, the measurement of the final velocity of both fragments A and B (from which the relative kinetic energy E_k can be readily calculated) will give a unique identification of the final quantum states of A and B, if ϵ is well defined. Moreover, because the energy spacing of the vibrational and rotational states of the molecule AB is in general much smaller than the energy difference between the various electronic states of the fragments A and B, one can also deduce the initial (distribution of) state(s) of AB prior to dissociation.

14.3 EXPERIMENTAL PRINCIPLES

Based on the above simple description of the molecular dissociation processes, the goal of an imaging experiment is to measure the final three-dimensional velocities of each of the fragments originating from a single dissociation event.

Two basic experimental geometries are used: The first is when the molecule AB is originally (i.e. prior to dissociation) at rest in the laboratory frame of reference (molecular target), and the second is when the molecule AB is moving with a velocity v_0 relative to the laboratory frame of reference (molecular beam). In the following, we will discuss the characteristics of each of these two geometries.

14.3.1 Dissociation of fast molecular beams

A beam of molecules (or molecular ions) with an energy E_0 is directed toward the center of an imaging detector. At some distance L from the detector (usually from half a meter to few meters), dissociation is induced by interaction with an electron, photon, single atom, or a thin solid film (see Fig. 14.3). For simplicity, we will assume that the molecule is diatomic and homonuclear.

As the molecule dissociates, both fragments gain some velocity in the center-of-mass frame. Since the imaging detector is located in the LAB frame of reference, we need to derive the transformation between the measured values in the LAB frame to the center-of-mass frame. If we denote v_0 for the beam velocity, then the relative distance D between the two fragments after traveling the distance L is given by (see Fig. 14.3):

$$D = \frac{1}{2} v_{\perp} \left[\frac{L}{v_0 + (1/2)v_{\|}} + \frac{L}{v_0 - (1/2)v_{\|}} \right] \tag{14.6}$$

where $v_{\perp}$ and $v_{\|}$ are the perpendicular and parallel components of the mutual velocity of the fragments, relative to the beam direction. Because the initial velocity of the beam v_0 is much larger than the parallel components of the relative velocity $v_{\|}$, one can reduce the previous equation to:

$$D = L \frac{v_{\perp}}{v_0}. \tag{14.7}$$

With the same approximation, the flight time difference is given by

$$\Delta t = t_2 - t_1 = L \frac{v_{\|}}{v_0^2} \tag{14.8}$$

where t_1 and t_2 are the times-of-flight from the dissociation point to the detector for the particles, Thus, the kinetic energy release E_k in the center-of-mass, which is the value of interest, can be expressed as a function of D and Δt only:

$$E_k = \frac{E_0}{4L^2} \left[(v_0 \Delta t)^2 + D^2 \right]. \tag{14.9}$$

The general formula for E_k for a heteronuclear molecule with mass $M = m_1 + m_2$ is:

$$E_k = \left(\frac{E_0}{L^2} \right) \left(\frac{m_1 m_2}{M^2} \right) \left[(v_0 \Delta t)^2 + D^2 \right] \left(1 + 2 \frac{m_1 - m_2}{M} \frac{v_0 \Delta t}{L} \right) \tag{14.10}$$

For a polyatomic molecule, when more than two fragments are produced in the process of dissociation, the relevant quantities are the velocities of each fragment in the center-of-mass, and not only the total kinetic energy release. The Cartesian components of these velocities v_x^i, v_y^i and v_z^i can be easily expressed as a function of the impact coordinates x^i and y^i for each particle i on the surface of the detector and by the time-of-flight t^i relative to the center-of-mass time of flight t_{cm}:

$$v_x^i = v_0 \frac{x^i}{L} \tag{14.11}$$

$$v_y^i = v_0 \frac{y^i}{L} \tag{14.12}$$

$$v_z^i = v_0 \frac{(t_{cm} - t^i)}{t_{cm}} \tag{14.13}$$

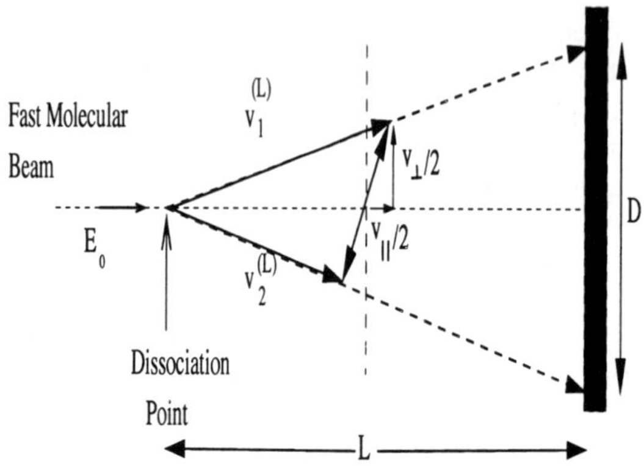

FIGURE 14.3 Schematic view of the dissociation of a molecule from a fast diatomic molecular beam. $v_1^{(L)}$ and $v_2^{(L)}$ are the velocities of the fragments in the laboratory frame of reference.

For the diatomic case, (see Eq. 10) the energy resolution depends mainly on the spatial and time resolutions. Of these, the time resolution is the most critical. Although the required values are a function of the beam velocity and the energy release, it is useful to take an example. Let us suppose that a beam of H_2^+ (M=2) with an energy of 1 MeV dissociates (e.g. after electron stripping in a thin foil) into two protons at a distance L=2 m from a detector. In this case, assuming a pure Coulomb repulsion between the two protons, with an initial internuclear distance of 1.06 Å, the energy release is about E_k=15 eV. Based on Eqs. 7 and 8, one obtains for the maximum relative distance on the surface of the detector D_{max}=15 mm, and the maximum flight time difference is Δt_{max}=1.6 ns. Thus, a resolution of 100 μm will result in a relative error which is better than 1% for D. On the other hand, a resolution of 100 ps (which is about the present limit) yields a relative error of about 7% on Δt. Hence, in most cases, the time resolution will generally be the limiting factor as far as resolution is concerned. Other factors, such as the spread in the initial beam energy are found to be negligible in most cases.

From the above description of the dissociation kinematics, it is clear that in order to measure the final velocity of each of the fragments, the detector should be able to measure the position and relative time of arrival of each particle of a given molecule, one event at the time. This is called three-dimensional (3-D) imaging, while 2-D imaging refers to to the measurement of the position only, i.e. the transversal velocity components only. 1-D imaging usually refers to the measurement of one velocity component only, generally the time-of-flight. It is interesting to point out that, in the case of 3-D imaging, only the relative time of arrival is needed (in addition to the spatial transverse coordinates), and not the absolute time-of-flight from the dissociation point to the detector.

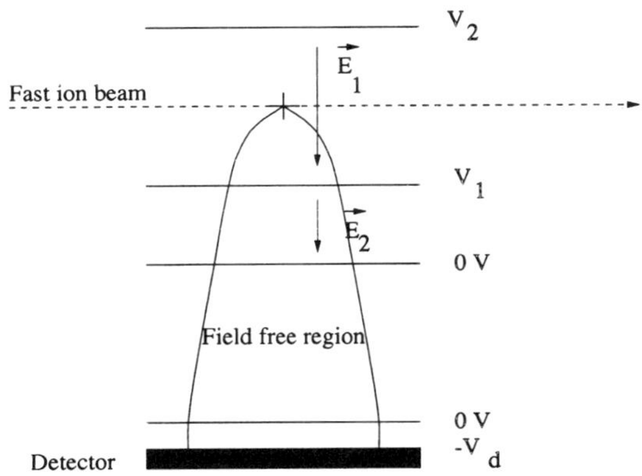

FIGURE 14.4 Molecular target configuration.

Another point is the comparison between 2-D and 3-D imaging: as 2-D imaging is much simpler than 3-D imaging, it might be important to consider wether a 2-D imaging system does yield sufficient information on the dissociation process, so that a 3-D imaging detector is not required. Very often, a 2-D imaging detector can be built as a sub-system, or a starting point of a full 3-D imaging system.

14.3.2 Dissociation of molecular targets

In the molecular target configuration, molecules in the gas phase can be dissociated using fast beams such as electrons, photons and ions. A scheme of a typical experimental set-up is shown in Figure 14.4 [10–12]. In this type of experiment, a fast well-defined ion beam is directed towards a thin gas jet of molecules which are either in thermal equilibrium with the surroundings, or even better, are cooled to a few degrees K using a special cooling setup [13,14]. In order to measure the final velocities of the fragments induced by the dissociation of the molecule, the products are extracted by a uniform electrostatic field which consists of one or two stages of acceleration. The two acceleration-stage set up has the advantage of minimizing the influence of the initial beam width over the time of flight of each of the dissociating fragments, from the target to the imaging detector. This is achieved by tuning the ratio between the two acceleration fields $\vec{E_1}$ and $\vec{E_2}$ (see Fig. 14.4), a technique called space focusing [15].

A drift region with no electrostatic field, from a few centimeters up to a few tens of centimeters, is located between the acceleration stages and the imaging detector. The main role of this drift region is to enhance the time and space separation between the dissociated fragments. The imaging detector is usually based on microchannel plate technology (see Section 14.4.2). As these detectors have a low efficiency for slow fragments, an additional short acceleration stage is provided after the drift region in order to enhance their efficiency. An acceleration field of 2 kV for protons, and up to 7 kV [16] for heavy particles is usually sufficient.

The kinematics of the dissociation imaging in the present case is different from to the one described in the previous section. The approximation used to get Eq. 7 is no longer valid, since the the fragments' parallel and perpendicular velocities ($v_{||}$ and $v_{\perp}$ respectively) are of the same order of magnitude. Thus, in this configuration, the distances and time of arrival on the detector plane will depend on the spectrometer geometry and electrostatic fields.

For any fragment i with mass m^i and charge q^i, the component of the velocity along the electrostatic field is

$$v_{||}^i = f_{\sqrt{\frac{m^i}{q^i}}}(t^i - t^0) \tag{14.14}$$

The function $f_{\sqrt{\frac{m^i}{q^i}}}$ is a universal function which depends directly on the geometry and the electrostatic fields of the apparatus and parametrically on the square root of the ratio m^i/q^i. From Eq. 14, $v_{||}$ is a function of the difference between the ion time of flight t^i and t^0 which is the time of flight of a fragment with a zero initial parallel-velocity component. The perpendicular velocities v_x and v_y (perpendicular to the field direction) are given by

$$v_x^i = (x^i - x^0)/t^i \tag{14.15}$$

$$v_y^i = (y^i - y^0)/t^i \tag{14.16}$$

where x^i and y^i are the position on the detector plane and x^0 and y^0 are the positions of an ion with a zero initial perpendicular velocity component. The value t^0, x^0 and y^0 can be easily measured separately using an atomic target for calibration.

Caution should be taken when the initial position of the molecular target is not well defined i.e. when the projectile beam is wide. In this case, a more detailed calculation should be done, and the position and time of the center-of-mass deduced from momentum conservation for each individual event should be used instead of the values t^0, x^0 and y^0.

When compared to the imaging of fragments from a fast molecular beam, several advantages of the molecular target setup can be pointed out. First, in the present geometry, the parallel component of the fragment velocities ($v_{||}$) is generally measured with a better resolution since the time of flight differences are of the order of several tens of nanoseconds. This also permits a simpler timing scheme for the design of the anode of the detector as well as the related electronics. Another advantage is that the molecules are initially in their ground vibrational state, and in low rotational states. This is generally not the case for the fast molecular beam geometry, as the molecules need to be ionized before acceleration. This yields a beam of molecular ions which may be in a wide distribution of vibrational states, depending on the Franck-Condon factor between the neutral and ionic electronic states.

14.4 DETECTION PROCEDURE

Various schemes for the measurement of the asymptotic velocities for each of the fragments of a dissociating molecule have been designed. Although the technical requirements for measurements from a fast molecule and from a molecule which is initially at rest are slightly different, some common characteristics can be described:

- Simultaneous determination of the fragment positions and time, independent of where the fragments actually strike the detector
- The active area of the detector must be larger than the diameter of the "velocity cones"
- A large dynamical range, i.e., the simultaneous detection of fragments with different masses
- Good position resolution (depending on the specific case, but generally better than 1 mm)
- Good time resolution (generally much better than 1 ns in the case of a fast molecular beam)
- High detection efficiency - especially for polyatomic molecules

The above list is not complete, and clearly can vary from one type of experiment to another. In the following, we will review a few types of detectors that have been used for the imaging of molecular dissociation. We will concentrate mainly on detectors used in 3D imaging, although some examples will be given for 2D imaging. Most modern detectors are based on Micro-Channel Plates (MCP), and these will be described in more detail than the earlier types, the gas-filled wire chamber.

14.4.1 Gas detector

Gas detectors (or multiwire chamber) have been used for many years in nuclear Physics, and on a much larger scale in high energy Physics. These detectors are based on the production of a localized electron avalanche which develops along the track of the particle when it penetrates a low pressure gas volume. These electrons are detected by either a series of pickup electrodes, or by a segmented anode, yielding (fast) signals which can be analyzed to reveal both time and position of impact. An example of such detector a is shown in Fig. 14.5.

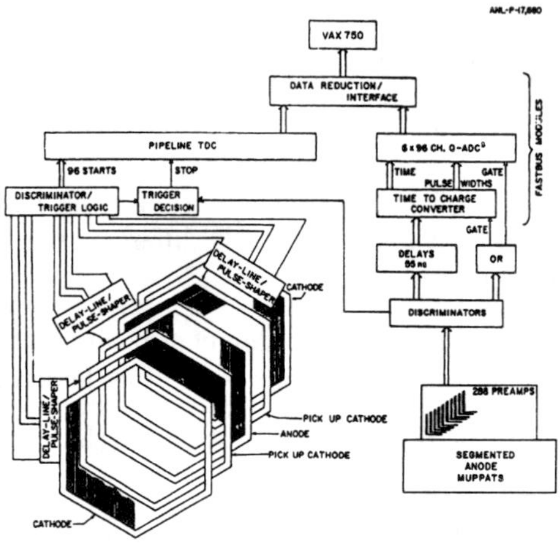

FIGURE 14.5 The MUPPATS detector. Reprinted from W. Koenig, A. Faibis, E. P. Kanter, Z. Vager and B.J. Zabransky, "A multiparticle 3D imaging technique to study the structure of molecular ions", NIM B10/11, 259 (1985), with kind permission of Elsevier Science - NL, Sara Burgerhartstraat 25, 1055 KV Amsterdam, The Netherlands.

The MUltiParticle Position And Time Sensitive (MUPPATS) [17] detector (see Fig. 14.5) has been developed at Argonne National Laboratory for the measurement of the final velocity of ion fragments originating from the Coulomb explosion of fast (MeV) molecular ions. The heart of the system is a low pressure multi-wire avalanche detector, consisting of 5 wire planes, where an anode at positive potential ($\sim$ 350 V) is sandwiched between two cathodes (each 5 mm apart) at a negative potential ($\sim$ -350 V). In addition, two pick-up cathodes at ground potential are arranged close to the anode. The wire spacing is 1 mm for the anode and 0.4 mm for the two cathodes. The pick-up cathodes consist of pairs of wires 0.4 mm apart. The whole volume of the detector is contained by a window foil made of stretched polypropylene of $\sim$ 80 μg/cm^2 thickness.

When an ion passes through the active volume electron avalanches develop towards the anode. The low pressure, together with the large electron drift velocity in isobutane (the filling gas of the detector) results in short signal rise times ($\sim$ 4-5 ns) allowing a time resolution of a few 100 ps. A position resolution of about a hundred microns can be achieved by interpolating between wires according to the pulse height.

The main feature of the MUPPATS detector is its ability to detect several particles belonging to the same event, independent of their relative position.

One major problem is that two sets of orthogonal wires are not sufficient to determine the position of more than one particle. A set of of N x- and y-coordinates can be combined in N^2 different ways, resulting in an (N-1)N-fold ambiguity. In the MUPPATS detector, this ambiguity is resolved by taking the position information from three planes rotated by 120° with respect to each other. More details can be found in Ref. [17].

For data readout, signal from the wires of the pick-up cathodes are pipelined along a delay line with 5 ns delay between neighboring pairs of wires. Therefore, within one event, several pulses appear at both ends of the delay line, which are connected to preamplifiers. The position in time of the leading and trailing edges of each signal (as measured by leading edge discriminators) are recorded in a so-called 'pipeline time to-digital converter (TDC)". The common stop signal is used for a simple trigger decision, requiring that at least one particle is seen by each pick-up cathode. More details concerning the electronic setup, data readout and reduction can be found in the literature [17]. The SAM (Segmented Anode MUPPATS) detector [18], also developed at Argonne National Laboratory, is based on the same principle as the MUPPATS detector, except that the pickup electrodes are replaced by a segmented anode, on which the electron charge is collected. The anode is made of interleaved strips, each being connected to its own preamplifier.

An advantage of these detectors is the large active area (MUPPATS has a diagonal length of 34 cm, and SAM is 25 cm by 50 cm), whichs allow the measurement of a large energy release (as in Coulomb explosion), or the placement of the detector at a large distance from the dissociation point, which is advantageous for relative spatial and time resolution. On the other hand, the data readout and electronics of such a detector are quite cumbersome. Also, the window foil needed to contain the gas in the detector volume, sets a limit on the minimum energy of the fragments, as they need to penetrate into the detector with enough energy to produce an electron avalanche. Thus, this class of detector is useless in the case of molecular target dissociation (see Section 14.5.3) as the energy of the fragments is much too low.

14.4.2 Micro-Channel-Plates detector

Micro-Channel-Plates (MCP) detectors have the ability, when used with a suitable anode, to yield excellent position and time information for the impact of particles over a wide range of energies. MCPs are basically electron amplifiers which create an electron avalanche through narrow channels close to the position where the particle hits the surface. A single MCP creates about 10^4 electrons per impact. Better amplification can be obtained by using a two stage geometry (Chevron) (about 10^6 electrons per impact) or even a three stage (Z-Stack). By detecting these electrons on an anode located behind the MCP, one can easily measure the time as well as the position of impact of a single particle. Position resolution for the MCP itself (i.e. not including the anode) is of the order of 40 μm, and time resolution of the order of 100 ps can be achieved. The most important part of such a detector for the purpose of imaging, is the design of the anode. Among the various schemes are the resistive anode [19], the segmented anode [20], and the phosphor screen coupled to a CCD camera [21]. As this last method is now becoming more and more popular, we will describe it in more detail after reviewing the basic principles of the first two techniques.

Resistive anodes have been used to measure the 2D position of impact of a single particle [11]. The resistive anode uses a charge division principle. The anode itself is a square distributed resistive network with electrodes at the four corners. When a charged particle strikes the input micro-channel plate, a pulse of a few tenths of a picoCoulomb is deposited on the anode. The position of the pulse is then computed electronically from the ratio of the charge collected at each of the four corners [19]. Since the pulse from the resistive anode is relatively slow ($\sim$ ns) it is impossible to measure the position (and time) of impact of two or more coincident fragments using a single resistive anode. Only if the time difference between the impact of each fragment is longer than a few ns, can the resistive anode be used for multi-particle detection.

The segmented anode method (sometimes called multi-anode) [20,22] has also been very much used to measure the 2D position of impact of a single particle. It has the advantage, as compared to the resistive anode, of producing faster signals, resulting in better time resolution. Various types and design of segmented anodes exist for single particle detection. An example of a segmented anode used for the simultaneous detection of two particles in the study of the dissociation of fast molecular beams is shown in Fig. 14.6 [22,23]. The anode pattern consits of concentric, circularly shaped anodes that alternatively increase and decrease - linearly with the angle - in thickness from one extreme to the other. All the anode strips are interconnected by capacitors and resistors. The time resolution amounts to 2 ns, the position resolution to 0.3 mm radially, and 0.05 rad azimuthally. This anode takes advantage of the geometry of the dissociation (cylindrical symmetry) which reduces the number of readouts. The detector itself is made of two separate anodes as depicted in Fig. 14.6, allowing for the detection of two simultaneous particles.

Another design, which uses an etched "crossed-wire" anode structure, has been used in a multi-particle position and time detector for molecular target dissociation studies [10]. The anode consists of two separated layers of wires strained in the x- and y- direction. These wires are in fact an etched structure made of 256 squares (16×16), each 1.5 × 1.5 mm^2 in size (see Fig. 14.7). If an electron cloud from the channel plates hits at a crossing of two wires, coincident pulses in x- and y- wire are registered in the corresponding channel of a Time-to-Digital Converter (TDC) system. By independent determination of the x- and y-coordinates of the impinging particle a significant reduction in complexity and cost of electronic equipment was achieved. The time resolution of such a system is about 2 to 3 ns, which is sufficient in the case of recoil

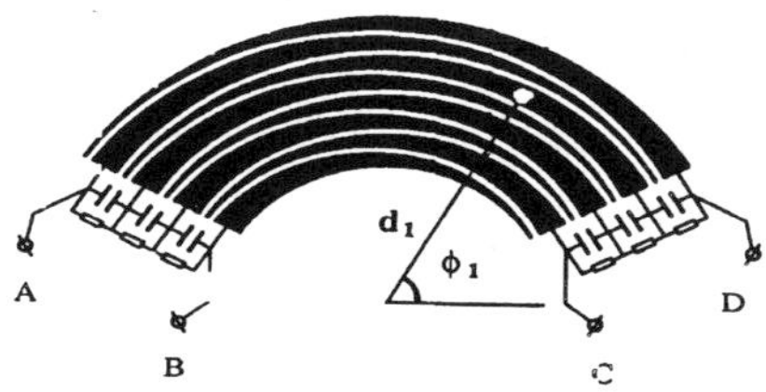

FIGURE 14.6 The FOM anode design. Multianode pattern allowing for simultaneous determination of two fragment positions in two dimensions: a distance from the detector center d_1 and an angle around that center ϕ_1. Distance and angle follow trivially from the total charges registered at A, B, C and D. The detector is made of 2 MCP's and two separated anodes. Reprinted from *Surface Science* **280** (1993) 145, with kind permission from Elsevier Science - NL, Sara Burgerhartstraat 25, 1055 KV Amsterdam, The Netherlands.

fragments (i.e. in a molecular target configuration), as their time of flight difference is relatively large. The main problem of this setup is that when two particles arrive within the chosen time resolution, which defines a single particle in the x- and y- direction, the correlation between the x- and y- wires is lost. Such an event is processed as if four particles hit the detector in a symmetrical arrangement.

During the last decade, imaging systems based on the use of a phosphor screen as an anode for the MCP, have gained considerable popularity. The reason is that this method allows for the measurement of the position of a large number of particles from a single event, with good efficiency, thus making the design of a 2D imaging system relatively simple.

In this method, the cloud of electrons exiting the MCP channels are accelerated toward a metalized phosphor screen located behind the MCP. The impact of these electrons on the phosphor produces a visible light spot. The position of these spots can then be recorded by a standard CCD camera (see Fig. 14.8), coupled to a frame grabber [21], with a resolution which is basically limited by the MCP and phosphor screen. Resolution of better than 100 μm is possible. One of the advantages of the phosphor screen - CCD system is that if the region of interest (i.e. the area on the screen where the events hit the detector) is smaller than the MCP, one can use a simple lens to focus the whole CCD area on this region, thus increasing the number of pixels per unit of length. In any case, with a standard CCD camera with more than 512 pixels per line and a 40 mm diameter MCP, the resolution is better than 100 μm per pixel. The main drawback of the system is the relatively low rate of digitization (the number of frames per second of a standard CCD camera is 50 (60 in the US)). Fast cameras are now available [24], with readout rate as

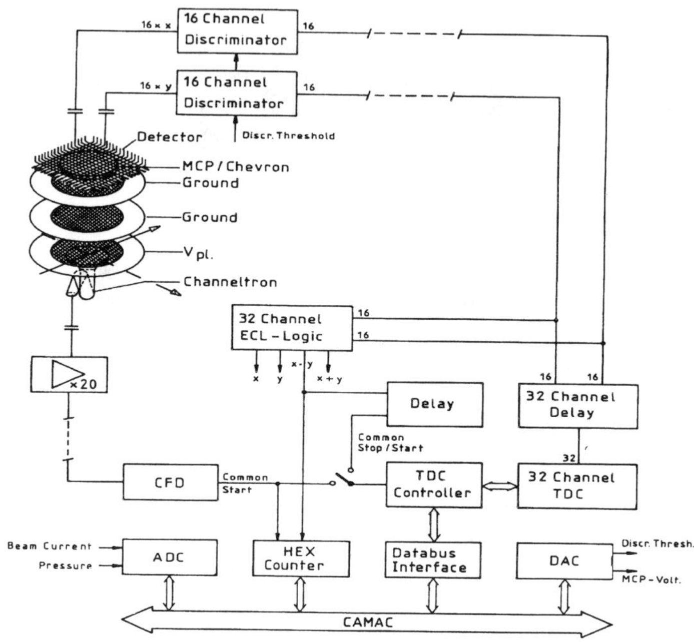

FIGURE 14.7 The "crossed-wire" anode structure, together with the related electronic, from the university of Bielefeld. The detector is capable to measure the positions and times of many slow molecular fragments. Reprinted from J. Becker, K. Beckord, U. Werner and H.O. Lutz, "A system for correlated fragment detection in dissociation experiments" [10], with kind permission of Elsevier Science - NL, Sara Burgerhartstraat 25, 1055 KV Amsterdam, The Netherlands.

high as 500 frames/s, although usually the resolution (number of active pixels) is reduced. Another important problem is the measurement of correlated events: as the integration time of a frame is 20 ms (16.7 ms in the US) for a standard CCD camera, all events occuring during this time will be recorded on the same frame. Thus, in order to record the position of only correlated events (i.e. fragments originating from the same molecule), one needs either to generate the events in phase with the camera timing cycle (one event per frame) [21], or to "switch off" the detector once an event has been detected [4]. Another possibility is the use of a gated camera, which can reduce the effective integration time.

The extraction of the time information for each particle impact is more cumbersome. A method which has been successfully used for the Coulomb Explosion Imaging experiment at the Weizmann Institute is based on para-

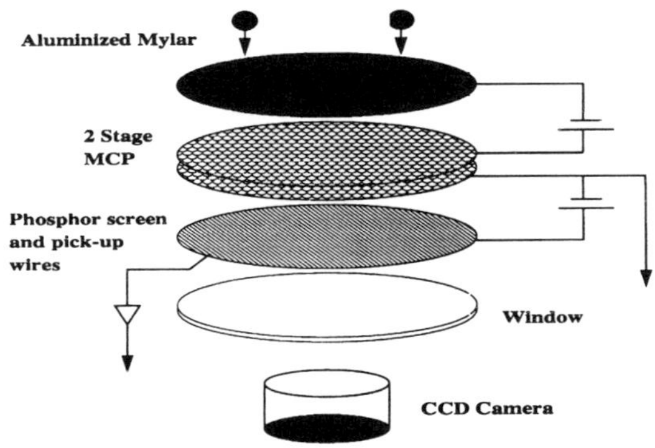

FIGURE 14.8 Schematic view of the CCD detector used at the Weizmann Institute of Science for the Coulomb Explosion Imaging studies. This detector is capable to measure the positions and times of impact of many fast fragments.

sitic pickup from wires stretched very close to the phosphor screen [21] (see Fig. 14.8). As the electrons are accelerated toward the phosphor screen, they (or their impact on the screen) induce a charge on the closest wire(s). Each wire is connected to a charge amplifier, and the output signal is fed into a constant fraction discriminator, which outputs a charge signal (Time to Charge convertor) which is proportional to the time of arrival relative to a common trigger signal (see Fig. 14.8). The charge is analyzed by a charge-sensitive converter ADC(Q-ADC). Thus, the output of the Q-ADC is a number which is proportional to the time of impact of each particle on the detector, relative to a common reference obtained from the MCP itself. The relative time of impact between each particle (which is the only quantity of interest (see Section 14.3)) can then be extracted by subtracting the values obtained from the Q-ADC, after calibration. Time resolutions of the order of 60 ps have been obtained with such a design. More details of this detector, related electronics and software can be found in the literature [21].

14.5 EXAMPLES

14.5.1 Coulomb Explosion of fast molecular beams

Coulomb Explosion Imaging (CEI) is a technique that has been used during the last several decades for the study of molecular structure [6,7,25,26]. The basic idea behind the experimental technique is the following: as a fast molecular ion ($>$ 100 keV/amu) passes through a thin solid film ($\sim$ 100 Å), its valence electrons are stripped away very rapidly ($\sim 10^{-16}$ s) initiating a fast dissociation process. This process has been termed the "Coulomb explo-

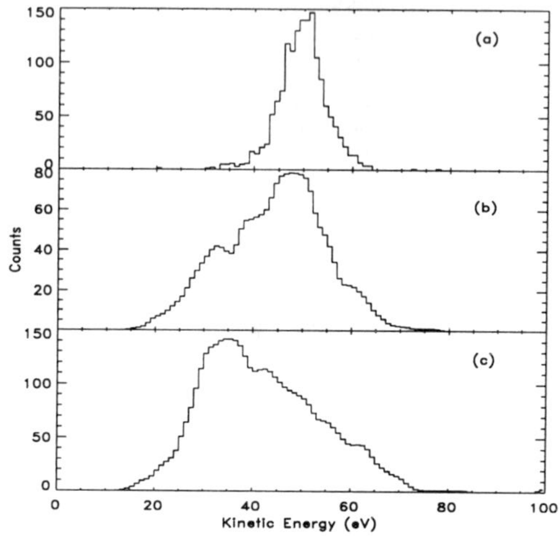

FIGURE 14.9 Kinetic energy release due to the Coulomb explosion of a fast He_2^+: (a) vibrationally cold (mainly $\nu = 0$), (b) vibrationally excited ($\nu = 0-5$), and (c) vibrationally excited ($\nu = 5 - 10$).

sion" because the electrostatic repulsion between the remaining ions is the dominant term in the resultant potential. For a molecule containing N atoms, measurements of the 3×N velocity components after the explosion provides information on the 3×N spatial components within the original molecule. Since the time scale for electron stripping is very short compared to the characteristic vibrational and rotational motions in the molecules, the Coulomb explosion of each individual molecular ion gives a "snapshot" of the relative positions of the nuclei within that molecule. Combining such measurements for a large ensemble of molecules yields the fully-correlated many-particle density distribution describing the nuclear coordinates for that ensemble. As an example of a simple CEI experiment [27], Fig. 14.9 shows various distributions of the final kinetic energy release of the fragments in the center-of-mass as measured for the Coulomb explosion of He_2^+ for a vibrationally cold (a) and a vibrationally hot (b and c) beam. The vibrational distribution was controlled by using a supersonic expansion ion source in the high voltage terminal of the accelerator [28]. As explained in Section 14.2.1, the kinetic energy release distribution shown in Fig. 14.9 is directly related to the distribution of bond length in the molecule.

The shift toward lower kinetic energy in the distributions shown in Fig. 14.9(b) and (c) is a result of the anharmonicity of the He_2^+ molecular potential. As the high-lying vibrational states become populated, there is a higher probability for the molecule to be found at large inter-nuclear distances (R). Because of the dominant 1/R Coulomb interaction, larger R values correspond to lower final kinetic energies. Further, the increased width of the distribution at high temperatures is consistent with the broader vibrational state population.

The same cooling effect was observed for C_3 – a very floppy linear molecule – using a different setup [21]. The C_3 molecules were produced by laser photodetachment of a C_3^- beam produced using either a Cs sputter source or a laser vaporization source which utilizes a supersonic expansion to produce cold molecules. The negative molecular ions were accelerated to an energy of 11 MeV, photodetached ($h\nu$=2.3 eV) to produced neutral C_3 which impinges on a thin stripping foil, producing three carbon ions with various charge states. A magnet located after the stripping foil selected a particular charge state. In the present case, only the ions exiting the thin foil with a charge q=+3 were steered toward the detector. The distance between the target and the multi-particle position and time detector is 2.2 m. The time and position information from the detector, with a resolution of 100 ps and 100 μm respectively, is digitized and collected by a computer. As both beam velocity and target-detector distance are known, the time and position coordinates are transformed into 3 velocity components for each fragment of each molecule. Each event is analyzed separately and the total distribution is investigated. The 3 velocity components of each carbon ion were used to compute the relative velocities v_{12}, v_{13} and v_{23} between the fragments. In order to analyze the data in internal coordinates, the following symmetry coordinates were used:

$$S_2 = (v_{12} - v_{23})/\sqrt{2} \tag{14.17}$$

$$S_3 = (2v_{13} - v_{23} - v_{12})/\sqrt{6} \tag{14.18}$$

where v_{ij} are the relative velocities between atom i and j. Because of the identity of the three carbons, all possible permutations of i and j are included when plotting the distribution of these coordinates. In this coordinate system, the origin (S_2=S_3=0) represents an equilateral triangular structure since in this case, if all the carbon fragments have the same charge state one can easily show that v_{12}=v_{13}=v_{23}. On the other hand, for a linear structure, one can show that v_{12}=v_{23}=$v_{13}/2$=v_c, where the central carbon atom in the chain was tagged with the index 2 for the permutation shown in Eq. 17 and 18. Such a symmetry would produce a distribution which peaks at (S_2,S_3)=$(0,2v_c/\sqrt{6})$,$(-v_c/\sqrt{2},-v_c/\sqrt{6})$, $(v_c/\sqrt{2},-v_c/\sqrt{6})$ when all the permutations are included.

Fig. 14.10 shows the data in the (S_2,S_3) coordinates: In Fig. 14.10(left), the ions were created using the Cs sputter source, while in Fig. 14.10(right) they were produced using the cold ion source [29]. In Fig. 14.10(left), the density distribution peaks at (0,0), which correspond to a triangular shape. This is due to the fact that a highly vibrationally excited linear molecule such as C_3 –when produced by the "hot" Cs sputter source– spends most of its

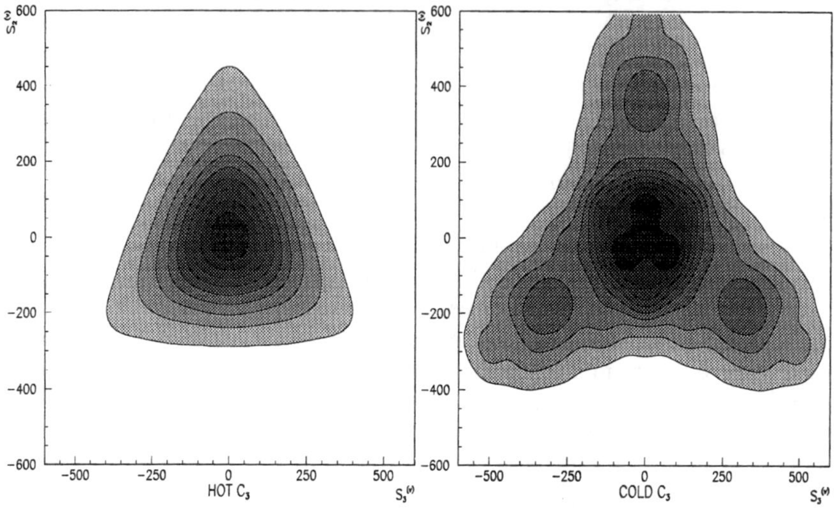

FIGURE 14.10 Coulomb Explosion Imaging of C_3 for two different vibrational distribution: (left) vibrationally hot and (right) vibrationally cold. The coordinates S_2 and S_3 are described in the text.

time in a non linear configuration, and thus is "seen" with a triangular configuration. In Fig. 14.10(right), the three symmetric peaks which correspond to a linear structure start to appear, as a result of the cooling process. It is important to point out that since the ion produced in the ion source is C_3^-, the photo-detachment process, at an energy which is ~ 0.2 eV above the threshold, populates a series of vibrational states of the neutral cluster, even if the negative ion is produceed in the vibrational ground state. This is the main reason for the central peak in the "cold" data of Fig. 14.10(right). Using advanced data analysis technique [30,31,9], these data sets can be transformed to the original configuration space (not the velocity space in which the data is plotted as in Fig. 14.10); it can be shown that then both "hot" and "cold" data peak around a linear configuration, the hot data being much more spread than the cold one, as expected from the difference between the hot Cs sputter ion source and the supersonic expansion source.

14.5.2 Dissociative recombination of molecular ions

Interest in dissociative recombination (DR) arises out of the need to understand the basic properties of laboratory and astrophysical plasma, such as planetary ionosphere. [32–34]. In this process, a free electron excites the ion and is simultaneously captured into a doubly excited molecular state. This

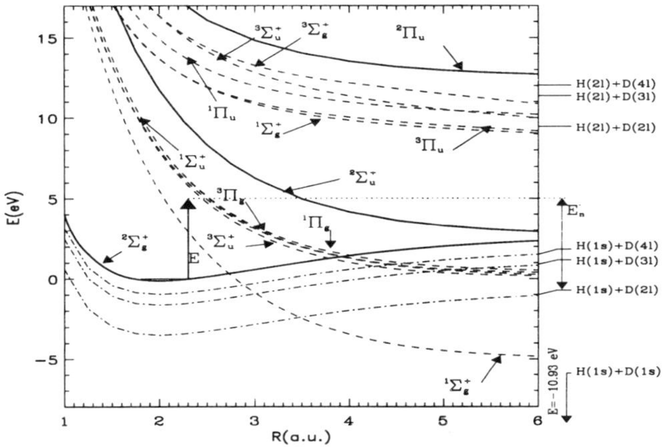

FIGURE 14.11 Relevant potential curves for HD^+ (thick solid lines) and HD (dashed and dash-dotted lines), showing the ground state and the two excited states of HD^+, the lowest singly excited Rydberg states of HD (dash-dotted) and the lowest doubly excited states of HD (dashed) below the ${}^2\Sigma_u^+$ and ${}^2\Pi_u$ states of HD^+. The asymptotic energies for various final states of the fragments (D can be replaced by H and vice-versa) are drawn on the right-hand side of the picture. The relation between the energy release E_n' and the electron energy E is illustrated.

state then dissociates along a repulsive potential curve, yielding fragments with translational energy and, possibly, internal excitation depending on the energy balance of the reaction. A schematic description of this process is shown in Fig. 14.11 for HD^+. The two most important questions in such studies are: (a) What is the total cross section for the DR of a molecular ion in a well defined vibrational level and (b) What are the final states of the neutral fragments? The answer to (a) was difficult to obtain for most of the molecular ions of interest, until substantial progress in the production of relaxed (cold) molecular ions was achieved by using the heavy-ion storage ring technique [35], which enables the storage of fast molecular ion beams for a time long enough to allow relaxation through spontaneous radiative decay. The answer to question (b) was obtained recently by using 2D imaging techniques on fully relaxed HD^+ [4]. In this experiment, a beam of HD^+ was accelerated to an energy of 2.08-MeV and stored in the Test Storage Ring (TSR) located at the Max-Planck-Institut für Kernphysik, Heidelberg. Once vibrationally cold, the molecular ions were merged with a quasi-monochromatic electron beam (the electron cooler) over a length of 1.5 m. An 80-mm diam. Chevron MCP detector was mounted straight ahead of the electron cooler section at a distance of 6 m to detect the neutralized particles as produced in the DR reaction. This reaction can be described as

$$HD^+ + e \rightarrow HD^{**} \rightarrow H(nl) + D(n'l') \qquad (14.19)$$

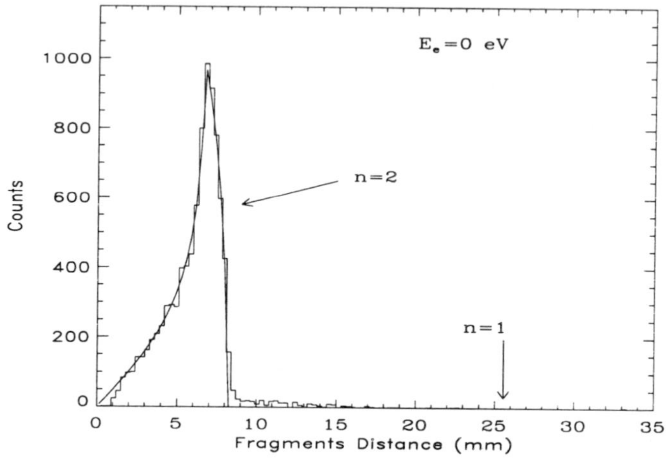

FIGURE 14.12 Projected distance distribution for the DR of HD^+ at E_e=0. The solid line is the calculated projected distance assuming that one of the fragment is in the n=2 state. The n=1 arrow shows the expected position if both fragments were in the n=1 state.

with HD** denoting the doubly excited dissociating state.

Each impact on the MCP produced a light spot on a phosphor screen located 1 mm behind the MCP. The image generated by these spots was digitized using a CCD camera, coupled to a fast frame grabber device [21]. On analysis, the positions of the light spots were determined using a peak finding procedure, and the relative projected distance between any two fragments on the plane of the detector was deduced. The detector was operated in a trigger mode in order to suppress random coincidences between single hits due to two unrelated single fragments produced by collisions with the residual gas. The main problem with such a setup is the long overlap between the electron cooler and the molecular beam. In the present case, the final energy resolution for the kinetic energy release is smeared by the fact that the dissociation comes from an extended source. Thus, the projected distance D observed on the surface of the detector represents an average over the longitudinal extension of the interaction region and over the molecular orientation (this last part would have been resolved using 3D imaging). The projected separation distribution function P(D) can be computed analytically and derivation of this function for isotropic as well as anisotropic distribution of the fragments in the center of mass can be found in the literature. Fig. 14.12 shows the measured projected-distance spectrum for the reaction depicted in Eq. 19, for an electron energy E_e=0. At this energy, the only asymptotic states which are energetically accessible for HD^+ ions in their vibrational ground state are those for which at least one atom is in the ground state and the other one is either in the $n = 1$ state [H(1s) + D(1s)] or the $n = 2$ state [H(1s) + D(2s)

or D(1s) + H(2s)]. Theory predicts that at this energy DR occurs through the lowest lying $^1\Sigma_g^+$ state. This curve crosses the singly excited HD Rydberg states at inter-nuclear distances of 2.5–3.5 a.u. (see Fig. 14.11) but does not connect to the $n = 1$ dissociation limit. Thus, if $^1\Sigma_g^+$ is the dominant dissociating state, only the $n = 2$ asymptotic states should be formed. The results are in perfect agreement with this theoretical prediction, as shown in Fig. 14.12 by the solid line which was calculated for an angular isotropic fragmentation and an energy release of $E_2' = 0.726$ eV; the arrow indicates the expected maximal projected distance of DR events involving the $n = 1$ asymptotic state (energy release $E_1' = 10.93$ eV), where in fact no events above the noise level could be observed. Moreover, this shows that no significant vibrational excitations are present in the stored beam: The additional energy release of 0.288 eV from the first vibrational level would be easily resolvable in the projected-distance spectrum. The branching ratio for the final state has also been measured at higher electron energy and the results can be found in the literature [4].

14.5.3 Dissociation of molecular targets

Molecular target dissociation by fast ion beams has been a subject of intense research at the Cyclotron Institute at Texas A&M University. In this experiment, a fast (97 MeV) Ar^{14+} is directed toward a well-collimated jet of neutral CO (see Fig. 14.13). In this collision, many electrons are ejected from the neutral molecules, and the fragments start to dissociate along various dissociative states of the ionized molecules [12].

The molecular fragments (ions) are extracted from the interaction region using a two stage extraction field (see Section 14.5.3). The fragments are directed toward an imaging detector made of a double stage MCP coupled to a phosphor screen. The images are recorded by a CCD camera and analyzed by a fast frame grabber [12]. For every event, the time-of-flight between the first arriving fragment on the surface of the MCP detector and an Ar ion projectile is recorded too. The time-of-flight difference between the C and O fragments is obtained by a second consecutive signal from the MCP detector itself. The values of these two time of flights is used to define a unique combination of mass and charge for each fragment hitting the detector.

The bottom part of Fig. 14.14 shows the 2-D image of many pairs of C^+ and O^+ as produced by the dissociation of CO^{++}. In the upper part of Fig. 14.14 is shown the 2-D image of the position distribution when the target is atomic Ar. In this case, there is no kinetic energy release (no fragmentation), and the residual width reflects the original spread of the gas jet. Also, it shows that the momentum transfer from the projectile to the target is negligible. From the above spectrum, together with the time-of-flight of each fragment, the 3-D velocity vector of each fragment can be calculated, as shown in Section 14.5.3 (See Eqs. 14-16). From these values, the final kinetic energy release from the dissociation can be calculated, as well as the angular correlation between the

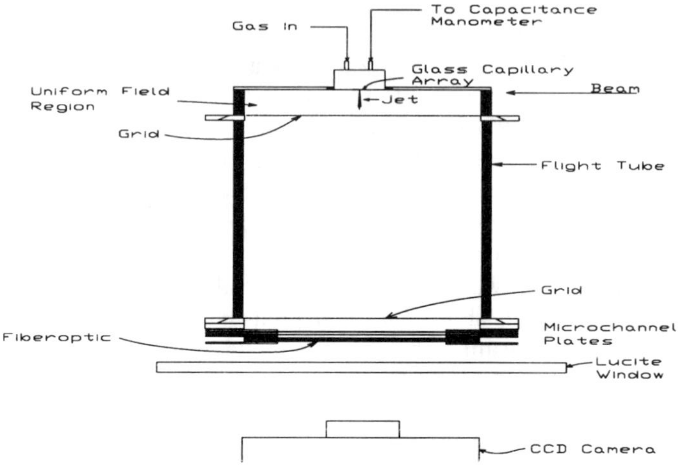

FIGURE 14.13 Schematic diagram of the gas jet, time-of-flight spectrometer, and optical detection system. Reprinted from V. Horvat, O. Heber, R.L. Watson, R. Parameswarn, and J.M. Blackadar, "3-D imaging of binary dissociation events" [12], with kind permission of Elsevier Science - NL, Sara Burgerhartstraat 25, 1055 KV Amsterdam, The Netherlands.

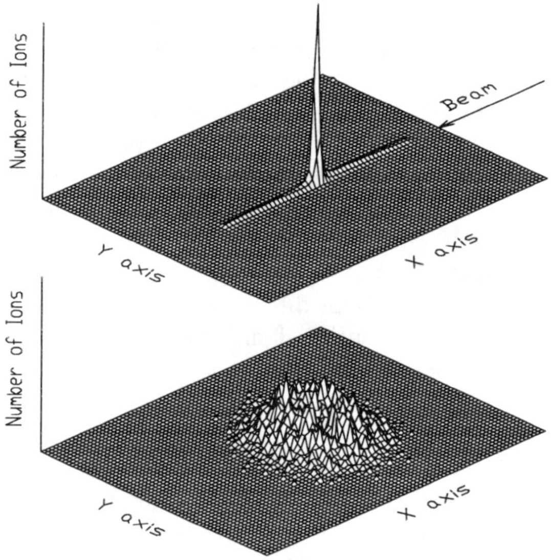

FIGURE 14.14 Position distribution for (top) Ar^+ ions and (bottom) $C^+ + O^+$ ion pairs from the dissociation of CO^{2+}. Reprinted from V. Horvat, O. Heber, R.L. Watson, R. Parameswarn, and J.M. Blackadar, "3-D imaging of binary dissociation events" [12], with kind permission of Elsevier Science - NL, Sara Burgerhartstraat 25, 1055 KV Amsterdam, The Netherlands.

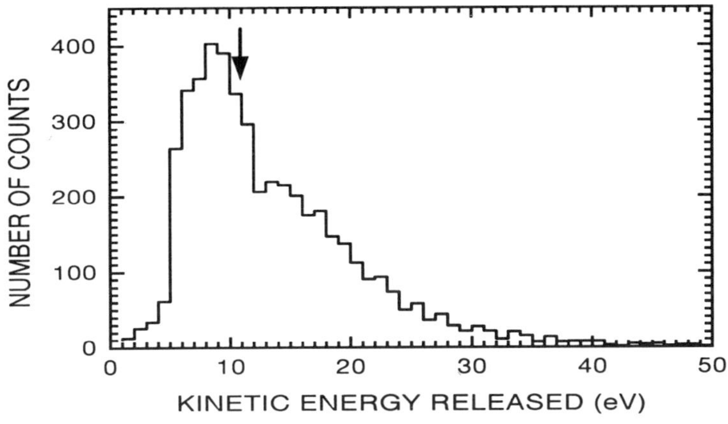

FIGURE 14.15 Kinetic energy release distribution for the dissociation of CO^{2+} into $C^+ + O^+$.

ion beam direction and the internuclear axis of the molecule upon collision. Fig. 14.15 shows the kinetic energy release when two singly charged fragments are produced in the collision. From this spectrum, it can be seen that the energy release might sometimes be larger than the Coulomb limit, which is indicated by an arrow. This indicates that the process of dissociation often occurs through highly excited states of the intermediate CO^{++} compound. More details can be found in Ref. [12]. The angular correlation indicates that the dissociation probability is maximum when the internuclear axis of the molecule is perpendicular to the beam direction.

14.6 SUMMARY

3-D as well as 2-D imaging is a very promising technique for the study of molecular dissociation. The two types of configuration described in this chapter each have their specific advantages, and the geometry is typically a function of the process to be studied. The technology used in the imaging detectors has progressed tremendously during the last decade making possible better time and space resolutions. These two parameters are still the main preocupation of the experimentalist.

On the physics side, there are still many processes, not described in this chapter, which can be studied using imaging technique. Among them one can cite dissociation by tunneling (see the chapter by B. Rosner et al. in this book) which allows measurement of the lifetime of specific vibrational and rotational levels in doubly charged ions. Also, the reaction path is a new aspect which is just emerging. The possibility to "image" the motion of the nuclei inside a molecule has been realized [25,26] and provides a new challenge for theoretical calculation of molecular structure.

ACKNOWLEDGMENTS

The authors would like to acknowledge Z. Vager, E. P. Kanter, R. L. Watson and V. Horvat for helpful discussions. Also, we would like to thank W. J. van der Zande and H. Lutz for providing us material for this work. During the writing, the authors were supported partially by the Minerva Foundation-Munich, by the German Israel Foundation (GIF) under Contract No. I-0208-202.07/91, and by the Israel Science Foundation.

REFERENCES

1. E. Herbst, in *Dissociative Recombination: Theory, Experiment and Applications*, Eds J. B. A. Mitchell and S. L. Guberman, pp. 403, World Scientific (1989).
2. W. G. Graham, in *Dissociative Recombination: Theory, Experiment and Applications*, Eds J. B. A. Mitchell and S. L. Guberman, pp. 194, World Scientific (1989).
3. H. Helm. D. P. de Bruijn, and J. Los, *Phys. Rev. Lett.* **53,** 1642 (1984); D. P. Bruijn and H. Helm, *Phys. Rev. A* **34,** 3855 (1986).
4. D. Zajfman et al, *Phys. Rev. Lett.* **75,** 814 (1995).
5. O. Heber, D. Zajfman, D. Kella, Z. Vager, R. L. Watson, and V. Horvat, *Nucl. Instr. and Meth.* **B99,** 90 (1995).
6. Z. Vager, R. Naaman, and E. P. Kanter, *Science* **244,** 426 (1989).
7. D. Zajfman, *Comments At. Mol. Phys.* **29,** (1994).
8. I. Gertner, I. Ben-Itzhak and B. Rosner, *Nucl. Instr. and Meth.* **B94,** 47 (1994).
9. J. Levin, D. Kella and Z. Vager, *Phys. Rev. A* **53,** 1469 (1996).
10. J. Becker, K. Beckord, U. Werner and H. O. Lutz, *Nucl. Instr. and Meth.* **A337,** 409 (1994).
11. S. Cheng, C. L. Cocke, V. Frohne, E. Y. Kamber, J. H. McGuire and Y. Wang, *Phys. Rev. A* **47,** 3923 (1993).
12. V. Horvat, O. Heber, R. L. Watson, R. Parameswarn and J. M. Blackadar, *Nucl. Instr. and Meth.* **B99,** 94 (1995).
13. J. Ullrich, R. Dörner, R. Olsen and H. Schmidt-Böcking, *Phys. Rev. Lett.* **63,** 147 (1989).
14. R. Dörner, V. Mergel, L. Spielberger, O. Jagutzki, S. Nüttgens, M. Unverzagt, H. Schmidt-Böcking, J. Ullrich, R. Olsen, K. Tokesi, W. E. Meyerhof, W. Wu and C. L. Cocke, *Nucl. Instr. and Meth.* **B99,** 111 (1995).
15. W. C. Wiley and I. H. McLaren, *Rev. Sci. Instr.* **26,** 1150 (1955).

16. T. LeBrun, H. G. Berry, S. Cheng, R. W. Dunford, H. Esbensen, D. S. Gemmell, E. P. Kanter and W. Bauer *Phys. Rev. Lett.* **72,** 3965 (1994).
17. W. Koenig, A. Faibis, E. P. Kanter, Z. Vager, and B. J. Zabransky, *Nucl. Instr. and Meth.* **B10/11,** 259 (1985).
18. A. Belkacem, A. Faibis. E. P. Kanter, W. Koenig, R. E. Mitchell, Z. Vager, and B. J. Zabransky *Rev. Sci. Instr.* **61,** 946 (1990).
19. C. Firmani, E. Ruiz, C. W. Carlson, M. Lampton, and F. Paresce, *Rev. Sci. Instr.* **53,** 570 (1982).
20. C. Martin, P. Jelinsky, M. Lampton, R. F. Malina, and H. O. Anger, *Rev. Sci. Instr.* **52,** 1067 (1981).
21. D. Kella et al., *Nucl. Instr. and Meth.* **A329,** 440 (1993).
22. D. P. de Bruijn and J. Los *Rev. Sci. Instr.* **53,** 1020 (1982).
23. J. M. Schims, R. B. Vrijen, W. J. van der Zande, and J. Los, *Surface Science* **280,** 145 (1993).
24. EG&G Reticon and Dalsa Inc.
25. Z, Vager, T. Graber, E. P. Kanter, and D. Zajfman, *Phys. Rev. Lett.* **70,** 3549 (1993).
26. Z, Vager, D. Zajfman, T. Graber, and E. P. Kanter, *Phys. Rev. Lett.* **71,** 4319 (1993).
27. D. Zajfman and Z. Amitay, *Phys. Rev. A* **52,** 839 (1995).
28. T. Graber, D. Zajfman, E. P. Kanter, R. Naaman, Z. Vager, and B. J. Zabransky, *Rev. Sci. Instrum.* **63,** 3569 (1992).
29. D. Zajfman, H. Feldman, O. Heber, D. Kella, D. Majer, Z. Vager, and R. Naaman, *Science* **258,** 1129 (1992).
30. D. Zajfman, G. Both, E. P. Kanter, and Z. Vager, *Phys. Rev. A* **41,** 2482 (1990).
31. D. Zajfman, E. P. Kanter, T. Graber, and Z. Vager, *Phys. Rev. A* **46,** 194 (1992).
32. Dissociative Recombination: Theory, Experiment and Applications, Eds: J. B. A. Mitchell and S. L. Guberman, World Scientific, 1989.
33. Dissociative Recombination: Theory, Experiment and Applications, Eds: B. R. Rowe, J. B. A. Mitchell and A. Canosa, Nato Asi Series, 1993.
34. Dissociative Recombination: Theory, Experiment and Applications, Eds: D. Zajfman, J. B. A. Mitchell, B. R. Rowe and D. Schwalm, World Scientific, in press.
35. D. Habs et al., *Nucl. Instr. and Meth.* **B43,** 390 (1989).

CHAPTER 15

Fragment-Fragment Coincidence Methods for Mean Lifetime Measurements of Doubly Charged Molecular Ions

B. Rosner*, I. Gertner*, and I. Ben-Itzhak†

The mean lifetimes of long-lived molecular ions formed in fast beams of charged particles can be determined by measuring the number of surviving molecular ions as a function of their flight time from the moment of creation. A few techniques for such measurements are described in detail for doubly charged molecular ions formed by charge stripping. In general, one varies the distance between the target cell and the analyzer (electrostatic or magnetic) and measures the number of molecular ions emerging from the analyzer. When the molecular ions dissociate before reaching the detector their number at the analyzer's exit can be determined from the number of fragment-fragment coincidences. These methods can be also used for mean lifetime measurements of other long-lived molecular ions.

15.1 INTRODUCTION

Doubly charged molecular ions containing noble gas atoms are of increasing interest, experimentally [1–13], as well as theoretically [14–28]. These molecular ions may have bound and metastable states even though noble gas atoms are not strongly active chemically. Furthermore, theoretical studies show that the bond strength does not always decrease as electrons are removed from the molecule as one would expect intuitively [14,15]. Some of these molecular ions decay by electronic transitions to lower repulsive states, while others dissociate by tunneling through the potential energy barrier. The main interest in this

*) Department of Physics, Technion, Haifa 32000, Israel

†) Department of Physics, Kansas State University, Manhattan, KS 66506, USA

"exotic" family of molecular ions stems from the weak binding, which makes their study a stringent test of molecular structure theory because the small depth of the potential well makes them very sensitive to small differences in the total energy of the system. It is, therefore, important to determine the long-lived states of such molecular ions. This may be sometimes accomplished by measuring the decay rate and the dissociation energy. Furthermore, the dynamics of the decay can be studied by measuring the mean lifetime and comparing isotopic effects, which are expected to be large especially for molecular ions dissociating by tunneling through the potential energy barrier.

Decay by Electronic De-excitation. Long-lived doubly charged molecular ions can be formed in either bound or metastable electronic states, both of which may decay by electronic transitions to lower electronic states, if such states exist. A typical example of such a molecular ion is HeH^{2+}. The large repulsion between the two nuclei, bound by one electron only, causes its $1s\sigma$ ground electronic state and most other states to be repulsive [16,17]. However, some excited states are still bound, the lowest of which is the $2p\sigma$ first excited state. It has a minimum 0.849 eV deep at R =3.89 a.u. that can sustain about 15 vibrational levels. Out of the lowest 20 electronic states two highly excited states are also bound, namely the $4f\sigma$ and $4f\pi$ states [17]. The latter states have their minima at much larger internuclear separation then the $2p\sigma$ state. The first excited state of HeH^{2+} can be produced in fast charge-stripping collisions between HeH^{+} and Ar atoms [12]. In these collisions a wide distribution of vibrational states, bound to the $2p\sigma$ first excited state, is populated by a vertical Franck-Condon transitions. This distribution is typically peaked around highly excited vibrational states because of the large difference between the equilibrium internuclear distance of the parent and daughter molecular ions, as shown in Fig. 15.1. These excited HeH^{2+} molecular ions later decay by dipole transitions to the repulsive $1s\sigma$ ground state. Surprisingly, the mean lifetimes of these bound-free transitions depend on the vibrational state, especially for highly excited states [13]. The decay rate of the first excited $2p\sigma$ is therefore not expected to give a single exponential decay curve because strictly speaking each vibrational level has its own mean lifetime [13,29]. The electronic potential energy curves for all HeH^{2+} isotopes are of course identical within the Born-Oppenheimer approximation, although, due to the difference in the reduced mass of the isotopes, the energies of the vibrational levels and their relative population may differ. The isotopic effect on the decay rates is usually small but can be pronounced if the highest vibrational states are involved [29], as shown in Fig. 15.2.

Decay by Tunneling. Doubly charged molecular ions in metastable states may also decay by tunneling through their potential energy barrier, in addition to the decay by electronic transitions to lower states, as discussed before. The population of such states by vertical Franck-Condon transitions from the parent singly charged molecular ion is rather similar to the previous case, but the properties of the decay by tunneling are completely different.

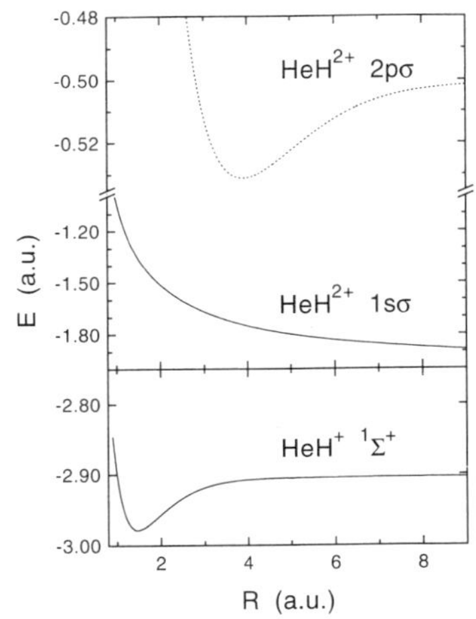

FIGURE 15.1 Potential energy curves of the electronic ground states of HeH^+ and HeH^{2+} (solid lines), and the first excited state of HeH^{2+} (dashed line).

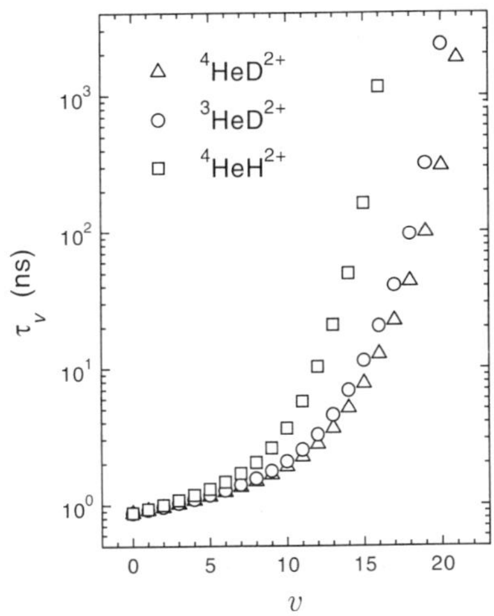

FIGURE 15.2 Theoretical mean lifetimes of the vibrational states bound to the $2p\sigma$ state of HeH^{2+} isotopes.

One significant difference is the large differences between the mean lifetimes of different vibrational states, typically a few orders of magnitude. For example, the metastable ground state of He_2^{2+} and the vibrational levels of its heteronuclear isotope are shown in Fig. 15.3. The mean lifetimes of the vibrational states differ from each other by about 4 to 6 orders of magnitude [23–26]. Furthermore, these mean lifetimes are very sensitive probes of the

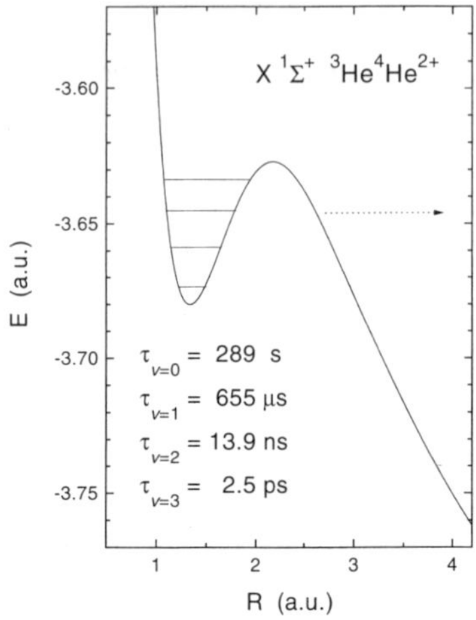

FIGURE 15.3 Potential energy curve of the He_2^{2+} electronic ground state and the vibrational states of its heteronuclear isotope.

potential energy curve, the energies of the rovibrational states, and the calculation method used. This can be clearly seen from the well-known WKB approximation decay rate formula [30]

$$\tau^{-1} = \lambda = f_v \; e^{-2\sqrt{2\mu} \; \int_a^b \; dR \; \sqrt{V(R)-E_v}}. \tag{15.1}$$

Another significant difference between tunneling decay rates and those of dipole transitions is the large isotopic effects expected in the former. Different isotopes of the same molecular ion have the same potential energy curve within the Born-Oppenheimer approximation. However, due to the differences in their reduced masses, μ, the vibrational energy levels are different from each other, lower energy values for the larger μ isotope, and thus their mean lifetimes are expected to differ significantly from each other. The large isotopic effect on the tunneling rates can come in handy when one tries to determine the decay mechanism of a long-lived molecular ion as discussed, for example, for $NeAr^{2+}$. It has been suggested that the long-lived molecular ion is associated with the decay by tunneling of its metastable electronic ground state [27] (see Fig. 15.4). It has been shown by Chen *et al* [28] that the calculated mean lifetime of the $v = 12$ vibrational state of the ${}^{20}Ne^{40}Ar^{2+}$ isotope is in reasonable agreement with the measured value [10]. Such highly excited vibrational states are expected to be populated by vertical transitions from the $NeAr^+$ parent as shown in Fig. 15.4. If the long-lived $NeAr^{2+}$ decays by tunneling, as suggested, then one would expect a large difference between the mean lifetimes of its different isotopes. Recent measurements, however, indicate that the mean lifetimes of the ${}^{20}Ne^{40}Ar^{2+}$ and ${}^{22}Ne^{40}Ar^{2+}$ isotopes are

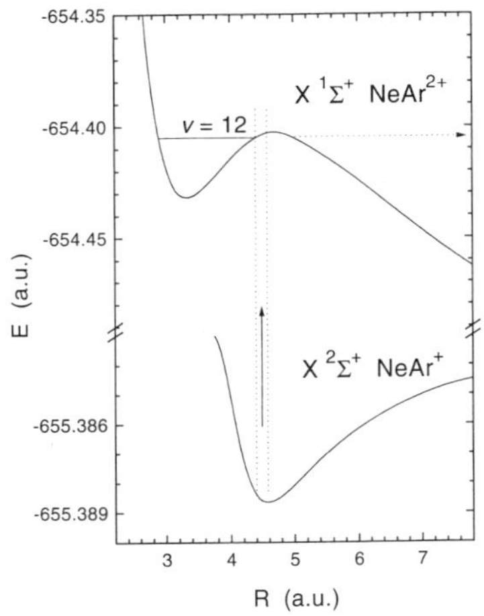

FIGURE 15.4 Potential energy curves of the electronic ground states of $NeAr^+$ and $NeAr^{2+}$ molecular ions.

rather similar to each other while the closest calculated mean lifetime for the dissociation by tunneling of the heavier isotope is more than a factor of 20 off the measured value [31]. Thus, suggesting that the long-lived $NeAr^{2+}$ molecular ion has some bound or metastable excited states that decay by electronic transitions. For such transitions much smaller isotopic effects are expected, as discussed for HeH^{2+} previously.

Kinetic Energy Release in the Dissociation. Measurements of the kinetic energy released in the dissociation can help one determine the molecular state (electronic, vibrational or rotational). In the case of tunneling, the kinetic energy released in the dissociation of a particular metastable molecular state is the difference between the total energy of the initial state and that of the two final fragments far away from each other. The width of this distribution is very narrow because the rovibrational states have narrow widths (see Fig. 15.3). On the other hand, states which decay by electronic transitions to lower states are expected to have broad kinetic energy distributions, because the kinetic energy release depends on the internuclear distance at the time of the vertical electronic transition. This significant difference between the kinetic energy distributions of the two decay mechanisms can also be used to determine which is the decay mechanism of a given molecular state (see Sec. 15.5). As an example of the use of kinetic energy release measurements let us return to the HeH^{2+} example. The possible long-lived states of this molecular ion are the $2p\sigma$ first excited state [16,17] and the highly excited $4f\sigma$ and $4f\pi$ states [17]. The measured kinetic energy distribution was found to be consistent with the one expected for the decay of the $2p\sigma$ state and much larger than the distribution expected for the highly excited states. Thus, the measurement of the kinetic energy release affirmed that the measured long-lived HeH^{2+} was formed as expected in its first excited state [12]. Due to the typical width of the kinetic energy distributions one can use them to determine the molecular state only in some cases.

Now that we have described the physics one can learn from such studies it is time to focus on the techniques used to conduct them. Fast beam techniques for studies of molecular ions provide the experimentalist with good control of the molecular ions under investigation as it is relatively easy to select the molecular ion of interest from any other ions (atomic or molecular) produced in the ion source. Among such studies one finds Coulomb "explosion" imaging of the stereometric structure of molecular ions [32,33] and dissociative recombination of molecular ions [34]. Such dissociative recombination experiments have been performed recently, using storage rings in which the molecular ions were allowed to cool down to their ground state [35–37]. Using the parent molecular ion beam one can produce more "exotic" molecular ions in collisions with a thin gas target: for example, triply charged diatomic molecules [38–40] or doubly charged rare gas dimers [1–13]. These molecular ions are usually very weakly bound and in many cases live just long enough to be detected. In the following we will describe our experimental studies of doubly charged diatomic molecular ions containing one or two noble gas atoms, and in particular measurements of their mean lifetimes. Even though our main interest is in phenomena related with this small family of molecular ions the techniques described in the following sections can be easily used to study many other long-lived molecular ions.

15.2 EXPERIMENTAL TECHNIQUES

In order to investigate the decay-mechanisms and -rates of long-lived doubly charged molecular ions one has to produce and separate them from other ions. An efficient method to accomplish this goal is to produce the molecular ion of interest in fast charge-stripping collisions between the parent singly charged molecular ion and a target atom, for example $HeH^+ + Ar \rightarrow HeH^{2+}$, and then separate the doubly charged molecular ion from the other ions using electric or magnetic fields. These charge-stripping collisions are efficient at high collision velocities where ionization dominates over electron capture. The experimental work was conducted using the Technion 1 MV Van de Graaff accelerator and the experimental setup shown in Fig. 15.5. The typical beam energies of 100-900 keV and the size of the experimental apparatus determine the range of mean lifetimes which we can study (about 1-500 ns). This range can be extended, in principle, by using lower and higher beam energies available at other accelerators. The general methods used are described in this section.

The mean lifetimes of molecular ions span a wide range, *i.e.* many orders of magnitude. For example, the ground state of $NeAr^{2+}$ has a mean lifetime longer than the universe while the highest vibrational level will decay within

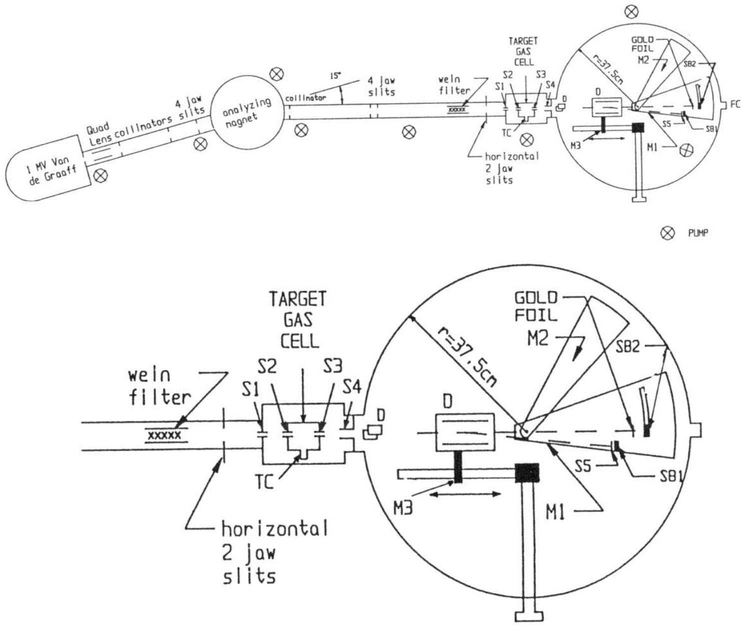

FIGURE 15.5 Schematic view of the beam line used for these studies. M1 and M2 are arms that can be rotated around the axis of the scattering chamber. M3 is a translation stage parallel to the beam axis on which the deflector is mounted. SB1 is the surface barrier detector used to detect the doubly charged molecular ions while SB2 is a similar detector used for normalization.

about 60 ps [28]. Thus, it is impossible to design an experimental setup and develop methods which will be applicable over the whole range, and one has to design the experiment for the case under study. In this article we describe our experimental system in which several methods have been used to determine mean lifetimes from about 1 to 1000 ns. Generally speaking, the methods are based on the determination of the number of molecular ions, dissociated and non-dissociated ones, as a function of the distance they have traveled. The need for several methods is due to the large time range covered, competing reactions and impurities in the beam. Furthermore, the kinetic energy imparted on the fragments during the dissociation process causes a deviation of the fragments from the trajectory they would have followed if they gained no energy in the dissociation (or if their transverse velocity remained unchanged). Due to the transverse velocities gained by both fragments, they are confined to move within cones around the trajectory followed by the center of mass of the molecular ion. The cone's apex angle α_1 for m_1 is given by:

$$\tan \alpha_1 = \sqrt{(m_2/m_1)(E_k/E)}\ , \tag{15.2}$$

where E_k and E are the dissociation energy and beam energy, respectively. This broadening of the spots produced by the fragments on the detector plane is one of the main reasons the techniques used for mean lifetime measurements of molecular ions differ somewhat from those used for Atomic beams, for which there is practically no broadening. In the next section we describe the experimental setup and the common features of all experiments involving long-lived molecular ions. In particular we will discuss how one distinguishes between the molecular ions of interest and beam impurities, a major problem when dealing with "exotic" molecular ions which are produced in very small rates relative to the beam intensity. In the following section we will describe the Fragment-Fragment coincidence method developed for studies of short-lived molecular ions, *i.e.* those which decay faster then their travel time to the detectors. In the fourth section we will dwell on a few specific methods for mean lifetime measurements of long-lived molecular ions, *i.e.* cases where both the molecular ions and their fragments reach the detector in measurable quantities. Finally, in the last section possible future developments will be suggested.

15.2.1 Production and selection of the parent molecular ion

The first step in the experiments described in this report is the production of the parent singly charged molecular ion, HeH^+ in the example above. This is accomplished by introducing a mixture of He and H_2 gases (typically 80 and 20 %, respectively) into the accelerator rf ion source, in which ionization is caused by radio frequency EM power, and extracting the singly charged molecular ions of interest (such molecular ions are produced also in some other ion sources). Beams of molecular ions which are produced by ionization of stable neutral molecules have typical currents similar to those of atomic species. In contrast "exotic" molecular ions like HeH^+ are created inefficiently in three body collisions in the ion source and thus have very weak currents. This creation mechanism is very sensitive to the pressure in the ion source and it is very helpful if a pressure gauge is connected as close as possible to it. Typical beam currents are about four orders of magnitude smaller than those of atomic beams through the same experimental setup. For example, the collimated HeH^+ beam was of the order of 10^6 particles per second, *i.e.* about 0.1 pA. For comparison, under the same conditions a current of about 1 nA of He^+ ions have been measured. Because of these weak currents one has to search for the beam with very sensitive current meters or with particle detectors. The beam currents, though typical, differ from one molecular ion to another.

The parent molecular ions of interest are then selected by a 15° analyzing magnet which is carefully calibrated each time for the slit settings of that measurement. We then check the energy spectrum of the direct beam with

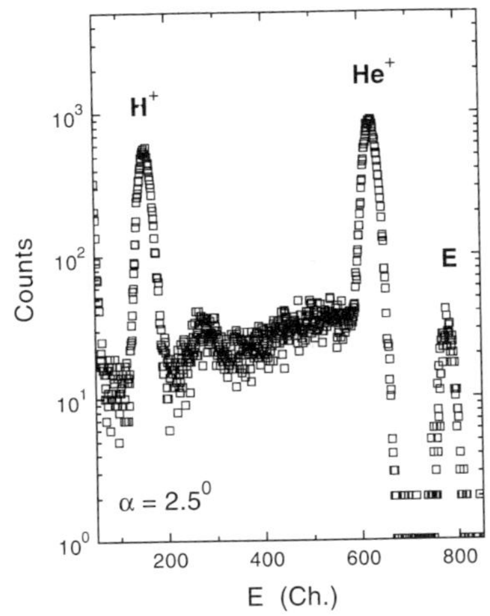

FIGURE 15.6 The energy spectrum measured when the detector is on the $^4HeH^{2+}$ trajectory, *i.e* a deflection angle of 2.5°. The full energy peak, labeled E, is associated with the *fragment-fragment* coincidences (see text).

one of the surface barrier detectors shown in Fig. 15.5 to assure that slit scattering is negligible and that the ions have the correct beam energy. Any additional peaks in the energy spectrum indicate contaminations in the beam. Then, some gas is introduced into the target cell to dissociate the incoming beam and verify that the weak beam is truely the parent molecular ion of interest. A heteronuclear diatomic molecular ion is expected to produce three peaks in the energy spectrum one for each fragment and one for the molecular ion. For example, $^4HeH^{2+}$ produces 3 energy peaks at $\frac{1}{5}$, $\frac{4}{5}$ and 1 of the full beam energy, as shown in Fig. 15.6 (the HeH^+ has a similar energy spectrum).

The selection of the particular molecular ion of interest, however, is not always a simple task because of the various possible impurity ions which may go with it through the analyzing system of the accelerator. Even when the impurity is hardly detectable in the main beam it might compete significantly with the doubly charged molecular ions because the cross sections for formation of the "exotic" states under study are orders of magnitude smaller than the cross sections for stripping an electron of atomic or molecular ions with strongly bound doubly charged states. For a specific example, if we are interested in the study of $^4He^{22}Ne^{2+}$ molecular ions produced by stripping an electron from $^4He^{22}Ne^+$ projectiles we may find small amounts of impurities of other atomic or molecular ions with the same mass number, for example Mg^+ and CN^+. Another source of contamination is higher-mass atoms, molecules and clusters at higher energies and charge states, which have the same trajectories in the analyzing magnet. These impurities are system-dependent and in our case they typically include H, C, N and O atoms. As discussed above it is essential to separate the beam of interest from the impurities. Very good vacuum throughout the system can significantly help to reduce the severity of the problem. However, it has been our experience that it is not always the cure, and additional analyzing components on the beamline are very helpful. Some of the impurities can be avoided by the use of electromagnetic lenses such as the strong focusing quadrupole doublet and the einzel lenses as they focus the main beam and the impurities at different points. A Wien filter (crossed magnetic and electric fields) placed after the analyzing magnet (see Fig. 15.5) is used to select the beam velocity and thus to eliminate impurities with the same mv/q (*i.e.* $\sqrt{2mE}/q$) as the main beam but a different velocity. (For example an ion with four times the mass, the same energy and twice the charge has the same mv/q but half the velocity, *e.g.* $^{20}Ne^{2+}$ will have the same trajectory as the $^4HeH^+$ beam but the two can be separated by the velocity selector).

Neutral molecules, fragments and atoms that are typically created by collisions with the remnant gas in the system and slit scattering can be an additional source of background counts, because they are not affected by the deflecting fields. Thus, it is important to guide the beam through the target cell on an "off center" trajectory using a "cleaning" deflecting field (either electrostatic or magnetic) as close as possible to the entrance of the target cell.

15.2.2 The gas target cell

The charge-stripping collisions in which the doubly charged molecular ions are produced take place in a differentially pumped target cell indexmolecular ion production!target gas cell (see figure 15.5). The length of this cell is not important if a single exponential decay curve is expected and most molecular ions of interest will not decay before reaching the analyzer. Using a relatively long cell of 5 cm reduces the load on the pump in the differentially pumped region for a given target thickness. The entrance and exit apertures of the target cell should be as small as possible (diameters of about .5 and .7 mm, respectively). These small apertures make it possible to increase the target gas pressure with only a minute effect on the differentially pumped region (1:1000 ratio in our case). Furthermore, they define a beam spot that is small relative to the spot produced by the dissociating fragments (see subSec. 15.2.6). The exit aperture has to be slightly larger than the entrance to allow all fragments from the dissociation within the target cell to exit. The vertical position and orientation of the target cell is adjusted mechanically to align the entrance and exit apertures along the beam axis in order to maximize the beam current and reduce slit scattering.

The gas pressure needed in the target cell depends on the particular ex-

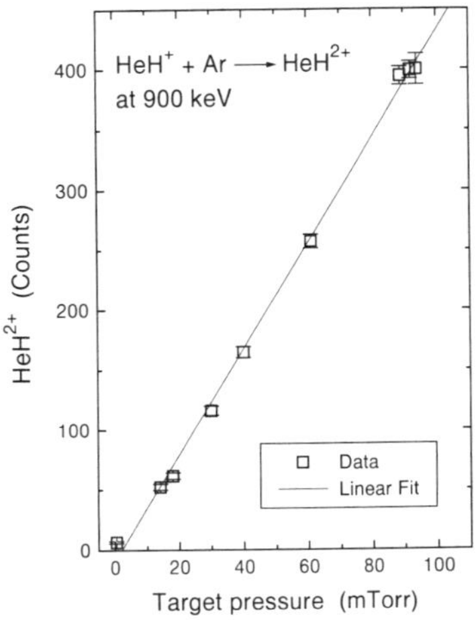

FIGURE 15.7 The number of HeH^{2+}, *i.e.* full energy counts, normalized to the HeH^+ rate as a function of the pressure in the target cell.

periment. At low pressures the yield of the doubly charged molecular ions produced by the interaction of the beam with the target gas increases linearly with the pressure. At higher pressures this linear trend will break as many of the doubly charged molecular ions, as well as a significant fraction of the main singly charged molecular beam, will dissociate. Typical values for the gas target pressures are of the order of a few mTorr. We typically used argon as the target gas, although, any other many-electron atom or molecule can be used. A typical target gas pressure of 5 mTorr in the 5 cm long cell provides a target thickness of about 10^{14} atoms/cm^2. Under these conditions the pressure in the system (scattering chamber and beam line) increases to about 1×10^{-6} Torr from a base pressure of 5×10^{-7} Torr.

For short mean lifetimes the length of the target cell must be as small as possible to allow most of the short-lived molecular ions produced to reach the analyzer. Furthermore, if the decay curve is not a single exponent, the exact location where the interaction took place is very important. In such cases (for example HeH^{2+}), a 3 mm long target cell is used and the gas pressure is increased accordingly. The doubly charged molecular ion production cross section can be evaluated from their normalized yield measured as a function of the target gas pressure, as show in Fig. 15.7. However, if the mean lifetime of the doubly charged molecular ion of interest is comparable to the flight time through the analyzer this information yields only an apparent cross section, from which the production cross section can be evaluated once the mean lifetime is determined [12,13].

15.2.3 Analysis of the reaction products

Quite a few atomic and molecular ions emerge from the target cell following the molecular-ion atom interactions. For example, the $NeAr^+$ + Ar collision results in the following reaction products: singly and doubly charged molecular ions, both having the full beam energy; Ne^{q+} and Ar^{q+} fragments in various charge states having one third and two thirds of the beam energy, respectively. The molecular ions will proceed along the original beam trajectory, while the fragments will spread out a little due to the energy released in the dissociation, contained in a cone whose apex angle is given in Eq. 15.2. These numerous reaction products must be identified and spatially separated from each other, and in particular from the doubly charged molecular ion, before its decay rate can be measured. This is done by the use of an electrostatic analyzer for molecular ions with longer mean lifetimes, and by a small strong permanent magnetic analyzer for those with short mean lifetimes.

The electrostatic analyzer is a parallel plate deflector 12 cm long with 2 cm spacing between the plates, operating typically with potential difference V up to 20 kV. This analyzer separates the reaction products according to their charge-to-energy ratio. The $NeAr^+$ molecular ion, for example, is deflected 15 mm from the neutral fragments 450 mm down stream from the exit of the deflector for V=10 kV. Under the same conditions, the $NeAr^{2+}$ molecular ions are deflected by 30 mm, the Ar^+ fragments are deflected by 22.5 mm, and the Ne^+ and Ar^{2+} ions are deflected by 45 mm. The spot of the Ar^+ fragments at this distance is a couple of mm. Thus, the doubly charged molecular ion is well separated from the other ions. The analyzer is mounted on a movable stage so that the distance between the deflector exit and the target cell can be varied to allow measurements of mean lifetimes, as described in Sec. 15.4. Obviously, when changing the deflector position its voltage needs to be adjusted such that the ions will hit the same spots on the detector plane.

For short-lived molecular ions, *i.e.* those which travel only a few millimeters during their mean lifetime at 1 MeV, an experimental setup much smaller than the one described above is needed. Such a system was developed for the study of one of the most interesting doubly charged molecular ions, HeH^{2+}, which consists of a proton and a helium nucleus bound by a single electron. The compact analyzing system is based on a small strong permanent magnet 12.7 mm in diameter with a 0.8 T field in between the magnet poles. This analyzer separates the reaction products according to their momentum-to-charge ratio. The HeH^+ molecular ion, for example, is deflected 23.4 mm from the neutral fragments 1070 mm down stream from the exit of the magnet. Under the same conditions, HeH^{2+} is deflected by 46.8 mm while the He^+ and He^{2+} fragments are deflected by 37.4 and 74.9 mm, respectively. The H^+ fragments are deflected way off the HeH^{2+} trajectory; thus, the doubly charged molecular ion is well separated from the other ions. This analyzer is placed right next to a 3 mm long target cell, so that a significant fraction of the HeH^{2+} molecular ions produced survives through the analyzer. Furthermore, the decay curve of this molecular ion has more than one mean lifetime, and thus it is essential to know the point of its production as discussed in Sec. 15.3. The distance between the exit of the magnetic analyzer and the target cell can be varied by moving the small target cell, which is mounted on a translation stage, along the beam axis. Because the analyzer's position remains fixed, the trajectories beyond it do not vary and the detector positions need not be adjusted during the experiment.

In order to determine experimentally the deflection angles of the various reaction products after passing through the (electrostatic or magnetic) analyzer and determine the width of the fragments distributions, an angular scan is performed with a narrow vertical slit in front of one detector while a second detector located at a fixed position is used for normalization by measuring either the beam current or the rate of neutral fragments. The analyzer field is increased and adjusted to avoid any overlap between the doubly charged molecular ions (including those that dissociated after the analyzer) and the adjacent fragments.

15.2.4 Detection system

Two similar surface barrier detectors are needed, one for measuring the reaction products, including the decaying doubly charged molecular ion, and one for normalization. These detectors have 100% detection efficiency at these energies and they produce signals proportional to the energy of the detected particles. Generally, there are three peaks in the energy spectrum, one for each fragment at an energy proportional to its mass (relative to the molecular ion mass), and one at the full energy of the beam produced by molecular ions and by pairs of fragments that hit the detector in coincidence. The energy spectrum measured when the detector was on the HeH^{2+} trajectory is shown in Fig. 15.6. The use of energy-sensitive detectors is of great importance in such studies for two main reasons: (a) fragment-fragment coincidence events are detected as full energy counts because all the beam energy is deposited in the detector (see Sec. 15.3 for a detailed discussion), and (b) some impurities can be identified and sometimes separated using the energy spectrum, for example an impurity with twice the mass, energy and charge can only be separated by using the fact that it appears at twice the full energy of the beam in the energy spectrum. It is useful to use an energy-sensitive detector for normalization in order to assure that normalization is done on the beam of interest excluding impurities as discussed in the next subsection.

Both detectors are mounted in a large vacuum chamber on arms (M1 and M2 in Fig. 15.5) that can be rotated externally without breaking vacuum, thus allowing the needed angular scans described in the previous subsection and small adjustments in their positions to match the "off center" beam trajectory.

15.2.5 Normalization

In order to determine the decay rate of a long-lived (or short-lived) molecular ion one measures the number of doubly charged molecular ions as a function of their flight time from the moment of production. This usually requires repeating the measurement for different flight paths, thus one has to normalize all measurements to the same number of parent singly charged molecular ions. The intensity of the main beam of singly charged molecular ions after collimation and alignment is typically below the intensities that can be normalized using Faraday cup readings; thus, techniques based on particle counting are necessary. Furthermore, because of the wide range of cross sections for the charge-stripping collisions, in which the doubly charged molecular ions are produced, different normalization procedures must be used in specific cases. In cases were the cross section is very small but an intense beam can be obtained, Rutherford back scattering from a thin gold foil onto a surface barrier detector at 90° to the beam trajectory is a good normalization procedure, since the number of fragments counted is proportional to the beam intensity (see Fig. 15.10). On the other hand, if only a low intensity beam is available, another reaction channel can be used for normalization. The most convenient channel in such cases is the dissociation into neutral fragments, which can be counted by a surface barrier detector with the appropriate aperture, mounted on the beam axis (see Fig. 15.9). This reaction channel is preferred because it is not affected by changes of the analyzing field. Furthermore, for long measurements where the gas pressure in the target cell may fluctuate, normalization based on the measurement of reaction products will yield a welcome compensation as long as single collision conditions remain valid for both channels. The disadvantage of the use of the neutrals' rate for normalization is that it might include some neutrals caused by slit scattering. This is a problem especially when the target cell is moved along the beam axis because the alignment is not likely to be perfect, and thus slit scattering might depend on the target cell position. However, Rutherford back scattering normalization is well suited for this case because the trajectories of the reaction products do not change after the analyzer.

If the main beam contains some impurities, they might be revealed in the energy spectrum of the normalization detector as signals in the wrong energy range. By setting a proper window on the energy spectrum one can avoid contributions from these impurities and assure that normalization is done relative to the intensity of the parent molecular ion alone.

15.2.6 Collecting all dissociation fragments

The energy released in molecular dissociation causes an additional complication in the measurement of mean lifetimes of molecular ions as compared to similar measurements of atomic ions. The trajectories of fast atomic ions are not affected by de-excitation, even if an electron is ejected, because of the large mass difference. In contrast, the fragments of a dissociating molecular ion will move significantly away from each other (a few mm in our studies). Thus, special care is needed to ascertain that *all the dissociation products are detected*, otherwise one can mistakenly associate changes in the number of counts due to the detection solid angle with the mean lifetime of the molecular ion. If the kinetic energy released in the dissociation channel of interest is known, then one can calculate the radius of the spot that contains all the fragments. If this information is not available one can make some estimates using the Coulomb "explosion" model, *i.e.* $E_k \sim q_1 q_2 / R_0$, where R_0 is the equilibrium internuclear distance of the parent molecular ion. However, a better approach is to directly determine the spot size for each fragment and for the fragment-fragment coincidence channel. The first can be determined from the width of the peaks in the angular scan described previously (see subSec. 15.2.3). The width of the fragment-fragment coincidence spot can not be determined from an angular scan with a narrow vertical slit because the two fragments are deflected away in opposite directions from the trajectory of the doubly charged molecular ion due to conservation of linear momentum in the dissociation. The radius of this spot can be determined by using an iris aperture placed on the detector which in turn is positioned along the trajectory of the doubly charged molecular ion. When the iris aperture is just slightly larger than the beam spot most full energy counts are associated with doubly charged molecular ions that have survived all the way to the detector (or more precisely to very close in front of it). As the iris aperture is opened, the number of full energy counts increases due to the collection of more pairs of fragments, and it levels off once the aperture reaches a diameter beyond which all fragment pairs are collected, thus determining the minimum aperture needed to ensure that all the fragments of the doubly charge molecular ions which dissociated in flight between the analyzer exit and the detector are accounted for. This is demonstrated in Fig. 15.8. This measurement has to be done, of course, for the largest possible distance between the detector and the exit of the analyzer.

15.2.7 Collisions with the residual gas

Collisions between the molecular ions and residual gas atoms and molecules can cause systematic errors in mean lifetime measurements. *Dissociation caused by collisions* between the doubly charged molecular ions and the residual gas between the exit of the analyzer and the detector will reduce the number of molecular ions. The amount of the reduction gets smaller when the distance between the analyzer exit and the detector is reduced because the residual target thickness is smaller. Thus, if significant, these collisions make the mean lifetime *seem shorter*. Second, *charge-stripping of the singly charged molecular ions* in collisions with the residual gas can create more dou-

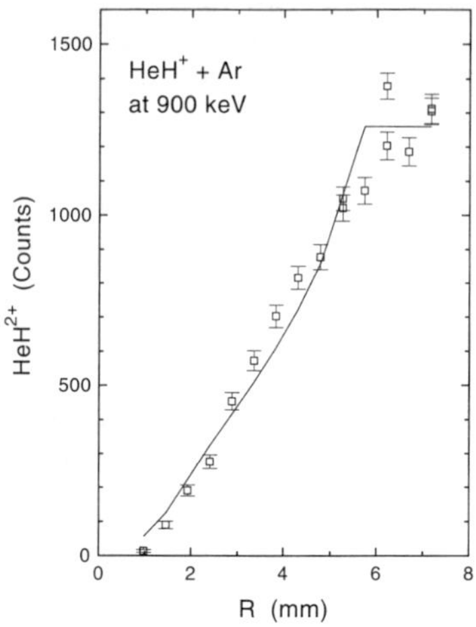

FIGURE 15.8 The number of HeH^{2+} normalized to the number of H^0 fragments as a function of the radius of the iris aperture in front of the detector. The solid line is the expected distribution for a kinetic energy release of 6 eV.

bly charged molecular ions than the number emerging from the target cell. If the method used for the measurement of the mean lifetime is based on changing the distance between the target cell and the analyzer then the doubly charged molecular ions produced over the added distance might make the mean lifetime *seem longer*. Fortunately, the cross sections for collisional dissociation and charge-stripping are typically small, and in our setup they cause negligible effects as long as the pressure in the system is kept below a few μTorr. For example, if one assumes a geometrical cross section for collisional dissociation, $\sigma_{coll} \sim \pi R_0^2 \sim 10^{-15}$ cm^2, then for a 17 cm translation length of the deflector and a pressure of 10^{-6} Torr the fraction of doubly charged molecular ions dissociated because of collisions is less then 0.06 %, *i.e.* much smaller then other errors in the measurement. In a similar way one can estimate the number of $NeAr^{2+}$ molecular ions produced in collision with the residual gas atoms. Assuming that the charge-stripping cross section is similar to the one measured for Ar atoms, $\sigma_{CS} \sim 3\times10^{-18}$ cm^2 [10], then for the same added path length and residual gas pressure about 4 $NeAr^{2+}$ molecular ions are produced relative to about 10,000 measured, *i.e.* a relative correction of less then 0.04 %. Both of these corrections are clearly negligible.

However, it is not guaranteed that the above corrections will always be m=6negligible; furthermore, usually the information needed in order to estimate them is not available. Repeating the mean lifetime measurements at a few different values of the base pressure in the vacuum system is a simple, direct approach verifying that corrections due to collisions with the residual gas are negligible.

15.2.8 Homonuclear *vs* heteronuclear molecular ions

The methods described above, and discussed in further detail using explicit examples in the next two sections, can be used only for heteronuclear doubly charged molecular ions. This is because the angular separation between the doubly charged molecular ions and their singly charged fragments resulting from dissociation before the analyzer is large enough that the spots of the fragments do no overlap with the aperture needed to collect all fragment pairs. *Unfortunately, this is not the case for homonuclear molecular ions.* Doubly charged homonuclear molecular ions follow the same trajectory as their singly charged fragments except for some broadening of the fragments around the trajectory. The fact that the energy of the molecular ions is two times larger than that of the fragments can not help either, because many of the fragments, whose rate is typically orders of magnitude larger, hit the detector simultaneously and thus also produce full energy signals. Reducing the beam current helps only partly as some of these events are true coincidences associated with the direct dissociation of transient doubly charged molecular ions in the target cell, for example $^4He_2^+ + Ar \rightarrow {}^4He^+ + {}^4He^+$. One can reduce the aperture in front of the detector to approximately the beam spot size, and thus reduce only the fragments yield, except for fragmentation of molecular ions oriented along the beam direction. This "trick" can be used sometimes to show the existence of doubly charged homonuclear molecular ions [5,8]. However, it cannot be used for mean lifetime measurements because these measurements require that all the dissociating fragment pairs be collected.

15.2.9 Kinematical coincidences

When the angular resolution between the doubly charged molecular ion and its neighboring fragments on both sides is marginal, special care is needed to avoid kinematical coincidences between the fragments, *i.e.* coincidences between two fragments from the direct dissociation in the target cell. This scenario is typical for "almost" homonuclear molecular ions (*i.e.* when the masses of the two atoms are similar), like $^3He^4He^{2+}$, and the analyzer field is not strong enough to prevent the detection of any $^3He^+$ and $^4He^+$ fragments, *i.e.* that their spots on the detector plane do not overlap with the aperture in front of the detector). Under such conditions, the direct dissociation in the target cell, $^3He^4He^+ + Ar \rightarrow {}^3He^+ + {}^4He^+$, can be detected in coincidence when the molecular ion is aligned along the analyzer field. In this case, due to conservation of linear momentum if one fragment moves toward the trajectory of the doubly charged molecular ion the other one will do so too, because their velocities are in opposite directions in the center of mass system (see Fig. 15.9). Thus, if one will be detected it is very likely that the other one will also be detected. As these events are true coincidences, though not the ones of interest in mean lifetime measurements, their rate will not be reduced

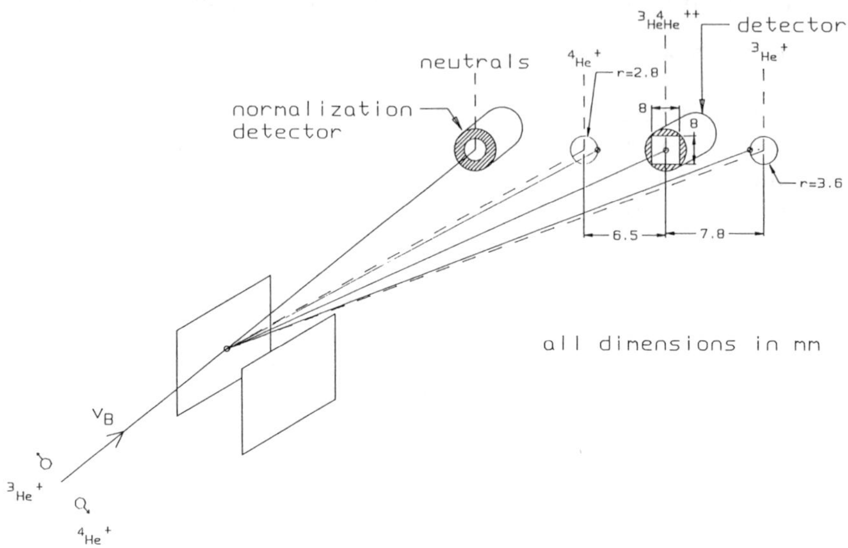

FIGURE 15.9 A schematic view of the spots of the $^3He^+$ and $^4He^+$ fragments of the dissociating $^3He^4He^{2+}$ molecular ion, about 500 mm after the exit from the analyzer.

relative to the dissociation of the doubly charged molecular ion if the beam current is reduced. To avoid such kinematical coincidences one has to use a strong enough analyzing field and verify that the singly charged fragments can not be detected. (In practice it is sufficient to verify that one of the fragments can not be detected.)

15.3 SHORT-LIVED MOLECULAR IONS

15.3.1 The Fragment-Fragment Coincidence Method

This method is most suitable for the determination of short mean lifetimes, *i.e.* in cases where none of the molecular ions reach the detector. It is based on knowing the number of the doubly charged molecular ions that managed to reach the exit of the analyzer. Both fragments of the short-lived molecular ion that dissociated in flight after the analyzer are detected in coincidence if the acceptance angle of the detector is sufficiently large (they are detected as events with the full energy of the beam). On the other hand, any fragments from the dissociation before (or within) the analyzer will be deflected away. An angular scan around the deflection angle associated with the doubly charged molecular ion is done in order to verify that these fragment-fragment coincidences are true events associated with the molecular ion of interest.

An electrostatic analyzer is not suitable for this purpose because the length needed to provide the needed angular dispersion is too long to allow adequate resolution. Furthermore, it can not be placed close enough to the target cell because of voltage breakdown. On the other hand, a small strong permanent magnet is most suitable especially for the study of light molecular ions such as the HeH^{2+}. It is of course necessary to have a very short gas target close to the analyzing magnet so that a measurable fraction of the short-lived molecular ions produced may reach the analyzer exit.

The mean lifetime of the short-lived molecular ion is then determined from the number of fragment-fragment coincidences normalized to the molecular ion beam as a function of the distance between the target cell and the analyzer exit. This distance is varied by moving the target cell along the beam axis and thus keeping the trajectories of all ions fixed beyond the analyzer. The mean lifetime measurements of the HeH^{2+} molecular ion are described below as an example [13,29,41].

The smallest mean lifetime values that can be measured using the system shown in Fig. 15.10 is limited not only by its size ($\sim$ 15 mm), but of course, also by the velocity of the projectiles. For a 1 MeV molecular ion with M=5 amu the velocity is about 6 mm/ns, so that $\sim$1 ns is the lowest measurable limit. For heavier molecular ions which move slower, there is a somewhat higher limit. The beam energy can be adjusted in order to match the expected mean lifetime of the molecular ion to the flight time through the setup. However, if a lower beam energy is needed then one might be limited by the energy resolution of available detectors for low energy heavy ions, and if a higher beam energy is needed then one might be limited by the angular resolution achievable for a given analyzing field strength.

15.3.2 Decay rate measurements of short-lived HeH^{2+}

The experimental method described in this subsection was developed for the study of one of the basic doubly charged molecular ions, HeH^{2+}, which consists of a proton and a helium nucleus bound by one electron. As discussed in the introduction this molecular ion has a few bound excited electronic states and a repulsive ground state. The first excited $2p\sigma$ state is the most likely bound state to be populated by charge-stripping collisions of the parent HeH^{+} molecular ion, and its mean lifetime is of the order of 1 ns. The HeH^{2+} has a velocity of 5.85 mm/ns at 900 keV (*i.e.* about 6 mm in one mean lifetime). Thus, the analysis of the reaction products must be completed within a very short distance, of the order of a couple of centimeters. Furthermore, no molecular ions are expected to reach the detector because the flight time to the detector is about 200 ns. This demand dictates the dimensions of the experimental setup shown in Fig. 15.10. The system consists of a 3 mm long target cell with small circular entrance and exit apertures (0.3 and 0.5 mm diameter, respectively), in which the HeH^{+} + Ar collisions take place. The target cell is situated in a differentially pumped region and operates with

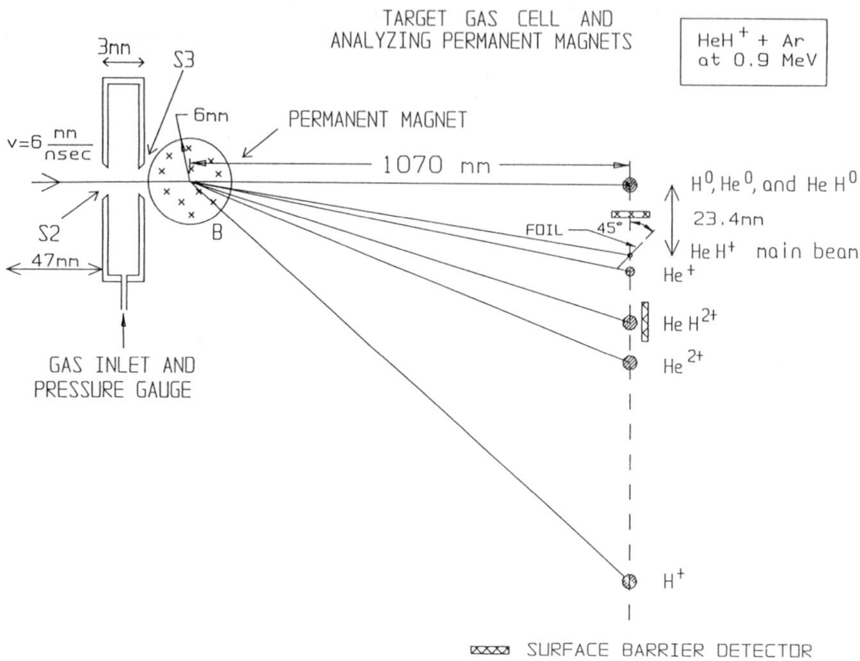

FIGURE 15.10 Schematic view of the experimental setup used for mean lifetime measurements of HeH^{2+}. The target cell and magnetic analyzer are mounted in the differentially pumped region labeled target gas cell in Fig. 15.5. The detectors are mounted in the scattering chamber in a similar way to the ones shown in Fig. 15.5.

pressures of about 10 mTorr, while the background pressure in the rest of the system is below 10^{-6} Torr. The ions produced in the target cell are analyzed according to their momentum-to-charge ratio, by a 12.7 mm long ~0.8 T magnetic field produced by small permanent magnets. The target cell is mounted on a translation stage so that the distance from its center to the exit of the magnetic analyzer can be varied from 15.5 to 62.5 mm. The trajectories of the ions passing through this analyzer are determined within a few ns from the instant of their production, and they remain unchanged throughout the measurement. It can be seen from Fig. 15.10 that HeH^{2+} is well resolved from the other ions. The analyzed ions are detected by a surface barrier detector 107 cm down stream. A 6 mm wide aperture in front of the detector defines the angular resolution of the system to be 0.16°, and is large enough to collect all the fragments from the HeH^{2+} dissociation, as can be seen from Fig. 15.8. The detector has a larger acceptance angle in the vertical direction (~ 19 mm) in order to make beam alignment easier. These measurements were normalized to the number of Rutherford back scattered projectiles to 90° from a 900 Å thick Au foil. (Some measurements were normalized to the H^0 rate,

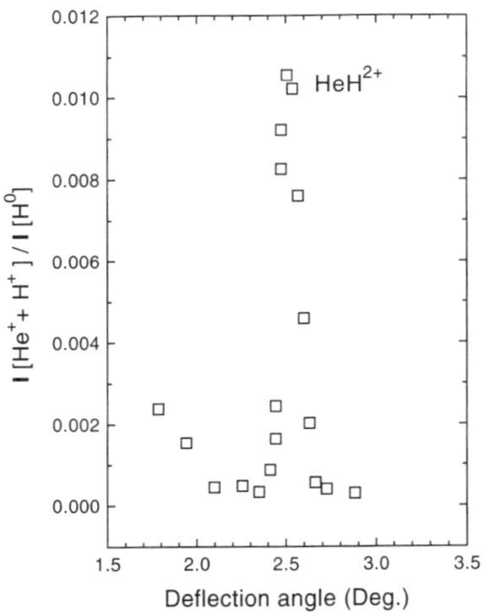

FIGURE 15.11 The number of $He^+ + H^+$ coincidences normalized to the number of H^0 fragments as a function of the deflection angle. HeH^{2+} are deflected by 2.5°.

while others were normalized to the HeH^+ beam rate using the RBS method, as indicated in the figure captions.)

Only HeH^{2+} molecular ions that have passed the magnetic field before their dissociation are deflected by an angle of 2.5°. The $He^+ + H^+$ fragments of the HeH^{2+} detected in coincidence along the trajectory associated with the 2.5° deflection angle are a clear signature of the survival of this short-lived molecular ion through the analyzer. These "fragment-fragment" coincidence events are detected as counts under the full energy peak in the energy spectrum shown in Fig. 15.6. For a true short-lived molecular ion that has dissociated in flight after the analyzer these counts have a well defined peak at the expected deflection angle, as shown in Fig. 15.11, and random coincidences are negligible around this deflection angle. One might be tempted to use the H^+ fragment yield at the deflection angle associated with the HeH^{2+} trajectory as the signature that the HeH^{2+} was created and decayed after the analyzer. However, the rate of H^+ fragments from collisional dissociation in between the magnet poles is much larger and one has to use the "fragment-fragment" coincidence technique. The pressure outside the target cell is still low enough to have negligible effect on the HeH^{2+} rate, but this pressure gradient before and after the target cell increases the effective length of the target and thus adds to the uncertainty of the apparent decay rate.

The mean lifetime of the bound $2p\sigma$ state of HeH^{2+} is determined from the number of $He^+ + H^+$ coincidences normalized to the HeH^+ beam as a function of their time of flight from the target cell and the exit of the analyzing magnet, as plotted in Fig. 15.12. These coincidence events are the number of HeH^{2+} molecular ions which survived all the way through the analyzer. This yield

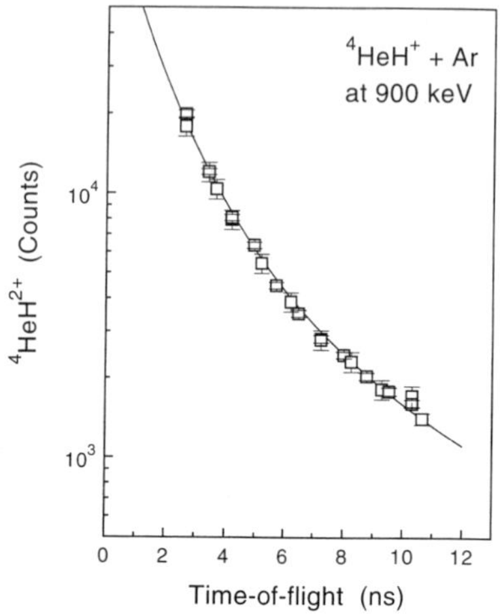

FIGURE 15.12 The number of He^+ + H^+ coincidences normalized to the HeH^+ rate as a function of the HeH^{2+} time-of-flight at a deflection angle of 2.5°.

decreases with increasing flight time, however, a single exponential decay curve does not fit the data. This is a typical result for the decay of a system which is a superposition of a few similar decay rates. For the HeH^{2+} molecular ions the mean lifetimes of the different vibrational states differ from each other significantly, especially for highly excited states as shown in Fig. 15.2. The decay curve of a wide distribution of vibrational states produced by Franck-Condon transitions from the HeH^+ parent, assumed to have a Boltzmann distribution of vibrational states, is in good agreement with the data shown in Fig. 15.12 (see Ref. [13,29] for details).

15.4 LONG-LIVED MOLECULAR IONS

Long-lived molecular ions are those which dissociate significantly during their flight time through the system, some of them making it all the way to the detector. This definition is of course system dependent, but so are the methods described below for measuring their mean lifetimes. In all methods the doubly charged molecular ions of interest follow a separate trajectory than all other ions. The mean lifetime is then determined from: (subSec. 15.4.1) their number as a function of flight time from the target cell; (subSec. 15.4.1, 15.4.1) the ratio of dissociated to non-dissociated molecular ions reaching the detector; and (subSec. 15.4.1) the number of one of the fragments dissociated after the analyzer as a function of their flight time from the target cell. Typically, a single exponential decay is assumed. However, the molecular ions might be formed in a few states with different mean lifetime. In particular, mean lifetimes much longer than the flight time through the system do not contribute to the dissociation. The effect of such very long-lived states is discussed in detail for the first method and it should be considered no matter which method is being used.

15.4.1 Specific techniques for decay rate measurements of long-lived molecular ions

The Deflector Translation Method

In the *deflector translation* method the distance between the electrostatic analyzer and target cell is varied by known amounts. The number of counts in the full energy peak, which contains molecular ions as well as ones which dissociated in flight, is measured as a function of the distance l_1 between the target cell and the analyzers' exit. For an exponential decay this number is given by

$$N(l_1) = N_0 \; exp(-t_1/\tau) \tag{15.3}$$

where N_0 is the number of doubly charged molecular ions produced in the target cell, τ is their mean lifetime and t_1 is their flight time given by $t_1 = l_1/v$, where v is the beam velocity. For the evaluation of the mean lifetime using this method it is important that all the long-lived molecular ions which have passed the deflector will be counted even if they dissociated in flight. Molecular ions which decay in flight do not follow exactly the original trajectory because of the kinetic energy released in the dissociation. The aperture in front of the detector must be large enough so that both fragments may hit it simultaneously and produce a full energy count (see subSec. 15.2.6). For example, the number of $NeAr^{2+}$ molecular ions as a function of their flight time from the target cell is shown in Fig. 15.13. The fit of Eq. 15.3 to the data yields a mean lifetime of $\tau = 275 \pm 25$ ns [10]. However, *this is not a unique interpretation of the data*, it is possible that some of the $NeAr^{2+}$ molecular ions are produced in states with negligible decay during their flight time through the system, designated N_{Long} (*i.e.* states with $\tau \gg t_1$). If this is the case, the number of full energy counts is given by

$$N(l_1) = N_{Long} + N_0 \; exp(-t_1/\tau) \; . \tag{15.4}$$

Typically, the additional free parameter in the equation above yields a shorter mean lifetime for the same data. For example, Eq. 15.4 can be fitted to the data in Fig. 15.13 with mean lifetimes as short as 175 ns. Thus, if nondissociating states can not be excluded the mean lifetime of $NeAr^{2+}$ is $\tau = 275 \;^{+25}_{-100}$ ns [28].

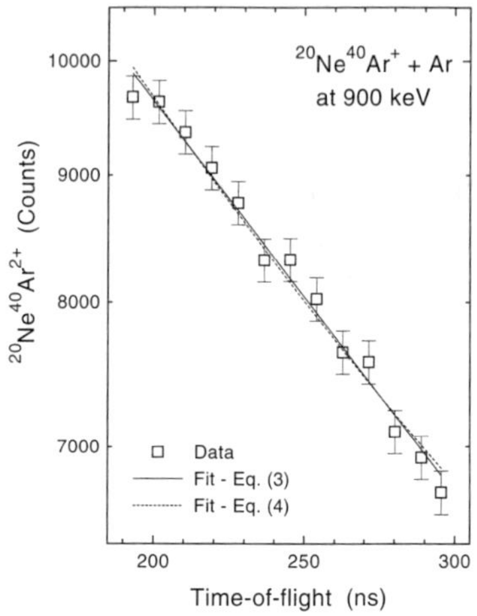

FIGURE 15.13 The number of $NeAr^{2+}$ molecular ions normalized to the number of Ne^0 fragments as a function of their flight time from the target cell to the exit of the electrostatic analyzer.

The Mesh Method

This method can be used for systems in which the the distance between the target cell and the deflector can not be varied, and is based on evaluating the fraction of molecular ions that survive after traveling a given distance. First it is necessary to verify that all the fragments of molecular ions that dissociated in flight between the analyzers' exit and the detector are detected (see subSec. 15.2.6). The separation of the counts in the full energy peak into molecular ions and fragment-fragment coincidences is based on the different transmission probabilities of molecules and both fragments of a dissociated molecule through a mesh with transmission T [42,34]. The number of non-dissociated molecules is reduced by a factor of T, while the number of fragment pairs passing simultaneously through the mesh is reduced by a factor of T^2. This is due to the fact that upon arrival at the mesh, the average distance between the two fragments is much larger than the mesh spacing so that their transmission through the mesh can be treated as two independent events. The number of counts under the full energy peak is measured twice, once with no mesh and once with a mesh with a transmission probability T in front of the detector (with both measurements normalized to the same beam current). The first measurement yields $M = [NeAr^{2+}] + [Ne{+}Ar]$, where the first term is the number of molecular ions and the second term is the numbers of Ne+Ar fragment pairs in this full energy peak. The second measurement yields $N = T[NeAr^{2+}] + T^2[Ne{+}Ar]$. Assuming a single exponential decay the mean lifetime of the molecular ion is given by,

$$\tau = -\frac{l_m}{v\ ln(\frac{(\frac{N}{M}-T^2)}{(T-T^2)})}\ , \qquad (15.5)$$

where l_m is the distance from the exit of the analyzer to the mesh. Using a 20 % transmission mesh with a spacing of s = 0.03 mm and l_m = 406 mm the number of counts under the full energy peak is reduced to $N = 931$ from the M = 8099 measured without the mesh. Substituting these values into Eq. 15.5 a mean lifetime of $\tau = 315 \pm 52$ ns is determined. The error is evaluated by repeating the measurement a few times and calculating twice the standard deviation of the sample. The assumption that the transmission of two fragments can be considered as independent events introduces an additional systematic error because it is not valid for molecules which dissociated in flight near the mesh as the distance between them caused by the dissociation is smaller or comparable to the mesh spacing, s. Using the "Coulomb explosion" model to estimate the kinetic energy released in the $NeAr^{2+}$ dissociation yields $E_k = \frac{q_1 \cdot q_2}{R_0} = \frac{1}{3.3} = 0.30$ a.u. $= 8.2$ eV. Thus, molecules which dissociate up to 6 mm in front of the mesh pass through a single hole and as a result the effective distance l_m is somewhat shorter than the real distance and if not corrected the mean lifetime is slightly over estimated, by about 10 % for the example above [43]. Another issue which can affect the interpretation of the measurement is the validity of the assumption of a single exponential decay as discussed for the previous method. *This is the main problem in evaluating the mean lifetime from the measurement.* If one assumes the existance of a state with a mean lifetime much longer than the flight-time to the detector in addition to the state which dissociates in flight, instead of the single exponentially decaying state, then the mean lifetime is given by

$$\tau = -\frac{l_m}{v\ ln(\ \frac{(\frac{N}{M}-T^2)}{(T-T^2)}\ (1+F)\ -\ F\ )}\ , \qquad (15.6)$$

where F is the ratio of very long-lived states (with $\tau \gg l_d/v$) to dissociating states formed in the target cell, *i.e.* $F = N_{long}/N_0$. Obviously, having a single measurement and two unknowns, τ and F, makes it impossible to determine the mean lifetime exactly. For the $NeAr^{2+}$ used as an example, a fraction F=0.5 yields $\tau = 150$ ns, however this fraction can reach a maximum value of 0.88 for which only an lower limit of $\tau \geq 33$ ns can be established. A mean lifetime much shorter than this limit is inconsistent with the fact that these $NeAr^{2+}$ molecular ions reach the exit of the analyzer. Thus, mean lifetimes from about 30 to 300 ns are consistent with the data. This large range can be reduced by repeating the measurement for another distance between the deflector exit and the mesh, l_m. (A similar problem exists in the method discussed in the following subsection).

The Iris Aperture Method

This method is similar to the previous one, in that both are based on the evaluation of the ratio of non-dissociated to dissociated doubly charged molecular ions measured for a fixed geometry. This ratio is determined from the number of counts in the full energy peak measured with an open and closed iris aperture. With the large opening both molecular ions as well as their fragments are counted whereas with the small opening, approximately as small as the beam spot, mainly the nondissociated molecular ions contribute.

For example, the ratio of non-dissociated to dissociated $NeAr^{2+}$ molecular ions detected can be evaluated directly from the number of counts under the full energy peak as a function of the radius of the iris aperture (see Fig. 15.8). At large openings the number of counts is $N(r > 3\text{mm}) = [\text{NeAr}^{2+}] + [\text{Ne+Ar}] = 5120 \pm 120$. While for the smallest iris aperture, which is slightly bigger than the beam spot ($\sim 0.2 \times 0.2$ mm), the number of counts is reduced to $N(r = 0.5\text{mm}) = [\text{NeAr}^{2+}] + \alpha[\text{Ne+Ar}] = 2748 \pm 74$, where α is the fraction of fragment pairs that hit the detector simultaneously. When $R_{Iris} \ll R_{Spot}$ then $\alpha \ll 1$, because only molecular ions which were aligned approximately parallel to the beam trajectory are detected in addition to the small contribution of molecular ions that dissociated close to the iris aperture. Assuming a single exponential decay the mean lifetime is given by

$$\tau = -\frac{l_I}{v\ ln(\frac{\frac{N(r=0.5)}{N(r>3)} - \alpha}{1-\alpha})}, \tag{15.7}$$

where $l_I = 515$ mm is the distance between the iris aperture and the exit from the analyzer. The determination of α is the main problem of this method. An upper limit for the mean lifetime can be easily calculated assuming that the fragment contribution for the minimum aperture is negligible, *i.e.* $\alpha = 0$, this yields $\tau_{max} = 490$ ns. A more realistic value of τ can be obtained using a simple estimate for α. The value of α depends on the kinetic energy released in the dissociation, ~8.2 eV for the $NeAr^{2+}$ dissociation (see subSec. 15.4.1). Given this dissociation energy the speed of the light fragments is $v_{Diss} = 0.0072$ mm/ns and the radius of their spot is $R_{spot} = \frac{v_{Diss}}{v}x = 0.00427x$ for a 900 keV beam energy, where x is the distance from the point of dissociation to the iris. Out of these ions a fraction $\alpha(x)$ will go through the collimator at its smallest radius $R_{Iris} = 0.5$ mm. For an isotropic dissociation, *i.e.* a uniform density on a sphere with radius R_{spot}, we get $\alpha(x) = 1 - \sqrt{1 - R_{Iris}^2/R_{spot}^2}$ if $R_{Iris} < R_{Spot}$ and 1 otherwise. The average value of α is 0.33 if one assumes for simplicity a uniform rate of dissociation. Using this value the minimum mean lifetime is estimated to be $\tau_{min} = 260$ ns. A better value of $\tau = 350$ ns can be evaluated by averaging over the exponential decay along the trajectory using τ_{max} which yields $\alpha = 0.20$. Therefore, the mean lifetime is $\tau = 350 \pm 100$ ns. The large error can be further reduced if the dissociation energy is measured and

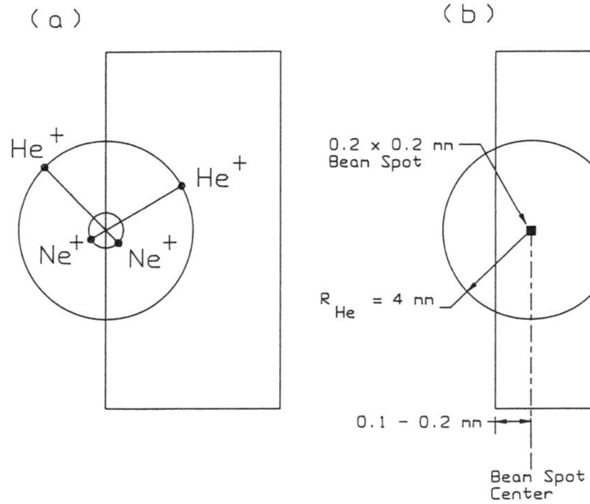

FIGURE 15.14 A schematic view of the beam spots relative to the detector and the mask covering half of its active area.

a detailed simulation of the fragments' trajectories is performed. Comparing such simulations to the intensity of counts under the full energy peak as a function of the radius of the iris aperture yields a more precise value of τ but it is time consuming and some of the information needed, like the distribution of dissociation energy is not always known. Another issue which can affect the interpretation of the measurement is the validity of the assumption of a single exponential decay as discussed for the previous method (see subSec. 15.4.1).

The "Masked Detector" Method

In this method the mean lifetime of the doubly charged molecular ion of interest is determined by measuring one of its fragments instead of both of them in coincidence and the molecular ions themselves as in the methods described previously. This main difference makes this method useful in the following cases: (a) If the molecular ion can not be separated from another atom or molecular ion by the analyzing system, for example, the $HeNe^+$ and C_2^+ molecular ions. In this case the $HeNe^{2+}$, C_2^{2+} and C^+ follow similar trajectories, and thus, methods based on the measured full energy counts can not be used [11,43]. (b) If one wants to focus on the decaying state and avoid any additional states that live much longer than the flight time through the system. Such long-lived states do not contribute to the production of fragments, and thus, will not affect the measurement [31].

In order to measure only one of the two fragments from the dissociation of a molecular ion one has to block half of the detector. The doubly charged molecular ions of interest are directed toward this mask such that none of them will be detected. Under these conditions only one fragment can be detected, either the He or the Ne from the $HeNe^{2+}$ used for example, because the two fragments move away from the beam spot in opposite directions, due to conservation of linear momentum in the dissociation (see Fig. 15.14). The fact that these fragments peak around the deflection angle of the $HeNe^{2+}$ indicates that this molecular ion is produced in the target cell and dissociates after passing through the analyzer. The He fragment is a better choice for the mean lifetime measurement because it has a larger spot on the detector, so uncertainties caused by alignment are smaller than for the heavier Ne fragment. The mean lifetime is determined from the measured number of He fragments as a function of the flight time of the $HeNe^{2+}$ from the target cell to the exit of the analyzer, as shown in Fig. 15.15. (This flight time is varied by moving the electrostatic deflector, for example.) This number can be written as

$$N(He) = \frac{1}{2} N_0 \left[\, e^{-x/v\tau} - e^{-l_d/v\tau} \, \right] \tag{15.8}$$

assuming a single exponential decay, where x and l_d are the distances from the target cell to the deflector exit and the detector, respectively. The best fit of the relationship above to the data shown in Fig. 15.15 yields $\tau = 180 \pm 30$ ns for $HeNe^{2+}$ [11]. In this method the spot of the fragment used for the mean lifetime measurement has to be much larger than the beam spot. Otherwise the molecular ions dissociating parallel to the edge of the mask, which are not counted (see Fig. 15.14), are a significant fraction of the dissociation and the factor of $\frac{1}{2}$ in Eq. 15.8 above overestimates the fraction of events measured. Under the conditions of the $HeNe^{2+}$ mean lifetime measurement [11] the deviation from 50% was negligible. This deviation is approximately given by $2R_{He}\delta/\pi R_{He}^2$, where $\delta \sim 0.1 - 0.2$ mm is the distance between the center of the beam spot and the mask edge, and R_{He} is the radius of the He spot.

15.5 FUTURE DEVELOPMENTS

In the near future our main effort will be focused on two major issues: (a) Developing a method for mean lifetime measurements of homonuclear doubly charged molecular ions, and (b) Improving the accuracy of the measurements of the kinetic energy released in the dissociation (see for example the section by D. Zajfman and O. Heber in this book). The first will enable us to measure many additional doubly charged molecular ions and study isotopic effects in their decay, for example by measuring the decay rates of all He_2^{2+} isotopes which decay by tunneling from their electronic ground state. The latter improvement will help us determine the decay mechanism of the molecular ion of interest and the rovibrational state which is decaying [44]. The knowledge of the quantum state, electronic and nuclear, of the decaying molecular ion is crucial for the comparison between the measurements and theory.

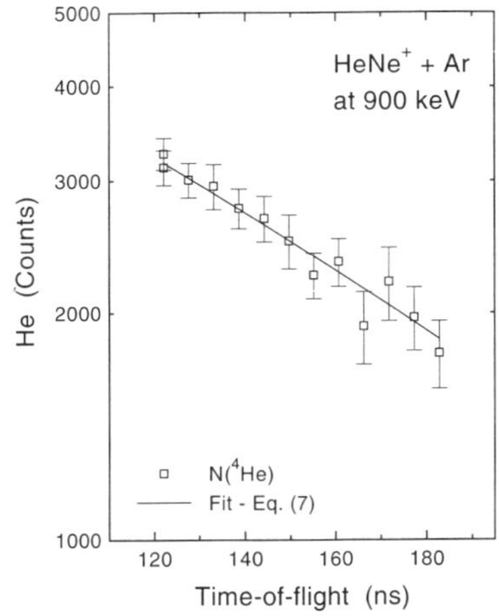

FIGURE 15.15 The number of He fragments of $HeNe^{2+}$ normalized to the number of He^0 fragments as a function of their flight time from the target cell to the exit of the electrostatic analyzer.

ACKNOWLEDGMENTS

We wish to thank J.P. Bouhnik, O. Heber and D. Zajfman for their contributions to some of the experiments, J. Saban, H. Shemi, and A. Shterenberg for their invaluable technical assistance, Z. Chen, B.D. Esry, G. Frenking, C. Heinemann, W. Koch, and C.D. Lin for their theoretical support, and K.D. Carnes, V. Krishnamurthi, and O.L. Weaver for many useful discussions. This work was supported in part by the Foundation for Promotion of Research at the Technion and in part by the Division of Chemical Sciences, Office of Basic Energy Sciences, Office of Basic Energy Research, U.S. Department of Energy.

REFERENCES

1. R. Johnsen and M.A. Biondi, Phys. Rev. A **20**, 87 (1979).
2. H. Helm, K. Stephan, T.D. Mark and D. L. Huestis, *J. Chem. Phys.* **74**, 3844 (1981).
3. K. Stephan, T.D. Mark and H. Helm, *Phys. Rev. A* **26**, 2981 (1982).
4. M. Guilhaus, A.G. Birenton, J.H. Beynon, M. Rabrenovic and P. Von Raque Schleyer, *J. Phys. B. At. Mol. Phys.* **17**, L605 (1984).
5. O. Heber, I. Ben-Itzhak, I. Gertner, A. Mann and B. Rosner, *J. Phys. B. At. Mol. Phys.* **18**, L201 (1985).
6. P. Jonathan, R.K. Boyd, A.G. Brenton, and J.H. Beynon *Chemical Physics* **110**, 239 (1986).

7. A. Belkacem, E.P. Kanter, R.E. Mitchell, Z. Vager, and B.J. Zabransky, *Phys. Rev. Lett.* **63**, 2555 (1989).
8. I. Ben-Itzhak, I. Gertner, D. Bortman and D. Zajfman, *Phys. Rev. A* **41**, 6548 (1990).
9. D. Zajfman, E.P. Kanter, Z. Vager, and J. Zajfman, *Phys. Rev. A* **43**, 1608 (1991).
10. I. Ben-Itzhak, I. Gertner, and B. Rosner, *Phys. Rev. A* **47**, 289 (1993).
11. I. Ben-Itzhak, I. Gertner, O. Heber, and B. Rosner, *Chem. Phys. Lett.* **212**, 467 (1993).
12. I. Ben-Itzhak, I. Gertner, O. Heber, and B. Rosner, *Phys. Rev. Lett.* **71**, 1347 (1993).
13. I. Ben-Itzhak, Z. Chen, B.D. Esry, I. Gertner, O. Heber, C.D Lin, and B. Rosner, *Phys. Rev. A* **49**, 1774 (1994).
14. W. Koch and G. Frenking, *J. Chem. Phys.* **86**, 5617 (1987).
15. D.L. Cooper and S. Wilson, *Molecular Physics* **44**, 161 (1981).
16. D.R. Bates and T.R. Carson, *Proc. Roy. Soc. A* **234**, 207 (1956).
17. T.G. Winter, M.D. Duncan, and N.F. Lane, *J. Phys. B At. Mol. Phys.* **10**, 285 (1977).
18. G. Frenking, W. Koch, D. Cremer, J. Gauss, and J.F. Liebman, *J. Phys. Chem.* **93**, 3397 (1989).
19. M.C.B. Montabonel, R. Cimiraglia, and M. Persico, *J. Phys. B: At. Mol. Phys.* **17**, 1931 (1984).
20. N.N. Penkina and T.K. Rebane, Opt. Spectrosk. **62**, 514 (1987), and *Opt. Spectrosk.* **46**, 454 (1979).
21. N.A. Gushchina and V.K. Nikulin, *Opt. Spectrosk.* **71**, 432 (1991), and private communication.
22. H. Yagisawa, H. Sato, T. Watanabe, *Phys. Rev. A* **16**, 1352 (1977).
23. A. Metropoulos and C.A. Nicolaides, *Chem. Phys.* **114**, 1 (1987).
24. N.C. Bacalis, Y. Komninos, and C.A. Nicolaides, *Phys. Rev. A* **45**, 2701 (1992).
25. J. Ackermann and H. Hogreve, *J. Phys. B. At. Mol. Phys.* **25**, 4069 (1992).
26. J.F. Babb and M.L. Du, *Chem. Phys. Lett.* **167**, 273 (1990).
27. W. Koch, G. Frenking, and A. Gobbi, *Chem. Phys. Lett.* **203**, 205 (1993).
28. Z. Chen, I. Ben-Itzhak, C.D. Lin, W. Koch, G. Frenking, I. Gertner, and B. Rosner, *Phys. Rev. A* **49**, 3472 (1994).
29. I. Ben-Itzhak, J.P. Bouhnik, B.D. Esry, I. Gertner, O. Heber, and B. Rosner, accepted for publication in *Phys. Rev. A* (1996).
30. E. Merzbacher, *Quantum Mechanics* (John Wiley & Sons, New-York 1961).
31. I. Ben-Itzhak, J.P. Bouhnik, Z. Chen, I. Gertner, C. Heinemann, W. Koch, C.D. Lin, and B. Rosner, *Phys. Rev. A* **52**, R3401 (1996).
32. Z. Vager, R. Naaman, E.P. Kanter, *Science* **244**, 426 (1989).
33. Z. Vager, D. Zajfman, T. Graber, and E.P. Kanter, *Phys. Rev. Lett.* **71**, 4319 (1993).
34. J.B.A. Mitchell, J.L. Forand, C.T. Ng, D.P. Levac, R.E. Mitchell, P.M. Mul, W. Claeys, A. Sen, and J.Wm. McGowan, *Phys. Rev. Lett.* **51**, 885 (1983).

35. P. Forck, M. Grieser, D. Habs, A. Lampert, R.Repnow, D. Schwalm, A. Wolf, and D. Zajfman, *Phys. Rev. Lett.* **70**, 426 (1993).
36. M. Larsson, H. Danared, J.R. Mowat, P. Sigray, G. Sundström, L. Broström, A. Filevich, A. Källberg, S. Mannervik, K.G. Rensfelt, and S. Datz, *Phys. Rev. Lett.* **70**, 430 (1993).
37. T. Tanabe, I. Katayama, N. Inoue, K. Chida, Y. Arakaki, T. Watanabe, M. Yoshizawa, S. Ohtani, and K. Noda, *Phys. Rev. Lett.* **70**, 422 (1993).
38. T.D. Märk, *Int. J. Mass Spectrom and Ion Proc.* **55**, 325 (1983/1984).
39. L. Morvay and I. Cornides, *Int. J. Mass Spectrom and Ion Proc.* **62**, 263 (1984).
40. D.L. Weathers, F.D. Macdanials, S. Matteson, J.L. Duggan, J.M. Anthony, and M.A. Douglas, *Nucl. Instrum. and Meth.* **B56/57**, 889 (1991).
41. I. Ben-Itzhak, J.P. Bouhnik, I. Gertner, O. Heber, and B. Rosner, *Nucl. Instrum. and Meth.* **B99**, 127 (1995).
42. K.H. Berkner, T.J. Morgan, R.V. Pyle, and J.W. Steams, in *Proceedings of the Seventh International Conference on the Physics of Atomic Collisions, Amsterdam 1971, Abstracts,* edited by J. Kistemaker (North-Holland, Amsterdam, 1971), p-422; T.J. Morgan, K.H. Berkner, and R.V. Pyle, *Phys. Rev. Lett.* **26**, 602 (1971).
43. I. Gertner, B. Rosner, and I. Ben-Itzhak, *Nucl. Instrum. and Meth.* **B94**, 47 (1994).
44. I. Ben-Itzhak, J.P. Bouhnik, Z. Chen, I. Gertner, O. Heber, C.D. Lin, B. Rosner, and D Zajfman, in preparation (1996).

CHAPTER 16

Angular Distribution of Auger Transitions in Oriented Molecular Ions

R. M. Wood* and A. K. Edwards†

We report the measurement and analysis of the angular distribution of Auger electrons emitted by homonuclear diatomic molecules such as nitrogen. The distributions are measured relative to the orientation of the internuclear axis of the molecule. The initial inner-shell vacancies are produced by bombardment of the target by fast projectiles. The ensuing Auger transitions proceed to dissociative states of a doubly charged molecular ion. The experimental technique involves the coincident detection and analysis of one of the charged molecular fragments and an Auger electron. The measured angular distributions are analyzed in terms of a two-center effective potential using prolate spheroidal coordinates. We discuss corrections to the measurements made necessary because of the thermal translational and rotational motion of the target molecules, and a further correction which accounts for the angular resolution of the electrostatic analyzers and detectors.

16.1 INTRODUCTION

16.1.1 Review of Auger electron angular distributions

The angular distribution of Auger electrons in atomic systems has been discussed by Mehlhorn and co-workers [1,2]. If the initial state of the transition has an equal population of M_L (or M_J) levels, i.e., the initial state is not aligned, then the subsequent transition will be isotropic. This is not the case for diatomic molecules. The Auger transition in diatomic molecules is a two-center phenomenon and does not have the spherical symmetry of the atomic system. The internuclear axis serves as an axis of quantization in reference

*) University of Georgia.
†) University of Georgia.

to which M_L is defined. States of $M_L = 0$ are Σ states, states of $M_L = 1$ are Π states, etc., and the symmetry of each should be manifested in the angular distribution of the Auger electron.

Dill and co-workers [3] have investigated theoretically the angular distribution of Auger electrons following K-shell photoionization. The results are expressed in terms of an asymmetry parameter associated with the photoabsorption process and a factor that depends on the Auger decay. Calculations were made for the asymmetry parameter for N_2 and CO. Synchrotron radiation experiments were performed [4,5] with the polarization vector serving as the axis of quantization for photoexcitation. The expected angular distributions were not observed. Several reasons were given for the lack of agreement between theory and experiment. Photoexcited states continue to be of interest, both theoretically [6] and experimentally [7], in the study of Auger transitions in molecules.

16.1.2 Molecular dissociation

The Auger studies which are the subject of this report all proceed to a final state which is a dissociative state of a doubly charged molecular ion. The experimental technique requires detection and analysis of the dissociation products, and it is therefore useful to consider the dissociation process and the information accessible to the experimentalist. For purposes of example, consider the direct double ionization of N_2 leading to $N_2^{2+}(1\pi_u^{-2})^1\Sigma_g^+$. Figure 1 illustrates the process. The potential curves for the ground state [8] and dissociative state [9] are taken from the literature.

Suppose a doubly ionizing collision takes place in a time interval so short that the internuclear separation does not change. The transition to the dissociative final state of the molecule can be described as a vertical arrow. The base of the arrow is located at the energy of the N_2 ground state, and at the separation coordinate at the moment of the ionizing collision. The tip of the arrow lies on the excited state potential curve. Once the dissociative state is formed, the nuclei begin to separate, picking up kinetic energy as they go. In the case of the $N_2^{2+}(1\pi_u^{-2})^1\Sigma_g^+$ state chosen as the example, the separated atom limit is $N^+(^3P) + N^+(^3P)$. The total kinetic energy of the fragments is the difference between the potential energy at the moment of formation and potential energy in the separated atom limit. The dissociation fragments share the available kinetic energy according to the dictates of the conservation of momentum principle. In this case the two identical N^+ ions have equal energy of approximately 4 eV.

The actual kinetic energy spectrum of the N^+ fragments is determined by the shape of the potential curve for the dissociative state and the ground state wave function. The reflection approximation [10] is frequently used to calculate energy spectra [11,12]. The spectrum predicted for the channel diagrammed in Fig. 1 is shown in Fig.2. Before the predicted spectra can be

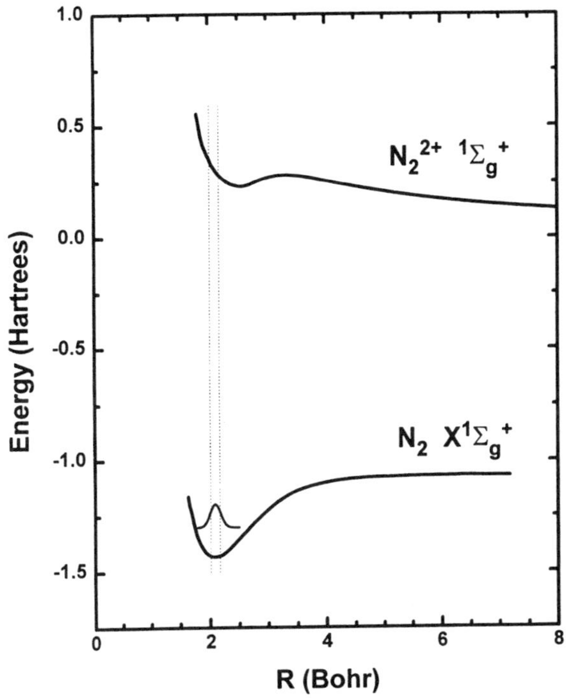

FIGURE 16.1 Potential energy curves for the ground state of N_2 and the dissociative $^1\Sigma_g^+$ state of N_2^{2+}. The energy scale has been adjusted so that the asymptotic energy of the $^1\Sigma_g^+$ state is zero. The vertical dotted lines indicate the classical turning points of the ground state wave function and define the Franck-Condon region. The ground state wave function is also shown.

compared to measured data, some corrections must be applied to the calculations. First there is a thermal broadening correction which has a small effect [13]. Secondly, the energy resolution function of the analyzer system used to collect the data has an additional broadening effect.

16.1.3 Axial recoil approximation

In the laboratory frame diatomic molecules are rotating with an angular velocity that can be estimated from the rotational energy of a rigid rotator with energy kT, where k is Boltzmann's constant and T is the Kelvin temperature. At room temperature the rotational period is usually very long compared to the time of dissociation, so the molecule dissociates approximately along the line of the internuclear axis. A requirement of the approximation is that the potential curve of the dissociating state has a steep slope in the

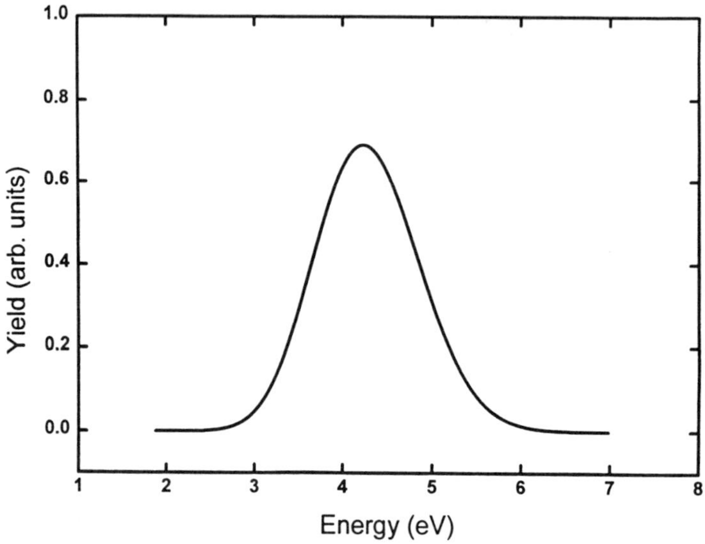

FIGURE 16.2 The reflection approximation prediction of the shape of the energy spectrum produced by ionization and subsequent dissociation through the channel $N_2^{2+}(1\pi_u^{-2})^1\Sigma_g^+ \longrightarrow N^+(^3P)+N^+(^3P)$. Thermal and instrumental broadening have been included.

Franck-Condon region. Ignoring the rotational motion allows the assumption of knowledge about the orientation of the internuclear axis when fragment ions are detected following dissociation. Zare [14] presented this axial recoil approximation in a study of the photodissociation of H_2^+. Since then, the axial recoil approximation has been used in studies such as, orientation effects in the excitation of H_2 [15] and the angular distribution of photoelectrons [16].

16.1.4 The Auger process

The Auger process for diatomic molecules is considered to be a multi-step process. First, the collision occurs to produce the inner-shell vacancy. This is followed by the Auger transition which ejects a fast electron and leaves the molecule in a doubly ionized state. If the final state is dissociative, then the Auger transition is followed by dissociation with the ejection of two fragment ions, each of several-eV energy.

A coincidence experiment can be performed in order to measure the angular distribution of Auger electrons from diatomic molecules. When the molecule is left in a dissociative state following a radiationless transition, it is assumed that the fragment ions produced in the dissociation travel along the line of the internuclear axis, i.e., the axial recoil approximation holds. The detection of ions being emitted at some angle relative to the beam direction yields the orientation of the internuclear axis at the time of the transition. The detection of the Auger electron in coincidence with the ion then gives the angle of emission of the electron relative to the internuclear axis.

Energy analysis is performed on both the Auger electron and the fragment ion. The coincident count-rate is very slow so that only moderate energy resolution is used. Energy analysis of the fragment ion gives some added selectivity to the measurements, in particular, if two peaks in the electron spectrum are not well resolved but the final states dissociate to different atomic states, then energy measurements of the ion does not allow coincident detection of one of the transitions. Or, if one of the final states does not dissociate then the coincidence requirement eliminates one of the transitions.

16.2 EXPERIMENTAL TECHNIQUE

16.2.1 Apparatus

Collision region

A schematic diagram of the collision region is shown in Fig. 3. The N^+ analyzer is set at angle θ_B, and the Auger electron analyzer at angle θ_A with respect to the incident electron beam direction. Both analyzers are free to rotate independently in the plane defined by the projectile beam and analyzers. During the measurements of the angular distributions, the angle θ_B is fixed and the angle θ_A is varied so that the angle θ, defined as the angle between the direction taken by the Auger electron and the internuclear axis, is varied from 0° to 90°.

The collision region is a differentially pumped volume that is formed by the gap between two opposing, vertically mounted tubes. The opening between the tubes allows for the unobstructed view of the interaction volume by the detectors as they are moved in angle. The diameter of the tubes is about 6 mm and the spacing between them is about 6 mm. The larger the diameter, the more uniform the target density is over the target volume. The target gas enters the interaction region from above and below through the tubes which are mounted in small cylindrically shaped chambers. These chambers are maintained at low pressure by a leak valve. A capacitance manometer is used to monitor the pressure in the small chambers to assure a constant flow of gas to the collision region. The diameter of the tubes is large and the pressure is low so that there are no effects as found in gas jets. Measurements show that the target can be considered to be a static gas.

The target density has been calibrated relative to the pressure in the cylindrical chambers by measuring the Rutherford scattering of protons from argon. With the analyzers at angles in the range of 30° to 150° the target gas pressure is constant over the very small length of beam (about 1.75 mm at 90°)

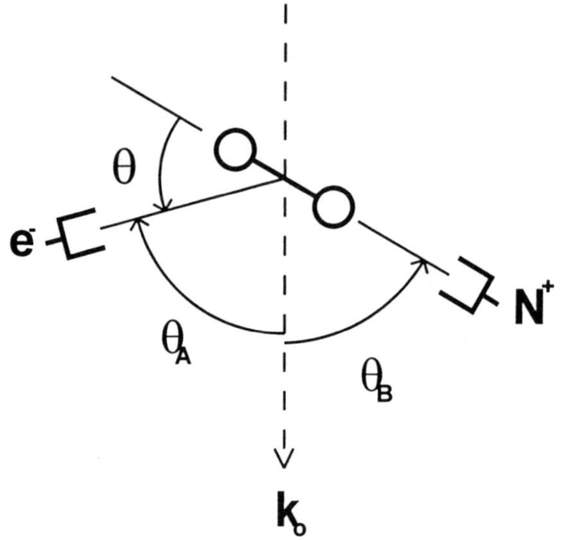

FIGURE 16.3 A schematic diagram of the scattering geometry (not to scale). The dashed line represents the path of the incident beam. The N^+ detector is set at angle θ_B, and the Auger e^- detector at θ_A with respect to the incident beam direction. Both detectors can be rotated in the plane defined by the projectile beam and detectors.

that contributes counts to the detectors. At angles below 30° or beyond 150° the analyzers look at a beam length that extends into a region where the gas pressure has decreased due to the differential pumping. This is manifested by a change in the rate of increase of the count rate with detector angle, i.e., it no longer follows an approximate $1/\sin\theta$ behavior as θ approaches 0° or 180°. The Rutherford scattering measurements can be used to determine corrections for the changes in target densities as a function of angle as well as to determine a pressure calibration.

An attempt was made to calculate the target density by a proper extrapolation of the procedure described by Brinkmann and Trajmar [17]. However, the calculated values of the target density fell off much more rapidly with distance from the target center than the measured values. The Rutherford scattering measurements indicated a more uniform density distribution in the gap between the gas-inlet tubes.

Analyzers and focusing elements

The hemispherical, electrostatic analyzers used in this work are commonly used devices and their properties have been reviewed by Rudd [18]. The electrostatic focusing elements and design considerations using the matrix

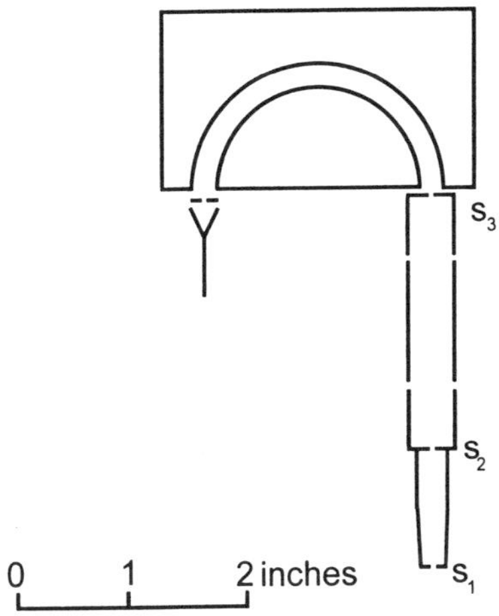

FIGURE 16.4 Hemispherical analyzer and its associated focusing elements. Dimensions are given in text.

method have been described by Kuyatt [19] and Dahl [20]. Numerical tables for electrostatic lenses by El-Kareh and El-Kareh [21] are useful in constructing a focusing system from tubes. A program called SIMION is available and can be used in the design of analyzers and focusing systems.

The energy analyzers for the positive ions and electrons are identical 180°-hemispherical analyzers illustrated in Fig. 4. The apparatus was originally designed to measure two low-energy fragment ions in coincidence from the dissociation of diatomic molecules. For such an experiment high resolution was not of primary concern. However, it is possible to use one of the analyzers to measure high energy electrons under special conditions that are discussed below. The inner-hemisphere radius is 2.22 cm and the outer-hemisphere radius is 2.86 cm. The relative positions of the inner and outer hemispheres are set by sapphire balls placed in the mounting hardware. The analyzers are constructed of aluminum in order to keep them light in weight. They are coated with Aquadag to prevent problems of surface contamination and contact potentials.

Each analyzer is constructed with equal width entrance and exit slits. If aberration terms are neglected, such an analyzer has a transmission function which is triangular in shape. The energy width (FWHM) of the triangle can be written as

$$\Delta E = E_0 \frac{Aw}{2R_0}, \tag{16.1}$$

where w is the slit width, R_0 is the radius of the ion path, and E_0 is the energy of the ion as it leaves the target region. The multiplicative factor A

is determined by the lens system which precedes the analyzer. There are two observations to be made: The counting rate increases as A and ΔE increase, and the energy resolution improves as A and ΔE decrease. Because the lens systems described in the next paragraph were operated with a magnification less than one, the size of the image of the interaction region is smaller than the width of the entrance slit, and the transmission function is a trapezoid rather than a triangle. However, the essential behavior of the analyzer remains the same. In the work reported here, the energy resolution of the N^+ analyzer system was $0.15E_0$, and the resolution of the electron analyzer was $0.0075E_0$.

The focusing elements in front of each analyzer consist of a set of three equi-diameter tubes spaced at 0.1 times the diameter of the tubes. The two gaps between the three tubes make up two lenses of adjustable focal length. These focusing elements serve the dual purpose of focusing either positive ions or electrons onto the entrance of the analyzer and accelerating or decelerating the ions or electrons prior to energy analysis. A pair of slits collimate particles entering the focusing system so that the particles are never more than one-half a tube radius from the axis of the lenses. This minimizes the effects of aberrations of the tube lenses. The focus elements are designed and operated so that the interaction volume is focused onto the entrance of the analyzer with a magnification of about one-half.

The kinetic-energy spectrum of fragment ions consists of many broad overlapping peaks and has no fine structure. Therefore fine resolution is not required, and the ions can be accelerated prior to analysis in order to increase the count rate. On the other hand, fine resolution is desired for the electron spectrum, so the electrons are decelerated by the focusing system prior to entering the hemispherical analyzer. Favorable focal conditions for our apparatus occur when N^+ ions are accelerated to ten times their original energy, and when electrons are slowed down by a factor of one-half. The desired factor of acceleration or deceleration must be considered in the design of the focusing elements.

The voltages on the inner and outer hemispheres, V_1 and V_2, respectively, are given by:

$$V_1 = \frac{E_0}{q}(1 - A\frac{R_2}{R_1}), \tag{16.2}$$

and

$$V_2 = \frac{E_0}{q}(1 - A\frac{R_1}{R_2}), \tag{16.3}$$

where E_0 is the initial energy of the particle in eV, q is the charge state of the particle being analyzed (e.g., +1 for N^+, -1 for electrons), A is the factor by which the energy is changed (10 for ions, 0.5 for electrons), and R_1 and R_2 are the radii of the inner and outer hemispheres, respectively. The potential at the radius half-way between R_1 and R_2 is $E_0(1 - A)/q$.

The voltages on the focus elements for the fragment ion analyzer are determined by the required focal lengths. The potentials used are: $V_{mid} = -4E_0$, and $V_0 = -9E_0$, where V_{mid} is the voltage on element 2, and V_0 is the voltage on element 3. The first element is kept at ground potential so that stray fields do not perturb the interaction region. The energy of the ion is increased by a factor of five by the first lens and another factor of two by the second lens. The voltage of the last focusing element matches the central potential of the energy analyzer. Typical ion energies are 1 to 10 eV.

The voltages on the focusing elements for the electron detector are set so that the first lens decelerates the electrons to an energy of $0.2E_0$ and the second lens accelerates them to increase their energy by a factor of 2.5. Consequently, V_{mid} is set at $-0.8E_0$ volts and V_0 is set at $-E_0/2$ volts. For example, a 100-eV electron would see element 2 at -80 volts and slow down to 20 eV. It would then be accelerated across the second gap from -80 V to -50 V, or an increase of 30 eV to 50 eV. The voltages on the hemispheres would be proper to pass 50-eV electrons. Calculations using the matrix method show that these conditions give an image of the interaction region near the entrance of the analyzer with a magnification of about one half.

When the analyzers are operated in the mode of sweeping the energy, the voltages on the focusing elements are swept at a constant ratio to the analyzer voltages and to each other so that the focal lengths of the lenses remain fixed. All of the voltages for a given system are derived from a common source that can be swept, and all of the voltages are set by potentiometers.

Common acceptance

In front of the focusing elements are two slits that collimate the particles being detected and define the effective beam length and solid angle observed by the analyzers. All particles that pass through the slits and have a post-acceleration or post-deceleration energy that matches the transmission energy of the analyzer are focused onto the entrance of the analyzer. The acceptance of the analyzing system is set by the geometry of the slits and not the focusing voltages.

With the electron analyzer and the ion analyzer set at different angles there are regions of the interaction volume which can contribute counts to one detector but not the other. Therefore, it is necessary to correct the coincidence count rate for the changes in the effective beam lengths common to both detectors as the electron detector is changed in angle while the ion detector is not moved. We have done a first-order correction for our apparatus and found it to be about 24% at its largest value.

In order to understand the correction procedure, it is helpful to visualize the slit system and possible ion paths or rays which leave the target volume, successfully pass through the slits, and reach the analyzer. The projectile beam path, the slit system, and four particularly significant rays are shown in

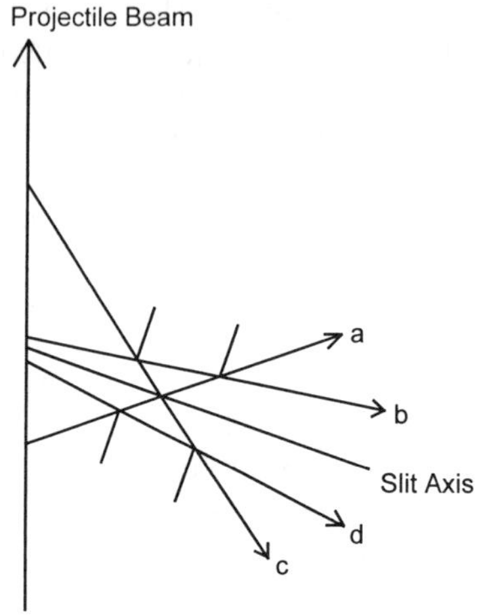

FIGURE 16.5 Schematic of beam length as observed by an analyzer (not to scale).

Fig. 5. Each ray in the figure can be characterized by two coordinates. The first coordinate is the position of the point of origination of the ray measured from where the slit axis intersects the projectile beam path. The second coordinate is the slope of the ray relative to the slit axis.

The information gathered in Fig. 5 is used to plot the one-dimensional acceptance [20] of the slit system. Each ray in Fig. 5 – indeed, each of the infinite number of rays successfully passing through the slits – is represented by a single point in the acceptance plot. The ordinate of the point is the second coordinate (slope) mentioned in the last paragraph, and the abscissa is the first coordinate (position).

Figure 6 is a superposition of the one-dimensional acceptance plots for the N^+ analyzer set at 60°, and the electron analyzer set at 90°. Each plot is very nearly a quadrangle. At 90° the quadrangle is a parallelogram. At other angles the sides of the quadrangles are actually bowed slightly, but are approximated as straight lines for the sake of argument. As the electron analyzer is moved in angle the set of rays that pass through the slits changes, causing the quadrangle to change.

Only points of emittance along the beam axis that are common to both detectors can contribute coincident counts. For each of these points only certain ranges of slopes can contribute. The slope x' can be written in terms of the slit widths, slit separation, and distance from the beam to the first slit, while the distance along the beam axis depends on these quantities plus the angle of the lens axis relative to the beam direction (which is the angular position of the electron analyzer). A weighting factor for each setting of the electron detector is found that is proportional to the product of the small areas of the quadrangles of each detector in the region x_i to $x_i + \delta x$ and

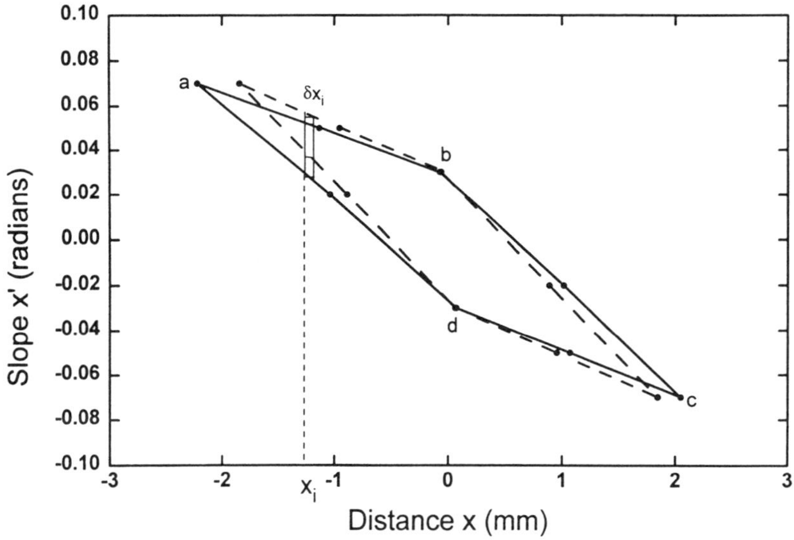

FIGURE 16.6 The phase diagrams for detectors set at 90° (dashed line) and 60° (solid line) relative to the projectile beam. Points a - d correspond to the limiting rays of Fig. 5. The common acceptance observed by the two detectors is described in text.

summed over x_i. Each small area of each quadrangle represents the product of an interaction length δx and acceptance angle $\delta\theta$ needed to measure cross sections. For coincidence work this product is needed for both detectors. The data are normalized to the measurements with the electron detector at 90° where the interaction length of the electron detector is a minimum.

16.2.2 Procedures

Auger electron spectra

While the angular distribution data are being collected, the voltages on both analyzers are held fixed so that electrons and N^+ ions of appropriate energy are accepted. The first step in selecting the proper analyzer settings is to sweep the voltages on the electron analyzer and measure the energy spectrum of all the Auger electrons emitted in the energy range of interest without reference to any positive ions. Figure 7 shows the spectrum obtained when N_2 is bombarded with 1634 eV electrons. The spectrum is calibrated by comparison to the measurements of Moddeman et al. [22]. The overall features of our Auger electron spectrum are in agreement with the results of Moddeman et al. except for the fact that their energy resolution was 0.09% FWHM compared to our 0.75% FWHM. Our initial interest is in the study

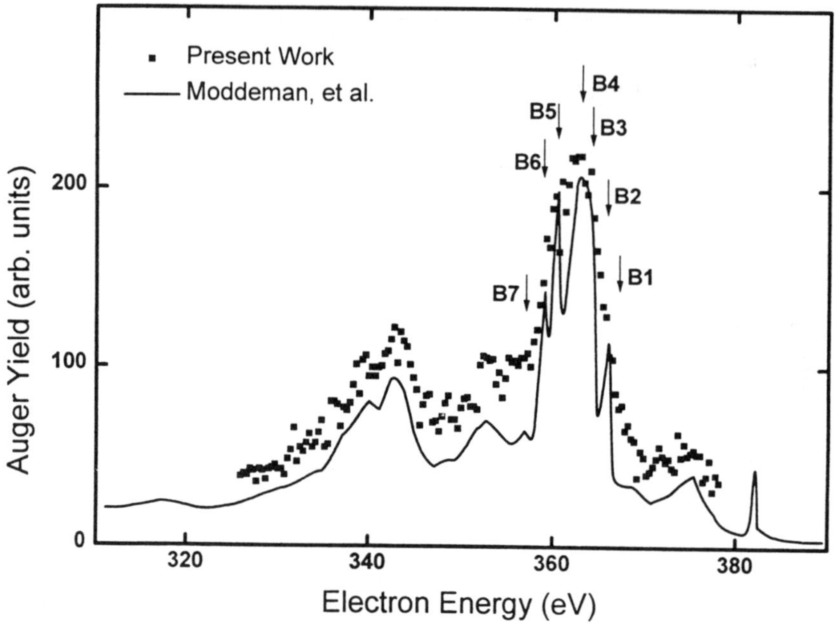

FIGURE 16.7 The molecular nitrogen K-shell Auger electron spectrum. The highest peak is the B-4 peak at 362.5 eV. The solid curve is the experimental Auger spectrum of Moddeman, et al.

of the peaks labelled B4, B5, and B6. While it is true that improved energy resolution would help to isolate an individual peak, it is also true that the counting rate would decrease proportionately. The selected resolution is a compromise between the two effects.

N^+ spectrum

The energy spectrum of positive ions without reference to any electrons is shown in Fig. 8. The voltage settings on the positive ion analyzer were swept in order to sample all energies of interest. Since the true shape of the energy spectrum is unimportant for the purposes of this experiment, the data are not corrected by taking into account the energy-dependent efficiency of the analyzer. The correction is described elsewhere [18]. The N^+ spectrum is a superposition of many broad, overlapping peaks produced by the dissociation channels resulting from the bombardment of N_2. The channels involving Auger transitions are expected to make up a small fraction of the total yield which is dominated by direct double ionization. Examination of the singles spectrum alone does not allow us to set the N^+ analyzer voltages correctly. Coincidence

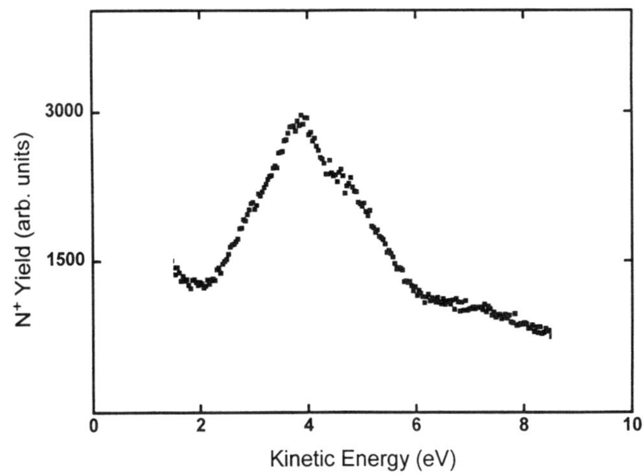

FIGURE 16.8 The measured kinetic energy spectrum on N^+ ions without a coincidence condition. The N_2 target was bombarded with 1634 eV electrons.

measurements are needed in order to make those adjustments.

Channel isolation

Our first angular distribution measurements were made with the electron analyzer voltages fixed on the peak labelled B4. The energy of the Auger electrons in that channel is 362.5 eV. The reflection approximation [10] and the N_2^{2+} potential curves calculated by O'Neil [9] were used to predict the energy range of the N^+ ions coming from that dissociation channel. The N^+ analyzer was set to accept ions of the predicted energy and the electron-N^+ coincidence yield was measured. The voltage settings on the N^+ analyzer were then varied to maximize the coincidence yield. Figure 9 is a plot of the coincidence yield shown as a function of the kinetic energy of the positive ions. The error flags indicate statistical uncertainties only. The presence of the peak near 4.3 eV in the coincidence spectrum infers the existence of the dissociation channel $N_2^{2+} \rightarrow N^+ + N^+ + 8.6$ eV. The 8.6 eV energy release is consistent with the predicted values from the dissociation of the $N_2^{2+}(1\pi_u^{-2})^1\Sigma_g^+$, $N_2^{2+}(1\pi_u^{-2})^1\Delta_g$ and $N_2^{2+}(2\sigma_u^{-1}1\pi_u^{-1})^3\Pi_g$ states. The solid curve in the figure represents the reflection approximation prediction of the profile of the energy spectrum calculated from the dissociation of the $N_2^{2+}(1\pi_u^{-2})^1\Sigma_g^+$ state using O'Neil's N_2^{2+} potential-energy curves. It is the same curve shown in Fig. 2. Thermal and instrumental broadening effects have been included in the calculation. Figure 9 shows that the measured coincidence spectrum is in good agreement with the reflection approximation prediction. As a result of the procedure just de-

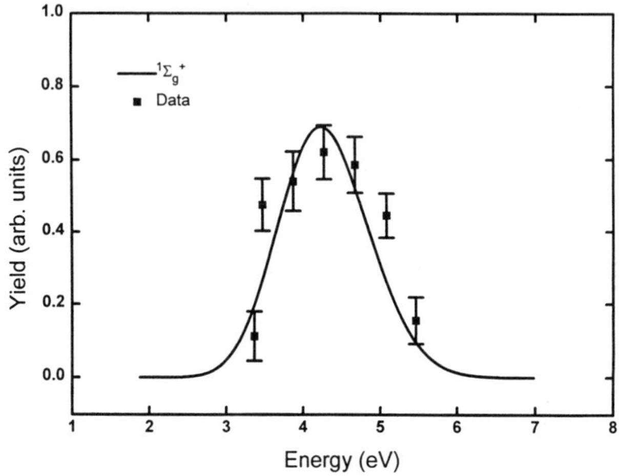

FIGURE 16.9 N^+ ion and 362.5-eV Auger electron coincidence yield as a function of N^+ ion energy. The solid curve represents the reflection approximation prediction of the shape of the energy spectrum produced by ionization and subsequent dissociation through the channel $N_2^{2+}(1\pi_u^{-2})^1\Sigma_g^+ \longrightarrow N^+(^3P)+N^+(^3P)$. Thermal and instrumental broadening have been included. The solid curve has been normalized to the experimental data at the position of the coincidence peak.

scribed, the voltages of the N^+ analyzer are set for the peak of the coincidence yield.

If greater accuracy is desired the measurements can be repeated by first stepping the voltages on the electron analyzer to find a new maximum, and then stepping the N^+ analyzer once more. It should be noted that some care must be taken in the selection of the angle between the two analyzers. If the axial recoil approximation is strictly true, only Σ final states will produce coincidence yield at zero degrees relative to the internuclear axis. In the case of the B4 measurements, the angle θ was set to 30°.

Electronics

Standard time-of-flight techniques are used to record electron-ion coincidences. The time of flight of the Auger electrons from their point of origin in the target volume to the detector is of the order of 10 ns. The time of flight of the massive positive ions is of the order of 10 μs. An electronic delay is used to delay the signal from the electron detector before it starts a time-to-amplitude converter (TAC). After shaping, the signal from the N^+ detector stops the TAC. The height of the output pulse from the TAC is proportional to the length of time elapsed between the start and stop inputs. If the TAC is started but does not receive a stop pulse within 40 μs, the TAC is reset,

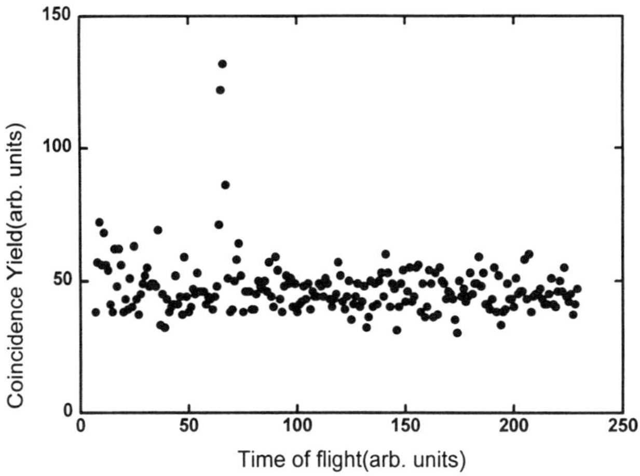

FIGURE 16.10 A time of flight spectrum showing the time correlation of electron and N^+ detection. The spectrum was recorded with the electron and N^+ analyzers settings at 60° with respect to the projectile beam direction. Full scale on the time axis is approximately 40 μs.

and awaits another start pulse. The TAC signal is processed by an analog-to-digital converter and recorded in a multichannel analyzer. A typical time of flight spectrum is shown in Fig. 10. The time correlated events appear as a peak superimposed on a flat background. The flat background is caused by accidental coincidences between electrons and N^+ ions which emerge from two independent events in the target.

The coincidence counting rate is very low: typically one count in two hours. The typical N^+ counting rate is 10 counts per second, and the electron counting rate is 5 counts per second. One could increase the count rate in both detectors by increasing either the projectile beam current or the target gas pressure, but doing so will decrease the signal to noise ratio. The reason for this is that the accidental coincidence counting rate is proportional to the product of the counting rates in the two detectors and therefore the square of the beam current, while the true coincidence counting rate is proportional to the first power of the beam current. In this work the operating parameters are chosen to permit reasonable count rates with acceptable statistical uncertainties in the calculated true coincidence rate. Values of beam current and target gas pressure in the interaction region are typically 50 nA and 10^{-4} Torr. The background pressure in the absence of target gas is less than 3×10^{-7} Torr.

Individual measurements of coincidence yield often last for several days. In order to reduce the effects of target gas pressure and temperature, and projectile beam focus and current, measurements are normalized to the same number of N^+ counts without a coincidence requirement. Beam current is collected in a Faraday cup and monitored. The number of N^+ counts and electron counts per unit charge are determined for each run.

16.3 RESULTS AND ANALYSIS

16.3.1 Measurements

The three Auger transitions studied so far in this work are the ones labelled B4, B5, and B6 in Fig. 7. Consider the B4 transition first. The *ab initio* configuration interaction calculations of Wetmore and Boyd [23] and of O'Neil [9] show that the $(1\pi_u^{-2})^1\Sigma_g^+$, $(1\pi_u^{-2})^1\Delta_g$, and $(2\sigma_u^{-1}1\pi_u^{-1})^3\Pi_g$ states of N_2^{2+} are in the proper range of energies for the transition. The bar diagram in Fig. 11 shows the predicted peak positions for the various possible final states of N_2^{2+} alongside the measured values and peak assignments due to Moddeman.

The potential-energy curves for the three excited states mentioned in the last paragraph and the ground state of N_2^{2+} all calculated by O'Neil are shown in Fig. 12. The vertical dotted line in the figure indicates the equilibrium separation of the N_2 molecule and therefore marks the center of the Franck-Condon region. According to O'Neil's calculations, the kinetic energy of the Auger electrons leading to the triplet state $(2\sigma_u^{-1}1\pi_u^{-1})^3\Pi_g$ is about 1 eV higher than the kinetic energies of those leading to the singlet states $(1\pi_u^{-2})^1\Sigma_g^+$, and $(1\pi_u^{-2})^1\Delta_g$. The curve shapes show that all three excited states are dissociative in the Franck-Condon region. Since the energy resolution of the electron analyzer used in the B4 measurements is 2.72 eV at 362.5 eV, all three states will contribute to the electron yield we measure. Similarly, the potential curves show, and the reflection approximation confirms, that the N^+ kinetic energy spectra from the three dissociation channels are very nearly the same. That is, since the vertical dotted line in Fig. 12 crosses the three excited state curves at nearly the same energy, and since the asymptotic final state is identical for each of the three curves, the kinetic energy release in each case is essentially the same. Consequently, the N^+ analyzer cannot distinguish between N^+ ions from the three channels. The conclusion to be drawn from these arguments is that our angular distribution measurements for the Auger transition labelled B4 represent a sum over three channels. The model calculations used to fit the data reflect this summation.

Figure 13 presents Auger electrons!N_2 the angular distribution measurements for the B4 transition. Recall that the angle θ is measured with respect to the internuclear axis. The fits to the data are discussed in a later paragraph. The error bars indicate statistical uncertainties only. The data were obtained with the N^+ analyzer set at $\theta_B = 60°$ relative to the incident beam.

We began these experiments with the expectation that the angle between the internuclear axis of the target molecule and the direction of the projectile beam has no influence on the angular distributions we measure. To test this hypothesis we measured angular distributions for the B4 channel for θ_B set

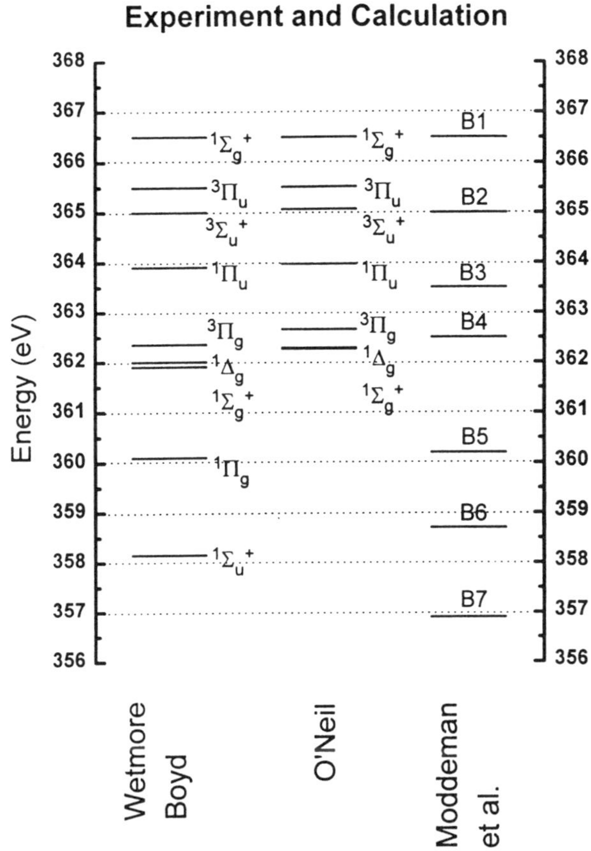

FIGURE 16.11 Assignments of peaks appearing in the valence Auger spectrum of N_2 by different authors. The diagram shows that the B-4 peak is likely to be a mixture of transitions to the ${}^1\Sigma_g^+(1\pi_u^{-2})$, ${}^1\Delta_g(1\pi_u^{-2})$, and ${}^3\Pi_g(2\sigma_u^{-1}1\pi_u^{-1})$ states. We use these three states as final states to compute the angular distribution function of the B-4 peak. The B-5 peak should be a transition to the ${}^1\Pi_g(2\sigma_u^{-1}1\pi_u^{-1})$ state.

equal to 90° and 120°. These data were less complete than the data in Fig. 13 because of spatial restrictions in the way we could position the analyzers in the scattering chamber. Nonetheless, the additional data is consistent with the data presented here and the hypothesis just mentioned. In addition, while the angle θ_B was set to 90°, the angular distribution was measured as the electron detector was moved from one side of the internuclear axis to the other in six degree steps. It was found that the results were symmetric within statistical errors about the line defined by $\theta = 0°$.

Now shifting attention to the B5 transition, the bar chart in Fig. 11 shows

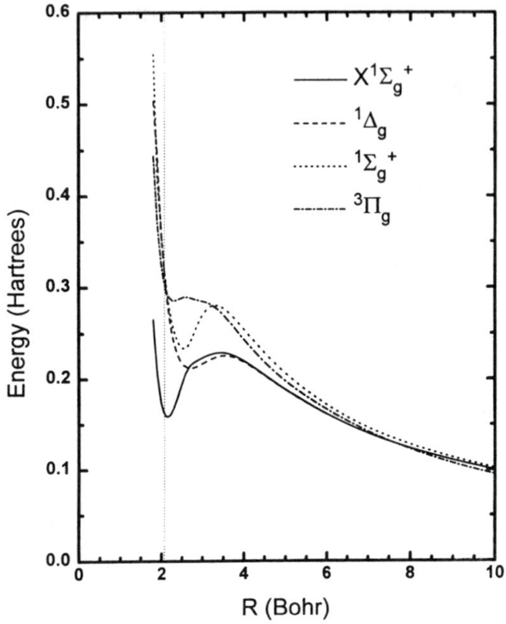

FIGURE 16.12 Potential-energy curves for selected molecular ion states of N_2^{2+} predicted by the theoretical calculations of O'Neil.

that the energy of the Auger electrons is 360.2 eV. Furthermore, Wetmore and Boyd [23] suggest that the B5 transition takes place to the $N_2^{2+}(2\sigma_u^{-1}1\pi_u^{-1})^1\Pi_g$ final state. Before work on the B5 peak was begun, the apertures at the entrance and exit of the electron analyzer were reduced in size so that the energy resolution of the analyzer was improved to 0.375% which amounts to 1.35 eV FWHM at the settings for the B5 channel. The improved resolution made contributions to the measured data from the $^1\Sigma_g^+$ and $^1\Delta_g$ states negligible. The reflection approximation was applied to the potential-energy curve for the $^1\Pi_g$ state calculated by Wetmore and Boyd [23] and served as the starting point for the search to determine the appropriate energy settings for the N^+ analyzer. The procedures described earlier determined that the peak of the kinetic energy distribution occurs at 5.4 eV. Figure 14 presents the measured angular distribution for the B5 transition. The fits will be discussed later.

The B6 transition provides an interesting test of experimental procedures as well as theory. The Wetmore and Boyd calculations [23] of the potential curve for the $^1\Sigma_u^+$ final state show a well in the Franck-Condon region so that the N_2^{2+} molecular ion should not dissociate. Our measurements support this result in that no coincidence yield above background could be detected. The full search procedure described in an earlier section of this chapter was employed.

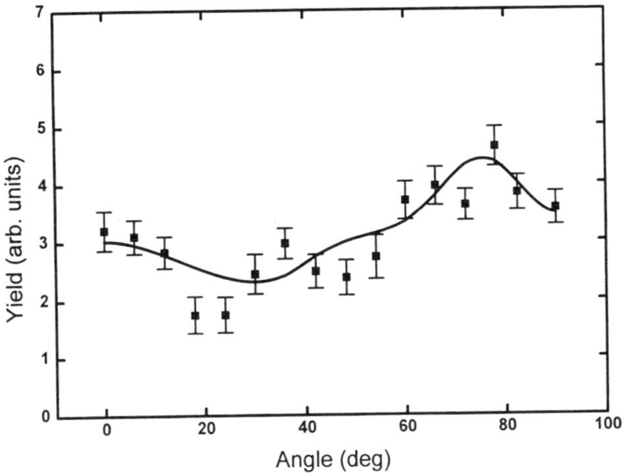

FIGURE 16.13 The distribution of Auger electrons versus the angle θ between the direction of emission and the line defined by the internuclear axis for the B-4 transition. We assume that $1\sigma_g$ and $1\sigma_u$ holes are equally formed in the primary projectile-target collision. The least-square fitting procedure allows the relative amount of the transitions to the $N_2^{2+}(1\pi_u^{-2})^1\Delta_g$, $^1\Sigma_g^+$, and $N_2^{2+}(2\sigma_u^{-1}1\pi_u^{-1})^3\Pi_g$ final states to be varied. The plot shows the fit to the experimental data including the effects of the rotation of the N_2^{2+} ions, the finite angle correction, and the correction for the thermal translational motions of the N^+ ions.

16.3.2 Analysis

Symmetry arguments can be made that will give a general description of the expected angular distribution. In order for the transition matrix element not to vanish, the direct product of the irreducible representations of the final state of the molecular ion and Auger electron must contain the irreducible representation of the initial state. For example, in the Auger transition in N_2,

$$1\sigma_g^{-1}(^2\Sigma_g^+) \longrightarrow 1\pi_u^{-2}(^1\Delta_g) + e^-, \tag{16.4}$$

the direct product of the irreducible representations of the Auger electron with the Δ_g final state must yield the Σ_g^+ of the initial state. This means that the irreducible representation of the electron is δ_g. Of course, the spins of the final state and Auger electron must couple to match that of the initial state. From this argument it is possible to determine the representation and basis for the Auger electron. Once this is known, general statements can be made about the expected angular distribution.

For example, if the Auger electron has σ symmetry then ejected electrons are expected along the internuclear axis at $\theta = 0°$. If $m > 0$ ($\pi, \delta, \ldots$, electrons) then no electrons will be ejected in that direction. Furthermore, if the Auger wave has gerade symmetry then a maximum is expected at 90° relative to the

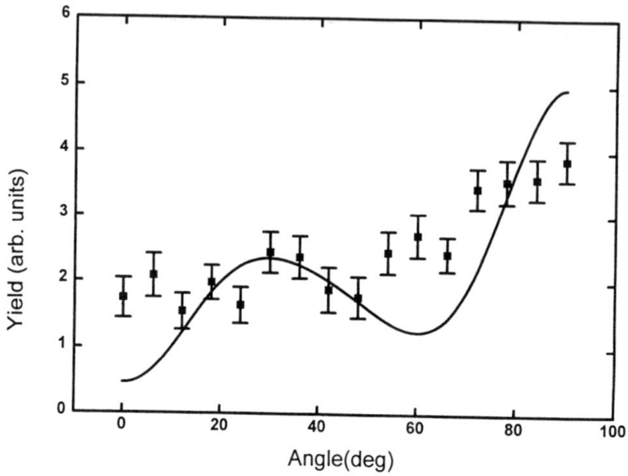

FIGURE 16.14 The distribution of Auger electrons versus the angle θ between the direction of emission and the line defined by the internuclear axis for the B-5 transition. We assume that $1\sigma_g$ and $1\sigma_u$ holes are equally formed in the primary projectile-target collision. The plot shows the fit to the experimental data including the effects of the rotation of the N_2^{2+} ions, the finite angle correction, and the correction for the thermal translational motions of the N^+ ions.

internuclear axis while ungerade symmetry yields a minimum. These general features can be inferred from the behavior of the associated Legendre functions from which the Auger wave function can be formed.

However, in order to determine a detailed angular distribution, it is necessary to evaluate the transition moments. Unlike the atomic case where the angular dependence can be treated entirely separately from the radial dependence, the two-center problem requires solutions of both the radial and angular parts. This occurs because of the form of the volume element in the chosen coordinate system (see below).

The Auger transition occurs from the Coulombic interaction between the pair of electrons involved in the transition. Therefore, the transition moment has the form

$$\langle \psi_f(1,2)|r_{12}^{-1}|\psi_i(1,2)\rangle \tag{16.5}$$

where $\psi_f(1,2)$ and $\psi_i(1,2)$ are antisymmetrized two-electron wave functions. A description on how to evaluate this matrix element in prolate-spheroidal coordinates by numerically integrating Schroedinger's equation for homonuclear molecules has been published [24,25].

A potential of the form Z_{eff}/r_1 and Z_{eff}/r_2, where r_1 and r_2 are the distances of an electron from the two nuclei, yields a separable hamiltonian of a form that can be solved numerically. Energies for bound state orbitals are taken from photoionization measurements [26] and wave functions are obtained by numerical integration while adjusting Z_{eff} so that the wave matches the proper boundary conditions. These bound state wave functions are used to predict the Z_{eff} seen by the Auger electron, which is used, in turn, to find a set of partial waves and their phase shifts for the Auger electron. An expansion [27] of r_{12}^{-1} is used along with the newly generated wave functions to evaluate the transition moment.

This procedure does not readily work for heteronuclear diatomic molecules. A term involving the differences of effective charges of the two nuclei $(Z_1 - Z_2)$, which does not occur for homonuclear molecules, appears in the angular part of the separated hamiltonian. An assumption would have to be made about this difference in order to solve for a separation constant that is needed for a solution of the radial part.

16.3.3 Corrections

Rotational correction

The calculated angular distributions just presented apply to an ideal experiment in which the axial recoil approximation is strictly true and the slit widths which determine the angular resolution of the particle analyzers are negligibly small. In fact, a rotational correction is needed because the dissociating molecules do rotate during the dissociation process. A simple semiclassical model has been developed to account for the effects of rotations [28]. Briefly, the model assumes that the molecule is a rotating object consisting of two point nuclei separated by an internuclear distance which increases as the molecule dissociates. Angular momentum conservation dictates that the molecule's angular velocity will decrease as the nuclei move apart. The initial angular velocity of the N_2^{2+} ion when it is just formed by the Auger transition is determined by the ion's rotational energy. The rate at which the nuclei separate is determined by the shape of the potential-energy curve for the N_2^{2+} final state. By following the separation process through many small steps, the total angle through which the molecule turns can be calculated. What one calculates is a distribution of angles since the initial rotational energy depends on the value of the quantum number J, and the probability that molecules have a given J is described by the Boltzmann distribution function.

In the present work the target gas enters the interaction region through tubes which are 6.35 mm in diameter which means there is no appreciable rotational cooling. Consequently, the rotational temperature of the target gas was taken to be 300 K. For the N_2 target in this work the most probable value of J is 7. It is interesting to note that the angle turned through is very sensitive to the shape of the repulsive potential curve in and slightly beyond the Franck-Condon region. In the case of the three final states involved in the B4 transition the greatest rotation takes place for the $^3\Pi_g$ state, and the least for the $^1\Delta_g$ state. Using the potential curve for the $^3\Pi_g$ state calculated by O'Neil [9], the model predicts that it takes approximately 385 fs for the dissociating N_2^{2+} to reach an internuclear separation of 100 bohr. During that time an N_2 molecule having J = 7 rotates through an angle of 6.2°. In the case of the $^1\Delta_g$ state the corresponding time is 363 fs, and the angle turned through is 3.8°.

Figure 15 illustrates the scattering geometry and the significance of the angle of rotation. The N^+ analyzer is positioned on the negative z-axis, and the electron analyzer is placed in the z-y plane at an angle θ with respect to the z-axis. If the axial recoil approximation were strictly true, the detection of an N^+ ion would signal that the dissociating molecular ion was oriented with its internuclear axis parallel to the z-axis. The coincident detection of an N^+ ion and an Auger electron would mean that the Auger electron was emitted at an angle θ with respect to the internuclear axis of the parent molecule. If on the other hand, the molecule rotates through angle δ during dissociation, then an N^+ ion detected on the z-axis was not oriented along the z-axis at the time of Auger electron emission. The unit vector **A** in Fig. 15 indicates one possible orientation of the molecular axis. It is drawn for the case wherein the molecule will rotate through angle δ during dissociation, and one of the N^+ fragments will be detected on the z-axis. The vector **A** does not necessarily fall in the z-y plane. The family of possible vectors **A** which rotate through angle δ describe a cone of half-angle δ centered on the z-axis. The actual angle at which the Auger electron is emitted is α, and α depends on θ, δ and the azmuthial angle ϕ.

Reference [28] describes how the predicted angular distributions are modified so that they are expressed in terms of θ and compared directly to experiment. In essence, for each value of θ a weighted sum is performed over the range of possible values of δ and ϕ. The weighting factor for each term includes the predicted intensity at the angle α , and a probability factor describing the likelihood of the occurance of a particular value of δ.

Translational correction

A second correction is needed in order to account for the thermal distribution of translational velocities of the N_2 target molecules. The important factor in calculating this correction is the component of the thermal translational velocity approximately perpendicular to the line from the target to the analyzer. The resultant direction of the N^+ fragment is determined by the sum of the thermal velocity component and the velocity acquired in the basic dissociation process. The net effect is to allow some N^+ fragments which were not originally directed toward the N^+ analyzer to reach the analyzer and be detected. The detailed analysis of the translational correction is similar to that for the rotational correction. In particular, the geometry illustrated in Fig. 15 is the same for both corrections.

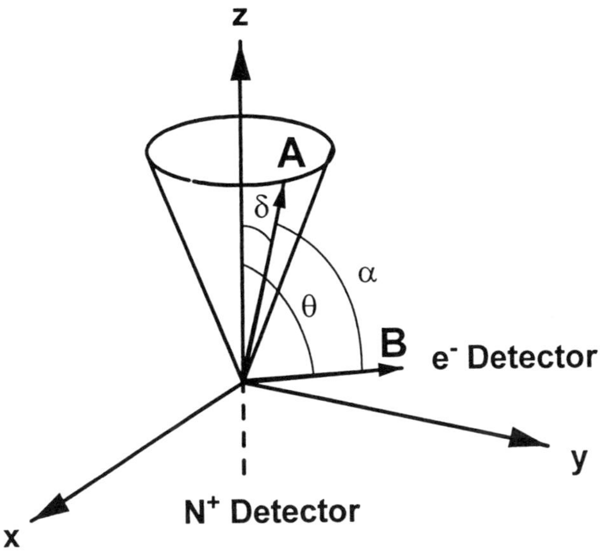

FIGURE 16.15 The directions and angles of interest in accounting for the rotational effect in angular distribution of the Auger electrons. The N^+ detector is set on the negative z-axis. The vector $\vec{B}$ in the y-z plane points in the direction of the electron detector which makes an angle θ relative to the positive z-direction. The vector $\vec{A}$ makes an angle δ to the positive z-direction and indicates one possible orientation of the molecular axis. The angle between $\vec{A}$ and $\vec{B}$ is denoted as α.

Instrumental correction

A third correction is needed because of the finite size of the slits in front of the positive ion and electron analyzers. This allows for a spread in the angle of orientation of the internuclear axis and the angle of emission of the Auger electron. Assuming that the axial recoil approximation holds, the nominal angle of ejection of the electron relative to the internuclear axis for a molecule located at the center of the interaction region is $\theta_{nom} = 180° - (\theta_A + \theta_B)$ (see Fig. 3). For a molecule located at an arbitrary point x_i along the beam axis, its orientation can have a spread in angle that can be determined from a diagram such as that shown in Fig. 6. In that figure one of the quadrangles refers to the positive ion detector, and the other to the electron detector. Combining the spread in acceptance angles in each of the two detector systems gives a spread in the angle θ_{nom}. At each selected x_i the angular distribution $I(\theta_i)$ is calculated for equal steps sizes from the minimum to the maximum values of allowed angles of θ_i and then summed. The sum is continued over the length of the beam axis that is common to both detectors by taking equal step sizes. Finally, the sum is divided by the total number of steps. This procedure is carried out for each setting θ_A of the electron detector.

Comparisons

The relative importance of the three corrections was determined by applying each of them individually to a rectangular step function arbitrarily placed at 45°. Each correction smoothed the sharp step. In order to make quantitative comparisons, the angular range over which the corrected distribution changed from 10% to 90% of the step size was recorded. The rotational correction, using the numbers associated with the $^1\Sigma_g^+$ state, yielded an angular range of 11°. The translational correction and the instrumental correction both yielded angular ranges of 8°.

16.3.4 Comparison of calculations and measurements

The calculations of the angular distribution of the Auger electrons comprising the B4 transition are discussed in detail in Ref. [25,28]. The solid curve in Fig. 13 represents the best fit to the data and includes all three corrections. The fit involves only three adjustable parameters. Similarly, the solid curve in Fig. 14 is the result of the calculation for the B5 transition. Because only one final state is thought to be involved in the B5 transition, there is only one adjustable parameter in the fit. It is a scaling factor. The details of the B5 calculations and measurements are as yet unpublished.

16.4 FUTURE MEASUREMENTS

16.4.1 Other targets

Several intriguing questions arise when one considers studying Auger transitions in heteronuclear diatomic molecules. We have just begun experiments with a CO target . CO was chosen because it is isoelectronic with N_2 and has been studied extensively [29,30]. Measurements of the kinetic energy spectra of the positive ions produced in the dissociation of CO following fast ion bombardment were measured in this laboratory and reported in the literature [31].

One major difference between N_2 and CO is that the CO will display two sets of Auger transitions. One set will derive from a K-shell hole on the carbon atom, and the other from a K-shell hole on the oxygen atom. The two sets will lie about 250 eV apart in transition energy. What makes this interesting is that the experimental technique can identify which atom possessed the initial hole. In addition, detection and identification of a positive ion to be either a C^+ or alternatively an O^+, permits us to know the orientation of the CO molecule at the moment of ionization. In the case of the N_2 target we could only determine the orientation of the molecular axis since the nitrogen atoms are indistinguishable. In the case of the CO target, we can determine the orientation of the axis and which end of the molecule is the carbon or oxygen end.

Reconsideration of Fig.3 will help the discussion. Suppose the positive ion detector is placed at angle θ_B and set to detect C^+ ions of the energy appropriate to the Auger transition being studied. The consequence of the set up is that the target molecule must have been positioned so that carbon atom was on the end of the molecule facing the detector. It is important to note that the angular distributions measured do not need to be symmetric about $\theta = 90°$. There may well be interesting screening effects which occur when θ approaches 0° or 180°. The probability density of the bound-state electrons is presumed to be greater near the oxygen atom, and this asymmetry of the electron cloud is expected to affect the angular distribution.

We also intend to examine the possibility that the cross section for the creation of the initial K-shell hole depends on the angle θ_B. Suppose the apparatus is arranged so that $\theta_B = 0°$, and the positive ion detector is again set to detect a C^+ ion. Then the incoming ionizing projectile "sees" the oxygen atom screening the carbon atom. Will this naive idea mean that the ionization cross section will be significantly different at $\theta_B = 0°$ and $\theta_B = 180°$?

16.4.2 Experimental considerations

The experiments with heteronuclear molecules place new demands on the apparatus. As a result a new larger, more versatile scattering chamber has been constructed. The new chamber accommodates four analyzers each of which can be positioned at any angle on either side of the axis defined by the projectile beam. The finite size of the analyzers means that any of the analyzers intercept the projectile beam if placed too close to either 0° or 180°. In order to complete the measurements discussed in the previous section of this report, a parallel plate analyzer is being designed which will be drilled so that the projectile beam can pass through the analyzer when it is placed at 0°, and positive ions can still be detected. The increased number of analyzers in the new chamber permits data to be collected more rapidly.

REFERENCES

1. Cleff B., and Mehlhorn W.,*J. Phys. B: Atom. Molec. Phys.* **7**,593(1974).
2. Mehlhorn W.,*Atomic Inner-Shell Physics* , New York: Plenum Press, 1985, p. 119.
3. Dill D., Swanson J. R., Wallace S., and Dehmer J. L., *Phys. Rev. Lett.* **45**,1393(1980).
4. Truesdale C. M., Lindle D. W., Kobrin P. H., Becker U. E., Kerkhoff H. G., Heimann P. A., Ferrett T. A., and Shirley D. A., *J. Chem. Phys.* **80**, 2319(1980).
5. Lindle D. W., Truesdale C. M., Korbin P.H., Ferrett T. A.,. Heimann P. A, Becker U., Kerkhoff H. G., and Shirley D. A., *J. Chem. Phys.* **81**, 5375(1984).
6. Chandra N., and Sen S., *J. Chem. Phys.* **98**, 5242(1993).

7. Schaphorst S. J., Caldwell C. D., Krause M. O., and Jimnez-Mier J., *Chem. Phys. Lett.* **213**, 315(1993).
8. Gilmore F. R., *J. Quant. Spectrosc. Radiat. Transfer* **5**, 369(1965).
9. O'Neil S. V., private communication.
10. Coolidge A. S., James H. M., and Present R. D., *J. Chem. Phys.* **4**, 193(1936).
11. Dunn G. H., and Kieffer L. J., *Phys. Rev.* **132**, 2109(1963).
12. Edwards A. K., Wood R. M., Davis J. L., and Ezell R. L., *Phys. Rev. A* **42**, 1367(1990).
13. Misakian M., and Zorn J. C., *Phys. Rev. A***6**, 2180(1972).
14. Zare R. N., *J. Chem. Phys.* **47**, 204(1967).
15. Edwards A. K., Wood R. M., Mangan M. A., Ezell R. L., *Phys. Rev. A***46**, 6970(1992).
16. Shigemasa E., Adachi J., Oura M., and Yagishita A., *Phys. Rev. Lett.* **74**, 359(1995).
17. Brinkmann R. T., and Trajmar S., *J. Phys. E: Sci. Instrum.* **14**, 245(1981).
18. Rudd M. E., *Low Energy Electron Spectrometry*, New York: Wiley- Interscience,1972, p.17.
19. Kuyatt C. E., private communication.
20. Dahl P., *Introduction to Electron and Ion Optics*, New York: Academic Press, (1973).
21. El-Kareh A. B. and J. El-Kareh J. C., *Electron Beams, Lenses, and Optics*, New York: Academic Press, 1970, Vol. 1.
22. Moddeman W. E., Carlson T. A., Krause M. O., Pullen B. P., Bull W. E., and Schweitzer G. K., *J. Chem. Phys.* **55**, 2317(1971).
23. Wetmore R. W. and Boyd R. K., *J. Phys. Chem.* **90**, 5540(1986).
24. Zheng Q., Edwards A. K., Wood R. M., and Mangan M. A., *Phys. Rev. A* **52**,3940(1995).
25. Zheng Q., Edwards A. K., Wood R. M., and Mangan M. A., *Phys. Rev. A* **52**,3945(1995).
26. Siegbahn K., Nordling C., Johansson G., Hedman J., Hedn P. F., Hamrin K., Gelius U., Bergmark T., Werme L. O., Manne R., and Bear Y., *ESCA, Applied to Free Molecules*, Amsterdam: North-Holland, 1971.
27. Rüdenberg K., *J. Chem. Phys.* **19**, 1459(1951).
28. Wood R. M., Edwards A. K., Zheng Q., and Mangan M. A., Manuscript in preparation.
29. Lablanquie P., Delwiche J., Hubin-Franskin M.-J., Nenner I., Morin P., Ito K., Eland J. H. D., Robbe J.-M., Gandara G., Fournier J., and. Fournier P. G, *Phys. Rev. A* **40**,5673(1989).
30. Larsson M., Olsson B. J., and Sigray P., *Chem. Phys.* **139**,139,457(1989).
31. Wood R. M., Steuer M. F., and Edwards A. K., *J. Chem. Phys.* **69**,4487(1978).

CHAPTER 17

Radiative-Lifetime Measurements on Highly Charged Ions

Elmar Träbert*

Transition probabilities reflect the dynamics of atomic systems. In order to determine these in multiply charged ions, usually the mean lives of excited levels are measured using foil-excited swift ion beams, after the atomic structure has been established from spectroscopic studies. In some cases, however, time-resolved spectra and lifetime measurements come first, providing clues for the analysis of atomic structure.

The historical development from beam-foil spectroscopy to fast-beam spectroscopy is reviewed as a background to the concepts of specific instrumentation used in the field. The instruments range from target chambers, standard and tailored spectrometers of various working principles, equipped with a variety of detectors, to specific accelerator designs like ion storage rings. Specific properties of experimental arrangements and particular systematic errors are discussed, as well as hardware and software techniques to counter the latter.

17.1 INTRODUCTION: WHY MEASURE ATOMIC LIFETIMES?

Atomic physics deals with the structure and dynamics of atomic systems. Atomic structure and discrete lines, observed in the spectra of radiation emitted from atoms, are explained by Bohr's postulates of stable, discrete energy levels and the emission and absorption of energy by atoms in transitions from one level to another. However, not every combination of energy levels results in observable spectral lines, and not all spectral lines are equally intense. Apparently selection rules are at work which govern the dynamics of the atomic system and relate to symmetries, and quantum mechanics tells us of transition probabilities. Transition probabilities result in relative line intensities in spectra; such relative line intensities in transition multiplets can yield very instruc-

*) Ruhr-Universität Bochum, D-44780 Bochum, Germany.

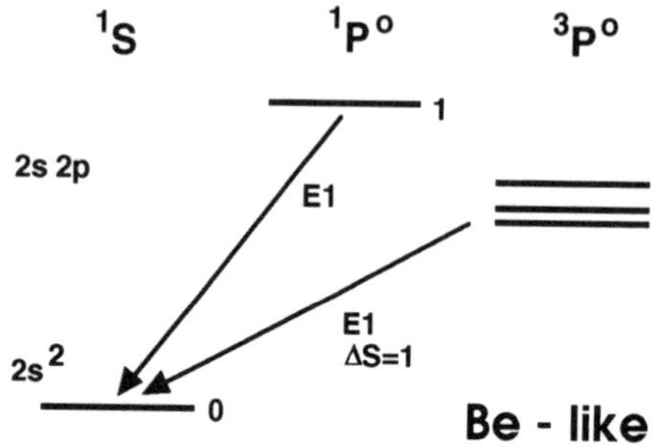

FIGURE 17.1 Schematics of the level structure of a two-electron system (*Be*-like ions have two electrons outside of the closed $n = 1$ shell), with resonance and intercombination transitions from $J = 1$ levels.

tive information on the electron-electron coupling in multi-electron atoms. For example, in the three-electron spectra of the B and Al isoelectronic sequences, relative intensities of the $ns^2\ np\ ^2P^o_{1/2,3/2}$ - $ns\ np^2\ ^2S_{1/2}$ ground state transitions tell about intermediate coupling and wave function interferences which make the $np\ ^2P^o_{3/2}$ - $^2S_{1/2}$ line (the stronger one of the line pair for low ionic charge) gradually disappear (near $Z = 26$ for the B sequence) and reappear along the isoelectronic sequence [1,2].

The concept of isoelectronic sequences (ions of a given number of electrons, but of different nuclear and thus ionic charge) has been developed in the late 1920's. According to Hylleraas [3,4], many atomic structure parameters can be expanded in powers of the nuclear charge Z. The energies of excitation levels or the transition rates of certain types of transitions can thus be parameterized in simple ways. In order to study the physics of, say, a five-electron system, one follows the sequence of five-electron ions from the neutral atom, boron, to highly charged ions, and disentangles the roles of basic structure effects (hydrogen-like atomic-shell gross structure, energy level splittings from electron-electron interaction within a given shell, transition rates between shells or inside a shell) from those of relativity and higher-order effects etc. with their different Z dependences.

Intercombination transitions connect term systems of different multiplicity (Fig. 1), and their knowledge helps to put all excitation energies onto a common footing. These transitions, however, require a spin change, and that reduces their transition probability in comparison to transitions between similar levels of the same multiplicity. The result is a difference in lifetime of otherwise comparable levels, which is exploited in plasma diagnostics: if the collision frequency in a plasma is in between the radiative decay rates of the upper levels of resonance and intercombination transitions, the relative intensity of this line pair is an electron density indicator. For this application it is obviously helpful to know the two transition rates. The ratio of the two transition rates, besides some transition energy correction, depends on the spin-orbit mixing matrix element and thus on a physics entity not trivially contained in the pure spectral information. A lifetime measurement then can be used to invert the problem and to gain information on this "physics" part [5,6].

Along the isoelectronic sequences, the coupling of the electrons generally develops from the LS-coupling limit toward the jj-coupling limit. Systematic measurements of both the structure (wavelengths) and the dynamics (relative intensities, lifetimes) are necessary to test the detailed understanding atomic structure theory has reached, with numerical results usually accomplished by approximations. Whereas there is an optimization rule for the calculation of energy levels (Ritz' principle), there is no such guideline for transition probabilities. The reason for this is that the transition probability depends on the square of the radial wavefunctions, and these need not be orthogonal. Therefore lifetimes provide a test of the wavefunctions which is quite different from that obtained by atomic energy levels alone. Of course, perfect wavefunctions will yield both correct excitation energies and level lifetimes.

Due to the increasing magnitude of relativistic effects, the fine structure components of what are line multiplets at low Z move apart in energy for high Z, and the individual wavefunctions of fine structure states suffer different relativistic distortions. These details of the wavefunctions are important for the computation of transition rates, as the overlap of the (non-orthogonal) radial wavefunctions of initial and final states matters, as well as the transition energy that is observed and measured by spectroscopy. On top of this, quantum electrodynamics (QED) causes appreciable level shifts in highly charged ions of high atomic number (high-Z). The transition energy enters the expression for lifetime in the third power, and level shifts (for example, due to QED) by a few percent therefore relate to quite measurable lifetime effects.

Examples will be given throughout the text. The references cited in the following are often not to the original work, but to later papers or reviews with their own reference lists which are more suitable for further information retrieval. Most of the subsequently discussed measurements are being done by fast-beam spectroscopy. A recent review of spectroscopy on fast ion beams is to be found in [7]; for atomic lifetime measurements in general, see also [8].

17.2 HISTORICAL NOTE

In the 1920's, W. Wien observed the optical emission from canal rays and tried to study both the spectra and the intensity distribution along the flight path. Poor vacuum was a limiting factor, and thus the method of spectroscopy on fast ion beams fell dormant for another four decades. Since the 1930's, accelerators were developed with the aim of giving atoms sufficient energy so that in close collisions the Coulomb repulsion between the nuclei might be overcome and nuclear reactions be initiated. In such studies often a fast

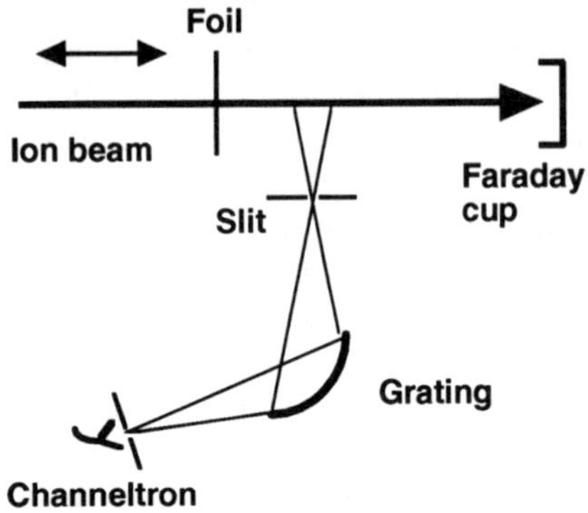

FIGURE 17.2 Schematics of beam-foil spectroscopy. Shown is a grating spectrometer with a channeltron detector for the vacuum-ultraviolet spectral range.

ion beam strikes a thin target, and the effect of energy loss of the projectiles in the target as well as a modification of the projectile ion charge state by the interaction with a material target was realized and measured. Because of the small geometric cross sections of the nuclei, most projectiles do not interact with the target nuclei, but rather with the screened Coulomb field of the nuclei (Rutherford scattering) and with the ubiquitous target electrons. Consequently most of the projectile ions continue their trajectories with little change in velocity vector, and - from a nuclear physics view - are considered as wasted. It is exactly these projectile ions that are of use for the atomic physics experiments discussed here, whereas those projectiles which undergo collisions resulting in a large angle of deflection or experience a considerable energy loss cause systematic errors and spoil otherwise clean atomic physics measurements.

Normally detectors for particles or γ-rays would be used to see whether fast ions (guided inside a vacuum vessel from the accelerator to the experimental site) had hit a nucleus in a thin foil so that it would break up or otherwise tell of nuclear reactions. In 1962/63, independently of each other, two nuclear physicists working at accelerators dared to look with their own eyes. These two people, L. Kay at Manchester (England) and S. Bashkin at Tucson (Arizona, USA), realized that the incoming ions have electronic shells outside the nucleus, and that after the violent passage of the ions through the foil target these shells must rearrange and possibly emit light in the process (Fig. 2). This was the birth of beam-foil spectroscopy (bfs) which later developed into fast-beam spectroscopy (fbs) [9–14].

Among the features of the new light source was the light intensity that behind the exciter foil dropped along the excited ion beam. This was immediately recognized as a signature of the exponential decay of excited atomic levels. Some of the early lifetime measurements gave good results. Many others, however, failed in the attempt of determining meaningful radiative lifetimes of specific atomic levels, for lack of linear detectors, poor spectral resolution and under-appreciation of the complex decay chains activated by the non-selective ion-foil interaction. These points need to be adressed in order to obtain reliable lifetime data.

17.3 PROPERTIES OF THE BEAM-FOIL LIGHT SOURCE

17.3.1 Advantages

Adjustable charge state. The beam-foil light source provides a charge-state equilibrium as a function of incident ion energy [15]. With ever larger accelerators, any charge state has become accessible, in any (naturally abundant) element, up to uranium ions with only a single electron left (hydrogen-like). This last electron can be stripped off as well, but then nothing is left to radiate light, unless one studies the radiative electron capture by a bare ion - which is being done nowadays.

Efficient excitation. The excitation process is non-thermal and reaches levels not reached in any other light source, including multiply-excited levels, electron configurations with several electrons not in the ground state. Most of the multiply (and core-) excited ions will autoionize; that is, they will emit electrons to stabilize. However, there are selection rules reducing autoionization probabilities for high-angular momentum and high-spin states in particular, which then may stabilize preferentially by emission of radiation. Both electron spectroscopy and light spectroscopy thus face a wide field of work in the quest for understanding the structure of multiply excited ions. Beam-foil spectroscopy has shown the existence of negative ions, of excited levels in negative ions, and even of radiative transitions between bound levels of negative ions (in the case of Li^-; see discussion in [16]). Thus beam-foil spectroscopy deals with ions from H^-, He^-, Li^- and Be^- up to U^{91+}.

Unperturbed decay. The spectra emitted by foil-excited ions sometimes look rather different from those of other laboratory light sources, because of the different excitation process and the inherent time resolution. One of Bashkin's first papers on the subject dealt with the similarities of features in the beam-foil spectrum with the spectrum of a Nova, an astrophysical phenomenon [10]. Many later studies followed up on astrophysical leads, or made comparison with solar spectra; one of the decisive factors being the low particle density in the ion beam which lets the excited ions decay without any notable interaction with each other or with the rest gas in the target chamber: at typical ion beam currents (of order 1 μA), the average distance between projectile ions along the beam trajectory is of the order of 1μm, while the ions are distributed laterally over a beam cross section of the order of 1 mm^2 to 1 cm^2. This "cleanliness" has made possible lifetime studies not plagued by the systematics of many other laboratory light sources, including some measurements that are helpful for solar elemental abundance determinations.

Sudden excitation. Passage of swift ions through typical exciter foils of areal mass density 10 $\mu g/cm^2$ takes of order $10^{-15}s$. If one takes the end of the excitation process, that is the passage through the back layer of the exciter foil, as a measure, excitation is defined to better than 10^{-15} s. Although the ions loose some energy in the interaction with the exciter foil and also suffer some energy and angular straggling, the trajectories and velocities of the ions are rather similar to each other. Via the velocity v, the distance x of the ions from the foil is linearly connected to time t after excitation, $x = vt$, and the spatial decrease of intensity directly relates to the decrease with time. Measuring this decrease by either moving the detector along the ion beam or (more generally done, Fig. 2) by moving the foil in the opposite direction, permits the recording of decay curves from which the lifetimes of the excited levels could be extracted - in principle of any level, of any ion, of any element. Ion energies of order 0.5 MeV/amu correspond to velocities of 1 cm/ns (about 3% of the speed of light), and this, together with the dimensions of practical experimental chambers, indicates the lifetime range accessible: Foil displacements in micrometer increments relate to picosecond lifetimes as the lower limit, maximum displacements of order 1 m (limited by practical sizes of vacuum chambers and by ion beam optics) yield an upper limit of order 100 ns. Slow ion beams and storage rings move the upper limit by several orders of magnitude.

These advantages can be exploited only if a number of severe disadvantages are dealt with: Doppler effect, low light intensity, and non-selective excitation.

17.3.2 Disadvantages

Doppler effect

The high velocity of fast ion beams results in a wavelength shift which depends on the angle θ between the ion beam trajectory and the line of observation (first order Doppler shift) and on relativity (second order Doppler shift, γ factor), and a Doppler broadening due to the in-beam-plane component (of the finite solid angle of observation Ω)(Fig. 3). Obviously an observation along the ion beam suffers the largest Doppler shift (and requires the most precise determination of the ion beam velocity), but can be done with the smallest Doppler broadening. This is the geometry of choice for many experiments working with lasers and ion beams, in particular for hyperfine structure measurements and for recent precision experiments measuring QED effects on the structure of almost bare ions [17]. For time-resolved measurements, however, this geometry necessitates additional measures, like the localized Doppler matching of the ion beam velocity to the available laser light, which may be achieved, for example, by retardation or acceleration of the ions in an electric

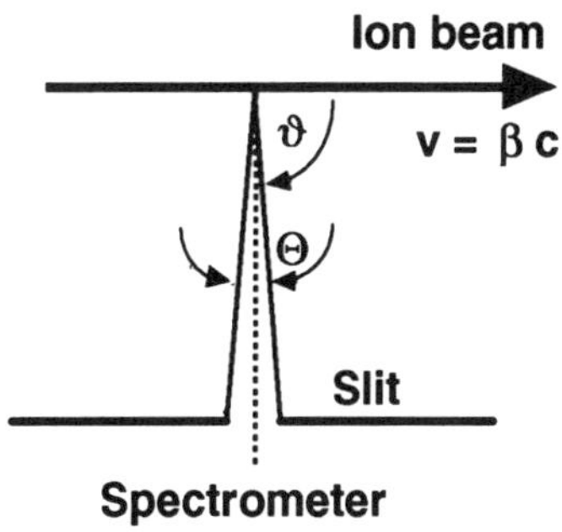

FIGURE 17.3 Geometry for Doppler shift and broadening.

field pointing along the ion beam axis [18,19].

For not-too-fast ion beams, observation at right angles offers an approximation to a vanishing first-order Doppler shift, at the cost of maximum Doppler broadening. At higher velocities $\beta = v/c$ (specific energies of many MeV/amu), it may be useful to have the first and second order Doppler shifts cancel, by observing at an angle $\beta/2$ slightly forward of 90°. For very high (relativistic) ion velocities, the emission pattern points progressively forward, and most light will be detected under forward angles. This is exemplified by the use of granular detectors at the ion storage ring ESR, where the granularity permits one to simultaneously record a set of individual spectra with different Doppler shifts but relatively small Doppler broadening, whilst the placement of the detector assembly takes the relativistic emission pattern into account [20].

Low intensity and high spectral line density.

The efficiency of the interaction of ion beam and exciter foil (or gas target) is rather low in the sense of the fraction of ions to be found in excited states after excitation. Measurements of the population of low-lying, long-lived (often called "metastable") levels typically indicate a fraction of, maybe, a few percent of the total number of available ions. In order to gather sufficient signal, efficient detection is necessary, but may not be achieved by simply enlargening the solid angle of detection, for the adverse effect on line widths which stems from the Doppler broadening discussed above. With the high cost of high-power accelerators, multichannel detection (see below) is a very important tool for saving on accelerator time, as it increases the integration time per spectral feature compared to observation by a scanning spectrometer.

At the same time the interaction of ion beam and exciter foil is rather efficient in the sense that the interaction is so violent that practically all levels, including multiply excited levels and displaced terms, may be reached. The excitation is largely nonselective, and therefore the available fraction of excited ions is distributed over many levels. Consequently the resulting spectrum is very rich in lines, most of which are individually weak. In order to study such spectra properly and to isolate lines of interest for lifetime studies, sufficient spectral resolution in the face of Doppler broadening is required, but not easily achieved.

17.4 SPECTROSCOPIC TECHNIQUES

A number of techniques have been developed over the years to deal with the special properties of the fast ion beam as a (feeble) light source, aiming at higher spectral resolution and at more precise wavelength measurements. Laser-excitation (mostly restricted to visible light and low-charge state ions) and granular detectors (for x-ray spectroscopy on relativistic, highly charged ions) have already been mentioned as means for precision spectroscopy. Certain spectrometers for visible or ultraviolet light can be modified to achieve high speed combined with good resolution, and thus a more generally applicable tool becomes available.

17.4.1 Refocusing

Light from different positions along the ion beam reaches the entrance slit of the spectrometer under different observation angles and thus with different Doppler shifts. This results in a spectrally broadened image of the entrance slit in the exit focal plane. The procedure of refocusing can turn this problem, Doppler line broadening, into a useful entity, by reducing the observed line width and improving the light gathering power of a spectrometer simultaneously. The latter gain results from collecting light from a longer section of the ion beam than would be advisable otherwise.

Stoner-Leavitt refocusing. In a regular normal-incidence spectrometer, Doppler shifted light from an extended, moving light source is imaged in a location which is different from that for unshifted light. The imaging can be treated as if the spectrometer had a different focal length for Doppler-dispersed light [21], and by moving the exit slit (or the diffraction grating) appropriately inward or outward along the optical axis, a rather good (displaced) focus can be found. The displacement of this new focus from that for a stationary light source is a linear function of wavelength. Many normal-incidence spectrometers can be modified (for example, by an adjustment to the cam which anyway provides for a wavelength- dependent displacement of the grating during the scanning motion of the spectrometer, or by displacing the exit slit under servo control), so that the refocusing condition can be met for extended regions of the spectrum.

Bergkvist refocusing. Another type of refocusing first invented for magnetic spectrometers was later on adapted to the fast ion beam [22]. A lens in front of the spectrometer collects "pencils" of parallel light from sections of the ion beam and focuses the light into the entrance plane of the spectrometer. This Doppler-dispersed light pattern (from a range of observation angles) is then diffracted and imaged by the spectrometer onto the exit plane. If the two dispersions are matched in magnitude (adjusted by the lens and the ion velocity), but of opposite sign (achieved by setting the spectrometer on either the left or the right side of the fast ion beam, or operating the spectrometer in negative orders of diffraction), the result can be rather narrow spectral lines.

The signal yield achieved with both methods can be higher by one to two orders of magnitude compared to a non-refocused spectrometer [23], in addition to the gain of information which results from the improved spectral resolution. The technique of refocusing has, for example, been essential in achieving sufficiently high resolution for the observation of lifetime-broadened lines in few-electron spectra, from radiative transitions between an upper level of relatively long lifetime and a lower, very short-lived, autoionizing level [24,16].

17.4.2 Reference wavelengths

For slow ion beams, it is well possible to obtain reference wavelengths from stationary light sources (like a hollow cathode lamp) mounted to the same observation chamber and observed alternatingly with the ion beam. If refocusing schemes are employed, the situation is more difficult, because a spectrometer refocused for the observation of fast ions will be defocused for the observation of a stationary light source and may suffer a change of dispersion as well. One way out of this dilemma is the use of reference lines emitted from the same ion beam (in-beam calibration), for example of Rydberg transitions which are often prominent in the spectra of foil-excited ion beams and may be calculated to some precision. Bergkvist refocusing employs parallel light from the ion beam; the light from a calibration light source may be made parallel by suitable optics and fed into the same light path [25]. In any case, the referencing of wavelengths measured on a fast ion beam to those measured on stationary light sources (see the chapter by R. Deslattes) requires a detailed knowledge of observation angles, and that is often difficult to achieve. One of the many problems arises from the fact that the efficiency of a diffraction grating or an extended detector may change locally as a function of wavelength [26], and thus the light path in the experiment may be different from that used when setting up with complex optical techniques using visible light, and polarization effects may play a role, too.

17.4.3 Time resolved spectroscopy

The fast ion beam intrinsically provides time resolution. This is a unique feature. Looking close to the exciter foil, the (prompt) spectrum usually is full of lines, richer in lines than almost any other light source. These lines include transitions from the multitude of core-excited levels, a system so complicated that no thorough theoretical analysis has been reported yet. Moving the foil by a few millimeters brings the observation zone away from excitation to times later by a few hundred picoseconds. This often is enough to cause dramatic changes (Fig. 4). The lines from extremely short-lived levels are exhausted, whilst the decays of slightly more long-lived levels still show. Moving the foil even farther reduces the overall intensity of the spectrum the more, but decays of long-lived levels will now persist and may be recognized and measured. Such delayed spectra are a unique tool to identify levels by their differential longevity and thus to assign intercombination transitions [27,28] or forbidden lines which otherwise appear only in low-density solar spectra. Decay curves of individual lines observed in any such spectra may then be analyzed with the aim of finding out the transition probabilities for a given transition, or at least the lifetime of a given upper level.

Atomic structure (as seen by the line positions in a spectrum) and atomic dynamics (relative line intensities and lifetimes) are, however, intertwined: For two-electron ions, lifetime measurements have been done to study the influence of hyperfine mixing between spectroscopically unresolved levels which becomes measurable by opening the otherwise forbidden decay channel from the $1s2p\ ^3P^o_0$ level to the $1s^2\ ^1S_0$ ground state [31]. In He-like U^{90+}, a lifetime measurement on the same level ($\tau = 54.4 \pm 3.4\ ps$) was interpreted in terms of the $1s2s\ ^3S_1$ - $1s2p\ ^3P^o_0$ term difference, which contains the Lamb shift and thus reveals quantum electrodynamical effects [32].

17.5 LIFETIMES

Atomic level lifetimes are the inverse of the sum of the transition probabilities for decays from a given level. Their atomic structure description involves overlap integrals over initial and final state wave functions (the so- called "physics" part), plus the energy difference between the participating levels. Lasers can be used to populate or depopulate levels selectively, and thus very clean decay curves, unaffected by cascades, can be obtained and evaluated with little uncertainty (sometimes as low as a fraction of a percent). Laser-interaction with such atoms or ions in fast beams is helped by the kinematic compression: the absolute velocity distribution of particles in the source is superimposed on a large common velocity component from the acceleration process, resulting in a low relative energy spread along the beam trajectory. Because of the steep increase of excitation energies with charge state, multiply charged ions cannot (presently) be excited by laser from the ground state. In some cases foil-excited beams of multicharged ions have been used, and fine structure intervals have been determined, often with a view to the Lamb shift, by laser spectroscopy starting from excited levels. Lifetime measurements utilizing the laser - ion-beam interaction are covered elsewhere (see C.E. Tanner, following chapter); therefore only a few notes are given here.

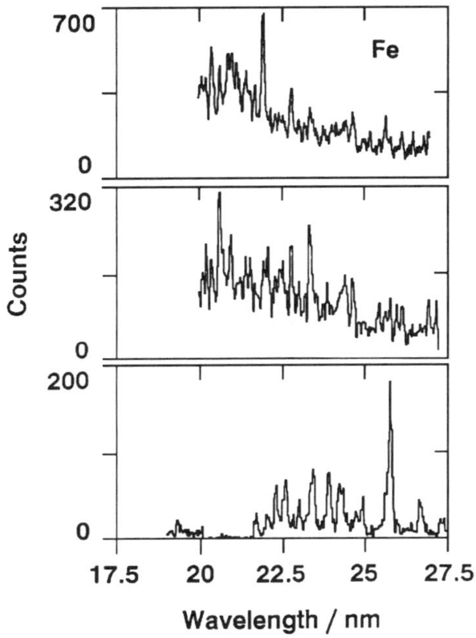

FIGURE 17.4 Spectra of foil excited *Ni* ions at an ion energy of 15 *MeV*, recorded at three foil positions. Top spectrum: Observation close to the foil (prompt spectrum) yields a spectrum most similar to those obtained with laser-produced plasmas. Middle spectrum: Foil displacement 2 mm, delay after excitation 300 *ps*. Bottom spectrum: Foil displacement 10 *cm*, delay after excitation 14 *ns*. This spectrum contains mostly intercombination transitions in *S*- and *Cl*-like ions of *Ni* (spectra *Ni XIII* and *Ni XII*) and relates to solar corona spectra [29,30]. The integration time is much longer for the delayed spectra (and the spectrometer slits are set somewhat wider, for more detected light at the cost of spectral resolution); on a common intensity scale the lines of the delayed spectra would be hidden in the signal of the prompt spectrum. See [29] for a corresponding set of Fe spectra.

17.5.1 Laser techniques

Gas - Laser. The matter density encountered by an ion beam in crossing a dilute gas target is rather low; therefore the effects of energy loss and straggling are not severe. The collisions of the ions with the target gas atoms, however, can excite the projectile ions to metastable levels much higher in excitation energy than any practical laser could reach from the ground state. The interaction with the laser light can then be used to reach further excited levels. This technique has been pioneered and pursued at Kaiserslautern [34].

If the ions capture an electron from the gas, they neutralize and form a fast atomic beam. On such beams, selective laser excitation has been used to provide extremely precise lifetime data (uncertainties of much better than

1%) of Li, Na [33] and Cs [35]. The experimental lifetime data on Li and Na invalidated many early calculations, and for more than 15 years they were at variance with advanced and involved theoretical calculations on these seemingly simple atoms having only a single electron outside a closed shell. At long last, improved measurements at Kaiserslautern [36] superseded those earlier results (as did some quite different work [37]) and vindicated the best calculations. Part of the experimental improvement is due to better geometrical control of ion beam trajectory - this certainly is a problem at the reported 0.15% level of lifetime uncertainty - and to a detector arrangement which encircles the particle beam.

Foil - Laser. Interaction of the ion beam with even a thin foil causes a more massive energy loss and straggling than a gas target and reduces the fraction of ions which might interact with a narrowband laser. Undeterred of this prospect, the Liège group has managed to do laser spectroscopy on multiply charged foil-excited ion beams (in contrast to spectroscopy on neutral or singly charged atoms done elsewhere) [38]: Laser light is used to selectively influence the population of a given level after foil excitation. Neither the decay curve with "laser off" nor a curve with "laser on" can be evaluated easily, due to the non-selective population mechanism of ion-foil interaction. The difference signal, however, can be a rather clean single exponential.

17.5.2 Foil excitation

Foil excitation is the most widely applicable method for lifetime measurements in highly charged ions, and the technical steps involved merit a more detailed discussion (below). Typical lifetime measurements using exciter foils proceed by stepping the foil along the ion beam, across and beyond the field of view of the detection system, which normally looks side-on. The technique is well suited for lifetimes in the range from 1 *ps* to 100 *ns*.

The ion-foil interaction is not yet well understood. Inside the foil, very rapid collisions of the projectile ions with the multitude of electrons in the solid bring about processes of ionization, excitation, electron capture, and loss. For geometrical reasons (radii of the electronic orbits in the projectile ions vs. interatomic distances in the target), only the innermost and most tightly bound electrons of the incoming ions may be considered as travelling with the projectiles, whereas for the outer electrons the screening by the target atoms will be complete (except for some collective effects). If ionization rates exceed capture rates, the charge state of the travelling ion will increase, and vice versa. For the relative populations of excited Rydberg levels, massive calculations regarding transport inside the material finally yielded reasonable results for specific ions [39], but for the present purpose we need only a qualitative picture. On leaving the foil, the ion collects a swarm of accompanying electrons and restructures its electron cloud in order to get rid of excess energy. Surface fields are essential in redistributing the level population.

This messy excitation and redistribution process affects the intensity distribution in spectra as well as the shape of decay curves, that is, the signal at a given wavelength as a function of time after excitation. Instead of a single exponential decay component as often realized in laser excitation, the curves will represent superpositions of the wanted decay with many exponentials from feeding transitions. This complexity, coupled with imperfect signal statistics, limits the amount of information one can extract from a decay curve (see section on decay curve analysis).

17.5.3 Techniques for short lifetimes

There are several techniques that aim at very short lifetimes. The optical measurement of spectral line broadening due to the rapid autoionization of a lower level in few-electron ions [16] has already been mentioned above. The following two techniques exploit the interactions of swift ions with the material during foil passage.

The first technique deals with inner-shell vacancies in projectile ions while travelling in a foil. Ionization, excitation and electron capture are likely to yield an equilibrium fraction of inner-shell vacancy ions for sufficiently thick foils. Before reaching the equilibrium, the x-ray yield is a function of the target thickness. A superposition of the emission from inside the foil (which varies) and from outside the foil will be observed, and by making observations with foils of different thicknesses (mass per unit area) the lifetime of the inner-shell vacancies (of order $10^{-15}s$ to $10^{-14}s$) can be obtained [40]. The second technique exploits the different probability of survival of excited states of highly charged ions travelling in a solid. The level of interest, for example a fine structure level of the $1s2p\ ^3P$ term in a highly-charged helium-like ion, is excited in a first foil. The ion beam next passes through a very thin second foil placed at a very short distance after the first (Fig. 5). If the original level decays in between the foils, for example to the $1s2s\ ^3S_1$ level, the latter level may survive the second foil and can afterwards be observed by its x-ray decay to the $1s2\ ^1S_0$ ground state. Those of the original ions which did not decay, are most likely stripped of their outer electron in the second foil. Therefore a variation of the distance between the foils results in a signal variation of the x-ray observation channel (and of the charge state of the projectile) which can be evaluated for the upper level lifetime. Cheng *et al.* in this way determined the lifetime of the $1s2p\ ^3P_2$ level in Kr^{34+} to $\tau = (9.5 \pm 0.9)ps$ [41].

17.5.4 Techniques for long lifetimes

With typical ion energies available at accelerators, it is not possible to measure certain atomic level lifetimes with the above methods, because they are so long that the associated decay path lengths (velocity · lifetime) exceed the dimensions of any reasonable laboratory. The long lifetimes are of physics

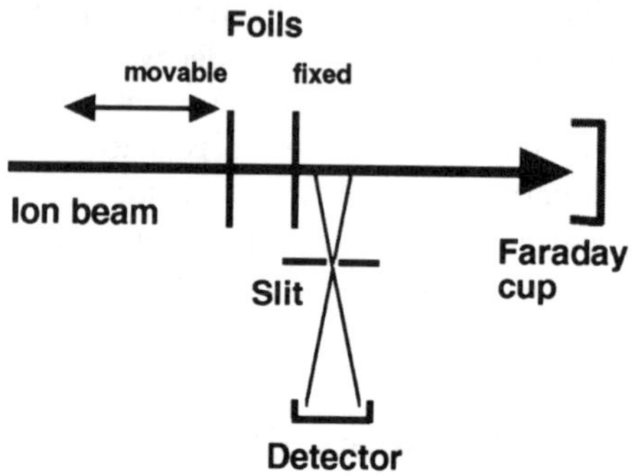

FIGURE 17.5 Two-foil technique for measuring very short lifetimes in highly charged ions [41].

interest, because they may result from structural peculiarities or relate to higher-order effects and decay channels which only become notable in the absence (or non-availability) of the regular electric dipole (E1) transitions. Such decay modes may be: spin-forbidden electric dipole (intercombination) transitions of astrophysical or plasma diagnostic interest, which are sensitive to the spin-orbit coupling matrix element; magnetic dipole (M1) transitions (as for the decay of the lowest triplet level, $1s2s\ ^3S_1$, in He-like ions, a purely relativistic effect, or the so-called forbidden transitions between fine structure levels of a given term or configuration; these latter are of great interest in the diagnostics of terrestrial and solar plasmas); or electric quadrupole (E2) transitions, and so forth.

In order to measure such long lifetimes on ion beams, one has two options: Either one curls the ion trajectory around - as is being done in ion storage rings - or one decouples the ion production from the formation of an ion beam. The latter requires an ion source for multiply charged ions. There now are a number of such ion sources available, with the electron cyclotron resonance (ECR) ion source not reaching charge states as high as in the electron beam ion trap (EBIT)(which can be operated as an ion source, too), but delivering rather intense ion beams. (For such devices, see the contributions by M. Stöckli, G. Melin, A. Girard and H. Winter in this volume.) It is, however, difficult to extract a very low energy, low-energy-spread ion beam from such sources (for μs lifetimes one would want an ion beam energy of order 1 keV and below, with an energy spread of less than 10 eV). Such extraction is under development and will be a desirable future option. At present another type of ion source already works at those properties, that is a source based on recoil ions. We will discuss this source of slow ion beams (for the μs lifetime range) as well as storage rings (for the ms lifetime range) in the following. The EBIT ion trap is also suited to measure lifetimes of some energetically isolated, long-lived levels in multiply charged ions [42], but this is beyond our present scope.

Recoil ion beams

Multiply charged energetic ions colliding with the atoms of a gas target will most likely do so at large impact parameters. The large impact parameter will cause only small momentum transfer, while highly charged projectiles may well sweep away a large number of target electrons in a single collision. Thus we find multiply charged recoil ions at very low velocities (sub-eV energies), which can be exploited for detail studies of the collision process (see chapter by H. Schmidt-Böcking). The recoil ions can be extracted from the gas target and formed into a beam of slow, multicharged ions. Some long-lived levels might survive for longer than the time needed for extraction, charge-state selection, and beam-formation.

This assumption was tested in an experiment [43] on the $1s2s\ ^3S_1$ level in He-like Ar^{16+} ions (Fig. 6). A differentially pumped argon gas target was bombarded with 1.4 MeV/amu ions of Fe^{18+} and La^{24+}. After extraction by a weak electric field (in order to keep the energy spread due to the extraction low) the ions were accelerated by a 1 kV potential drop, passed across the field-of-view of a 10 cm long position-sensitive x-ray detector with a Soller collimator in front, were decelerated again and then analyzed for their charge state. The Soller collimator would let only those X rays reach the detector which had been emitted sideways; thus the position information was preserved from the emitter to the detector. A coincidence of x-ray photon and ion of proper charge filtered out most of the noise and unwanted signal. The result was a clean single-exponential curve with a decay length of about 6 cm, yielding a lifetime of $\tau = (203 \pm 12)\ ns$. This is much closer to theory (best prediction of $\tau = (209.4 \pm 0.4)\ ns$ [44]) than the results of much earlier fast-beam experiments at Berkeley, which had to deal with massive systematic errors because of the immense decay length there (of order 8 m) [45,46]. The lifetime of this particular level has meanwhile been studied from He $(Z = 2)$ to Xe $(Z = 54)$, with a lifetime range of 15 orders of magnitude, using laser absorption (He), storage ring techniques (C, N) [47], an electron beam ion trap (Ne, Mg, Si) [42], recoil ions (Ar) [43] and foil-excited fast ion beams ($P - Xe$) [48,49]. Although the experimental precision has been increasing, theory has developed as well and now claims uncertainties of less than 0.5% [44], which will be hard to match by experiment. There are other levels, however, mostly the upper levels of intercombination (spin-changing E1) transitions in low charge state ions that would be studied best on recoil ion beams: the charge states (for example $q = 4+$ to $6+$ in Ne and Ar) are low enough for astrophysical interest (and the calculational problems dominated by correlation effects), yet higher than what can be achieved by electron excitation in a radiofrequency (rf) ion trap, whilst the lifetimes (1 to 100 μs) are too long for fast-beam work and too short for regular rf ion trap or storage

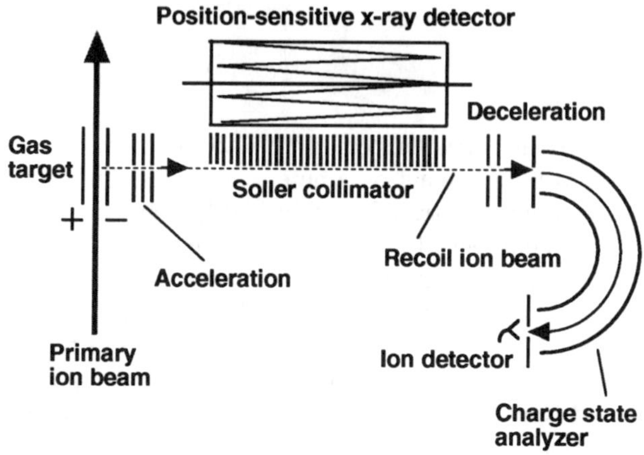

FIGURE 17.6 Experimental arrangement for the x-ray lifetime measurement on a beam of Ar^{q+} recoil ions [43].

ring work.

Storage rings

Storage rings are equivalent to very long beam lines with detector positions at regular intervals. Economizing on vacuum equipment, the beam tube is curled into a ring, and the same detector position is being passed by the circulating ions again and again. This opens up several ways of measuring lifetimes of excited levels in multicharged ions. Some of these will be covered elsewhere (see chapter on storage rings by A. Müller and A. Wolf in this volume). For example, laser resonance can be employed to influence the level population of ions in such a ring. An example is the transition between the hyperfine components of the ground state in Bi^{80+} [50], the high-nuclear charge equivalent to the radioastronomical $21 - cm$ line. The following discussion, however, will be limited to the extensions of lifetime work with straight ion beams as outlined above.

For the lifetime measurements of interest here, the ion production and excitation takes place in the injector (for example, in the foil or gas stripper of a tandem accelerator serving as a pre-accelerator), and the storage ring serves to store the ions under optimum conditions in order to facilitate data acquisition even at long times after excitation. Ion beams regularly need to be stabilized (stacked and cooled) in the storage ring, which may take 0.1 to 1 ms. The ion cycle time in the ring is of order 1 to $10 \mu s$; an observation of any decay over several round trips in the ring thus corresponds to a minimum measurable lifetime of order 0.1 to 0.2 ms. Multicharged ions can regularly be stored for times of seconds and longer, and this storage time determines

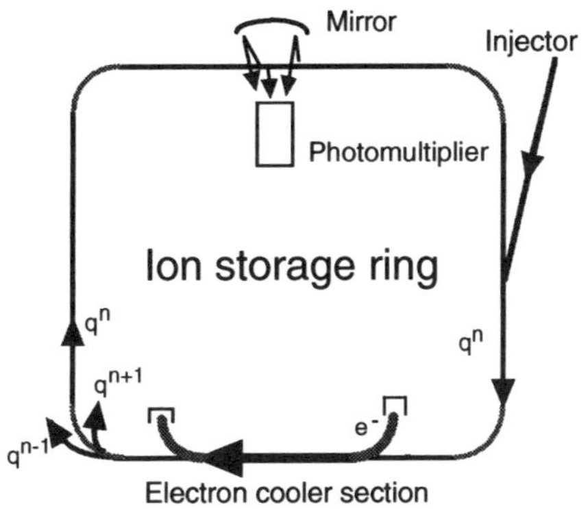

FIGURE 17.7 Schematics of an ion storage ring with atomic lifetime experiments. Bottom section: Detection of ions after recombination or ionization in the ring. Top section: Photon detection.

the upper physical limit of lifetimes to be measured. Detector noise in the face of low detection efficiencies and other factors may suggest an optimum working range between 1 and 20 ms for a storage ring like TSR [47]. For such lifetimes the ring is operated with fast cycling (filling, cooling, time for detection, ejection of the leftover beam), which at TSR is possible at rates up to 5 Hz. For nuclear lifetime studies or experiments with laser- excitation of the ions inside the ring, such cycling can be much slower and may take advantage of the many hours of storage time achieved with bare and almost bare ions.

There are two detection techniques which are presently being explored; particle detection and optical detection. Particle detection involves, for example, dielectronic recombination (DR, the inverse of the Auger effect) as a probe for the surviving excited levels: in the electron cooler section of the storage ring, the electron energy (after the cooling cycle) is switched to match a DR resonance on the level of interest, and some of the ions will capture an electron and stabilize radiatively. They thus have changed charge state and can be detected at the next bend of the storage ring, because the recombined ions will follow a different trajectory outside the main beam (Fig. 7, bottom). The DR cross section is so low that the ion beam is hardly affected by this probe; particle detection of the energetic recombined ion, however, is possible with 100% efficiency. The actual lifetime measurement relates to the electronic timing of the events after injection into the storage ring. This DR and particle-detection technique yielded precise lifetimes on the $1s2s\ ^3S_1$ level in the two-electron ions (He-like) of C and N [47].

Measurements on four-electron ions, however, encountered problems with the DR process which are not yet understood [51]. In the latter experiment, decay curves were also obtained on ions which were further ionized (by collisions with the rest gas). The curves showed a decay behavior that may be linked to the preferential destruction of ions in excited, long-lived levels, because for these the ionization cross sections are larger than for ground-state ions.

For many-electron ions with complex ground state configurations, the DR spectrum is expected to be very complicated, and it may be difficult to achieve state-selective radiative stabilization. The option of choice then is optical detection (Fig. 7, top), which is state-selective. Problems arise from the small signal rate: An observed section of 5 cm length of the ion beam corresponds to only 10^{-3} of the storage ring circumference (TSR: 55 m). An optical coupling system (lenses or mirror) is needed to collect as much light as possible (solid angle of order 10^{-2} of 4π steradian) and concentrate this light on the photocathode of a photomultiplier mounted outside of the ring vacuum. Spectral filtering may be necessary to reject stray light from vacuum gauges and ion getter pumps in the ring vessel, so that a total detection efficiency of order 10^{-7} to 10^{-8} can be expected, which is comparable to that of grazing-incidence spectrometers used on straight, foil-excited ion beams. A photon signal of order 10 counts per second has been obtained from 30 μA of C^{2+} ions stored in the TSR ring, at a wavelength of 190 nm [52], resulting in a very clean, single-exponential decay curve. At an overall uncertainty of about 0.3%, this is one of the most precise lifetime measurements on multiply charged ions ever achieved. It resolved a long-standing discrepancy of data from radiofrequency ion traps and involved theoretical treatments in favor of the latter. For a sufficient signal-to-background ratio it was necessary to use a really low-noise photomultiplier and thorough (double) shielding of the detector and data cables against the ubiquitous stray signals present in a storage ring environment. However, even keeping the optimum spectral range of low-noise photomultipliers in mind (wavelengths between 120 nm (MgF_2 window transmission cut-off) and 300 nm ("solar blind" detector response)), there are many intercombination transitions as well as M1 and E2 transitions in the ground complexes of multicharged ions with, say, 4 to 14 electrons, which are of astrophysical interest, and for which calculations have never been tested by experiment.

17.6 EQUIPMENT

17.6.1 Detectors

Single-channel detection

The low brightness of foil-excited ion beams regularly requires detectors of high efficiency, but very low dark rate, so that single-photon counting can be done. Photoelectric detectors have a linear response over a wide range of signal dynamics, which is needed to record properly the decay curves from which then lifetimes are to be extracted. For the different spectral regions these detectors come with different physical properties. Photomultipliers need a suitable cathode material, with a very low dark rate; they usually are cooled when intended for the visible and near-UV ranges. Solar-blind photomultipliers (with considerable rejection against visible stray light) are available for the near-vuv range; channeltrons (single-channel continuous dynode open photomultipliers) are used for the VUV and EUV; and proportional counters and solid-state devices are used for the x-ray range.

Multiplex detectors

Any mechanical drive system for the foil or a scanning spectrometer introduces problems. Ideally for spectroscopy, one would use a system with two-dimensional detection: a (stigmatic) spectrometer would produce a linear wavelength spectrum along one axis of the detector, and an image of the ion beam along the other. This finely grained multichannel detector would be linear both in position read-out and in response to signal at any position, and this arrangement would permit the simultaneous study of spectral features and their time dependences. An elegant solution to this wish for spectrally separated, distortion-free decay curves recorded with linear, multiplex detectors has been demonstrated for visible light [53]. In this system a stigmatic Ebert-Fastie spectrometer is illuminated by a lens system which provides telecentric imaging of the ion beam (parallel light from the ion beam is focused onto the spectrometer), and the detector is a charge-coupled device (CCD) in the exit focal plane. The sensitivity of the CCD is extremely high compared to photographic film, the granularity of the detector (pixel size) is very fine, and detection is photoelectrically linear. The two-dimensional detector permits extraction of spectral slices and decay curves. The latter are practically free of systematic errors otherwise caused by the detection system. Even such clean recording, however, does not solve the intrinsic problems of decay curve analysis (see below).

For other spectral ranges, no such near-perfect solutions exist (yet). Photographic plate provides an extended detector which covers a rather wide wavelength range; the simultaneous exposure of the full plate takes care of fluctuations of the ion beam intensity which in a scanning instrument might affect the measurement of relative intensities. Optical scanners equipped with length gauges or interferometric control permit to read the developed plates with high positional precision. However, even photographic glass plates are not perfectly linear in positional information, the handling of the thin glass plates is cumbersome, the spectral response is quite non-uniform, and the local response to photon numbers (blackening) is very non-linear.

Photodiode arrays mostly work in the visible and near-uv ranges, with recent developments towards the vacuum-uv (VUV) and extreme uv (EUV) ranges. Microchannelplates (MCP), arrays of miniature continuous-dynode photomultipliers, work in the VUV, EUV and soft-x-ray ranges. They can be sensitized for improved detection of photons in the low-energy part of the spectrum, and by shaping the front plate (funneling), the angular response can be widened. This angular response is important, because for optimum imaging qualities one would like to work with a detector that follows the Rowland circle circumference [54]. This would imply a (slightly) curved-surface MCP; because of the low detection efficiency in grazing-incidence geometry one regularly tilts the MCP so that it cuts through the Rowland circle [55]. This tilt causes distortions in the spectrum and increases the instrumental line widths for positions outside the point of intersection. The multiplex spectra thus available offer additional advantages for lifetime measurements (see below), but suffer from the need for extensive off-line calibration and data reduction work.

A single MCP amplifies a first photoelectron signal by a factor of 10^3 to 10^4, and in a stack of two or three MCPs amplification reaches saturation, which improves the pulse height distribution so that electronic noise can be recognized and discriminated against. Saturation also helps to obtain amplification which is fairly uniform across the MCP (non-uniformities in the few percent range). However, the amplified signal (an electron cloud) needs to be collected and evaluated. For the wanted spatial information this can be done using a resistive or a structured (wedge-and-strip, crossed wire grids, and similar designs) anode, both of which permit a positional read-out in one or two dimensions. Unfortunately, both anode types need amplification and intercomparison of small signals, which introduces electronic non-linearities and instabilities on top of the distortions that are intrinsic properties of each anode type [56]. An alternative read-out is the use of a phosphor screen which lights up when struck by electrons (and thus also shifts the original VUV or EUV photon signal into the visible range), light transport via optical fibre where necessary, and detection of the light pattern by a photodiode array or a CCD. The output of the last stage can be fed into the data acquisition system and processed for evaluation. Typical MCPs are of order 20 to 40 mm diameter, and this corresponds to spectral ranges of 2 nm to 10 nm covered simultaneously in standard grating spectrometers of focal length 1 to 2 m, using diffraction gratings of 300 to 1200 $lines/mm$. Position-sensitive x-ray detectors have also been suggested for decay curve measurements [57]; an example is discussed in the section on recoil ion beams.

17.6.2 Spectrometers

A variety of spectrometers has been deployed at fast ion beams. Almost without exception these are commercial spectrometers intended for use with stationary light sources, modified for attachment to target chambers. As the ion beam is a feeble light source and the light intensity changes with the ion beam current, one needs to monitor the data collection time for a given (generally equidistant) spectrometer setting. As an important technical measure, all of these spectrometers were converted from synchronous drive to step motor drives. Furthermore, the gear boxes with their additional periodic errors are regularly bypassed and the motor coupled directly to the lead screw. Many of the instruments are also equipped with provisions for direct positional read-out. This immediate control over the actual position of optical elements relaxes the need for mechanical expert help in constructing spectrometers and opens ways for home-built, purpose-designed devices. Since precision mechanics is very expensive, whereas computer control is dropping in price, the design of spectrometers based on precision translation stages and complete stepping-motor control of the complex arrangements of grating and slits (necessary for scanning a spectrum in some geometries) appears increasingly attractive.

Normal-incidence spectrometers

The most widely used size of instrument is of one meter focal length, with gratings of line densities 600 to 2400 *lines/mm*. Such instruments are regularly mounted inside a vacuum tank, which can be coupled directly to the vacuum chamber with the foil target. The working range of these spectrometers (depending on the grating) is from 35 *nm* (limit of normal-incidence reflective optics) to about 600 *nm* (geometrical limit of the most common designs, with angles of 10° to 15° between incoming and outgoing light). Depending on the wavelength range of interest, any of the aforementioned open or closed photoelectric detectors can be used.

Normal-incidence spectrometers regularly have been refocused for the observation of the fast-beam light source. The short-wavelength limit of normal-incidence optics is presently being shifted to much shorter wavelengths by introducing multilayer structures [58–60]. Such multilayer coatings may considerably enhance the reflectivity of a surface (for illumination near to the surface normal), over a wavelength interval of about 20% of the mean wavelength. Bundles of small spectrometers optimized for different wavelengths have been flown on rockets to observe the sun in the soft x-ray range [61,62]. The larger collection angle compared to grazing incidence spectrometers (see below) makes such multilayer structures attractive for work on fast ion beams, too.

Grazing-incidence spectrometers

Grazing-incidence spectrometers are needed for studies in the wavelength range between the cut-off for normal-incidence optics ($\lambda \geq 35$ nm) and the range of good x-ray crystals ($\lambda \leq 3$ nm). Ground-complex transitions in all highly charged ions fall into this range, and lifetime measurements on these low-lying levels are of interest for relativistic and QED effects (atomic structure) as well as for the diagnostics of some hot plasmas.

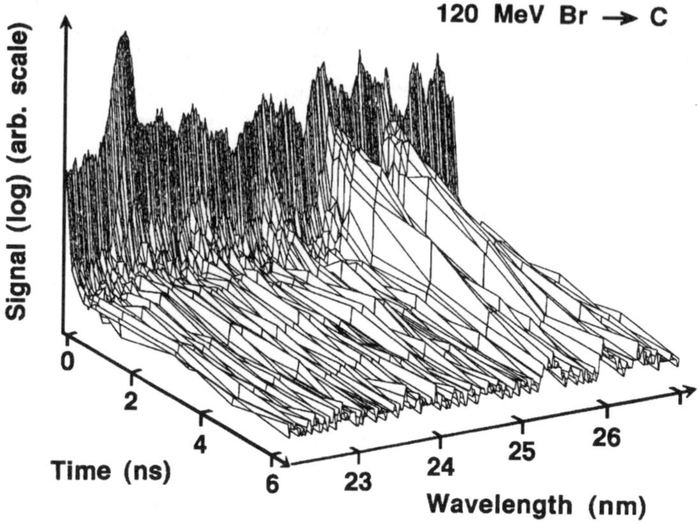

FIGURE 17.8 Set of multichannel spectra (from a grazing-incidence spectrometer equipped with a position-sensitive (multiplex) detector based on a MCP with a resistive anode. The spectra refer to various foil displacements and thus map the spectrum as a function of time after excitation. Two spectral lines stand out by their longevity and are readily identified with intercombination transitions in *Mg*- and *Al*-like ions of *Br* [63]. (Figure adapted from [6,7])

Because of the small acceptance angle and the low grating reflectivity in the extreme ultraviolet (EUV) region, these spectrometers are rather inefficient. The small acceptance angle, however, keeps Doppler broadening at bay. Grazing incidence of light onto the diffraction grating implies that the grating normal will normally be pointing along the ion beam, and in that geometry none of the refocusing schemes is known to work.

Most such spectrometers in use at fast ion beams have a Rowland circle diameter of 2.2 m, with the notable exception of a 5 m instrument at GSI Darmstadt [64]. Several of the instruments have been equipped with position-sensitive (multichannel) detectors. The grazing-incidence geometry renders the dispersion across these detectors rather non-linear, and there is a lack of suitable reference lines both to calibrate the spectra and to study the properties of the multiplex detectors. This is of lesser importance for lifetime studies, but the low efficiency is important: a single set of delayed multiplex spectra from which one can construct decay curves (see Fig. 8) may well require a full day of beam time - which is a very expensive enterprise at large accelerators.

X-ray spectrometers

Both single-channel (scanning) and multiplex-detector x-ray instruments have been used on fast ion beams. Of particular interest for lifetime studies, however, are energy-dispersive detectors, like $Si(Li)$ diodes, because of their much higher detection efficiency (at the cost of lower spectral resolution than is achievable with a crystal spectrometer). The efficiency is high enough to permit coincidence measurements of the two photons from those levels which decay by 2E1 decay, like the $2s_{1/2}$ level in H-like (one-electron) ions and the $1s2s\ ^1S_0$ level in He-like (two-electron) ions. As the two continua of these ions overlap, an additional coincidence with the projectile-ion signal after charge state separation (in a magnetic spectrometer) is desirable. Very precise measurements of rather short lifetimes in a series of highly charged few-electron ions have been performed with such detectors [65].

Multiple spectrometers

Multiple spectrometers may be used for lifetime measurements in different ways:

a) one spectrometer may serve as a monitor for light emitted from the ion beam at one position, while the other spectrometer travels along the ion beam to obtain a decay curve.

b) spectrometers set to different wavelengths may simultaneously record decay curves which are to be entered into an ANDC analysis (Arbitrarily Normalized Direct Cascades, [66]) or which otherwise merit a correlated analysis. For example, the fine structure levels of the $1s2p\ ^3P^o$ term in He-like ions (and similarly the $2s3p\ ^3P^o$ term in Be-like ions) have different lifetimes because all three levels can decay to the $1s2s\ ^3S_1$ level, but (for relatively low Z) only the $^3P^o_1$ level can also decay via spin-changing intercombination decay to the singlet ground state. The transitions in the same shell can be spectrally resolved. By measuring the individual lifetimes of the fine structure levels, one can obtain the transition probability of the additional decay branch [67]. In order to minimize systematic errors in the measurement, it would be desirable to measure the decay curves simultaneously, and this has been done at Oxford: The VUV spectrometer was equipped with a double exit slit matching the separation of the two spectral lines of interest, and a channeltron detector was mounted behind each slit. For the measurement, the foil was moved as usual. The desired transition probability resulted from the "decay curve" which was obtained by dividing the signal in one channel by the signal in the other [25].

c) One (fixed) spectrometer may observe light from one transition in a cascade chain, whereas the other (travelling) spectrometer records the signal from a subsequent step in the same decay chain. By evaluating only coincident signals (with the proper delay accounting for the time of flight of the ions), a

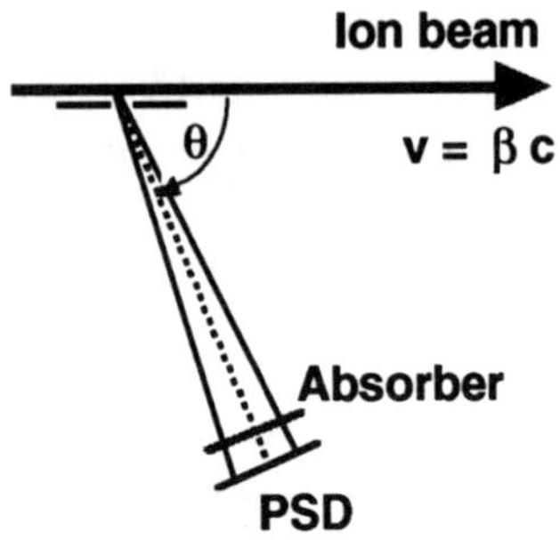

FIGURE 17.9 Schematics of a Doppler-tuned spectrometer equipped with a position-sensitive detector (PSD).

cascade-free measurement can be performed. Such coincidence measurements on neutral atoms were done for quite some time, before lasers offered the better method of selective excitation. For multiply charged ions, where such a technique would be wanted for lack of suitable lasers, the transitions of interest generally are in the UV and VUV ranges. Filters are not available for the latter range, and the solid angle of suitable spectrometers is rather small. An experiment has been set up which uses two sets of seven identical miniature grating spectrometers, each set serving a single detector, one for the start signal and one for the stop signal. The precision reached in tests on the well-known systems of neutral Li and on Ba^+ ions was about 5%, and an application to Na-like Si^{3+} has been tried [68]. Apparently the high level of random coincidences and the rather short lifetimes in multiply charged ions discouraged continuation of this experiment.

Doppler-tuned spectrometers

Observation of radiation at varying angles between ion beam and detector element is at the heart of Doppler-tuned spectrometers (Fig. 9), which exploit the angle-dependent Doppler shift to match emission from fast ions with absorption features in a stationary absorber [69]. The absorber provides an edge filter (ideally, though not in practice - absorption features have lots of structure) which lets pass all radiation of energy below, say, the K- or L-absorption edge of the absorber material. The decay curve of a single spectral line would be obtained as the difference of two decay curves obtained at wavelength settings slightly above and below the spectral feature of interest [70]. This Doppler-tuned spectrometer technique has been used in one of the pace-making experiments measuring QED contributions to the structure of high-Z ions, by measuring the wavelength of a $2s - 2p$ transition in Li-like U^{89+} [71].

Electron spectrometers

Of course, multiply excited ions may stabilize not only by photon emission, but also by autoionization.

Electron spectra recorded close to the exciter foil are often overwhelmingly complex, as are the photon spectra. However, the same technique of delayed observation also works with electrons, and spin-forbidden and similar decays may be identified readily in delayed spectra [72]. These long-lived levels are amenable to lifetime measurements as well, by displacing either the foil or the electron spectrometer along the ion beam trajectory.

17.7 LIFETIME MEASUREMENTS WITH FOILS: TECHNICAL PROBLEMS

17.7.1 Foil life, structural changes, geometry changes

Although *Be* and *Al* foils (and some other materials) have been used over the years, carbon is the most widely used foil material, for reasons of manufacture, sturdiness and price. Foil thicknesses typical for work with ions up to ion energies in the *MeV* per nucleon range are of order 10 $\mu g/cm^2$. This is a factor of 800 thinner than ordinary writing paper, and such carbon foils are very transparent. The structure of a freshly prepared foil is most likely one of the very many modifications of graphite, for example of the class of turbostratic graphite. In this, granules or neighbourhoods of ordered graphite are interspersed with voids.

Foils exposed to ion beams (Fig. 10) suffer radiation damage, develop strain and break or rupture. Apparently the most damaging effect of ion bombardment is dislocation of carbon atoms from their ordered place in those graphite grains, brought about by close collisions with projectile ions. As these are related to the nuclear stopping power, formulae have been developed that permit estimation of foil life based on the accumulated nuclear energy loss.

What happens in the foil in detail, however, is not clear. Dislocations by collisions may be partly balanced by annealing due to the heating of the foil by intense ion beams. Laser-induced heating prior to ion irradiation has resulted in somewhat longer lives in certain kinds of foils made by evaporation of carbon. An even longer foil life has been reached by foil production from cracking hydrocarbon molecules in a discharge. In this case the foil may be expected to consist not of (rather) pure carbon, but of carbon-hydrogen compounds, the balance of evaporation under ion bombardment, sputtering and cracking of rest gas molecules resulting in a prolonged life. What is important for using such foils in atomic lifetime measurements is the fact that the mass density and areal density of foils may well vary with time, and should be monitored when aiming at utmost precision (see below).

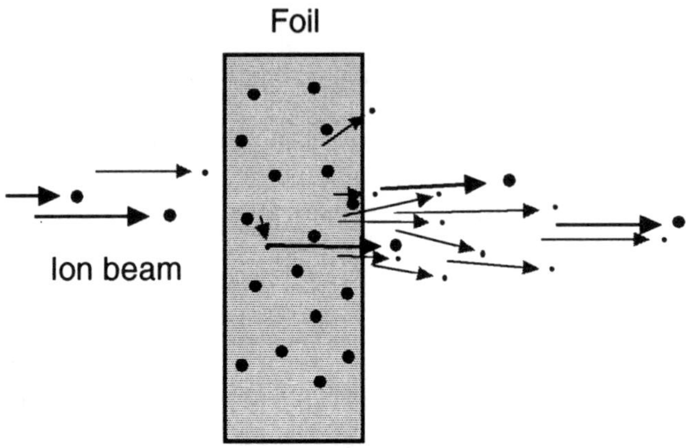

FIGURE 17.10 Illustration of the ion-foil interaction. Most ions will continue after only electronic energy loss with almost the same trajectory and speed, accompanied by a swarm of electrons. Only projectiles colliding with the ion cores of the atoms in the foil (black dots) may experience larger-angle scattering, which goes along with a notable momentum transfer and nuclear energy loss.

17.7.2 Foil chamber design

Foil carriage. Precise foil displacement is usually and easily achieved by rotating a precision lead screw by means of a stepping motor; the foil holders are mounted on a carriage which is coupled to the lead screw by ball bearings. Although the pitch of such lead screws is often well defined, most screws, gears and bearings display periodic errors. It is thus desirable to monitor the actual displacement of the foil carriage (if not of the foil), for example by a moiré fringe length gauge. Such devices are readily available, with measuring ranges of a few centimeters and reading with μm precision [73].

Signal normalization. For signal normalization, a number of techniques are available. The simplest technique is the collection of the ion beam on a Faraday cup, determining equal amounts of incident beam by a preset amount of charge. The Faraday cup needs a (negative) suppression voltage on an entrance aperture in order to avoid misreadings due to secondary electron emission under ion bombardment of the Faraday cup material. However, the charge state distribution of the ion beam after passage through a foil will regularly have a mean value that differs from the single incident-ion charge state, and therefore there is no strict relationship between the measured charge and the number of incident ions. Under ion irradiation, foils change their internal structure, suffer sputtering of the foil material and deposition of cracked foreign materials, develop voids and microscopic holes. Consequentially the charge state distribution after the foil may change with time, and with a broken foil the Faraday cup still collects charge, even as foil-excitation no longer takes place.

One solution is an optical monitor, for example collecting light (preferably of a spectral line emitted by an ion in the desired charge state) from the ion beam close to the exciter foil and guiding it to a monitor photomultiplier via flexible fiber optics. Such a system can detect the breakage of a foil by the associated loss of excitation. However, breaking foils often develop spots and edges which emit bright light in the visible and thus counteract the intention of a monitor working in this spectral region. On top of this, even the light yield of a given foil varies with the radiation damage of the foil.

Although the above measures can be used for lifetime measurements with precision at the five to ten percent level (discarding all data runs during which any irregularities are noted), a proper solution needs a much more complex experimental approach. The survival of the foil as well as the average foil thickness may, for example, be monitored (even on-line) by a particle attenuation, of a well-collimated a particle beam crossing the ion beam at the position of the foil. Rutherford scattering of projectile ions on target atoms can at the same time provide an absolute measurement of the projectile particle current [74].

The same Rutherford scattering process causes the so-called nuclear energy loss and an increase in the angular divergence of the ion beam. Both are more important at low ion energies (up to a few hundred keV) than at high beam energies, but the angular scattering causes a systematic error in lifetime measurements (as long as the Faraday cup does not travel with the foil) and thus merits mentioning: After the foil, the ion beam widens because of the scattering. Thus part of the ion beam may bypass the field of view of a detection system which is tailored to match the ion beam cross section close to the foil. The progressive lack of observed ions results in a typical kink and increased slope in some decay curves, and the lifetimes obtained when overlooking this effect are always too short. However, a Faraday cup that is too small will also see only a fraction of the ion beam which depends on the spatial separation between foil and Faraday cup and may contribute an error of opposite sign. Furthermore, ions which are deflected by close collisions in the foil will travel a longer path to the Farady cup (if they reach it), and such a slanted trajectory also smears out the time scale. Although these problems are long since known, they are not yet fully solved: the charge state distribution and electronic excitation of particles which are deflected notably by Rutherford scattering (with an associated longer path through the foil material and a different energy loss) may be quite different from the mainstream ions which experienced electronic collisions only. Fortunately, with heavy ion projectiles onto light element foils and at high ion velocities, these particular problems may be largely (and fairly safely) neglected.

Foil positioning. In a decay curve composed of several components with different time constants, the relative amplitudes are a function of time after excitation. If one wants to intercompare results obtained from decay curves (as is necessary in correlated analyses like in the ANDC scheme) or wants to obtain meaningful initial population or intensity parameters, the time zero of a decay curve needs to be established. This is most easily done by letting the exciter foil pass the field-of-view of the detection system during the measurement, for example by beginning a decay curve measurement with the foil positioned downbeam of the optical axis and afterwards moving it towards the accelerator (Fig. 2). Then some data points show the true background level (Fig. 11), without any light from the excited ion beam reaching the detector. Next comes a section in which the curve rises steeply, because the brightest section of the decay curve moves into and then across the field-of-view. After the foil has moved out of view, the decay curve represents a superposition of exponential components without further distortions. A curve analysis which takes the window shape into account then easily determines the data point which relates to the foil position meeting the optical axis of the detection system, that is, the true time zero for the observed decay curve.

Time resolution. When measuring lifetimes, it is certainly important to have appropriate time resolution. As seen in Fig. 2, the detection system has a field-of-view of finite width at the position of the ion beam, the so-called window. The details of this window, its shape parameters, can be determined from the geometry of the two apertures involved (here: spectrometer slit and grating mask) and from grating reflectivity properties, either synthetically [75] or working backwards from the distortions of observed decay curves [76]. The Lund group meanwhile has introduced a higher level of complexity by three-dimensional modeling of this problem [77]. Using such techniques, a number of lifetimes down to the range of a few *ps* have been measured.

Time resolution is not determined by the total width of the window (which is governed primarily by the distance of the spectrometer from the ion beam), but by the smallest defined structure of the window, that is by the steepness of the flanks or wings of the (approximately trapezoidal) window function. The latter depend on the spectrometer slit width and on the distance from the ion beam. However, the observed decay curve results from a convolution of the true decay curve with the window function (Fig. 11). This convolution involves integration of the decay curves over the width of the window, and decay components of different time scale are being weighted differently in this process. A narrow window function is optimum for the observation of short-lived decays, keeping the contribution from slow components (only a small fraction of which decays within the length of the window) small. A wider window function does not notably increase the amplitude observed on fast decay components, but it does increase the amplitude of slow components (by detecting a larger fraction of a decay length). Although a spectrometer farther away from the ion beam will span a smaller solid angle and thus loose signal, the imaging conditions of grazing-incidence spectrometers, for example, with their severe astigmatism, partly compensate this loss. Also, screening against ion-beam related noise may sometimes be easier for a spectrometer entrance

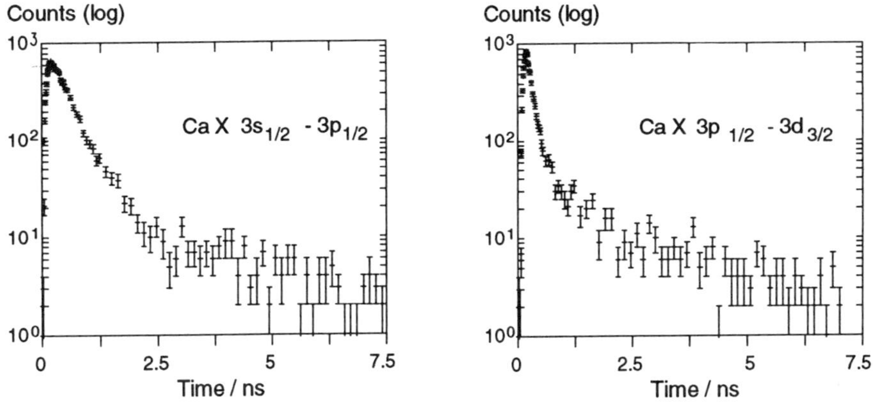

FIGURE 17.11 Decay curves of one of the resonance lines (left) and of its strongest cascade (right) in sodium-like Ca^{9+}. Both levels are repopulated by cascades and thus give rise to non-linear curves in these logarithmic plots. The distortion of the steep initial rise of both curves is due to the finite spatial resolution of the detection system. For a proper lifetime analysis, the left-hand curve has to be evaluated in correlation with the right hand one; otherwise a systematic error of order 30% is likely.

slit not so close to the ion beam (see below). There is some range of play, and optimization is possible for different experimental aims. For maximum time resolution, an auxiliary slit positioned near the ion beam may be a good option, although it reduces the window width and thus the collectible signal.

Ion velocity calibration. In order to establish the time scale applicable to decay curves, the velocity of the ions after interaction with the target must be known. The simplest way is to rely on the properties and settings of magnets in the beam transport system of the accelerator. Under favorable conditions, a precision of the velocity determination close to 10^{-4} may be achieved, with the magnets being calibrated by ion-beam - induced nuclear reactions. With the larger apertures common at high-energy machines this method is of rather limited precision. With the pulsed ion beams typical of linacs and synchrotrons, however, time-of-flight measurements are possible. With flight path lengths of order 10 m (time of flight of order 1 μs and less), sub-nanosecond timing is necessary to achieve a precision of better than 10^{-3}. Both methods measure the ion velocity before interaction with the target and thus require a correction of the actual energy by reference to tabulations on energy loss in matter [78,79]. As the target thickness often is not precisely known (manufacturers' quoted uncertainties of mass densities of commercial foils are typically of order 10 to 20%, standard laboratory measures by quartz frequency during evaporation often suffer systematic errors of order 30%), this is not a satisfactory situation. Fortunately the relative energy loss decreases for higher energy ions, and the problem thus becomes less important. For low energy ions this is a severe problem which, however, can be solved by using an electrostatic energy analyzer (see below).

Beam-foil excitation is so sudden that fine structure levels are excited coherently, and their decay will show this coherence by displaying the so-called quantum beats, with a frequency which directly relates to the energy separation of the coherently exited levels. If the fine structure interval is known sufficiently well, the beat pattern can be exploited as an internal clock, and the observed spatial beat frequency, via the known time frequency, translates into the desired particle velocity. Exploiting this phenomenon in neutral He, for which a precise calculation predicted the fine structure intervals in the $1s3p\ ^3P$ term (and precise microwave data are also available), a most precise (0.26%) lifetime measurement became possible [80]. In fact, the precision is practically the same as reached in the best ion-beam laser experiments of the time. It was already noted that lasers cannot regularly excite levels in multicharged ions. Although it uses foil excitation, the quantum beat technique also fails for higher ionic charges, because the much larger fine structure intervals correspond to unresolvably fine beat patterns, and the beat amplitudes usually are very low (low degree of non-statistical population).

The same quantum beats have been used to calibrate an electrostatic analyzer at Berlin, which in turn was employed to measure the velocity of other ion beams (after ion-gas or ion-foil interaction). Practical limits on electric field strengths in such analyzers (90° sectors of a cylindrical capacitor) restrict their use to ions in the few-hundred keV energy range.

Background reduction. With the low light levels encountered in much of fast beam spectroscopy, it is very important to keep the ion-beam related background level of the signal low. This involves radiation shielding of the detector against neutrons and γ rays released from ions striking apertures and the Faraday cup. Also, the light path in the detection system needs to be baffled properly against stray light, stray ions and electrons set free by ion impact or stray light. It helps, for example, to block the zero order light in VUV spectrometers so that it does not strike a surface close to a particle-sensitive detector like a channeltron. Screening meshes in front of such detectors can be used against slow particles of one polarity, and the front end of a (negatively biased) channeltron is best run near ground potential so as not to attract stray ions. Further mechanical measures include an obstacle close to the entrance slit of the spectrometer (Fig. 12) which shields the slit jaws against the many electrons ejected from the foil under ion bombardment; these electrons have a spatial distribution which points forward and may otherwise cause a background (by photon emission when hitting the slit) which becomes notable for some ranges of foil displacement, and which cannot be blocked inside the spectrometer. Instead of a mechanical obstacle a magnetic field (small permanent magnets mounted onto the foil wheel) can be used to deflect these electrons away from the spectrometer slit.

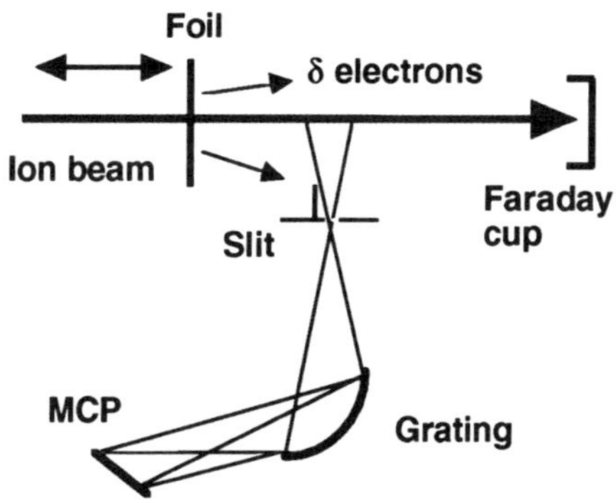

FIGURE 17.12 Schematics of beam-foil spectroscopy using a multichannel detector (MCP). A piece of sheet metal close to the entrance slit of the spectrometer blocks fast electrons emerging from the foil from striking the slit jaws.

A second, electronic, measure exploits the fact that most ion beams at high-energy accelerators (with their higher rate of nuclear reactions along the ion beam path) are pulsed. The flight time of photons from the ion beam to the detector is known, and the signal circuitry can be gated to admit only those signal pulses which are detected within a specified time window adjusted to the time structure of the ion beam. In this way neutrons produced at the target (because of their lower velocity) and γ rays from far-away apertures or the Faraday cup (because of the larger distance) can largely be rejected via their travel time.

17.8 DECAY CURVE ANALYSIS

Fitting a spectral line to obtain a line position is aided by the fact that, in many cases, a symmetric profile is expected. The position of the symmetry axis can thus be determined rather well even with imperfect trial functions. In decay curve analysis, however, the problem is much more difficult: the curve shape must be determined. A typical decay curve is a superposition of exponential functions, and the number of components may be very large. The art of lifetime analysis is to extract sensible information from a decay curve or a number of correlated decay curves, which often cannot be analyzed in their entirety.

The complexity arises from the fact that most excitation mechanisms are not selective and thus do not lead to a single, exclusively populated excited level. The spontaneous decay of a single level can be described by a single exponential. For the level population N of level k which at time $t = 0$ had a population N_k^0 we have $N_k = N_k^0 exp(-t/\tau)$, the level mean life t being the inverse of the sum of the transition probabilities A_{ki} for transitions to other levels i:

$$\tau = 1/\Sigma A_{ki} \tag{17.1}$$

The observed signal is proportional to the decay rate,

$$Signal = \epsilon \, \frac{d}{dt}(N_k) = \epsilon A_{ki} \; N_k^0 \; exp(-t/\tau), \tag{17.2}$$

with the detection efficiency ϵ for a transition at a given wavelength. While selective excitation of each individual level is a dream that comes true only for a number of levels in neutral atoms or a few levels of some ions in very low charge states, the normal case involves many levels which are excited simultaneously and may decay, not only to the ground state, but to other excited levels as well. Consequently we find decay *chains* and speak of *cascades*, the latter referring to the multitude of decays that may, via decay chains, replenish the population of a level of interest.

This cascade problem is also known in nuclear lifetime analysis [81]. While the range of lifetimes encountered in the latter may be as wide as that in the electron shells of multiply charged ions, the detection efficiency usually is much higher for nuclear radiation, because physically large, energy-dispersive detectors are available. What makes atomic lifetime measurements attractive in this comparison is the easier comparison with theory; theory is better able to describe a few-body Coulomb field problem than the complex nuclear interactions and excitations. As soon as many electrons are involved, however, this difference may appear less clear cut.

As each individual level decays spontaneously, its decay is described by an exponential function. Solving the differential equation for a level which decays and is replenished by other decays yields a sum of exponential functions in which the individual lifetimes of the levels are preserved, but in which the amplitudes of the decay components depend on population factors and on differences and ratios of transition probabilities [82]. For an analysis of a decay curve, not only the amplitudes, but also the time constants of the components have to be varied, and that makes the fitting problem complex and non-linear. Because of the mathematical similarity of the decay components and the unavoidable statistical limitations of the data, only a few components can be fitted sensibly, some of which will represent a multitude of components that cannot be analyzed individually.

Although such multi-exponential fitting can yield useful results, it has been shown that there are many cases in which the solutions show systematic errors, and that there are cases in which the procedure is bound to fail or in which a fit algorithm cannot decide on a best solution. Hence, a blind treatment will usually meet with only limited success, even as non- (observer-) biased automatic fitting routines are available which also vary the number of components of the trial function [83]. However, by utilizing prior knowledge of the problem - analyzing the atomic structure of the system studied, or making measurements on cascade transitions - one can often clarify the fitting problem and from physics arguments decide upon alternative solutions. This open-eyed analysis, with auxiliary information on (physically less interesting and often better calculable) cascades often taken from theory, may seem observer-biased. However, by varying the assumptions used in the analysis, one sometimes can identify the role of observer bias and thus reduce it. Anyway, none of these analyses make sense without the intricate framework of atomic structure knowledge that physicists have assembled over time.

There is one measurement and evaluation strategy which avoids most of the above ambiguities and delivers reliable lifetime data. It requires the measurement not only of the decay of the level of interest, but also of the decay properties of all levels which decay directly towards this level. These cascade decay curves contain all the information on contributing higher lying levels, and thus the knowledge on the cascade repopulation is complete. The fit procedure then consists of correlating all decay curves of direct cascades with the decay curve of interest (ANDC scheme [66]), by adjusting the relative amplitudes of the cascades. (As the lifetime of a level relates to the sum of all transition probabilities, one does not even have to measure the direct cascade itself, but may choose a decay branch of the same level which might be better amenable to measurement. Optimum conditions for ANDC are given if all lines of interest can be measured under comparable conditions; for this purpose, a facility has been set up at Lund which comprises several spectrometers viewing the same ion beam section.) This, then, is a linear fit problem with is solvable with much more mathematical stability and reliability than the non-linear fitting problem of multi-exponential analysis. Tackling this integro-differential problem of cascade replenishment, the various ANDC codes proceed via numerical integration or differentiation of the primary decay and the cascade data curves, checking the consistency of the results. In order to reduce the influence of individual data point statistical fluctuations, the most widely used code, CANDY [84], represents the data curves by multi-exponential smoothing functions.

17.9 ACHIEVEMENTS AND OUTLOOK

Beam-foil spectroscopy has demonstrated amply that the lifetimes of many levels in many ionization stages of many elements can be measured. Lifetime measurements on highly charged "one-electron" systems have reached Li-like Kr^{33+} [85], Na-like Au^{68+} [86] and Cu-like U^{63+} [87], whereas in two-electron systems the lifetime measurements reach up to U^{90+} [32]. Beyond the sheer demonstration of lifetime measurements is the quest for precision that permits the testing of theory, and hence our understanding of atomic structure and dynamics. Here many measurements of single ionic species have been superseded by systematic studies along isoelectronic sequences, both for structure (best done by spectroscopy of laser-produced plasmas and, increasingly and particularly worthwhile for high-Z ions, electron beam ion traps (EBIT) [7])

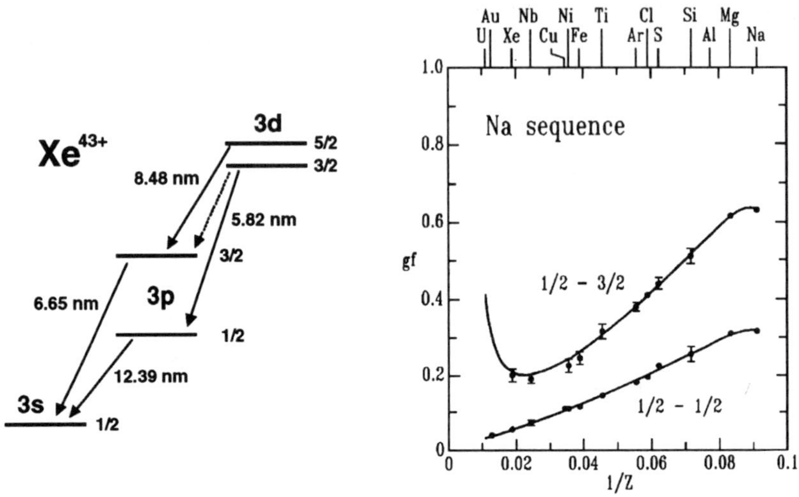

FIGURE 17.13 Left: Level and transition scheme of $n = 3$ levels in sodium-like Xe^{43+}. Right: Oscillator strengths f of the resonance transitions in Na-like ions. All of the experimental data have been obtained with ion beams. The measurement on Na used a neutralized ion beam and laser excitation, the one on Mg^+ used laser excitation of an ion beam; all other data are from foil-excited ion beams and involved suitable cascade correction techniques. The solid lines represent the results of the best calculations available (see [86] for references). (Figure adapted from [89])

and for lifetimes. So what? What has been achieved and learned? A few examples may highlight the physics:

Using ANDC for most ions, $3s - 3p$ and $3p - 3d$ transitions in Na-like ions have been studied up to Au^{68+} (at GSI Darmstadt, using ion energies up to 13 MeV/amu). For such heavy ions, relativity has a major effect on the level structure: The fine structure interval of the $3p$ term (upper level of the "sodium D lines") is about as big as the $3p$ excitation energy (Fig. 13, left). Consequently the lifetimes of the two fine structure levels differ by a factor of about seven. The precision of the lifetime measurement in Xe^{43+} is good enough to make the uncertainty comparable to the size of the relativistic modification of the transition matrix element, and another factor of two improvement seems possible [86]. For a time, the spectra and transition probabilities of the Na-like ions seemed easily tractable by theory. However, several sets of measurements along the Na isoelectronic sequence showed up insufficiencies of theory. Meanwhile theory has caught up with experiment again and is now in good overall agreement with the experimental data (Fig. 13, right).

If for reasons of spectral range or equipment the cascades cannot be measured, cascade modeling and simulation of the decay curve of interest may be of some help. The simulated data can be evaluated by standard techniques, and systematic errors be recognized. An example is shown in Fig. 14. Here the decay curve of the $2p^2P_{3/2}$ level in Li-like Kr^{33+} ($\lambda = 9.10nm$) was studied [85], but the cascades along the chain $2p - 3d - 4f - 5g - 6h$ etc. (so so-called yrast cascades) could not be measured directly. The cascade lifetimes, however, are reliably obtainable from a simple hydrogenic approximation (treating the system as a one-electron ion of core charge $q = 34+$). Although at face value the decay curve (which on a logarithmic scale looks rather straight) may be taken for a single-exponential curve, the simulation indicates some slight, but significant, distortion which can only be recognized in data of high statistical quality. The cascade model analysis [85] brought about a systematic change of the lifetime result, compared to earlier experiments, by about 6% and halved the error bar. The new result is in excellent agreement with the results of several theoretical calculations.

The above lifetime measurements on the $2p_{3/2}$ level in Kr^{33+} and on the $3p_{3/2}$ level in Na-like Xe^{43+} [86] have uncertainties comparable to the size of relativistic corrections to the transition matrix element. Any significant improvement would actually test this physical concept, and a measurement in heavier ions would encounter larger contributions from these interesting effects. The experimental limit, however, is not the ionic charge, but the short lifetime (of a few ps) and the spectral range: in those isoelectronic sequences the $np_{1/2}$ level is longer lived and remains measurable up to rather high nuclear charges, but it is the less interesting level as it develops rather regularly. The interesting level is the short-lived $np_{3/2}$ level which is so short-lived because of the increasing "fine structure" interval (relativity); the associated decay moves towards the most difficult part of the EUV range where grazing-incidence spectrometers fail and x-ray spectrometers are not very efficient either. Recent soft x-ray diode developments as well as multilayer x-ray optics might help a long way for such systems. Lifetime values precise to 1% are the present goal, which may be in reach only for specific cases. Theory, after its recent advances, is difficult to beat in the race for precision on these more fundamental systems.

In the aforementioned examples the systems were relatively simple and had a single, dominant decay channel. The situation, however, is often much more complex, which is a challenge for theory and experiment. Near-neutral high-Z ions, for example, feature a multitude of electrons in a rather weak Coulomb field, and relativity, correlations etc. are very important. Because of cancellations of the wavefunctions and because configuration interaction influences particular decay channels, the $Hg\ II$ $7p$ fine structure levels - which in regular LS-coupling would be of equal mean life - have lifetimes which differ by more than a factor of ten. This has been verified by a beam-foil measurement [92].

A different kind of problems is found in few-electron, core-excited (multiply excited) ions, where electrons are in different open shells, and several, very different, decay channels are available: in core-excited three-electron ions (Li

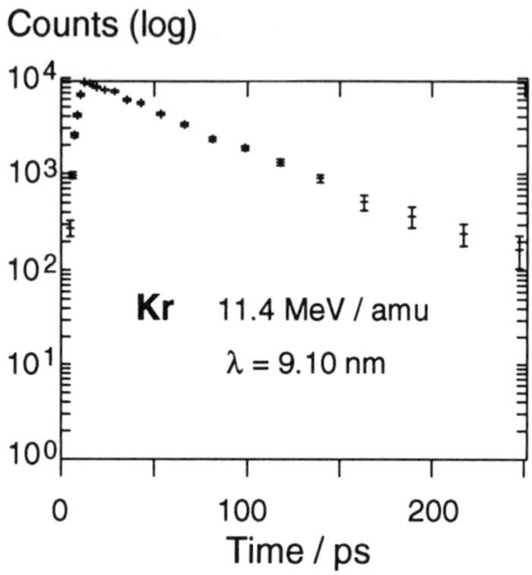

FIGURE 17.14 Decay curve of the Kr^{33+} $2p\ ^2P^o_{1/2}$ level. These data are from spectra obtained with a multiplex detector [85].

isoelectronic sequence), the $1s2p^2\ ^4P$ levels lie high above the first ionization limit. These levels nevertheless are long-lived, because a number of selection rules hinder autoionization. However, three different decay channels are open which scale differently with nuclear charge Z and give the levels a different longevity: one decay mode is via E1 radiation to the $1s2s2p\ ^4P^o$ levels, and the transition probability for this decay branch scales linearly with Z. The second decay mode, autoionization, proceeds via spin-spin and spin-other-orbit interaction (roughly with the 3rd to 4th power of Z), while spin-orbit-induced autoionization is suppressed. The autoionization rate is highest for the $J = 5/2$ level. Last, but not least, there is a spin-changing intercombination transition to the doublet ground term, which also scales with a high power of Z. Observing the first of these decay branches (a transition within the same shell, producing UV light), the fine structure can be resolved, and level-specific lifetimes can be measured. By such lifetime measurements reaching from neutral Li to Si^{11+} ions it was possible to disentangle and study the roles of the individual decay channels, and shortcomings in the prediction for the UV transition were found and corrected [93].

In singly-excited systems with more electrons the situation is different again. Electron-electron interaction and relativity need to be taken into account simultaneously, and this causes severe problems for theory. There are a number of fairly simple spectroscopic systems (with two, three or four electrons outside closed shells) that show an interesting (predicted) behavior of the intercombination transitions along the isoelectronic sequences and which never have been tested outside the low-Z range - where the shortcomings of theory were obvious even at the 10 to 20% level of experimental lifetime uncertainty [6]. Within limits, calculated lifetimes may be "fine-tuned" by adjusting the theoretical level structure with an eye on experimental data [88], using correction factors based on basic theory. Sometimes this is quite sufficient to obtain agreement among calculated data or of predicted and measured lifetimes, but in other cases it is not. This then is the basis for tests of the different wavefunctions used in the calculations, proving the value and merit of lifetime measurements which reach beyond the information obtained from energy level spectroscopy.

Under favourable conditions, lifetime uncertainties of 5% and less are within reach for many of such systems. The expected lifetimes of certain levels are long enough for measurements up to U, and many of the associated wavelengths are in reach of existing spectrometers. Some of the lifetimes show isoelectronic trends not easily accommodated in the low-Z Hylleraas expansions [6,90]. Furthermore, hotter and hotter fusion test plasmas will contain such emissions from heavier and more highly charged ions.

At the other, lower, end of the isoelectronic sequences there are good prospects for accelerator-based atomic lifetime measurements, too. Ion storage rings will be used to probe intercombination and forbidden transitions of interest for both astrophysical and terrestrial plasma diagnostics, exploiting the recently demonstrated percent range precision [52], in some productive competition with electron beam ion trap work. However, the latter devices work much better with higher charge state ions, so that in the low- to intermediate-Z range the flexibility of accelerators and the purity of ion beams will be major factors in favor of storage rings and of slow ion beams with good energy definition over any ion traps.

17.9.1 Suggested readings

The development of the field of Beam-foil / Fast-beam spectroscopy may be followed by looking up the series of beam-foil conferences, a series that started out at Tucson (1967) and continued at Lysekil 1970, Lyon 1971, Tucson 1972, Gatlinburg 1975, Lyon 1978, and Laval 1981. The field was last formally incorporated in the subtitle of the Oxford meeting on the Physics of Highly Charged Ions in 1984, by which time other devices besides fast ion beams had become available for the study and application of such ions. This conference in turn was followed by a string of meetings on that wider subject of highly charged ions (HCI) (Groningen 1986, Grenoble 1988, Giessen 1990, Manhattan 1992, Vienna 1994). Many of the proceedings of these conferences appeared in Nucl. Instr. Meth. (among others Vols. 90, 110, 202, B9, B23, B31, B98). Some of the considerable number of reviews on specific, important subfields of fast-beam spectroscopy are included in the following list of references, under nos. [14,16,6,7], and in the contributions by L.J. Curtis and E.H. Pinnington to ref. [8]. For the analysis of isoelectronic sequences, Edlén's Handbook article [91] gives many examples and practical tools. Ref. [5] is a book based on lecture notes and designed to introduce students to the field of spectroscopy by realistic examples (including some from the accelerator-based physics background of the authors), but also to yield information on where to look for more detailed explanations in the more specialized literature. Spectroscopic instruments, their workings and designs, light sources and so on are comprehensively covered in [54].

REFERENCES

1. Doerfert, J., and Träbert, E., *Physica Scripta* **47**, 524 (1993).
2. Engström, L., Kirm, M., Hellborg, R., Martinson, I., Bengtsson, P., Maniak, S.T., Curtis, L.J., Träbert, E., Granzow, J., Doerfert, J., Heckmann, P.H., *Physica Scripta* **52**, (1995).
3. Hylleraas, E.A., *Z. Physik* **54**, 347 (1929).
4. Hylleraas, E.A., *Z. Physik* **65**, 209 (1930).
5. Heckmann, P.H., Träbert, E.: *Introduction to the Spectroscopy of Atoms*, Amsterdam: North-Holland, 1989.
6. Träbert, E., *Physica Scripta* **48**, 699 (1993).
7. Träbert, E., *Nucl. Instrum. Meth. Phys. Res.* B **98**, 10 (1995).
8. Drake, G.W.F. (ed.): *Atomic, Molecular, and Optical Physics Reference Book*, Woodbury: AIP Press 1996
9. Kay, L., *Phys. Lett.* **5**, 36 (1963).
10. Bashkin, S., Meinel, A.B., *Astrophys. J.* **139**, 413 (1964).
11. Bashkin, S., *Science* **148**, 1047 (1965).
12. Kay, L., *Nucl. Instrum. Meth.* B **9**, 544 (1985).
13. Bashkin, S., *Nucl. Instrum. Meth.* B **9**, 546 (1985).
14. Andrä, H.J., in: Hanle, W., Kleinpoppen, H. (eds.): *Progress in Atomic Spectroscopy*, New York: Plenum Press, 1978, Part B, pp. 829.
15. Shima, K., Kuno, N., Yamanouchi, M., Tawara, H., *At. Data Nucl. Data Tables* **51**, 173 (1992).
16. Mannervik, S., *Physica Scripta* **40**, 28 (1989).
17. Liesen, D., Beyer, H.F., Finlayson, K.D., Bosch, F., Jung, M., Klepper, O., Moshammer, R., Beckert, K., Eickhoff, H., Franzke, B., Nolden, F., Spädtke, P., Steck, M., Menzel, G., Deslattes, R.D., *Z. Physik* D **30**, 307 (1994).
18. Gaillard, M.L., Pegg, D.J., Bingham, C.R., Carter, H.K., Mlekodaj, R.L., Cole, J.D., *Phys. Rev.* A **26**, 1975 (1982).
19. Short, R.T., Mannervik, S., Larsson, M., Sigray, P., Sonnek, D., *Phys. Rev.* A **39**, 3969 (1989).
20. Mokler, P.H., Stöhlker, Th., Kozhuharov, C., Moshammer, R., Rymuza, P., Stachura, Z., Warczak, A., *J. Phys. B: At. Mol. Opt. Phys.* **28**, 617 (1995).
21. Stoner, jr., J.O., Leavitt, J.A., *Optica Acta* **20**, 435 (1973); Leavitt, J.A., Robson, J.W., Stoner, jr., J.O., *Nucl. Instr. Meth.* **110**, 423 (1973).
22. Bergkvist, K., *J. Opt. Soc. Am.* **66**, 837 (1976).
23. Jelley, N.A., Silver, J.D., Armour, I.A., *J. Phys.* B **10**, 2339 (1977).
24. Berry, H.G., Désesquelles, J., Dufay, M., *Phys. Lett.* **36**A, 237 (1971); *Phys. Rev.* A **6**, 600 (1972); *Nucl. Instrum. Meth.* **110**, 43 (1973).

25. Myers, E.G., Silver, J.D., Klein, H.A., Pinnington, E.H., Träbert, E., (unpublished work, Oxford, England, 1979).
26. Hutley, M.C., Hunter, W.R., *Appl. Opt.* **20**, 245 (1981).
27. Kauffman, R.L., Woods, C.W., Hopkins, F.F., Elliott, D.O.,Jamison, K.A., Richard, P., *J. Phys.* B **6**, 2197 (1973).
28. Träbert, E., Heckmann, P.H., Hutton, R., Martinson, I., *J. Opt. Soc. Am.* B **5**, 2173 (1988).
29. Träbert, E., Brandt, M., Doerfert, J., Granzow, J., Heckmann, P.H., Meurisch, J., Martinson, I., Hutton, R., Myrnäs, R., *Phys. Scripta* **48**, 580 (1993).
30. Jupén, C., Isler, R.C., Träbert, E., *Mon. Not. R. Astron. Soc.* **264**, 627 (1993).
31. Birkett, B.B., Briand, J.-P., Charles, P., Dietrich, D.D., Finlayson, K., Indelicato, P., Liesen, D., Marrus, R., Simionovici, A., *Phys. Rev.* A **47**, R2454 (1993).
32. Munger, C.T., Gould, H., *Phys. Rev. Lett.* **57**, 2927 (1986).
33. Gaupp, A., Kuske, P., Andrä, H.J., *Phys. Rev.* A **26**, 3351 (1982).
34. Schmoranzer, H., Roth, H., Volz, U., Marger, D., *J. Phys. B: At. Mol. Opt. Phys.* **24**, 595 (1991).
35. Tanner, C.E., Livingston, A.E., Rafac, R.J., Serpa, F.G., Kukla, K.W., Berry, H.G., Young, L., Kurtz, C.A., *Phys. Rev. Lett.* **69**, 2765 (1992).
36. Volz, U., Majerus, I., Liebel, H., Schmitt, A., Schmoranzer, H.: *Phys. Rev. Lett.* **76** , 2862 (1996).
37. McAlexander, W.I., Abraham, E.R.I., Ritchie, N.W.M., Williams, C.J., Stoof, H.T.C., Hulet, R.G., *Phys. Rev.* A **51**, R871 (1995).
38. Baudinet-Robinet, Y., Garnir, H.P., Dumont, P.D., El Himdy, A., *Physica Scripta* **39**, 221 (1989).
39. Kemmler, J., Reinhold, C.O., Burgdörfer, J., *Phys. Rev.* A **44**, 2993 (1991).
40. Betz, H.-D., Bell, F., Panke, H., Kalkoffen, G., Welz, M., Evers, D., *Phys. Rev. Lett.* **33**, 807 (1974).
41. Cheng, S., Berry, H.G., Dunford, R.W., Gemmell, D.S., Kanter, E.P., Kurtz, C., Rehm, R.E., Zabransky, B.J., *Phys. Rev.* A **50**, 2197 (1994).
42. Wargelin, B.J., Beiersdorfer, P., Kahn, S.M., *Phys. Rev. Lett.* **71**, 2196 (1993).
43. Hubricht, G., Träbert, E., *Z. Physik* D **7**, 243 (1987).
44. Lindroth, E., Salomonson, S., *Phys. Rev.* A **41**, 4659 (1990).
45. Schmieder, R.W., Marrus, R., *Phys. Rev. Lett.* **25**, 1245 (1970).
46. Marrus, R., *Nucl. Instrum. Meth.* **110**, 333 (1973).
47. Schmidt, H.T., Forck, P., Grieser, M., Habs, D., Kenntner, J., Miersch, G., Repnow, R., Schramm, U., Schüssler, T., Schwalm, D., Wolf, A., *Phys. Rev. Lett.* **72**, 1616 (1994).
48. Cheng, S., Dunford, R.W., Liu, C.J., Zabransky, B.J., Livingston, A.E., Curtis, L.J., *Phys. Rev.* A **49**, 2347 (1994).
49. Marrus, R., Charles, P., Indelicato, P., de Billy, L., Tazi, C., Briand, J.P., Simionovici, A., Dietrich, D., Bosch, F., Liesen, D., *Phys. Rev.* A **39**, 3725 (1989).
50. Klaft, I., Borneis, S., Engel, T., Fricke, B., Grieser, R., Huber, G., Kühl, T.,

Marx, D., Neumann, R., Schröder, S., Seelig, P., Völker, L., *Phys. Rev. Lett.* **73**, 2425 (1994).
51. Doerfert, J., Träbert, E., Granzow, J., Wolf, A., Kenntner, J., Habs, D., Grieser, M., Schüssler, Th., Schramm, U., Forck, P., *Nucl. Instrum. Meth. Phys. Res.* B **98**, 53 (1995).
52. Doerfert, J., Träbert, E., Granzow, J., Wolf, A., Uwira, O., (work in progress 1996).
53. Kay, L., Sadler, D.A., *Physica Scripta* **51**, 459 (1995)
54. Samson, J.A.R.: *Techniques of Vacuum Ultraviolet Spectroscopy*, New York: Wiley 1967.
55. Demarest, J.A., Watson, R.L., *Physica Scripta* **38**, 670 (1988).
56. Martin, C., Jelinsky, P., Lampton, M., Malina, R.F., Anger, H.O., *Rev. Sci. Instrum.* **52**, 1067 (1981).
57. Mowat, J.R., *IEEE Trans.* **NS-28**, 1563 (1981).
58. Barbee, Jr., T.W., Mrowka, S., Hettrick, M.C., *Appl. Opt.* **24**, 883 (1985).
59. Kozhevnikov, I.V., Vinogradov, A.V., *Physica Scripta* T **17**, 137 (1987).
60. Barbee, jr., T.W., Bixler, J.V., Dietrich, D.D., *Physica Scripta* **41**, 740 (1990).
61. Bruner, M.E., Haisch, B.M., Brown, W.A., Acton, L.W., Underwood, J.H., *J. Phys. (Paris) Colloque* **49**, C1-115 (1988).
62. Catura, R.C., *Physica Scripta* T **31**, 208 (1990).
63. Träbert, E., Suleiman, J., Cheng, S., Berry, H.G., Dunford, R.W., Kanter, E.P., Kurtz, C., Livingston, A.E., Kukla, K.W., Serpa, F.G., Curtis, L.J., *Phys. Rev.* A **47**, 3805 (1993).
64. Kraus, B., Schartner, K.-H., Folkmann, F., Livingston, A.E., Mokler, P.H., *Proc. SPIE* **1159**, 217 (1989).
65. Dunford, R.W., Berry, H.G., Cheng, S., Kanter, E.P., Kurtz, C., Zabransky, B.J., Livingston, A.E., Curtis, L.J., *Phys. Rev.* A **48**, 1929 (1993).
66. Curtis, L.J., Berry, H.G., Bromander, J., *Phys. Lett.* A **34**, 169 (1971).
67. Engström, L., Jupén, C., Denne, B., Huldt, S., Meng, W.T., Kaijser, P., Ekberg, J.O., Litzén, U., Martinson, I., *Phys. Scripta* **22**, 570 (1981).
68. Becker, B., Andrä, H.J., *Nucl. Instrum. Meth.* B **9**, 650 (1985).
69. Schmieder, R.W., *Rev. Sci. Instrum.* **45**, 687 (1974).
70. Varghese, S.L., Cocke, C.L., Curnutte, B., *Phys. Rev.* A **14**, 1729 (1976).
71. Schweppe, J., Belkacem, A., Blumenfeld, L., Claytor, N., Feinberg, B., Gould, H., Kostroun, V.E., Levy, L., Misawa, S., Mowat, J.R., Prior, M.H., *Phys. Rev. Lett.* **66**, 1434 (1991).
72. Rødbro, M., Bruch, R., Bisgaard, P., *J. Phys. B: At. Mol. Phys.* **12**, 2413 (1979).
73. Heidenhain, Traunreuth (Germany) is one of the manufacturers. After drilling a venting hole into the gauge head box, their devices work inside high vacuum.
74. Garnir, H.P. (unpublished work, Liège 1995.)
75. Träbert, E., Winter, H., Heckmann, P.H., Buttlar, H.v., *Nucl. Instrum. Meth.* **135**, 353 (1976).
76. Baudinet-Robinet, Y., Garnir, H.P., Dumont, P.D., Livingston, A.E., *Physica Scripta* **14**, 224 (1976).

77. Engström, L., Bengtsson, P., *Phys. Scripta* **43**, 480 (1991).
78. Andersen, H.H., Ziegler, J.F.: *The Stopping and Ranges of Ions in Matter*, New York: Pergamon Press 1977.
79. Hubert, F., Bimbot, R., Gauvin, H., *At. Data Nucl. Data Tables* **46**, 1 (1990).
80. Astner, G., Curtis, L.J., Liljeby, L., Mannervik, S., Martinson, I., *Z. Physik* A **279**, 1 (1976).
81. Fossan, D.B., Warburton, E.K.: *Lifetime Measurements*, in J. Cerny (Ed.): *Nuclear Spectroscopy and Reactions, Part C*, p. 307(374, New York: Academic Press 1974.
82. Curtis, L.J., *Am. J. Phys.* **36**, 1123 (1968).
83. Provencher, S.W., *J. Chem. Phys.* **64**, 2772 (1976).
84. Engström, L., *Nucl. Instrum. Meth.* **202**, 369 (1982).
85. Träbert, E., Doerfert, J., Granzow, J., Büttner, R., Brauckhoff, J., Nicolai, M., Schartner, K.-H., Folkmann, F., Mokler, P.H., *Phys. Lett.* A **202**, 91 (1995).
86. Träbert, E., Doerfert, J., Granzow, J., Büttner, R., Staude, U., Schartner, K.-H., Rymuza, P., Mokler, P.H., Engström, L., Hutton, R., *Phys. Lett.* A **188**, 355 (1994).
87. Träbert, E., Büttner, R., Schartner, K.-H., Mokler, P.H., (unpublished work, Bochum / Giessen / Darmstadt 1994).
88. Fleming, J., Hibbert, A., Stafford, R.P., *Physica Scripta* **49**, 316 (1994).
89. Träbert, E., Pinnington, E.H., Kernahan, J.A., Doerfert, J., Granzow, J., Heckmann, P.H., Hutton, R., *J. Phys. B: At. Mol. Opt. Phys.* **29**, (1996). (in print)
90. Träbert, E., Doerfert, J., Granzow, J., Büttner, R., Staude, U., Schartner, K.-H., Rymuza, P., Engström, L., Hutton, R., *Z. Physik* D **32**, 295 (1995).
91. Edlén, B., *Atomic Spectra*, in S. Flügge (Ed.): *Handbuch der Physik*, Bd. XXVII, Heidelberg: Springer 1964.
92. Maniak, S.T., Curtis, L.J., Irving, R.E., Martinson, I., Hellborg, R., *Physics Letters* A **182** 114 (1993).
93. Träbert, E., *Nucl. Instrum. Meth. Phys. Res.* B **23**, 287 (1987).

CHAPTER 18

Laser Excitation of Fast Atomic Beams for Precision Lifetime Measurements

C. E. Tanner*

Precision measurements of excited-state lifetimes play important roles in many areas of physics. Measurements of atomic lifetimes are used to determine fundamental interaction parameters, test atomic and molecular theory, and place constraints on models of stellar evolution and big-bang nucleosynthesis. There are a variety of techniques for determining atomic lifetimes, however, this chapter focuses on measurements by laser excitation of fast atomic beams. The unique features of several experiments are discussed.

18.1 INTRODUCTION

Throughout many fields of research there are a variety of motivations and techniques for measuring atomic and molecular lifetimes. Lifetime measurements in general provide fundamental information about interactions and/or initial and final state wave functions. A number of techniques for measuring lifetimes are described in the excellent review article by Imhof and Read [1]. Many of these techniques can be applied to both atoms and molecules. This chapter specifically focuses on measurements of atomic lifetimes by laser excitation of fast atomic beams. A number of very precise measurements have been made using this technique, and these measurements have important implications for atomic theory. The relevant formulas for comparing experimental and theoretical results can be found in Reference [2].

*) Department of Physics, University of Notre Dame, Notre Dame, IN 46556, USA

18.1.1 Motivations

Lifetimes of neutral atoms and ions play an important role in the interpretation of data from many areas of physics. In astrophysics, accurate oscillator strengths (or f-values) are needed to interpret interstellar and stellar spectroscopic data. Elemental abundances are determined relative to that of hydrogen by observing stellar emission spectra, and the data are interpreted using oscillator strengths. Often relative values of oscillator strengths are measured precisely within a particular atomic species, but absolute values are needed. One of the most accurate methods for normalizing the relative measurements uses precisely measured atomic lifetimes [3]. For example in recent studies of the Hyades cluster, the authors emphasize the importance of lithium in models of stellar evolution, in determining primordial abundances, and in understanding big-bang nucleosynthesis [4]. In this work, the lithium 2p $^2P_{1/2,3/2}$ state lifetimes are used to interpret the stellar spectra. Many singly and doubly ionized species are also of great astrophysical interest. The applications of lifetimes to the interpretation of data in plasma research and geophysics are similar. Various schemes have been developed to calculate oscillator strengths to meet these needs. However, precision measurements are required to test the accuracy of atomic theory.

Other motivations come from atomic physics. The interpretation of parity nonconservation (PNC) experiments requires accurate knowledge of the associated atomic structure including transition matrix elements. Several neutral atomic systems have been studied in a variety of PNC experiments [5–10]. Precise lifetime measurements also provide information important to theoretical development in many electron systems [11,12]. The alkali and alkali-like atoms with a single electron outside of a closed shell represent the simplest open shell systems for evaluating the calculational capabilities of atomic theory applied to many electron systems [13]. Many experimental techniques have been applied to lifetime measurements of the low lying excited states in the neutral alkalis. However, the versatile fast-beam-laser technique can be applied to both ions and neutrals and yields lifetime results with fractions of a percent precision.

18.1.2 Exponential decay

The finite lifetime of an excited atomic state is revealed in a variety of phenomena. Most relevant to the fast beam technique is the characteristic exponential decay of the population of a single excited state. The decay of a population, N, as a function of time, t, is given by Equation 18.1, where τ is the lifetime.

$$N(t) = N(0)e^{-t/\tau} \tag{18.1}$$

A typical experiment is designed to generate a population in the excited state, and then time-resolve the decay of this population. These experiments usually involve allowed decays with lifetimes from fractions of a nanosecond to several microseconds and tend to fall into two classes.

In one class of experiments, the atoms or molecules being studied are "at rest" in the laboratory, and the delayed coincidence between excitation and emission is observed. The excited state can be populated by pulsed excitation or by cascade from another excited state. Pulsed excitation may be accomplished with a variety of techniques which include pulsed lasers or pulsed electron beams. A single decay is often observed by detecting the emission of a photon, and the time delay between excitation and emission is resolved electronically. The process is repeated to obtain an exponential distribution of times. Similarly, if the state under study is populated by cascade, then the time delayed coincidence between the emission of the cascade photon and the decay photon is measured. Photomultiplier tubes (PMTs) are typically used as single-photon detectors. These time resolved techniques have some sources of uncertainty in common, such as calibration and linearity of the time scale. Calibration usually entails detailed evaluation of fast-coincidence electronics including a time-to-amplitude converter and pulse height analyzer. Extracting a decay constant from the observed coincidence spectrum may involve deconvolving the spectrum with an instrument response function. With pulsed electron excitation, the selectivity of the excited level is limited allowing cascades from unwanted levels to contribute. With laser excitation, the selectivity of the excited state is much improved, and cascading is eliminated. However, high peak-power short-pulsed lasers can introduce other problems. The high power produces scattered light leading to large signals at short times that can produce nonlinearities in the detectors. In cascade delayed coincidence experiments, signal rates are much lower, but scattered light is typically very low and can be filtered away. In either case, often the sample under study is contained in a fluorescence cell or thermal beam at densities where quenching collisions and radiation trapping may become apparent. A thermal velocity distribution can also limit the selectivity of laser excitation.

Another class of experiments involves the use of fast atomic beams. Ions are produced, accelerated into a beam, and velocity selected. The ions can be neutralized by charge exchange, if desired. The excitation takes place during passage through a foil, gas target, or laser beam. The exponential decay is observed by detecting the number of photons emitted as a function of position, and the speed is used to convert the distance scale to a time scale. In these fast-beam experiments, the time scale is determined by the precision of the position and velocity measurements. However, ion beam experiments have their own set of difficulties. Variations in ion-beam current and target excitation efficiencies are common problems. Excitation of residual gas in the vacuum system leads to an observable background. Beam-foil excitation is a widely used and versatile technique and is often the only means available to excite very highly-ionized systems (see Chapter 17. However, the nonselective excitation from a foil or gas target can induce cascades from unwanted excited states. Curtis *et al.* have reported that cascade effects can be eliminated in some cases [14]. Another disadvantage is that foil excitation can increase the angular divergence of the ion beam which can influence the light collection efficiency as a function of position. The narrow velocity distribution and low density in a fast beam are advantages that minimize unwanted collisional and radiation-trapping effects. Laser excitation of a fast beam offers additional advantages including selectivity of excitation, high signal rates due to large excited state populations, and the elimination of cascade problems. However, lasers tend to introduce a scattered light background.

In both pulsed-laser excitation and fast-beam laser experiments, the finite excitation time transforms into a spread in excitation energy leading to coherent excitation of closely spaced energy levels. The coherence introduces the possibility of quantum beats that can cause systematic shifts in the measurements. Quantum beats between coherently populated hyperfine levels can be suppressed by carefully selecting observation angles and polarizations. Residual magnetic fields cause quantum beats between magnetic sublevels, but these can be minimized by shielding or cancellation of fields and selecting observation angles and polarizations.

18.2 GENERATION OF FAST ATOMIC BEAMS

18.2.1 Ion sources

Precision measurements of atomic lifetimes are performed in many different species; therefore no one type of ion source is suitable for all needs. There are three types of sources commonly used for accelerator based atomic physics. These are the radio-frequency-discharge source, the duoplasmatron, and the hollow cathode source. This section describes the general features of these sources. A number of specific details of these and other sources are described in Ref. [15].

RF discharge ion source

The RF discharge source was originally developed [16,17] as a source of protons generated from hydrogen and is still used in many Van de Graaff accelerators. A commonly used design [18] is shown in Figure 18.1. An RF discharge ion source is usually constructed from a Pyrex vessel with a small opening at one end and a small tungsten probe at the other end. A metallic base fits over the output end and protrudes through the opening in the glass. A hole drilled into the base serves as the exit channel. The metal base functions as the cathode and the probe as the anode. Either an inert buffer gas or the gas under study fills the glass envelope at a pressure of a few Pascal. An oscillator supplies the RF energy necessary to ignite a discharge in the gas with either an inductively coupled magnetic field or a capacitively coupled electric

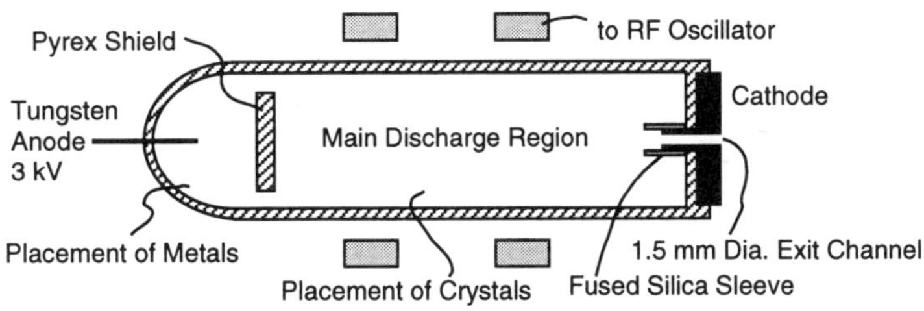

FIGURE 18.1 The RF discharge ion source is constructed from a cylindrical Pyrex tube. The size is approximately 10 cm long by 3 cm wide.

field. RF frequencies of tens to hundreds of MHz are typically used with an input power of approximately 100 Watts. An axial magnetic field can be used to confine the electrons, effectively increasing the ionization probability and reducing the RF power requirements. A potential difference of about 3 kV applied between the probe and the base accelerates the ions toward the base where some ions pass through the exit channel. The loss of ions due to surface recombination and the formation of molecules is reduced by keeping the amount of metal surface exposed to the plasma to a minimum. A Pyrex shield or sidearm hides the metal anode from the discharge region, and a fused-silica sleeve over the cathode reduces the exposure of the metal. Secondary electrons, which are created when ions strike the cathode, accelerate toward the probe end; therefore the anode region of the tube may require cooling to prevent excessive heating.

A simple modification to the RF source makes it possible to produce selected metallic ions [19]. The exit channel is lined with the material to be studied, and ions from a buffer gas sputter material from the exit channel lining. The buffer gas is often an inert gas such as argon or neon. The sputtering rate can be enhanced by taking advantage of chemical interactions. In Reference [20], the authors mixed oxygen with the buffer gas thereby oxidizing the iron exit channel and increasing the sputtering rate of iron ions compared to the buffer gas alone. Other non-gaseous materials can be studied [21] by placing dry solids into the ion source bottle where they are heated by the discharge, or by placing metals into the anode region which is away from the main discharge.

Duoplasmatron ion source

The duoplasmatron ion source [22] in Figure 18.2 incorporates a variety of different techniques. A heated filament functions as the cathode which produces electrons by thermionic emission. An advantage of thermionic emission is that the voltage required to ignite a discharge is considerably reduced from

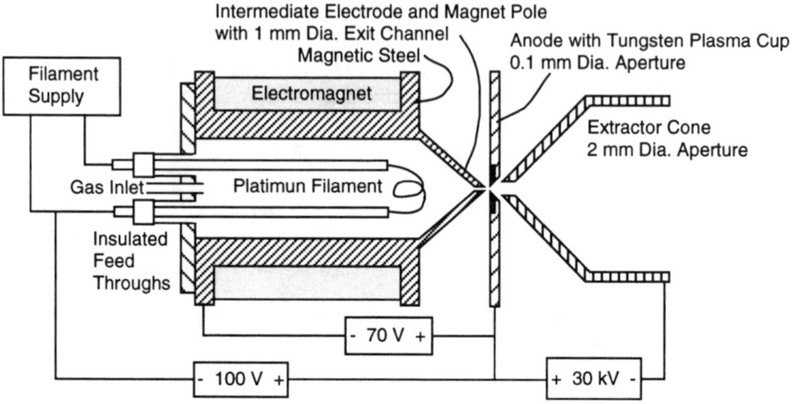

FIGURE 18.2 The duoplasmatron ion source uses a heated filament for efficient electron emission. The conical intermediate electrode increases the plasma density by compression, and the axial magnetic field supresses diffusion of the electrons to the walls. The inner dimensions are about 8 cm long by 3 cm in diameter.

10-50 kV for a cold cathode to about 300 V for a heated cathode. However, the lower voltages tend to reduce the ion density in the plasma thus limiting the extractable current. The design of the duoplasmatron ion source addresses this limitation in two ways. First, the conical shape of the intermediate electrode narrows the plasma and increases the ion concentration near the extraction channel. Second, a magnetic field along the axis of the discharge forces the electrons to move in helical orbits. Diffusion of the electrons toward the walls is reduced and the path length of the electrons in the discharge is increased. Thus, the electrons have a greater probability of colliding with and ionizing a gas particle which tends to increase the ion concentration. The magnetic field only slightly influences the motion of the ions because of their greater mass. The overall effect of the magnetic field is similar to that of increasing the gas pressure in an RF discharge source. Thus, the duoplasmatron can function over a wide range of gas pressures from 0.5 to 8 Pascal where the optimum pressure depends on the desired gas.

Hollow cathode ion source

The hot filament hollow cathode ion source is now one of the most commonly used sources because of its versatility and reliability. Sidenius developed the hollow cathode ion source as a universal source for isotope separation [23–25]. A typical configuration is shown in Figure 18.3. A cylindrical electron emitter is contained within a hollow cathode. In practice, the electron emitter is a coiled tungsten filament. A cylindrical anode sits behind the cathode. The

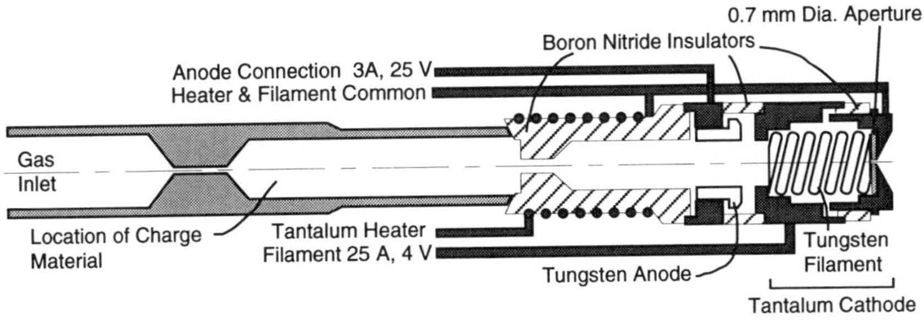

FIGURE 18.3 The hot filament hollow cathode ion source can be used with both solids and gases. The cylindrical cathode is about 1 cm long by 1 cm in diameter.

plasma column extends from the anode into the cylindrical cathode, and is essentially a neutral mixture of electrons and positive ions. Nearly the entire anode voltage appears across the boundary between the electron emitter and the plasma column. In the presence of this high electric field, the electrons are extracted perpendicularly to the cylindrical axis of the emitter, travel through the center of the plasma, and are reflected by the field on the opposite side. Thus, the electrons are trapped within the cathode while diffusing slowly to the anode through scattering and due to the small field within the plasma. If a magnetic field is present, it will prevent the electrons from passing through the center of the plasma column which reduces the ionization efficiency. Therefore, the magnetic field due to the filament current is canceled with an electromagnet. Sidenius designed the electromagnet such that the system is insensitive to changes in filament resistance when the magnet is operated in series with the filament current. The opening in the anode is used for charging the source with the material under study. The hollow anode has the additional advantage that the discharge current is distributed over a large area which reduces wear.

18.2.2 Acceleration

Low energy ion accelerators have been used in a wide variety of research and industrial applications. There are many variations in design, but most have the same basic components. References [26,27] discuss the details of low energy accelerator design. In low energy (less than 500 kV) applications, electrostatic acceleration is usually employed. The ion source is held at the desired energy. A high voltage platform electrically isolates the ion source from the experimental apparatus which is usually at earth ground. For the safety of personnel and sensitive detection equipment, a grounded cage surrounds the high voltage platform. An example of a typical system is the Atomic Physics Accelerator Laboratory at Notre Dame (APALaND) shown in Figure 18.4.

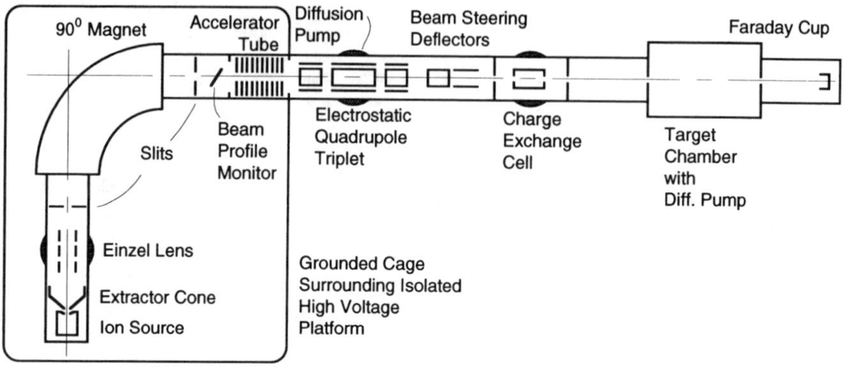

FIGURE 18.4 The Atomic Physics Accelerator Laboratory at Notre Dame houses a 10-200 kV ion accelerator with a hollow cathode ion source.

A hollow cathode ion source is the first stage of the system. An extractor cone is located just outside of the ion source and held at a potential of about −30 kV relative to the ion source. Next an *einzel* lens collimates the ions into a beam. The fast ion beam then passes through a 90 degree bending magnet with an entrance and exit slit for charge state selection. The ion beam is focused at the entrance of the bending magnet and is refocused at the exit slit after passing through the magnet. The final acceleration is accomplished with a stack of ring electrodes each attached to one segment of a precision voltage divider. The ion beam leaves the accelerator stack at ground potential. During acceleration, the ions can interact with background gases or strike the electrodes producing electrons which are potentially problematic. Care must be taken to prevent the electrons from being accelerated backwards through the accelerator stack because the back streaming electrons can produce X rays when they strike the electrodes or walls of the vacuum system. To prevent this problem, an electron trap is used to extract the electrons from the accelerator. In some systems, the final acceleration is performed before the analyzing magnet. This configuration is potentially more dangerous because the ion beam load on the accelerator stack is higher, thus increasing the potential for an X-ray hazard.

After the final acceleration, the ion beam usually passes through focusing elements resulting in a typical beam diameter of a few millimeters. Quadrupole magnets are often used to focus the ion beam into the experimental apparatus. The accelerator in Figure 18.4 uses electrostatic quadrupoles for focusing, and vertical and horizontal deflection plates to steer the ion beam through the experimental chamber.

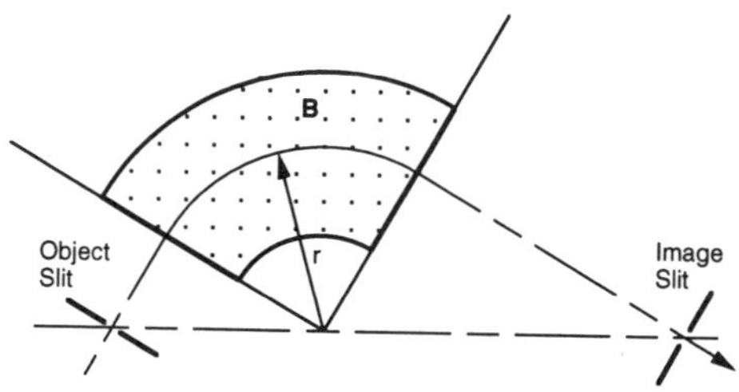

FIGURE 18.5 The magnetic sector analyzer images the entrance slit to the point which lies on a line through the center of the sector. Often the angle is 90 degrees and the slits are placed symmetrically before and after the magnet.

18.2.3 Charge state selection

A magnetic sector analyzer (Fig. 18.5) performs the charge state and mass selection as in a mass spectrometer. The main component is a DC electromagnet that produces a uniform magnetic field for deflecting the charged particles perpendicular to their momentum. Entrance and exit slits are used to restrict the main trajectory through the magnet to a particular radius of curvature, and the potential difference between the source and the magnet equals the energy per unit charge of the particles in the magnetic field region. Thus in the nonrelativistic limit, the charge-to-mass ratio of the transmitted particles is given by Equation 18.2

$$\frac{q}{m} = \frac{2V}{r^2 B^2} \tag{18.2}$$

where V is the potential difference between the ion source and the magnet, r is the radius of the main trajectory, and B is the magnetic field in the center of the magnet.

The sector magnet is also referred to as being double focusing. In the plane of the main trajectory (Fig. 18.5), focusing is easily understood by noting that the particles following a path slightly longer than the main trajectory spend more time in the magnetic field and are deflected through a slightly larger angle. Similarly, those particles following a path shorter than the main trajectory are deflected through a slightly smaller angle. In this plane, the image of the entrance slit lies on a line that passes from the entrances slit through the center of curvature of the sector to the other side of the magnet (Barber's Rule) [28]. The exit slit is usually placed at the location of this image. Focusing in the direction perpendicular to the plane of the main trajectory also occurs. This effect is due to the fringing fields of the magnet and occurs only for those trajectories that enter the magnet at an angle to and off the axis of the main trajectory. Details of charged particle optics can be found in References [26,27].

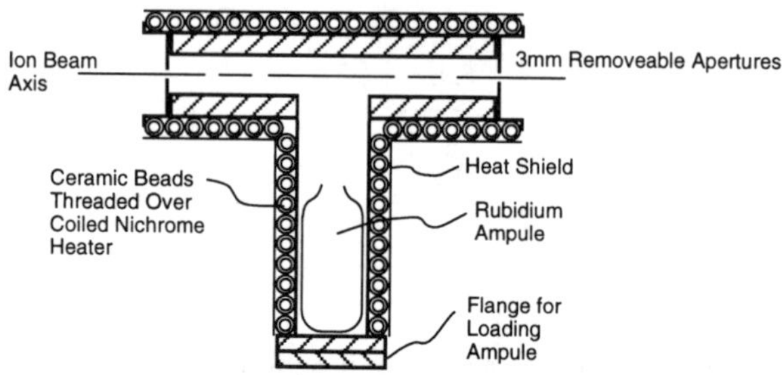

FIGURE 18.6 This simple charge exchange cell produces Rb vapor for near resonant neutralization of a 50 keV CsII beam. The interaction length is 7.5 cm.

18.2.4 Neutralization of ions

Many fast beam experiments are performed on neutral atoms. In these cases, the accelerated positive ions can be neutralized by passing the fast ions through a gas target. The interaction between a fast ion and an atom in the gas can result in an exchange of an electron from the gas atom to the fast ion. The fast ion emerges as a fast neutral atom and the formerly neutral atom is ionized. A near resonant charge exchange cross section is of order 10^{-16} to 10^{-15} cm^2 at an energy of about 50 keV [29]. In our work [30,31], a fast CsII ion beam is neutralized by charge exchange with Rb vapor.

$$\mathrm{CsII}[+50\ \mathrm{keV}] + \mathrm{Rb} \rightarrow \mathrm{CsI}[+50\ \mathrm{keV}] - 0.28\mathrm{eV} + \mathrm{RbII} \tag{18.3}$$

Although this process takes place between two different atoms, the interaction is resonantly enhanced because the ionization potentials of the two neutral species differ by only 0.28 eV. Figure 18.6 shows a charge exchange cell constructed for our lifetime measurements of the 6p $^2P_{1/2,3/2}$ states of neutral cesium. In this experiment, the 50 keV CsII ion beam was more than 90% neutralized over the cell length of 7.5 cm where the density of rubidium was about 10^{14} $/\mathrm{cm}^3$ at a temperature of 340 K.

18.3 LASER EXCITATION SCHEMES

Each of the experiments discussed in this section demonstrates a unique application of the use of laser excitation in measurements of atomic lifetimes. The details depend upon the atomic system under study, but each experiment benefits from the selectivity of resonant laser excitation. The simplest configuration is single-photon laser excitation from the ground state to the desired excited state. In some examples, the desired state may not be accessible from the ground state; therefore a combination of excitation schemes is employed.

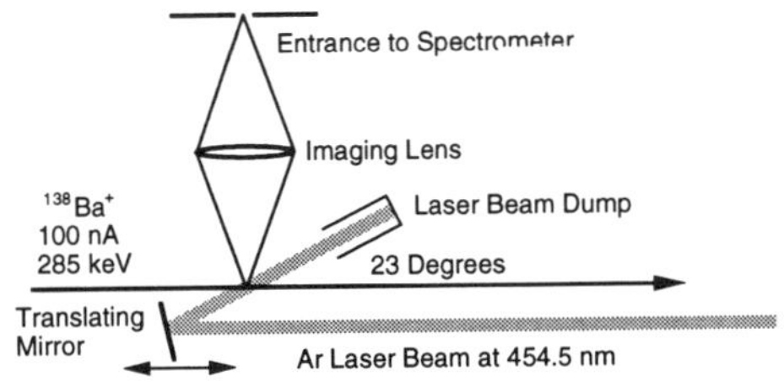

FIGURE 18.7 The first fast-beam laser lifetime experiment utilized a stationary detector system and a foil translator to move a laser mirror under vacuum.

By combining resonant laser excitation with excitation by a foil or a gas target, highly excited states can be studied. The selective advantage is retained when the laser is used for the last step in the excitation process.

18.3.1 Crossed laser-atomic beam excitation

Precision lifetimes in BaII, Li, and Na by Andrä et al.

Andrä *et al.* introduced laser excitation of fast beams as a new technique for measuring atomic lifetimes with their report of the BaII 6p $^2P_{3/2}$ state lifetime in 1973 [32]. In this work, the authors took advantage of the properties of fast beams that were discovered through the development of beam foil spectroscopy (for a more detailed discussion, see the preceding chapter by E. Träbert) and coupled these with the emerging technology of lasers. The authors noted that in a fast beam, the emission is unaffected by collisions, self–absorption, and induced emission because of the low relative velocities, densities, and pressures found in a typical ion beam apparatus.

Taking advantage of the Doppler effect, an argon ion laser (454.5 nm) was angularly tuned into resonance with the BaII 6s $^2S_{1/2}\rightarrow$6p $^2P_{3/2}$ transition at 455.4 nm. The argon ion laser entered the vacuum system parallel to the ion beam direction and reflected from a mirror that was held fixed at the required angle. A position resolution of 0.4 mm was achieved by mechanically translating the mirror along the beam direction with the mechanism that was usually employed to move a foil (Fig. 18.7). In this arrangement, the collection optics remained fixed. A lens collected light emitted at an angle of approximately 90 degrees to the ion beam direction and imaged it at the entrance slit of a spectrometer. With a resolution of 0.6 nm, the fluorescent light at 90 degrees was effectively filtered from the scattered laser light. At constant laser power, the fluorescence signal was normalized by integrating the ion current to a preset value with a Faraday cup.

The data consisted of photon counts as a function of mirror position. The results exhibited the functional form of a single exponential with a flat background convoluted with excitation and detection functions. Starting at positions well after the excitation terminated, the data was fit to a single exponential plus a constant background. This procedure eliminated the necessity of knowing the excitation function. The exponential decay constant was turned into a lifetime using the velocity as inferred from the accelerator voltage. Further development of the fast-beam laser technique led to the precise result of 6.312(16) ns for the BaII 6p $^2P_{3/2}$ state lifetime [34].

Fractions of a percent precision were also reported for the lowest lying excited states of Li 2p $^2P_{1/2}$ and Na 3p $^2P_{1/2}$ [35]. In this experiment, a singly charged fast beam was neutralized in a gas cell. A tunable dye laser aligned perpendicularly excited the atoms at a fixed position. A movable detector was mounted on a translation stage in the vacuum system, and monitored the decay as a function of position. Rectangular apertures defined the solid angle viewed by the detector determining the spatial resolution along the atomic beam. A second detector monitored the interaction region, and this signal was used for normalization. Baffles around the laser entrance and exit windows prevented most of the scattered laser light from reaching the detectors. The authors carefully controlled and measured backgrounds due to stray light, collisional excitation of the atomic beam, and detector dark counts. An electrostatic energy analyzer was used to determine the energy of the ions from which the energy of the neutral atoms was inferred. The results of this work placed the lifetimes of the Li 2p $^2P_{1/2}$ and Na 3p $^2P_{1/2}$ states at 27.29(4) ns and 16.40(3) ns respectively. For a number of years, this work has set the standard of precision for subsequent lifetime measurements. However, atomic theory and recent measurements in these atoms have shown that the accuracy of these results may have been influenced by systematic errors.

Pulsed-laser excitation by Pinnington et al.

The transition wavelengths for many ionic systems of astrophysical interest are too short for excitation by most CW lasers. Beam-foil excitation may be the only way to excite some levels, but with nonselective excitation cascading severely limits the precision. To avoid this problem, Pinnington *et al.*, explore the many possibilities of high power pulsed-laser excitation. In experiments on the CaII 4p levels [36], the ion beam was transversely excited from the 4s ground state with an excimer-pumped dye laser. The laser typically produced light near 395 nm at a power density of 200 kW/cm^2 in a 5 ns pulse at a repetition rate of 200 Hz. The measurement technique went through various stages of development and finally achieved an ultimate precision of 1%. One important feature of the experiment was a target chamber designed specifically to minimize the scattered light level. Another important feature included monitoring a branch transition to the 3d levels with narrow-bandpass filters which provided additional discrimination against scattered light. This approach required a PMT with high quantum efficiency at 850 nm in the near infrared. This combination of techniques resulted in a signal to background ratio of 1000 to 1 at the peak of the decay curve.

Measurements of the MgII 3p state lifetimes with precisions of 1% were performed using frequency doubled laser radiation [37]. The 3s-3p transition was used to populate the excited state with 280 nm light intersecting the ion beam at 45 degrees. The 280 nm light was generated with a frequency doubled excimer-pumped dye laser. In this work, the authors also explored two-step and two-photon excitation of MgII 3d levels. The two-step excitation 3s $^2S_{1/2}$→3p $^2P_{1/2}$→3d $^2D_{3/2}$ was accomplished with the laser intersecting the beam at 45 degrees then retro-reflecting the laser for a second pass. With the accelerator voltage tuned appropriately, the first step was Doppler tuned into resonance on the first pass and the second step on the second pass. The authors were also able to demonstrate two photon excitation of the 3d $^2D_{3/2}$ level from the 3s $^2S_{1/2}$ ground state.

Pinnington *et al.* also demonstrated laser excitation of ions from metastable states which are produced in the ion source [38]. This technique extends the fast–beam laser technique to states that would not normally be accessible by laser excitation from the ground state. Lifetimes of the 4f and 5p levels of SrII and the 4f, 6p, and 7p levels of BaII have been recently measured using this approach [39]. In each case, the lowest nd ^2D states were excited with a single photon to the desired excited state, and the basic experimental procedure was similar to earlier measurements.

Multi-step excitation

Demtröder *et al.* also used a two-step excitation process that included short lived intermediate states by combining a gas cell with laser excitation [40]. The fast ion beam passed through a differentially pumped gas cell. Collisions with the target gas produced highly excited atomic states of both ions and neutrals. Shortly after the gas cell, a laser aligned perpendicularly to the atomic beam was tuned to resonantly drive optically allowed transitions between two excited states. Fluorescence of the upper level was observed as a function of position after the laser excitation. This technique was first demonstrated with a 0.9% measurement of the Li 3d ^2D state lifetime. Schmoranzer *et al.* have made several important additions and improvements to this apparatus which are thoroughly described in Reference [41]. A unique improvement to this apparatus is that the ion beam now passes through a CW laser resonator. The CW dye laser is pumped with an Ar ion laser and tuned to the appropriate wavelength with a three plate birefringent filter. An advantage of this scheme is that the intra-cavity power is typically several Watts, many times higher than the power available through the output coupler of a typical dye laser. Dumont *et al.* also use intra-cavity laser excitation in combination with beam-foil excitation to measure lifetimes in neutral hydrogen [42].

Precision lifetimes in neutral Na by Volz et al.

Recent lifetime measurements of the Na 3p $^2P_{1/2,3/2}$ states [43] seem to resolve a long standing discrepancy between experiment and theory [44]. *Ab initio* theoretical calculations of the 3s-3p transition matrix elements in neutral sodium differ from the precise experimental results reported by Gaupp *et al.* in 1982 [35]. New measurements of the Na 3p lifetimes employ the apparatus developed by Schmoranzer *et al.* for combined beam-gas laser excitation.

The gas target in the Na experiment provided 30% neutralization of a 170 keV NaII ion beam. A linearly polarized CW dye laser excited the neutral Na as the beam passed perpendicularly through the resonator cavity. The laser was tuned to one of 3s-3p resonance transitions. A translating detector system measured the emitted fluorescence as a function of distance from the interaction region. The translating detector system consisted of a conical arrangement of optical fiber bundles which collected and transmitted the fluorescence to a monochromator and cooled PMT. A similar detector system located closer to the interaction region remained fixed and provided a normalization signal. The fluorescent light was collected at angles other than 90 degrees so that the Doppler shift could be used to discriminate against laser scattered light. Both the laser and atomic beam were chopped on and off to provide additional background subtraction.

The Na data consisted of 21 position scans with 30 equally spaced points for each fine structure state. The total translation along the atomic beam was 58 mm. At the first data point, the count rate was about 100,000 counts/s and corrections due to detector saturation were small. The exponential decay as a function of position was converted to a lifetime with the beam velocity. The velocity was determined from the kinetic energy of the ion beam. The accelerator voltage was measured with a calibrated voltage divider, and an electrostatic energy analyzer was used to characterize the dependence of the ion kinetic energy on the source conditions. A small negative shift between the accelerating voltage and the ion kinetic energy was measured. The energy loss due the charge exchange process was also measured by introducing a second gas cell to re-ionize the neutral atoms. Corrections to the accelerating potential were made to account for both of these shifts resulting in a velocity precision of 0.04%.

Various systematic effects were also considered most of which contributed less than or equal to 0.05% each to the uncertainty. One effect was measured to have the largest influence on the decay: the divergence of the atomic beam along length of the translation affected the measurement through the solid angle of the fiber optic system, because the collection efficiency decreases as a function of distance from the central axis. This effect was also measured and taken into account. The final results for the Na 3p $^2P_{1/2,3/2}$ state lifetimes are 16.299(21) ns and 16.254(22) ns respectively.

There are also recent measurements of the Na 3p lifetimes by other techniques. Tiemann *et al.* [45] observed the vibrational spectrum of Na_2 in a molecular beam where they were able to measure the C_3 coefficient for long range dipole-dipole interactions. Oates *et al.* [46] measured the linewidth of the 3s $^2S_{1/2}$→3p $^2P_{3/2}$ transition in a sample of laser-cooled Na atoms. All three results agree to within 0.1%.

Precision lifetimes in neutral Cs with diode laser excitation by Tanner et al.

The lifetimes of low-lying electronic states of neutral cesium are being studied because of their importance to the interpretation of parity nonconservation (PNC) in the 6s $^2S_{1/2}$→7s $^2S_{1/2}$ transition. PNC measurements [6] are approaching a sensitivity that can test the standard model of electroweak interactions at the level of radiative corrections making it possible to look for new physical effects [47]. In order to obtain precise weak interaction information from PNC experiments, the atomic radial matrix elements between low-lying states must be known to an accuracy approaching the 0.1% level.

Our recent lifetime measurement of the 6p $^2P_{3/2}$ state in atomic cesium represents the first application of the use of diode lasers in a fast beam experiment [30]. A diode laser resonantly excited the Cs 6s $^2S_{1/2}$→6p $^2P_{3/2}$ (D2) transition at 852 nm. Following a series of improvements, the systematic uncertainties were reduced to a level that is comparable with the statistical uncertainties, and the measurements were extended to include the 6p $^2P_{1/2}$ state. Current lifetime results by the fast beam laser technique for the 6p $^2P_{1/2,3/2}$ states are 34.934(94) ns and 30.499(70) ns respectively [31]. The precision of these results test many-body calculations of the 6s $^2S_{1/2}$→6p $^2P_{1/2}$ D1 and 6s $^2S_{1/2}$→6p $^2P_{3/2}$ D2 transition matrix elements to better than 15 parts in 10^4.

Our technique employed a fast atomic beam crossed with an excitation laser-beam and subsequent detection of the decay in flight of the atomic beam fluorescence. CsII ions were accelerated through a potential of 50 kV, velocity selected, and neutralized in a rubidium charge exchange cell. Light from a diode laser was directed perpendicularly to the collimated fast beam of cesium atoms and tuned to populate the desired state. The decay of fluorescence was monitored with a translatable fiber optic collection system that views light emitted at 90 degrees to the atomic beam. A similar stationary detector system monitored the beginning of the decay for normalization. (See Figure 18.8.) Narrow-bandwidth interference filters suppressed non-resonant light background, and baffles physically shielded the fiber inputs from scattered light. Detector dark counts were minimized by lowering the temperature of the photomultiplier tubes. Atomic beam-induced scattered light was measured by using a laser beam shutter to collect both fluorescence and background data. The atom beam velocity was determined by measuring the Doppler shift of the 6s $^2S_{1/2}$→6p $^2P_{3/2}$ transition when excited collinearly.

In our measurements [31], the data consisted of about 30 position scans

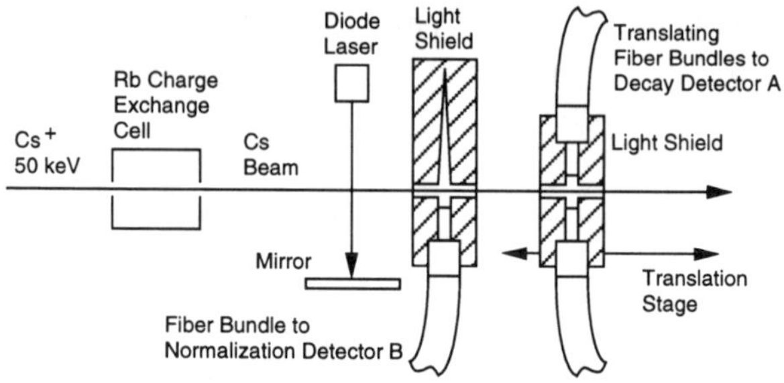

FIGURE 18.8 The apparatus for measuring Cs lifetimes employed fiber optic bundles to collect and couple the fluorescent light through the vacuum chamber to cooled PMTs. With perpendicular alignment, the diode laser excited the atoms on both the first pass and the second pass after reflection.

for each fine structure component monitored at 31 separate positions over about 5 decay lengths. The data acquisition sequence included laser-on and laser-blocked pairs of photon-count signals in both the translating and the normalization detectors. These signals were accumulated at each point of the decay curve for approximately ten seconds. Diode lasers have an output power that is stable to better than 1 part in 10^3, thus it was not necessary to monitor the laser power in this experiment. Decay data were acquired for both directions of translation. Following six pairs of decay measurements, a pair of control runs was acquired with the atomic beam blocked in order to verify the stability of the scattered laser light and detector dark counts. Detector dark counts and scattered laser light contributions were typically less than 100 Hz. Scattered light produced by the atomic beam was less than 2 kHz. For each detector, counts due to all backgrounds were recorded during data acquisition, and their contributions to each signal were subtracted during analysis. The fluorescent signal rate for the D2 transition at 852 nm was 70 kHz. For the D1 transition at 895 nm the available diode laser power was only about 15% of that for the 852 nm laser, and the detector quantum efficiency was about 1/3 of that at the lower wavelength, resulting in fluorescence rates of about 4 kHz. The background-corrected and normalized decay data were fit to the function of a single exponential plus constant background. Nearly all the decay fits resulted in backgrounds consistent with zero, as expected from the data acquisition and treatment scheme. The reproducibility (standard deviation of the mean) of the fitted lifetimes was 0.14% for the 6p $^2P_{3/2}$ state and 0.22% for the 6p $^2P_{1/2}$ state. The estimated systematic uncertainties are 0.05% due to solid angle, 0.14% in atomic beam velocity, and less than 0.05% due to mechanical nonlinearities, temperature, and pressure effects. Residual quantum beat effects in the decays contributed a negligible uncertainty for the 6p $^2P_{1/2}$ state and less than 0.1% for 6p $^2P_{3/2}$. The sum in quadrature of the uncertainties yields 0.23% for the 6p $^2P_{3/2}$ state and 0.27% for the 6p $^2P_{1/2}$ state.

18.3.2 Longitudinal excitation

Most of the fast beam lifetime experiments discussed in this chapter were performed by observing the fluorescence decay in flight after excitation of the atomic beam by a crossed laser beam. Gaillard *et al.* introduced an alternative approach called rapid Doppler switching where the laser and ion beam are aligned collinearly [48]. This technique takes advantage of the velocity bunching that occurs along the accelerating direction which results in narrow resonance lines. A homogeneous electric field is applied along the beam direction to either accelerate or decelerate the ions. At each position in this region, the ions travel at a different velocities. By tuning the frequency, the laser can be brought into resonance with the ions at different locations within the electric field region. With a detector fixed at one position, tuning the laser is equivalent to moving the excitation point along the ion beam direction. Gaillard *et al.* demonstrated this technique by measuring the BaII 6p $^2P_{3/2}$ state lifetime which compared favorably with the results of Reference [34].

Precision lifetimes in ArII and CaII by Jin and Church

Jin and Church implemented the velocity switching technique in a slightly different way for their precision measurements in ArII and CaII [49,50]. In these experiments the laser frequency was held fixed, and the incident velocity of the ions was adjusted before they entered the additional uniform electric field region. (See Figure 18.9.) The fluorescence detector viewed a narrow length of the ion beam at a fixed position. For different incident velocities, the ions were in resonance with the fixed laser at different positions resulting in different time intervals between excitation and detection. The fluorescence signal was recorded as a function of accelerator voltage. The conversion from accelerating potential to time is given by Equation 18.4

$$t_1 - t_2 = \frac{(2m/qe)^{1/2}}{E}(U_1^{1/2} - U_2^{1/2}) \tag{18.4}$$

18.4 DETECTION

Techniques of detection vary from one experimental group to the next. There is a great deal of variety in the design of the collection optics and in obtaining position sensitivity. Most experimenters employ PMTs as the final stage of photon detection. However, there are some important differences here as well.

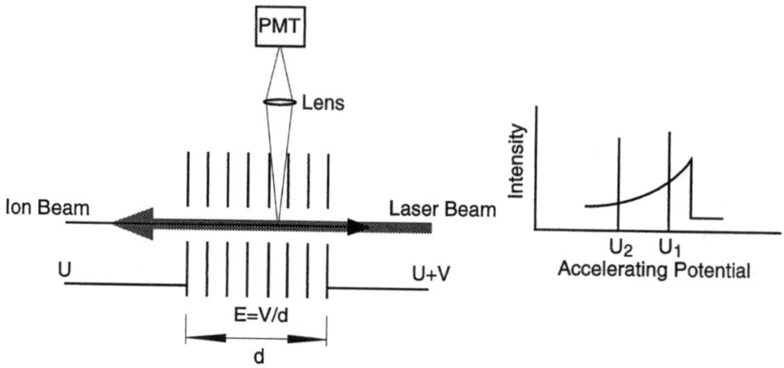

FIGURE 18.9 The diagram shows the main features of the velocity switching technique. The laser frequency is held fixed, and the ions have a different velocity at each position in the uniform electric field region. A change in the accelerating potential changes the incident beam velocity which brings the ions into resonance with the laser at a different position relative to the observation region.

18.4.1 Position sensitivity and collection optics

In fast–beam lifetime experiments, the time resolution is achieved by either translating the point of excitation or the point of detection. Each of these approaches has a different criterion for the design of collection optics. Imaging lenses efficiently collect light from a very localized position both transversely and longitudinally relative to the atomic beam direction. This type of system is very sensitive to changes in collection efficiency resulting from misalignment of the translation system, motion of the ion beam, or the divergence of the ion beam. An imaging system is better suited to applications where the detection system remains fixed and the point of excitation is moved [49,32].

Translating the detection system is the other alternative. Usually, a stepper motor drives a precision lead screw on a translation stage that moves the collection optics in the vacuum system. To avoid systematic changes in collection solid angle, the optical system views the greatest solid angle possible without picking up too much scattered light. In the later work of Andrä [34,35], a movable PMT was placed in the vacuum system with the photocathode viewing the atomic beam at 90 degrees. Light baffles limited the angular view along the beam direction to provide adequate position resolution. Transverse to the atomic beam direction, the acceptance angle was larger so that the collection efficiency was insensitive to alignment between the translation stage and the beam direction.

Demtröder *et al.* used a conically shaped fiber optic bundle to collect the fluorescent light [40]. At the other end of the bundle, the fiber ends were arranged into a narrow rectangle that uniformly illuminated the entrance slit to a spectrometer. The bundles viewed a 2 mm length of beam at an angle of 30 degrees from the axis. Since the fibers viewed the entire 360 degrees of azimuthal angle, systematic changes in collection solid angle due to misalignment of the translation stage only contributed to second order. Another advantage to this system is that the fluorescence emitted at 30 degrees was Doppler shifted relative to the scattered laser light which could then be filtered away with the spectrometer. Later Schmoranzer *et al.* modified this technique by making the half-angle of the cone 54.7 degrees [51]. This 'magic angle' suppresses quantum beats and polarization effects due to laser induced alignment of the internal atomic states.

For our measurements in Cs, the collection optics consisted of a pair of fiber optic bundles that viewed the atomic beam from opposite sides (Fig. 8). This configuration reduced systematic uncertainties due to possible misalignments between the translation of the fiber inputs and the atomic beam direction. A similar single fiber optic bundle was used to provide a normalization signal. The fibers collected light within an acceptance angle of 60 degrees. Each bundle contained thousands of fibers. The diameter of each bundle was about 5 mm, approximately twice that of the atomic beam. The bundles were positioned about 1 1/2 times their diameter from the beam center. This configuration provided a compromise between maximizing the collection of fluorscence and minimizing the collection of scattered laser light. The fiber-optic bundles were coupled through the vacuum wall and into cooled PMT housings.

18.4.2 Detectors

In most fast beam lifetime experiments, PMTs are used in a single-photon-counting mode. A counting channel usually consists of a PMT coupled to a preamp, amplifier, and discriminator. PMTs with low dark currents are most desirable. Dark count rates of order 10 counts per second are easily achieved with high quality tubes in refrigerated housings. Commercially available housings using thermoelectric coolers can typically keep the detectors at a temperature of −30 C. With single-photon counting, the maximum signal must be limited to count rates of less than 1 MHz to avoid uncertainties due to pulse-pile-up corrections. This problem is especially significant when pulsed lasers are used because the intensity of each pulse can be very high, of order tens of kW/cm^2, in a pulse of width less than 10 ns. With such high intensities, there can be a large number of scattered photons reaching the detectors within a short time interval making it necessary to apply dead time corrections. In a typical pulsed laser experiment, the overall count rate in the PMTs must be less than 1 photon/pulse when using single-photon counting.

An alternative approach was taken by Pinnington *et al.* as described in Reference [37]. The output of the PMT used for detecting the laser induced fluorescence was recorded using gated charge integration. With this approach, a wide dynamic range was achieved. For example, the authors recorded signals over the range from 10 photons/pulse down to 0.03 photons/pulse. For comparison, their uncertainty due to pulse-pile-up corrections became significant at 0.3 photons/pulse in the single-photon counting mode. Thus the gated integration technique increased their dynamic range by a factor of 30. To perform the charge integration, the output of the PMT was passed through a conventional spectroscopy amplifier which integrates the PMT output current. The output of the amplifier was a shaped pulse with an amplitude proportional to the number of photons arriving during the time resolution of the PMT. A gated analog-to-digital converter was used to digitize the integrated signal.

18.5 VELOCITY MEASUREMENT TECHNIQUES

In several of the experiments discussed, the lifetimes are measured by observing the fluorescence decay as a function of position. Conversion of the position coordinate to time requires the measurement of the mean atomic velocity.

18.5.1 Electrostatic measurements of velocity

Ion velocities are usually measured with electrostatic techniques. The simplest approach uses the accelerator voltage to obtain the velocity of the ions. The accelerator voltage can be determined with a calibrated resistive voltage divider that provides a lower measurable voltage. This technique can be used to obtained a precision of about 0.1% in the voltage. For voltages much greater than 100 kV, a divider may not be practical. In this regime, the accelerator voltage can be calibrated with known nuclear reactions [52]. Andrä *et al.* [32] employed this technique in their early work to a precision of 1% in the voltage. However, the kinetic energy of the ions can be affected by the operating conditions of the source.

Another approach uses an electrostatic energy analyzer (Fig. 18.10). The most common geometry uses concentric cylindrical electrodes to produce a radial electric field along a 90 degree arc [53]. This type of analyzer has advantages for lifetime experiments since it is sensitive to the kinetic energy of the charged particles which could in principle differ from the accelerating voltage. The electrostatic analyzer also reduces the voltage to be measured by a factor of about 100, however the geometry of the electrodes must be known. Accurate knowledge of the geometry limits the calibration to a precision of about 0.2%. In later work [34], Andrä *et al.* take advantage of the precisely measured HeII 3p ^{3}P J=1 to 2 fine structure splitting to calibrate the electrostatic analyzer by observing quantum beats in the 3p ^{3}P decay, resulting in an ultimate precision of 0.07% in the calibration factor [35].

Neutral atom velocities can be inferred from the accelerating potential of the ions, but care must be taken to measure systematic effects. As discussed

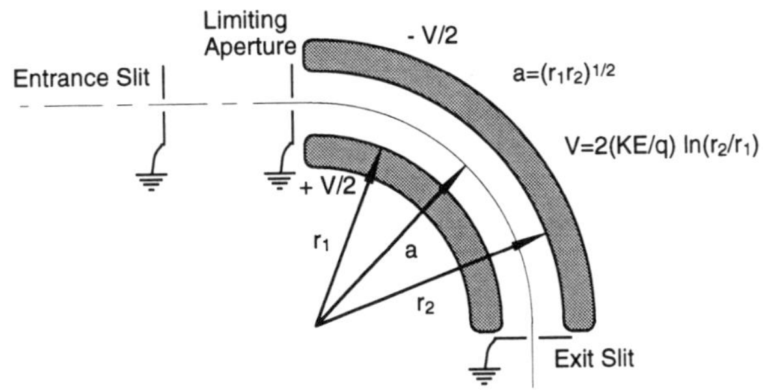

FIGURE 18.10 The electrostatic energy analyzer measures the kinetic energy of the incident ion beam. For a given ion kinetic energy, the analyzer voltage, V, is tuned to allow the ions to traverse the path of radius, a, determined by the apertures.

previously, fast neutral atomic beams are usually produced by charge exchange with a neutral gas target. The charge exchange processes can result in a reduction in the kinetic energy of the fast neutrals relative to the ions. In addition, there may be differences between the ion kinetic energy and the accelerating potential. In the recent precision lifetime measurements of Volz *et al.* [43], the authors carefully evaluated the difference between the ion beam energy and the accelerating potential with an electrostatic energy analyzer and placed an upper limit on the energy loss due to charge exchange. Thus they were able to determine the velocity of the neutralized atoms to a precision of 0.04 % from the accelerator voltage.

18.5.2 Doppler shift measurements of velocity

Another approach to measuring the velocity of fast atoms involves observing the Doppler shift of a resonance transition [54]. The technique requires an atomic state that can be optically coupled to another state where the transition energy or wavelength is known. A single frequency tunable laser is best suited to drive the resonance, and the fluoresced photons are detected. This approach also requires a precise means for measuring the laser wavelength. The Doppler shift of the laser in the rest frame of the atom is given by the expression:

$$\omega_L' = \omega_L \frac{1 - \frac{v}{c} cos\theta}{\sqrt{1 - \left(\frac{v}{c}\right)^2}} \tag{18.5}$$

where ω_L is the laser frequency in the laboratory frame, ω_L' is the laser frequency in the rest frame of the atom, v is the velocity of the atoms in the laboratory frame, c is the speed of light, and q is the angle between the atomic velocity and the direction of the laser. Preferably the laser is directed either parallel or antiparallel to the atomic velocity where the measurement is least

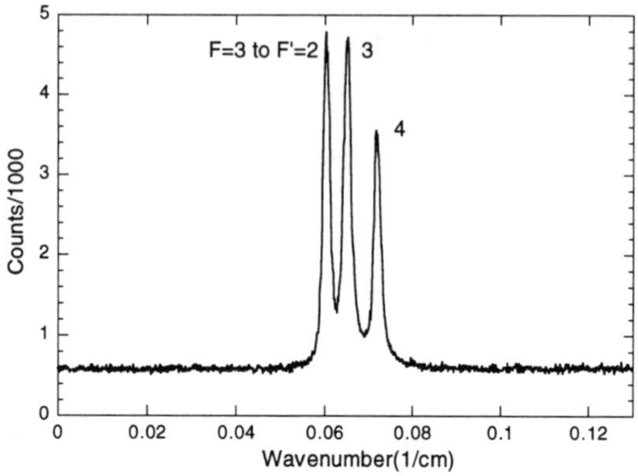

FIGURE 18.11 This laser scan of the Cs 6s $^2S_{1/2}\rightarrow$6p $^2P_{3/2}$ resonance transition at 852 nm in a 50 keV beam shows the narrow resonance lines that can be obtained with a diode laser aligned antiparallel to the fast-beam direction. The hyperfine structure of the transition is clearly resolved. In this experiment, measurements of velocity were limited by the wavemeter employed, not the longitudinal velocity distribution.

sensitive to the uncertainty in angle. When the laser is tuned such that it drives a resonance transition in the atomic frame, the number of scattered photons is a maximum. Figure 18.11 shows data from the experiment described in Reference [31]. Using a wavemeter to measure the laser wavelength in the laboratory frame, one can determine the atomic velocity to a precision of 0.01% or better depending on the wavemeter. This technique has the advantage that it yields a direct measurement of the velocity for the atoms under study, and it can be applied to both ions and neutrals.

The velocity distribution along the beam direction is also important for time resolution. In the source, the velocity distribution of the ions is roughly determined by the temperature of the plasma. A fortuitous feature of fast beams is that electrostatic acceleration compresses the velocity distribution along the beam direction [55]. The compression occurs because the fastest atoms spend the least time in the accelerator and the slowest atoms spend the most time in the accelerator. By scanning the laser frequency over the resonance transition, laser excitation can be used to map out the velocity profile in both the longitudinal (Fig. 18.11) and transverse directions. The velocity is also affected by the stability of the accelerating potential and the operating conditions of the ion source. To reduce these effects, Koch *et al.* [54] used the Doppler shifted resonance fluorescence to stabilize the atomic velocities by feeding an error signal back to the accelerating potential.

Pinnington *et al.* take an alternative approach [56] to measuring atomic beam velocities using the Doppler shift. The technique involves measuring the wavelength of the fluorescence observed at an angle other than 90 degrees to the atomic motion. However, their particular implementation of this technique eliminates sources of systematic errors associated with alignment and collection solid angle. A prism with a precisely known apex angle of 100 degrees is aligned such that one facet faces up stream and the other facet faces down stream along the beam direction. Thus, both the blue shifted and red shifted fluorescence are symmetrically reflected into a direction perpendicular to the ion beam and are viewed with a spectrometer. The wavelength difference between the two shifted fluorescence peaks equals twice the Doppler shift. The uncertainty depends on the apex angle of the prism, but is only second order in the angular misalignment of the prism and the viewing angle. Velocity measurements of 0.5% precision have been reported with this approach.

18.6 DATA COLLECTION AND ANALYSIS

An important part of data collection is identifying what needs to be measured and recorded for analysis. In the typical lifetime experiment, the output of the fluorescence detector recorded as a function of position (or in some cases voltage) contains the information about the desired lifetime. However, it also contains backgrounds from various sources, and all of these contribute to the noise. Our approach for measuring and controlling the backgrounds has evolved with our desire for higher precision. Using our current data collection scheme as an example, we identify the translating detector output as A and a stationary normalizing detector output as B, and record each PMT signal for several seconds at each position. The components that contribute to each signal (Eqs. 18.6, 18.7) are identified as f for laser induced fluorescence, l for laser scattered light, b for atomic beam background, and d for detector dark counts or offset. Ideally, the fluorescence makes the largest contribution to each signal.

Translating Detector Signal:

$$A(x,t) = f_A(x,t) + l_A(x,t) + b_A(x,t) + d_A(x,t) \tag{18.6}$$

Stationary Detector Signal:

$$B(t) = f_B(t) + l_B(t) + b_B(t) + d_B(t) \tag{18.7}$$

The procedure for recording data and measuring the background components must take into account the time scale of variation for several experimental parameters. Two experimental parameters that can vary rapidly in time are the intensity of the ion beam and the laser power (for some types of lasers) both of which affect the fluorescence, f. By recording the A and B signals simultaneously, these effects can be normalized away provided that the l, b, and d backgrounds can be realistically subtracted. In our case, the detector dark counts vary slowly on the time scale of hours because the cooled

PMT housings have some sensitivity to the temperature of the room. With the atomic beam on, we record the signals with the laser both on and shuttered off at each position. We use these data to subtract the slowly varying dark counts and the majority of the beam background. Then with the atomic beam blocked, we again record the signals for the laser on and off. These data are used to determine the laser scattered light as a function of position. However, the scattered light, l, and the beam background, b, can vary between laser on and laser off data, and between atomic beam on and off data. We found that by adding two more detectors we could more effectively subtract the scattered light and the beam background. The detector signal, C, (Eq. 18.8) is optimized to monitor the atomic beam background before the laser interaction region; b_C is proportional to the atomic beam intensity and l_C is the scattered light in this detector. Another detector signal, D, (Eq. 18.9)monitors the laser power, l_D. These detectors can also have a background in the absence of an input referred to as d.

$$\text{Atomic Beam Monitor:} \qquad C(t) = b_C(t) + l_C(t) + d_C(t) \tag{18.8}$$

$$\text{Laser Power:} \qquad D(t) = l_D(t) + d_D(t) \tag{18.9}$$

At each position, x_i signals from all four detectors are integrated over several seconds for each of the four combinations of ion beam on and off, and laser beam on and off. The measured signals are all combined as in Equations 18.10 and 18.11 to obtain the decay fluorescence from signals A and B. The first subscript refers to the ion beam, 1 for on, and 0 for off; the second subscript refers to the laser, 1 for on, and 0 for off. The first term in parenthesis subtracts dark counts; the second term subtracts laser scattered light where the ratio involving signal D is used to scale the atomic beam off data for variations in laser power. The third term involving signal C subtracts the atomic beam background with scaling factors to account for variations in beam intensity and laser power.

$$f_A(x_i) = (A_{11} - A_{10}) - \left[\frac{D_{11} - D_{10}}{D_{01} - D_{00}}\right](A_{01} - A_{00})$$
$$- \left[\frac{A_{10} - A_{00}}{C_{10} - C_{00}}\right]\left\{(C_{11} - C_{10}) - \left[\frac{D_{11} - D_{10}}{D_{01} - D_{00}}\right](C_{01} - C_{00})\right\} \tag{18.10}$$

$$f_{B_i} = (B_{11} - B_{10}) - \left[\frac{D_{11} - D_{10}}{D_{01} - D_{00}}\right](B_{01} - B_{00})$$
$$- \left[\frac{B_{10} - B_{00}}{C_{10} - C_{00}}\right]\left\{(C_{11} - C_{10}) - \left[\frac{D_{11} - D_{10}}{D_{01} - D_{00}}\right](C_{01} - C_{00})\right\} \tag{18.11}$$

The ratio of these quantities results in the normalized fluorescence as a function of position, and for most cases discussed here, we expect the behavior of a single exponential as in Equation 18.12

$$\frac{f_A(x_i)}{f_{B_i}} = \frac{f_A(0)}{f_B}e^{-x/v\tau} \tag{18.12}$$

where x is the position, v is the atomic beam velocity, and τ is the desired lifetime. We obtain our lifetime results by performing a least squares fit of this ratio to the three parameter function:

$$a_1 e^{-a_2 x} + a_3 \tag{18.13}$$

The quantity a_2 is converted into a lifetime using the known position scale and the measured atomic beam velocity. A scheme similar to this was used for the analysis of the data in Reference [31]. For nearly all of the runs, the constant background a_3 was statistically consistent with zero indicating the effectiveness of such a data collection and subtraction scheme.

18.7 SUMMARY

Precision measurements of atomic lifetimes are important to the analysis of data from many areas of physics and provide fundamental atomic structure information. Scientists in the fields of astrophysics, geophysics, and plasma fusion all depend on oscillator strengths to determine the relative abundances of elements. Often relative values of oscillator strengths are measured precisely, but accurate atomic lifetimes are needed to obtain absolute values. In addition, the interpretation of parity nonconservation experiments requires accurate knowledge of atomic structure including radial matrix elements. Some of these scientific needs are addressed theoretically. However, precision lifetime measurements also play the important role of assessing the accuracy of many-electron atomic structure calculations and guiding further theoretical development. Several experimental groups have addressed these needs with laser excitation of fast atomic beams.

The generation of fast atomic beams is based on well established ion source and accelerator technology making the fast beam a practical laboratory tool. The low relative velocity distribution within a fast beam implies that the emission is free of the systematic effects associated with collisions. With the selectivity of laser excitation, the effects of cascades can be effectively eliminated from observations. Many of the low lying states of neutral atomic systems are accessible with the variety of CW lasers available today. High power pulsed lasers, frequency doubling, and multiphoton excitation open up much of the near UV spectrum to beam-laser techniques. When laser excitation is combined with beam-foil or beam-gas excitation, atomic states that are not normally accessible with a laser can also be studied. Fast-beam lifetime measurements depend on the atomic velocity, and a variety of precise techniques are available for velocity determination. The experiments discussed in this chapter demonstrate that laser excitation of a fast atomic beam is an extremely versatile technique, which can provide precision measurements of excited-state lifetimes in both neutral and ionic systems.

ACKNOWLEDGMENTS

I thank the editors for the invitation to contribute to this work. In addition, I thank my colleagues A. E. Livingston and H. G. Berry for their assistance and helpful discussions. I thank R. J. Rafac and D. DiBerardino for helpful conversations and their attention to experimental needs during the preparation of this manuscript. I also thank E. H. Pinnington for providing information on his work, and K. W. Kukla for assistance with formatting the equations. Partial support is acknowledged for C. E. T. from the Department of Energy and the National Science Foundation.

REFERENCES

1. Imhof, R. E. and Read, F. H., *Rep. Prog. Phys.* **40**, 1-104 (1977).
2. Tanner, C. E., "Precision Measurements of Atomic Lifetimes", in *Proceedings of the Fourteenth International Conference on Atomic Physics*, Boulder, CO, 1994, pp. 130-145.
3. Lawler, J. E., Bergeson, S. D. and Wamsley, R. C., *Physica Scripta T* **47**, 29-35 (1993).
4. Thorburn, J. A., Mobbs, L. M., Deliyannis, C. P., and Pinsonneault, M. H., *Astro. J.* **415**, 150-173 (1993).
5. Bouchiat, M. A., Guena, J., Hunter, L., Pottier, L., *Phys. Lett. B* **117**, 358-364 (1982).
6. Noecher, M. C., Masterson, B. P., Wieman, C. E., *Phys. Rev. Lett.* **61**, 310-313 (1988).
7. Drell, P. S., and Commins, E. D., *Phys. Rev. A* **32**, 2196-2199 (1985).
8. Meekhof, D. M., Vetter, P. A., Majumder, P. K., Lamoreaux, S. K., and Fortson, E. N., *Phys. Rev. Lett.* **71**, 3442-3445 (1993).
9. Vetter, P. A., Meekhof, D. M., Majumder, P. K., Lamoreaux, S. K., and Fortson, E. N., *Phys. Rev. Lett.* **74**, 2658-2661 (1994).
10. Edwards, N. H., Phipp, S. J., Baird, P. E. G., and Nakayama, S., *Phys. Rev. Lett.* **74**, 2654-2657 (1995).
11. Blundell, S. A., Johnson, W. R., and Sapirstein, J., *Phys. Rev. A* **43**, 3407-3418 (1991).
12. Dzuba, V. A., Flambaum, V. V., Sushkov, O. P., *Phys. Lett. A* **142**, 373-377 (1989).
13. Johnson, W. R., Liu, Z. W., Sapirstein, J., Atomic and Nuclear Data Tables (to be published).
14. Curtis, L. J., Berry, H. G., Bromander, J., *Phys. Lett. A* **34**, 169-170 (1971).
15. Valyi, L., *Atom and Ion Sources*, London: John Wiley, 1977.
16. Thonemann, P. C., *Nature* **61**, 158 (1946).
17. Rutherglen, J. G. and Cole, J. F. I., *Nature* **160**, 545-546 (1947).
18. Moak, C. D., Reese, Jr., H., and Good, W. M., *Nucleonics* **9**, 18-23 (1951).
19. Oona, H., and Bickel, W. S., *Rev. Sci. Instrum.* **47**, 517-518 (1978).

20. Dumont, P. D., Garnir, H. P., and Baudinet-Robinet, Y., *Nucl. Instrum. Meth. Phys. Res. B* **164**, 193 (1979).
21. Kernahan, J. A., and Pinnington, E. H., *Nucl. Instrum. Meth.* **166**, 313-314 (1979).
22. Moak, C. D., Banta, H. E., Thurston, J. N., Johnson, J. W., and King, R. F., *Rev. Sci. Instrum.* **30**, 694-699 (1959).
23. Sidenius, G., *Nucl. Instrum. Meth.* **38**, 19-22 (1965).
24. Sidenius, G., "The Hollow-Cathode Ion Source and Its Application to On-Line Separations", in Proceedings of the International Conference on Electromagnetic Isotope Separators and the Techniques of Their Application, Marburg, September 7-10, 1970, pp. 423-432.
25. Sidenius, G., *Nucl. Instrum. Meth.* **151**, 349-362 (1978).
26. Wilson, R. G. and Brewer, G. R., *Ion Beams with Applications to Ion Implanation*, New York: John Wiley, 1973, pp. 189-257
27. Holmes, A. J. T., "Beam Transport," in *The Physics and Technology of Ion Sources* ed. by Brown, I. G., New York: John Wiley, 1989, ch. 4, pp. 53-70
28. More, J. H., Davis, C. C., Coplan, M. A., *Building Scientific Apparatus*, 2nd. ed., Addison-Wesley, Inc., 1989, Ch. 5, p. 333.
29. Barnett, C. F., "Atomic Collision Properties," in *Physics Vade Mecum*, ed. Anderson, H. L., New York, AIP Press, 1981, Ch. 4, pp. 90-91.
30. Tanner, C. E., Livingston, A. E., Rafac, R. J., Serpa, F. G., Kukla, K. W., Berry, H. G., Young L., and Kurtz, C. A., *Phys. Rev. Lett.* **69**, 2765-2767 (1992).
31. Rafac, R. J., Tanner, C. E., Livingston, A. E., Kukla, K. W., Berry, H. G., and Kurtz, C. A., *Phys. Rev. A* **50**, R1976-R1979 (1994).
32. Andrä, H. J., Gaupp, A., and Wittmann, W., *Phys. Rev. Lett.* **31**, 501-504 (1973).
33. Bickel, W. S., *Appl. Opt.* **6**, 1309 (1967) and **7**, 2367 (1968).
34. Andrä, H. J., "Laser Excitation in Fast Beam Spectroscopy," in *Proceedings of the Conference on Beam-Foil Spectroscopy*, ed. by Sellin, Ivan, A. and Pegg, D. J., Vol. 2, pp. 835-851, 1976.
35. Gaupp, A., Kuske, P., and Andrä, H. J., *Phys. Rev. A* **26**, 3351-3359 (1982).
36. Gosselin, R. N., Pinnington, E. H., and Ansbacher, W., *Phys. Rev. A* **38**, 4887-4890 (1988).
37. Ansbacher, W., Li, Y., Pinnington, E. H., *Phys. Lett. A* **139**, 165-169 (1989).
38. Guo, B., Ansbacher, W., Pinnington, E. H., Ji, Q., and Berends, R. W., *Phys. Rev. A* **46**, 641 (1992).
39. Pinnington, E. H., Berends, R. W., and Lumsden, M., *J. Phys. B: At. Mol. Opt. Phys.* **28**, 2095-2103 (1995).
40. Schulze-Hagenest, D., Harde, H., Brand, W., and Demtröder, W., *Z. Physik A* **282**, 149-156 (1977).
41. Schmoranzer H., and Volz, U., *Phys. Scripta* **14**, 42-48 (1993).
42. Dumont, P. D., Garnir, H. P., Baudinet-Robinet, Y., and El Himdy, A., *Nucl. Instrum. Meth. Phys. Res. B* **35**, 191-195 (1988).
43. Volz, U., Majerus, M., Liebel, H., Schmitt, A., and Schmoranzer, H., *Phys.*

Rev. Lett. **76**, 2862-2865 (1996), and references there in.
44. Wiese, W. L., *Physica Scripta* **35**, 846-850 (1987).
45. Tiemann, H., Richling, and Knockel, H., (to be published).
46. Oates, C. W., Vogel, K. R., Hall, J. L., *Phys. Rev. Lett.* **76**, 2866-2869 (1996).
47. Marciano, W. J., and Rosner, J. L., *Phys. Rev. Lett.* **65**, 2963-2966 (1990).
48. Gaillard, M. L., Pegg, D. J., Bingham, C. R., Carter, H. K., Mlekodaj, R. L., and Cole, J. D., *Phys. Rev. A* **26**, 1975-1983 (1982).
49. Jin, J. and Church, D. A., *Phys. Rev. A* **47**, 132-136 (1993).
50. Jin, J. and Church, D. A., *Phys. Rev. Lett.* **70**, 3213-3216 (1993).
51. Schmoranzer, H., Roth, Volz, U., and Marger, D., *J. Phys. B: At. Mol. Opt. Phys.* **24**, 595-604 (1991).
52. Marion, J. B., *Rev. Mod. Phys.* **33**, 139 (1961).
53. Herb, R. G., Snowdon, S. C., and Sala O., *Phys. Rev.* **75**, 246-259 (1949).
54. Koch, P. M., *Opt. Com.* **20**, 115-118 (1977).
55. Kaufman, S. L., *Opt. Com.* **17**, 309-312 (1976).
56. Ansbacher, W., and Pinnington, E. H., *Nucl. Instrum. Meth. Phys. Res. B* **4**, 207-209 (1984).

CHAPTER 19

Generation of Slow Positron Beams with the Oak Ridge Electron Linear Accelerator (ORELA)

Lester D. Hulett, Jr., T. A. Lewis, and Jun Xu*

This paper describes a practical technique for generating slow positron beams with an electron linac. The linac that has been adapted to this function is the Oak Ridge Electron Linear Accelerator (ORELA). The positron converter-moderator has been designed and positioned so as not to impair the ability of the ORELA to supply photoneutrons of defined energies for nuclear cross section measurements. Thus, the accelerator has been upgraded in ability: It can supply particles simultaneously for two independent experiments, with no appreciable increase in operating cost. The moderation technique for converting fast positrons to the monoenergetic state is described, and a practical guide for preparing moderators is given.

19.1 INTRODUCTION

In recent years slow positron spectroscopy has undergone a spectacular rise in its application to applied and fundamental problems. It is especially valuable for the non-destructive evaluation of materials. Linear electron accelerators offer an especially convenient avenue for generating slow positron beams.

The gamma bremsstrahlung from the impact of the electron beam on a heavy element target produces fast positrons, which are subsequently moderated to low energies. The first group to accomplish this was Groce *et al.* [1], using the electron linac at Lawrence Livermore National Laboratory. This was done before the positron moderation process was thoroughly understood. At a much later date, Howell *et al.* [2] started generating practical intensities of slow positrons for applications to materials science and other problems. This technique has since proliferated to many other institutions such as the

*) Chemical and Analytical Sciences Division Oak Ridge National Laboratory

University of Giessen [3], The University of Tokyo, The University of Mainz, the University of Osaka, KEK at Tsukuba, and several others. This paper is a summary of the facility at ORNL, describing its structure and principles for the non-specialist, and offering an introduction and overview to the techniques of positron moderation and brightness enhancement for those who might be considering the use of the method.

As will be shown below, the ORELA slow positron facility has a significant advantage with respect to economy of operation. Both neutron and positron spectroscopy can be performed simultaneously. On occasion, gamma spectroscopy experiments are also done. This means that two or more research programs can be supported with no significant increase in accelerator operating costs.

19.2 GENERATION AND TRANSPORT OF THE POSITRON BEAM

An electron beam, having an energy of 10 MeV or higher, directed onto a heavy metal target will produce a usable intensity of gamma photons of energy above the positron-electron pair formation threshold of 1.02 MeV. The gamma photons are forward scattered, and if they are intercepted by a secondary heavy metal target, the electrostatic fields of the high-Z nuclei will promote their conversion to pairs. To enable the positrons to escape without being converted back into gamma photons by annihilation, the secondary target must be fabricated with a high surface-to-volume ratio.

Such an arrangement is shown in Figure 19.1. A 150 MeV electron beam impacts the ORELA primary target, which consists of a layered array of tantalum plates inside a water cooling baffle. Gamma bremsstrahlung is produced, most of which is consumed in the ejection of photoneutrons from the tantalum and the water. Approximately 15% of the power escapes the primary target and is forward scattered to the tungsten converter-moderator plates.

The converter-moderator target has three functions: (1) the electrostatic fields of its high-Z nuclei induce the disproportionation of the gamma photons into positron-electron pairs; (2) its high electron density thermalizes a large fraction of the positrons and electrons ; (3) its specially-prepared surfaces allow the negative work function to take effect, which causes some of the thermalized positrons to be expelled with well defined energies, of the order of 2-3 eV. The emission of the low energy positrons, the moderator properties that enable this process, and the preparation of efficient positron moderators are discussed below.

The electron beam from the ORELA is delivered to the primary tantalum target in short bursts of 6-25 nanoseconds duration. The emission of the forward scattered gamma photons, their conversion to positrons, and the moderation and emission of the slow positrons, all occur in less than 10 picoseconds. The slow positron emission thus tracks the ORELA electron beam

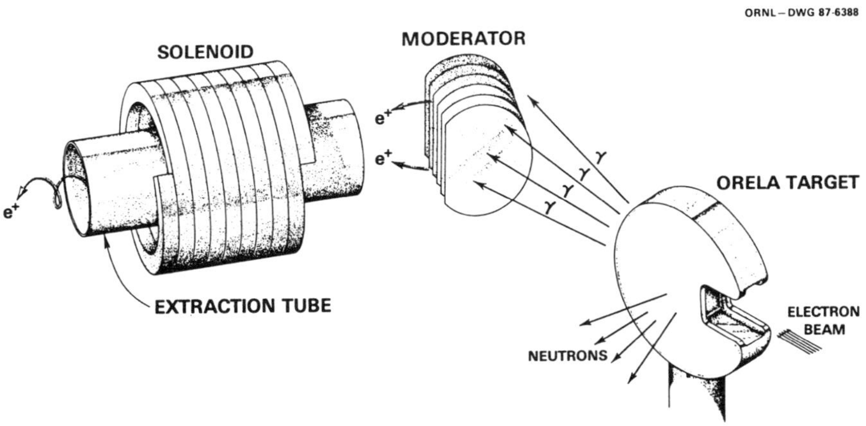

FIGURE 19.1 Configuration of ORELA primary target for producing neutrons and secondary target for positron-electron pair production and positron moderation.

pulses.

As the slow positrons emerge from the moderator they are extracted and transported along the magnetic field of a long solenoid leading out of the target room. In Figure 19.1 the open end of the extraction tube is shown facing the moderator plates. Figure 19.2 shows the extraction tube running coaxially with the solenoid. Outside the target room, in an access area, the extraction tube goes through a ferrite torus and then connects to ground. A negative pulse of 3-4 kV is applied to the tube at a point on the target-side of the ferrite torus. The high inductive reactance of the ferrite torus prevents the negative pulse from going directly to ground. The negative charge is propagated to the end of the tube facing the positron moderator, arriving simultaneously with the emergence of the burst of slow positrons from the moderator plates.

The positrons are accelerated into the tube and are conducted by the B-Field (0.005 T) along the axis of the solenoid. They are transported out of the target room to an adjacent experiment room. The negative charge on the end of the extraction tube builds up over a period of a few nanoseconds and is reflected back to the other end. The polarity of the charge changes to positive and it is then returned to the moderator end. The length of the extraction tube is about 3 meters, so the half-period of the negative-positive charging of the end of the extraction tube near the moderator is about 25 nanoseconds. After the positrons enter the extraction tube they are shielded from changes in the potential on its end and they continue to travel with the kinetic energy initially imparted to them. The charge on the extraction tube is damped to zero in a few hundred nanoseconds as it reflects back and forth. This technique of extraction has the advantage of leaving the tungsten moderator at ground potential. Other workers have operated with the positron moderators above

ORELA FACILITY FOR GENERATING NEUTRONS AND SLOW POSITRONS

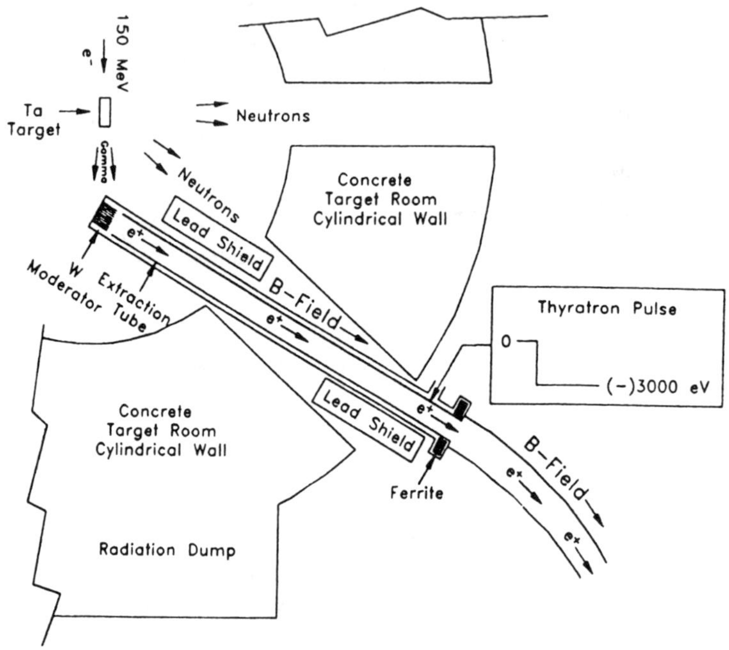

FIGURE 19.2

ground. This leads to significant problems involving glow discharge.

Initially, the primary function of the ORELA was to generate neutrons for nuclear cross section measurements. For purposes of clarity, only one neutron beam line is indicated in Figure 19.2. Several more are actually present. Also, to allow clear labeling of the components, the diagram in Figure 19.2 is not drawn exactly to scale. The width/length ratio of the solenoid going into the target room is exagerated. The neutrons are emitted with approximately equal intensities in all directions and are delivered by line-of-sight collimation tubes to experiment stations. The positron beam line was installed in one of the neutron collimation tubes leading obliquely away from the tantalum target. Both positrons and neutrons can be simultaneously delivered to experiments, without mutual interference.

The prevention of neutron transmission to the positron experiment station is a practical problem that was overcome by using the magnetic fields of the solenoids to conduct the positron beam along an indirect route. This allows neutron barricades to be placed so that there is no direct line-of-sight path from the tantalum target to the positron station. Note that the positron solenoid is at a skew angle with respect to the radius of the cylindrical target room. This allowed the introduction of lead shields to stop the neutrons. After the positron beam exits the extraction tube, passing beyond the ferrite torus, the solenoids guide it through three bends having radii of about 1 meter. The 0.0050 T magnetic fields are enough to retain the 3 keV positrons on an axial path as they traverse the bent solenoids.

With the ORELA operating at a power of 30 kilowatts, 25 nanosecond pulse width, 800 Hz repetition rate, an average current slightly greater than 1×10^8 slow positrons per second was obtained. Constraints on the current are partly due to the parasitic mode of operation. To prevent interference of the positron apparatus with the neutron delivery, it was necessary to offset the tungsten target from the tantalum target about 300 mm. This allows the gamma beam to diverge appreciably before it impacts the tungsten moderator, causing a significant decrease in the brightness of the positron beam.

19.2.1 The positron moderation process; brightness enhancement

At present, energy moderation is the only known method by which monoenergetic ('slow') positrons can be produced. We must start with fast positrons, generated by pair production or emitted from radioisotopes. The energy distributions of the positrons from both of these sources range over many hundreds of keV, and because of the law of conservation of brightness (Liouville's theorem), it is not practical to use electrostatic retardation to reduce the energies to a low range. It becomes necessary to use non-conservative forces to circumvent the constraints of Liouville's theorem. This is done by injecting the fast positrons into solids so that the inelastic collisions with electrons will dissipate their high energies to levels that will allow them to be focused and transported as beams. We must, of course, retrieve the positrons from the solids before they annihilate, and this has required the development of the science of positron moderation. Before 1972, positron moderation was more art than science. In 1974 Pendyala *et al.* [4] were the first to measure distinct spectral peaks, in the range of 1-3 eV, of positrons emitted from metal surfaces. We now know these to be due to negative work function effects, which have since been quantitatively measured and elucidated by Mills *et al.* [5]. In 1980 Dale *et al.* [6] published methods for preparing high efficiency moderators from polycrystalline tungsten, molybdenum, and iridium. In 1986 Mills and Gullikson published preparation methods for solidified inert gas moderators [7].

When a fast positron enters a solid, it loses energy by exciting core electron ejections, interband transitions, and plasmon oscillations. In times of less than 10 picoseconds it is thermalized and enters a band state. For certain metals the potential energy of the positron in the band state is positive with respect to the vacuum level. The metal thus exhibits the negative work function effect, such that, if the positron is able to diffuse to the surface before it annihilates, it may be spontaneously emitted.

It is not surprising that positron work functions are negative for some solids. Using first-approximation reasoning, one might expect all solids to exhibit negative work functions for positrons. The negative potential energies of electrons in the band states of solids are the result of the attractive forces of the atomic nuclei being greater than the repulsive forces of the other electrons. Thus one would expect positive potential energy for a positively charged electron. This simplistic model must be modified by considering the correlation potentials due to the tendency of electrons to be attracted to the positron. As pointed out by Brandt [8], and by Nieminen and Hodges [9], the correlation energy is usually the overriding term. There are a few metals, however, such as tungsten, molybdenum, iridium, nickel, copper, chromium, and platinum, for which the correlation potentials do not override the repulsion of the atomic nuclei for the positrons. Of the metals, tungsten has the most negative work function, -2.7 eV. The most negative work function reported to date is for diamond, - 3.7 eV, measured by Brandes *et al.* [10].

In 1986 Mills and Gullikson [7] reported that solid neon, argon, krypton and xenon have moderation efficiencies higher than tungsten. The moderation mechanism for these materials is apparently not due to a negative work function effect. Solidified rare gases have large band gaps, making energy loss processes less than 10 eV improbable. It is believed that the loss processes within the solids do not reduce the positrons to a thermal state, but rather they achieve epithermal kinetic energies below 10 eV. Annihilation probability for epithermal positrons is low and eventually they are reflected outside the solid, where they can be extracted and focused into a beam. Moderation efficiencies greater than 0.5% have been reported for solidified rare gases. An attractive advantage of the rare gas moderators is the ability to easily renew them by evaporating them off of their substrates and recondensing new films. After a moderator has been in use for an extended period of time, its surface will become contaminated and its efficiency will decrease. The energy distributions of slow positrons from rare gas moderators range from 1-10 eV in width, not as well defined as for those from metal surfaces. The full-width-half-maximum of the slow positron peak from tungsten is of the order of 0.2 eV at room temperature.

In addition to the material and its perfection, the efficiency of positron moderation depends on other factors, such as the energy distribution of the fast positrons and the configuration of the moderator. Most efficiency measurements have been made using radioisotope sources, such as ^{22}Na , ^{58}Co, or ^{64}Cu, for which the average energy of the fast positrons is of the order of 250 keV. For single crystal tungsten moderators, prepared *in situ* under high vacuum, efficiencies as high as 0.3% have been reported for radioisotopes. For polycrystal tungsten vane moderators, which have been allowed to have air exposure after annealing, the efficiencies are usually of the order of 0.05%. For the rare gas moderators, condensed on the inside of cylinders, the efficiency is of the order of 0.5%.

The use of brightness enhancement is essential for effectively manipulating positron beams. Without this technique we would not be able to perform most of the useful experiments with slow positrons. The concept of brightness enhancement was first explained in detail by Mills [11]. The practicality of the method has been demonstrated by Brandes, Canter and Mills [12] in their very impressive experiments with positron emission microscopy. Brightness is defined as follows:

$$\text{Brightness} = n/(\alpha \times a \times \varepsilon) \tag{19.1}$$

Where n is the current of positrons, α is the angular divergence of the beam, a is the cross sectional area of the beam, and ε is the kinetic energy. The Liouville theorem states that the brightness-quantity is constant and cannot be increased by means of conservative forces, such as those of magnetic or electrostatic lenses. If one attempts to make either α, a, or ε smaller by focusing with lenses or retarding fields, the current of positrons, n, is decreased.

Brightness enhancement is a method of circumventing the Liouville theorem by the use of non-conservative forces to reduce the energy, angular divergence, and cross sectional areas of positron beams. The initial production of the slow positron beam by the moderation of fast positrons is a form of brightness enhancement, since it involves the inelastic collisions of positrons in a solid. But in most cases the area of the moderator surface is large, which results in a broad slow positron beam of low brightness. Also, if the moderator is in the form of parallel vanes, the positrons will be emitted perpendicular to the optical axis, producing a large angular divergence. Thus, it becomes desirable to re-moderate, but in the second stage one usually has more control on the size of the beam and it can be condensed to a minimum area as it enters the moderator.

Figure 19.3 shows schematically the use of a thin film moderator to enhance the brightness of a slow positron beam. A broad beam is shown approaching from the left. By means of either an electrostatic or magnetic lens, the beam is condensed to a cross-over point coincident with the surface of the film. When the beam enters the thin film it is thermalized, and the positrons begin to diffuse in all directions. Those positrons that reach the forward surface are spontaneously emitted due to the negative work function. They emerge with most of their momenta in the forward direction, thus greatly reducing the divergence angle. The area over which they are emitted and their kinetic energies are reduced by factors of 10 and 1000 respectively. The number of positrons emerging in the forward direction will be lowered by over 50%, due to backward and lateral diffusion in the film, but this is overridden by the lower divergence angle and surface area.

For effective brightness enhancement, the moderator film must be smooth, i.e. as wrinkle-free as possible, and the surface should be clean and annealed

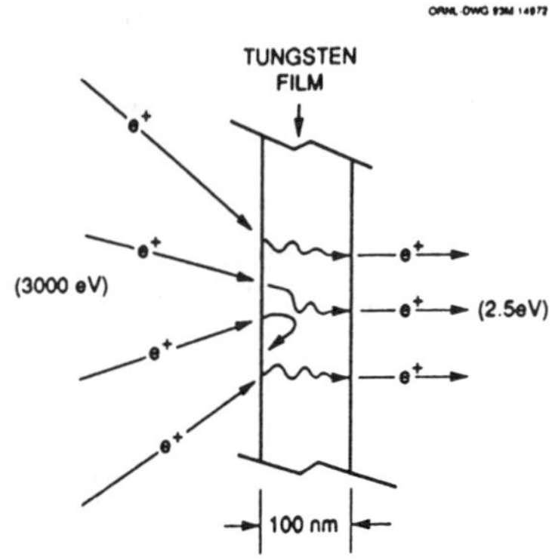

FIGURE 19.3 Brightness enhancement of positron beam by re-moderation with a tungsten film.

in situ so as to have a negative work function. If tungsten films are used, their thicknesses should be about 100 nm, and the energy of the incoming positron beam should be in the range of 3-4 keV. The thickness of the film and the energy of the incoming beam should be adjusted so that most of the positrons thermalize as close to the forward surface as possible.

The efficiency of positron recovery for brightness enhancement remoderator films has been reported to be as high as 40% [17]. This is no doubt due to the controlled energy (3-4 keV) of the positron beam that is injected into the film. The narrow distribution of the depth of penetration of the positron beam into the film allows a high probability of recovery for all of the positrons. Positron recovery in the primary moderation process has never been reported higher than 0.5 %. This is due to the extremely large energy ranges of fast positrons. A large fraction of the positrons have energies so high that they penetrate the moderator to depths too large to allow escape.

Canter *et al.* [13] have shown that brightness enhancement can also be done in the reflection mode, such that the positrons emerge from the same surface that they enter. In fact, the positron recovery efficiency is probably better than that obtained with films, due to cleaner and flatter surfaces. The energy of the incoming beams is about the same as that for films.

19.2.2 A Practical Perspective on Positron Moderator Preparation

Presently, there are many well-known techniques for producing high-efficiency positron moderators. For the generation of slow positrons by pair-production with electron linacs, the metallic tungsten moderator appears to be the most practical. As indicated above, the high-Z nuclei induce the pair production process, which is especially fortunate. Tungsten metal has another highly advantageous property, however, in that its surfaces are very easily cleaned, and they maintain their cleanliness even under poor vacuum conditions.

Dale, Hulett and Pendyala [6] were the first to develop reliable techniques for preparing metallic tungsten positron moderators. In this same publication, preparations for two more very useful moderators, metallic molybdenum, and metallic iridium, were also described. The value of molybdenum and iridium as positron moderators, and their ease of preparation, has been overlooked by most workers in the field.

In addition to negative work functions for positrons, tungsten, molybdenum and iridium metals all have the fortunate property of higher melting points than their respective oxides. This allows them to be prepared by simply annealing in vacuum. The oxides are melted and volitilized, leaving clean metallic substrates, free of gross contamination. If the annealing is done at temperatures in excess of the oxide melting/volatilization points one will always obtain functional moderators; however, the temperatures of the metals should be raised to levels as least as high as 50% of their absolute melting points. At these temperatures the vacancies in the metals become mobile and are greatly reduced in concentration. Dale *et al.* [6] have shown that, for the case of tungsten, it is highly desirable to heat to temperatures sufficiently high to anneal vacancies and dislocations. This greatly increases the probability that the thermalized positrons will escape to the surface of the metal and be ejected without becoming trapped in these defects and annihilated. For tungsten moderators to be used in linac sources, Akahane [14] reports that it is profitable to anneal at temperatures in excess of 2000°C. Functional moderators can be prepared under vacuum conditions as poor as 10^{-5} Torr, but better results are obtained with lower pressures.

It is significant that the positron work functions are negative for all three of the easily-prepared moderators (W, Mo, Ir). Their respective moderation efficiencies correlate with the magnitudes of their work functions. Nickel [12], chromium [15], and copper [16] are efficient moderators, and their calculated work functions are also negative.

It appears that the highest moderation efficiencies can be achieved by *in situ* annealing of the moderators, without allowing air contact afterwards [17], but this is not practical for linac-based slow positron sources. Fortunately, air contact reduces the efficiency of the tungsten moderator by no more than a factor of 4. This is a significant loss of efficiency, of course, but not enough to destroy the effectiveness of linacs as slow positron sources. For the preparation of brightness enhancement films, however, in situ annealing is practical and should be performed if circumstances permit. Techniques for brightness enhancement using metal film moderators are still under development. Both molybdenum and tungsten should be further explored in our efforts to improve brightness enhancement methods. Films of molybdenum are significantly easier to prepare than those of tungsten. The development of single crystal tungsten and molybdenum moderator films was first done by Chen *et al.* [18]. The authors have obtained single crystal films of these materials from the University of Arhus [19].

One should not overlook the future possibility of using solidified rare gas moderators in linac slow positron sources. At present there is the practical constraint of the difficulty in maintaining cryogenic temperatures under the heavy heat load (> 1 kw) generated by the impact of the linac beam. In future designs, however, this problem may be overcome.

19.2.3 Future Research Programs

Slow positron spectroscopy has a proven record of worthwhile applications in chemistry, physics and materials science. These were reviewed and elaborated in the proceedings of a recent workshop, sponsored by the United States Department of Energy [20]. The applications to materials science, which were the principal focus of the workshop, are especially broad. Yet, there are many fundamentals that are not understood. Workers in the field therefore enjoy the opportunity to do worthwhile research in both the applied and fundamental sciences. Much of the future work with the ORELA slow positron facility will be done in materials science. The measurement of positron lifetimes, using positrons of variable energy, and positron stimulated Auger electron spectroscopy will be major areas of endeavor.

Positrons injected into solids are repelled by atomic nuclei and are attracted to regions of potential energy minima, such as vacancies, voids, free volume interstices and surfaces. Upon entering such regions, they often undergo inelastic collisions and become trapped. The reduced electron densities cause annihilation rates to be lower. Positron lifetime spectra are therefore quite useful for determining defect natures of solids. Such measurements have been done for many years, using ^{22}Na sources of fast positrons, but the large energy range caused the depths of penetration to be undefined. Applications to surface problems have been quite limited. Using slow positron beams of controlled energy, however, the depths of penetration can be better defined, allowing surfaces and interfaces to be delineated from the bulk. In a recent review, Asoka-Kumar, Lynn and Welch [21] have described the application of slow positron beams to problems involving defects in Si and at Si-SiO_2 interfaces. Such work has obvious importance in microelectronics technology. By injecting slow positron beams of variable energy into the specimens and measuring their lifetime spectra, it will be possible to determine the types of defects present and their concentration as functions of depth. Recently Hulett *et al.* [22] showed that positron spectroscopy can be used to determine the depths of penetration of weathering for protective polymer coatings. Positron lifetime measurements will be used to expand this work.

Positron-induced Auger electron spectroscopy (PAES) has recently been shown by Weiss *et al.* [23] to be more surface sensitive than the conventional mode of Auger electron spectroscopy, induced by electron bombardment. The creation of the core electron hole that induces the Auger electron emission is done by annihilation with positrons, rather than by a collisional hole. The positron is at rest on the outermost atomic layer of the solid when the annihilation takes place. The method is therefore sensitive exclusively to the outermost atomic layer.

ACKNOWLEDGMENTS

Funding for this work was provided by the Chemical Sciences Division of the U.S. Department of Energy, Office of Basic Energy Science, Contract No. DE-AC05-84OR21400 with Lockeed-Martin Energy Systems, Inc.

REFERENCES

1. D. E. Groce, D. G. Costello, J. W. McGowan, and D. F. Herring, *Bull. Am. Phys. Soc.* **13**, 1397 (1968); D. G. Costello, D. E. Groce, D. F. Herring, and J. W. McGowan, *Phys. Rev. B* **5**, 1433 (1972).
2. R. H. Howell, M. J. Fluss, I. J. Rosenberg, P. Meyer, *Nuclear Instrum. and Methods in Physics Research* **B10**, 373 (1985).
3. F. Ebel, W. Faust, C. Hahln, S. Langer, M. Ruckert, H. Schneider, A. Singe, and I. Tobein, *Nuclear Instrum. and Methods in Physics Research* **A272**, 626-628 (1988).
4. S. Pendyala, D. Bartell, F. E. Girourard, J. W. McGowan, *Phys. Rev. Lett.* **33**, 7 (1974).
5. A. P. Mills, Jr., P. M. Platzman, B. L. Brown, *Phys. Rev. Lett.* **41** 1076 (1978); A. P. Mills, Jr, Appl. Phys. Lett., **37** 667 (1980).
6. J. M. Dale, L. D. Hulett, Jr., S. Pendyala, *Surface and Interface Analysis* **2,** 199 (1980).
7. A. P. Mills, Jr., E. M. Gullikson, *Appl. Phys. Lett.* **49**, 1121 (1986).
8. W. Brandt: in *Radiation Effects on Solid Surfaces, Advances in Chemistry Series*, Vol 158, ed. by M. Kaminsky (Washington, D.C.), p 219.
9. R. M. Nieminen and C. H. Hodges, *Solid State Commun.* **18**, 1115 (1976).
10. G. R. Brandes, A. P. Mills, Jr., D. M. Zuckermann, *Positron Annihilation* Vol. 105-110 (1993) ed. by Zs. Kajcsos and Cs. Szeles, Trans Tech Publications Ltd., Aedermannsdorf, Switzerland.
11. A. P. Mills, Jr., *Applied Physics* **23**, 189 (1980).
12. G. R. Brandes, K. F. Canter, A. P. Mills, Jr., *Phys. Rev. Lett.* **61**, 492 (1988).
13. K. F. Canter, T. Horsky, P. H. Lippel, W. S. Crane, and A. P. Mills, Jr., *Positron (Electron)-Gas Scattering*, W. E. Kauppila, T. S. Stein, and J. M. Wadehra, eds., World Scientific Press, Singapore, p. 202, 1986.
14. T. Akahane, National Institute for Research in Inorganic Materials, Tsukuba, Japan, Private Communication.
15. W. Cherry, Ph.D. Dissertation, Princeton University, University Microfilms, Inc., Ann Arbor, MI.

16. A. P. Mills, Jr., *Proceedings of the International School of Physics (Enrico Fermi)*, Vol. 83, ed. by W. Brandt and A. Dupasquier, July 14-24, 1981, Italian Society of Physics, North Holland Publishing Co., New York, p471.
17. M. R. Poulsen, M. Charlton, J. Chevallier, B. I. Deutch, F. M. Jacobsen, G. Laricchia, page 597 of *Positron Annihilation*, Proceedings of the 8th International Conference on Positron Annihilation, Gent, Belgium, 29 Aug. - 3 Sept., 1988, Ed. by L. Dorikens-Vanpraet, M. Dorikens, and D. Segers, World Scientific, Singapore.
18. D. M. Chen, K. G. Lynn, R. Pareja, and B. Nielsen, *Phys. Rev. B* **31**, 4123 (1985).
19. J. Chevallier, Institute of Physics, Arhus University, Dk-8000 Arhus C, Denmark.
20. L. D. Hulett, Jr., R. Gonzalez, B. L. Brown, A. L. Denison, R. H. Howell, Y. C. Jean, P. L. Jones, W. J. Kossler, K. G. Lynn, A. P. Mills, Jr., S. Okada, D. M. Schrader, L. C. Smedskjaer, A. H. Weiss, R. N. West, Y. Chen, *Application of Positron Spectroscopy to Materials Science*, Advanced Materials Research Volume 3, Scitec Publications, Trans Tech Publications Ltd., CH-4714 Aedermannsdorf, Switzerland, 1994.
21. P. Asoka-Kumar, K. G. Lynn, D. O. Welch, *Applied Physics Reviews*, in *J. Appl. Physics*, **76**(9), 4935 (1994).
22. L. D. Hulett, Jr., S. Wallace, Jun Xu, B. Nielsen, Cs. Szeles, K. G. Lynn, J. Pfau, Ann Schaub, *Applied Surface Science* **85**, 334-338 (1995).
23. A. Weiss, R. Mayer, M Jibaly, C. Lei, D. Mehl, K. G. Lynn, *Phys. Rev. Lett.* **61**, 2245-2248 (1988).

CHAPTER 20

Ion Bombardment Induced Light Emission

D. Ghose* and R. Hippler†

Energetic ions penetrating a solid lose their energy by inelastic electronic and elastic nuclear collision processes. The elastic collision results in sputtering of target atoms, while the inelastic part of the interaction may lead to emission of particles in an excited or ionized state. In this article, we consider mainly the sputtered excited particles that are short-lived (lifetimes $< 10^{-6}$ s) and decay radiatively away from the target surface. The various mechanisms of atomic excitation in sputtering are discussed. The experimental studies on secondary photon emission yields are briefly reviewed. Finally, the use of sputtering-induced optical emission as a sensitive tool for surface analysis is discussed.

20.1 INTRODUCTION

It is a common experience when working with ion-solid interaction phenomena in an ion beam system or an ion accelerator that the impact region is seen to glow, often with a faint blue light. This collision induced photon emission is not confined to the visible region of the spectrum but can extend to the ultraviolet (UV) and soft X-ray region. The different sources of light emission phenomena are illustrated in Fig. 20.1 [1]. First, the incident ions that are scattered back from the solid may be in an excited state. Second, some of the sputtered target particles that leave the surface, consisting of atoms, ions, molecules, and clusters, are also in an excited state. All these may de-excite radiatively in front of the target surface. Third, photons can be emitted by the solid itself due to, for example, excitation of electrons within the solid or

*) Saha Institute of Nuclear Physics, Sector I, Block 'AF', Bidhan Nagar, Calcutta 700 064, India

†) Institut für Physik, Ernst-Moritz-Arndt-Universität Greifswald, Domstraße 10a, D-17487 Greifswald, Germany

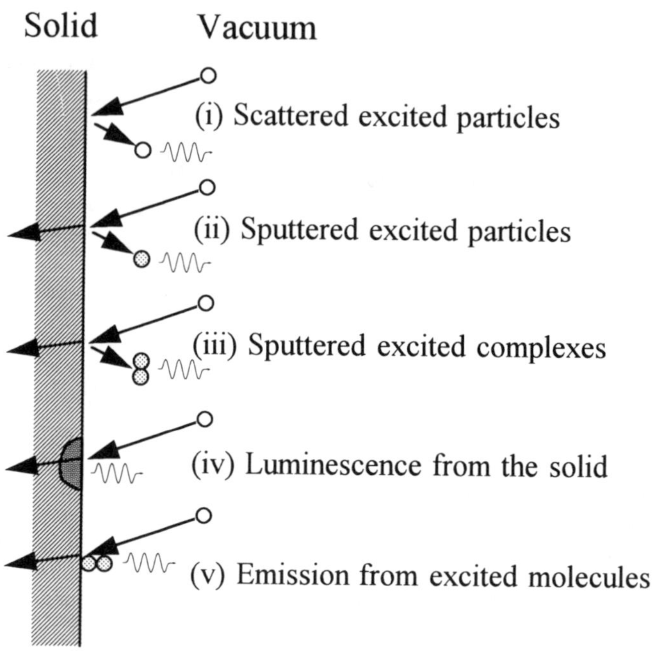

FIGURE 20.1 The various sources of light emission phenomena. Reprinted from E.W. Thomas, "Formation of Excited States by ion impact on surfaces" [1], with kind permission of Elsevier Science Ltd, The Boulevard, Langford Lane, Kidlington 0X5 1GB, UK.

radiative recombination of collision-induced electron-hole pairs. This results in a continuum luminescence and is particularly evident in the bombardment of alkali halide crystals. Finally, molecules contaminating the surface may become excited by ion bombardment leading to molecular band emission. In the present article we concentrate on the light emission phenomena from sputtered atoms which in many cases represent the major contribution to the emitted light. A brief account of the other light emitting sources is also included.

The *bombardment-induced light emission (BLE)* received rather limited attention up to 1970; however, the interest in this field has grown since then. The present article is not intended to review the field. Rather, it is an attempt to outline the general picture with reference to the excited sputtered particles as it has emerged over the years.

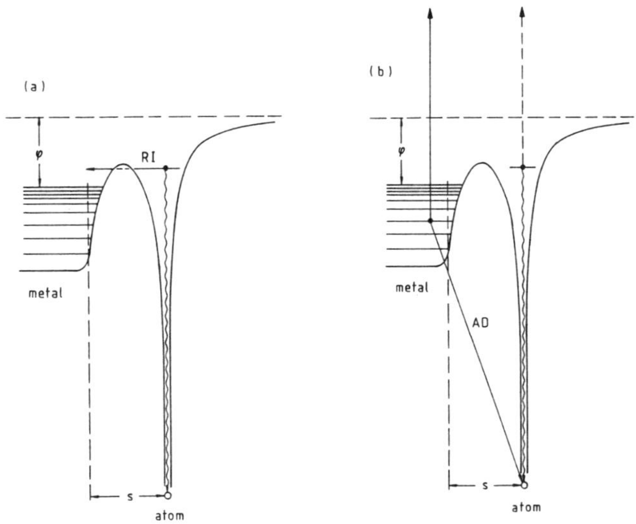

FIGURE 20.2 Schematic representation of the energy levels of an atom in front of a metal surface. Two non-radiative de-excitation processes, (a) resonance ionization (RI) and (b) Auger de-excitation (AD) are indicated.

20.2 MECHANISMS OF LIGHT EMISSION PHENOMENA FROM SPUTTERED ATOMS

20.2.1 Electron tunneling model

The electron-tunneling or band-structure model was first proposed by van der Weg and Rol [2] following the early work of Hagstrum in ion neutralization spectroscopy measurements [3]. It was assumed that the populations of sputtered excited atoms are intimately connected to the surface band structure. Fig. 20.2 shows the potential energy diagram of a solid metal and an atom separated by a small distance s. The metal surface is characterized by a square potential well of depth ϵ_0 filled with electrons up to the Fermi level ϵ_F. The Fermi level lies below the vacuum level by the work function φ. Two types of electron transitions can be distinguished. In the resonance ionization (RI) process, the electron from an upper excited level of the atom tunnels through the surface potential barrier to an empty state in the conduction band of the solid.

$$X^{0*} \overset{(RI)}{\longrightarrow} X^{+} + e_s^{-} \tag{20.1}$$

where X^{0*} is the excited atom, X^{+} is the resulting ion and e_s^{-} the transfer electron. Since the potential energy of the electron does not change during

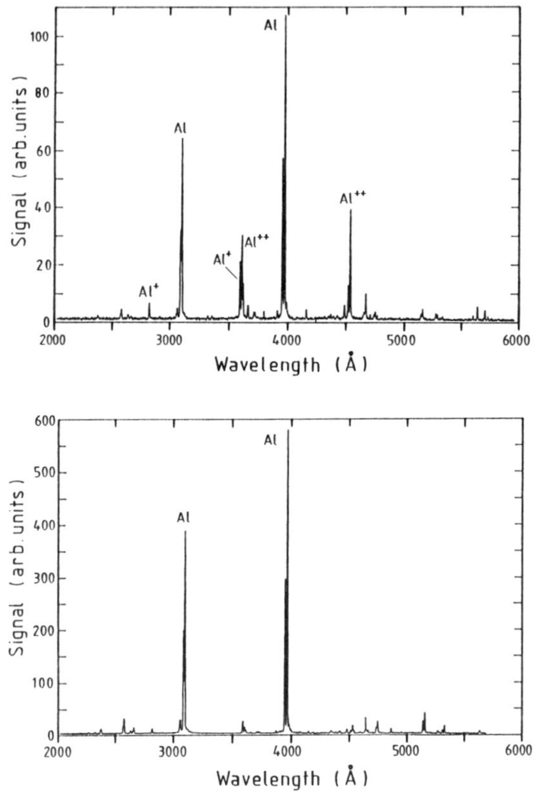

FIGURE 20.3 Light emission spectra from (a) a clean and (b) an oxygen-covered aluminum surface following 300 keV Ar^+ bombardment. Note the different intensity scales; the same arbitrary scale factor applies in both cases. Reprinted from Stefan Reinke, Dieter Rahmann and Rainer Hippler, "Light emission from sputtered atoms following ion bombardment of polycrystalline Al and Cu targets" [4], with kind permission of Elsevier Science Ltd, The Boulevard, Langford Lane, Kidlington 0X5 1GB, UK.

the transition, it is called a resonant process. In the two-electron Auger de-excitation (AD) process, electrons from occupied states of the solid tunnel to energetically lower vacant states of the atom; the energy released in that transition is used to liberate the excited atomic electron,

$$X^{0*} + e_s^- \xrightarrow{(AD)} X^0 + e_A^- \tag{20.2}$$

where e_A^- is the ejected Auger electron. These two radiationless de-excitation processes compete with radiative de-excitation by photon emission. The transition rate for optical emissions is of the order of 10^8/s, which is several orders of magnitude smaller than that of non-radiative transition processes ($\sim 10^{15}$/s). This leads to a significant reduction of the number of slow excited particles in the case of a clean metal surface. On the other hand, when the excited level of an atom lies opposite to a forbidden energy gap as, e.g., in the case of insulators, the tunneling of the electron is impossible and de-excitation always takes place by photon emission.

The probability that an excited atom will escape from the target without undergoing a radiationless de-excitation event is approximately given by

$$P = \exp(-A/av_{\perp}) \tag{20.3}$$

where $v_{\perp}$ is the velocity component of the excited atom perpendicular to the surface and the quantity A/a is sometimes called the survival parameter. This equation simply expresses that only atoms with relatively high energies can leave a metal surface in an excited state and de-excite at large distances by photon emission.

Fig. 20.3 shows a comparison of two optical spectra obtained from 300 keV Ar^+ bombardment of an Al target [4]. The measurements were performed for a clean surface (Fig. 20.3a) and with a partial oxygen pressure of $\approx 5 \times 10^{-5}$ mbar inside the target chamber (Fig.20.3b). An oxygen enhancement of the most prominent spectral lines (4s $\rightarrow$ 3p, 3d $\rightarrow$ 3p, 4d $\rightarrow$ 3p and 5d $\rightarrow$ 3p) is observed. It is assumed that the oxygen exposure creates a band gap at the surface which suppresses resonance tunneling between escaping particles and the surface (Fig. 20.4). There are several drawbacks of the band-gap model. Here RI is thought to be the dominant non-radiative de-excitation process when energetically allowed. Thomas and de Kluizenaar [5] deliberately lowered the work function of Al and Cu by Cs vapor deposition which should lead to an enhancement of the emission of the 3092Å Al line and the 3247Å Cu line due to the blocking of electron tunneling. To the contrary, a 20% decrease in the intensities was actually observed. Second, this model favours the depopulation of the levels of excited neutrals that lie within the energy range of the empty conduction band. It cannot explain the oxygen enhancement of some excited ionic lines whose levels lie deeper inside the solid [6]. Inoue *et al.* [7] have pointed out the importance of AD process in order to explain the variation of photon yield with the Cs-induced work function changes of a GaAs target.

A further major drawback of the electron-tunneling model is that merely the de-excitation mechanism is considered while no satisfactory explanation for the processes leading to the production of excited atoms is provided.

20.2.2 Bond-breaking model

This model was developed on the basis of similarities observed in oxygen enhancement of excitation and ionization [8–11]. It is assumed that at the end of a collision cascade process, a pair of sputtered metal (M) and oxygen (O)

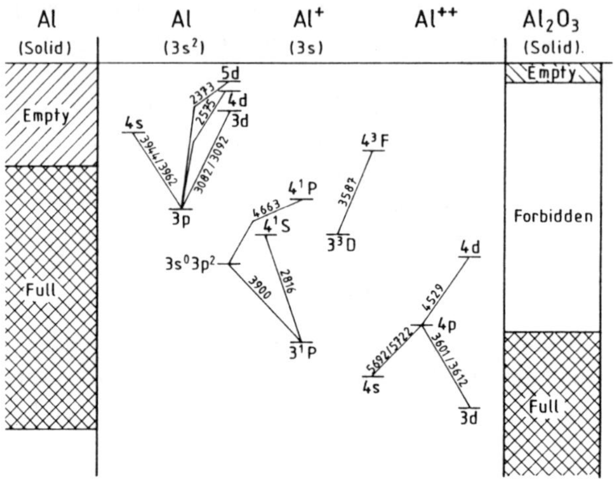

FIGURE 20.4 The atomic energy levels of several Al, Al^+, and Al^{2+} states in comparison to the band structure of metallic (Al) and oxidized (Al_2O_3) aluminum. The wavelengths (in Å) of some transitions are indicated. Reprinted from Stefan Reinke, Dieter Rahmann and Rainer Hippler, "Light emission from sputtered atoms following ion bombardment of polycrystalline Al and Cu targets" [4], with kind permission of Elsevier Science Ltd, The Boulevard, Langford Lane, Kidlington 0X5 1GB, UK.

atoms forms a transient molecule. Provided these atoms are approaching each other close enough, the metal atom can become excited through level-crossing processes in the transiently formed molecule (Fig. 20.5).

20.2.3 Local thermodynamic equilibrium (LTE) model

This model was developed by Andersen and Hinthorne [12] for a quantitative interpretation of secondary ion yields and has been applied to also describe excitation processes. This theory assumes that the sputtering region resembles a dense plasma where the majority of sputtered atoms, ions, molecules and electrons are in local thermodynamic equilibrium. In such a situation, the population of atoms in the excited state is given by Boltzmann distribution. However, this model does not gain popularity primarily on the grounds that the concept of a plasma at LTE conditions is difficult to reconcile with the present state of theoretical knowledge of the basic sputtering process. In addition, some experiments show that the existence of LTE plasma may not be physically justifiable. For example, Snowdon *et al.* [13] showed that relative level populations of Zn I lines can be fitted to a Boltzmann-like distribution, whether the levels are excited by $Ar^+ \rightarrow Zn$ sputtering or $Zn^+ \rightarrow Ar$ gas phase single collisions. The latter certainly cannot be described in LTE terms.

In an another approach Kelly *et al.* [14,15] assumed that the excitation pro-

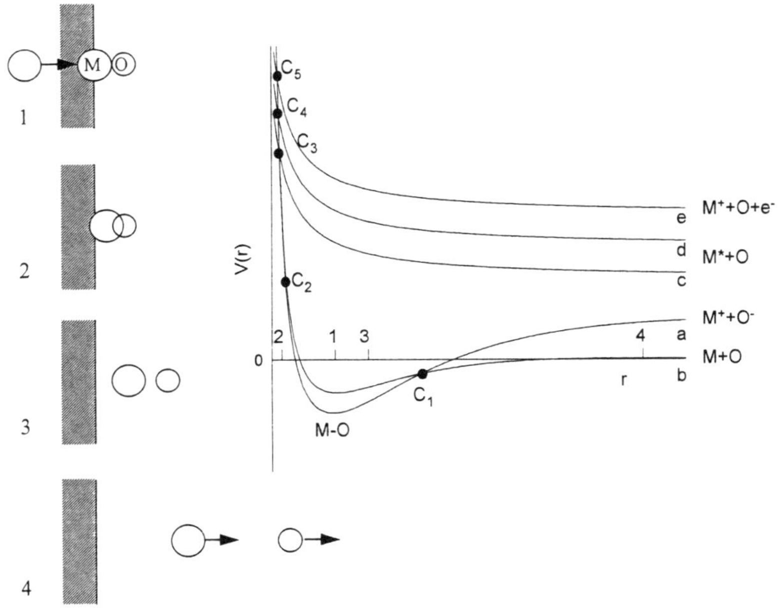

FIGURE 20.5 The potential energies V(R) of the metal-oxygen (M - O) system during its formation and the following dissociation as a function of the internuclear separation R are shown. At high collision energies, the system may cross onto curves c, d, or e at points C_3, C_4 or C_5, respectively, and emission of excited sputtered atoms becomes possible. At low energy collisions the system may follow the curves a or b and ground state metal and oxygen ions or atoms will be sputtered. After Ref. [10], with kind permission of Springer - Verlag, Berlin.

cess takes place in LTE conditions while de-excitation occurs in the free space outside the plasma region. However, the "temperatures" obtained from the LTE plots are found to be independent of ion or target properties contrary to what should be expected from considerations of energy transfer and transport processes within the solid.

20.2.4 A brief discussion of other excitation models

(i) Kelly [16] proposed a statistical model in which electronic excitation takes place through random inelastic energy transfer during energetic collisions. In this model, the population of level i is proportional to the statistical weight factor g_i, provided that the excitation energy is less than the inelastic energy transfer.

(ii) The molecular orbital electron promotion model [17] assumes that excitation is the result of energetic binary collisions among the recoiling target atoms. During the collision the two electron shells overlap and a short-lived quasimolecule is formed. At the crossing of the energy levels, an electron has the possibility of making a transition or being "promoted" to an empty level. According to Veje [18,19], this mechanism is sensitive to the incident ion angle.

(iii) The surface polarization model was put forward by Tsong and Tsuji [20] in connection with the oxygen dependence of the photon yield. It is postulated that the recoil-implanted oxygen atoms act as the electron-emissive sites at the surface. The metal ions, which are supposed to form during the metal-oxygen bond breaking process, pick up electrons through interaction with these polarized surface sites, thereby creating more excited neutrals and giving rise to an enhanced light emission.

(iv) The quantum mechanical model of Schroeer *et al.* [21] assumes excitation of the sputtered atom to be a result of the perturbation imposed by the metal surface. The wavefunction of the valence electron initially belonging to the metal conduction band slowly changes with time as the sputtered atom leaves the surface. The excitation probability was calculated using the adiabatic approximation of the perturbation theory.

20.3 EXPERIMENTAL ARRANGEMENT

The apparatus for a typical bombardment induced light emission experiment [22] is shown in Fig. 20.6. A well-defined beam both with regard to ion species and energy is collimated and directed onto the surface of the solid under consideration. The projectile energy may be varied from as low as a few hundreds of eV to as high as several hundreds or thousands of keV. The light emitted from the target surface is allowed to pass through a quartz window and is subsequently focussed by a quartz lens onto the entrance slit of a grating monochromator to disperse the light beam. The optical signal is recorded by a photomultiplier tube. Single photon pulse counting techniques with computer-controlled wavelength scanning and data acquisition into a multichannel analyzer are usually used. Analog detection can also be performed if the signals are not too weak. In passing, it may be mentioned that if some well-separated specific spectral lines are of interest to study, narrow-band interference filters in place of the monochromator can be used.

In most of the experiments relative line intensities are measured, which gives only qualitative information about the excitation process. However, in order to measure the absolute photon yield it is necessary to know the sensitivity or quantum yield of the optical equipment, i.e., the mean output signal per incoming photon at the relevant wavelengths. The most straightforward calibration method is to use an absolute standard source such as synchrotron radiation [23] or blackbody radiation [24] which emits a known number of photons per second. Andersen *et al.* [25] have discussed in detail various possible ways of determining the overall quantum efficiency with an optical monochromator.

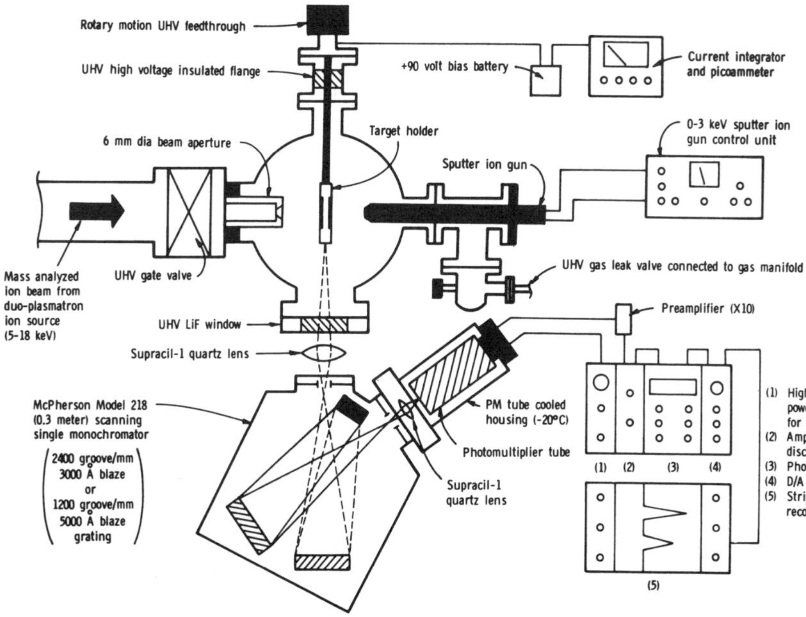

FIGURE 20.6 Experimental setup (schematic). Reprinted from R. B. Wright and D. M. Gruen, "Secondary photon emission studies of ion bombarded beryllium" [22], with kind permission of Elsevier Science - NL, Sara Burgerhartstraat 25, 1055 KV Amsterdam, The Netherlands.

In light emission studies it is often also necessary to know the spatial distribution of the line intensity, i.e., how the intensity varies with distance from the target surface. The extent of the light emitting volume ensures whether the signals refer to all the emissions in the observation region. Yusuf and Tsong [26] employed an intensified silicon intensifier target (ISIT) detector followed by an optical multichannel analyzer (OMA) for direct imaging of the light emitting region. The system resembles a vidicon camera covering the spectral range from ultraviolet to infrared. A similiar setup was used by Arlinghaus and Bispinck [27]. The special features of OMA are the high spatial resolution and real time measuring facility in contrast to a conventional monochromator which detects only the spatially averaged light intensity through a defined slit.

Recently, a digital image processing system has been introduced for light emission spectroscopy especially in plasma diagnostics (e.g., Ref. [28]). It includes an optical spectrograph and a charge coupled device (CCD) camera. The CCD camera combines three important features, namely, the ability to detect low level signals, extreme short shutter times of several ns and digital image acquisition. The CCD sensor holds the image information in a matrix of separated image elements (pixels). The photoelectron charge of each pixel is then read out and digitized by a computer. A typical read-out sensitivity is about 1 camera count per 13 electrons which, for a typical (wavelength-dependent) quantum efficiency of ~ 50 % corresponds to about 1 count per 25 photons. To further improve the sensitivity of the spectrograph/CCD system, a light intensifier is frequently coupled to the CCD camera. The intensifier device consists of a photocathode followed by a single or a tandem microchannel plate electron amplifier and a phosphor screen. It effectively converts a single photon with a photon yield comparable to that of a typical photomultiplier to an avalanche of electrons which in turn produce many photons when hitting the phosphorous screen.

Since the light emission is found to be highly sensitive to the chemical state of the surface, the other important requirements for BLE studies are a sufficient ultra-high-vacuum in the target chamber and an atomically clean target surface. The time for a monolayer of adsorbate formation is few seconds at 10^{-6} mbar of residual gas pressure, while it takes almost an hour when the gas pressure is 10^{-9} mbar assuming all molecules arriving at the surface stick and are incorporated. It may be mentioned that also at high- vacuum conditions one can maintain a clean bombardment condition if the target temperature is kept sufficiently high [29,30]. The target chamber may also be equipped with a small quadrupole mass spectrometer providing an analysis of residual gas composition as well as being useful as a leak detector.

20.4 EXPERIMENTAL OBSERVATIONS

20.4.1 Transients

It is well known that the intensity of spectral lines changes abruptly with time after interrupting the incident beam or due to a sudden change of ion beam current or oxygen pressure. The transient, i.e., the variation of line intensity as a function of time is, generally, explained as due to a change of oxygen coverage on the surface. Transients are quite often observed right after the beginning of the ion bombardment (e.g., Ref. [31]). 2516 Å line at base The light intensity first rises sharply and then declines exponentially to an equilibrium level corresponding to the intensity for the clean surface. Such an effect is not only caused by oxygen and oxygen-containing molecules but, for example, N_2 and NH_3 adsorptions are also known to affect the photon yield in a similar way [32,33].

Oxygen-pressure-induced transients have been studied most frequently. Introduction or withdrawl of oxygen leads to an instantaneous rise or fall, respectively, in the photon signal. The corresponding time constants, however, may vary to a large extent depending upon the nature of the reactions that take place on the surface, e.g., adsorption at random or accompanied by recoil implantation. Betz [32] showed that the existence of transients is related to the development of an oxide-like or nitride-like band structure at the surface. Van der Weg and Lugujjo [34] noted that elements whose d-shells are filled or nearly filled are susceptible to strong transient effects caused by oxygen. The transient measurement due to incremental changes of oxygen pressure is quite interesting. This is shown in Fig. 20.7 for the Si 2882 Å line following 300 keV SF_5^+ ion bombardment of Si. In this experiment [35], a small amount of high-purity oxygen gas was introduced into the target chamber through a pulsed valve and a leak valve in series. The pulsed valve operates for 150 μs for single pulse generation. The pressure first rises to about 5×10^{-5} mbar almost instantaneously and then gradually decreases to the base pressure ($\sim 10^{-9}$ mbar). The line intensity, on the other hand, shows an initial increase, passes through a maximum and then decreases exponentially to an equilibrium level. It is interesting to note that there was a sharp rise in photon intensity at times when the oxygen pressure was rapidly falling down. This is not surprising if one keeps in mind that the degree of oxygen coverage is the consequence of two competing effects, namely, the adsorption of oxygen atoms from the ambient oxygen atmosphere and the sputter removal of oxygen atoms by the incident ion beam. Thus in the rising part of the photon yield, the rate of adsorption exceeds the rate of erosion. The yield proceeds to a maximum corresponding to an instantaneous equilibrium oxygen coverage, and finally falls down due to both sputtering of oxygen atoms and excessive declination of the rate of adsorption.

The measurement of line intensity, thus, requires sufficient time delay to allow for the disappearance of transient regime. Hippler and Reinke [36] studied equilibrium photon intensities of several atomic and ionic aluminum lines as a function of oxygen pressure for 300 keV Ar^+ bombardment of an Al target. The results are shown in Fig. 20.8. A pronounced oxygen dependence is observed for the lines from neutral aluminum atoms; the yield, at first, increases with oxygen pressure and then reaches a plateau at high pressure. The ionic lines are less affected by the oxygenated surface. This indicates that either the suppression of the non-radiative transitions is less complete for deeper energy levels (ionic lines) or that the production mechanisms of excited ions and atoms are different. The increase in the photon yield of atomic lines has been interpreted as being caused by the increasing oxygen coverage of the surface. The oxygen adsorption on the target surface is governed by the rate of arrival of the oxygen atoms Φ_o multiplied by the sticking probability c_s and the rate of removal of oxygen atoms due to sputtering by the incident beam. The equilibrium oxygen coverage Θ is given by [37]

$$\Theta = \frac{\Phi_o}{\Phi_o + \frac{Y_o}{c_s}\Phi_i}, \tag{20.4}$$

where Y_o is the sputtering yield of oxygen and Φ_i the flux density of the incident ion beam. Eq. 20.4 implies a random adsorption of oxygen atoms without any subsequent arrangement or incorporation process. Θ, hence, ranges from

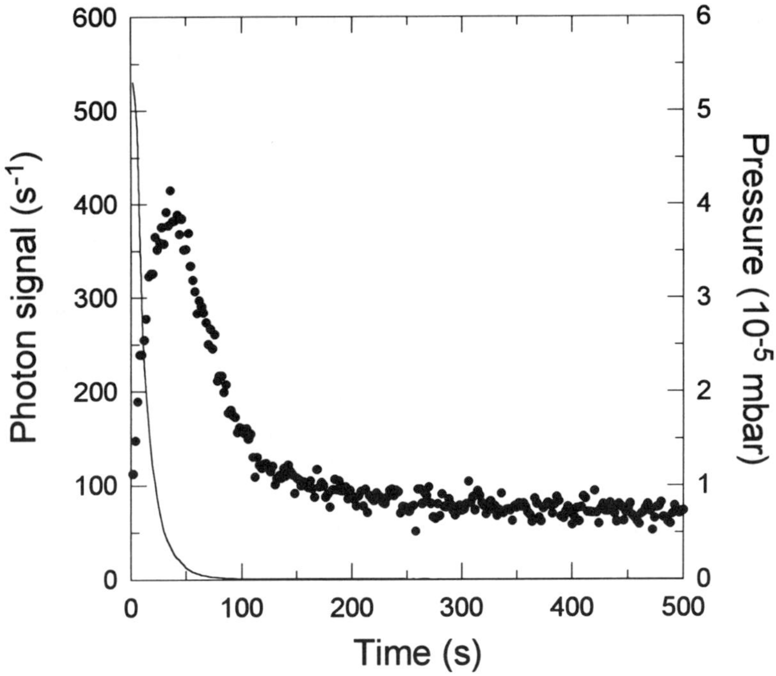

FIGURE 20.7 Transient in the photon intensity of the Si 2882 Å line (•) following 300 keV SF_5^+ bombardment of a silicon target when a small amount of oxygen is introduced by a pulsed valve into the vacuum chamber momentarily; the solid curve shows the oxygen pressure variation with time (after Ref. [35]).

zero (for a clean surface) to unity (complete coverage with oxygen). The photon signal I(Θ) for a given equilibrium coverage Θ may be expressed as a sum of two contributions I_c and I_o from the clean and from the oxygen-covered part, respectively, of the surface

$$I(\Theta) = I_o\Theta + I_c(1 - \Theta). \tag{20.5}$$

Combining Eqs. 20.4 and 20.5, the pressure dependence of photon intensity can be predicted provided Y_o and c_s are known. Vice versa, a fit to the predicted pressure dependence (Eq. 20.5 in combination with Eq. 20.4) of the experimental data will yield the ratio Y_o/c_s. From Fig. 20.8, this ratio is obtained as 30 at an ion incidence angle of 45^0. Assuming $c_s = 0.1$ [37], the sputtering coefficient of adsorbed oxygen atoms on an Al surface is about 2.1 atoms/ion for normal ion incidence.

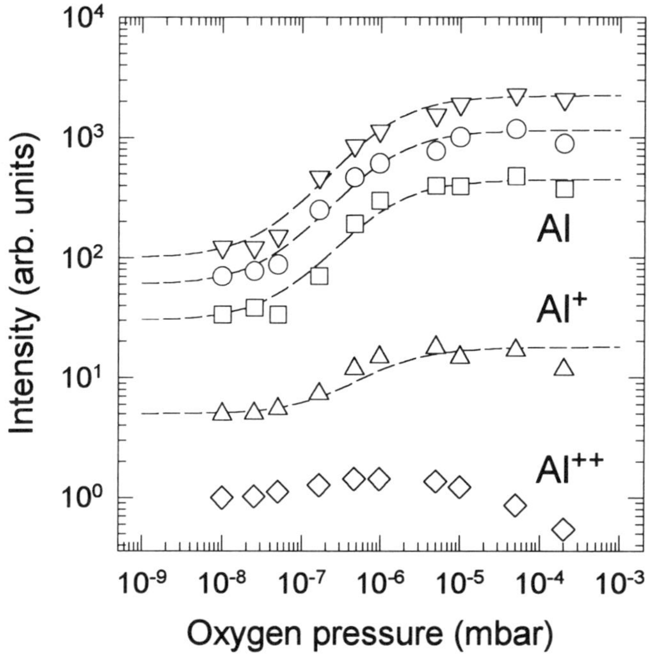

FIGURE 20.8 Light intensity *versus* oxygen pressure for 300 keV Ar^+ bombardment of an aluminum target and for various Al (∇, $\circ$, $\square$), Al^+ ($\triangle$), and Ar^{2+} ($\diamond$) transitions. The dashed lines are corresponding calculations according to Eqs. 20.5 and 20.4. Reprinted from Rainer Hippler and Stefan Reinke, "Oxygen dependence of light emission from sputtered atoms and ions following ion bombardment of polycrystalline metal targets" [36], with kind permission of Elsevier Science - NL, Sara Burgerhartstraat 25, 1055 KV Amsterdam, The Netherlands.

20.4.2 Dependence on ion energy and ion incidence angle

Since a fraction of the sputtered particles leaves in some excited state and participates in the photon emission, it may be thought that a variation of the projectile energy and/or of the angle of ion incidence would affect the photon yield in the same way as the sputtering yield. We have studied the photon emission due to Xe^+, SF_5^+, Kr^+, Ar^+ and Ne^+ ion bombardment of Al, Si and Cu targets as a function of the primary ion energy [38,39]. In a first approximation, the sputtering yield should be proportional to the nuclear stopping power of the projectile ion [40]. Therefore, the data are presented (e.g., Fig. 20.9) as the reduced photon yield, Y^*, plotted as a function of reduced projectile energy, ϵ, as defined by Lindhard *et al.* [41]. In general, the energy dependence of the photon yield is similiar to that of the sputtering yield

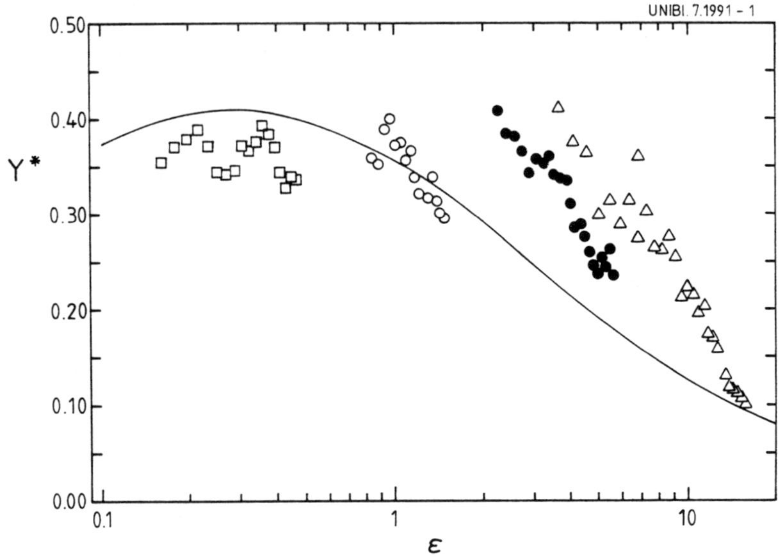

FIGURE 20.9 Reduced photon yield Y^* of the neutral Al 3962 Å line following Ne^+ (Δ), Ar^+ (•), Kr^+ (○), and Xe^+ (□) bombardment of aluminum *versus* reduced energy ϵ. The solid line is the reduced sputtering yield [40], to which the experimental data were normalized. Reprinted from [38] with kind permission of Elsevier Science - NL, Sara Burgerhartstraat 25, 1055 KV Amsterdam, The Netherlands.

but significant differences are also noted. For example, large discrepancies have been observed for the Si I line intensity following Xe^+ and Kr^+ bombardment at small reduced energies. However, in the case of Al III photon yield, the experimental data agree fairly well with the theoretical prediction [38].

Wright and Gruen [42,43] showed that the secondary photon emission yield is essentially determined by the product of the survival probabilities of all possible electron exchange processes between excited particle and the surface. The photon yield Y_{if}^* of a given emission line $i \rightarrow f$, can be expressed as:

$$Y_{if}^* \;\propto\; J \cdot Y \cdot \exp[-(A/a)_i/v_m], \tag{20.6}$$

where J is the ion current density, Y the sputtering yield and v_m the velocity corresponding to the maximum energy transferred between the incident ion and the target atom. $(A/a)_i$ is the effective nonradiative de-excitation parameter, or survival parameter, for the state i.

The angular dependence of the photon yield curve is found to be qualitatively similar to that of the neutral sputtering yield, i.e., the yield first rises monotonically with ϕ, the angle of ion incidence with respect to surface normal, passes through a maximum at $\hat{\phi}$ which is typically at $60 - 80^o$ and then decreases sharply as ϕ approaches 90°. When the target surface is tilted the enhancement of photon yield is, mainly, the result of two effects: the increase

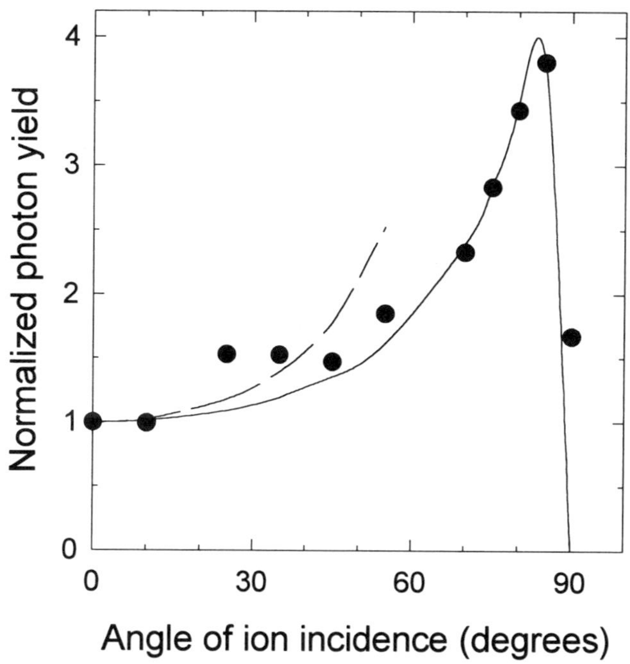

FIGURE 20.10 Normalized photon yield (Si I 2882Å) as a function of the angle of ion incidence for 150 keV Ar^+ bombardment of Si. • - experimental data points. The dashed line is the theoretical calculation of Sigmund [40]. The solid line represents the fit curve to Yamamura's empirical formula with the parameters $f = 1$ and $\Sigma = 0.1$. Σ and f are related by the relation $\hat{\phi} = \cos^{-1}(\Sigma/f)$ (after Ref. [39]).

of (i) the total sputtering yield and (ii) the number of high velocity sputtered atoms. For not-too-oblique ion incidence, Sigmund [40] predicted a $(\cos\phi)^{-f}$ dependence, where $1 < f < 2$, rather than the normally expected $(\cos\phi)^{-1}$ trend. More recently, Yamamura [44] proposed an empirical formula containing two adjustable parameters for the angular dependence of sputtering yield covering the whole angular range. One of the parameters corresponds to Sigmund's f, while the other is related to $\hat{\phi}$. Fig. 20.10 shows the angular variation of the Si 2882 Å line intensity for 150 keV Ar^+ bombardment of a Si target [39]. Larsen and Veje [45] noted that the actual increase depends on the elements and the levels studied. For example, they found stronger angular variations of Be $2p^2$ triplet and $2p$ singlet P levels than for the other Be levels. An electron-promotion process is suggested for such selective excitation.

20.4.3 Doppler broadening of spectral lines

The velocity component of an excited particle in the direction of the monochromator gives rise to a Doppler shift of the detected wavelength. Because of the velocity spread of the moving excited atoms and the different directions in which they move, there is a considerable Doppler broadening of the emission line profiles. Van der Weg and Bierman [46] have measured the wavelength shifts and line profile widths of the Cu 3247 Å line for different angles of ion incidence. The largest Doppler shift occurs at grazing ion incidence, where the line profile is also found to be considerably broader. Their results demonstrate that high kinetic energy particles originating from two-particle collisions at the surface contribute significantly to the line emission. For instance, the observed maximum shift of 0.6 Å corresponds to ~ 1 keV excited Cu atoms.

Hippler *et al.* [47] performed a systematic study of the line shape measurements of neutral 3247.5 Å Cu, 4810.5 and 3075.9 Å Zn and 3092.8 Å Al emission lines during 300 keV Ar^+ bombardment of Cu, Zn and Al. They analyzed the line shape following a similar approach used by van der Weg and Bierman [46], i.e. a model based on binary collision sputtering component combined with non-radiative de-excitation processes by electron tunneling. A limited agreement between the measured and calculated line profiles leads them to conclude that contributions from the collision cascade should also be taken into account. The values of the survival coefficient A/a obtained from a fitting procedure were, 3×10^6 cm/s for Cu, 5×10^6 cm/s for Zn, 5×10^4 cm/s for ZnO, and 5×10^5 cm/s for Al. The above value of A/a for Zn was found for the 4810.5 Å emission line originating from an upper level above the Fermi energy. For the 3075.9 Å line of Zn, which has the excited level below the Fermi level, the coefficient was found to be one order of magnitude smaller.

Betz [48] carried out a comparative study between line profiles at normal and oblique ion incidences, respectively. The results showed that at normal incidence the photon emission is contributed mainly by slow atoms resulting from the collision cascade, while at grazing incidence the contribution is largely due to fast excited atoms resulting from a few violent collisions at the surface. Loxton [49] observed a reasonable "blue" shift (i.e., the shift towards shorter wavelengths) in the intensity peak as well as pronounced asymmetry for the Si 2882 Å line profile at large incidence angle. Another characteristic feature of grazing incidence collisions is that there is also a "red" shift (i.e., the shift towards longer wavelengths) due to reflected component of the observed emission causing a double peak structure in the line profiles [47–50].

The Doppler broadened line shapes can be well-fitted to Thompson's velocity distribution of sputtered atoms [51] provided that the surface binding energy, E_b, is treated as a fitting parameter. The energy (or velocity) spectrum of the excited atoms is determined by the corresponding distributions of all emitted particles and the survival probability, i.e.,

$$\Phi(E, v_\perp) \propto \frac{E}{(E+E_b)^k} \exp(-A/av_\perp) \tag{20.7}$$

where the exponent $k = 3$ means the usual Thompson's formula. E is the kinetic energy of the sputtered target atoms. The results of Bissesur and Tsong [52] showed that the full-width at the half-maximum (FWHM) of the line shape, however, increases consistently with distance away from the surface indicating the presence of both slow and fast particles.

Reinke *et al.* [4] found that a better fit to Thompson's distribution, Eq. 20.7, is obtained if the exponent of the denominator is chosen to be $k = 2.6$ rather than 3. They studied the line shape of the neutral Al 3093 Å transition as well as that of the ionic Al^{2+} 4529 Å transition under 300 keV Ar^+ bombardment, for a clean and an oxygen-covered Al surface (Fig. 20.11). In the case of Al 3093 Å line, the profile for the oxidized surface is narrower than that for the clean surface. In the case of the Al^{2+} 4529 Å line, on the other hand, the line width changes little under exposure of oxygen. This is not unexpected since owing to the energetic position of the involved Al^{2+} state with respect to the band energies of the Al bulk (Fig. 20.4), the non-radiative de-excitation processes cannot take place. However, the most striking feature is that the broadening of the ionic line is found to be almost double than that of the neutral atomic line. Moreover, when the ionic line profile is fitted to Eq. 20.7, the parameter E_b yields a very high value of 500 eV [53]. All these observations imply that an appreciable fraction of the excited Al^{2+} ions has kinetic energies of more than 1 keV. This is considerably larger than the kinetic energies with which the excited neutral Al atoms leave the surface and may be taken as evidence of different production mechanisms for excited Al atoms and Al^{2+} ions. Presumably, the Al^{2+} ions are produced in close and, hence, more violent encounters than the neutral atoms which are believed to result from the sputtering cascade.

20.4.4 Velocity distribution of sputtered excited atoms

The velocity or energy distributions of sputtered excited atoms can conveniently be determined from Doppler broadened line profile measurements. For example, Bissessur and Tsong [54] measured the line shape of the K I 7665/7699 Å emissions and analyzed their results in terms of Thompson's energy distribution formula [55]. The kinetic energies of the excited atoms are found to be rather small and around 10–40 eV.

A second method to determine the velocity distribution is by measuring the light intensity as a function of the perpendicular distance from the surface. The intensity decay curve can be described by the following relation

$$I(x) = I_0 \exp(-x/v_\perp \tau) \tag{20.8}$$

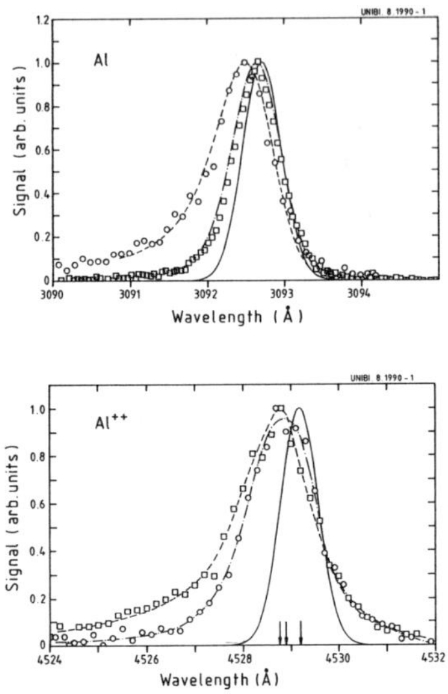

FIGURE 20.11 Line shape of (a) the $3s^23d \rightarrow 3s^23p$ transition ($\lambda = 3092.7$ Å) in neutral Al and (b) of the $3d \rightarrow 3p$ transition ($\lambda = 4529$ Å) in ionic Al^{2+} following 300 keV Ar^+ bombardment of clean (○) and an oxidized (□) aluminum target. The solid line represents a reference line measured with a hollow-cathode lamp. In (a), dashed and dash-dotted lines are corresponding calculations based on the resonance tunnel model. In (b), dashed and dash-dotted lines are to guide the eye only. Reprinted from S. Reinke, D. Rahmann and R. Hippler, [4], with kind permission of Elsevier Science Ltd, The Boulevard, Langford Lane, Kidlington 0X5 1GB, UK.

where *I(x)* is the line intensity for a particular transition at a distance x from the target surface, I_0 the initial intensity close to the surface, $v_\perp$ the velocity component normal to the surface and $\tau = 1/\sum A_{if}$, where A_{if} is the transition rate from an initial level i to a particular final state *f*, is the lifetime of the decaying level i.

The experimental decay curves, however, often show departures from linearity on a semi-logarithmic plot, especially at large distances from the surface. For many levels the plots show a pronounced curvature which is due to the velocity spread of the sputtered particles and is also caused by cascading transitions from higher-lying levels. Ignoring the latter effect, Gritsyna *et al.* [56,57] and Braun *et al.* [58] have analyzed their data as the sum of two exponential decay curves thereby resulting in two distinct groups of sputtered excited atoms, one at ($10^2 - 10^3$ eV) and the other at ($10^3 - 10^4$ eV). Later, Mladenov and Braun [59] simulated their light-versus-distance data incorporating the effect of cosine angular distribution and E^{-2} kinetic energy distribution of excited particles. Fitting the decay curve to several exponentials, they found contributions of 50 eV, 100 eV, 500 eV, 1 keV, 5 keV and 10 keV particles.

Dzioba *et al.* [60] examined the problem from a different point of view. They assumed a minimum kinetic energy E^* for the formation of an excited state and defined the probability of excitation *P(E)* as the step function given by

$$P(E) = \begin{cases} 0 \text{ for } E < E^* \\ 1 \text{ for } E \geq E^* \end{cases} \tag{20.9}$$

With an E^{-2} energy distribution of the sputtered particles and including the radiative and non-radiative de-excitation terms in their usual forms, Dzioba *et al.* [60] were able to derive both the integral photon yield in front of the target and the differential photon yield when the emission is viewed through a narrow slit, respectively. The results are shown in Fig. 20.12. The experimentally measured yield values are first converted to z using Fig. 20.12 and then the values of z are plotted against x, the slope of which gives the value of E^*. Here z is a dimensionless distance variable, given by

$$z = (a + \sum A_{if}\, x\, \sqrt{m/2})/\sqrt{E^*} \tag{20.10}$$

where a is a constant and m is the mass of the excited atom. Later Dzioba and Kelly [61] used a variety of different excitation functions $P(E)$ in their calculations. The values of E^* were, however, found to be largely independent to the choice of *P(E)* and to lie mainly in the interval 10–1000 eV [60–67], which is rather large when compared to that of sputtered ions and neutrals [61].

In this context, we should distinguish among the kinetic energy parameters derived from different measurements. The first one is the E^* introduced in the decay curve model of Dzioba *et al.* [60,61]. It is considered as the threshold energy for excitation with the equivalent meaning if excitation were via a "curve-crossing" process. However, such a sharp cut-off in the energy distribution of excited atoms has been criticized by Sigmund [68]. The second one is $E_\perp$ which is the average energy component perpendicular to the surface. It is defined through Eq. 20.8, where the experiment is performed at normal incidence bombardment and photon emission is observed at 90^0 to the beam. The third energy parameter is associated with the line profile measurements. It is the average energy with component parallel to the observation direction. Experimentally, it has been found that all these energy parameters yield comparable values [64,65].

One disadvantage of the light-versus-distance method is the need to know the lifetime of the excited state. Also, due to use of "defining slits", the spatial resolving power is generally not sufficient to identify excited particles with low kinetic energy. However, the results obtained from a slitless approach using OMA by Yusuf and Tsong [64], where the spatial resolution was considerably improved, are found to be not that different from those of "defining slits" experiments. The difficulty to properly account for low energy particles is also present in Doppler-broadening experiments. One should note that there may

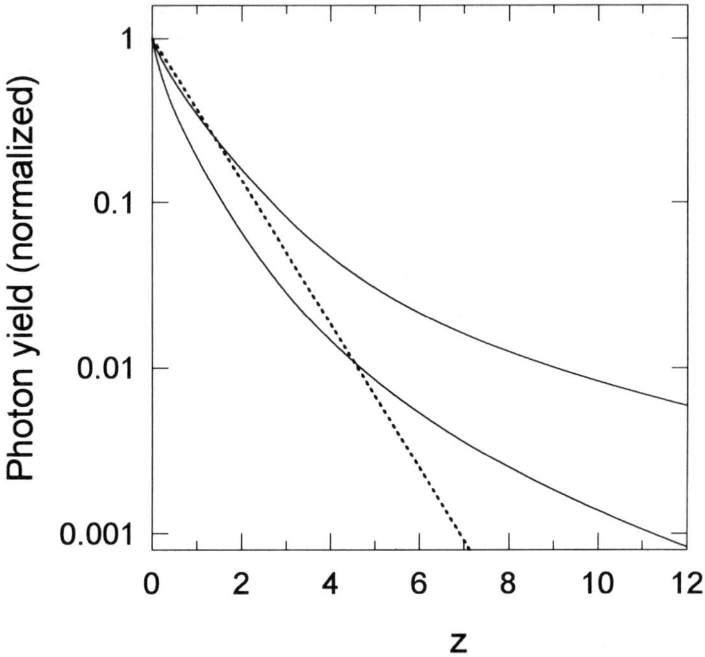

FIGURE 20.12 The normalized photon yield functions in integral, Y^{int}, and differential, Y^{diff}, forms versus the dimensionless distance variable z (Eq. 20.10). The dotted curve represents the exponential decay according to Eq. 20.8 Reprinted from Steven Dzioba, Orlando Auciello and Roger Kelly, "On the kinetic energies of sputtered excited particles I. Atoms sputtered from Li, LiF and NaCl," [60]) with kind permission from Gordon and Breach Publishers.

be additional line-broadening effects due to the interaction of the excited atom with the surface, which will introduce an error in the estimation of the velocity. Finally, although intensity decay measurements are influenced by cascading from higher levels which tend to give high energies at large distances, the line profile technique is essentially free from cascading.

20.4.5 Effect of ion charge on light emission

A multiply charged ion is characterized by its high potential energy, which is the sum of the successive ionization energies. For example this energy is ~ 14 keV for Ar^{18+} and ~ 500 keV for U^{92+}. When a slow multiply charged ion approaches the surface, it strongly interacts with the surface before actually touching it. The incident ion captures electrons from the surface into high-lying Rydberg states via resonance neutralization followed by de-excitation steps by Auger transitions filling the lower-lying levels. This results in a depletion of electrons near the surface. If the electronic relaxation time of the solid is long enough (e.g. in insulators and semiconductors) the target particles are ejected via mutual electrostatic repulsion to an amount roughly proportional to the ion's potential energy. This phenomenon, dubbed "Coulomb explosion" or "electronic" sputtering, was first proposed by Parilis [69] and the idea was extended by Bitenskii *et al.* [70]. However, controversy exists whether there is any additional sputtering due to this mechanism. For example, de Zwart *et al.* [71] studied the sputtering of Si by 20 keV Ar^{q+} ions ($1 \leq q \leq 9$). They found that while the secondary ion yield increases by a factor of 2 when q changes from 4 to 9, the total sputtering yield remains constant up to $q = 9$. They concluded that the emission of neutral particles is essentially a "kinetic" process, while the "electronic" process leads to the emision of particles in an ionized or excited state. More recently Neidhart *et al.* [72] observed an enhancement of the total sputter yield in proportion to the potential energy stored in the projectile during impact of slow ($\leq$1 keV) Ar^{q+} ions ($q \leq 9$) on LiF. Most of the sputtered particles are found to be neutrals having thermal energies. It has been shown that the ejection of particles is related to the formation of additional holes in the $F(2p)$ valence band during neutralization processes of the incident ion followed by decay of electron-hole pairs rather than Coulomb sputtering.

Tribble *et al.* [73–75] measured the photon yield of the Na D lines from sodium atoms sputtered from a NaCl target by 48 keV Ar^{q+} ions ($q \leq 12$). The light yield from sputtered atoms that decayed at least 0.5 mm away from the target surface appears to be insensitive to the charge state of the projectile. The light emission at or very close to the surface, on the other hand, shows an increase of about 25 % , when q increases from 4 to 12. The authors speculated that the large energy required to neutralize a highly charged projectile may cause an increase of electronic excitation of the surface atoms but not of the ejected atoms. Also, they thought that the lattice defects may have some role in the excitation. The results of another experiment during Ar^{q+} ($q =$ 4–11) bombardment of Ag by the same group [76] showed no charge state dependence of the light yield (Ag I 3383 Å), presumably because of short electronic relaxation times of metals.

excited Si atoms

Xe ion charge states.

20.4.6 A brief account of the light emission resulting from other sources

(a) Continuum emission

Ion bombardment of solids often exhibits unresolved broad-band or continuum radiation, the first kind of which is generally observed for transition metals with unfilled nd-shells (Mo, Ta, W, Ni, Nb and Cr) [77] Although there are a number of speculations about the origin of this continuum, it is believed that the continua are due to the radiative decay of excited metal oxide molecules or clusters ejected as a result of collision processes occurring in the near surface region.

The second type of continuum arises during ion bombardment of a solid with energetic inert gas ions [78]. The projectile ions are implanted and trapped in the target material and subsequently released as excited molecules. These broad band emissions occur in the vacuum ultraviolet (VUV) spectral region and are attributed to the transitions from a transient bound excited state to the repulsive ground state of a rare gas dimer.

The third kind of radiative continuum occurs when insulating targets like alkali halides are bombarded with energetic ions [79]. A similar continuum luminescence is also observed in metals, e.g., Al during medium energy (10–30 keV) H^+ and He^+ bombardment [80]. Very recently, Suchańska *et al.* [81] studied ion-beam-induced luminescence in a number of pure and impurity-doped alkali halide crystals, namely, NaCl, KCl, KI, KBr, LiF and NaF. The results show that apart from activated luminescence bands, there exist intrinsic luminescence bands which are more prominent and characteristic of pure crystals. The band positions and half-widths of the intrinsic luminescence spectra remain invariant with the change of ion beam parameters and also with target temperature. However, a blue shift of the peak intensity occurs for crystals with wider band-gap energies. It has been proposed that the ionoluminescence is caused by the radiative recombination of electrons from the conduction band with holes at the V_2 center. A large number of such centers is formed for ion doses larger than $10^{16}/\mathrm{cm}^2$ and these are found to be stable at room temperature. At high temperatures (> 300 K) or doses ($> 10^{17}/\mathrm{cm}^2$), the V_2 centers tend to agglomerate and transform to V_3 centers. These changes are reflected in the decrease of the luminescence intensity.

(b) Light emission from excited projectiles

A fraction of the projectiles backscattered from the surface may also be in an excited state either as scattered ions or as neutrals, which may decay radiatively. As with sputtered excited particles radiationless processes such as RI and AD can significantly alter the proportion of the light emitting species when the departing particles are in close proximity to the surface. However, the mechanisms of excitation are likely to be different from those for sputtered particles. Veje [82] studied the energy dependence of some excited triplet levels of neutral helium backscattered from Au, Tl and Pb targets. In the projectile energy range 20–110 keV, the yield first rises to ~ 40 keV and then reaches a plateau. It was concluded that excited particles are predominantly fast particles and the excitation results from the direct capture of electrons from the top of the valence band when the projectile is leaving the surface. The excited levels are found to be populated according to n^{-4} dependence, where n is the principal quantum number.

There are several interesting investigations of projectile excitation in grazing ion-surface collisions [83]. For example, Martin *et al.* [84] observed circular polarization of H_α Balmer line during glancing collisions of 50 keV H_2^+ with a Nb surface. The values of the normalized Stokes parameters are found to be closely dependent on the degree of surface oxygen coverage.

Quite recently, Schiwietz *et al.* [85] studied photon and particle emissions by highly charged Xe^{q+} ion impact on SiO_2. The specific energy was 2.2 keV/q and the incident charge q was varied between $q = 30$ to 50. No spectral analysis was performed for the emitted photons, which were detected with an open channelplate detector sensitive to the vacuum ultraviolet (VUV) part of the electromagnetic spectrum. A pronounced rise of the photon yield with increasing charge state was observed while at the same time the total sputtering yield displayed a significantly weaker q-dependence. The absolute photon yield was about 1 to 10 photons per incident ion. The physical origin of the emitted photons did not become clear from this investigation, although it was expected that the major fraction could result from backscattered projectiles rather than from sputtered excited atoms.

20.5 APPLICATIONS TO SURFACE ANALYSIS

The phenomenon of light emission from sputtered atoms serves as the basis of surface chemical analysis, where the species present at the surface can be identified by their characteristic line emissions. White *et al.* [86–88] proposed the acronym SCANIIR (surface composition by analysis of neutral and ion impact radiation) for this surface analytical technique, which has the following advantages:

(i) The presence of a relatively small number of lines simplifies the identification of elements without complications of line interference.

(ii) It has the ability of depth profiling, i.e., one can study not only the composition at the surface but also as a function of depth. Since layer by layer of the target surface is successively sputtered away, the depth distribution of elements can be found from the variations of the analytical signal with time. This is particularly useful in the investigation of thin films and layered structures in which the location of interfaces can also be determined.

(iii) It is also capable to detect hydrogen [89].

(iv) Finally, this technique is suitable for the analysis of insulating targets. Not only is the sensitivity greatly enhanced because of the increased photon yield in the presence of a large forbidden energy band gap, but also the charge buildup at the surface does not affect the kinetic energy of the light-emitting excited sputtered neutrals.

The above advantages are, however, offset by the lack of quantitative understanding of photon emission processes. Nevertheless, it may be possible to do quantitative analysis using a standard [88]. SCANIIR also provides photon imaging for element mapping across the sample surface. Following the early works of Mériaux *et al.* [90], Inoue *et al.* [91] recently, developed a scanning SCANIIR microscope employing a photon counting technique and a ratemeter for lateral and depth composition analysis of solids. In this system, the primary beam current is kept low enough to minimize the damage of the specimen surface. The authors estimate the spatial and depth resolutions as 30 μm and 40 nm, respectively.

20.6 FUTURE PROSPECTS

At least qualitatively, we do now understand some of the basic processes of BLE. Nevertheless, there are not enough data on basic parameters such as the dependence of the photon yield on the incident ion energy, the spatial distribution of the emitted light, the effect of adsorption of reactive species on the yield, etc., in well-defined conditions. Since the surface cleanliness and the chemical nature of the contaminants strongly affect the observations, it is highly desirable to perform the measurements in ultra-high-vacuum systems.

Though the electron tunneling model has had some success in the interpretation of experimental results, strictly speaking it does not provide so much on the excitation mechanism itself. However, the relative importance of one-electron process or two-electron processes during the interaction of the excited particle with the surface has not yet been made clear. Therefore, it will be of great interest to look for Auger electrons from the AD processes simultaneously with photon emission analysis. It is worthwhile to compare the ion-solid experiments directly with those of isolated ion-gas atom collisions in order to shed some light on the debate whether excited particles are produced during the first few collisions at the surface or at the end of the collision cascade formation. The excited ion-photon and Auger electron-photon coincidence experiments of Yurasova [92,93] suggest that a significant fraction of excited sputtered particles is produced by energetic collisions between the primary ion and target atoms as well as between target atoms through inner-shell excitations. The study of secondary photon emission under polyatomic or cluster ion bombardment is quite untouched so far as the review article of Andersen [94] is concerned. Such experiments may provide answers whether the excitation depends on the energy-deposition density within the collision cascade.

The vacuum ultraviolet (VUV) part of the emission spectrum is not properly explored so far. It is known that most of excited multicharged ions have strong emission lines in the VUV region and so also for the elements of several group III - V compound semiconductors, e.g., GaAs, which are important in microelectronics industry. Moreover, the resonance lines of the contaminants, like H_2, N_2, O_2 and C that are usually present on the target surface, lie in VUV rather than in the visible part of the spectrum. Thus, direct monitoring of contaminants during the experiment via their VUV emission should be favourable and possible. For example, Schartner *et al.* [95] detected strong spectral lines from excited oxygen atoms and ions in the region 800–1000Å under 300 keV Ar^+ impact of Si. These lines provide alternative means of studying the mechanism of oxygen enhancement of photon yield as well as monitoring surface oxygen coverages. Schartner *et al.* [95] also noted that current models of excited state formation are inadequate to explain the excitation of oxygen atoms.

There is little progress on light emission phenomenon due to bombardment by multiply charged ions. The "potential sputtering" via "Coulomb explosion" may be verified using ultra-slow very highly charged ions interacting with insulator targets [96]. It will also be interesting to investigate large band gap metal oxides, e.g., Al_2O_3, which are somewhat electrically conducting and hence facilitate the measurements. Parilis [96] proposed that one of the possible ways to obtain evidence for Coulomb sputtering is to use a high-temperature superconducting target. In that case, one can look at abnormal changes in the photon yield, if any, around the critical temperature. Photon emission study from highly charged ions at grazing ion-surface collisions will be highly interesting especially from the point of view of circular polarization of the emitted light [97]. The measurements of Stokes parameters would provide information concerning the non-statistical populations of the various magnetic sublevels, which could serve as a signature of the charge transfer processes at the surface.

Finally, less research has been carried out to examine the potentiality of BLE as an analytical tool. Although this technique has several limitations, as do all analytical methods, it can be combined with other techniques to provide a detailed characterization of the surface. It should be noted that since the line intensity is matrix and surface sensitive, a quantitative analysis is not possible unless there is a well-formulated theory to describe the excited-state formation.

We may conclude, as Yu [98] remarked, that "this field is still wide open, with exciting opportunities".

REFERENCES

1. E. W. Thomas, Prog. Surf. Sci. **10** (1980) 383.
2. W. F. van der Weg and P. K. Rol, Nucl. Instr. Meth. **38** (1965) 274.
3. H. D. Hagstrum, Phys. Rev. **96** (1954) 336.
4. S. Reinke, D. Rahmann and R. Hippler, Vacuum **42** (1991) 807.
5. G.E. Thomas and E. E. de Kluizenaar, Nucl. Instr. Meth. **132** (1976) 449.
6. J.P. Mériaux, R. Goutte and C. Guillaud, Appl. Phys. **7** (1975) 313.
7. M. Inoue, Y. Sugiyama, S. Nishigaki and T. Noda, Surf. Sci. **209** (1989) L157.
8. G. Blaise, Surf. Sci. **60** (1976) 65.
9. G. E. Thomas, Rad. Effects **31** (1977) 185.
10. I.S.T. Tsong, in: *Inelastic Particle-Surface Collisions*, (E. Taglauer and W. Heiland, Eds.), Springer-Verlag, Berlin (1981), p. 258.
11. M. L. Yu in: *Secondary Ion Mass Spectrometry SIMS VI* (A. Benninghoven, A. M. Huber and H.W. Werner, Eds.) Wiley, New York (1988), p.41.
12. C.A. Andersen and J. R. Hinthorne, Anal. Chem. **45** (1973) 1421.

13. K.J. Snowdon, G. Carter, D. G. Armour, B.Andresen and E. Veje, Surf. Sci. **90** (1979)429.
14. R. Kelly, C. J. Good-Zamin, M. T. Shehata and D. B. Squires, Nucl. Instr. Meth. **149** (1978) 563.
15. C. J. Good-Zamin, M.T. Shehata, D. B. Squires and R. Kelly, Rad. Effects. **35** (1978) 139.
16. R. Kelly, Phys. Rev. **B25** (1982) 700.
17. M. Barat and W. Lichten, Phys. Rev. **A6** (1972) 211.
18. E. Veje, Nucl. Instr. Meth. **B2** (1984) 520.
19. P. Larsen and E. Veje, Phys. Rev. **B28** (1983) 5011.
20. I.S.T. Tsong and S. Tsuji, Surf. Sci. **94** (1980) 269.
21. J.M. Schroeer, T.N. Rhodin and R. C. Bradley, Surf. Sci. **34** (1973) 571.
22. R. B. Wright and D. M. Gruen, Nucl. Instr. Meth. **170** (1980) 577
23. J. A. R. Samson, *Methods of Experimental Physics*, Vol. 13, Part A (D. Williams, Ed.), Academic Press, New York (1976).
24. J. van den Bos, G. J. Winter and F. J. de Heer, Physica, **40** (1968) 357.
25. N. Andersen, K. Jensen, J. Jepsen, J. Melskens and E. Veje, Appl. Opt. **13** (1974) 1965.
26. N. A. Yusuf and I.S.T. Tsong, Surf. Sci. **108** (1981) 578.
27. H. Arlinghaus and H. Bispinck, Surf. Sci. **134** (1983) 567
28. M. Mimura and K. Sato, *The Transactions of the IEICE*, **E73** (1990) 1361.
29. L. Gabla and M. Suchańska, J. Phys. : Condens. Matter **3** (1991) 6159.
30. C. Assad and R. E. Tribble, Nucl. Instr. Meth. **B93** (1994) 427.
31. D. Ghose, U. Brinkmann and R. Hippler, Surf. Sci. **327** (1995) 53.
32. G. Betz, Nucl. Instr. Meth. **B27** (1987) 104.
33. R. S. Bhattacharya, J. F. Van der Veen, C. B. W. Kerkdijk and F.W. Saris, Rad. Effects **32** (1977) 25.
34. W. F. Van der Weg and E. Lugujjo, In: *Atomic Collisions in Solids*, Vol. 2 (S. Datz, B. R. Appleton and C. D. Moak, Eds.), Plenum Press, New York (1975), p. 511.
35. D. Ghose, U. Brinkmann and R. Hippler, to be published.
36. R. Hippler and S. Reinke, Nucl. Instr. Meth. **B68** (1992) 413.
37. R. Kelly and C. B. Kerkdijk, Surf. Sci. **46** (1974) 537.
38. S. Reinke and R. Hippler, Nucl. Instr. Meth. **B67** (1992) 620.
39. D. Ghose, U. Brinkmann and R. Hippler, to be published.
40. P. Sigmund, Phys. Rev. **184** (1969) 383.
41. J. Lindhard, V. Nielsen and M. Scharff, Mat. Fys. Medd. Dan. Vid. Selsk. **36** (1968) No. 10.
42. R. B. Wright and D. M. Gruen, Nucl. Instr. Meth. **170** (1980) 577.
43. R. B. Wright and D. M. Gruen, J. Chem. Phys. **73** (1980) 664.
44. Y. Yamamura, Rad. Effects **80** (1984) 57.
45. P. Larsen and E. Veje, Phys. Rev. **28** (1983) 5011.
46. W. F. Van der Weg and D. J. Bierman, Physica **44** (1969) 206.
47. R. Hippler, W. Krüger, A. Scharmann and K. -H. Schartner, Nucl. Instr. Meth. **132** (1976) 439.

48. G. Betz, Nucl. Instr. Meth. **B27** (1987) 104.
49. C.M. Loxton, Ph.D. Thesis, Australian National University (1981).
50. P.G. Fournier, J. Fournier and B. Bellaoui, Nucl. Instr. Meth. **B67** (1992) 604.
51. M.W. Thompson, Phil. Mag. **18** (1968) 377.
52. V. Bissessur and I.S.T. Tsong, Nucl. Instr. Meth. **B52** (1990) 129.
53. S. Reinke, Diplomarbeit, Universität Bielefeld (1991).
54. V. Bissessur and I. S. T. Tsong, Nucl. Instr. Meth. **B52** (1990) 129.
55. M. W. Thompson, Phil. Mag. **18** (1968) 377.
56. V. V. Gritsyna, T. S. Kijan, R. Goutte, A. G. Koval' and Ya. M. Fogel', Bull. Acad. Sci. USSR **35** (1971) 530.
57. T. S. Kijan, V. V. Gritsyna and Ya. M. Fogel', Nucl. Instr. Meth. **132** (1976) 435.
58. M. Braun, B. Emmoth and I. Martinson, Physica Scripta **10** (1974) 133.
59. G. M. Mladenov and M. Braun, phys. stat. sol. (a) **53** (1979) 631.
60. S. Dzioba, O. Auciello and R. Kelly, Rad. Effects **45** (1980) 235.
61. S. Dzioba and R. Kelly, Surf. Sci. **100** (1980) 119.
62. S. Dzioba and R. Kelly, Nucl. Instr. Meth. **182/183** (1981) 207.
63. I. S. T. Tsong and N. A. Yusuf, Nucl. Instr. Meth. **170** (1980) 357.
64. N. A. Yusuf and I. S. T. Tsong, Surf. Sci. **108** (1981) 578.
65. C. M. Loxton and R. J. MacDonald, Surf. Sci. **110** (1981) 339.
66. S. V. Teplov and V. P. Shestov, Sov. Phys. Journal **28** (1985) 182.
67. A.I. Bazhin, V.V. Kozel, V.P. Shestov, M. Suchańska and S.V. Teplov, phys. stat. sol. (a) **112** (1989) 437
68. P. Sigmund, in: *Inelastic Particle-Surface Collisions* (E. Taglauer and W. Heiland, Eds.), Springer-Verlag, Berlin (1981), p. 251.
69. E.S. Parilis, In : *Atomic Collision Phenomena in Solids*, (D.W. Palmer, M.W. Thompson and P.D. Townsend, Eds.), North-Holland, Amsterdam (1970), p.324.
70. I.S. Bitenskii, M.N. Murakhmetov and E.S. Parilis, Sov. Phys. Tech. Phys. **24** (1979) 618.
71. S.T. de Zwart, T. Fried, D.O.Boerma, R.Hoekstra, A.G.Drentje and A.L. Boers, Surf. Sci. **177** (1986) L 939.
72. T. Neidhart, F. Pichler, F. Aumayr, H.P. Winter, M. Schmid and P. Varga, Phys. Rev. Lett. **74** (1995) 5280.
73. R.E. Tribble, M.H. Prior and R.G. Stokstad, Nucl. Instr. Meth. **B44** (1990) 412.
74. C. Assad and R.E. Tribble, Nucl. Instr. Meth. **B93** (1994) 427.
75. C. Assad and R.E. Tribble, Nucl. Instr. Meth. **B99** (1995) 152.
76. C. Assad, W. Liu and R.E. Tribble, Nucl. Instr. Meth. **B62** (1991) 201.
77. e.g., C. W. White, N. H. Tolk, J. Kraus and W. F. Van der Weg, Nucl. Instr. Meth. **132** (1976) 419.
78. e.g., M. Braun and B. Emmoth, Nucl. Instr. Meth. **170** (1980) 585.
79. e.g., N. H. Tolk, D. L. Simms, E. B. Foley and C. W. White, Rad. Effects **18** (1973) 221.

80. M. Zivitz and E. W. Thomas, Phys. Rev. **B13** (1976) 2747.
81. M. Suchańska, A. I. Bazhin and E. I. Konopelko, phys. stat. sol. (b) **182** (1994) 231.
82. E. Veje, Nucl. Instr. Meth. **B2** (1984) 218.
83. H. Winter and R. Zimny, in : *Coherence in Atomic Collision Physics*, (H. J. Beyer, K. Blum and R. Hippler, Eds.), Plenum Press, New York (1988), p. 283.
84. P. J. Martin, L. Berzins and R. J. MacDonald, Surf. Sci. **95** (1980) L 277.
85. G. Schiwietz, D. Schneider, M. Clark, B. Skogvall, D. DeWitt and J. McDonald, Rad. Effects and Def. Solids **127** (1993) 11.
86. C. W. White, D.L. Simms and N.H. Tolk, Science **177** (1972) 481.
87. C.W. White, D. L. Simms and N.H. Tolk, In : *Characterization of Solid Surfaces* (P.F. Kane and G.R. Larrabee, Eds.), Plenum, New York (1974), p. 641.
88. N.H. Tolk, I.S.T.Tsong and C.W. White, Anal. Chem. **49** (1977) 16A.
89. I.S.T. Tsong and R.B. Liebert, Nucl. Instr. Meth. **149** (1978) 523.
90. J.P. Mériaux, R. Goutte and C. Guillaud, J. Radioanal. Chem. **12** (1972) 53.
91. M. Inoue, S. Kunitomo, S. Nishigaki and T. Noda, Jpn. J. Appl. Phys. **31** (1992) L 200.
92. V. E. Yurasova and I. F. Urazgil'din, Rad. Effects and Def. Solids **117** (1991) 99.
93. D. V. Ledyankin, I. F. Urazgil'din and V. E. Yurasova, Sov. Phys. JETP **67** (1988) 2442.
94. H. H. Andersen, Mat. Fys. Medd. Dan. Vid. Selsk. **43** (1993) 127.
95. K. -H. Schartner, H. J. Flaig, D. Hasselkamp and A. Scharmann, Nucl. Instr. Meth. **168** (1980) 419.
96. E. S. Parilis, Z. Phys. D **21** (1991) S 127.
97. H.J. Andrä, Phys. Lett. **54 A** (1975) 315.
98. M. L. Yu, in: *Sputtering by Particle Bombardment III*, (R. Behrisch and K. Wittmaack, Eds.), Springer-Verlag, Berlin (1991), p.91.

CHAPTER 21

Accelerator-Based Rydberg Atom Collision Experiments

Brett DePaola*

The use of Ryberg atoms as a target in ion-atom collisions experiments is becoming more common. In addition, the study of Rydberg projectile ions formed in, for example beam-foil experiments, is an area of growing interest. In this section we discuss various methods for producing Rydberg atom targets as well for analyzing projectile Rydberg ions.

21.1 INTRODUCTION

The last two decades have seen a tremendous rise in the number and variety of accelerator-based ion-Rydberg atom collisions experiments. The processes under study are electron loss, electron capture, and electron excitation/de-excitation, specifically n and l-changing collisions. These collision processes are characterized as having very large cross sections, some measured [1] capture cross sections being greater than 10^{-10} cm^2, with l-changing cross sections estimated as being even greater by 2 orders of magnitude.

In principle, this sort of collision system is very simple: It may be thought of as a pure 3-body system consisting of a projectile of charge q, a target nucleus of charge 1, and a target electron; there are no "correlation" effects to complicate the tasks of either experimentalist or theorist. In fact, however, the system is not so simple as one may think. The target electron is most likely captured (or excited) into a very high-lying state of the projectile (or target), making "exact" quantum mechanical treatment problematic. Classical treatments such as the Bohr-Lindhard model and classical trajectory Monte Carlo calculations are certainly tractable, but there is no guarantee that their results are correct. Part of the motivation for studying these "simple" systems

*) Department of Physics, Kansas State University, Manhattan, KS 66506

is therefore to provide a check on classical treatments, as well as to provide a data base against which future quantum mechanical results may be compared.

Though not as mature as for ground state atoms and molecules, the history of using Rydberg atoms in accelerator experiments goes back at least 2 decades. In 1975 Koch and Bayfield published [2] the results of the first accelerated ion-Rydberg atom charge transfer experiment. In this pioneering work, the target Rydbergs were hydrogen atoms, collisionally excited to a band of high n's from 44 to 50. In the 20 years since this landmark experiment, techniques have been continuously refined, leading to a variety of target preparation and projectile detection methods. In this chapter some of these methods will be discussed and compared.

We begin our treatment of Rydberg atom collisions by discussing various means by which Rydberg atoms or ions are prepared. We next describe how target Rydberg production/destruction may be monitored. Following this we present various techniques for detecting projectile Rydberg ions/atoms. Finally, we speculate on future directions which studies of collisions involving Rydberg atoms/ions may take.

21.2 RYDBERG ATOM AND ION PREPARATION

Three major schemes are currently in use for the laboratory preparation of Rydberg atoms and ions. These are:

$$A + e^- \longrightarrow A(nl) + e^- \tag{21.1}$$

$$A^{q+} + B \longrightarrow A^{(q-1)+}(nl) + B^+ \tag{21.2}$$

$$A + h\nu \longrightarrow A(nl) \tag{21.3}$$

In addition, there is a fourth method, which is actually a combination of the last two: an ion beam captures an electron to an excited state and is then subsequently excited to a high Rydberg state. This will be briefly discussed in the section on "Charge Transfer to Rydberg States".

21.2.1 Electron impact excitation

The first method above, electron impact excitation, has yet to be used to produce a Rydberg target for use in ion-atom collisions experiments. Although simple enough to implement, this is an inefficient, "brute force" technique which is not state selective in the Rydbergs it produces. the cross section for electron impact excitation of Rydberg atoms goes as $1/n^3$, with the total cross section being [3] of order 1 Å^2. Thus, the cross section for production of He($n = 20$) is of order 10^{-4} Å^2. In practice, atoms in the electron beam collisions region are allowed to drift into a separate "applications" region, typically for spectroscopic studies. Because radiative decay is more probable for low values of l, this technique favors production of high angular momentum

Rydberg states by the time the atom has drifted into the applications region. Though, as mentioned, this technique has not yet been used to create Rydberg targets in ion-atom collisions experiments, one might expect it to eventually be used, particularly in experiments where the exact state of the Rydberg target is unimportant.

21.2.2 Charge transfer

An important method of producing beams of Rydberg atoms (or ions) involves charge transfer between an ion beam and an atomic (or molecular) or solid foil target. This technique was first demonstrated [4] by Riviere and Sweetman, and has been used to produce beams of Rydberg atoms/ions in elements ranging from hydrogen [5] to uranium [6].

In its most common form, the technique consists of directing a beam of ions into a cell containing a charge exchange gas or foil. The exiting beam is then charge state analyzed. The process is not state selective, with the cross section for capture into a particular n state going as $1/n^3$; little is known [7] about the distribution of l and m quantum numbers. In principle, without some sort of filtering process, the broad distribution of states produced in the capture process would make the resulting projectile beam useless for any sort of quantitative studies. Bayfield and Koch [5] developed a method by which the projectile beam can be made usable for (the first) experiments on microwave ionization of Rydberg hydrogen atoms. Basically, following charge state separation, the hydrogen beam was passed between two parallel plates. The electric field between these plates was modulated between two values, E_1 and E_2, which were strong enough to field ionize all Rydberg atoms having principal quantum numbers greater than n_1 and n_2, respectively. (The details of Rydberg field ionization will be discussed in sections 21.3 and 21.4.) By looking at difference signals in their experiments downstream from this "filter", Bayfield and Koch were able to deduce information about microwave ionization of the Rydberg states having principal quantum numbers between n_1 and n_2.

In summary, the charge transfer technique for producing beams of Rydberg atoms or ions is an extremely useful one, especially when the desired final state of the atom or ion is not accessible to direct laser excitation (the topic to be discussed in the following section).

21.2.3 Laser excitation

By far the most common technique used to produce atomic Rydberg targets in ion-atom collisions experiments involves the use of lasers to excite the target to well-defined Rydberg states. As in electron-impact excitation, the cross section for laser excitation goes as $1/n^3$. The biggest advantage of laser excitation over other techniques is that specific Rydberg states may be produced.

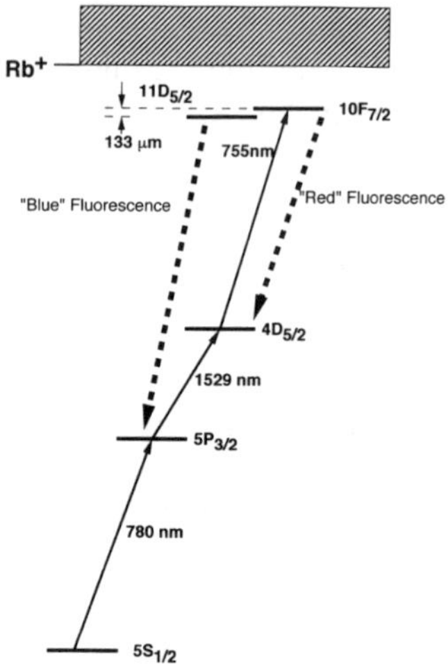

FIGURE 21.1 Simplified energy level diagram of rubidium, showing the 3-photon excitation scheme. For the geometry discussed in the text, the 10F and 11D states are coupled by mirrorless maser activity (see the text for details). From [17].

That is, states having a specified n, l, m, (and therefore even orientation) may be created. In section 21.5 an extreme level of control, involving the production of the nearly classical "elliptical states", is discussed.

Although some collisions experiments [8] have used single photon excitation to a high Rydberg state, most have used resonant two or three photon excitation schemes. This latter approach has the disadvantage of requiring more lasers, but has the advantage that the laser wavelengths are longer, and the required laser power is lower. A sample excitation scheme used in recent collisions experiments [1] is shown in Fig. 21.1. Here rubidium is excited to the 10F state using a cw three photon pumping scheme. The first two transitions, 5S - 5P, and 5P - 4D have large oscillator strengths, thus the lasers easily saturate those transitions even at the few mW per cm^2 level. The third transition, 4D - 10F, is driven by a laser having an intensity of several tens of mW per cm^2. In this excitation scheme, all the lasers are in the infra-red.

Other excitation schemes make use of pulsed dye lasers. In those studies, a laser such as a Nd:YAG is used to pump two or three dye lasers, whose outputs are in the visible or near infra-red. Generally, most of the pump laser power is directed to the dye laser driving the final transition (which usually has the smallest oscillator strength).

To date, all collisions experiments using purely laser-excited Rydberg targets have been done on alkali vapors, including lithium, sodium, and rubidium. The reason for this is two-fold. First, the alkalis have an easily accessible (in terms of both wavelength and oscillator strength) S - P transition available for

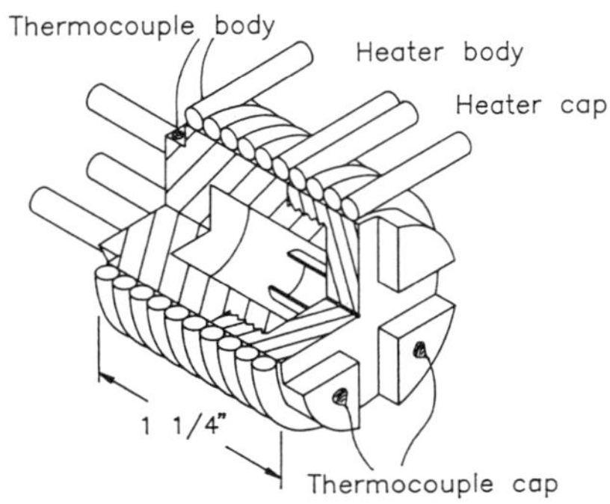

FIGURE 21.2 A typical effusive oven for the production of alkali beams which can be laser excited to high Rydberg states. In this case, the oven is heated by coils of coaxial wire. Other designs make use of "insertion heaters".

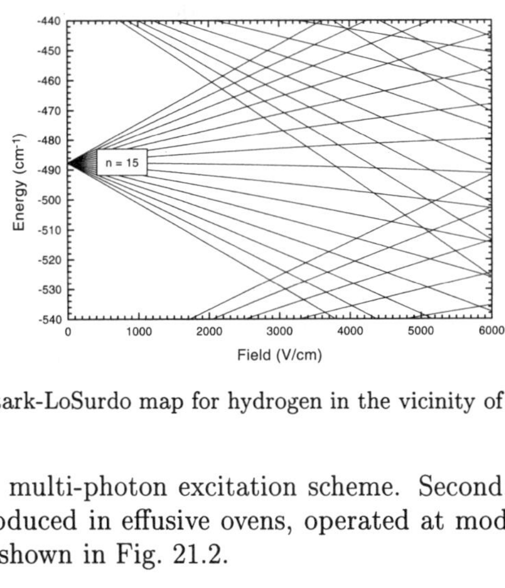

FIGURE 21.3 Stark-LoSurdo map for hydrogen in the vicinity of $n = 15$, for $m = 3$.

the first step of a multi-photon excitation scheme. Second, alkali vapor targets are easily produced in effusive ovens, operated at modest temperatures. A sample oven is shown in Fig. 21.2.

21.3 TARGET RYDBERG DETECTION

After having created a target of Rydberg atoms, it is important to be able to detect them, both to monitor their production and to measure any changes due to (for example) collisions with a beam of projectile ions. By far the most common methods make use of selective field ionization (SFI). As this is also an important feature in projectile Rydberg detection (section 21.4) we will discuss in some detail how Rydberg atoms/ions are field ionized by an external electric field. For simplicity we limit our discussions to systems having only one active electron.

For hydrogen or a 1-electron ion, the Stark-LoSurdo "map" looks essentially like that shown in Fig. 21.3. In general, in an electric field, l is no longer a good quantum number. Technically speaking, there are no stationary states in even weak electric fields due to the finite probability of ionizing by tunneling. However, if the field is not too strong, the parabolic quantum numbers n_1, n_2, and m are still approximately good. These are related to the spherical quantum numbers by the relation $n = n_1 + n_2 + |m| + 1$. Additionally, the pseudo quantum number k is often used, where $k \equiv n_1 - n_2$. Thus, one may uniquely describe a state using the set of "mixed" quantum numbers n, k, and m. The energy of a Rydberg atom in an electric field F may then be expressed to second order in F by:

$$E = \frac{-Z^2}{2n^2} + \frac{3}{2}\frac{Fnk}{Z} - \frac{F^2}{16}\left(\frac{n}{Z}\right)^4 \left[17n^2 - 3k^2 - 9m^2 + 19\right] \quad , \tag{21.4}$$

where Z is the nuclear charge. In this expression, the first term on the right is simply the Coulomb term; the second term is the linear Stark-LoSurdo effect. Note that the m quantum number doesn't appear until the quadradic Stark-LoSurdo term. Taking this equation to first order in F, we find that the critical field at which the lowest level in a manifold crosses the highest level in the next lower manifold is given by

$$F_{\text{cr}} = \frac{Z^3}{3n^5} \quad , \tag{21.5}$$

where it was assumed $n \gg 1$ and the extreme values of k are $k = \pm(n-1)$, which implies that $m = 0$. Note also that only in hydrogen or 1-electron ions is there a true crossing of levels; in all other systems, including quasi-1-electron systems, the crossing are "avoided". We discuss the implications of this crossing later in this section.

If we consider an electron moving in the combined potentials of a nucleus of charge Z and an applied electric field in the z-direction, we find a saddle point will be formed at $z = -\sqrt{\frac{Z}{F}}$, having a value $V = -2\sqrt{ZF}$. If the atom or ion has an energy given by the first term of Eq. 21.4, one then finds a "classical" threshold for field ionization given by:

$$F_{\text{th}} = \frac{Z^3}{n^4} \quad . \tag{21.6}$$

Implicit in this derivation was the the assumption that $m = 0$; a non-zero value of m leads to a centrifugal term in the potential. Taking this into account leads [9] to a correction

$$\frac{\Delta F}{F} = \frac{|m|}{2n} \quad . \tag{21.7}$$

Even with the correction of Eq. 21.7 the formulation of Eq. 21.6 suffers from two major deficiencies. The first is that it comletely neglects tunneling.

However, it has been shown experimentally [10] that ionization by an applied electric field shows a very sharp threshold, indicating that tunneling effects are minimal. The other deficiency is more serious. The effective quantum number k may be considered as an indication of the polarization of the Rydberg atom, with positive values corresponding to the electron lying upstream of the nucleus (with respect to the electric field) and therefore more easily ionized. Similarly, negative values of k indicate the electron lying downstream of the nucleus, making it less easily ionized. Thus, one might expect the threshold field to be a strong function of k; this is not reflected in the above equations. Experimentally, one sees [3] about a factor of 2 in F_{th} for the extreme states in a given Stark-LoSurdo manifold.

A closed form, semi-empirical expression developed by Damburg and Kolosov [11] is often used to estimate Γ, the electric field ionization rate. This formula,

$$\begin{aligned}\Gamma = \quad & \left(\tfrac{4}{Fn^3}\right)^{2n_2+m+1} \exp\left[3k - \tfrac{2}{3Fn^3}\right] \tfrac{1}{n^3 n_2!(n_2+m)!} \\ \times & \left(1 - \tfrac{n^3}{8}\left[36nk - 21k^2 + 17n^2 + 68n_2^2 + 68n_2 m \right.\right. \\ & \left.\left. +92n_2 + 5m^2 + 46m + \tfrac{163}{3}\right] F + O(F^2)\right) \quad , \end{aligned} \tag{21.8}$$

in which Γ expressed as a function of all the parabolic quantum numbers, has been found to agree well with experiment [10] except for a slight offset in F. ("Exact" numerical calculations [12] by the same group are in excellent agreement with experiment.) Unfortunately, Eq. 21.8 is difficult to generalize to hydrogen-like ions.

A semi-empirical expression based on the assumption that

$$F_{\text{th}} = \frac{Z^3}{n^4} \Phi_{n,k,m} \tag{21.9}$$

has been developed [13] and seems to be consistant with experimental [10] results for hydrogen, while allowing the calculation of threshold fields for hydrogen-like ions. The semi-classical dependance of F_{th} on Z and n from Eq. 21.6 is modified by the semi-empirical function Φ which is weakly dependant on n, k, and m. While direct comparisons with experiment have not yet been made for hydrogen-like ions (due to experimental difficulty in exciting known n, k, and m states in ions) Eq. 21.9 is in agreement with more sophisticated perturbative calculations.

Unfortunately, although through the use of Eq. 21.9 one can accurately determine the threshold field for ionization, the inverse is not true. That is, in general, a given threshold field does not *uniquely* determine a particular set of n, k, and m quantum numbers. The impact of this is profound: one must always view results based on electric field ionization to determine final quantum states with caution.

Most experiments using laser-excited Rydberg targets are done with alkali atoms. Although in many ways they are similar to hydrogen, they do differ

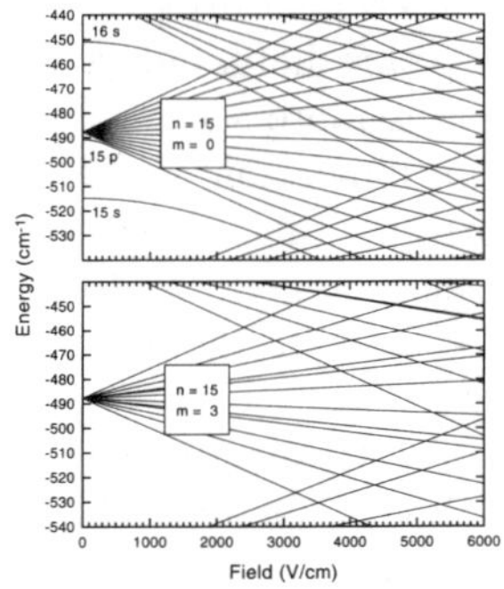

FIGURE 21.4 Stark-LoSurdo manifolds for lithium in the vicinity of $n = 15$. In the upper spectrum, $m = 0$, while in the lower spectrum $m = 3$. In the upper spectrum one can clearly see the avoided crossings; in the lower spectrum the "crossings" are still avoided, but they come much closer than in the upper spectrum.

significantly in two ways. The first is that there is a non-zero quantum defect in the alkalis. The second (and in terms of electric field ionization, more important) difference is the avoided crossings between adjacent Stark-LoSurdo manifolds. In Fig. 21.4 we show two calculated energy spectra for lithium in the vicinity of $n = 15$. In the upper spectrum, $m = 0$, while in the lower spectrum $m = 3$. In the upper spectrum one can clearly see the avoided crossings; in the lower spectrum the "crossings" are still avoided, but they come much closer than in the upper spectrum. One may understand this qualitatively by noting that the higher the value of m, the less core penetration takes place, and the more hydrogen-like the alkali atom becomes.

The reason all this is relevant – indeed, critical – to electric field ionization, is that as the ionizing field is ramped up, the electron may – or may not – make a level crossing [14], the probability for making such a transition being proportional the field slew rate and inversely proportional to the "crossing" separation. Because of this, electrons in higher m states are more likely to "cross", making them follow an ionization path very similar to that of hydrogen. For low m states, depending on the slew rate of the field, the state follows the "adiabatic", or non-crossing path. Thus, the already complicated and somewhat ambiguous method of field ionization becomes even more difficult to interpret. Having said this, there is one sense in which the field ionization data from non-hydrogen-like atoms/ions is actually *simpler* to interpret than the hydrogen-like case. That is, under adiabatic conditions the normally diverging levels in a manifold stop diverging and at $F \gtrsim F_{\rm cr}$ have an energy approximately independent of F. For the adiabatic case, then, Eq. 21.6 becomes a fairly good indicator of n. With all this in mind, we now proceed to review the various techniques in use for detecting target Rydberg atoms.

The simplest method for monitoring target Rydberg population is through

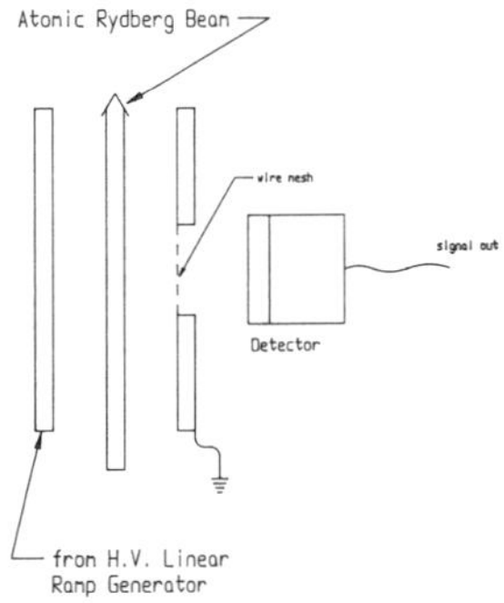

FIGURE 21.5 Simplified schematic of Selective Field Ionization (SFI) apparatus.

the application of a pulsed electric field. A proto-typical setup is shown in Fig. 21.5. A pair of parallel plates, one fitted with a grid, is placed around the Rydberg target. Behind the grid is placed a particle multiplier, for example a pair of micro-channel plates. When the electric field between the plates is 0 (or small), no signal comes from the detector. When a field greater than F_{th} is placed between the plates, through application of a voltage pulse to one of the plates, the Rydberg atoms are ionized and accelerated to the particle multiplier. This method is a simple and reliable means of monitoring relative Rydberg atom populations, either as a function of time, or for laser shot-to-shot. Note that by capacitively coupling the applied pulse to the plate, one may maintain a weak electric field in the region between the plates in order to, for example, explore [15] Stark-LoSurdo manifolds.

With just a bit more effort the setup in Fig. 21.5 may be used to monitor Rydberg atoms, differential in the principal quantum number n, *within the uncertainties expressed in the preceding section.* This is done using the method of Selective Field Ionization (SFI). SFI is based on the assumption that different quantum states have different thresholds for electric field ionizations. If one applies an electric field between the two plates of Fig. 21.5 which varies linearly [16] with time, and then observes the output of the particle multiplier as a function of time, one could, in principle, determine the distribution of quantum states in the region between the plates. In Fig. 21.6 we show the SFI "spectrum" arising from the application of a linearly ramped electric field. In this case the state being observed is a circular state having $n = 25$. The "sidebands" on either side of the main peak correspond to $n = 24$, $m = 24$, and $n = 26$, $m = 26$, respectively. (These sidebands were produced by the stimulated emission and absorption, respectively, of blackbody radiation from the main circular state represented by the center peak.) The small structure located at about 26 μsec is due to the adiabatic ionization of the 26s state which was produced through laser excitation at the same time as the circular state. For a more thorough treatment of adiabatic/diabatic paths to

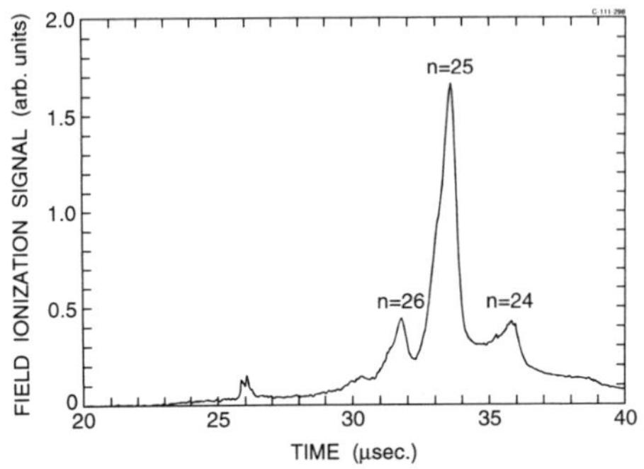

FIGURE 21.6 Applied electric field and SFI "spectrum" as a function of time. The species being observed is lithium which was prepared in an $n = 25$ circular state.

ionization, the reader is referred to reference [14].

Although most techniques for monitoring target Rydberg populations are some variation on the field ionization method just mentioned, other approaches based on absorption or emission of photons are also possible. In a recent experiment [1] involving a rubidium Rydberg target, the principal quantum number was too low for field ionization methods to be applied conveniently. A partial energy level diagram for rubidium is shown in Fig. 21.1. In the experiment Rb(10F) was produced in a c.w. 3-photon excitation scheme (described in section 21.2.3). For the geometry in the experiment, the 10F state is connected [17] to the 11D state via mirror-less maser activity. Strong fluorescence is observed at 508 nm due to the decay of the 11D to the 5P state. It was found that under conditions of constant rubidium oven temperature the ratio of fluorescence from the 10F to 4D transition to the 11D to 5P transition was essentially a constant. Thus, a measurement of the 508 nm fluorescence was taken to be an indication of the relative population of the *total* (10F + 11D) Rydberg population.

The use of absorption or emission to deduce relative populations of Rydberg atoms is particularly important in c.w. laser pumping since these signals can be fed back into locking circuits to keep the lasers from drifting off resonance. Field ionization would not be convenient to accomplish this since the feedback would be in the Hz regime, and not in the kHz regime available from simple detection of photon signals.

21.4 PROJECTILE RYDBERG DETECTION

In charge capture collisions involving Rydberg targets, the projectile is left in a Rydberg state which is centered about the final principal quantum number n_f, which is expected [18] to be given by

$$n_f = \left(\frac{Z_p}{Z_t}\right)^{\frac{3}{4}} n_i \quad , \tag{21.10}$$

where Z_p and Z_t are the effective core charges of the projectile and target, respectively, and n_i is the initial principal quantum number of the target.

The detection of projectile Rydberg atoms or ions is thus often a critical part of a collisions experiment. In most experiments involving ion-Rydberg target collisions, the projectile is singly charged; the resulting charge capture product is therefore a neutral Rydberg atom which must be detected in total, or differentially with respect to final state. Recently, collisions between highly charged projectiles and Rydberg atoms have been [1,8] studied. In this case the resulting Rydberg projectile is an ion and the detection techniques can be quite different. In the next two sections we describe various experimental approaches to detecting neutral and charged Rydberg projectiles.

21.4.1 Neutral projectiles

In charge capture experiments involving Rydberg atom targets one generally measures the projectiles which have captured a single electron from the Rydberg target. Similarly, in experiments in which the Rydbergs under study are in a beam one performs some sort of analysis on the projectiles in the beam (to measure, for example, l-changing cross sections in collisions with neutral ground state targets). In this section we discuss various techniques for detecting neutral beams of Rydbergs; in the following section we review techniques for detecting beams of Rydberg ions.

Homogeneous Field Strippers Early experiments on beams of Rydberg atoms [2] made use of simple flat plates which produced a uniform electric field parallel to the beam axis. These "stripper plates" are preceded and followed by electrostatic deflectors. The upstream deflectors remove all ions from the beam. The downstream deflectors direct the ions produced in the stripper region onto a detector. In operation, the field between the stripper plates is incrementally changed, and the signal, $S(F)$, from the detector is recorded as a function of this field. An approximation to the projectile n-distribution can then be deduced by taking the derivative $\mathrm{d}S(F)/\mathrm{d}F$. The technique is simple and efficient, and suffers only from the usual deficiencies inherent in using field ionization as an indication of principal quantum number (as already discussed in section 21.3. There is also, of course, the slight disadvantage that one only sees a "final result" after taking all the data and then taking a derivative.

Inhomogeneous Field Stripper In an effort to improve on the beam stripper approach, MacAdam and Rolfes developed [19] a "stripper analyzer" which is inherently differential in electric field. In this device, and others like it [20] the projectile beam is presented with a spatially varying electric field, and associated potential. A schematic is shown in Fig. 21.7. Here, a potential difference, V_B is placed across two concentric semi-cylinders. The potential

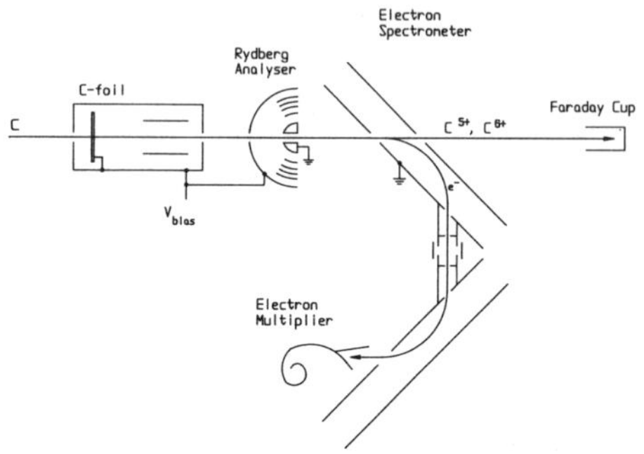

FIGURE 21.7 Schematic of the MacAdam-type stripper analyzer. See text for details.

and electric field magnitude in the annular region are then given by

$$V(r) = \frac{\ln(r/\mathrm{r}_1)}{\ln(\mathrm{r}_2/\mathrm{r}_1)} V_B \tag{21.11}$$

and

$$F(r) = \frac{V_B}{r \ln(\mathrm{r}_2/\mathrm{r}_1)} \tag{21.12}$$

where r_1 and r_2 are the radii of the inner and outer cylinders, respectively, and r is measured from the common axis of the cylinders. An atom (or ion) which is field ionized in the annular region at some position r will then gain (or lose, depending on the sign of V_B) an amount of energy given by $eV(r)$.

The ions are then electrostatically energy analyzed. The energy of the ion determines r; r gives F; and F is used to estimate n, using Eq. 21.6. Thus, within the uncertainties of electric field ionization, and within the resolution limits of the ion energy analyzer, the n-distribution of the projectile beam is determined. In a variation of this method electrons have been energy analyzed [20,21] instead of ions. The method is simple, elegant, and could even be modified to allow the continuous (non-scanned) energy analysis of ions (or electrons) by using a position sensitive detector on the output of the ion (or electron) energy analyzer.

Total Projectile Rydberg Detection: In experiments where the total rate of projectile Rydberg production is to be measured (for example total capture from a Rydberg target), one would like to reduce uncertainties by directly measuring total rates (rather than integrating the partial rates given by the previously mentioned techniques).

A device has been developed [22] which detects beams of Rydberg atoms with high efficiency (nearly independent of n, for n large) while having very

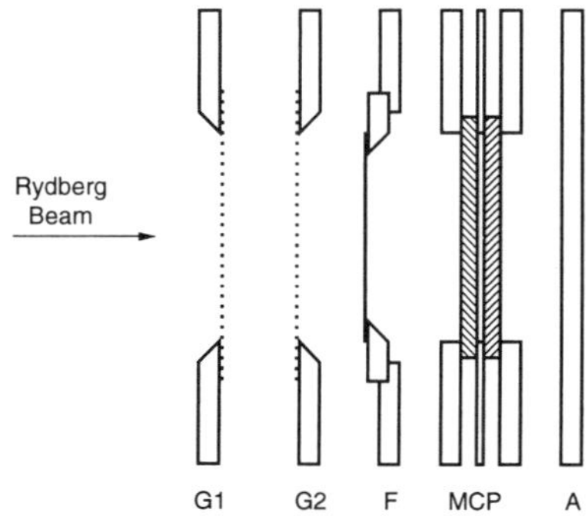

FIGURE 21.8 Schematic of total Rydberg detector for neutral projectile Rydberg atoms. See the text for for details.

low efficiency for non-Rydbergs. The detector is shown in Fig. 21.8. It consists of a grounded grid followed by a thin carbon foil at a high negative potential, followed by a channel plate electron multiplier with anode. When the Rydberg projectile enters the region between the grid and foil it is immediately ionized and accelerated *through* the foil whereupon it strikes the channel plate detector, producing a signal. Non-Rydberg projectiles are not ionized and therefore do not pick up enough energy to penetrate the foil. The technique is limited to use with neutral Rydberg projectiles which do not have enough energy to penetrate the foil in the absence of a stripping/accelerating field. Also, care must be taken to ensure that the stripper field is sufficiently strong to ionize all projectile states under consideration.

The overall efficiency of the detector is given [23] by

$$\epsilon = T_{G1}\, T_{G2}\, T_{foil}(E_{in})\, \epsilon_{mcp}(Et \quad , \tag{21.13}$$

where T_{G1} and T_{G2} are the grid transmissions, T_{foil} is the foil transmission at a given incident energy E_{in}, and ϵ_{mcp} is the detection efficiency of a transmitted particle with energy E_t

In a recent set of experiments [23,24] total projectile Rydberg production (*via* capture from elliptical Rydberg states) was measured using such a device. For typical projectile conditions of 0.8 - 3.0 keV Na^+, a foil potential of -12 kV (resulting in an ionizing field of 12 kV/cm), foil thickness of 3 μgm/cm^2, and $T_{G1} = T_{G2} = 88\%$, the overall efficiency for detecting Rydberg atoms was estimated to be about 0.4, while that for non-Rydberg projectiles was estimated to be below 10^{-4}.

In order to determine final l distributions following electron capture from a Rydberg target, Lundeen *et al* employed a technique which utilized a clever combination of laser excitation and electric field ionization. The collision system was S^+ incident on Rb(10F). Following charge capture, the neutral projectiles are electrostatically separated from the ions and then subjected to

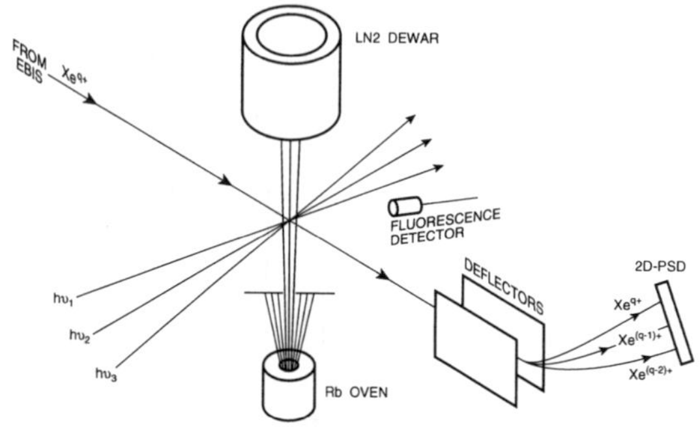

FIGURE 21.9 Apparatus for total charge capture cross sections for slow, highly charged (q = 8-40) xenon ions incident on rubidium Rydberg targets. From [1].

radiation from a single mode CO_2 laser. The neutral sulfer beam then passes through a region having an electric field. The field is not strong enough to ionize the sulfer atoms *unless* they had been laser excited. The CO_2 laser radiation seen by the sulfer beam is Doppler tuned and the ionization signal is measured as a function of laser frequency. The result, an ionization signal only when the CO_2 laser is on resonance, can be converted into a final l distribution in the sulfer *without the uncertainties of field ionization techniques.*

21.4.2 Charged projectiles

The detection of charged Rydberg projectiles often requires the use of different techniques then for neutral projectiles. This is because until ionized by a stripper field, a neutral projectile beam remains unperturbed by it. A charged projectile beam, on the other hand, depending on its rigidity, may become distorted by stripper fields well before any ionization takes place. In this section we discuss some of the techniques which have been used to study projectile Rydberg ions.

In recent experiments [1] total charge capture cross sections were measured for slow, highly charged (q = 8-40) xenon ions incident on rubidium Rydberg targets. To measure these *total* cross sections simple charge separation was sufficient. The experiment is shown schematically in Fig. 21.9. Although the deflection (and, therefore, charge separation) was in the horizontal plane, a two-dimensional position sensitive detector (2D-PSD) was used in order to help verify that the projectile beams were not partially missing the detector in the vertical direction.

The 2D-PSD was of the resistive anode type, but a "wedge and strip backgammon" anode has also been used with similar results. A typical 2-D output is shown in Fig. 21.10. Note the log scale on the vertical axis. Generally, to determine capture fractions it was convenient to integrate over

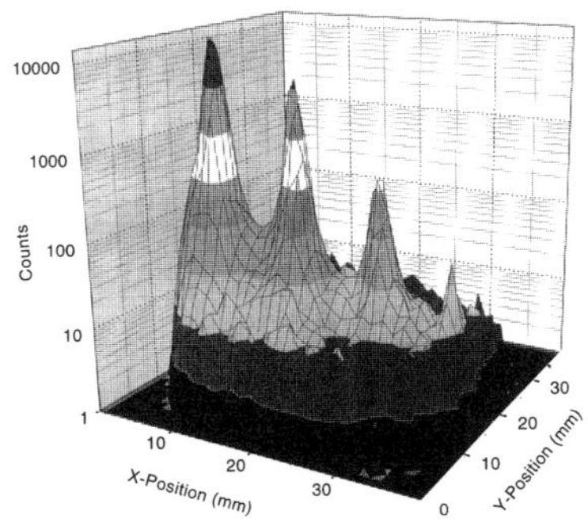

FIGURE 21.10 A typical result from the experiment shown schematically in Fig. 21.9. Here, $q = 40$, and $v_p roj = 0.2$ atomic units. Note the log scale in the vertical axis.

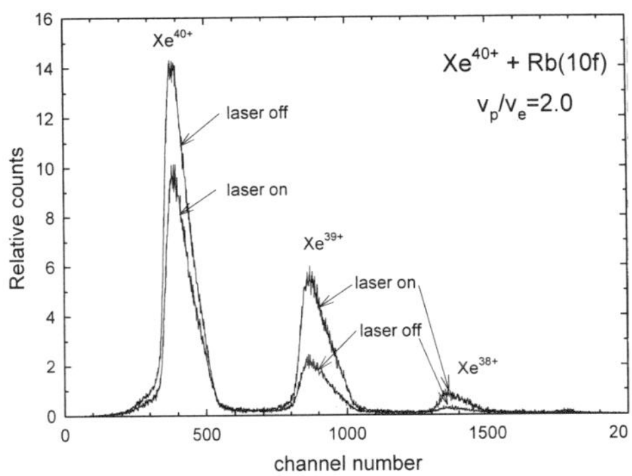

FIGURE 21.11 Data from Fig. 21.10, integrated over y and gated with the lasers on and off. From [1].

the vertical dimension of the detector to produce data like that in Fig. 21.11.

The ease of use of this sort of detection scheme is readily apparent. There are, however, three drawbacks to this approach. The first is that the detection is not final state selective. Secondly, one must excercise caution that the fields in the deflector are not sufficiently high to cause field ionization of the high-n wings of the projectile Rydberg distribution. The latter problem is endemic to projectile Rydberg detection in general, and may be addressed by using a sufficiently long set of deflector plates and beam pipe. One may then use a relatively weak deflection field and still achieve the desired charge state separation. The third drawback is potentially the most severe: This method becomes difficult (or impossible) for systems in which the probability of capturing an electron from a ground state (or background) target is much greater than capture from the Rydberg target. This is usually the case [25] since in most experiments the Rydberg target thickness is much less than the ground state target thickness.

A clever final state sensitive system based on field ionization has been used [8] recently in collisions of Kr^{8+} with Rb(17p). The cross sections in this case were not absolute but were differential to within the uncertainty inherent in field ionization methods (as discussed in section 21.3). A schematic of the detection system has not been published as of the printing date of this manuscript, but is described [8] in the literature. Briefly, the krypton ions which have undergone capture are electrostatically separated from the primary beam and directed into a cylindrically symmetric electrostatic field ionizing lens system of 11 electrodes. The arrangement of electrodes and applied potentials result in a uniformly decreasing potential $U(z) = -cz^2$ towards the center of the stripper lens. This results in a stripping field $F(z) = 2cz$, uniformly increasing from 0 V/cm at the lens entrance to a maximum at the lens center, and then back to 0 V/cm at the lens exit. Ions which are further ionized with some field F_I are then "tagged" by having a well-defined energy given by $E_i - U_I$, where E_i and is the initial energy and U_I is the magnitude of the electric potential at the point where field ionization occurred. Upon leaving the stripper lens the ions are energy analyzed. We see then, that the stripper lens operates on essentially the same principle as the MacAdam stripper analyzer, but with an emphasis on controlling the beam optics – essential for slow, highly charged projectiles.

21.5 NEW TRENDS: ELLIPTICAL RYDBERG PRODUCTION

An exciting new development in the study of collisions with Rydberg atoms is the efficient production of Rydberg "elliptical targets". A coherent elliptical state (CES) is characterized by n, the expectation values of the angular momentum $< \mathbf{L} >$, and the Runge-Lenz-Pauli vector $< \mathbf{A} >$, and may be defined as a state in which the electron spatial density distribution lies, with a minimum in quantum fluctuations consistant with the uncertainty principle, on an ellipse with the nucleus at one of the foci. In momentum space as well, the density distribution is given by that of a particle following a classical elliptical orbit, with a minimum in quantum fluctuations consistant with the uncertainty principle. The prospect of using a CES target in ion-Rydberg collision studies is very exciting since the experimentor may control (a) the energy (by choosing n, the principle quantum number), (b) the eccentricity, and (c) the spatial orientation of the target. This allows great flexibility to the experimentor in investigating various collisions processes, and thus allows ever more stringent checks on theory.

Elliptical states are produced [26] using the adiabatic switching method. In this technique extreme Stark states are populated by laser excitation in a moderately strong (order of 150 V/cm) electric field. The electric field is then adiabatically reduced to something of order of a few V/cm (the "ellipse field") and a perpendicular magnetic field is simultaneously adiabatically switched on. This leaves [27,28] the Rydberg atom in an elliptical state with eccentricity, ϵ, given by

$$\epsilon = \frac{1}{\sqrt{1+\frac{\omega_Z^2}{\omega_S^2}}} = \frac{1}{\sqrt{1+\left(\frac{2.18B[\text{Gauss}]}{3nF[\text{V/cm}]}\right)^2}} \quad , \tag{21.14}$$

where B is the applied magnetic field in Gauss, F is the "ellipse field", n is the principle quantum number, and ω_Z and ω_S are the Larmor and Stark frequencies, respectively. In actual practice, the magnetic field (of order a few tens of Gauss) is always on, and it is only the electric field which is changed; when the "Stark field" is on, the relatively weak magnetic field is negligible.

The result is a target in an elliptical state, characterized by principle quantum number n and eccentricity ϵ. The orientation vectors $<\mathbf{L}>$ and $<\mathbf{A}>$ are given in atomic units by

$$<\mathbf{L}>=(n-1)\sqrt{1-\epsilon^2}\frac{\mathbf{B}}{\mathrm{B}} \quad ,<\mathbf{A}>=-n(n-1)\epsilon\frac{\mathbf{F}}{\mathrm{F}} \quad . \tag{21.15}$$

Note that the eccentricity, and therefore the orientation vectors $<\mathbf{L}>$ and $<\mathbf{A}>$ are continuous variables, determined by the ratio of electric to magnetic fields. Also note that at the two extremes, $F=0$ and $F\to\infty$, the eccentricity goes to 0 and 1, respectively.

Because n and the eccentricity do not specify the electron's orbital direction, a sign convention has been adopted in which, for projectiles traveling perpendicular to the electron angular momentum axis, ϵ is positive when the electron at perikarpion [29] is moving parallel to the projectile direction, and negative otherwise.

The experimental setup for producing elliptical Rydberg states is surprisingly simple. An example is shown schematically in Fig. 21.12. In this case [30] lithium elliptical states were produced using 3-photon excitation. The lasers were linearly polarized, with the polarization axis parallel to the Stark field. In the figure, the star indicates where the lasers intersect the atomic beam, which enters the apparatus from below. In theis region the electric field is strong enough to produce strongly Stark-split levels. After being excited to the uppermost level in the (for example) $n=25$ manifold, the lithium beam drifts into a variable field region. It is in this region in which the Rydberg atoms evolve into elliptical states of the desired eccentricity. It is also the region in which the projectile ion beam, in this case Na^+, intersects the elliptical target at 90°. The lithium beam continues to drift up into a selective field ionizer, which, as will be discussed in section 21.3 is used to monitor Rydberg production.

A schematic of the complete collisions experiment is shown in Fig. 21.13. After leaving the collision region, the sodium beam is charge state analyzed and the neutrals, which have captured an electron from either a ground state

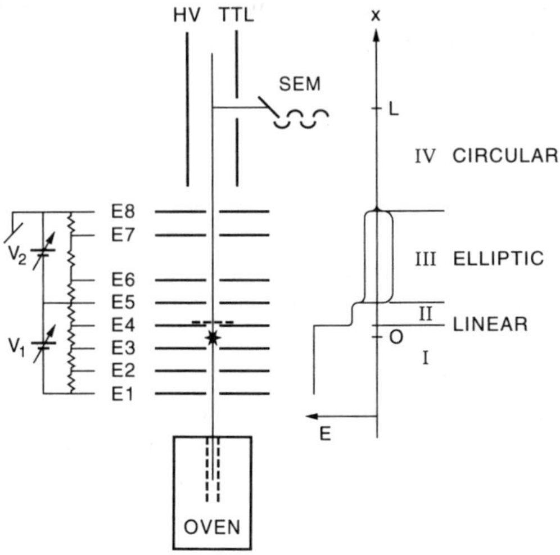

FIGURE 21.12 The experimental setup for producing elliptical Rydberg states. The star indicates where the lasers intersect the atomic beam, which enters the apparatus from below. See the text for details.

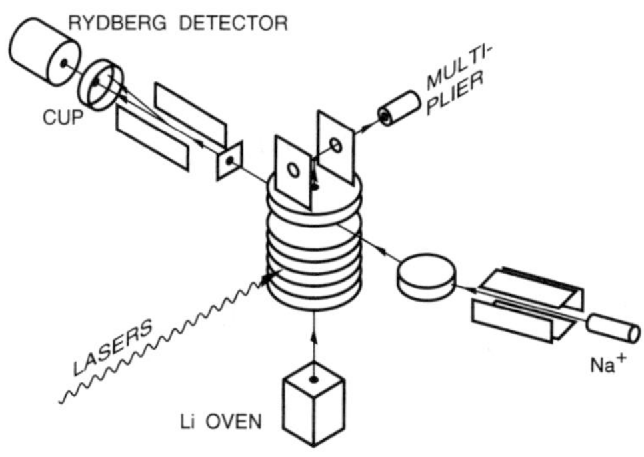

FIGURE 21.13 A schematic of a collisions experiment is involving an elliptical Rydberg atom target. After leaving the collision region, the sodium beam is charge state analyzed and the neutrals, which have captured an electron from either a ground state or Rydberg target, enter a projectile Rydberg detector.

or Rydberg target, enter a projectile Rydberg detector. This detector, as will be discussed in section 21.4 is only sensitive to neutral projectile Rydbergs. Since only the electrons captured from the Rydberg target end up in Rydberg states of the projectile, this is an efficient means of determining the Rydberg capture fraction.

A typical experimental result is shown in Fig. 21.14. Here, the charge capture cross section (in arbitrary units) is plotted *versus* target eccentricity.

The projectile was 2.5 keV Na^+, and the target was oriented such that the projectile trajectory was parallel to the plane of the elliptical orbit. The cross section is clearly asymmetric with respect to the eccentricity. A simple classical explanation could account for this behavior. At the projectile energy used, the scaled velocity is $v_p/v_e = 1.65$, where v_p is the projectile velocity and v_e is the velocity an electron would have in a circular Bohr orbit. For this particular orientation of the target, the projectile and electron move on parallel paths of the target only while the electron is at perikarpion. For only one eccentricity would an electron in a classical elliptical orbit have a velocity at perikarpion which exactly matches the projectile velocity, thus maximizing the time that it and the projectile can interact. It is at this eccentricity that the cross section is a maximum. Studies with projectiles of different energies always yield the same result: the cross section is *always* a maximum when the eccentricity of the target is such that electron's velocity at perikarpion matches the projectile velocity.

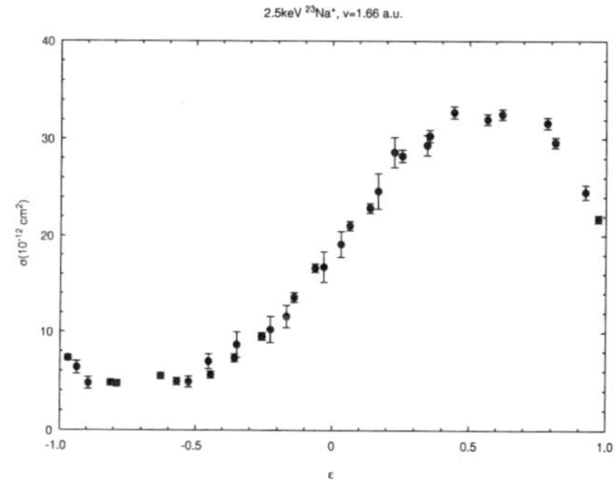

FIGURE 21.14 A typical experimental result. Here, the charge capture cross section (in arbitrary units) is plotted *versus* target eccentricity. The projectile was 2.5 keV Na^+, and the elliptical target was oriented such that the projectile trajectory was parallel to the plane of the elliptical orbit.

In addition to the fascinating possibilities opened up through the use of elliptical states, there are other areas in which advances will hopefully be made. One of these is in final state determination following capture from a Rydberg, or excitation/de-excitation of a Rydberg. As has been discussed in some detail in this chapter, electric field ionization by itself does not uniquely determine the state of a Rydberg atom or ion. However, if a tunable laser is first used to excite an atom/ion from n, k, m to n', k', m', then field ionization of the new state may yield enough information to determine n, k, and m. For example, if a linearly polarized laser is scanned in the wavelength region necessary to excite a hydrogen atom from n to n', a subsequent electric field will show evidence of n' only when the laser is resonant withe the n – n' transition. Knowledge of the laser wavelength at resonance then uniquely determines n (and n'). If the electric field was aligned to the laser polarization axis, then $m = m'$. Now, given n', and using the stripper analyzer method described in section 21.4.1, it could be possible to determine k' and m'. Thus, at least in principle, it should be possible to uniquely determine the complete set of quantum numbers for a system. Some preliminary progress has been made in this area [31], but probably several years' effort will have to be made to turn this into a reliable technique.

REFERENCES

1. DePaola, B. D., Huang, M.-T., Winecki, S., Stöckli, M. P., Kanai, Y., Lundeen, S. R., Fehrenbach, C. W., and Arko, S. A., *Phys. Rev. A* **52**, 2136 (1995).
2. Koch, P. M. and Bayfield, J. E., *Phys. Rev. Lett.* **34**, 448 (1975).
3. Gallagher, T. F. in *Rydberg Atoms* (Cambridge University Press, Cambridge, 1994).
4. Riviere, A. C. and Sweetman, D. S., in *Atomic Collision Processes*, ed. McDowell, M. R. C. (North Holland, Amsterdam, 1964).
5. Bayfield, J. E. and Koch, P. M., *Phys. Rev. Lett.* **33**, 258 (1974).
6. Schneider, D., private communication, (1990).
7. DePaola, B. D., Kanai, Y., Richard, P., Awaya, Y., Schöne, H., Sanders, J. M., Giese, J. P., and Zouros, T. J. M., *J. Phys. B: At. Mol. Opt. Phys.* (1996).
8. Pesnelle, A., Trainham, R., Pascale, J., Monnand, E., and Andrä, H.J., *Phys. Rev. Lett.* **74**, 4169 (1995).
9. Cooke, W. E. and Gallagher, T. F., *Phys. Rev. A* **17**, 1226 (1978).
10. Koch, P. M. and Mariani, D. R., *Phys. Rev. Lett.* **46**, 1275 (1981).
11. Damburg, R. J. and Kolosov, V. V., *J. Phys. B: At. Mol. Opt. Phys.*, **12**, 2637 (1979).
12. Damburg, R. J. and Kolosov, V. V., *J. Phys. B: At. Mol. Opt. Phys.*, **9**, 3149 (1976).
13. Banks, D. and Leopold, J. G., *J. Phys. B: At. Mol. Opt. Phys.*, **11**, 37 (1978).
14. Jeys, T. H., Foltz, G. W., Smith, K. A., Beiting, E. J., Kellert, F. G., Dunning, F. B., and Stebbings, R. F., *Phys. Rev. Lett.* **44**, 390 (1980).
15. Zimmerman, M. L., Littman, M. G., Kash, M. M., and Kleppner, D., *Phys. Rev. A* **20**, 2251 (1979).
16. Fuqua III, W. L. and MacAdam, K. B., *Rev. Rev. Sci. Instr.* **56**, 385 (1985).
17. Fehrenbach, C. W., Lundeen, S. R., and Weaver, O. L., *Phys. Rev. A* **51**, R910 (1995).
18. Hickman, A. P., Olson, R. E., and Pascale, J., in *Rydberg States of Atoms and Molecules*, edited by Stebbings, R. F., and Dunning, F. B., (Cambridge University Press, Cambridge, 1983).
19. MacAdam, K. B. and Rolfes, R. G., *Rev. Sci. Instr.* **53**, 592 (1982).
20. DePaola, B. D., Axmann, W. J., Parameswaran, R., Lee, D. H., Zouros, T. J. M., and Richard, P., *Nucl. Instrum. Meth. in Phys. Res.* B**40/41**,187 (1989).

21. Lee, D. H., Zouros, T. J. M., Sanders, J. M., Shinpaugh, J. L., Tipping, T. N., Varghese, S. L., DePaola, B. D., and Richard, P., *Nucl. Instrum. Meth. in Phys. Res.* B**40/41**,1229 (1989).
22. Hansen, S. B., Gray, L. G., Horsdal-Pedersen, E., and MacAdam, K. B., *J. Phys. B: At. Mol. Opt. Phys.*, **24**, L315 (1991).
23. Hansen, S. B., *Electron Capture from Oriented Circular Rydberg Atoms*, PhD thesis, Aarhus University, unpublished (1993).
24. Hansen, S. B., Ehrenreich, T., Horsdal-Pedersen, E., MacAdam, K. B., and Dubé, L. J., *Phys. Rev. Lett.* **71**, 1522 (1993).
25. For the target used in the experiment of Fig. 21.10, this was *not* the case, as the Rydberg target was extremely thick [1] and the projectile velocity was close to the matching velocity of the Rydberg electron.
26. Delande, D. and Gay, J. C., *Europhys. Lett.* **5**,303 (1988).
27. Day, J. C., Ehrenreich, T., Hansen, S. B., Horsdal-Pedersen, E., Mogensen, K. S., and Taulbjerg, K., *Phys. Rev. Lett.* **72**, 1612 (1994).
28. Mogensen, K. S., Ehrenreich, T., Horsdal-Pedersen, E., and Taulbjerg, K., *Phys. Rev. A* **51**, 4038 (1995).
29. Perikarpion, with "karpion" coming from the Greek for nucleus, is the point of closest approach between an electron in its elliptical orbit and the nucleus.
30. Ehrenreich, T., Day, J. C., Hansen, S. B., Horsdal-Pedersen, E., MacAdam, K. B., and Mogensen, K. S., *J. Phys. B: At. Mol. Opt. Phys.*, **27**, L383 (1994).
31. M. Troike and B. D. DePaola, private communication (1995).

CHAPTER 22

Collision Spectroscopy of Multiply Charged Recoil Ions in Collisions With Atomic Targets

Emanuel Y. Kamber*

Doubly differential (in energy and angle) cross sections for state-selective single-electron capture by low-energy multiply charged ions from atomic and molecular targets have been studied at laboratory impact energies between 30 and 600 eV and scattering angles between 0° and 6°, by means of translational energy-gain spectroscopy. The studies utilized ion beams from a recoil-ion source that was pumped by a beam of fast ions from the Western Michigan University 6 MV tandem Van de Graaff accelerator, along with a translational energy spectrometer to characterize the reaction channels produced in the collisions. This chapter will concentrate on collision spectroscopy of multiply charged ions in ion-atom collisions.

22.1 INTRODUCTION

Collision spectroscopy consists of measuring cross sections, doubly differential in translational energy-gain/loss and projectile scattering angles, using the energy-gain/loss technique. Such a technique is usually utilized to measure the total change in internal energy of the colliding system that takes place during the collision (sometimes called the energy gain/loss or energy balance of the collision) in order to understand the collision dynamics at small scattering angles. This energy balance of the capture reaction is the most important parameter in determining the levels which are most strongly populated in the process. The measured translational energy spectrum will reveal spectroscopic information on the states of the participating particles, as well as detailed information on the scattering process, such as its dynamics, collision

*) Department of Physics, Western Michigan University
Kalamazoo, Michigan 49008

cross sections, and populations of the projectile states. Doubly differential studies of the type presented contribute significantly to the understanding of important inelastic scattering processes that occur in low-energy multiply charged ion-atom collisions and provide substantially more information about the nature of the interaction than do measurements of total cross sections. In addition to which, differential measurements can provide a test of applicability of various theoretical models and the possibility of relating the scattering angle to a classical impact parameter. Furthermore, the energy gained by the projectile is tied kinematically to both the change in electronic energy for the reaction and the projectile scattering angle, so that simultaneous measurement of both parameters is desirable.

Among the principal topics for research in atomic physics is the study of collisions involving atomic and molecular particles. Three fundamental processes may occur as a result of ion-atom/molecule collisions: excitation, ionization, and electron capture, i.e. charge transfer. Combinations of these three processes may also result. Of particular interest is single-electron capture by a projectile from a target atom since this is a predominant reaction for low-energy collisions.

The importance of understanding electron capture can be seen from the variety of physical phenomena that involve this process. Electron capture processes involving multiply charged ions are of interest in the study of astrophysical plasmas where they have been found to modify the ionization structure of plasmas and exert a major influence on plasmas created by the absorption of high frequency radiation. Another important application is toward the development of techniques for controlled thermonuclear fusion. Several previous experimental measurements on electron capture by multiply charged recoil ions from different target atoms have been reported using the translational energy-gain spectroscopy technique [1–4]. In this work we have studied doubly differential cross sections for single-electron capture from neutral targets by low-energy multiply charged recoil ions of argon and neon at impact energies of 30 to 600 eV [5,6].

22.2 EXPERIMENTAL TECHNIQUE

The differential energy-gain spectrometer used to study energy-gain spectra of low-energy multiply charged ions undergoing single-electron capture consists of a recoil ion source followed by a mass analyzer, a collision gas cell, a 90° double focusing electrostatic analyzer, and a position sensitive detector. In this work, a fluorine ion beam from the Western Michigan University (WMU) tandem Van de Graaff accelerator was used in the production of recoil ions that were subsequently used as a projectile for low-energy ion-atom collision studies.

22.2.1 Apparatus

The experimental apparatus which we have used to study electron capture processes involving low-energy multiply charged ions is shown in Fig. 1 [5]. Fast ions (sometimes called the "pump" or "hammer" beam) from the WMU tandem accelerator, a 25 MeV F^{4+} beam in our case, was poststripped to a median charge state of $\langle 7+ \rangle$ by a ten $\mu g/cm^2$ carbon foil to improve its stripping power, then collimated with a set of four-jaw slits. After collimation, the beam enters into a recoil ion source, which was designed and built at WMU, where highly charged projectile ions are produced. This source consists of a collision chamber, a pusher, and two lenses (L1 and L2).

The fast beam passed through the collision chamber which has entrance and exit apertures of 2.5 and 3 mm diameter, respectively. The fast beam was collected by a Faraday cup after it passes through the interaction region in the recoil ion source. Typical beam currents collected by the Faraday cup were between 0.5 and 1 μA to optimize the recoil beam current. Target gas atoms come into the source through the collision chamber; the pressure, which is held constant, is monitored by noting an increase in the background pressure of the main chamber from about 1.2×10^{-6} Torr to about 2.5×10^{-6} Torr. Recoil ions formed in the collision chamber were extracted perpendicular to the fast beam with an acceleration voltage ($V_{acc} = V_{pusher}$) through a 2.5 mm diameter aperture in lens L1, under the action of a potential field set up by voltages applied to the pusher and lenses L1 and L2. The pusher was kept at the same potential as the collision chamber. The voltage V_{pusher} (the voltage on the pusher) and V2 (the voltage on the L2) were set so as to optimize the resolution and transmission of the detected ions. Typical potentials for the pusher and lenses L1 and L2 were +50, +49, and +40 volts, respectively.

Once extracted, the recoil ion beam was focused with an Einzel lens onto the entrance aperture of a 180° double focusing magnet, of radius 11.5 cm, to be mass analyzed. Here, the various charge states of the recoiling projectile ions were separated due to their different mass-to-charge ratios by varying the magnetic field of the analyzing magnet. Figure 2 shows a charge-state spectrum for argon recoils. The reason for the smaller size of the peak for charge state 1+, is that the detector reached saturation for this charge state during data collection. Following mass analysis, the ion beam was collimated by an aperture of 12.5 mm and directed with horizontal and vertical deflection plates (D1 and D2) toward the 3mm-long collision gas cell. The entrance and the exit apertures of the cell were 1 and 2 mm, respectively.

The target gas pressure in the cell (monitored by noting an increase in the background pressure in the main chamber from about 2.5×10^{-6} Torr to about 3.5×10^{-6} Torr) was sufficiently low to ensure single-collision conditions. Ions scattered through a nominal angle θ into a solid angle $\Delta\Omega$ of about 3×10^{-3} sr were energy analyzed by a 90° double focusing electrostatic analyzer (ESA) (mean radius 35 mm). The scattering angle θ is selected by means of an aperture A1 (1mm diameter) in front of the ESA. The analyzed ions were then

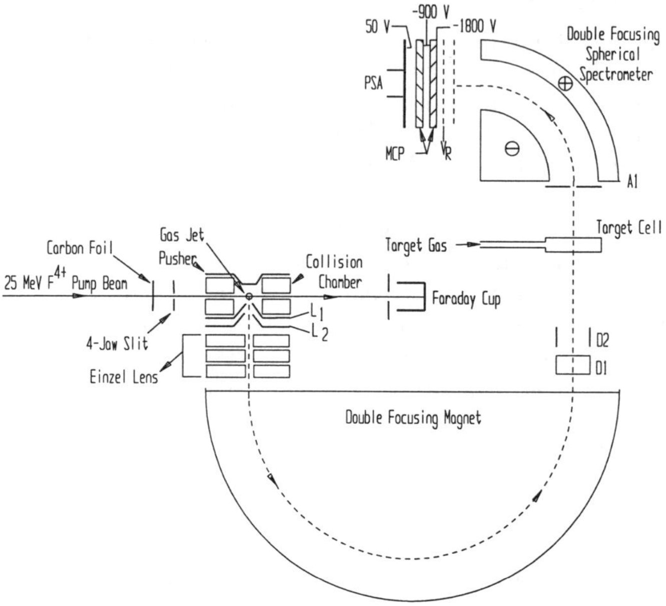

FIGURE 22.1 Schematic diagram of the doubly differential energy-gain spectrometer. From [5].

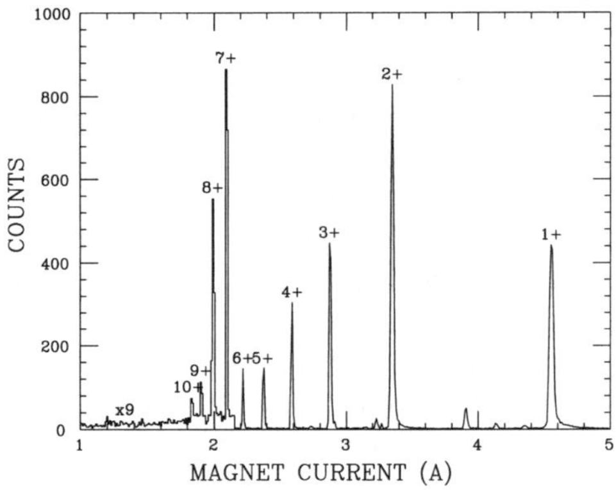

FIGURE 22.2 Typical charge-state spectrum for argon recoil ions produced in the recoil ion source. From [5].

detected by a one-dimensional position sensitive microchannel plate detector, which is located at the focal plane of the ESA. The detector device consists of two 3.5 cm diameter microchannel plates (MCP) and a one-dimensional position-sensitive anode encoder (PSA). The potential across each plate was

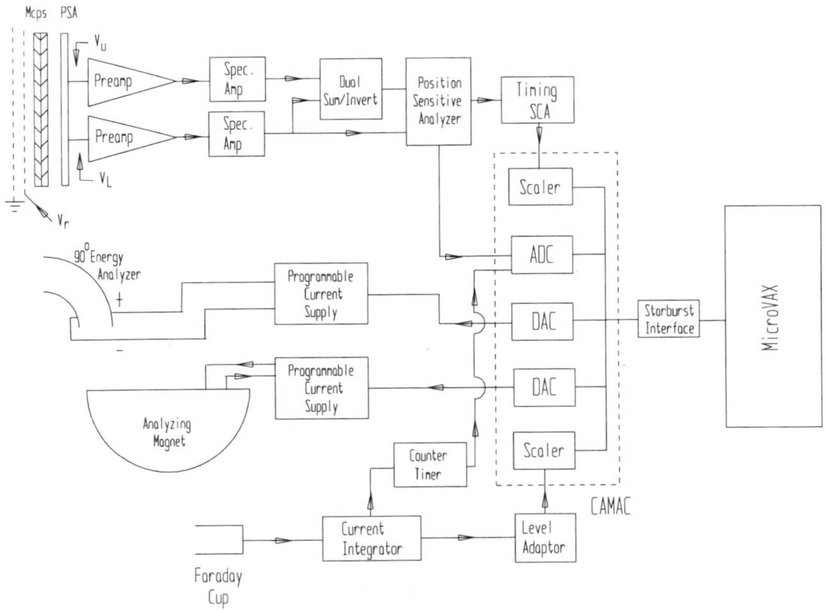

FIGURE 22.3 Electronics block diagram for the energy-gain spectrometer.

maintained at approximately -900 V, for a total of approximately -1800 V across the entire configuration. The output electrons generated by the MCPs were accelerated towards and collected by the PSA. The charge gathered by the PSA was used to determine the position of events along a single dimension. Electrodes on either side of the resistive anode collect fractions of the total charge deposited. The larger the fraction of the charge collected at an electrode, the closer the event location was to that electrode.

The relative position of an event is determined by the ratio between the resulting voltage pulse at one electrode and the sum of the voltage pulses of both electrodes, that is

$$X = \frac{V_L}{V_U + V_L} \tag{22.1}$$

Here, X is the relative position along the PSA axis and V_U and V_L are the voltage pulse heights at the upper and lower electrodes, respectively.

Data were collected using the electronics setup shown schematically in Fig. 22.3. Signals from each electrode of the position-sensitive anode (PSA) were equally amplified by preamplifiers followed by spectroscopy amplifiers, and then summed using a dual sum/invert. The signal, along with the signal pulse from the lower electrode, were then guided into the energy and position inputs of the position sensitive detector analyzer (PSD), respectively.

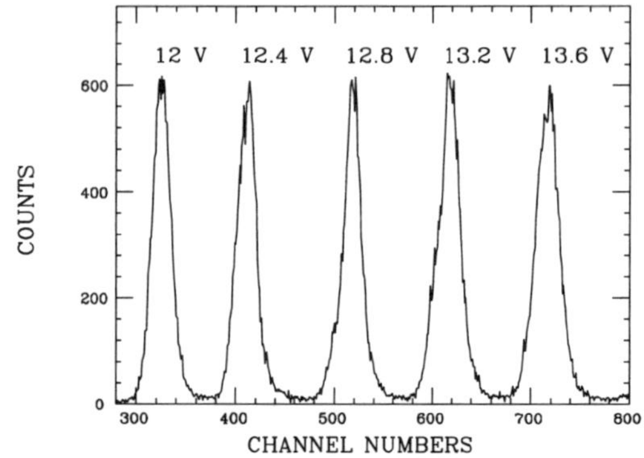

FIGURE 22.4 Position of the projectile peaks as a function of the ESA voltages.

Within the PSA, the division in Equation (22.1) takes place producing a voltage signal whose height is proportional to the location of the incident ion's position on the PSA. The analog signals of the position output of the PSD were directed to an analog-to-digital converter (ADC), which was gated by the counter/timer, preset for some maximum number of counts of integrated fast beam current accumulated by the Faraday cup. Digital position signals from the ADC go to the input module of the CAMAC crate. This information is accessed through the STARBURST interface to the microVAX computer system enabling conversion from position distribution information to energy distribution information.

In order to minimize distortion due to small changes in linearity near the edges of the PSA, only the central portion of the PSA was used. The detector was periodically calibrated by positioning the direct beam at different locations by varying the applied voltage to the ESA. Plotting the measured voltages versus the channel number produced a line whose slope and voltage per channel number allowed for conversion from channel numbers to voltages (see Fig. 22.4). In order to correct for contributions due to background gas in single-electron capture spectra, data were taken both with and without target gas in the collision cell. After normalizing each spectrum to the total amount of fast beam charge collected, the background spectrum was subtracted from the spectrum with target gas to produce a spectrum corresponding to the translational energy distribution of the projectile capturing a single electron from the target gas.

22.2.2 Energy analysis

The initial kinetic energy E_0 of the ion beam of atomic charge q passing through the ESA is given by

$$E_o = qV_{acc} = qKV_0 \tag{22.2}$$

where V_0 is the voltage applied to ESA, and K is the ESA constant, given by

$$K = \frac{2R_i R_0}{R_0^2 - R_i^2} \tag{22.3}$$

where R_i and R_0 are the radii of the inner and outer electrodes of the ESA, respectively. Measured values of R_i and R_0 are 27.6 and 42.4 mm respectively. This yields a K of 2.26. The experimentally determined K was 2.4 ± 0.1.

After a collision, the final kinetic energy E of the charge-changed projectile with charge q' where $q' = q - 1$,

$$E = q'KV = q'(\frac{V_{acc}}{V_0})V \tag{22.4}$$

where V is the voltage applied to the ESA.

The energy gain (Q) of the projectile following single-electron capture is

$$Q = E - E_0 = (\frac{qV}{qV_0} - 1)V_{acc}q \tag{22.5}$$

Errors in the measured Q-values are mainly due to the least-squares fit to the calibration curve and the stability of the voltage applied to the ESA. Under these conditions errors in the translational energy scale, as calibrated against the scattered ions, were always less than $\pm$ 1 eV. The measured energy resolution $\left(\frac{\Delta E}{E_0}\right)$ of the primary beam ranged from 0.7 to 0.85%, where ΔE is the full width at half maximum (FWHM) of the energy distribution of the main peak (primary beam) and E_0 is the projectile energy at which the analyzer is tuned.

22.3 THEORETICAL CONSIDERATIONS

Single-electron capture (as opposed to excitation and ionization) is known to be the most probable event as the result of low-energy multiply charged ion - atom/molecule collisions. This process of charge transfer, by which an electron is captured by a multiply charged ion, A, from a neutral target, B, can be represented by the expression,

$$A^{q+} + B \rightarrow A^{(q-1)+} + B^{+} + \Delta E \tag{22.6}$$

where q is the charge state of the projectile, and ΔE represents the energy defect of the reaction. Furthermore, the projectile and target products may be formed in excited states. In exothermic reactions, distinguished by a positive ΔE, single-electron capture is dominated by the diabatic (curve crossing) transitions [14,16] within a finite number of relatively well-separated diabatic potential energy curve crossing distances. The ion-atom interaction in the

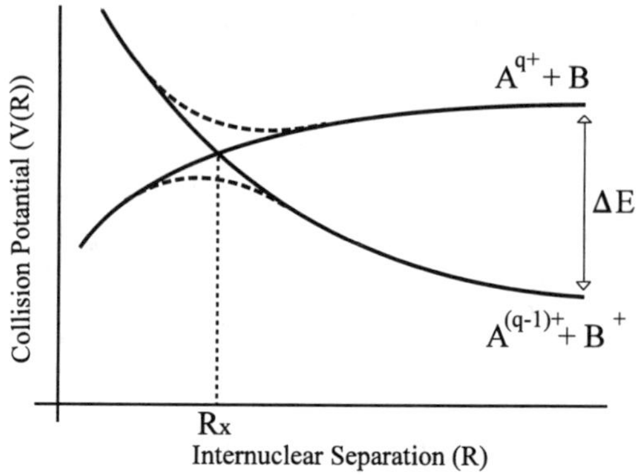

FIGURE 22.5 Schematic representation of the diabatic (solid lines) and adiabatic (dashed lines) potential energy $V(R)$ curves for the entrance ($A^{q+} + B$) and exit ($A^{(q-1)+} + B^{+}$) channels near a curve crossing R_x versus Internuclear separation R.

entrance channel ($A^{q+} + B$) is dominated by the weak attractive polarization potential, whereas the interaction in the exit channels ($A^{(q-1)+} + B^{+}$) is dominated by Coulombic repulsion (see Fig. 22.5). The important curve crossings between the diabatic potential energy curves associated with the entrance channel and various exit channels are those which occur at moderate curve crossings, where the probability for single-electron capture is large. This intermediate range of crossings is commonly referred to as the reaction window for the electron capture process [7–9]. The location of the reaction window for any collision system depends mainly on the impact energy of the projectile. In this section, the focus will be on those aspects which are directly related to the prediction of the final state populations of the captured electrons and the model used for determining the angular distributions for single-electron capture, without great emphasis upon the formal calculations or the approximations made.

22.3.1 Kinematics

Using classical two-body dynamics, the final kinetic energy E of an ion undergoing inelastic scattering differs from the initial kinetic energy of the projectile ion E_o by

$$Q = E - E_o = \Delta E - \Delta K \tag{22.7}$$

where ΔK is the translational energy given to the target, ΔE is the energy defect of the reaction, and Q is the energy gained by the projectile ion during a collision. Commonly, one defines the energy-gain as the Q-value of the reaction. This value is significant since it can be determined experimentally, as when translational energy spectroscopy is employed. The energy defect ΔE is calculated according to the formula,

$$\Delta E = I_P(A^{(q-1)+}) - I_P(B) - E_x \tag{22.8}$$

where $I_P(A^{(q-1)+})$ and $I_P(B)$ are the ionization potentials of the projectile product ion $A^{(q-1)+}$ and target atom B with the target atom being assumed to be in its ground state and the captured electron as the most loosely bound. E_x is the excitation energy of the xth level of projectile product ion $A^{(q-1)+}$ or target product ion B^+.

The translational energy ΔK given to the target can be represented through the following equation [10]:

$$\Delta K = \frac{m_p}{m_p + M}(1 - \cos\theta_p)[\frac{2ME_0}{m_p + M} - \Delta E] + \frac{m_p(\Delta E)^2}{4ME_0}\cos\theta_p \tag{22.9}$$

where m_p and M are, respectively, the projectile and target masses, E_0 is the impact energy of the projectile, and Θ_p is the final laboratory scattering angle of the projectile. For zero scattering angle, this energy reduces to

$$\Delta K = \frac{m_p(\Delta E)^2}{4ME_0} \tag{22.10}$$

A typical electron capture reaction at low collision energy in the laboratory frame of reference is shown in Fig. 22.6, in which the projectile ion is scattered through an angle θ_p while the recoil target ion is scattered through an angle θ_r from 90°. Here, the multiply charged ions move in a single direction through atomic target gas that is essentially at rest. At some critical distance, at which the projectile ion approaches the neutral target, an electron is transferred and the projectile ion continues its motion past the newly formed target ion. Coulomb repulsion between the two charged ions ($A^{(q-1)+}$ and B^+) will then lead to an increase in the kinetic energy of the projectile ion, i.e., the projectile ion will gain some energy Q during the collision.

22.3.2 Landau-Zener model

In the Landau-Zener model [11,12], the single-crossing probability at each potential energy crossing R_x, that is, the probability that the system remains on a diabatic potential curve when traversing a curve-crossing, can be expressed [13]

$$p = \exp(\frac{-2\pi H_{12}^2}{v_r \Delta F}) \tag{22.11}$$

where H_{12} is the relevant coupling matrix element and v_r is the radial velocity at the curve-crossing, R_x, and can be given by

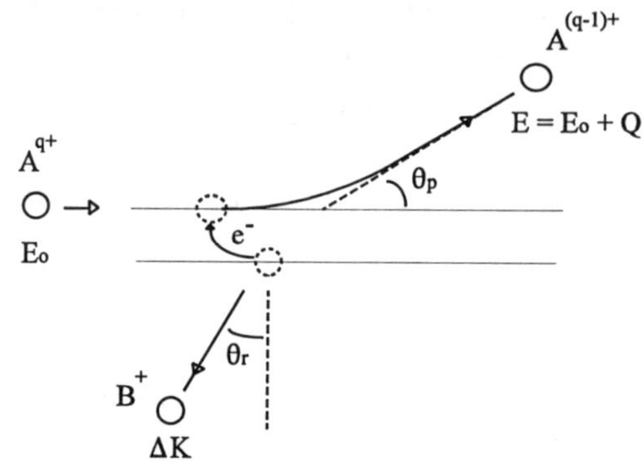

FIGURE 22.6 Typical electron capture reaction in the laboratory frame.

$$v_r = v_0[1 - (\frac{b}{R_x})^2]^{\frac{1}{2}} \tag{22.12}$$

where v_0 is the collision velocity and b is the impact parameter. The quantity ΔF is the difference in slope of the diabatic potential energy curves at the curve crossing (see Fig. 22.5), and is given by

$$\Delta F = [\frac{d}{dR}(U_{11} - U_{22})]_{R=R_x} \tag{22.13}$$

where U_{11} and U_{22} are the entrance and exit potential energies. Substituting for the energies used, coulombic for exit channel and zero for entrance channel, Equation (22.13) becomes

$$\Delta F = \frac{q-1}{R_x^2} \tag{22.14}$$

The crossing radius R_x can be calculated in terms of the energy gain, in which its value characterizes the particular state to which capture occurs, from the condition,

$$U_{11}(R_x) = U_{22}(R_x) \tag{22.15}$$

where $U_{11}(r) = 0$ and $U_{22}(r) = \frac{q-1}{r} - Q$.

The resulting expression for the crossing radii is

$$R_x = \frac{(q-1)}{Q} \tag{22.16}$$

To study state-selective single-electron capture by partially-stripped projectile ions, a correction term f_{nl} suggested by Taulbjerg [9] has been added to the analytical formula for evaluating the coupling matrix element H_{12}^{OS} presented by Salop and Olson [14] for bare projectiles on atomic hydrogen. Thus, the corrected coupling matrix can be given as

$$H_{12} = f_{nl} H_{12}^{OS} = 9.13 f_{nl} q^{-1/2} \exp[-1.324 R_x (\frac{2I_t}{q})^{1/2}] \tag{22.17}$$

where I_t is the ionization potential of the target (in atomic units) and q is the charge state of the projectile ion. The correction term (f_{nl}) depends on the quantum numbers n and l of the state into which the transferred electron is captured and is given by [9]

$$f_{nl} = (-1)^{n+l-1}(2l+1)^{\frac{1}{2}}\Gamma(n)[\Gamma(n+l+1)\Gamma(n-1)]^{-1/2} \tag{22.18}$$

Generally, p is defined to be the probability that upon encountering a curve crossing, the system behaves diabatically and remains on the same potential energy curve. Then $(1-p)$ is the probability that an adiabatic transition occurs and the system changes potential energy curves at the curve-crossing. Since the system actually traverses the curve-crossing twice during the collision, the total transition probability at a given impact parameter for two states and a single curve crossing (assuming that there is no interaction between the other paths) is given by

$$P = 2p(1-p) \tag{22.19}$$

Here we used the single crossing Landau-Zener model to interpret the final state populations by calculating the location of the reaction window in terms of the Q-value of the final states, i.e., the location of the curve crossings where the probability of single-electron capture is large. The reaction windows were calculated based on the Taulbjerg factors $f_{nl} = 1$ and f_{nl} for the dominant reaction channel.

The Landau-Zener formula for two states and a single crossing can also be generalized to a multichannel system, called the multichannel Landau-Zener (MCLZ) model, when the transition regions are well separated, assuming that there is no effect associated with the rotation of the internuclear axis and the crossings occur only between the entrance channel and exit channels each pertaining to a particular state of the projectile product. The total probability P_n ($n = 1, 2, \ldots, N$) for capture into the nth final state, calculated by summing over the possible paths necessary to make the transition from the entrance to the nth exit channel and assuming that there is no interference between different paths leading to a particular channel, is given by [15]

$$\begin{aligned} P_n = p_1 p_2 \ldots p_n (1-p_n)[1 + (p_{n+1}p_{n+2}\ldots p_N)^2 + \\ (p_{n+1}p_{n+2}\ldots p_{N-1})^2(1-p_N)^2 + (p_{n+1}p_{N-2})^2 + \\ (1-p_{N-1})^2 + \ldots + p_{n+1}^2(1-p_{n+2})^2 + (1-p_{n+1})^2] \end{aligned} \tag{22.20}$$

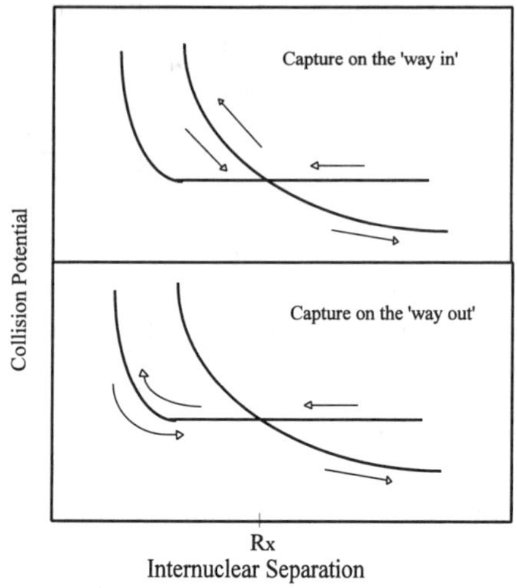

FIGURE 22.7 Two-state picture of trajectories for capture on the way-in (above) and capture on the way-out (below).

The cross section σ_n for capture into a particular final state n is given by

$$\sigma_n = 2\pi \int_0^{R_x} P_n b \, db \tag{22.21}$$

22.3.3 Semi-classical curve-crossing model

This model is called semi-classical because the trajectories of the collision reactants are explained by classical mechanics, while the electron's behavior is described by quantum mechanics. The Born-Oppenheimer approximation, which applies to slow collisions, where the velocities of the nuclei are much lower than those of the electrons, allows for treating, separately, the motion of the electrons and that of the nuclei. Therefore, deflection functions and differential cross sections are determined classically, while electron capture probabilities are determined using Landau-Zener formula [16–18].

The semi-classical two-state scattering model has been used to estimate the half Coulomb scattering angle or the critical angle θ_c, which corresponds to capture at an impact parameter equal to the crossing radius, by assuming that capture occurs at a localized curve-crossing between the potential energy curves for entrance and exit channels. The position of the critical angle in angular distribution spectra is considered a threshold angle for electron capture, i.e. charge transfer cannot occur until the collision system has reached the cross radius, R_x. For small scattering angles, the critical angle may be expressed as

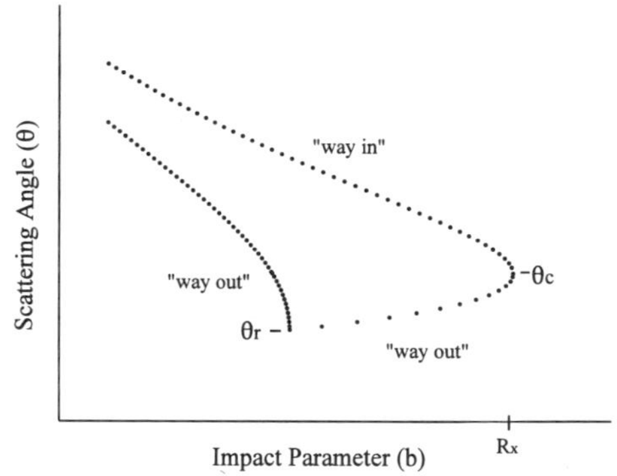

FIGURE 22.8 Typical deflection function for a two-state model.

$$\theta_c = \frac{Q}{2E_0} \tag{22.22}$$

where Q is the energy-gain and E_0 is the initial kinetic energy of the projectile.

For a two-state channel, i.e., one entrance and one exit channel, there are two possible paths that the system may follow for charge transfer which result in a double-branched deflection $\theta(b)$ (see Fig. 22.7). The upper branch of the deflection function relates to the projectile trajectory followed for capture on the way in, while the lower branch pertains to the trajectory followed for capture on the way out. The two branches meet at a critical angle θ_c. Electron capture on the way-out usually results in small scattering angle, whereas capture on the way-in typically extends to large angles.

At smaller internuclear distances, inside the crossing radius, there commonly exist reaction channels with higher Q-value than the active one which behave adiabatically at crossings with the entrance channel. Therefore, the entrance channel will not remain flat (i.e., zero interaction energy) inside the active crossing, but rather will be promoted to a repulsive potential energy curve. This promotion causes the lower branch of the deflection function to rise at small angles producing a rainbow angle θ_r, which corresponds to the smallest angle for which two impact parameters result in the same scattering angle (see Fig. 22.8). The result of this discontinuity, i.e., the discontinuity in the lower branch of the deflection function at θ_r as a result of a change in functional form from zero potential to pure Coulombic potential, is the presence of a maximum in the angular distribution spectra, known as a rainbow effect. For a particular collision system, we calculate a differential cross section for each path leading to a particular state, multiply each by its corresponding transition probability, and then sum them to determine the total differential cross section for capture to a particular final state of the projectile product ion.

22.4 EXPERIMENTAL RESULTS

22.4.1 Translational energy-gain spectra

In this work, we have measured translational energy-gain spectra of projectile product ions resulting from single-electron capture processes in collisions between low-energy multiply charged ions and atomic targets at selected projectile scattering angles. In addition to pure single-electron capture, other processes are found to be important. These processes include single-electron capture with simultaneous excitation of the projectile core (i.e., core-varying single-electron capture) and capture accompanied by target excitation or ionization. To facilitate the investigation of electron capture, we have selected a few typical collision systems that illustrate the main features of the above-mentioned collisional processes.

Figure 22.9 shows the translational energy spectra for single-electron capture by 250 eV Ar^{5+} ions from Ar at different laboratory scattering angles [5]. The outgoing reaction channels observed in the collision process were identified by comparing the measured Q-values with the calculated values from Equations 22.7 and 22.8. The energy level used in calculating the energy defect (Equation 8), were taken from Bashkin and Stoner [19]. The observed zero-angle spectrum shows that the dominant reaction channel is due to capture into the 4d state of Ar^{4+}, with a smaller contribution from capture into the 4p state. As the projectile scattering angle is increased, single-electron capture into the 4d state of Ar^{4+} remains dominant, but the relative importance of capture into the 4p state is strongly increased. This indicates that the 4d state produced in the single-electron capture process is strongly forward scattered, while the angular distribution for capture into the 4p state is not maximum at 0° but instead peaks at a greater angle. The broad peaks, centered around 22 eV, are contributions due to transfer ionization (TI) resulting from double-electron capture into autoionizing states that emit electrons before detection and thus appear to have captured only a single electron. The contribution of TI slowly increases with increasing projectile scattering angle, as would be expected.

Fig. 22.10 shows the probabilities for electron capture into the $4d$ and $4p$ states taken from the observed spectra as function of the scattering angle [5]. The electron capture probability is found from the ratio of the intensity of each reaction channel to the total signal, i.e., $P = I_{nl}/I_{total}$. In general, at the smaller angles, the larger impact parameter collisions play an important role in the electron capture process since the avoided crossings at small internuclear separation (large Q-value) cannot be reached, and make no contribution. As the angle is increased, contributions from successively smaller internuclear separation regions make their appearance. This feature has been observed in all the collision systems we have studied.

Figure 22.11a shows the translational energy-gain spectrum for single-electron capture by 150 eV Ne^{3+} ions from He at a scattering angle of 0°.

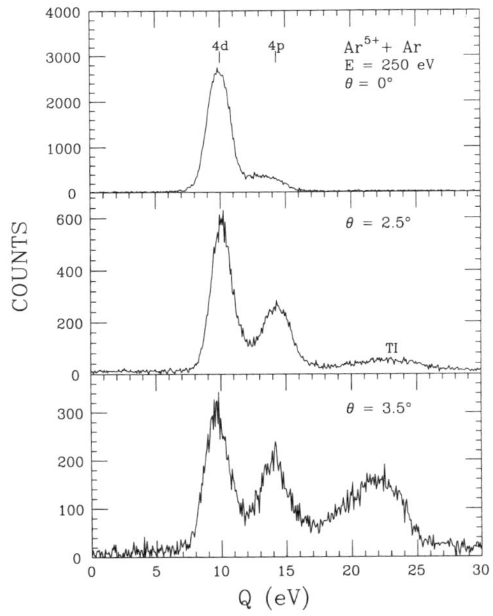

FIGURE 22.9 Translational energy-gain spectra for single-electron capture by 250 eV Ar^{5+} ions from Ar at different laboratory scattering angles. From [5].

Three peaks (I, II and III) are clearly resolved [6]. The dominant reaction channel (peak II) is due to single-electron capture from the $2p^3\ ^2P$ metastable state of Ne^{3+} into the $2s\ 2p^5\ ^3P$ state of Ne^{2+}, with an avoided crossing at $R_x = 5.44$ au. A small peak (peak III) at $Q = 13$ eV is due to capture from ground state Ne^{3+} $(2p^3\ ^4S)$ into the $2s\ 2p^5\ ^3P$ state of Ne^{2+}. The other peak (peak I) at around 6 eV is due to capture into the $3s\ ^{5,3}S$ states of Ne^{2+} from metastable state Ne^{3+} $(2p^3\ ^2P)$, with an additional contribution from capture into the $2s\ 2p^5\ ^1P$ state of Ne^{2+} from metastable state Ne^{3+} $(2p^3\ ^2D)$. Peaks II and III can only be explained due to core-varying single-electron capture transitions since final Ne^{2+} $(2s\ 2p^5)$ states cannot be described by single configuration states [20].

Figure 22.11b shows the translational energy-gain spectrum of product Ne^{2+} following single-electron capture in Ne^{3+} - Ne collisions at a collision energy of 150 eV and scattering angle of 0°. In this collision system, capture from the ground state of Ne^{3+} into the ground state of Ne^{2+} accompanied by excitation of the target product, dominates the single-electron capture processes. Peak C involves the formation of Ne^{2+} $(2p^4\ ^3P)$ from ground state incident Ne^{3+} $(2p^3\ ^4S)$ accompanied by excitation of the target product into the $2s\ 2p^6\ ^2S$ state of Ne^+. Peak B arises from capture into the $2p^4\ ^1S$ state of Ne^{2+} by the Ne^{3+} $(2p^3\ ^2D)$ metastable state component in the incident beam

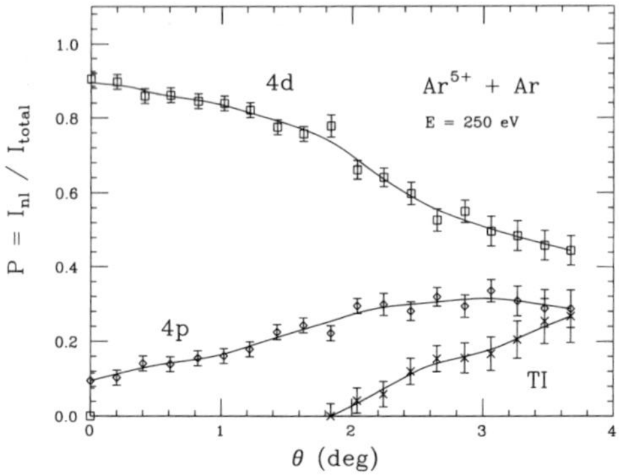

FIGURE 22.10 Probability of single-electron capture as a function of projectile laboratory scattering angle for 250 eV Ar^{5+} - Ar collisions, □, $4d$; ◇, $4p$; ×, transfer ionization. Smooth lines are drawn to guide the eye. From [5].

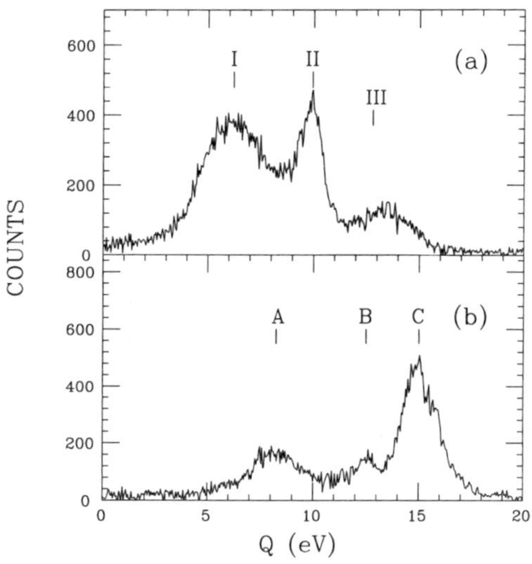

FIGURE 22.11 Translational energy-gain spectra of outgoing Ne^{2+} ions scattered at 0° from: (a) 150 eV Ne^{3+} - He, (b) 150 eV Ne^{3+} - Ne collisions.

accompanied by excitation of the target product. The broad peak (peak A) at about 8 eV is due to capture into the $3s\ ^{5,3}S$ states from the metastable state ($2p^3\ ^2D$) of Ne^{3+} with contributions from capture into the $2p^4\ ^1S$ of the Ne^{2+} from ground state incident Ne^{3+} accompanied by excitation of the target product.

In Ne^{4+} - He collision system, capture into the Ne^{3+} ($3s\ ^4P$) state occurs dominantly over the entire collision energy studied at 0° scattering angles,

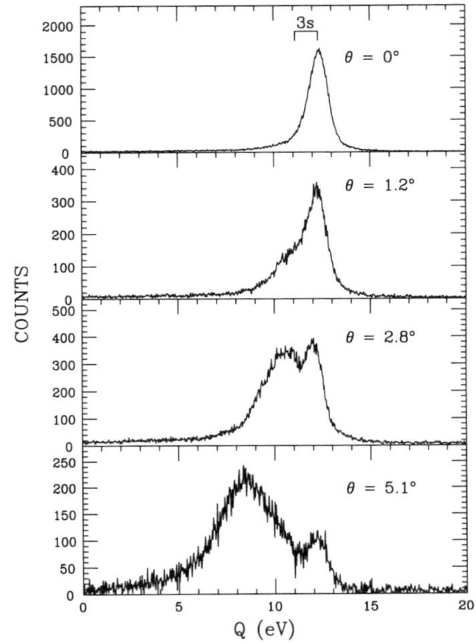

FIGURE 22.12 Translational energy-gain spectra for single-electron capture by 100 eV Ne^{4+} ions from He at different laboratory scattering angles. Reprinted from "Experimental study of single-electron capture by low-energy Ne^{q+} recoil ions (q=3-6) from He, Ne and Ar using translational energy spectroscopy" [6], with kind permission of IOP Publishing.

with an unresolved contribution involving capture into the $3s\ ^2P$ state. Figure 22.12 shows the translational energy-gain spectra of product Ne^{3+} for Ne^{4+} - He collisions at an impact energy of 100 eV and different laboratory scattering angles [6]. As the scattering angle is increased, contributions from capture into the $3s\ ^2P$ state increases relative to the dominant peak, and its maximum shifts toward smaller Q-values in contrast to the dominant peak which stays at approximately same position (same Q-value). Calculated kinematics [6] and experimental values for the observed reaction channels are shown in Fig. 22.13, where $Q(\theta_P)$ functions are shown. It is interesting to note that at large projectile scattering angles where translational energy of the recoil ions (ΔK) becomes important, the Q values decrease as would be expected from kinematics effects. However, the experimental Q values for capture into $3s\ ^4P$ stays constant with increasing projectile scattering angles, i.e. impact parameter-independent function. The reason for this might be due to oscillations in the differential cross sections reflected in the translational energy-gain spectra, i.e. an effect of angular scattering.

Tunnell et al [21] measured the differential cross sections for single-electron capture by Ne^{4+} ions from He at collision energies between 172 and 692 eV,

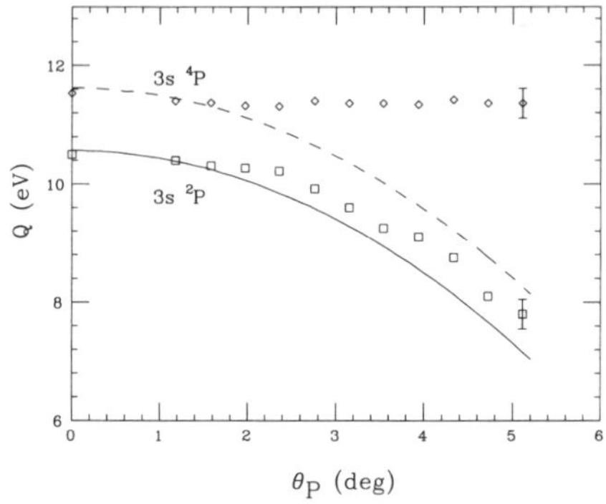

FIGURE 22.13 Experimental and calculated energy gain (Q) values for capture into $3s\ ^4P$ and $3s\ ^2P$ states of the Ne^{3+} ion from ground-state incident Ne^{4+} ($2p^2\ ^3P$) ions versus projectile laboratory scattering angle θ_p. Reprinted from "Experimental study of single-electron capture by low-energy Ne^{q+} recoil ions (q=3-6) from He, Ne and Ar using translational energy spectroscopy" [6], with kind permission of IOP Publishing.

and laboratory scattering angles between 0° and 3.44°. Their data show a strong forward peak close to 0° and an oscillatory structure with two discernable secondary maxima for angles greater than 0°. The separation in reduced scattering angles ($\tau = E\theta$) between the forward peak and the first maximum stays nearly constant with the impact energy E, while the separation between the first maximum and the next one increases with E. They suggested that the first feature is to be associated with the appearance at a particular impact parameter b of some distinguishing property of the deflection function, whereas the second maximum is dependent on an energy-dependent phase accumulation.

22.4.2 Differential cross sections

Studies of differential cross sections sometimes provide certain details on the collision process that are often obscured by total cross sections. Differential cross sections for single-electron capture processes in the collision of low-energy multiply charged ions with atomic targets have been studied experimentally by a number of groups, using different types of apparatus. The differential cross sections presented in this section have been obtained using two types of spectrometers. First, a differential energy-gain spectrometer, described previously in section 22.2.1, was used to measure double differential cross sections. Second, an angular distribution apparatus similar to that described by Tunnell et al [21] was used.

Briefly, the ions were produced in a recoil source by using 25 MeV F^{4+} ions from the WMU tandem Van de Graaff accelerator as a pump beam. The ions

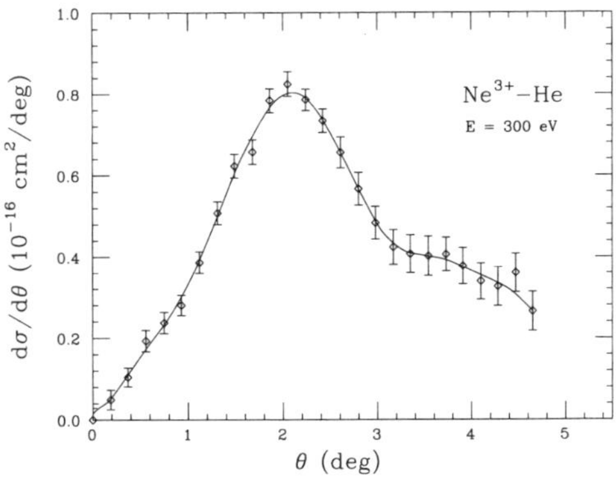

FIGURE 22.14 Differential cross sections for single-electron capture by 300 eV Ne^{3+} ions from He. Smooth line is drawn to guide the eye. Reprinted from "Experimental study of single-electron capture by low-energy Ne^{q+} recoil ions (q=3-6) from He, Ne and Ar using translational energy spectroscopy" [6], with kind permission of IOP Publishing.

were extracted at right angles to the pump beam direction, charge analyzed by the magnet, and passed through the target collision cell. Charge state selection was done with an electrostatic retarding grid located in front of the one-dimensional position-sensitive channel plate detector. The detector was collimated with a bow-tie shaped aperture to convert the radial distribution to an approximately linear one.

Figure 22.14 displays the measured differential cross sections for single-electron capture by Ne^{3+} ions from He at a collision energy of 300 eV [6]. The differential cross sections were found, using the translational energy-gain technique, by calculating the area under the peaks in energy-gain spectra at different scattering angles using a curve fitting program. The data show that the distribution for single-electron capture is peaked at an angle of about 2.1°. The data also show that the projectile products are distributed mostly at angles larger than $\theta_c = 1.02°$, which corresponds to the $2s\ 2p^5$ capture channel at an impact parameter equal to the crossing radius, indicating that capture took place on the way-out.

Figure 22.15 shows the measured differential cross section for single-electron capture by Ar^{8+} ions from He at a collision energy of 3 keV, obtained with the angular distribution apparatus [22]. The data show a broad distribution with a maximum near a scattering angle of about 8 mrad. The measurements also show that the projectile products are distributed near a scattering angle $\theta_c = 5.4$ mrad which corresponds to $(4f,4d)$ capture channels at an impact parameter equal to the crossing radius.

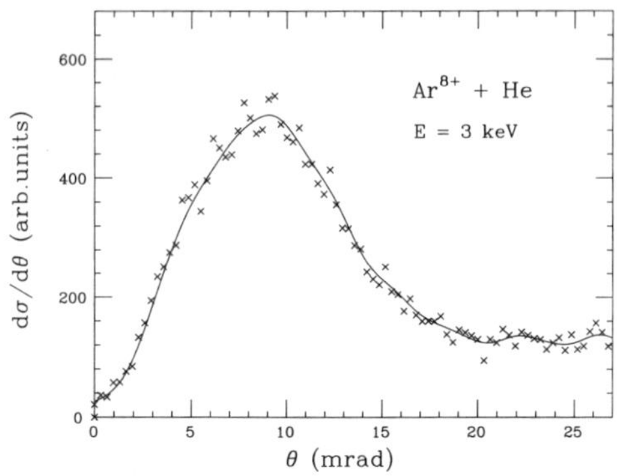

FIGURE 22.15 Experimental differential cross sections for single-electron capture by Ar^{8+} ions from He at 3 keV. Solid curve indicates the best fit to the experimental data. [22]

22.4.3 Comparison to theory

The single-crossing Landau-Zener model has been used to interpret the final-state populations by calculating the location of the crossing radii where the probability of single-electron capture is large, in terms of the energy-gain values of the final states, i.e. reaction windows. Figure 22.16 shows the translational energy-gain spectra for single-electron capture by 250 eV Ar^{5+} ions from He and Ar at 0° scattering angles [5]. Also shown are reaction windows calculated on the basis of a single-crossing Landau-Zener model using two values of H_{12}. Peak values have been normalized to the observed dominant peak value in the energy-gain spectrum. It can be seen that the reaction windows based on the factor 5.48 are in better agreement with the dominant reaction channels, while the reaction windows based on the factor 9.13 favors smaller energy-gain values than observed.

Also the MCLZ model has been used to calculate cross sections for single-electron capture into individual channels using the probability accumulation procedure which has been discussed in detail in section 22.3.2. The results of these calculations using the Taulbjerg expression (Equation 22.17) for H_{12} are shown as vertical lines in Figure 22.16, with the value of largest cross section being normalized to the height of the dominant peak observed in the energy-gain spectrum. For the Ar^{5+} - He collision system, the MCLZ calculations correctly predict the position of the dominant reaction channel and overestimate the contribution of the 3d reaction channel relative to the dominant reaction channel. For the Ar^{5+} - Ar collision system, the MCLZ theory predicts relative cross sections which are in good agreement with the measurements.

The differential cross sections for single-electron capture are interpreted using a simple semi-classical curve-crossing model based upon classical differen-

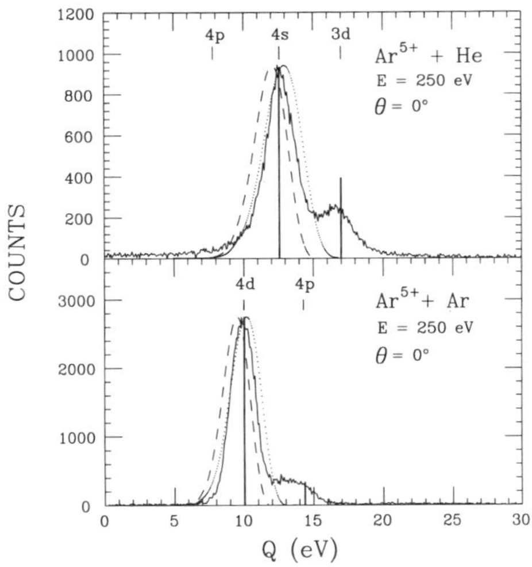

FIGURE 22.16 Translational energy-gain spectra for single-electron capture by 250 eV Ar^{5+} ions from He and Ar. Reaction windows are calculated on the basis of a single-crossing LZ theory using values of H_{12} with factors of 9.1 (broken curve) and 5.5 (dotted curve). The vertical lines show results of MCLZ calculations. From [5].

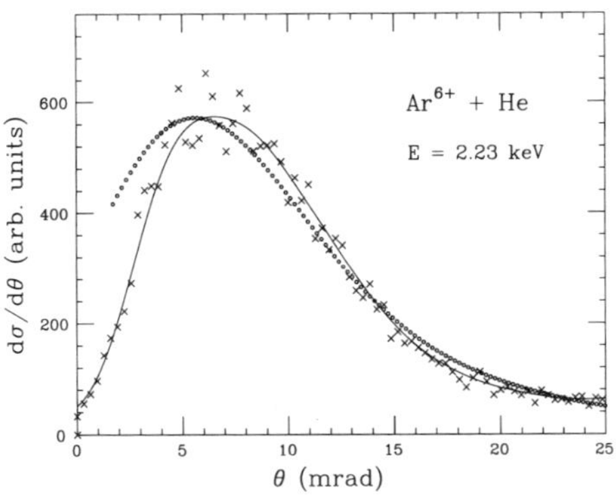

FIGURE 22.17 Experimental and calculated differential cross sections for single-electron capture by 2.23 keV Ar^{6+} ions from He. Data appear as ×'s, smooth curve as best fit to the data and theoretical calculation, folded with experimental angular resolution, as ∘'s [22].

tial cross sections coupled with Landau-Zener probabilities. The experimental differential cross sections and the theoretical calculations, folded with the experimental resolution, for single-electron capture by Ar^{6+} ions from He at collision energy of 2.23 keV are shown in Fig. 22.17 [22]. The value of the largest calculated cross section has been normalized to the height of the peak observed in the spectrum. The experimental data show a main peak centered at 7 mrad, outside the critical angle (θ_c) of 4 mrad. This distribution represents contribution from capture on the "way-in" branch of the deflection function. The calculation is performed assuming the $4d$ and $4p$ states of the Ar^{5+} ion to be open channels, with capture into the $4p$ state as the dominant reaction channel and the $4s$ channel taken to be the promotion channel. The theoretical predictions show close agreement with the measured results, but overestimate the contributions for capture on the "way-out" branch for angles less than $\theta_c = 4$ mrad.

22.5 CONCLUSIONS

There were two principal objectives of this chapter. The first objective was to describe an experimental apparatus capable of measuring state-selective differential cross sections for ion-atom collisions involving electron capture. The second purpose was to investigate the doubly-differential cross sections for single-electron capture processes by low-energy multiply-charged recoil ions from neutral targets.

The translational energy-gain spectra show that only a few final states are populated. The transition probabilities depend strongly on the charge state of the projectile ion and the ionization potential of the target atom, and only weakly on the impact energy. The reaction window obtained from the single-crossing LZ model using the factor 5.48 provides the best general description of the observed features. The MCLZ calculations correctly predict the positions of the dominant reaction channels that are observed.

The experimental angular distribution spectra contain a main peak lying near a critical angle, θ_c, corresponding to capture at an impact parameter equal to the crossing radius of the primary capture channel. The results are in qualitative agreement with calculations made using a semi-classical model employing Coulomb potential energy curves.

22.6 ADDITIONAL READING

1. Janev, R.K., in *Review of Fundamental Processes and Applications of Atoms and Ions*, edited by C.D. Lin, (World Scientific, Singapore, 1993) pp. 1-59.

2. Heil, T.G., *Nuclear Instruments and Methods in Phys. Research*, **B23**, 222 (1987).

3. Cocke, C.L., in *Review of Fundamental Processes and Applications of Atoms and Ions*, edited by C.D. Lin, (World Scientific, Singapore, 1993) pp. 111-161.

4. Janev, R.K., Winter, H., *Phys. Reports* **117**, 265 (1985).

5. Cocke, C.L., and Olson, R.E., *Phys. Reports* **205**, 153 (1991).
6. Kamber, E.Y., and Cocke, C.L., in *Physics of Ion Impact Phenomena*, edited by D. Mathur, (Springer-Verlag, Berlin, Heidelberg, 1991), pp. 91-122.
7. Hasted, J.B., (ed.) *Physics of Atomic Collisions*, Butterworth, 2nd Edition, London, 1972).
8. Hamdan, M., and Brenton, A.G., in *Physics of Ion Impact Phenomena*, edited by D. Mathur, (Springer-Verlag, Berlin, Heidelberg, 1991) pp. 165-196.
9. Brenton, A.G., *J. of Mass Spectrometry* **30**, 657 (1995).

REFERENCES

1. Nielsen, E.H., Andersen, L.H., Bárány, A., Cederquist, H., Heinemeier, J., Hvelplund, P., Knudsen, H., MacAdam, K.B., Sørensen, J., *J. Phys. B* **18**, 1789 (1985).
2. Giese, J.P., Cocke, C.L., Waggoner, W., Tunnell, L.N., Varghese, S.L., *Phys. Rev. A* **34**, 3770 (1986).
3. McCullough, R.W., Wilson, S.M., Gilbody, H.B.: *J. Phys. B* **20**, 2031 (1987).
4. Biedermann, C., Cederquist, H., Andersen, L.R., Levin, J.C., Short, R.T., Elston, S.B., Gibbons, J.P., Andersen, L., Lilijeby, L., Sellin, I.A., *Phys. Rev. A* **41**, 5889 (1990).
5. Yaltkaya, S., Kamber, E.Y. and Ferguson, S.M., *Phys. Rev. A* **48**, 382 (1993).
6. Said, R., Kamber, E.Y., Yaltkaya, S., and Ferguson, S.M., *J. Phys. B* **27**, 3993 (1994).
7. Winter, H., *Phys. Scr. T* **3**, 159 (1983).
8. Kimura, M., Iwai, T., Kaneko, Y., Kobayashi, N., Matsumoto, A., Ohtani, S., Takagi, S., Tawara, H., and Tsurubuchi, S., *J. Phys. Soc. Jpn.* **53**, 2224 (1984).
9. Taulbjerg, K., *J. Phys.* **B9**, L369 (1986).
10. Cooks, R.G. (ed.): *Collision Spectroscopy* (Plenum, New York), pp. 252, (1978).
11. Landau, L.D., *Phys. Z. Sowetunion* **2**, 46 (1932).
12. Zener, C., *Proc. Roy. Soc. London A* **137**, 696 (1932).
13. Janev, R.K., Presnyakov, L.P., Shevelko, V.P. (eds.): *Physics of Highly Charged Ions*, (Springer-Verlag, Berlin, Heidelberg, 1985).
14. Olson, R.E., Salop, A., *Phys. Rev. A* **14**, 579 (1976).
15. Salop, A., and Olson, R.E., *Phys. Rev. A* **13**, 1312 (1976).
16. Olson, R.E., Smith, F.T., *Phys. Rev. A* **3** 1607 (1971).
17. Cocke, C.L., Tunnell, L.N., Waggoner, W., Giese, J.P., Varghese, S.L., Kamber, E.Y., and Pedersen, J.O.K., *Nucl. Instrum. Methods* **B24/25**, 97 (1987).
18. Danard, H., and Barany, A., *J. Phys. B* **10**, 3109 (1986).

19. Bashkin, S., and Stoner, J.O., *Atomic Energy Levels and Grotrian Diagrams.* North-Holland, Amsterdam, 1978.
20. Winter, H., *Comment At. Mol. Phys.* **27**, 91 (1991).
21. Tunnell, L.N., Cocke, C.L., Giese, J.P., Kamber, E.Y., Varghese, S.L., and Waggoner, W., *Phys. Rev. A* **35**, 3299 (1987).
22. Edens, J.E., M.A. Thesis, Western Michigan University, 1995.

CHAPTER 23

Cold Target Recoil Ion Momentum Spectroscopy

H. Schmidt-Böcking*, M. Achler*, I. Ali*, H. Bräuning*, C. L. Cocke†, R. Dörner*, O. Jagutzki*, T. Kambara¶, Kh. Khayyat*, V. Mergel*, R. Moshammer*, M. H. Prior§, L. Spielberger*, W. Schmitt†, K. Ullmann-Pfleger*, M. Unverzagt*, J. Ullrich†, and W. Wu‡

A newly developed spectroscopy technique is presented to study the kinematics of many-particle atomic collision processes with high momentum resolution and high detection efficiency. These goals are achieved by detecting the low-energy recoil ion in coincidence with the emitted electrons or scattered projectiles. The recoil ions can be detected with nearly 100% efficiency by projecting them with a weak electrostatic field on a position-sensitive detector. The high momentum resolution for the recoil ion is obtained by using a cold super- sonic target jet and calculating the recoil-ion trajectory from the measured time-of-flight and position on the position-sensitive detection devices. The basic components of the experimental set-up for **COLTRIMS** (**COL**d **T**arget **R**ecoil **I**on **M**omentum **S**pectroscopy) and first results obtained with this technique are presented. Results for electron capture, target and projectile ionization in fast ion-atom collisions as well as photo ionization processes are described. Future applications of **COLTRIMS**, e.g. in the field of atomic physics for the investigation of inneratomic many electron dynamics or in the field of nuclear physics to study the kinetics of the nuclear beta-decay, are discussed.

---

*) Universität Frankfurt, Institut für Kernphysik, August-Euler-Str.6 60486 Frankfurt, FRG
†) GSI-Darmstadt, 64220 Darmstadt, FRG
¶) RIKEN, Wako, Saitama, Japan
§) Lawrence Berkeley National Laboratory, Berkeley, Ca, 94720, USA
‡) Kansas State University, Manhattan, Ks, 66506, USA

23.1 INTRODUCTION

High-resolution spectroscopy techniques in atomic collision physics in general suffer from the restriction of small detection efficiency. Thus, the coincident detection of several reaction products to study e.g. many-particle effects becomes extremely difficult. This is one of the reasons for the lack of systematic experimental data on the many-particle momentum exchange in atomic collision processes. Only very few data have been published so far in the literature for double target ionization [1–4] e.g. for (e,3e) and (γ,2e) reactions. Systematic data exist for single ionization processes, e. g. (e,2e) reactions (see e.g. [5]). In these experiments the traditional electron-electron-coincidence technique was used to measure the electron-electron momentum correlation in the final state. In case of three electrons in the continuum final state, triple e-e-e coincidence techniques yield extremely low coincidence detection efficiency, particularly if high momentum resolution is required. To our knowledge only one experiment has been reported in literature [4].

In the case of detecting the recoil ion instead of electrons, however, this ion can be detected with high resolution and a solid angle of nearly 100% and a total detection efficiency of about 60%. To achieve high momentum resolution the initial target motion (initial target temperature) has to be strongly reduced and the recoil-ion trajectory and its time-of-flight (TOF) must be measured precisely. Using standard super-sonic target devices, for helium as target gas an initial target temperature of a few mK can be obtained which is equivalent to a momentum distribution width of approximately $0.01 a.u.$, i.e. about $40 eV/c$. Depending on the size and the geometry of the recoil-ion spectrometer a relative resolution below 1% can be achieved for the recoil-ion momentum. Thus, for small recoil-ion momenta a resolution of $0.02 a.u.$ is feasible with a relatively small detection device. In the not-too-distant future Laser-cooled gas targets may provide even lower target temperatures and by using large scale detection devices a relative momentum resolution of a few 10^{-4} is achievable. The ultimate limit of momentum resolution obtainable with presently thinkable techniques may approach $0.001 a.u.$, which corresponds to $4 eV/c$.

Since the kinetic energy of practically all created recoil ions after the collision is mostly below $1 eV$ they all can be projected even by a weak electrostatic field onto a position-sensitive detection device. Thus, the recoil-ion momentum spectroscopy provides a much more efficient alternative technique for momentum spectroscopy in atomic many- particle collision processes than traditional electron spectroscopy. Particularly if two or more reaction partners in the final state have to be measured in coincidence it is usually sufficient to measure the recoil-ion momentum vector and only two momentum components e.g.the emission angle (i.e. the detection position) of the other ejected partners, to yield already complete information on the momentum exchange in such three-particle collisions.

For fast ion-atom collision processes, as will be shown below, the recoil-ion transverse (with respect to the incoming beam) momentum component is a good measure for the impact parameter, and its longitudinal component reflects the Q-value (inelasticity due to electronic excitation and ionization) and the mass exchange (number of transferred electrons) in such collisions. Thus, the measurement of the longitudinal momentum is equivalent to an energy-gain spectroscopy and it can provide for the projectile energy gain or loss in fast ion atom collisions a relative resolution even in the 10^{-9} regime. Therefore recoil-ion momentum spectroscopy is a tool which yields even in GeV ion-atom collisions precise information on the electronic transitions. Furthermore, also for (e,2e) or (e,3e) or (γ,2e) measurements **COLTRIMS** provides a very efficient way to study the complete momentum exchange between the electrons and nuclei in such collision processes. For a recoil-ion-electron coincidence an overall coincidence efficiency of 10 to 20% is easily feasible, therefore coincidence rates can nearly approach singles rates. Furthermore, **COLTRIMS** is not limited to any spatial direction therefore complete angular distributions are obtained.

In section 2 a short outline of the important kinematical equations for the recoil-ion spectroscopy is given and in section 3 the experimental technique is presented. In section 4 of this paper first results [6–12] for different collision processes are discussed and the power of the new technique is described.

23.2 RECOIL ION KINEMATICS

Since in all atomic collision processes energy and momentum are conserved, we can derive some general equations for the final momentum and energy state of the recoil ions. In the next two subsections we will discuss the equations for ion-atom collisions and photo absorption. Atomic units are used throughout ($e = \hbar = m_e = 1$).

23.2.1 Ion-atom collisions

The kinematical calculations for the recoil ion in fast ion-atom collisions become most simple if the momentum transfer between the collision partners is small compared to the initial momentum in the center-of-mass system. In this case the longitudinal momentum of the recoil ion depends only on the inelastic energy gain or loss of the collision process (i.e. electronic excitation, emission of Bremsstrahlung etc.), on the mass transfer in electron capture collisions, and on the longitudinal momenta of the electrons emitted to the continuum. The transverse momentum of the recoil ion in "close" encounters reflects the nuclear impact parameter. The condition of small momentum transfer is fulfilled in most of all atomic reactions of interest here. The three most important possibilities for target ionizing collisions are:

- capture from n_C target electrons to projectile bound states,

- ionization of m_T target electrons,
- ionization of l_P projectile electrons.

From energy- and momentum-conservation we obtain for the longitudinal momentum of the recoil ion in the final state ($p^f_{\|R}$):

$$p_{\|R}{}^f = \underbrace{\frac{Q}{v_P}}_{\text{Q-value}} \underbrace{- n_C \frac{v_P}{2}}_{\text{mass transfer}} \underbrace{+ \sum_{j=1}^{m_T} \left(\frac{E^f_{ej}}{v_P} - p^f_{\|ej} \right)^{\Sigma_{Lab}}}_{\text{target ionization}} + \underbrace{\sum_{j=1}^{l_P} \left(\frac{E^f_{ej}}{v_P} \right)^{\Sigma_P}}_{\text{projectile ionization}}$$

Q: Q-value (i.e. the sum of all electronic de- and excitation energies, v_P: projectile velocity, E^f_{ej}: final continuum energy of the jth electron, $p^f_{\|ej}$: final momentum of the jth electron, (given in the lab frame for electrons from the target and in the projectile frame for electrons emitted from the projectile). This equation can be simplified if we focus now on several special reaction channels.

For pure electron capture only discrete values of $p^f_{\|R}$ occur corresponding to the energy eigenvalues of the projectile and target electrons.

$$p^f_{\|R} = \frac{Q}{v_P} - n_C \frac{v_P}{2}. \tag{23.2}$$

This enables us to determine the energy levels of all excited states with high precision, often better than standard energy gain spectroscopy techniques can do.

For the emission of target electrons to continuum states we obtain:

$$p^f_{\|R} = \frac{Q}{v_P} + \sum_{j=1}^{m_T} \left(\frac{E^f_{ej}}{v_P} - p^f_{\|ej} \right)^{\Sigma_{Lab}} = -\frac{\Delta E_P}{v_P} - \sum_{j=1}^{m_T} p^{f\Sigma_{Lab}}_{\|ej}, \tag{23.3}$$

where ΔE_P is the kinetc energy change of the projectile. If $(Q + \sum_{j=1}^{m_T} E_{ej})/(v_P p^f_{\|R}) \ll 1$ is valid, i.e. the projectile velocity is much higher than the one of the outgoing electrons, the longitudinal momentum of the recoil ion reflects the sum of the longitudinal electron momenta. We obtain:

$$p^f_{\|R} \approx - \sum_{j=1}^{m_T} p^f_{\|ej}. \tag{23.4}$$

For the ionization of n_P electrons of the projectile it follows from equation 23.2.1:

$$p_{\|R}^{f} = \frac{Q}{v_P} + \sum_{j=1}^{l_P} \left(\frac{E_{ej}^{f}}{v_P} \right) \Sigma_P. \qquad (23.5)$$

For this case one can calculate the total energy loss/gain of the projectile directly from the measured $p_{\|R}$.

For all collisions the transverse momentum of the recoil ion reflects the sum of transverse momenta of projectile and electrons:

$$p_{\perp R}^{f} = - \sum_{j=1}^{l_P = m_T} p_{\perp ej} - p_{\perp P}. \qquad (23.6)$$

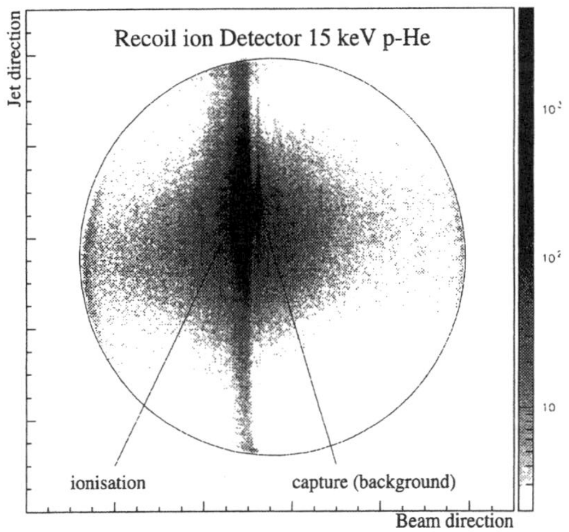

FIGURE 23.1 Image of He^{1+} ion on the recoil ion detector for $15keV$ p→He, measured in coincidence with an emitted electron. (see text)

In figure 23.1 the raw spectrum of one component of $p_{\perp R}^{f}$ and $p_{\|R}^{f}$ is shown for 15 keV protons on He showing capture and ionization, respectively. For electron capture the longitudinal recoil-ion momentum components (beam direction) are constant for all measured recoil-ion transverse momenta. For target ionization the recoil-ion longitudinal momenta as function of the longitudinal momentum of the emitted electron are presented in figure 23.2. They show the predicted dependence E (see equation 3). In figure 23.3 the projections of the recoil-ion longitudinal momenta are displayed. The energy-gain resolution of about 10^{-4} allows to separate the different reaction channels. Even the weak target ionization channel with simultanous excitation of the second He electron to the L shell is observable.

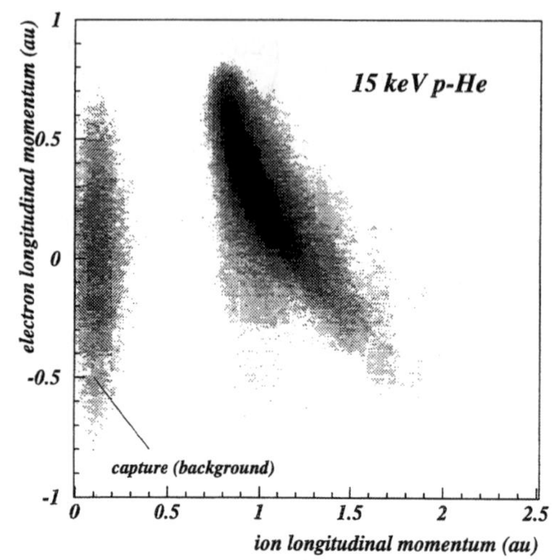

FIGURE 23.2 Correspondence of electron longitudinal momentum and He^{1+} longitudinal momentum for $15keV$ p + He → He^{1+} + e^-. The distribution has been integrated over the transverse momenta of all particles.
Since in this reaction the electron transverse momenta are small, one can see a structure as expected from equ. 23.2

23.2.2 Photo absorption

In the case of absorption of a single photon the approximation of small momentum is still fullfilled. But the electron can absorb all the photon energy E_γ leading to the momentum $p_e = \sqrt{2(E_\gamma - E_{Bind} - E_{excitation})}$. E_{Bind} is the binding energy of the emitted electron and $E_{excitation}$ the excitation energy of the residual ion. The recoil ion has to compensate this momentum. So the ions are distributed on a momentum sphere of a radius $p_R = p_e$ in the center-of-mass system, where E_γ is the photon energy. Because of the momentum of the incoming photon (p_γ) this sphere is shifted in the Lab system by $p_\gamma = E_\gamma/c$.

In figure 23.4 the measured recoil-ion momenta in the x-y-plane are plotted. x and y are the directions of the electric field vector and of the momentum of the linear polarized incoming photon, respectively. One can see two kinematic circles: the one with the larger radius corresponds to the emission of one K-electron, where the other electron remains in the He^{1+} ground state; the one with the smaller radius corresponds to ionization of one K-electron and simultaneous excitation of the second He electron to the L-shell.

These two cases presented are only a few examples for the variety of information one can extract from the momentum of the recoil ion in atomic collision processes.

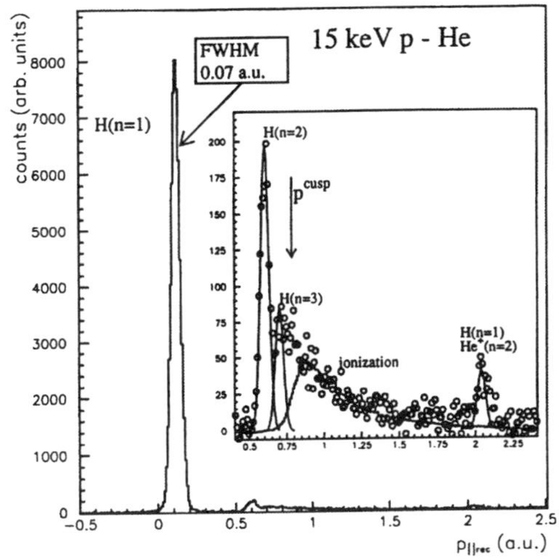

FIGURE 23.3 Longitutinal momentum distribution of He^{1+} ions from $15keV$ proton impact. The dominant peak is due to capture to the projectile ground state. The arrow indicates the position of the capture to the projectile continuum. The full line right of the arrow shows the momentum distribution for ionization only, which has been measured separately by detecting an electron in coincidence with the recoil ion. The momentum resolution is $\pm 0.035 a.u.$, equivalent to an energy gain of $\pm 0.7eV$ and a recoil energy of $\pm 4.5\mu eV$ [12].

23.3 EXPERIMENTAL SET-UP

23.3.1 Principle of COLTRIMS

The measurement of momentum transfer between different atomic reaction products is always performed by determining the final and the initial momenta of these products. Furthermore the accuracy of the determination of the momentum transfer is generally better, when the transferred momentum is of the order of the final and large compared to the initial momentum. The measurement of the initial momenta, however, becomes even obsolete when they are very small compared to the transferred momentum. These requirements are perfectly fullfilled in the COLTRIMS spectroscopy technique.

The main components of **COLTRIMS** are the very cold target gas jet and the recoil-ion spectrometer with a position-sensitive detector. In the center of the recoil-ion spectrometer the incoming projectiles collide with the cold target atoms.

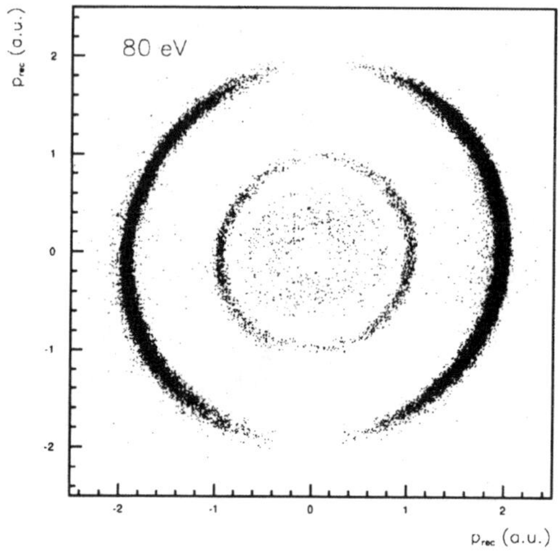

FIGURE 23.4 Momentum distribution of He^{1+} ion produced by $80eV$ photon impact. The x axis is the direction of the polarization vector of the linear polarized light, the y axis is the direction of the photon beam. The distribution has been integrated over a momentum range of $\pm 0.3a.u.$ in the z direction. For more detail see [10].

23.3.2 Super-sonic Gas Jet

In the **COLTRIMS**- systems presently used (see Figure 23.5) a supersonic helium gas jet, with an internal temperature of lower than $0.1K$, provides the rather cold localized gas target. The helium gas expands through a $30\mu m$ hole. The gas nozzle is mounted on the cold finger of a cryo pump and is cooled down to a temperature of 10 to $35K$. The gas jet is collimated at a distance of about $10mm$ from the expansion hole by a skimmer of approximately $0.3mm$ diameter. The gas jet leaves the collision chamber through a hole of about 1cm diameter into a jet dump, pumped by a small turbomolecular pump (260 or $360l/sec$) to reduce the helium residual gas pressure in the scattering chamber to a few $10^{-8}mbar$. Typical pressures are: $400mbar$ on the high pressure side of the $30\mu m$ hole and $5 \times 10^{-4}mbar$ in the source chamber (preskimmer stage). The source and the scattering chambers are separately pumped by two small (260 or $360l/sec$) turbomolecular pumps. At the collision region, $30mm$ above the skimmer, the helium gas jet has a diameter of 1 to $2mm$ and a local density of 1 to $5 \times 10^{11} atoms/\text{cm}^2$. The internal temperature is lower than $0.1K$ in the jet direction [13]. Perpendicular to the jet the atoms have a momentum spread of $\pm 0.02a.u.$ given by the velocity of the jet and the skimming geometry. Figure 23.5 shows the gas jet device with the target and source chamber.

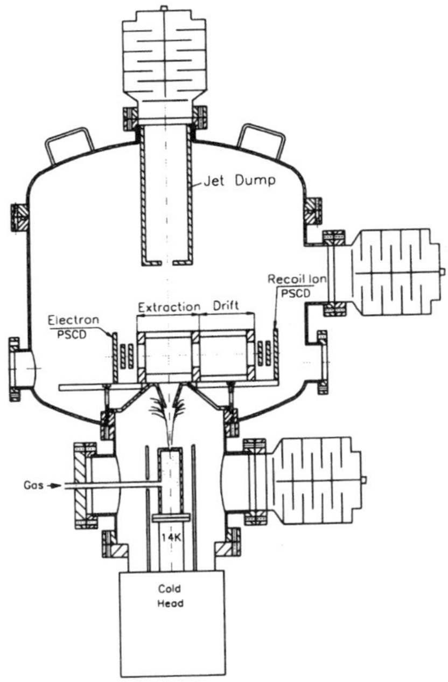

FIGURE 23.5 The upper part shows the target chamber with the recoil ion spectrometer: the lower part the source chamber where the gas jet is created by expanding through the $30\mu m$ nozzle, mounted on the cold head of a cryogenic system.

23.3.3 Recoil-ion spectrometer

The recoil ions created at the intersection point of the gas jet with the ion beam are extracted by a weak homogeneous electrostatic field varying between 0.1 to about $10V/cm$ depending on the expected recoil-ion momenta. After passing a field-free drift region they are post accelerated onto a position-sensitive channel-plate detector with a position resolution of $< 0.2mm$. Great care has been taken to assure proper field conditions in the spectrometer area. The time-of-flight of the recoil ions is measured by a coincidence with projectiles, electrons or the machine pulse, when using a pulsed beam. From the time-of-flight and the position on the channel-plate detector the three momentum components of the recoil ion can be calculated.

Different recoil-ion spectrometers have been developed in the last two years. For some applications the extraction of the recoil ions perpendicular to the projectile direction yields optimal detection and resolution for the investigated collision system, in other cases the recoil-ion extraction parallel to the projectile beam direction has substantial advantages. In the **COLTRIMS** spectrometer of [9] the recoil ions are extracted from the interaction region perpendicular to the beam. In these spectrometers (see figure 23.6) the drift region was separated from the extraction region by a woven mesh of $0.25mm$ mesh width. A stack of three of these meshes with 1mm spacing shields the drift region from the strong postaccelaration field just in front of the detector. On all parts, including the meshes, a thin layer of carbon is evaporated to avoid contact potentials. However, recent tests with improved spectrometers [12,10,14] have shown, that all meshes influence somewhat the recoil-ion trajectory and are thus limiting the resolution to about $0.05a.u.$. In these recoil-ion spectrometers a homogeneous field in the extraction region is obtained by shielding this area from external potentials e.g. with a carbon fiber. One thin carbon fiber of $7\mu m$ diameter and $10m$ length (or in some cases a thin metallic wire, see e.g. [15] is wound around four supporting germanium coated insulator screws. The fiber defines the potential in the extraction region and divides the voltage. Figure 23.6 shows a schematic of the recoil-ion spectrometer used by [10,14]. A sudden potential change in the extraction region provides a field geometry, which focuses recoil ions with the same momentum but created at different target positions onto the same position on the detector.

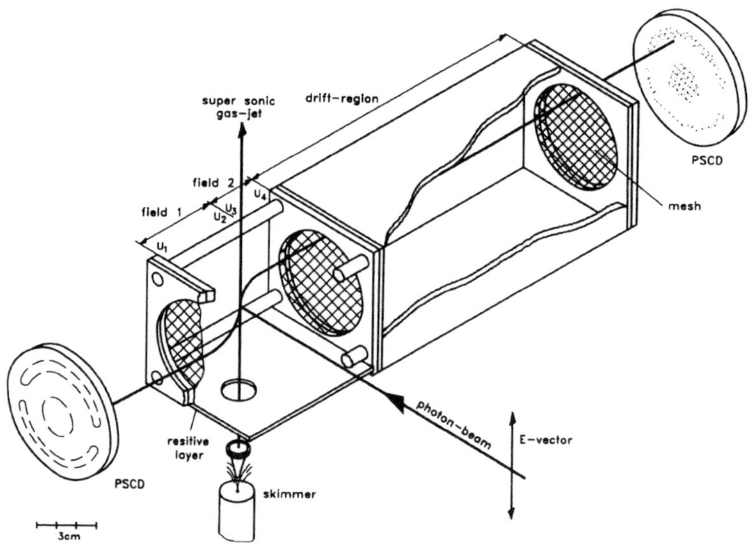

FIGURE 23.6 High-resolution recoil-ion momentum spectrometer with precooled supersonic gas jet target ([10,14]).

The resolution e.g. of the apparatus used in reference [10,12,14] for $p_{\|R}$ is $\pm 0.035a.u.$. This resolution is limited by the precooling target temperature of about $30K$. The momentum resolution in the direction of the extraction field is $\pm 0.02a.u.$, mainly restricted by the internal momentum spread of the jet perpendicular to the jet direction.

In the **COLTRIMS** spectrometer used by Moshammer and coworkers [8,16,17] a uniform electric field along the ion beam is generated by two parallel ceramic plates covered with burnt-in resistive layers. This spectrometer is shown in Figure 23.7. Recoil ions and electrons are accelerated into the forward and backward directions, respectively. The electric field is superimposed by a magnetic solenoid field which forces electrons on spiral trajectories thus yielding nearly 100% detection efficiency for electrons with transverse kinetic energies below $30eV$. In contrast to the spectrometers described above the recoiling ions are extracted in the longitudinal direction parallel to the ion beam. Thus very large longitudinal recoil-ion momenta (up to $160a.u.$) can be detected and the momentum resolution can strongly be improved. In [8] already a recoil-ion momentum resolution of $0.16a.u.$ (FWHM) was obtained. The theoretical limits for the momentum resolution of the recoil ion for the present super-sonic gas target are $< 0.02a.u.$ in transverse and $0.0005a.u.$ in longitudinal direction. Due to a specially designed burnt-in resitive layer structure in the spectrometer plates the electric field in the beam-target intersection region can be directed with in any direction. Thus the recoil-ion position on the channelplate can be chosen according to the requirements of the measurement. This spectrometer is perfectly suited for experiments carried out in ion beam storage rings. The system is described in detail [17].

An important component of the **COLTRIMS** system are the position-sensitive recoil-ion or electron detectors. The standard size detector is a $50mm$ diameter chevron or Z-stack channelplate electron multiplier with an active area of typically $47mm$ diameter. The position information is obtained either from charge division using "wedge-and-strip" anode structures or from time measurements using "delay-line" anode structures. Both position readout systems can yield a position resolution better than $0.1mm$ depending on the electronic moduls (preamplifiers, constant fractions, analog digital or time digital converters etc.) available. It might be worth to notice that the use of a Z-stack channelplate system yields, even for electron detection, signals far separated from noise pulses providing even with "standard electronic moduls" very good position resolution. In Figure 23.8 the "Delay-Line" detector and a typical position spectrum (figure 23.9) for electron detection are shown.

23.3.4 Coincidence experiments using COLTRIMS

In the recoil-ion spectrometers of reference [18] opposite to the detector for the recoil ions a second position-sensitive detector for electrons is located. For the same given momentum the electrons are much faster than the recoil ions due to the lighter mass. Thus, they are nearly unaffected by the weak electrostatic extraction field. The solid angle for electron detection is determined for the high energetic electrons by the geometry of this device and can reach for the present spectrometer about 15%, for low energetic electrons (less than

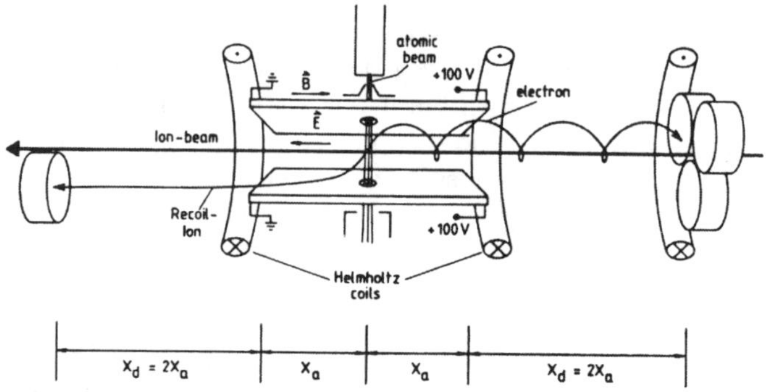

FIGURE 23.7 Recoil-ion spectrometer with solenoid field for collecting the electrons on a position-sensitive detector. The electric field is generated by carbon-coated ceramic plates parallel to the beam axis. [17]

about $5eV$ kinetic energy) the solid angle approaches nearly 100%.

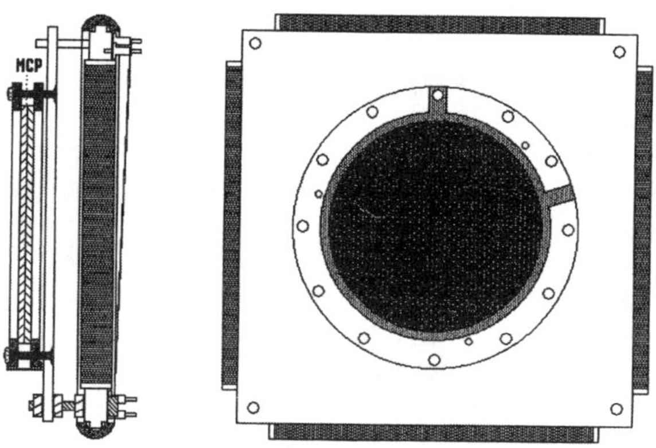

FIGURE 23.8 Scematic of a MCP detector with about 50mm active diameter and delay-line readout [25].

To improve the electron detection efficiency for high energetic electrons Moshammer and coworkers [8] have added to the electrostatic extraction field a weak magnetic solenoid field (see figure 23.7). The electrons are bent onto spiral-like trajectories by a magnetic field thus even for $100eV$ electrons a 4π solid angle can be obtained.

As discussed in detail in [17] for all electrons independent of their kinetic energy their Lamour frequency is constant, thus the electron detection position

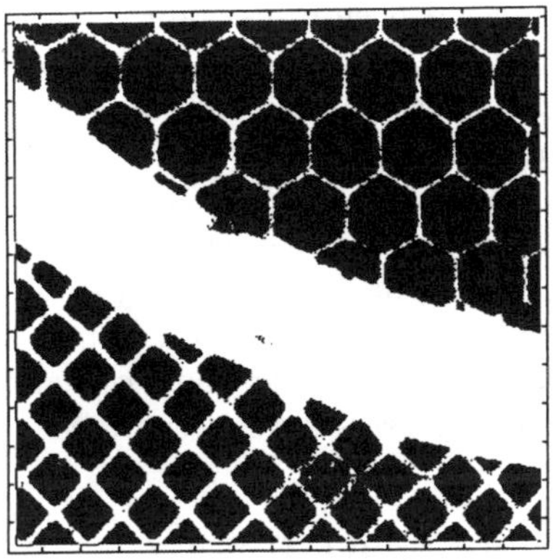

FIGURE 23.9 Shadow image of a wire mask. The MCP was homogeniously irradiated with an α-source. The picture shows an area of approximately $20mm \times 20mm$, the hexagonal wires have $0.2mm$ diameter. A position resolution of better than $50\mu m$ can easily be obtained

can be corrected for the deflection of the electron motion in the magnetic field. Using this spetrometer [8,16] have carried out several benchmark recoil-ion -electron coincidence experiments to investigate the dynamics of target ionization by fast highly charged ion impact.

23.4 EXPERIMENTAL RESULTS AND DISCUSSION

In this chapter the following ionization processes investigated by **COLTRIMS** will be discussed:

1. $He^{2+} + He \rightarrow He^{1+} + He^{1+}$

(electron capture)

2. $Ni^{24+} + He \rightarrow Ni^{24+} + He^{1+} + e$

(target ionization)

3. $He^{1+} + He \rightarrow He^{2+} + He^{1+} + 2e$

(projectile ionization)

4. $F^{8+} + He \rightarrow F^{9+} + He^{1+} + 2e$

(projectile ionization)

5. $\gamma + He \rightarrow \gamma' + He^{2+} + 2e$

(photo ionization)

The first reaction was measured at different projectile velocities of a few $a.u.$, and represents a single-electron transfer process, with two reaction partners in the final state [9,19]. The high resolution of **COLTRIMS** allows for the first time in such fast encounters the separation of electron capture into different excited states. In reaction 2 the single ionization of helium by $200MeV Ni^{24+}$ impact has been investigated by a recoil-electron-projectile triple coincidence [8]. The energy loss distribution of the projectile and the corresponding low-energy electron emission has been measured for the pure ionization channel. In reaction 3 and 4 [6,7] the projectile and the target are simultaneously ionized either by nucleus-electron or by electron-electron interaction. Both interaction processes yield different recoil-ion momentum distributions and can, indeed, be experimentally separated by **COLTRIMS**. In reaction 5 He-double ionization has been investigated for high-energy photon Compton scattering [11]. Measuring the recoil-momentum, the Compton scattering and photo-effect processes were separated.

23.4.1 Energy-gain spectroscopy

Reaction 1: In this single-electron transfer reaction only two particles are in the final state. Since the electron transfer can lead to an excited state there are 6 unknown momentum components plus one Q-value parameter in the final state. Because of energy and momentum conservation the measurement of the three recoil momentum components yields the complete determination of the full kinematics in this collision and thus provides fully differential cross sections for electron capture into different excited states. In Figure 23.10 for $0.25MeV He^{2+} + He$ the recoil-ion longitudinal momentum distribution is shown. Two separated peak structures can be seen. The peak at $p^f_{\parallel R} = -1.5a.u.$ results from electron transfer between the 1s-state; the peak at $-0.5a.u.$ shows the capture contributions, when one electron is excited into the L shell. As explained above in section 2, for electron capture the recoil ion receives a backward momentum transfer from the electron mass transfer between target and projectile. A positive Q-value (energy loss) gives the recoil ion a momentum transfer of Q/v_P in the forward direction. There is even a small contribution to be seen from forward directed recoil-ion emission, indicating a large positive Q value, i.e. simultaneous excitation of target and projectile. The recoil momentum resolution in the longitudinal direction obtained here is $0.26a.u.$, which yields for this collision system an energy-gain resolution of about $\Delta E/E = 1 \times 10^{-5}$.

In Figure 23.11 the differential capture cross sections as function of the transverse recoil-ion momentum $p_{\perp rec}$ are presented. The transverse momentum can be easily transformed into projectile scattering angle by dividing it by the incoming projectile momentum p_0. For $1MeV$ projectile energy the recoil transverse momentum of $1a.u.$ corresponds to a scattering angle of $4\times10^{-5}rad$. It can be seen that with increasing projectile energy the contribution of capture into the ground state increases. However, below $1MeV$ projectile energy capture into exited states dominates. For $0.25MeV$ bombarding energy, a dip

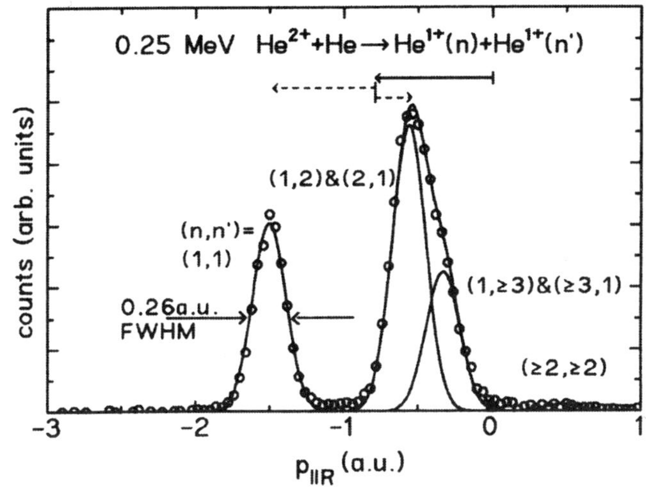

FIGURE 23.10 Longitudinal momentum distribution for the recoil ions from the reaction 0.25MeV and 1MeV He^{2+} + He → He^{1+} + He^{1+}. The momentum resolution is $0.26a.u.$ FWHM. The peaks correspond to the different states of projectile and target as indicated in the figure. The full arrow shows the contribution of the term $qv_P/2$ from equation 1, which is due to the mass transfer between target and projectile; the dashed arrows show the Q/v_P-term from equation 1 for the different states. (From [9].)

in the differential cross section for ground-state capture can be seen at a transverse momentum of about $1.5a.u.$ This structure is due to the interference of the relevant capture amplitudes on the in- and outgoing trajectories. A more detailed discussion of these data is given in [9,19]. Very recently Kambara and cooworkers [15] have used the **COLTRIMS** technique to investigate the capture process in fas O^{7+}-He collisions.

23.4.2 Target ionization in fast ion-atom collisions

Reaction 2: Using **COLTRIMS** [8] have recently investigated the low energy electron ejection in $200MeV\ Ni^{24+} + He \rightarrow Ni^{24+} + He^{1+} + e^-$ collisions, performing a recoil-electron-projectile coincidence. The apparatus they used is shown in Figure 23.7. Applying a weak magnetic solenoid field they obtained for electron energies below $50eV$ a nearly 30% triple coincidence detection efficiency and could measure for each event the complete momentum balance in these three-particle reactions. They could show that in these encounters the recoil ion and the ejected electron are strongly correlated in their longitudinal and transverse momenta (see figure 23.12). The projectile undergoes only a very small momentum change. The projectile energy-loss distribution is shown in Figure 23.13 as function of the correlated electron and recoil

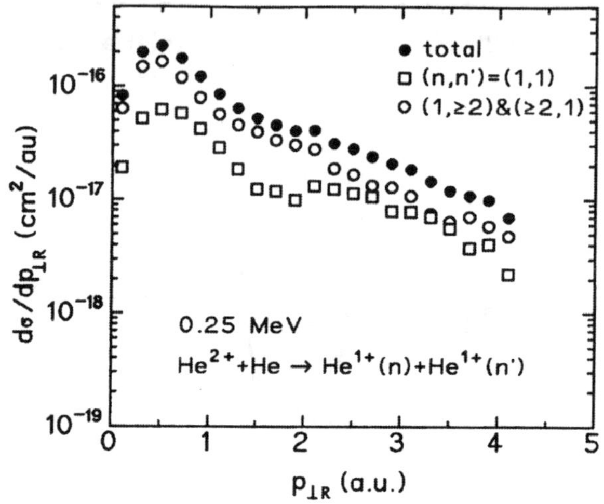

FIGURE 23.11 Differential state-selective capture cross sections. The x axis shows the transverse momentum of the recoil ion $p_{\perp R}$, which is equivalent to the projectile scattering angle ϑ_P ($\tan \vartheta_P = p_{\perp R}/p_0$, where p_0 is the incoming projectile momentum).
Full circles: sum over all states, squares: both partners are in ground state, open circles: one collision partner in some excited state. (From [9].)

longitudinal momentum. The measured correlated- momentum width reflects mainly the experimental resolution of $0.16a.u.$. This width corresponds to an projectile energy loss width of about $70eV$. The mean projectile energy loss is found to be about $30eV$ which corresponds to an energy gain/loss resolution of about 1×10^{-7}. The combined technique of weak electrostatic and magnetic fields provides important efficiency improvements for the **COLTRIMS** technique and is particularly suited for the investigation of ionization and capture processes in ion-beam storage rings. Details of this technique are described by [17].

Using **RIMS** Unverzagt and cooworkers [16] and Jardin and coworkers [27] have recently investigated also the multiple target ionization and have found strongly correlated electron emission into the forward direction.

23.4.3 e-e interactions in ion-atom collisions

Reaction 3 and 4: It is generally assumed that in fast ion-atom ionization processes the target ionization is due to a target-electron projectile-nucleus interaction. However, if the projectile is not fully stripped its remaining electrons can also contribute to the target ionization process via a pure electron-electron interaction [20,21,6,7]. In recent times the importance of these e-e contributions were discussed by comparing total ionization cross sections with

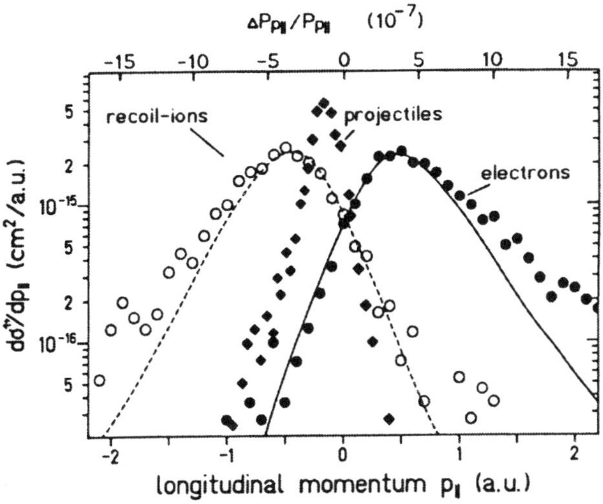

FIGURE 23.12 Longitudinal momentum distribution of electrons (full circles) and recoil ions (open circles) for single ionization of He by $3.6 MeV/u$ Ni^{24+} impact. The diamonds show the momentum change of the projectile. Full and dashed lines are results of nCTMC calculations.

theory [20,21]. So far it has been rather difficult to separate these different ionization mechanisms by traditional experimental detection techniques. **COLTRIMS** , however, provides a new experimental tool, to separate the different mechanisms by the very small differences in their final momentum states. In the case of a nucleus-electron interaction the projectile nucleus has to penetrate the target atom in a "close" collision yielding a transverse momentum exchange between the projectile and the target nuclei of more than $1 a.u.$. In case of e-e interaction only the two electrons in the projectile and target have to approach each other and the nuclei are mostly distant spectators. Thus, in these collisions a smaller transverse momentum is given to the recoiling target ion. Also the longitudinal momentum of the recoil ion in the case of e-e interaction should be smaller than for the nucleus-electron ionization process. In Figure 23.14 for He^{1+} on He collisions the measured transverse and longitudinal recoil-ion momentum distributions are shown for different bombarding energies. For experimental reasons the reaction channel with simultaneous target and projectile ionization was separated by a recoil-ion-projectile coincidence, since in this channel the e-e contribution is easily visible. Figure 23.14 shows that the distributions has two separate peak positions whose relative strengths vary with projectile energy. The peak near zero-recoil-momentum is indeed resulting from e-e interactions, whereas the peak at larger recoil momentum is due to nucleus-electron ionization processes. Both contributions show the expected projectile energy (E_p) dependence. The

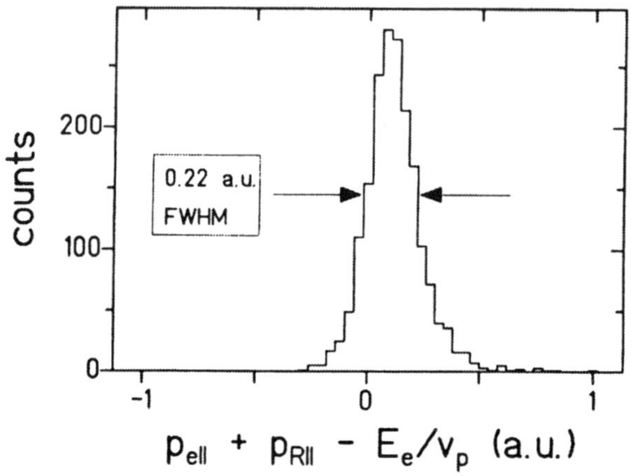

FIGURE 23.13 Projectile energy loss distribution as function of correlated longitudinal momenta of recoil ion and electron.

n-e interaction decreases with E_p^{-2}, whereas the e-e decreases with E_p^{-1} towards higher projectile energies. The e-e contribution is only observable above the expected threshold velocity. In Figure 23.14f the recoil-ion momentum distribution for $130eV$ electron impact on helium is also shown. This distribution agrees nicely with the e-e contribution at $1MeV$ Helium impact energy (i.e. the same projectile velocity). Using the same recoil detection technique Wu and coworkers [7] could recently separate for F^{8+} on helium the e-e and n-e contributions as well and determine the absolute cross sections for both ionization mechanisms.

23.4.4 Photo ionization

Reaction 5: **COLTRIMS** is a powerful technique for "complete" experiments of photon induced ionization processes. For example, $(\gamma\, 2e)$ reactions can very efficiently be investigated by recoil-ion-electron coincidence detection systems replacing the traditional low efficiency electron-electron coincidence technique. Since the absolute value of the recoil-ion momentum vector can be detected with nearly 100% efficiency and high resolution, only the emission angle of one electron has to be measured using large solid angle position-sensitive electron detectors. Thus a nearly 20% total coincidence efficiency is obtained. Vogt and coworkers (at Hasylab-Hamburg) [22] and Dörner and coworkers [18] (at the Advanced Light Source-Berkeley) have applied **COLTRIMS** recently, measuring the complete momentum balance in $(\gamma, 2e)$ reactions for different photon energies. They achieved a coincidence rate for the He-double ioniza-

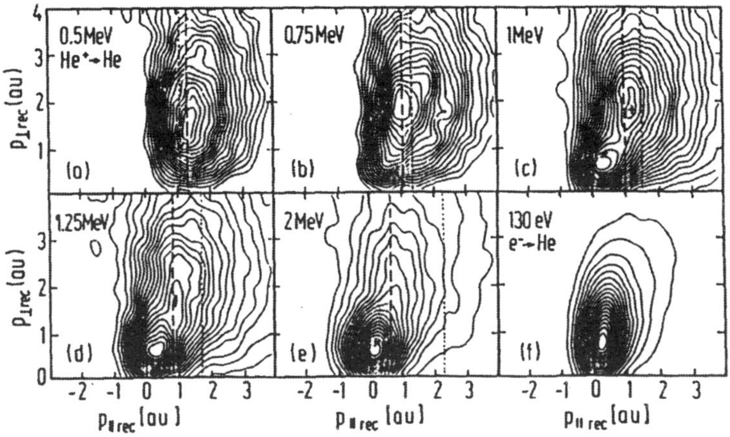

FIGURE 23.14 Doubly differential cross sections for reaction 3. The y axis shows the recoil transverse, the x axis the recoil longitudinal momentum. The contour lines represent a linear plot.

tion channel of about 100 coincidences per sec. In particular **COLTRIMS** will be a very powerful technique to study the angular distributions of the reaction products near the photo-ionization threshold. It can be applied to atomic as well to molecular targets. Spielberger and coworkers [11,23] have recently used **COLTRIMS** to measure the ratio of total cross sections of He double to single ionization by high-energy photon impact ($\hbar\omega = 9keV$ and $60keV$), respectively. **COLTRIMS** provides a clean method to separate the photo-effect and Compton-effect induced ionization processes. In case of photo-ionization the recoil-ion momentum is large and typically of the order of $\sqrt{2\hbar\omega}$ in $a.u.$. For a photon energy of $9keV$ the recoil momentum is approximately $25a.u.$. In case of Compton scattering the recoil ion is mostly a spectator and its final momentum is not very different from that in its initial state, and typically of the order of $1a.u.$. Using a "warm" super-sonic jet expansion [24] with a target density of about 2×10^{12}atoms/cm^2 the corresponding cross section ratios of single to double ionization of helium for photo-effect and Compton-effect was separately measured. Furthermore the Compton profile (*i.e.* the three-dimensional momentum distribution of the recoil ions for the Compton scattering process) was determined with very good statistics. In Figure 23.15a the density plot of the measured recoil-momentum distribution for He^{1+} ionization is presented. The outer ring distribution is due to photo-effect ionization and the inner spot is due to the Compton scattering process. Since in this experiment the photon beam had a broad energy distribution (parasitic experiment using a "white light" beam at the HASY-

LAB) the photo-effect distribution must show a large momentum width too. In figure 23.15b and 23.15c the projection of the distribution on the direction of the electric field vector is shown for single (He^{1+}) and double ((He^{1+}) ionization.

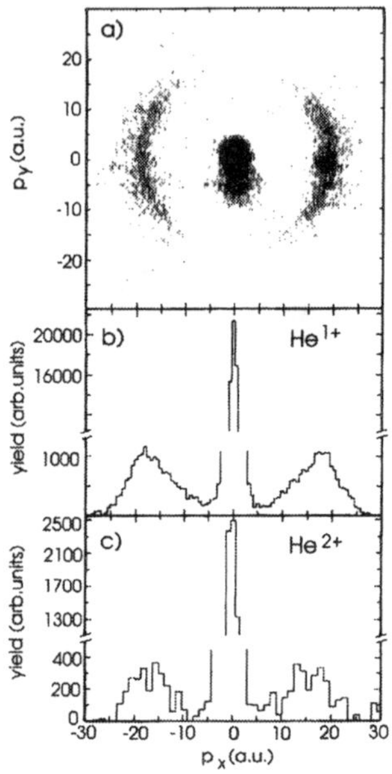

FIGURE 23.15 Measured He^{1+}-momentum distribution in the plane perpendicular to the photon-beam direction.

23.5 OUTLOOK AND PERSPECTIVES

Because of its momentum resolution and high detection efficiency for slow recoil ions **COLTRIMS** is an excellent technique to study the momentum exchange in many-particle atomic collision processes. It is well suited to investigate the dynamics of inner-atomic many-electron processes by electron-recoil coincidence methods particularly for photo-ionization processes. Its application is, however, not restricted to the field of atomic collision physics. It may open new detection windows for other fields in physics, *e.g.* nuclear physics, where *e.g.* β-decay can be investigated with this method. The angular correlation between electron and neutrino and even the neutrino mass can be determined with improved precision. Using super-sonic gas jets or in the near future even laser cooled target devices, a recoil momentum resolution of 0.01 or even $0.001 a.u.$ may be feasible, if the recoil-ion trajectory and the recoil-ion TOF can be measured with the required precision. This requires large position-sensitive detection devices which can be developed from existing technology. The momentum of the emitted electron has to be measured as well (emission angle and TOF) with equivalent precision. These coincidence measurements provide detailed information on the electron-neutrino angular correlation. With sufficiently large detection devices (several meters diameter) a neutrino-momentum resolution of a few eV/c seems feasible. One notes that in such a measurement the neutrino mass can, in principle, be derived from one single decay event. Using laser cooled targets even polarized target nuclei can be provided.

Last but not least, using very cold molecular targets, the Coulomb explosion of interesting molecules can be investigated [17,26]. The momenta of all emitted fragments can be measured with high precision by using multi-hit detection devices, if all (except one) of these fragments are ionized. Also here a momentum resolution of $0.02 a.u.$ seems feasible.

ACKNOWLEDGMENTS

This work was supported by DFG, BMBF(former BMFT), the Max-Planck-Forschungspreis of the Humboldt-Stiftung, Feodor-Lynen-Programm der Humboldt-Stiftung (R.D.) Graduierten Förderung des Landes Hessen (L.S.), Graduiertenkolleg-Schwerionenphysik (M.A.), Studienstiftung des Deutschen Volkes (V.M.), Willkomm-Stiftung, DAAD (I.A. and K.K.), and the Chemical Sciences Division, Office of Basic Energy Sciences, U.S. DOE. Furthermore we are indebted for many helpful discussions to our colleagues and friends R. Dreizler, H. J. Lüdde, J.Ast, S. Keller, R. E. Olson, M. Schulz, J. Burgdörfer, W. E. Meyerhof and U .Buck.

REFERENCES

1. O. Schwarzkopf, B. Krässig, J. Elmiger, and V. Schmidt. *Phys. Rev. Lett.*, **70**:3008, 1993.
2. O. Schwarzkopf, B. Krässig, V. Schmidt, F. Maulbetsch, and J. Briggs. *J. Phys.*, **B27**:L347–50, 1994.
3. A. Huetz, P. Lablanquie, L. Andric, P. Selles, and J. Mazeau. *J. Phys.*, **B27**:L13, 1994.
4. A. Lahmam-Bennani, C. Dupré, and A. Duguet. *Phys. Rev. Lett*, **63**:1582, 1989.
5. M.A. Coplan et al. *Rev.Mod.Phys.*, 66:985, 1994, and References therein.
6. R. Dörner, V. Mergel, R. Ali, U. Buck, C.L. Cocke, K. Froschauer, O. Jagutzki, S. Lencinas, W.E. Meyerhof, S. Nüttgens, R.E. Olson, H. Schmidt-Böcking, L. Spielberger, K. Tökesi, J. Ullrich, M. Unverzagt, and W. Wu. *Phys. Rev. Lett*, **72**:3166, 1994.

7. W. Wu, R. Ali, C.L. Cocke, V. Frohne, J.P. Giese, B. Walch, K.L. Wong, R. Dörner, V. Mergel, H. Schmidt-Böcking, and W.E. Meyerhof. *Phys. Rev. Lett*, **72**,:3170, 1994.
8. R. Moshammer, J. Ullrich, M. Unverzagt, V. Schmidt, P. Jardin, R.E. Olson, R. Mann, R. Dörner, V. Mergel, U. Buck, and H. Schmidt-Böcking. *Phys. Rev. Lett*, **73**:3371, 1994.
9. V. Mergel, R. Dörner, J. Ullrich, O. Jagutzki, S. Lencinas, S. Nüttgens, L. Spielberger, M. Unverzagt, C.L. Cocke, R.E. Olson, M. Schulz, U. Buck, E. Zanger, W. Theisinger, M. Isser S. Geis, and H. Schmidt-Böcking. *Phys. Rev. Lett.*, **74**, 1995.
10. R. Dörner, T. Vogt, V. Mergel, H. Khemliche, S. Kravis, C.L. Cocke, J. Ullrich, M. Unverzagt, L. Spielberger, M. Damrau, O. Jagutzki, I. Ali, B. Weaver, K. Ullmann, C.C. Hsu, M. Jung, E.P. Kanter, B. Sonntag, M.H. Prior, E. Rotenberg, J. Denlinger, T. Warwick, S.T. Manson, and H. Schmidt-Böcking. *Phys. Rev. Lett.*, 1996. accepted for publication.
11. L. Spielberger, O. Jagutzki, R. Dörner, J. Ullrich, U. Meyer, V. Mergel, M. Unverzagt, M. Damrau, T. Vogt, I. Ali, Kh. Khayyat, D. Bahr, H.G. Schmidt, R. Frahm, and H. Schmidt-Böcking. *Phys. Rev. Lett.*, **74**:4615, 1995.
12. R. Dörner, V. Mergel, L. Spielberger, O. Jagutzki, S. Nüttgens, M. Unverzagt, H. Schmidt-Böcking, J. Ullrich, R.E. Olson, K. Tökesi, W.E. Meyerhof, W. Wu, and C.L. Cocke. *AIP Conf. Proc. 360 (1995) Editors: L.J. Dube, J.B.A. Mitchell J.W. McConckey C.E. Brion*, AIP Press New York, 1995.
13. G. Brusdeylins, J.P. Toennies, and R. Vollmer. page 98, Perugia, 1989.
14. V. Mergel. *Phd thesis Universität Frankfurt*, 1996.
15. T. Kambara, J.Z. Tang, Y. Awaya, B.D. dePaola, O. Jagutzki, Y. Kanai, M. Kimura, T.M. Kojima, V. Mergel, Y. Nakai, H.W. Schmidt-Böcking, and I. Shimamura. Three-dimensional recoil-ion momentum analyses in 8.7 meV O^{7+}-He collisions. *J. Phys.*, 1995. accepted for publication.
16. M. Unverzagt et. al. *Phys. Rev. Lett.*, **76**, 1996.
17. R. Moshammer, M. Unverzagt, W. Schmitt, J. Ullrich, , and H. Schmidt-Böcking. *Rev. Sci. Instr.*, :submitted for publication, 1996.
18. R. Dörner, J. Feagin, C.L. cocke, O. Jagutzki, M. Jung, E.P. Kanter, H. Khemliche, S. Kravis, V. Mergel, M.H. Prior, H. Schmidt-Böcking, L. Spielberger, J. Ullrich, M. Unverzagt, and T. Vogt. *Phys. Rev. Lett.*, 1996. submitted for publication.
19. V. Mergel, R. Dörner, J. Ullrich, O. Jagutzki, S. Lencinas, S. Nüttgens, L. Spielberger, M. Unverzagt, C.L. Cocke, R.E. Olson, M. Schulz, U. Buck, and H. Schmidt-Böcking. *Nucl. Instr. Meth.*, **98**, 1995.
20. H.-P. Hülskötter, B. Feinberg, and W.E. Meyerhof. *Phys. Rev.*, **A44**:1712, 1991.
21. E.C. Montenegro, W.S. Melo, W.E. Meyerhof, and A.G. dePinho. *Phys. Rev. Lett*, **69**:3033, 1992.
22. T. Vogt. *Diploma Thesis, University Frankfurt 1995, to be published.*
23. L. Spielberger, O. Jagutzki, B. Krässig, U. Meyer, Kh. Khayyat, V. Mergel, Th. Tschentscher, Th. Buslaps, H. Bräuning, R. Dörner, T. Vogt, M. Achler,

J. Ullrich, D.S. Gemmel, and H. Schmidt-Böcking. *Phys. Rev. Lett.*, submitted for publication.

24. O. Jagutzki, L. Spielberger, R. Dörner, s: Nüttgens, V. Mergel, H. Schmidt-Böcking, J. Ullrich, R.E. Olson, and U. Buck. *Zeitschrift für Physik*, **D36**:5, 1996.
25. S. E. Sobottka and M. B. Williams, *IEEE Trans Nucl. Sci.*, **35** 348-351, 1988.
26. I. Ali, R. Dörner, O. Jagutzki, J. Ullrich, K. Ullmann, and H. Schmidt-Böcking. in preparation.
27. P. Jardin et al. to be submitted.

Index